Cellular Mechanisms of Conditioning and Behavioral Plasticity

Cellular Mechanisms of Conditioning and Behavioral Plasticity

Edited by

Charles D. Woody

Mental Retardation Research Center
Brain Research Institute
University of California, Los Angeles
Los Angeles, California

Daniel L. Alkon

National Institute of Neurological Communicative Diseases and Stroke
National Institutes of Health at the Marine Biological Laboratory
Woods Hole, Massachusetts

and

James L. McGaugh

Center for Neurobiology of Learning and Memory
University of California, Irvine
Irvine, California

Springer Science+Business Media, LLC

Library of Congress Cataloging in Publication Data

Cellular mechanisms of conditioning and behavioral plasticity.

Proceedings of the Symposium on the "Cellular Mechanisms of Conditioning and Behavioral Plasticity," held July 9-12, 1986, at the University of Washington, Seattle, Wash., as a satellite symposium of the International Union of Physiological Sciences 30th International Congress.

Includes bibliographies and index.

1. Conditioned response—Congresses. 2. Neuroplasticity—Congresses. 3. Learning—Physiological aspects—Congresses. 4. Neurophysiology—Congresses. I. Woody, Charles D. II. Alkon, Daniel L. III. McGaugh, James L. IV. Symposium on the "Cellular Mechanisms of Conditioning and Behavioral Plasticity" (1986: University of Washington) V. International Union of Physiological Sciences. Congress (30th: 1986: Vancouver, B.C.) [DNLM: 1. Behavior—physiology—congresses. 2. Conditioning (Psychology)—physiology—congresses. 3. Memory—physiology—congresses. 4. Neurons—physiology—congresses. WL 102.5 C3929 1986]

QP416.C45 1988 152′.3224 87-32891

ISBN 978-1-4757-9612-4 ISBN 978-1-4757-9610-0 (eBook)

DOI 10.1007/978-1-4757-9610-0

Cover illustration: Soma high Ca in response to iontophoretic glutamate stimulus. (Courtesy of Dr. John A. Connor.)

Originally published by Plenum Press, New York in 1988
Softcover reprint of the hardcover 1st edition 1988

Foreword

I would like first to thank Charles Woody and his organizing committee for arranging the symposium on the "Cellular Mechanisms of Conditioning and Behavioral Plasticity," which was also a satellite meeting of the International Union of Physiological Sciences 30th International Congress. The proceedings of this symposium are represented by the chapters that follow.

During the 1970s, Dr. Woody and co-workers were able to carry out a remarkable series of microelectrode studies, both intracellular and extracellular, of cortical nerve cells during conditioning of the eye-blink response to sound in the intact waking cat. He demonstrated enduring changes in excitability and membrane resistance in pericruciate cortical cells during associative conditioning of the eye blink, changes that are facilitated by ACh and cGMP and reinforced by stimulation of the hypothalamus (the latter confirming the original studies of Voronin). These findings have been of considerable importance in our attempt to understand the conditioning process at the cellular level.

In the early 1980s, Woody completed a review of the literature, citing 1150 references and presenting his own thoughtful views in a book entitled *Memory, Learning, and Higher Function: A Cellular View,* published by Springer-Verlag in 1982 (Woody, 1982a). In his prologue, Woody emphasizes the importance of understanding the molecular processes as substrates for memory, learning, and higher function, as in the molecular determinants of innate behavior. However, Woody adds, "Cellular views of behavior tend to focus on molecular processes, but the manner in which the involved cells are interconnected into a transmission network is also important. . . . We need an analytic methodology for handling the network as well as its adaptive elements." The importance of the network view of the cells and circuits involved in the conditioned eye blink and nictitating membrane response in the rabbit has since been illustrated by the work of Richard Thompson and associates (Thompson *et al.*, 1984) in their unicellular studies of hippocampal pyramidal cells, cells of the brainstem, and the dentate nucleus of the cerebellum. The cerebellar circuit was shown to be essential to the formation of the conditioned nictitating membrane response in the rabbit, though cortical and hippocampal circuits may play a role in the formation of other conditioned responses in vertebrate preparations. The hippocampus was essential for trace and reversal conditioning.

On completion of his own book, Woody proceeded to organize the first in this series of conferences on conditioning, with the help of some colleagues, particularly Dan Alkon, Richard Thompson, and David Cohen. Our first conference was held in October 1981 at the Asilomar Conference Center in Pacific Grove, California. It proved to be an exciting experience, resulting in the rapid publication of the proceedings, edited by Woody, by Plenum Press in 1982, with the title *Conditioning: Representation of Involved Neural Functions* (Woody, 1982b). This volume of 748 pages contained 46 chapters by most of the leading workers in the field studying the cellular and possible molecular bases of conditioning in a wide variety of animals, from marine mollusks to monkeys. The contributions of Voronin and associates were particularly interesting, containing numerous references to the Russian literature.

The second conference was hosted by Dan Alkon in his laboratories at the Marine Biological Laboratory in Woods Hole, Massachusetts, in November 1983. The emphasis here was on detailed biophysical and chemical mechanisms in postsynaptic cells or in photoreceptors of the marine mollusk *Hermissenda,* used by Alkon and associates for studies of associative conditioning in a "simple" model in which the importance of Ca^{2+}-dependent regulation of K^+ channels and intracellular protein metabolism was amply demonstrated. There was much discussion as to whether this model could be applied to mechanisms of conditioning in more complex vertebrate systems, in which more widespread circuits are involved, circuits not particularly involving primary sensory or motor nerve cells. This conference, edited by Alkon and Woody, was also published by Plenum Press in 1986 under the title *Neural Mechanisms of Conditioning* (Alkon and Woody, 1986). I was pleased to hear more emphasis being placed on changes in specific ionic conductances and intracellular metabolism of postsynaptic cells in contrast to the exclusively presynaptic mechanisms apparent in the studies of Eric Kandel and associates in *Aplysia* (Kandel, 1979).

The next symposium on neural mechanisms of conditioning was held on September 16–18, 1985, in Cambridge. It was organized largely by Lynn Bindman of University College, London, together with Gabriel Horn of Cambridge and Charles Woody and Dan Alkon. The Cambridge symposium was most important for introducing into our discussions the work and ideas of many English and European workers who have been studying mechanisms of learning and conditioning from somewhat different points of view.

The broadening scope of our discussion is reflected in the addition of "Behavioral Plasticity" to the title of the present symposium. We in this field will need increasingly to consider the importance of environment and experience in developmental neurobiology and the exciting new knowledge of the regenerative capacities of the brain, as summarized in a recent book entitled *Synaptic Plasticity,* edited by Carl Cotman and published by Guilford Press (Cotman, 1985).

Guilford Press has also published another important volume entitled *Neurobiology of Learning and Memory,* edited by Gary Lynch, James L. McGaugh, and Norman Weinberg (Lynch *et al.,* 1984). This book, with over 60 contributors, grew out of the conference held to inaugurate the Center for the Neurobiology of Learning and Memory at the University of California at Irvine. It is one of the most exciting detective stories I have read in a long time.

Finally, in my attempt to review current literature in preparation for this conference, I ran across an interesting article by Deborah Barnes in *Science* of March 14, 1986, on a European Conference on "Neural and Molecular Bases of Learning" held in West Berlin in December 1985. The actions of monoamine modulators of synaptic function, together with certain peptides, on ionic channels for Ca^{2+}, K^+, and Cl^-, with the activation of important intracellular enzymes involved in protein synthesis and phosphorylation, were emphasized in a number of models of the learning process. Changes in specific acetylcholine receptors by a Ca^{2+}-dependent regulator of protein synthesis were also described. A similar mechanism may be involved in the regulation of glutamate receptors in the hippocampus, previously described by Gary Lynch. While attempting to absorb and evaluate this interesting material, I read another critical and comprehensive review by Charles Woody, entitled "Understanding the Cellular Basis of Memory and Learning," which has just appeared in the latest volume of the *Annual Review of Psychology* (Woody, 1986). Once again, we are grateful to Dr. Woody for helping us to keep abreast of the current explosion of ideas and information on possible cellular and molecular mechanisms of conditioning.

Reflecting on the bewildering array of possible biophysical and biochemical mechanisms that may be involved in short-term learning and conditioning, and the possible morphological changes that may occur in the consolidation of long-term memory, I am reminded of an expression borrowed from developmental neurobiology, namely, "exuberance." It is used to describe the excessively redundant growth of nerve fibers and synapses during the early stages of development, with their death and degeneration later, according to a "Darwinian" principle of survival of the fittest. Certainly, there will remain multiple widespread mechanisms of importance to conditioning and learning, though some will be far more important than others in the learning process, as opposed to relatively transient sensorimotor mechanisms that are not specifically related to the storage of memories. The Darwinian principle will probably eventually come into play in the selection of those mechanisms most important for short- and long-term memory. We can all take part in this selection process by careful consideration of the following chapters.

Herbert H. Jasper

REFERENCES

Alkon, D. L., and Woody, C. D. (eds.), 1986, *Neural Mechanisms of Conditioning*, Plenum Press, New York.

Barnes, D. M., 1986, Lessons from snails and other models, *Science* **231**:1246–1249. (Report on Symposium "Neural and molecular bases of learning" 8–13 December 1985, West Berlin.

Cotman, C. W. (ed.), 1985, *Synaptic Plasticity*, Guilford Press, New York.

Kandel, E. R., 1979, Cellular insights into behavior and learning, *Harvey Lect.* **73**:29–92.

Lynch, G., McGaugh, J. L., and Weinberger, N. M. (eds.), 1984, *Neurobiology of Learning and Memory*, Guilford Press, New York.

Thompson, R. F., Clark, G. A., Donegan, N. H., Lavond, D. G., Lincoln, J. S., Madden, J. IV, Mamounas, L. A., Mauk, M. D., McCormick, D. A., and Thompson, J. K., 1984, Neuronal substrates of learning and memory: A 'multiple trace' view, in: *Neurobiology of Learning and Memory* (G. L. Lynch, J. L. McGaugh, and N. M. Weinburger, eds.), Guilford Press, New York, p. 528.

Woody, C. D., 1982a, *Memory, Learning, and Higher Function: A Cellular View*, Springer-Verlag, New York.

Woody, C. D. (ed.), 1982b, *Conditioning: Representation of Involved Neural Functions*, Plenum Press, New York.

Woody, C. D., 1986, Understanding the cellular basis of memory and learning, *Annu. Rev. Psychol.*, **37**:433–493.

Preface

This volume shows how cellular mechanisms of conditioning and behavioral plasticity can be studied in a variety of contexts using many different approaches. Planning for the conference from which this volume arose began several years ago at Woods Hole, was continued in Cambridge, England, and was concluded in Seattle. We thank all those who contributed their efforts so generously, and especially Judith Erstad, who assisted in the preparation of this volume.

Charles D. Woody
Los Angeles, California
Daniel L. Alkon
Woods Hole, Massachusetts
James L. McGaugh
Irvine, California

Contents

Part III—Conditioning

Part IV—Anatomy and Cable Properties

I

Some Different Perspectives on the Basic Mechanisms

1

The Cellular Basis for Short-Term Memory in Endocrine Systems

HOWARD RASMUSSEN, CARLOS ISALES, YOH TAKUWA, NORIKO TAKUWA, PAULA BARRETT, and WALTER ZAWALICH

1. INTRODUCTION

A common property of many endocrine systems is an anamnestic response. For instance, when pancreatic islets are reexposed to a standard glucose stimulus after a prior period of exposure to the same glucose challenge, their insulin secretory response is greater to the second than to the first stimulus (Gold *et al.*, 1982; Grill *et al.*, 1978, 1979; Grill and Rundfeldt, 1979; Grill, 1981; Grodsky *et al.*, 1969). A similar pattern is seen on repeated exposure of adrenal glomerulosa cells to the peptide hormone angiotensin II. Recent work in both of these systems indicates that protein kinase C plays a central role in mediating the sustained phase of each of these endocrine responses (Kojima *et al.*, 1984; Tanigawa *et al.*, 1982; Zawalich *et al.*, 1983, 1984). Work in our laboratories over the past 18 months has focused on the possible role of protein kinase C in this type of short-term cellular memory in bovine adrenal glomerulosa cells and isolated rat pancreatic islets.

2. THE CELLULAR AND MOLECULAR BASIS OF MEMORY IN ADRENAL GLOMERULOSA CELLS

From an analysis of the effects of angiotension II on phosphoinositide metabolism, calcium metabolism, and protein phosphorylation, we developed a two-branch model to account for the role of the Ca^{2+} messenger system in the regulation of aldosterone secretion from adrenal glomerulosa cells (Kojima *et al.*, 1984; Rasmussen and Barrett, 1984). In this model, the initial receptor-mediated events include activation of a phospholipase C, which is responsible for the hydrolysis of phosphatidylinositol-4,5-bisphosphate (Berridge, 1984; Nishizuka, 1984). This leads to the generation of two intracellular messengers, inositol-1,4,5-trisphosphate and diacylglycerol (DG) (rich in arachidonic acid).

HOWARD RASMUSSEN, CARLOS ISALES, YOH TAKUWA, NORIKO TAKUWA, PAULA BARRETT, and WALTER ZAWALICH • Departments of Internal Medicine and Cell Biology, School of Medicine and School of Nursing, Yale University, New Haven, Connecticut 06510.

Additionally, angiotensin II induces an immediate and sustained increase in the rate of Ca^{2+} influx across the plasma membrane (Kojima *et al.*, 1985). Whether this change in Ca^{2+} metabolism is a direct consequence of hormone–receptor interaction or an indirect consequence via the change in phosphoinositide metabolism is not yet known.

Inositol-1,4,5-trisphosphate ($InsP_3$) is a water-soluble messenger released into the cytosol, where it triggers the release of Ca^{2+} from a nonmitochondrial, intracellular pool (Streb *et al.*, 1983). This results in a transient rise in the Ca^{2+} concentration in the cell cytosol, $[Ca^{2+}]_c$, which, in turn, activates calmodulin-dependent enzymes including calmodulin-dependent protein kinases, thereby phosphorylating a specific subset of cellular proteins that mediate the initial phase of cellular response (Barrett *et al.*, 1986a,b).

The increase in the DG content of the plasma membrane, along with the $InsP_3$-induced transient rise in $[Ca^{2+}]_c$, leads to the conversion of protein kinase C from its Ca^{2+}-insensitive to its Ca^{2+}-sensitive, plasma-membrane-associated form (Takai *et al.*, 1984; Hannun *et al.*, 1986; May *et al.*, 1985; Wolf *et al.*, 1985). In this form, the activity of C-kinase is regulated by the rate of Ca^{2+} influx (or Ca^{2+} cycling) across the plasma membrane (Kojima *et al.*, 1984). An increase in its activity leads to the phosphorylation of a second specific subset of cellular proteins, which mediate the sustained phase of cellular response (Barrett *et al.*, 1986a,b).

In this model, Ca^{2+} ion plays a messenger function during both the initial and sustained phases of cellular response, but its cellular sites of action and its molecular targets differ (Rasmussen, 1986). During the initial phase of response, a change in $[Ca^{2+}]_c$ is the message, and calmodulin-regulated enzymes are the molecular targets: the calmodulin (CaM) branch. During the sustained phase, a change in Ca^{2+} cycling across the plasma membrane is the message, and the PM-associated C-kinase is the molecular target: the C-kinase branch. What is not clear is whether, during the sustained phase, C-kinase reads the message as a change in Ca^{2+} flux rate or as a change in $[Ca^{2+}]$ in a specific submembrane domain, $[Ca^{2+}]_{sm}$.

Validation of the model of events during the second phase of the response has come from studies that employ a pharmacological agent that increases Ca^{2+} influx rate (Bay K 8644) and one that activates C-kinase (1-oleoyl-2-acetylglycerol, OAG). Neither the addition of Bay K 8644, which leads to a doubling of Ca^{2+} influx rate, nor the addition of OAG, which leads to an association of C-kinase with the membrane, increases aldosterone production rate, but the combined addition of Bay K 8644 and OAG leads to a slowly developing yet sustained (submaximal) increase in aldosterone production rate (Kojima *et al.*, 1985).

With this model in mind, it is now possible to discuss our work on "memory" in this system. The basic observation is quite straightforward: readdition of angiotensin II to cells previously exposed to this peptide for 15–20 min leads to an aldosterone secretory response that is greater than the original response; i.e., the cells remember their prior exposure to angiotensin II (Barrett *et al.*, 1986a). In order for this memory response to occur, the cells must be treated with angiotensin II for at least 15 min during the initial period of exposure. Also, the magnitude of the memory response decreases as the time interval increases between the initial and the second periods of exposure to angiotensin II.

Of particular interest, the cells exposed to angiotensin II for an initial 20-min period display a different response to the calcium channel agonist Bay K 8644 than do naive cells. Addition of Bay K 8644 to naive cells, as noted above, induces a nearly twofold increase in Ca^{2+} influx rate but no change in aldosterone production rate. However, the addition of Bay K 8644 to cells previously exposed to angiotensin II for a 20-min period (and then treated with an angiotensin II antagonist) evokes an aldosterone secretory response. When the angiotensin II antagonist is added, the Ca^{2+} influx rate returns rapidly to its basal value, the aldosterone secretory rate declines, and the phosphorylation of the late-phase proteins is reversed. Addition of Bay K 8644 increases the Ca^{2+} influx rate, reinitiates aldosterone secretion, and reestablishes the phosphorylation of the late-phase proteins. Thus, the cells treated with angiotensin II for 20 min (and then treated with an angiotensin II antagonist) behave, when treated with Bay K 8644, exactly as naive glomerulosa cells treated with OAG.

From these data, we have concluded that on inhibition of angiotensin II action, Ca^{2+} influx rate returns rapidly to its basal value; as a consequence, the activity of C-kinase falls, and the aldosterone secretory rate declines, yet C-kinase remains in its Ca^{2+}-sensitive form. Thus, any agent that increases Ca^{2+} influx rate restimulates C-kinase activity and hence reactivates aldosterone secretion. A measure of the rate at which C-kinase relaxes back to its Ca^{2+}-insensitive state can be made by employing a system in which both an angiotensin II antagonist and Bay K 8644 are added simultaneously to cells exposed to angiotensin II for 20 min. As noted above, when the antagonist alone is added, Ca^{2+} influx rate falls promptly ($t_{1/2} \cong 2$ min), and aldosterone secretory rate falls nearly as rapidly ($t_{1/2} \cong 6$ min). However, when Bay K 8644 is added simultaneously with the antagonist, there is no change in Ca^{2+} influx rate. Under this condition the aldosterone secretory rate falls with a half-time of 45 min or more. We believe the slope of this decay in response is a measure of the time taken for the C-kinase to relax back from its membrane-associated, Ca^{2+}-sensitive form to its Ca^{2+}-insensitive form.

3. THE BASIS OF MEMORY IN THE PANCREATIC ISLET

Work in isolated rat pancreatic islets provides evidence for a similar type of memory mediated by protein kinase C. Yet, in the islet the particular arrangement of agonist-mediated insulin secretion is more complex. In order to describe the properties of this system, it is necessary to consider the actions of and interactions between four agonists: glucose, cholecystokinin 8S (CCK8S) (Holst *et al.*, 1980), forskolin, and tolbutamide (Table I).

When CCK8S acts on isolated rat islets, there is an increase in inositol trisphosphate production and a transient increase in the rate of Ca^{2+} efflux from an intracellular pool (Ahren and Linquist, 1981; Zawalich *et al.*, 1986, 1987; W. Zawalich, V. A. Diaz, and H. A. Rasmussen, unpublished data). Both of these changes are hallmarks of the activation of the Ca^{2+} messenger system via the hydrolysis of phosphatidylinositol-4,5-bisphosphate (PIP_2). Hence, it appears that CCK8S acts in islets much the same way that angiotensin II (AII) acts on adrenal glomerulosa cells. Complete validation of this statement requires measurement of the time course of change in the content of diacylglycerol (DG). For

TABLE I. Comparison of the Actions of Agonists on Islet Cell Function

Agonist	Ca^{2+} influx	cAMP	PIP_2 hydrolysis	IP_3 increase	Ca^{2+} efflux	Induce memory	Evoke memory
Glucose	Direct increase	Increase	Small	Small	Small	Yes	Yes
CCK8S	None	None	Large	Large	Large	Yes (only with glucose)	No
Tolbutamide	Direct increase	None	None	None	None	No	Yes

technical reasons this has not yet been achieved, but based on the fact that $InsP_3$ production goes up, one must assume that DG production rate also increases. Even so, there are important differences between the actions of CCK8S in islets and those of AII in adrenal glomerulosa cells. In the latter case, angiotensin II induces an immediate and sustained increase in the rate of Ca^{2+} influx across the plasma membrane, but in the former case, CCK8S has no apparent effect on Ca^{2+} influx rate. The other striking aspect of its action is that CCK8S stimulates PIP_2 hydrolysis, whether the glucose concentration is high (7 mM) or low (2.75 mM) but only induces a typical biphasic pattern of insulin secretion when the glucose concentration is high (Sakamoto *et al.*, 1982; Zawalich and Diaz, 1986): glucose determines whether the intracellular messengers generated by CCK8S–receptor interaction ($InsP_3$, Ca^{2+}, and DG) evoke an insulin secretory response or not.

In order to understand how glucose modulates the responsiveness of this tissue to CCK8S, it is first necessary to discuss the effects of glucose on islet cell function (Hedeskov, 1980; Henquin, 1985). A very well established and important effect of glucose is to increase Ca^{2+} influx rate. This effect is not mediated by an interaction of glucose with a surface receptor but depends on the metabolism of glucose. It now appears possible that the glucose metabolite of critical importance is ATP, because an ATP-dependent, tolbutamide-responsive K^+ channel has been discovered in the islet cell plasma membrane (Cook and Hales, 1984; Sturgess *et al.*, 1985, 1986). The properties of this channel are such that high ATP reduces the flux of K^+ through it. Hence, the first step in glucose action (via ATP) is that of reducing K^+ efflux. This change leads to the depolarization of the plasma membrane, the opening of voltage-dependent Ca^{2+} channels, an influx of Ca^{2+}, the opening of a Ca^{2+}-sensitive K^+ channel, an increase in K^+ efflux, and the repolarization of the membrane. The same sequence is then reinitiated; i.e., glucose induces a regenerative Ca^{2+} current in the islet cell plasma membrane. This is an effect that is not induced by CCK8S but is produced by tolbutamide.

A second effect of glucose is that of increasing the cAMP content of β cells (Henquin and Meissner, 1984). Two possible mechanisms by which it brings about this effect have been proposed (Pipeleers, 1984; Schuit and Pipeleers, 1985). One view is that the glucose-induced increase in Ca^{2+} influx rate activates adenylate cyclase directly via Ca^{2+}–calmodulin; the other is that glucose induces the release of glucagon (or other agents) from α cells, and this peptide interacts with specific receptors on the β cell membrane to activate adenylate cyclase in the classic manner.

Although there are numerous reports that glucose also activates PIP_2 hydrolysis (Best *et al.*, 1984; Turk *et al.*, 1986; Morgan *et al.*, 1985), this effect is small compared to that of CCK8S. Hence, it seems unlikely that glucose-induced increases in $InsP_3$ and DG are major mediators of glucose action. Nonetheless, it is apparent, as discussed below, that glucose has an additional, critically important effect on DG metabolism independent of any effect on PI metabolism. In order to discuss this effect, it is instructive to discuss the relationships between the actions of CCK8S and glucose. Before considering these relationships, it is necessary to introduce two other pharmacological agonists that have specific effects that allow one to dissect out CCK8S–glucose interactions (Table I). These are tolbutamide and forskolin. Tolbutamide has the specific effect of increasing Ca^{2+} influx rate in a manner similar to the action of glucose (Henquin, 1980) but has none of the other primary effects of glucose. When added to islets in the absence of glucose, tolbutamide induces only a submaximal first phase of insulin secretion but no second phase. Forskolin, on the other hand, is a specific activator of adenylate cyclase and in low concentrations increases the basal rate of insulin secretion to a slight degree. With these tools, it has been possible to explore the relationships between the actions of CCK8S and glucose.

As noted above, when islets are incubated in 2.75 mM glucose and then exposed to CCK8S, no insulin secretory response is seen even though CCK8S stimulates PIP_2 hydrolysis, $InsP_3$ production, and Ca^{2+} efflux. On the other hand, when islets are incubated in 7.0 mM glucose and then exposed to CCK8S, a typical insulin secretory response is seen even though CCK8S has the same effect on PIP_2 hydrolysis. Two of the changes in islet cell function induced by glucose that are not induced by CCK8S are an increase in Ca^{2+} influx rate and an increase in cAMP content. To determine whether either of these changes is important in the sensitizing action of glucose, islets were incubated in 2.75 mM glucose and exposed to forskolin and/or tolbutamide and then treated with CCK8S. Exposure of islets to either forskolin, tolbutamide, or the combination alters their responsiveness to CCK8S. In each instance, CCK8S induces an insulin secretory response, but this response is always confined to just a first-phase response. No second phase is seen. When both tolbutamide and forskolin are present, this first phase response is markedly greater than a normal first-phase response. The lack of a second- or sustained-phase response is not a result of depletion of insulin from the islets because the subsequent addition of glucose (7 mM) leads to an immediate second-phase response.

Only when glucose is present above some critical concentration (5.5–7.0 mM) does CCK8S induce a second-phase response. Hence, one must conclude that glucose has an additional action. This action is related to the metabolism of DG. The simplest models to account for this action of glucose are either that glucose changes the substrate specificity of the phospholipase, that glucose regulates the *de novo* synthesis of DG, or that glucose regulates the further metabolism of DG to phosphatidic acid (PA) such that the DG content increases; i.e., glucose either stimulates DG synthesis, inhibits DG kinase ($DG + ATP \rightleftarrows PA + ADP$), or activates PA phosphohydrolyase ($PA \rightleftarrows DG + P_i$). It is noteworthy that Dunlop and Larkins (1985) have reported that glucose stimulates the *de novo* synthesis of diacylglycerol from glucose via an acyldihydroxyacetone phosphate intermediate in cultured islets from neonatal rats. Regardless of which mechanism proves to be correct, from an operational point of view the consequences are postulated to be the same. When the glucose concentration is low, CCK8S increases DG production rate but does not increase DG content sufficiently to activate protein kinase C. When the glucose concentration is above the critical concentration, the CCK8S-induced increase in DG production rate leads to an increase in DG content and an activation of protein kinase C and hence a second phase of insulin secretion. In addition, glucose increases Ca^{2+} influx rate, which is the second necessary component of the second-phase response; i.e., glucose leads to the activation of protein kinase C and to an increase in Ca^{2+} influx rate across the plasma membrane, and this regulates the activity of the activated C-kinase, a situation similar to that seen in adrenal cells during the sustained phase of angiotensin-II-mediated aldosterone secretory response.

The response of the β cells in rat pancreatic islets displays a similar type of memory phenomenon as that seen in adrenal glomerulosa cells (Grill *et al.*, 1979; Grill and Rundfeldt, 1979; Grill, 1981; Grodsky *et al.*, 1969). An elevation of ambient glucose concentration (e.g., from 7 to 10 mM) leads to a biphasic insulin secretory response. When glucose concentration is lowered to 7 mM, the insulin secretory rate promptly returns to its basal value within 3–5 min. However, readdition of 10 mM glucose 10 min later leads to a greater insulin secretory response than that observed following the initial exposure to 10 mM glucose; i.e., the β cells remember their previous exposure to glucose.

If islets incubated in 7 mM glucose are exposed to CCK8S, they also respond with a biphasic pattern of enhanced insulin secretion (Zawalich *et al.*, 1987). If CCK8S is then removed and the glucose concentration is reduced to 2.75 mM, the insulin secretory response falls rapidly to basal rate. Addition of 10 mM glucose evokes a larger insulin secretory response in such islets than in islets not previously exposed. On the other hand, CCK8S will not evoke a greater than normal response in islets primed previously either with 10 mM glucose or with CCK8S in the presence of 7 mM glucose. Conversely, addition of tolbutamide, an agent known to increase Ca^{2+} influx rate across the plasma membrane of the β cell, will evoke a greater than normal secretory response in primed islets (Grill and Rundfeldt, 1979; Zawalich *et al.*, 1987; W. Zawalich, V. A. Diaz, and H. A. Rasmussen, unpublished data), but tolbutamide alone cannot prime the islets to the effects of either glucose or tolbutamide added subsequently (Grill and Rundfeldt, 1979).

On the basis of these results, one can divide agents involved in this memory phenomenon into those that can induce a memory-competent state and those that can evoke a response from memory-competent islets. Inducers of this state are glucose and CCK8S [only when glucose concentration is high enough (7 mM) to allow CCK8S to evoke a second phase of insulin secretion]. These two secretagogues have in common the ability to activate C-kinase. On the other hand, simply increasing Ca^{2+} influx rate with tolbutamide will not induce a memory-competent state in the β cell. Conversely, glucose and tolbutamide, both agents that stimulate Ca^{2+} influx rate, will evoke a magnified response from memory-competent islets, but CCK8S will not (Table I). Thus, the glucose-primed β cell behaves very similarly to the angiotensin-II-primed adrenal glomerulosa cell; it responds to agents that stimulate Ca^{2+} influx rate with a greater than expected response, because, we postulate, C-kinase has been activated by the prior exposure to primer (inducer) and remains in a Ca^{2+}-sensitive state for a considerable period of time after removal of the primer. In both cellular response systems, the memory appears to involve a changed state of protein kinase C.

4. PERSPECTIVE

These data in endocrine systems provide insights into the mechanism by which one type of short-term cellular memory is achieved. In both cases, this type of "memory" involves protein kinase C and is thought to be localized to a specific cellular site, the plasma membrane. Further, the contrasting fashions by which the Ca^{2+} messenger system is organized in adrenal glomerulosa cells and β cells of the islets of Langerhans provide new insights into several features of this system. In particular, the contrasting actions of CCK8S and angiotensin II illustrate the nature of the link between hormone receptor interaction and Ca^{2+} influx: even though each peptide hormone initiates PIP_2 hydrolysis in its particular target cell, one (AII) induces an increase in Ca^{2+} influx rate, but the other (CCK8S) does not. This means either that the effect of angiotensin II on Ca^{2+} influx rate is a direct consequence of hormone–receptor interaction (separate from PIP_2 hydrolysis) or that the Ca^{2+} channels in the plasma membranes of adrenal glomerulosa and islet cells differ, and PIP_2 hydrolysis is linked only to a particular type of Ca^{2+} channel, which exists in the glomerulosa but not the islet cell.

The other equally interesting postulate is that glucose can alter DG metabolism and thereby regulate C-kinase activation by a mechanism other than that of stimulating PIP_2 hydrolysis. This finding, in particular, may point the way to the discovery of mechanisms, other than PIP_2 hydrolysis, of C-kinase activation and by which a C-kinase-mediated memory response can be induced in a variety of target cells including those in the CNS.

ACKNOWLEDGMENTS. This work was supported by grants AM 19813, AM 34381, and AM 33001 from the National Institutes of Health and a grant from the Diabetes Research and Education Foundation.

REFERENCES

Ahren, B., and Linquist, I., 1981, Effects of two cholecystokinin variants CCK-39 and CCK8S on basal and stimulated insulin secretion, *Acta Diabetol. Lat.* **18:**345–356.

Barrett, P., Kojima, I., Kojima, K., Zawalich, K., Isales, C., and Rasmussen, H., 1986a, Temporal patterns of protein phosphorylation after angiotensin II, A23187 and/or TPA in adrenal glomerulosa cells, *Biochem. J.* **238:**893–903.

Barrett, P., Kojima, I., Kojima, K., Zawalich, K., Isales, C., and Rasmussen, H., 1986b, Short-term memory in the calcium messenger system: Evidence for a sustained activation of C-kinase in adrenal glomerulosa cells, *Biochem. J.* **238:**904–912.

Berridge, M. J., 1984, Inositol trisphosphate and diacylglycerol as second messengers, *Biochem. J.* **220:**345–360.

Best, L., Dunlop, M., and Malaise, W. J., 1984, Phospholipid metabolism in pancreatic islets, *Experientia* **40:**1085–1091.

Cook, D. L., and Hales, C. N., 1984, Intracellular ATP directly blocks K^+ channels in pancreatic β-cells, *Nature* **311:**271–273.

Dunlop, M. E., and Larkins, R. G., 1985, Pancreatic islets synthesize phospholipid *de novo* from glucose via acyl-dihydroxyacetone phosphate, *Biochem. Biophys. Res. Commun.* **132:**467–473.

Gold, G., Gishizky, M. L., and Grodsky, G., 1982, Evidence that glucose "marks" β cells resulting in preferential release of newly synthesized insulin, *Science* **218:**56–58.

Grill, V., 1981, Time and dose dependencies for priming effect of glucose on insulin secretion, *Am. J. Physiol.* **240:**E24–E31.

Grill, V., and Rundfeldt, M., 1979, Effects of priming with D-glucose on insulin secretion from rat pancreatic islets: Increased responsiveness to other secretagogues, *Endocrinology* **105:**980–987.

Grill, V., Adamson, U., and Cerasi, E., 1978, Immediate and time-dependent effects of glucose on insulin release from rat pancreatic tissue, *J. Clin. Invest.* **61:**1034–1043.

Grill, V., Adamson, U., Rundfeldt, M., Andersson, S., and Cerasi, E., 1979, Glucose memory of pancreatic B and A_2 cells, *J. Clin. Invest.* **64:**700–707.

Grodsky, G. M., Curry, D., Landahl, H., and Bennett, L., 1969, Further studies on the dynamic aspects of insulin release *in vitro* with evidence for a two-compartmental storage system, *Acta Diabetol. Lat. [Suppl.]* **1:**554–579.

Hannum, Y., Loomis, C. R., and Bell, R. M., 1986, Protein kinase C activation in mixed micelles, *J. Biol. Chem.* **261:**7184–7190.

Hedeskov, C. J., 1980, Mechanism of glucose-induced insulin secretion, *Physiol. Rev.* **60:**442–506.

Henquin, J.-C., 1980, Tolbutamide stimulation and inhibition of insulin release: Studies of the underlying ionic mechanisms in isolated rat islets, *Diabetologia* **18:**151–160.

Henquin, J.-C., 1985, The interplay between cyclic AMP and ions in the stimulus–secretion coupling in pancreatic β-cells, *Arch. Int. Physiol. Biochim.* **93:**37–48.

Henquin, J.-C., and Meissner, H. P., 1984, The ionic, electrical, and secretory effects of endogenous cyclic adenosine monophosphate in mouse pancreatic β cells: Studies with forskolin, *Endocrinology* **115:**1125–1134.

Holst, J. J., Jensen, S. L., and Morley, J. S., 1980, Neural regulation of pancreatic hormone secretion by the C-terminal tetrapeptide of CCK, *Nature* **284:**33–38.

Kojima, I., Kojima, K., Kreutter, D., and Rasmussen, H., 1984, The temporal integration of the aldosterone secretory response to angiotensin II occurs via two intracellular pathways, *J. Biol. Chem.* **259:**14448–14457.

Kojima, I., Kojima, K., and Rasmussen, H., 1985, Role of calcium fluxes in the sustained phase of angiotensin II-mediated aldosterone secretion from adrenal cells, *J. Biol. Chem.* **260:**9177–9184.

May, W. J., Jr., Sayhoun, N., Wolf, M., and Cuatrecasas, P., 1985, Role of intracellular calcium mobilization in the regulation of protein kinase C-mediated membrane processes, *Nature* **317:**549–551.

Morgan, N. G., Rumford, G. M., and Montague, W., 1985, Studies on the role of inositol trisphosphate in the regulation of insulin secretion from isolated rat islets of Langerhans, *Biochem. J.* **228:**713–718.

Nishizuka, Y., 1984, The role of protein kinase C in cell surface signal transduction and tumour promotion, *Nature* **308:**693–698.

Pipeleers, D., 1984, Islet cell interactions with pancreatic β-cells, *Experientia* **40:**1114–1126.

Rasmussen, H., 1986, The calcium messenger system, *N. Engl. J. Med.* **314:**1094–1101, 1164–1170.

Rasmussen, H., and Barrett, P. Q., 1984, Calcium messenger system: An integrated view, *Physiol. Rev.* **64:**938–984.

Sakamoto, C., Otsuki, M., Ohki, A., Yuu, H., Maeda, M., Yamasaki, T., and Baba, S., 1982, Glucose-dependent insulinotropic action of cholecystokinin and caerulein in the isolated perfused rat pancreas, *Endocrinology* **110:**398–402.

Schuit, F. C., and Pipeleers, D. G., 1985, Regulation of adenosine 3′,5′-monophosphate levels in the pancreatic β cell, *Endocrinology* **117:**834–840.

Streb, H., Irvine, R. F., Berridge, M. J., and Schultz, I., 1983, Release of Ca^{2+} from a non-mitochondrial store in pancreatic acinar cells by inositol-1,4,5-trisphosphate, *Nature* **306:**67–69.

Sturgess, N. C., Ashford, M. L., Cook, D. L., and Hales, C. N., 1985, The sulphonylurea receptor may be an ATP-sensitive potassium channel, *Lancet* **2:**474–475.

Sturgess, N. C., Ashford, M. L., Carrington, C. A., and Hales, C. N., 1986, Single channel recordings of potassium currents in an insulin secreting cell line, *J. Endocrinol.* **109:**201–207.

Takai, Y., Kikkawa, U., Kaibuchi, K., and Nishizuka, Y., 1984, Membrane phospholipid metabolism and signal transduction for protein phosphorylation, *Adv. Cyclic Nucleotide Prot. Phosphor. Res.* **18:**119–158.

Tanigawa, K., Kuzuya, H., Imura, H., Taniguchi, H., Baba, S., Takai, Y., and Nishizuka, Y., 1982, Calcium-activated, phospholipid-dependent protein kinase in rat pancreas islets of Langerhans, *FEBS Lett.* **138:**183–187.

Turk, J., Wolf, B. A., and McDaniel, M. L., 1986, Glucose-induced accumulation of inositol trisphosphates in isolated pancreatic islets, *Biochem. J.* **237:**259–263.

Wolf, M., Cuatrecasas, P., and Sahyoun, N., 1985, Interaction of protein kinase C with membranes is regulated by Ca^{2+}, phorbol esters, and ATP, *J. Biol. Chem.* **260:**15718–15722.

Zawalich, W., and Diaz, V. A., 1986, Asperlicin antagonizes the stimulatory effects of cholecystokinin on isolated perifused islets, *Am. J. Physiol.* **252:**E370–E374.

Zawalich, W., Brown, C., and Rasmussen, H., 1983, Insulin secretion: Combined effect of phorbol ester and A23187, *Biochem. Biophys. Res. Commun.* **117:**448–455.

Zawalich, W., Zawalich, K., and Rasmussen, H., 1984, Insulin secretion combined tolbutamide, forskolin, and TPA mimic action of glucose, *Cell Calcium* **5:**551–558.

Zawalich, W., Cote, S. B., and Diaz, V. A., 1986, Influence of cholecystokinin on insulin output from isolated perifused pancreatic islets, *Endocrinology* **119:**616–621.

Zawalich, W., Takuwa, N., Takuwa, Y., Diaz, V. A., and Rasmussen, H., 1987, Mechanisms of action of cholecystokinin and glucose in rat pancreatic islets, *Diabetes* **36:**426–433.

2

An Increased Basal Calcium Hypothesis for Long-Term Potentiation of Transmitter Release in Bullfrog Sympathetic Ganglia

K. KUBA, E. KUMAMOTO, S. MINOTA, K. KOYANO, K. TANAKA, and S. TSUJI

1. INTRODUCTION

Long-term potentiation (LTP) of synaptic transmission, a basis for learning and memory (cf. Tsukahara, 1981), occurs in response to conditioning stimuli in various neuronal elements at both central and peripheral synapses. In bullfrog sympathetic ganglia, there are two types of long-term potentiation of transmitter release, one induced by conditional tetanic stimulation of the preganglionic nerve through a Ca^{2+}-dependent mechanism (presynaptic LTP, pre-LTP; Koyano *et al.*, 1985), and the other generated by the action of epinephrine through a cAMP-dependent mechanism (epinephrine-induced LTP, adr-LTP; Kuba *et al.*, 1981; Kuba and Kumamoto, 1986). We describe here novel mechanisms of these LTPs in which a rise in the basal level of the intracellular free Ca^{2+} ($[Ca^{2+}]_i$) in the presynaptic terminal plays an important role.

2. EXPERIMENTAL METHODS

Intracellular recordings were obtained from the isolated ninth or tenth paravertebral sympathetic ganglia of bullfrogs. Fast excitatory postsynaptic potentials (fast EPSPs) were recorded in a low-Ca^{2+}, high-Mg^{2+} solution, and their quantal content and quantal size were analyzed as measures of the relative amounts of transmitter release from the presynaptic terminals and of the postsynaptic sensitivity to transmitter (acetylcholine; ACh), respectively. Spontaneous miniature excitatory postsynaptic potentials (MEPSPs) were recorded in normal Ringer or high-K^+ solution. The presynaptic terminal spike was extracellularly recorded concomitantly with the fast EPSP with an electrode inserted into

K. KUBA, E. KUMAMOTO, S. MINOTA, K. KOYANO, K. TANAKA, and S. TSUJI • Department of Physiology, Saga Medical School, Saga 840-01, Japan.

the postganglionic neuron. Other experimental procedures are described elsewhere (Kuba and Kumamoto, 1986; Kumamoto and Kuba, 1986).

3. HOMOSYNAPTIC AND HETEROSYNAPTIC LTP

In more than two-thirds of ganglion cells studied, conditioning tetanic stimuli (33 Hz, 5–10 sec) to the preganglionic nerve produced a long-term potentiation of the amplitude, quantal content, and/or quantal size of the fast EPSP, which became conspicuous after the subsidence of posttetanic potentiation, a synaptic potentiation much briefer than LTP (Fig. 1B). The long-term potentiation of quantal content (pre-LTP) decayed single-exponentially with a time constant of 39 min or lasted for more than 2 hr (Fig. 2B). The magnitude of the pre-LTP (10–50% of the control quantal content) increases with an increase in the duration of tetanus (33 Hz, 2–10 sec), but the duration of pre-LTP appeared to be independent of the tetanus duration. The pre-LTP strongly depended on Ca^{2+} influx during tetanus. A conditioning tetanus given in a Ca^{2+}-free solution did not produce pre-LTP (Fig. 2C), whereas a tetanus applied in normal Ringer resulted in a large pre-LTP (Fig. 2B). An interesting feature of pre-LTP is that it was enhanced by reduced temperature (Koyano *et al.*, 1985).

Treatment of the ganglion with epinephrine (0.1–100 μM) for more than 10 min resulted in long-term potentiation of the quantal content of the fast EPSP (50–75% of the control) with no change in quantal size. This adr-LTP lasted more than 3 hr with little tendency to decay (Fig. 1A). A similar potentiation was consistently induced by isoproterenol (10 μM) and, in some cells, by dopamine (10 μM). The generation of adr-LTP (induced by 10 μM epinephrine) was blocked by a β antagonist, propranolol (1 μM), but not by an α antagonist, phenoxybenzamine (1 μM), indicating that a β adrenoceptor mediates adr-LTP. Endogenous cAMP and a subsequent metabolic process seem to be involved in adr-LTP. Dibutyryl cAMP (0.8–1 mM) or cAMP (4 mM) induced LTP of quantal content like adr-LTP, although the former produced a larger LTP (Fig. 2A). Similarly, the phosphodiesterase inhibitors isobutylmethylxanthine (IBMX, 10 μM) and

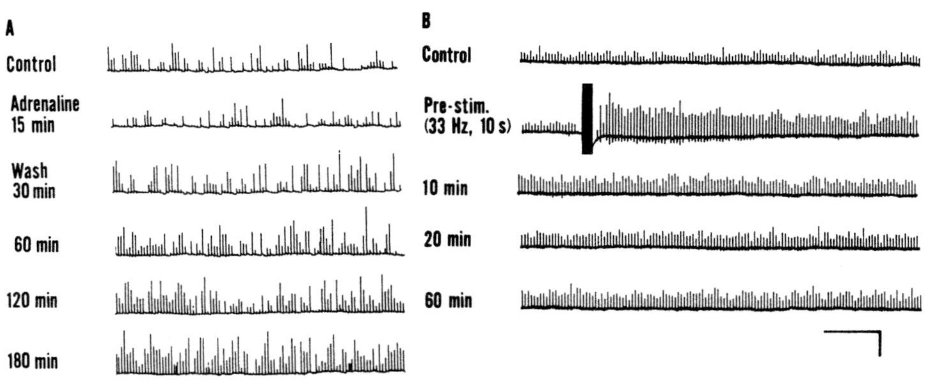

FIGURE 1. Long-term potentiations of the fast EPSP induced by the treatment with epinephrine (10 μM) for 30 min (A) and by a tetanic stimulation (33 Hz, 10 sec) to the preganglionic nerve (B). Fast EPSPS induced in a low-Ca^{2+}, high-Mg^{2+} solution every 3 sec were recorded on a pen-writing recorder (flat response up to 100 Hz). Calibrations are 1 min and 4 mV (A) and 5 mV (B).

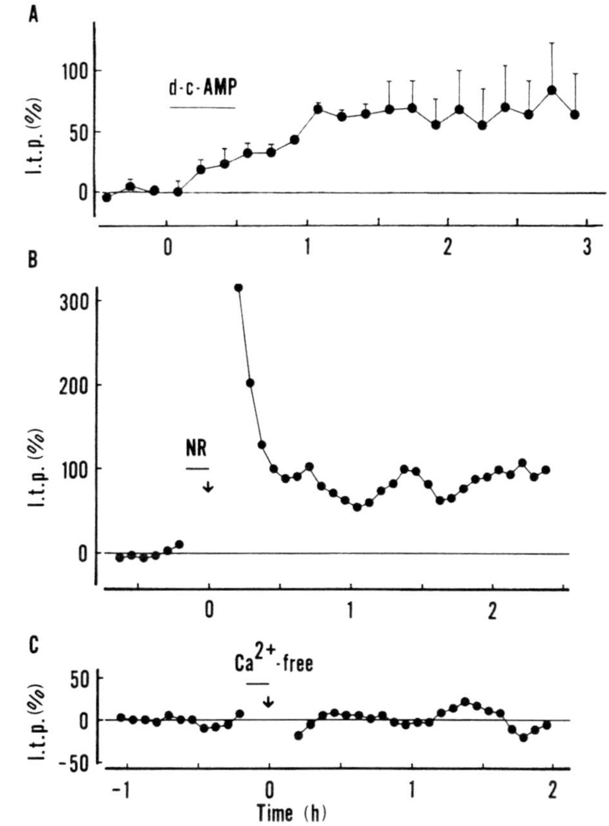

FIGURE 2. (A) LTP of quantal content induced by the exposure to dibutyryl cAMP (1 mM) for 30 min. (B,C) Pre-LTP produced by the tetanic stimulation applied in normal Ringer (B) and the lack of its generation by the tetanus given in a Ca^{2+}-free solution (C). The low-Ca^{2+}, high-Mg^{2+} solution was changed to a low-Ca^{2+}, high-Mg^{2+} solution containing dibutyryl cAMP (1 mM, A), normal Ringer (B), or a Ca^{2+}-free solution (C) during a period indicated by a horizontal bar. A tetanus (33 Hz, 10 sec) was applied at the downward arrow. Each point in A is the mean and S.E.M. of the data obtained from six cells, and data in B and C are obtained from a single cell. All the data are expressed as percentages relative to the control.

caffeine (2 mM) and an irreversible activator of adenylate cyclase (by ADP-ribosylation of a GTP-binding protein), cholera toxin (2 μg/ml), applied for 20–30 min all generated LTP of quantal content similar to adr-LTP. On the other hand, no LTP was seen when the ganglion was treated with AMP (4 mM) or adenosine (4 mM). In contrast to pre-LTP, adr-LTP was suppressed by lowering temperature (Kumamoto and Kuba, 1983).

There appear to be two regulatory mechanisms of adr-LTP. One is a mechanism involving endogenous cGMP. When dibutyryl cGMP (100 μM) was applied together with epinephrine (10 μM; Kuba and Kumamoto, 1986) or dibutyryl cAMP (1 mM; Kumamoto and Kuba, 1987), there was no potentiation of quantal content. However, when dibutyryl cGMP was given after the end of treatment with epinephrine, LTP of quantal content was sometimes seen. These results indicate that the inhibitory effect of dibutyryl cGMP is exerted on a process involving a rise in cAMP and a subsequent activation of cAMP-dependent protein kinase in the adr-LTP mechanism. Another regulatory mechanism involves desensitization of the β-adrenoceptor–adenylate cyclase system. An increase in the duration of treatment with epinephrine (10 μM) to a period longer than 20 min was not effective in augmenting the magnitude of adr-LTP (max. 75%), whereas repetition of a short-term exposure (20 min) to epinephrine with an interval longer than 20 min resulted in the summation of adr-LTPs (120%; Kumamoto and Kuba, 1983). On the other hand, the magnitude of LTP induced by dibutyryl cAMP (1 mM) increased with the duration of exposure. Thus, the desensitization appears to occur at a step before the production of cAMP in a cascade mechanism, presumably at the β-adrenoceptor.

4. MECHANISMS DIRECTLY INVOLVED IN THE POTENTIATION OF TRANSMITTER RELEASE

There are at least five possible mechanisms for an increase in the impulse-induced release of transmitter during adr-LTP and pre-LTP: (1) an increased Ca^{2+} influx caused by the broadening of the action potential as a result of blockade of K^+ conductance, (2) an enhancement of the Ca^{2+} conductance increase during a presynaptic spike, (3) an elevation of the basal level of $[Ca^{2+}]_i$ in the presynaptic terminals, (4) enhanced efficacy of excitation–secretion coupling, and (5) an increased rate of transmitter synthesis. These possibilities may be examined by observing changes in the following three phenomena regarding transmitter release during the generation of pre-LTP or adr-LTP: (1) rate of spontaneous release, (2) synaptic delay, and (3) short-term facilitation.

4.1. Spontaneous Transmitter Release

The frequency of miniature EPSPs would reflect the basal level of $[Ca^{2+}]_i$ in the presynaptic terminals, the efficacy of the transmitter release mechanism, and/or the amount of transmitter available for release. Epinephrine (2.5–160 μM) applied for 10–20 min potentiated miniature EPSP frequency by 1.5–3 times for more than 2 hr (Kuba and Kumamoto, 1986). Likewise, tetanic stimuli given to the preganglionic nerve (33 Hz, 5–30 sec) caused an increase in the frequency of miniature EPSPs that lasted over several tens of minutes with a slowly decaying time course (unpublished observations). These results indicate an increase in the basal level of $[Ca^{2+}]_i$ in the terminal (mechanism 3), the efficacy of the transmitter release mechanism (mechanism 4), or the amount of transmitter available for release (mechanism 5) during adr-LTP and pre-LTP.

4.2. Synaptic Delay

A cAMP-dependent mechanism activated by epinephrine may block K^+ channels at the presynaptic terminals and thereby cause broadening of the action potential, which should increase Ca^{2+} influx and enhance transmitter release. This was indeed observed in the potentiation of transmitter release at an *Aplysia* neuronal synapse (Kandel *et al.*, 1983). Under this condition, the synaptic delay would be lengthened because of elongation of a presynaptic spike (Fig. 3A; Kumamoto and Kuba, 1985). This possibility, however, is unlikely for the adr-LTP. The synaptic delay of the fast EPSP, measured from the positive peak of a presynaptic spike to the onset of the fast EPSP, remained unchanged throughout adr-LTP (Fig. 3B; Kumamoto and Kuba, 1986). Moreover, adr-LTP was generated in the presence of TEA (100 μM), which increased the quantal content about three times.

The hypothesis of K^+ channel blockade might be applicable to pre-LTP. Mallart (1985) reported the existence of a Ca^{2+}-dependent K^+ conductance at the motor nerve terminal that is inactivated for a period of 100 msec after its activation during a presynaptic impulse, presumably by a residual rise in $[Ca^{2+}]_i$ in the terminal (cf. Tokimasa, 1985). Thus, if the basal $[Ca^{2+}]_i$ in the presynaptic terminals rises continuously throughout the pre-LTP, it partially (if not entirely) inactivates the K^+ channel and potentiates transmitter release by an enhanced Ca^{2+} influx during an impulse. This possibility, however, can be ruled out for the mechanism of pre-LTP, since pre-LTP persists in the presence of tetraethylammonium (TEA; 100 μM), which blocks both the delayed rectifier K^+ channel and the Ca^{2+}-dependent K^+ channel at the presynaptic terminals (Mallart, 1985).

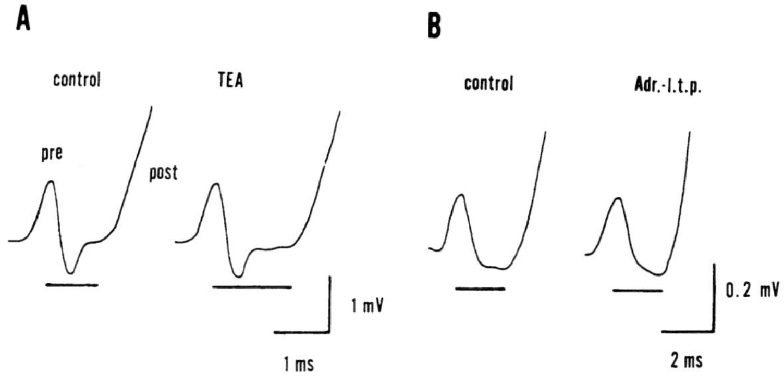

FIGURE 3. Elongation of synaptic delay by the blockade of the K^+ channel by TEA (100 μM) at the presynaptic terminal (A) and the unchanged synaptic delay during adr-LTP (B).

4.3. Short-Term Facilitation

A short-term facilitation of the fast EPSP (lasting several hundred milliseconds) induced by paired stimuli may be used to examine a possible change in the basal level of $[Ca^{2+}]_i$ in the terminal during LTP. This is based on the residual Ca^{2+} hypothesis, which is supported by substantial evidence in many preparations. The Ca^{2+} remaining in the presynaptic terminals after the first impulse would act to increase the total Ca^{2+} for the subsequent release. This effect would be amplified by the cooperative action of Ca^{2+} on the release mechanism, as shown below. The amount of transmitter release induced by the first impulse (m_1) in a low-Ca^{2+}, high-Mg^{2+} solution would approximately be expressed by

$$m_1 = k(\mathrm{Ca_I} + \mathrm{Ca_b})^n/K \qquad (1)$$

where $\mathrm{Ca_I}$ is a rise in $[Ca^{2+}]_i$ resulting from Ca^{2+} influx by an impulse, $\mathrm{Ca_b}$ is the basal $[Ca^{2+}]_i$, K is the dissociation constant for the Ca^{2+} action on a site essential for release in the terminal, n is the Hill coefficient, and k is the overall efficacy of excitation–secretion coupling. The amount of transmitter release by the second impulse (m_2) would be facilitated by the existence of residual Ca^{2+} ($\mathrm{Ca_r}$), which is added to $\mathrm{Ca_I}$ and $\mathrm{Ca_b}$ (Fig. 4A). Thus, assuming $\mathrm{Ca_I}$, k, and K remain constant for the second impulse,

$$m_2 = k(\mathrm{Ca_I} + \mathrm{Ca_b} + \mathrm{Ca_r})^n/K \qquad (2)$$

Accordingly, the magnitude of short-term facilitation (F_{control}) would be,

$$F_{\mathrm{control}} = (\mathrm{Ca_I} + \mathrm{Ca_b} + \mathrm{Ca_r})^n/(\mathrm{Ca_I} + \mathrm{Ca_b})^n - 1 \qquad (3)$$

If the basal level of $[Ca^{2+}]_i$ in the terminal is elevated to a higher level ($\mathrm{Ca_b'}$) during a LTP (Fig. 4B), the magnitude of short-term facilitation ($F_{\mathrm{LTP(basal)}}$) would alter to

$$F_{\mathrm{LTP(basal)}} = (\mathrm{Ca_I} + \mathrm{Ca_b'} + \mathrm{Ca_r})^n/(\mathrm{Ca_I} + \mathrm{Ca_b'})^n - 1 \qquad (4)$$

If $\mathrm{Ca_I}$ and $\mathrm{Ca_r}$ remain constant during LTP, $F_{\mathrm{LTP(basal)}}$ is obviously smaller than F_{control}, since the first term in the right side of eq. 4 is smaller than that of eq. 3 for $\mathrm{Ca_b'} > \mathrm{Ca_b}$.

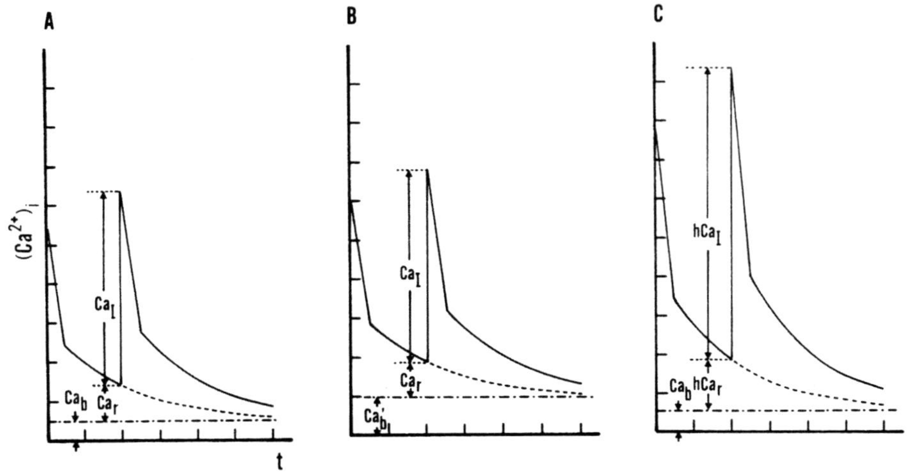

FIGURE 4. Schematic diagrams of alterations in the intraterminal Ca^{2+} during the course of short-term facilitation based on the residual Ca^{2+} hypothesis. It is assumed that the increased $[Ca^{2+}]_i$ decays at least in two phases: an initial rapid decline, which ceases completely within the refractory period of a spike, and a late single-exponential decrease. In these diagrams, the time course of the initial phase was assumed to be a simple linear function, although the exact function is not known. (A) Normal condition. (B) Increased basal $[Ca^{2+}]_i$. (C) Increased Ca^{2+} influx during an impulse. Scales in both ordinates and abscissas are arbitrary. Horizontal interrupted lines indicate the basal level of $[Ca^{2+}]_i$.

The predicted decrease in facilitation was 38–14% for $n = 1$–4 (Kumamoto and Kuba, 1986). The assumption of a constant Ca_I at an increased basal $[Ca^{2+}]_i$ would little affect the predictions because the resultant decrease in the electromotive force for Ca^{2+} across the terminal membrane would be too small to affect Ca^{2+} influx for a large preexisting Ca^{2+} gradient. (A twofold increase in Ca_b would cause only a few percent decrease in Ca^{2+} influx with a resting potential of -80 mV and the Ca^{2+} gradient of 10^5.)

On the other hand, if Ca^{2+} influx during each impulse for producing short-term facilitation is increased during adr-LTP or pre-LTP with the basal $[Ca^{2+}]_i$ being constant, the residual $[Ca^{2+}]_i$ would also increase proportionately (Fig. 4C). Accordingly, the magnitude of short-term facilitation would be expressed by,

$$F_{\mathrm{LTP(influx)}} = (hCa_I + Ca_b + hCa_r)^n/(hCa_I + Ca_b)^n - 1$$
$$= (Ca_I + Ca_b/h + Ca_r)^n/(Ca_I + Ca_b/h)^n - 1 \quad (5)$$

where h is a fraction of the increase in Ca^{2+} influx. Thus, facilitation will increase under this condition. However, the extent of the increase is expected to be small (2–8% for $n = 1$–4; see Kumamoto and Kuba, 1986), since a difference between Ca_b in eq. 3 and Ca_b/h in eq. 5 would be too small (when compared to Ca_I) to produce a large difference between $F_{\mathrm{(control)}}$ and $F_{\mathrm{LTP(influx)}}$. Furthermore, it is obvious that an increase in the efficacy of transmitter release or in the amount of transmitter available for release (both of which are included in a parameter k in eq. 1 or 2 does not affect the magnitude of short-term facilitation for the fast EPSPs of low quantal content, since the facilitation is independent of this parameter (see eq. 3).

Before these predictions for LTP are tested, this theory of analysis may be evaluated in several ways. First, the time course of short-term facilitation of the fast EPSPs induced by paired as well as triple or quadruple stimuli can be reconstructed by the equation derived from a more elaborate mathematical treatment of the residual Ca^{2+} hypothesis (Kumamoto and Kuba, 1986). Second, the residual Ca^{2+} hypothesis for the short-term facilitation predicts that an increase in Ca^{2+} influx during an impulse would affect only slightly the magnitude of short-term facilitation when the basal $[Ca^{2+}]_i$ is low. This was in fact observed. When the quantal content of the fast EPSP was increased by TEA (100 μM), undoubtedly through an increased Ca^{2+} influx during the presynaptic impulse, the magnitude of short-term facilitation remained unchanged (Fig. 5A). Third, the rate of an increase in miniature EPSP frequency after an impulse agreed well with that predicted by the residual Ca^{2+} hypothesis. Thus, the residual Ca^{2+} hypothesis seems to be valid for short-term facilitation in the bullfrog sympathetic ganglion.

Figure 5B shows a change in short-term facilitation during the course of adr-LTP. Short-term facilitation was significantly depressed during adr-LTP. Likewise, short-term facilitation was decreased in magnitude during pre-LTP (unpublished observations). These results are consistent with the idea that the basal level of $[Ca^{2+}]_i$ in the terminal is increased during the course of both adr-LTP and pre-LTP (mechanism 3) but not with the idea that an increase in the Ca^{2+} influx during an impulse (mechanisms 1 and 2), enhanced efficacy of transmitter release mechanism (mechanism 4), or increased synthesis of transmitter (mechanism 5) occurs during the generation of adr-LTP as well as pre-LTP.

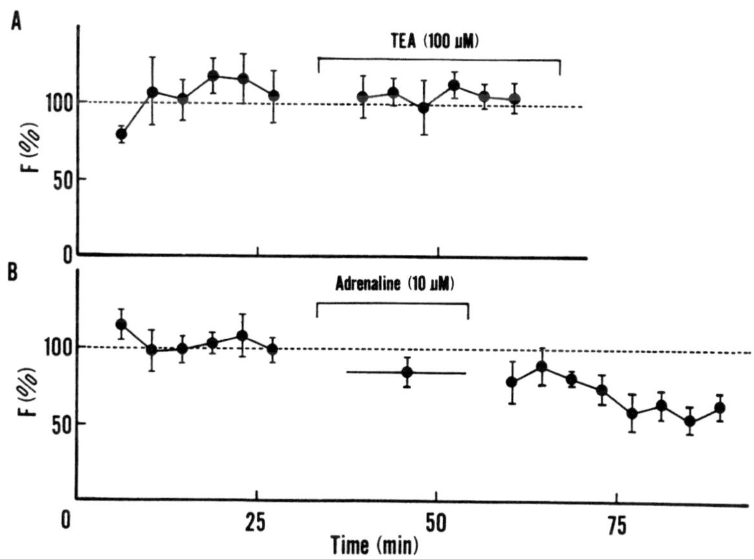

FIGURE 5. Time courses of the effects of TEA (100 μM, A) and epinephrine (10 μM, B) on the short-term facilitation [*F*(%)] of the fast EPSP induced at 50-msec intervals. Either TEA or epinephrine was added to the perfusing solution during the period shown by a horizontal bar. Each point, expressed as a percentage relative to control, represents the mean and S.E.M. of facilitations obtained from four (A) and six (B) cells and treated by the moving average method. The value in the presence of epinephrine indicates the mean and S.E.M. of data obtained during the period shown by the horizontal bar that were not treated by the moving average method.

5. MAINTENANCE OF POTENTIATION OF TRANSMITTER RELEASE

The experimental results and related discussions in the foregoing section suggest that a sustained rise in the basal $[Ca^{2+}]_i$ in the presynaptic terminals is a mechanism directly responsible for the enhanced release of transmitter during adr-LTP and pre-LTP. Then, a question arises how this elevated $[Ca^{2+}]_i$ is caused and maintained during both types of LTP.

The generation of LTPs of quantal content by cAMP, its analogue, or agents related to cAMP metabolism and the suppression of adr-LTP by lowering temperature suggest the following mechanism, which would lead to a rise in the basal $[Ca^{2+}]_i$ in the terminals (Fig. 6). The binding of epinephrine (diffused from either small intensely fluorescent cells or circulating blood) to a β-adrenoceptor at the terminal membrane activates adenylate cyclase through the activation of a GTP-binding protein. The resultant rise in cAMP in the terminals activates cAMP-dependent protein kinase, which phosphorylates a specific protein ether directly or indirectly involved in the active or passive transport mechanisms of Ca^{2+} in the terminals.

A constant magnitude of adr-LTP for a certain range of the extracellular Ca^{2+} concentration (which would indicate parallel increases in both Ca^{2+} influx during an impulse and the basal $[Ca^{2+}]_i$ by raising the extracellular Ca^{2+}) suggests that the increased basal $[Ca^{2+}]_i$ during adr-LTP is derived from an increase in resting Ca^{2+} influx. Thus, an increased basal $[Ca^{2+}]_i$ may arise from the suppression of active Ca^{2+} extrusion at, or the increase in the resting Ca^{2+} permeability of, the terminal membrane (cf. Kumamoto and Kuba, 1986), although a role of Ca^{2+}-storing organelles is not ruled out. The long-lasting nature of adr-LTP may be ascribed to the sustained phosphorylation but not to a sustained rise in cAMP concentration in the terminals, since a brief application of cAMP or dibutyryl cAMP produced a sustained potentiation of transmitter release even under

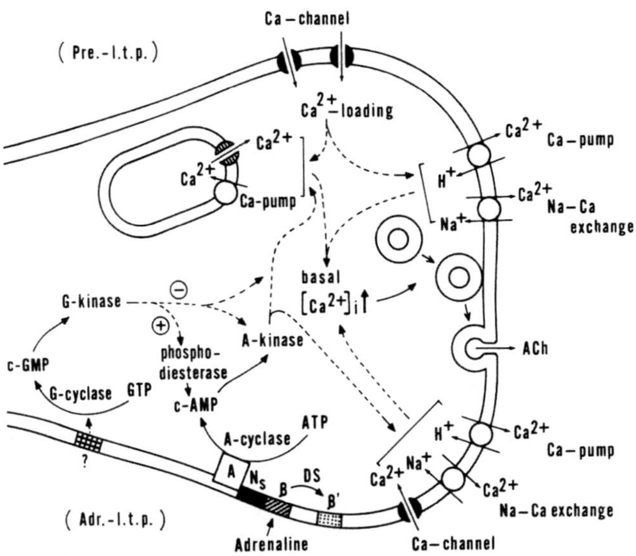

FIGURE 6. A hypothetical scheme for the mechanisms of pre-LTP and adr-LTP. See text for detailed explanation. β, β-adrenoceptor; Ns, GTP-binding protein for the activation of adenylate cyclase; A-cyclase, adenylate cyclase; G-cyclase, guanylate cyclase; A-kinase, cAMP-dependent protein kinase; DS, desensitization of β-adrenoceptor.

conditions in which the phosphodiesterase is active (Kuba and Kumamoto, 1986). This cascade mechanism of adr-LTP is regulated by two processes, one involving endogenous cGMP and the desensitization of β-adrenoceptors.

In contrast to adr-LTP, pre-LTP was enhanced by lowering the temperature. This finding suggests that the mechanism of the increased basal $[Ca^{2+}]_i$ during pre-LTP depends less on the metabolic energy, which is inconsistent with a protein phosphorylation mechanism. Alternatively, the enhancement of pre-LTP at low temperature could favor the possible involvement in the pre-LTP mechanism of the endoplasmic reticulum, whose Ca^{2+} transport systems are sensitive to lowering temperature (Kuba, 1980; Brattin and Waller, 1983). Thus, one of the most likely explanations would be that a massive loading of Ca^{2+} into the presynaptic terminals sets the basal $[Ca^{2+}]_i$ at a higher level by increasing the resting Ca^{2+} permeability or suppressing the Ca^{2+} transport system of the Ca^{2+}-storing organelle (Fig. 6). Such a high level of $[Ca^{2+}]_i$ may be maintained for a long time, thereby producing pre-LTP.

6. CONCLUDING REMARKS

Long-term potentiation has been found at many synapses of both vertebrate and invertebrate nervous systems (Bliss and Lømo, 1973; Yamamoto and Chujo, 1978; Andersen *et al.*, 1980; Brown and McAfee, 1982; Alkon, 1984). Although the presynaptic mechanism of LTP in the vertebrate central nervous system has not yet been clarified, the mechanism at invertebrate synapses has been well analyzed at the molecular level. For instance, both short-term and long-term potentiation of synaptic transmission involved in the gill-withdrawal reflex in *Aplysia* are explained by the blockade of a serotonin-sensitive K^+ channel through its phosphorylation by cAMP-dependent protein kinase at the presynaptic terminals (Kandel *et al.*, 1983). A similar phosphorylation-induced modulation of a K^+ channel involved in the associative learning of a phototaxic response was reported in another molusc (*Hemissenda*) (Alkon, 1984). Furthermore, many ion channels of invertebrate neurons are known to be common to those of vertebrate neurons. It is therefore expected that a similar channel-modulation mechanism would operate in the LTPs of vertebrate neurons.

Our findings, however, suggest different mechanisms for LTPs in bullfrog sympathetic neurons. In this context, the mechanism of increased basal $[Ca^{2+}]_i$ in the presynaptic terminal for adr-LTP and pre-LTP in sympathetic ganglia is quite novel. Furthermore, it is a mechanism consuming less metabolic energy than the channel modulation mechanism. A slight modification of the Ca^{2+}-buffering system by suppressing one of the active Ca^{2+} transport systems (leading to the reduction of ATP consumption) or increasing the permeability of a passive Ca^{2+} transport system produces a sustained potentiation of synaptic efficacy, whereas the modulation of K^+ channels induces a large Ca^{2+} influx during each impulse, which must be pumped out for the next synaptic transmission. The present findings also indicate the diversity of LTP mechanisms. This suggests that a mechanism clarified in the invertebrate can not necessarily be extrapolated to the LTPs of vertebrate neurons.

REFERENCES

Alkon, D. L., 1984, Calcium-mediated reduction of ionic currents: A biophysical memory trace, *Science* **226:**1037–1045.

Andersen, P., Sundberg, S. H., Sveen, O., Swann, J. W., and Wigström, H., 1980, Possible mechanisms for long-lasting potentiation of synaptic transmission in hippocampal slices from guinea-pigs, *J. Physiol. (Lond.)* **302:**463–482.

Bliss, T. V. P., and Lømo, T., 1973, Long-lasting potentiation of synaptic transmission in the dentate area of the anaesthetized rabbit following stimulation of the perforant path, *J. Physiol. (Lond.)* **232:**331–356.

Brattin, W. J., and Waller, R. L., 1983, Calcium inhibition of rat liver microsomal calcium-dependent ATPase, *J. Biol. Chem.* **258:**6724–6729.

Brown, T. H., and McAfee, D. A., 1982, Long-term synaptic potentiation in the superior cervical ganglion, *Science* **215:**1411–1413.

Kandel, E. R., Abrams, T., Bernier, L., Carew, T. J., Hawkins, R. D., and Schwartz, J. H., 1983, Classical conditioning and sensitization share aspects of the same molecular cascade in *Aplysia, Cold Spring Harbor Symp. Quant. Biol.* **48:**821–830.

Koyano, K., Kuba, K., and Minota, S., 1985, Long-term potentiation of transmitter release induced by repetitive presynaptic activities in bullfrog sympathetic ganglia, *J. Physiol. (Lond.)* **359:**219–233.

Kuba, K., 1980, Release of calcium ions linked to the activation of potassium conductance in a caffeine-treated sympathetic neurone, *J. Physiol. (Lond.)* **298:**251–269.

Kuba, K., and Kumamoto, E., 1986, Long-term potentiation of transmitter release induced by adrenaline in bullfrog sympathetic ganglia, *J. Physiol. (Lond.)* **374:**515–530.

Kuba, K., Kato, E., Kumamoto, E., Koketsu, K., and Hirai, K., 1981, Sustained potentiation of transmitter release by adrenaline and dibutyryl cyclic AMP in sympathetic ganglia, *Nature* **291:**654–656.

Kumamoto, E., and Kuba, K., 1983, Independence of presynaptic bimodal actions of adrenaline in sympathetic ganglia, *Brain Res.* **265:**344–347.

Kumamoto, E., and Kuba, K., 1985, Effects of K^+-channel blockers on transmitter release in bullfrog sympathetic ganglia, *J. Pharmacol. Exp. Ther.* **235:**241–247.

Kumamoto, E., and Kuba, K., 1986, Mechanism of long-term potentiation of transmitter release induced by adrenaline in bullfrog sympathetic ganglia, *J. Gen. Physiol.* **87:**775–799.

Kumamoto, E., and Kuba, K., 1987, Mechanisms regulating the adrenaline-induced long-term potentiation in bullfrog sympathetic ganglia, *Pflügers Arch.* **408:**573–577.

Mallart, A., 1985, A calcium-activated potassium current in motor nerve terminals of the mouse, *J. Physiol. (Lond.)* **368:**577–591.

Tokimasa, T., 1985, Intracellular Ca^{2+}-ions inactivates K^+-current in bullfrog sympathetic neurons, *Brain Res.* **337:**386–391.

Tsukahara, N., 1981, Synaptic plasticiy in the mammalian central nervous system, *Annu. Rev. Neurosci.* **4:**351–379.

Yamamoto, C., and Chujo, T., 1978, Long-term potentiation in thin hippocampal sections studied by intracellular and extracellular recordings, *Exp. Neurol.* **58:**242–250.

3

Sprouting as a Basis for Classical Conditioning in the Cat

FUJIO MURAKAMI, YOICHI ODA, and NAKAAKIRA TSUKAHARA

1. INTRODUCTION

Plasticity of synaptic transmission is thought to underlie behavioral plasticity in invertebrates. In mammals, circumstantial evidence has been accumulating to suggest that long-term potentiation of synaptic transmission contributes to behavioral plasticity (e.g., Morris *et al.*, 1986). Although long-term potentiation lasts for many days, some other mechanisms should be considered to explain behavioral modification that is retained for a much longer period.

Formation of new synaptic connections has been determined to occur in various loci of the mammalian central nervous system following various surgical procedures (Tsukahara, 1981). Formation of new synapses could be a good candidate for the cellular mechanism of long-term behavioral modification because (1) it has been established that the formation of new synapses occurs in various loci in the mammalian central nervous system, (2) it often occurs in parallel with behavioral modification (e.g., Tsukahara, 1981), and (3) the newly formed synapses are retained for more than several months.

We have obtained evidence to suggest, through two different and independent studies, that formation of new corticorubral synapses underlies classical conditioning in cats.

2. CLASSICAL CONDITIONING

Tsukahara, Oda, and their co-workers found that classical conditioning was established by pairing electrical shocks to the cerebral peduncle (CP) and forelimb skin of the cat in close temporal association (Tsukahara *et al.*, 1981). The conditioned stimulus (CS) applied to the cerebral peduncle was ineffective at the beginning of the training period. However, after 7–10 days' training period in which 100 CSs were paired every day with unconditioned stimuli (US) applied to the forelimb skin, the cat learned to flex its forelimb

FUJIO MURAKAMI, YOICHI ODA, and NAKAAKIRA TSUKAHARA • Department of Biophysical Engineering, Faculty of Engineering Science, Osaka University, Toyonaka, Osaka 560, Japan.

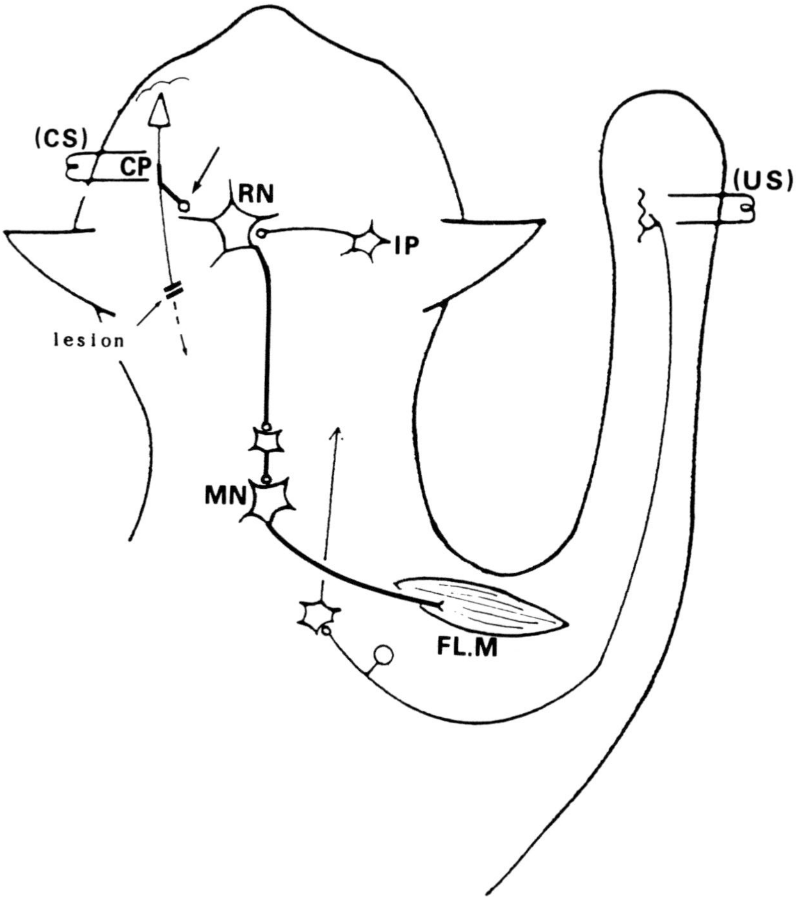

FIGURE 1. Schematic diagram showing the experimental diagram and the neuronal circuit involved in classical conditioning of the cat. RN, red nucleus; IP, interpositus nucleus; CP, cerebral peduncle; FL.M, flexor muscle; MN, motor neuron; CS, conditioned stimulus; US, unconditioned stimulus. All conditioned cats underwent pyramidotomy.

in response to the CS. They attributed the behavioral modification to some change in synaptic transmission at corticorubral synapses because (1) the pyramidal tract was lesioned below the level of the red nucleus (RN) to eliminate the participation of corticofugal descending systems other than the corticorubrospinal system for limb flexion and (2) no change in excitability was found along the interpositorubrospinal pathway after conditioning (see Fig. 1).

3. UNIT ACTIVITY OF RED NUCLEUS NEURONS

Analysis of rubral unit activity supported this view. Extracellular unit analysis showed that RN neurons tended to fire with higher frequency to stimulation of the CP in conditioned cats than in control ones, whereas it did not show any difference in responsiveness to stimulation of IP (Oda *et al.*, 1981).

4. DOES SPROUTING UNDERLIE THE BEHAVIORAL MODIFICATION?

Corticorubral synapses have been shown to form new collateral sprouts after partial denervation (Murakami *et al.*, 1977, 1982; Tsukahara *et al.*, 1974, 1975) or peripheral nerve cross reinnervation (Fujito *et al.*, 1982; Murakami *et al.*, 1984; Tsukahara and Fujito, 1976; Tsukahara *et al.*, 1982). The new synapses formed by the collateral sprouts were functionally more effective than the normally existing corticorubral synapses because they were situated on the soma–dendritic membrane much closer to the soma than normally existing corticorubral synapses. Therefore, it is possible that sprouting underlies this conditioning in the cat. Two different approaches were used to test the possibility of collateral sprouting in the conditioned cat.

4.1. Electrophysiological Study

In the two previous studies, new fast-rising corticorubral EPSPs were observed following lesion of the interpositus nucleus (IP) or peripheral nerve cross innervation. A preliminary study indicated that this holds true for animals that underwent classical conditioning (Tsukahara and Oda, 1981).

We recorded corticorubral EPSPs in conditioned cats and compared them with those recorded in control animals. Two groups of control animals were used; one consisted of normal (intact) cats, and another consisted of cats with pyramidal tract lesions. As exemplified in Fig. 2A, many of the EPSPs induced in conditioned cats by stimulation of the CP had rapidly rising components, which were uncommon in animals of control

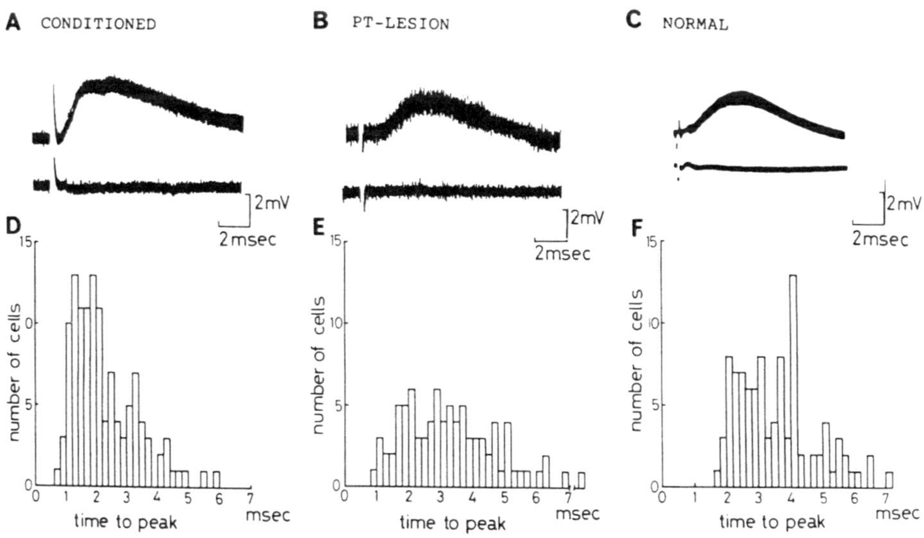

FIGURE 2. Corticorubral EPSPs in conditioned and control cats. (A–C) Specimen records of the EPSPs elicited by stimulation of the cerebral peduncle of a conditioned cat (A), a pyramidal-tract-lesioned cat (B), and a normal cat (C). Lower records show extracellular recordings. (D–F) Frequency histograms of the time to peak of the corticorubral EPSPs for conditioned cats (D), pyramidotomized cats (E), and normal cats (F).C and F from Tsukahara et al., 1975.

groups (Fig. 2B,C). Figure 2D–F shows frequency histograms of the time to peak of the corticorubral EPSPs in conditioned, pyramidal-tract-lesioned and intact animals, respectively. The time to peak of the corticorubral EPSP was significantly shorter than that with pyramidal-tract-lesioned or normal cats. The most plausible explanation for this change in the time to peak of the EPSP is that, as in the case of IP lesion or peripheral nerve cross reinnervation, new synapses were formed on the dendrites proximal to the recording site, i.e., soma.

Although similar changes could be induced by reduction of the dendritic electrotonic length, estimation of the electrotonic length revealed that it did not change after conditioning. Furthermore, one can see that the EPSPs shown in Fig. 2A have a second peak that approximately coincides with the peaks of the EPSPs of the control animals. The persistence of such a second peak, which was common in conditioned cats, suggests that the electrotonic length is as long as that of the control animals.

4.2. Morphological Study

The possibility of the formation of new synapses by conditioning was tested by a different and independent study. We analyzed the loci of the corticorubral synapses on the soma dendritic membrane of RN neurons. Since the diameter of dendrites of RN cells decreases approximately monotonically as a function of the distance from the soma, the diameter of the dendrites of RN cells can be used as a measure of the distance from the soma.

Therefore, we measured the diameters of the RN cell dendrites that received a corticofugal input. Corticorubral synaptic endings were identified by the degeneration method; i.e., ablation of the sensorimotor cortex a few days before tissue fixation elevated the density of synaptic endings. We identified the red nucleus by retrograde labeling with HRP, which was injected into the spinal cord. The area of the red nucleus that contained an HRP-labeled RN cell was dissected out, osmicated, and embedded in Epon. Then, ultrathin sections were made, and electron micrographs of the degenerating terminals synapsing on dendrites of RN cells were taken.

Figure 3A shows the relationship between the number of degenerating terminals on the RN neurons and the diameter of the dendrite contacted by them in the conditioned cats. The dendritic diameter was measured from the cross section of the dendrite that was contacted by the degenerating terminals. Histograms showing similar relationships in animals from the two control groups are shown in Fig. 3B and C. The first control group consisted of animals that underwent pyramidal tract lesion alone. There were few degenerating terminals caused by the pyramidal tract lesion when we observed the tissue, i.e., about 1 month after pyramidal tract lesion. The second control group consisted of intact animals.

One can see that a larger proportion of corticofugal axonal endings form synapses on the dendrites with large diameter and somata in conditioned cats than in control animals. This difference in the distribution of corticorubral synapses on the dendrites suggests that new synapses were formed on the proximal dendrites and somata of RN cells by conditioning of the cat.

Thus, the results of the physiological as well as the morphological studies suggest that new corticorubral synapses were formed on the proximal dendrites and somata in cats that acquired classical conditioning. Taken together with the results of the behavioral analysis, it seems likely that sprouting of corticorubral synapses is the cellular mechanism

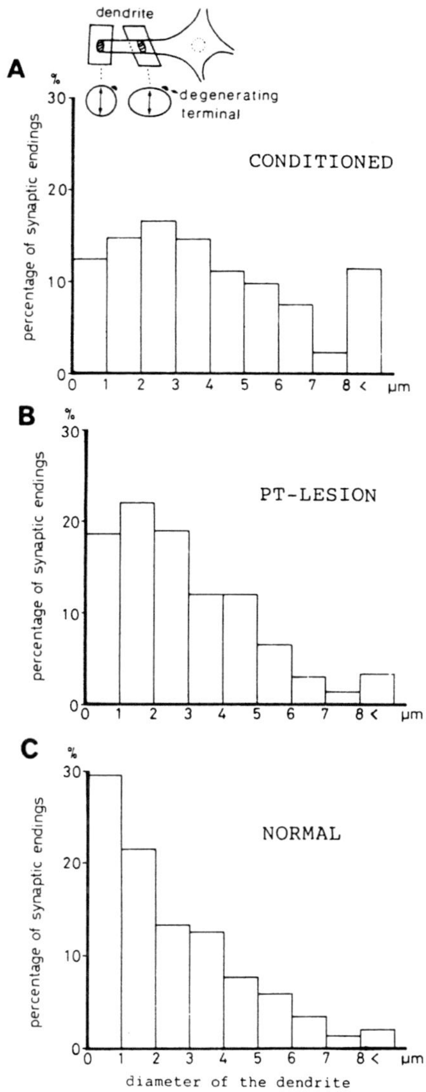

FIGURE 3. The relationship between the number of degenerating corticorubral terminals and the diameter of the dendrites. (A) The distribution of degenerating terminals in conditioned cats. Ordinate, number of degenerating synaptic endings as expressed by the percentage of the total observed terminals; abscissa, minor diameter of the dendrite contacted by degenerating terminals. (B and C) Similar distribution of degenerating terminals of pyramidotomized cats (B) and normal cats (C). Inset: A diagram illustrating that the minor diameter of the cross section of a dendrite expresses the actual diameter of the dendrite.

for conditioning described in Section 2. However, further study is needed to find whether this phenomenon fully explains the behavioral plasticity.

REFERENCES

Fujito, Y., Tsukahara, N., Oda, Y., and Yoshida, M., 1982, Formation of functional synapses in the adult cat red nucleus from the cerebrum following cross-innervation of forelimb flexor and extensor nerves. II. Analysis of newly appeared synaptic potentials, *Exp. Brain Res.* **45:**13–18.

Morris, R. G. M., Anderson, E., Lynch, G. S., and Baudry, M., 1986, Selective impairment of learning and blockade of long-term potentiation by an *N*-methyl-D-aspartate receptor antagonist, AP5, *Nature* **319:**774–776.

Murakami, F., Tsukahara, N., and Fujito, Y., 1977, Analysis of unitary EPSPs mediated by the newly-formed cortico-rubral synapses after lesion of the nucleus interpositus of the cerebellum, *Exp. Brain Res.* **30:**233–243.

Murakami, F., Katsumaru, H., Saito, K., and Tsukahara, N., 1982, A quantitative study of synaptic reorganization in red nucleus neurons after lesion of the nucleus interpositus of the cat: An electron microscopic study involving intracellular injection of horseradish peroxidase, *Brain Res.* **242:**41–53.

Murakami, F., Katsumaru, H., Maeda, J., and Tsukahara, N., 1984, Reorganization of corticorubral synapses following cross-innervation of flexor and extensor nerves of adult cat: A quantitative electron microscopic study, *Brain Res.* **306:**299–306.

Oda, Y., Kuwa, K., Miyasaka, S., and Tsukahara, N., 1981, Modification of rubral unit activities during classical conditioning in the cat, *Proc. Jpn. Acad. [B]* **57:**402–405.

Tsukahara, N., 1981, Synaptic plasticity in the mammalian central nervous system, *Annu. Rev. Neurosci.* **4:**351–379.

Tsukahara, N., and Fujito, Y., 1976, Physiological evidence of formation of new synapses from cerebrum in the red nucleus neurons following cross-union of forelimb nerves, *Brain Res.* **106:**184–188.

Tsukahara, N., and Oda, Y., 1981, Appearance of new synaptic potentials at cortico-rubral synapses after the establishment of classical conditioning, *Proc. Jpn. Acad. [B]* **57:**398–401.

Tsukahara, N., Hultborn, H., and Murakami, F., 1974, Sprouting of cortico-rubral synapses in red nucleus neurones after destruction of the nucleus interpositus of the cerebellum, *Experientia* **30:**57–58.

Tsukahara, N., Hultborn, H., Murakami, F., and Fujito, Y., 1975, Electrophysiological study of formation of new synapses and collateral sprouting in red nucleus neurons after partial denervation, *J. Neurophysiol.* **38:**1359–1372.

Tsukahara, N., Oda, Y., and Notsu, T., 1981, Classical conditioning mediated by the red nucleus in the cat, *J. Neurosci.* **1:**72–79.

Tsukahara, N., Fujito, Y., Oda, Y., and Maeda, J., 1982, Formation of functional synapses in the adult cat red nucleus from the cerebrum following cross-innervation of forelimb flexor and extensor nerves. I. Appearance of new synaptic potentials. *Exp. Brain Res.* **45:**1–12.

4

Is Conditioning Supported by Modulation of an Outward Current in Pyramidal Cells of the Motor Cortex of Cats?

CHARLES D. WOODY

1. INTRODUCTION

Layer V pyramidal cells of the motor cortex of cats are necessary for short-latency blink conditioning (Woody *et al.*, 1974). Increases in their excitability support production of this learned motor response (Woody *et al.*, 1970; Woody and Engel, 1972; Woody and Black-Cleworth, 1973; Brons and Woody, 1980). Early evidence suggested (Woody and Black-Cleworth, 1973) that a decreased postsynaptic conductance might support the excitability increase. Further studies indicate that the input resistance (R_m) of these cells is increased by acetylcholine (ACh), cGMP, and cGMP-dependent protein kinase (cGPK) but not by cAMP or Ca^{2+} (Woody *et al.*, 1978, 1986; Swartz and Woody, 1979, 1984; Wallis *et al.*, 1982; Bartfai *et al.*, 1985; Woody and Gruen, 1986b). The increase in R_m is persistent when depolarization-induced cell discharge accompanies application of ACh, cGMP, or cGPK. Recent studies using single-electrode voltage-clamp techniques *in vivo* now provide direct evidence that ACh and cGPK decrease a net outward current in these cells (Woody and Gruen, 1986a). We suggest that a persistent decrease in this current may mediate the excitability increase that supports short-latency conditioned blinking.

2. BEHAVIOR

A short-latency conditioned eyeblink response can be produced in the cat by pairing a click CS with a glabella tap US (cf. Woody and Brozek, 1969a; Woody, 1974). This conditioned reflex (CR) is characterized by (1) associative acquisition, depending on the order of CS–US presentation, (2) the performance of a specific motor response, i.e., a 20- to 40-msec onset latency eyeblink as opposed to some other movement (Woody and Engel, 1972; Woody *et al.*, 1974), and (3) the selective ability of a stimulus of conditioning

CHARLES D. WOODY • Mental Retardation Research Center, Brain Research Institute, University of California at Los Angeles, Los Angeles, California 90024.

significance, the CS, to elicit the CR as opposed to another, explicitly unpaired auditory stimulus (DS) of equal intensity (Engel and Woody, 1972; Woody *et al.*, 1974, 1976b).

Repeated presentation of click and glabella-tap stimuli in "random" temporal order and variable interstimulus interval (Rescorla, 1967) or in "backwards" order, US preceding CS, fails to lead to the development of conditioned blinking (Woody *et al.*, 1974, 1976b).

Extinction of the conditioned response can be accomplished by backwards pairing of US and CS or by presenting the CS alone (Engel and Woody, 1972). Extinction of the conditioned response by backwards pairing of US and CS has been evaluated with reference to (1) loss of both visually observed and myographically recorded conditioned movement (Woody and Brozek, 1969b; Woody *et al.*, 1970), (2) alteration in evoked potential activity at the facial nucleus (Woody and Brozek, 1969b), (3) alteration in evoked potential activity at the coronal–pericruciate (CPC) cortex (Woody, 1970), and (4) alteration in extracellularly recorded unit activity at the CPC cortex (Woody *et al.*, 1970) and in intracellularly recorded activity at the CPC cortex (Brons and Woody, 1980). Each of the above studies has been made with reference to concomitant series of observations made after conditioning and in naive cats. With the rapid conditioning paradigm described below, it is now possible to follow unit activity through conditioning, extinction, and retraining (Woody *et al.*, 1983).

The effects of associative and pseudorandom presentations of CS, US, and a discriminative stimulus (DS) have been studied behaviorally (Woody *et al.*, 1974) and electrophysiologically in relation to single-unit activity and excitability in caudal cortical areas (Woody *et al.*, 1976b). Discriminative learning of these motor responses to a particular CS has also been studied using paired click CS and tap US with a hiss DS randomly interspersed. These results were compared with those of presenting CS, US, and DS randomly (Woody *et al.*, 1974). Acquisition and performance of the conditioned response were specific for the CS and did not occur after pseudorandom presentation of the three stimuli (Fig. 3 of Woody *et al.*, 1974). Pseudorandom stimulus presentation produced unselective neural effects involving nonspecific increases in spontaneous unit activity and excitability compared with selective increases found after conditioning (Woody, 1977).

3. ANATOMY

Development of short-latency blink CRs depends on a functionally intact rostral cortex. If 25% KCl is applied to the rostral cortex to produce spreading depression in conditioned animals, the short-latency blink CR is abolished reversibly, returning later after recovery, whereas the unconditioned blink response to glabella tap is maintained throughout (Fig. 7 of Woody and Brozek, 1969b). This occurs in the rabbit as well as in the cat (Gutmann *et al.*, 1972). Bilateral ablation of the rostral cortex prevents acquisition of short-latency conditioned blinking (Fig. 1 of Woody *et al.*, 1974). The impairment persists despite extensive training for a period of 3 months after surgery. Control lesions of comparable size at more caudal regions fail to produce such deficits (Fig. 2 of Woody *et al.*, 1974).

The efferent pathway for mediation of the short-latency conditioned blink reflex runs from the motor cortex, by a possible interneuron, to the facial nucleus and finally to the orbicularis oculi muscles. The onset latency of the associatively conditioned eye blink evoked by a click CS in the click- and glabella-tap-trained cat is 20 msec at the periphery (Woody and Brozek, 1969a,b; Woody *et al.*, 1970). The afferent pathways for the short-latency blink CR elicited by the click CS remain to be investigated.

Using electrophysiological methods to investigate the conditioned eye blink acquired by pairing click CS with glabella tap US, we have shown the following. (1) The onset latency of the blink CR evoked by a click CS in the cat can be measured accurately by EMG activity in orbicularis oculi muscles (Woody and Brozek, 1969b). (2) The mean onset latency of the associated neural response at the rostral coronal–pericruciate cortex in the conditioned animal is 13 msec, and that at the facial nucleus is 17 msec (Woody and Brozek, 1969b; Woody, 1970; Woody *et al.*, 1970). Some units in the motor cortex are activated as early as 8–10 msec after click delivery. These units are primarily layer V pyramidal cells (Sakai and Woody, 1980). (3) Direct electrical stimulation of the rostral cortex via implanted electrodes is an adequate CS, when paired with glabella tap US, to produce a short-latency blink CR (Woody and Yarowsky, 1972).

4. PHYSIOLOGY

After conditioning, the excitability of single neurons to intracellularly (IC) injected depolarizing current has been found to be increased above naive levels in units of rostral cortical areas that project interneuronally to the target muscles of the CR (Woody and Black-Cleworth, 1973; Brons and Woody, 1980).

Increases in unit activity elicited by the CS after conditioning are recorded preferentially from motor areas that project interneuronally to target muscles of the conditioned response (Woody, 1970; Woody *et al.*, 1970; Woody and Engel, 1972). The changes in unit activity are associated with a reduction in the level of extracellular electrical stimulation required to produce an EMG response in the target muscles of the CR (Woody *et al.*, 1970). This has been shown for three different types of conditioned facial movements (Woody and Engel, 1972).

In awake blink-trained cats, significantly less intracellularly injected current is required to initiate action potentials in units of rostral cortex projecting ultimately to blink musculature than in adjacent units projecting elsewhere (Woody and Black-Cleworth, 1973). In cats trained to a different movement involving both eye blink and nose twitch, the observation that less intracellular current was required to discharge cortical units of target CR musculature projection was confirmed in cells with action potentials and resting potentions >40 mV. A comparison made against comparable units in naive animals showed that excitability was increased after conditioning. Further studies showed that the increase persisted with learning savings through an extinction procedure. Somewhat reduced excitability changes were found on applying the US alone, nonassociatively. Although these changes lasted several weeks, they did not last longer than a month unless associative CS–US presentations were given. Thus, magnitude and endurance of the changes are associatively dependent, whereas their induction is not (Brons and Woody, 1980).

Similar changes in neural excitability to those described above have been found in facial motoneurons after conditioning an eyeblink with click CS and tap US and after presentations of the US alone (Matsumura and Woody, 1982).

Pyramidal tract (PT) cells of the motor cortex preferentially mediate transmission of short-latency auditory messages comparable to those used as the CS in our model of eyeblink conditioning. On the basis of their short response latency, these cells constitute an "auditory-receptive cortex" within the cat motor cortex (Sakai and Woody, 1980).

Increases in excitability of CS-receptive units of other cortical areas, after conditioning, are thought to help mediate the discriminative aspect of this form of conditioning (Woody *et al.*, 1976b).

The number of pairings of click (CS) and glabella tap (US) needed to produce eye blink conditioning can be reduced from 1000 (Woody *et al.*, 1974) to fewer than 20 by adding hypothalamic stimulation (HS) 580 msec after the CS and 240 msec after the US (Kim *et al.*, 1983). The onset latencies of the major blink responses to CS, measured electromyographically, normally range between 80 and 320 msec when such an ISI is used. Further studies using the same stimuli at different interstimulus intervals demonstrate that short-latency (20–40 msec onset) discriminative CRs can be obtained equally rapidly. The CRs are discriminatively elicited by the CS and not by a hiss DS of comparable intensity and are associative, their emergence depending on the order of CS, US, and HS presentations. Changes in patterns of intracellularly recorded CS-elicited unit activity in the motor cortex are isomorphic with development and extinction of the CR. Short-latency cortical activation of layer V pyramidal cells by HS is predictive of loci of hypothalamic stimulation that will accelerate conditioning (Kim *et al.*, 1983; Woody *et al.*, 1983; Aou *et al.*, 1986).

Reduced afterhyperpolarization (AHP) and rapid activation of cortical cells can be produced by electrical stimulation of the lateral hypothalamus (*A* 18–20, *L* 2–4, *H* 1–3) in the monkey (*Macaca Fuscata*). Following electrical stimulation of the lateral hypothalamus with a four- or five-pulse train (100–500 μsec, 50Hz, 0.5–1.5 mA, bipolar), 14 of 23 cells in monkeys showed a reduction in both amplitude and duration of the AHP with little or no accompanying change in levels of spontaneous resting potential. The effect began 15 to 70 msec after stimulation and persisted for 50 to 300 msec after stimulation. Sometimes a decrease in the threshold level for spike generation accompanied the AHP reduction. This phenomenon could also be observed in neurons of the motor cortex of awake cats together with increases in input resistance. These findings provide evidence in two different mammalian species for shared commonalities in hypothalamocortical interactions that are of potential significance to the development of learned behavior (Aou *et al.*, 1985).

5. CELLULAR BASIS

The finding of increased excitability to intracellular current in the absence of increased spontaneous rates of discharge or detectable changes in resting membrane potential has been interpreted as suggesting a postsynaptic neural change (Woody and Black-Cleworth, 1973). If the membrane resistance of portions of the dendrites were increased in the cells, this would cause a change in the length constant, λ, governing the spread and weighting of synaptic inputs (as postsynaptic potentials) from these areas to regions of spike initiation (Rall, 1974). The increases in resistance would also be reflected by an increased ease of excitation by intracellularly injected current.

Increases in input resistance (R_m) can be produced artificially in cells of the pericruciate cortex by extracellular iontophoretic applications of acetylcholine (ACh) (Krnjevic *et al.*, 1971). We have confirmed Krnjevic's observations of this effect (Woody, 1970; Woody *et al.*, 1976a, 1978) and have found that the increases can be made to persist by pairing iontophoresis of ACh with intracellular injection of sufficient depolarizing current to discharge the neuron repeatedly. The increases in R_m do not occur if saline is substituted or if the neurons are simply discharged repeatedly with intracellularly injected current in the absence of extracellular iontophoresis (Woody *et al.*, 1976a). Further studies are described below.

Measurements were made of effects of ACh on intracellular K^+ ion concentrations in single neurons of the motor cortex of awake cats using ion-sensitive electrodes. Cells that responded to ACh with an increased rate of discharge showed small increases in intracellular concentrations of potassium. Cells that did not so respond did not show such changes. (The hyperpolarizing effect of the small increase in intracellular K^+ would not have been sufficient to overcome the depolarizing effect of the magnitude of associated decrease in potassium conductance predicted empirically.) Concentrations of ACh ten to 100 times higher than those applied iontophoretically were required to produce cross-reactive alterations in the potentials recorded through the potassium-ion-sensitive electrodes (Woody and Wong, 1981).

Horseradish peroxidase was injected intracellularly after studying the effects of ACh on single neurons of the motor cortex. Cyclic GMP was also applied intracellularly by pressure injection (Woody *et al.*, 1986). Pyramidal cells of layers V and VI were identified that responded to these agents with an increased resistance (Swartz and Woody, 1979; Woody *et al.*, 1986). The responsive neurons included those of layer V activated antidromically by PT stimulation. A comparison of the results of pressure-injected cGMP with those of intracellularly iontophoresed cGMP showed similar changes in resistance, but the increase in firing rate after the hyperpolarizing iontophoresis did not occur after pressure injection. The increase in firing rate following application of ACh appears to be a separate effect of this agent, apart from that supported by cGMP as a second messenger. This effect may arise from excitation of surrounding neurons presynaptic to the one recorded or from other, direct conductance effects of ACh binding at the neuronal receptors.

Calcium and cAMP produced decreases in input resistance when injected intracellularly into layer V pyramidal cells of the motor cortex of awake cats (Wallis *et al.*, 1982; Woody and Gruen, 1986b).

Effects of intracellularly applied calcium-calmodulin-dependent protein kinase (CaCMPK) were studied in neurons of the motor cortex of awake cats (Woody *et al.* 1984a). Intracellular iontophoretic application of CaCMPK was followed by a 30-sec application of steady depolarizing current (1.0 nA). These cells showed an increase in input resistance in comparison with a control group of 15 cells given depolarization only, without application of CaCMPK (Woody *et al.*, 1984a). Postiontophoretic measurements of input resistance in cells given CaCMPK alone were not increased, nor was input resistance increased in cells given equivalent negative currents through electrodes containing only KCl . The results indicate that intracellular injection of CaCMPK, followed by depolarization and depolarization-elicited impulse activity, increases input resistance of neurons of the motor cortex of cats transiently for a few seconds. A long-lasting increase of input resistance can be produced in the type B photoreceptor of *Hermissenda* by applying CaCMPK and sufficient depolarization paired with light to increase calcium conductance and internal calcium concentration (Alkon *et al.*, 1983).

Effects of intracellular antibodies to cGMP on responses of cortical neurons to extracellular application of muscarinic agonists were also studied (Swartz and Woody, 1984). Intracellular injection of specific antibody to cGMP (cGMP-Ab) produced substantial decreases in input resistance selectively in neurons of the motor cortex that had responded with increased resistance to prior application of muscarinic agents. Intracellular injection of nonspecific immunoglobulins (IgG) did not produce this effect. (Some nonspecific effects on spike production occurred in cells given IgG or cGMP-Ab.) The decrease in R_m may be interpreted as being consequent to a reduction in base-line levels of active cGMP from binding of cGMP with the injected antibody. In cells that demonstrated a prior increase in R_m following extracellular application of the muscarinic agonist aceclidine or ACh, injection of antibody to cGMP also resulted in suppression of the increase in R_m to subsequent applications of these mucarinic agents. Increases in firing rate to these agents continued to be observed after injection of cGMP-Ab. The results support the hypothesis that cGMP mediates effects of muscarinic neurotransmission on the conductances of neurons of the motor cortex of awake cats.

Intracellular injection of purified cGMP-dependent protein kinase (cGPK) produced increases in input resistance in neurons of the motor cortex of awake cats. Input resistances were measured with 1-nA, 40-msec rectangular, bridge-balanced, hyperpolarizing and depolarizing pulses. The mean input resistance increased within seconds (as rapidly as measurements could be made) after injection of cGPK and remained elevated for 2 min or longer when depolarization-induced discharge was associated. In these experiments the cGPK was incubated with 10 μM cGMP 30 min prior to filling the electrodes. Pressure injection of the cGPK without preincubation with cGMP caused smaller increases in R_m that were slower in onset, reaching a maximum value 60–90 sec after injection. "Control cells" injected with heat-inactivated cGPK, with or without preincubation with 10 μM cGMP, did not show such changes in R_m over the 2-min period of observation. The increases in input resistance after injection of activated cGPK were of a magnitude comparable to that observed on injection of cGMP. The results indicate that intracellular injection of the cGPK into neurons of the precurciate cortex of the awake cat can mimic the actions of extracellularly applied ACh and intracellularly applied cGMP (Bartfai *et al.*, 1985).

We have now identified a voltage-dependent, 4-aminopyridine-sensitive outward current in cortical neurons of awake cats using single-electrode voltage-clamp techniques *in vivo* (Woody *et al.*, 1985). Neurons of the precruciate cortex showed fast outward currents that increased with increasing positive-step voltage commands from holding potentials set between −60 and −80 mV. Preceding depolarizing pulses reduced the currents, whereas preceding hyperpolarizing impulses potentiated them. Cells given pressure injection of 4-aminopyridine showed reduction of the outward current. The degree of reduction varied from cell to cell.

The measurements of changes in current corresponded with previous *in vitro* measurements in other cortical cells (Gustaffson *et al.*, 1982) and in *Tritonia* (cf. Thompson, 1977). The cable properties of neocortical pyramidal cells are such that an actual voltage clamp is unrealizable with control of but 1–10% of a space constant estimated from previous modeling of theoretical cable properties in cells of known morphology. Also, the fast sodium currents are incompletely controlled. Further, the large number of active synaptic conductances along the cable length results in a kind of counterclamping toward the normal resting potential of the cell. Hence, the potentials are squeezed away from the normal resting potential and towards the desired holding and command potentials.

Nonetheless, the preparation affords qualitative examination of currents and time courses found in the actual *in vivo* state and semiquantitative measurements of sufficient precision and sensitivity to detect changes in outward currents on administration of pharmacological blocking agents. Net outward currents of the type described above were reduced after application of ACh or cGPK (Woody and Gruen, 1986a). We suggest that a long-lasting decrease in this current may mediate the excitability increase in layer V pyramidal cells that supports short-latency eyeblink conditioning.

Effects of local increases in membrane resistance on current spread in cortical pyramidal cell dendrites have been examined using a passive cable model for determining the transient potential in a dendritic tree of known geometry (Holmes and Woody, 1983). The morphology was obtained from a montage composed of photomicrographs taken at different overlapping areas within serial sections of an HRP-injected layer V pyramidal cell of the cat motor cortex. A passive cable model that could determine the transient potential in dendritic trees of arbitrary geometry was used to examine the efficacy of different loci of increased membrane resistance for given loci of current injection. The model used the passive cable equation (cf. Rall, 1962) to express the potential for each interbranch segment of the dendritic tree. With this model it was found that an increase in membrane resistance in the region immediately proximal to the point of current input was more effective in increasing soma potential than an increase in a comparable membrane area of a more proximal increase depending on the locus of current injection and the morphology of the dendritic tree. Tests with this model support the view that increases in membrane resistance could produce the increases in neural excitability found in these cells after conditioning and could account for the increase in activity of these neurons in response to the auditory CS (Holmes and Woody, 1983).

Additional studies (Woody *et al.*, 1984b) have assessed possible injury arising from cell penetrations. The response of penetrated neurons to repeated click stimuli was compared with that of unpenetrated (extracellularly recorded) units of the same cortical region. Responses obtained from penetrated neurons were separated into four groups according to the size of the recorded action potential. The magnitude of the response to click was much the same in cells with action potentials ranging between 50 and 60 mV, 40 and 50 mV, and 30 and 40 mV. The magnitude was slightly greater in the group with action potentials ranging between 20 and 30 mV (suggesting some slight depolarizing injury to some of these cells). The response profiles were comparable to those of extracellularly recorded units (Woody *et al.*, 1970; Woody and Engel, 1972). Studies using K^+ ion-sensitive microelectrodes indicated that "intracellular" recordings were in fact made intracellularly. It appears that whatever injury arose from the penetrations of these cells was minimal and was not sufficient to impair the ability of most cells to respond with spike activation to natural stimuli such as weak click.

Many of our recordings appear to be from penetrated dendrites. In further studies employing pressure injections of HRP, recordings characterized by action potentials of amplitudes smaller than the recorded resting potentials were correlated with recoveries of injected dendrites (Woody *et al.*, 1984b). Penetrations with dendritic recoveries had higher input resistances than did those with recoveries of both somas and dendrites. Increases in spike height during pressure injections were greater in recordings with dendritic recoveries than in recordings with recoveries of both somata and dendritic processes.

ACKNOWLEDGMENTS. This work was supported by AFOSR and NICHD grant HD 05958. I thank all of those (see references) who have collaborated with me in these studies.

REFERENCES

Alkon, D. L., Acosta-Urquidi, J., Olds, J., Kuzma, G., and Neary, J. T., 1983, Protein kinase injection reduces voltage-dependent potassium currents, *Science* **219**:303–306.

Aou, S., Woody, C. D., Chapman, C. D., Oomura, Y., and Nishino, H., 1985, Reduced afterhyperpolarization and rapid activation of cortical cells produced by electrical stimulation of hypothalamus in monkey and cat, *Soc. Neurosci. Abstr.* **11**:983.

Aou, S., Birt, D., and Woody, C. D., 1986, Activity and excitability of neurons of the cat pericruciate cortex after rapid acquisition of conditioned blink responses and during extinction, *Soc. Neurosci. Abstr.* **12**:555.

Bartfai, T., Woody, C. D., Gruen, E., Nairn, A., and Greengard, P., 1985, Intracellular injection of cGMP-dependent protein kinase results in increased input resistnce in neurons of the mammalian motor cortex, *Soc. Neurosci. Abstr.* **11**:1093.

Brons, J. F., and Woody, C. D., 1980, Long term changes in excitability of cortical neurons after Pavlovian conditioning and extinction, *J. Neurophysiol.* **44:**605–615.

Engel, J., Jr., and Woody, C. D., 1972, Effects of character and significance of stimulus on unit activity at coronal–pericruciate cortex of cat during performnce of conditioned motor response, *J. Neurophysiol.* **35:**220–229.

Gustaffson, B., Galvan, M., Grafe, P., and Wigstrom, H., 1982, A transient outward current in a mammalian central neurone blocked by 4-amino pyridine, *Nature* **299:**252–254.

Gutmann, W., Brozek, G., and Bures, J., 1972, Cortical representation of conditioned eyeblink in the rabbit studied by a functional ablation technique, *Brain Res.* **40:**203–213.

Holmes, W. R., and Woody, C. D., 1983, Effects on input currents of local increases in membrane resistance in cortical pyramidal cell dendrites explored using a passive cable model for determining the transient potential in a dendritic tree of known geometry, *Soc. Neurosci. Abstr.* **9:**603.

Kim, E. H.-J., Woody, C. D., and Berthier, N. E., 1983, Rapid acquisition of conditioned eye blink responses in cats following pairing of an auditory CS with glabella tap and hypothalamic stimulation, *J. Neurophysiol.* **49:**767–779.

Krnjevic, K., Pumain, R., and Renaud, L., 1971, The mechanism of excitation by acetylcholine in the cerebral cortex, *J. Physiol. (Lond.)* **215:**247–268.

Matsumura, M., and Woody, C. D., 1982, Excitability changes of facial motoneurons of cats related to conditioned and unconditioned facial motor responses, in: *Conditioning: Representation of Involved Neural Functions* (C. D. Woody, ed.), Plenum Press, New York, pp. 451–457.

Rall, W., 1962, Electrophysiology of a dendritic neuron model, *Biophys. J. (Suppl.)* **2:**145–167.

Rall, W., 1974, Dendritic spines, synaptic potency, and neuronal plasticity, in: *Cellular Mechanisms Subserving Changes in Neuronal Activity* (C.D. Woody, K. A. Brown, T. J. Crow, Jr., and J. D. Knispel, eds.), Brain Information Service, University of California, Los Angeles, pp. 13–21.

Rescorla, R. A., 1967, Pavlovian conditioning and its proper control procedures, *Psychol. Rev.* **74:**71–80.

Sakai, H., and Woody, C. D., 1980, Identification of auditory responsive cells in coronal–pericruciate cortex of awake cats, *J Neuropohysiol.* **44:**223–231.

Swartz, B. E., and Woody, C. D., 1979, Correlated effects of acetylcholine and cyclic guanosine monophosphate on membrane properties of mammalian neocortical neurons, *J. Neurobiol.* **10:**465–488.

Swartz, B. E., and Woody, C. D., 1984, Effects of intracellular antibodies to cGMP on responses of cortical neurons of awake cats to extracellular application of muscarinic agents, *Exp. Neurol.* **86:**388–404.

Thompson, S. H., 1977, Three pharmacologically distinct potassium channels in molluscan neurones, *J. Physiol. (Lond.)* **265:**465–488.

Wallis, R. A., Woody, C. D., and Gruen, E., 1982, Effects of intracellular pressure injections of calcium ions in morphologically identified neurons of cat motor cortex, *Soc. Neurosci. Abstr.* **8:**909.

Woody, C. D., 1970, Conditioned eye blink: Gross potential activity at coronal pericruciate cortex of the cat, *J. Neurophysiol.* **33:**838–850.

Woody, C. D., 1974, Aspects of the electrophysiology of cortical processes related to the development and performance of learned motor responses, *Physiologist* **17:**49–69.

Woody, C. D., 1977, Changes in activity and excitability of cortical auditory receptive units of the cat as a function of different behavioral states, *Ann. N.Y. Acad. Sci.* **290:**180–199.

Woody, C. D., and Black-Cleworth, P., 1973, Differences in excitability of cortical neurons as a function of motor projection in conditioned cats, *J. Neurophysiol.* **36:**1104–1116.

Woody, C. D., and Brozek, G., 1969a, Gross potential from facial nucleus of cat as an index of neural activity in response to glabella tap, *J. Neurophysiol.* **32:**704–716.

Woody, C. D., and Brozek, G., 1969b, Changes in evoked responses from facial nucleus of cat with conditioning and extinction of eye blink, *J. Neurosphysiol.* **32:**717–726.

Woody, C. D., and Engel, J., Jr., 1972, Changes in unit activity and thresholds to electrical microstimulation at coronal–pericruciate cortex of cat with classical conditioning of different facial movements, *J. Neurophysiol.* **35:**230–241.

Woody, C. D., and Gruen, E., 1986a, *In-vivo* effects of acetylcholine (ACh) and cGMP dependent protein kinase (cGPK) on outward currents of neurons of the motor cortex of awake cats, *Soc. Neurosci. Abstr.* **12:**725.

Woody, C. D., and Gruen, E., 1986b, Responses of morphologically identified cortical neurons to intracelluarly injected cyclic AMP, *Exp. Neurol.* **91:**596–612.

Woody, C. D., and Wong, B., 1981, Intracellular recording of potassium in neurons of the motor cortex of awake cats following extracellular applications of acetylcholine, in: *Ion-Selective Microelectrodes and Their Uses in Excitable Tissues* (E. Sykova and L. Vyklicky, eds.), Plenum Press, New York, pp. 125–132.

Woody, C. D., and Yarowsky, P. J., 1972, Conditioned eye blink using electrical stimulation of coronal–pericruciate cortex as conditional stimulus, *J. Neurophysiol.* **35**:242–252.

Woody, C. D., Vassilevsky, N. N., and Engel, J., Jr., 1970, Conditioned eyeblink: Unit activity at coronal–pericruciate cortex of the cat, *J. Neurophysiol.* **33**:851–864.

Woody, C. D., Yarowsky, P., Owens, J., Black-Cleworth, P., and Crow, T., 1974, Effect of lesions of cortical motor areas on acquisition of eyeblink in the cat, *J. Neurophysiol.*, **37**:385–394.

Woody, C. D., Carpenter, D. O., Gruen, E., Knispel, J. D., Crow, T. W., and Black-Clewroth, P., 1976a, *Persistent Increases in Membrane Resistance of Neurons in Cat Motor Cortex,* AFRRI Scientific Report (February, 1976), Bethesda, pp. 1–31.

Woody, C. D., Knispel, J. D., Crow, T. J., and Black-Cleworth, P., 1976b, Activity and excitability to electrical current of cortical auditory receptive neurons of awake cats as affected by stimulus association, *J. Neurophysiol.* **39**:1045–1061.

Woody, C. D., Swartz, B. E., and Gruen, E., 1978, Effects of acetylcholine and cyclic GMP on input resistance of cortical neurons in awake cats, *Brain Res.* **158**:373–395.

Woody, C. D., Kim, E. H.-J., and Berthier, N. E., 1983, Effects of hypothalamic stimulation on unit responses recorded from neurons of sensorimotor cortex of awake cats during conditioning, *J. Neurophysiol.* **49**:780–791.

Woody, C. D., Alkon, D. L., and Hay, B., 1984a, Depolarization-induced effects of Ca^{2+}-calmodulin-dependent protein kinase injection, *in vivo,* in single neurons of cat motor cortex, *Brain Res.* **321**:192–197.

Woody, C. D., Gruen, E., and McCarley, K., 1984b, Intradendritic recordings from neurons of the motor cortex of cats, *J. Neurophysiol.* **50**:925–938.

Woody, C. D., Nenov, V., Gruen, E., and Donley, P., 1985, A voltage-dependent, 4-aminopyridine sensitive, outward current studied *in vivo* in cortical neurons of awake cats by voltage squeeze techniques, *Soc. Neurosci. Abstr.* **11**:955.

Woody, C. D., Gruen, E., Sakai, H., Sakai, M., and Swartz, B., 1986, Responses of morphologically identified cortical neurons to intracellularly injected cyclic GMP, *Exp. Neurol.* **91**:580–595.

5

Conservation of Cellular Mechanisms for Models of Learning and Memory

DANIEL L. ALKON

1. INTRODUCTION

Obvious differences in the behavioral phenomenology of various learning preparations necessitate differences in underlying brain structure, the number and complexity of responsible neural networks, and the actual transformations and integration of information. The usefulness of a learning model must, therefore, be narrowly defined by real similarities between the model and what is modeled. Dissimilarities will aid discrimination of free-wheeling speculation from candidate mechanisms conserved during the course of evolution. A model, whether it be a mathematical description or a close behavioral parallel between one species and another, approaches uniqueness as more and more constraints are satisfied. The value of an interspecies model also depends on the degree to which physiology is conserved during evolution.

Review of behavioral features of major learning models reveals striking similarities for rabbit, cat, and *Hermissenda* Pavlovian conditioning. All three models show temporal and stimulus specificity, CS–UCS transfer, extinction, a requirement for contingency, savings, and duration for weeks or longer. The single "behavioral" feature shared by these three models with *Aplysia* sensitization and long-term potentiation, (LTP) is long duration. Long-term depression (LTD) shows some stimulus and temporal specificity but lasts only 1–2 hr. Review of cellular features of major learning models also reveals striking similarities. For *Hermissenda,* considerable evidence has been gathered for long-term reduction of specific K^+ currents (I_A and $I_{Ca^{2+}\text{-}K^+}$) having a causal role in the acquisition and retention of classically conditioned responses. Recently one of the K^+ currents, $I_{Ca^{2+}\text{-}K^+}$, was indirectly measured for CA1 neurons of hippocampal slices isolated from classically conditioned and control animals at least 1 day after training. As for *Hermissenda,* $I_{Ca^{2+}\text{-}K^+}$ of CA1 cells showed a conditioning-specific reduction. On the basis of these and other data, a Ca^{2+}/CaM- and C-kinase-mediated reduction of K^+

DANIEL L. ALKON • Section on Neural Systems, Laboratory of Biophysics, IRP, National Institute of Neurological and Communicative Disorders and Stroke, National Institutes of Health at the Marine Biological Laboratory, Woods Hole, Massachusetts 02543.

currents across postsynaptic membranes of Hermissenda has been shown to have considerable relevance to examples of vertebrate classical conditioning. This represents the first indication of the value of a "simple" associative learning model for elucidating conserved cellular mechanisms of learning of far more "complex" preparations.

2. MODELS

When is a model a model? And when is it an idle fantasy? When does a model capture some essence of what is its object and when does it provide such distortion that all reference, all context, is lost?

Most of us who study cellular mechanisms of learning and memory storage want to understand and describe these processes as we know them to function in ourselves. And therein lies our dilemma—how do we gain fundamental insight into our own learning and memory first as experiential phenomena, then as phenomena manifest by measurable behavior, then as function of neural systems, and finally as cellular and subcellular physiology? In varying degrees we are all in the same boat—we all must use models. A rat's ability to learn the spatial relationships of a maze shares common features with human ability to learn a route of travel form one place to another and, less directly, to learn the relationships of letters in a word or the numbers of a telephone exchange. A dog's ability to be classically conditioned to salivate at the sound of a bell resembles maze learning in one important respect, it is associative—i.e., it depends on the temporal relationship during conditioning of two specific stimuli: the bell and the meat.

Yet, the obvious differences in the phenomenologies of maze learning and classical conditioning almost necessitate differences in the underlying brain structures involved, the number and complexity of the responsible neural networks, and the actual transformations and integration of information required. Can classical conditioning be a model for maze learning? Only to a very limited degree, and with reference to precisely defined features. The two forms of learning have associative elements, increase as a function of practice, persist for months or years, can be extinguished, etc. Insofar as these features are concerned, there may be common mechanisms. Similarly, classical conditioning of a mollusk such as *Hermissenda* may share common mechanisms with classical conditioning of the rabbit nictitating membrane mechanisms that subserve stimulus and pairing specificity, retention, savings, extinction, etc. But the network encoding necessary for remembering spatial and temporal context, for generalization to related stimuli and/or contexts, are most certainly not available to the mollusk.

The usefulness of a model must, therefore, be narrowly defined by real similarities between the model and what is modeled. And the dissimilarities will help us to discriminate free-wheeling speculation from candidate mechanisms conserved during the course of evolution.

Long-term potentiation (LTP), as it has been analyzed until very recently, is not a model for associative learning—it is a model for long-term neural modification. Because it depends on electrical stimulation of axon bundles rather than sensory stimulation of sophisticated and sensitive neural systems, it does not, it cannot, utilize information processing within neuronal networks as occurs during maze learning or classical conditioning. Because it lasts for days or longer, it may, and this is only may, result from biophysical and biochemical transformations at the cellular level that bear some relationship to transformations of associative learning. Recent experiments in which single-cell neural modification is achieved by combining electrical stimulation of distinct axon bundles, or those including pharmacological treatments, incorporate stimulus associations but persist for many minutes rather than days. Such stimulation can still not be compared directly to natural stimulation of networks as occurs during learning and memory recall.

Similarly, we must examine what is common to nonassociative and associative learning to determine the usefulness of one as a model for the other. Typically, habituation and sensitization are transient in comparison to associative learning, although examples of sensitization lasting for weeks have been reported for both molluskan (Kandel and Schwartz, 1982) and vertebrate species. In fact, a potentially long (though not permanent) time course is one major feature found to be common in some cases to associative and nonassociative behavioral change. Both associative and nonassociative learning involve prolonged neural changes (Alkon, 1980, 1984; Kandel and Schwartz, 1982), but there is no evidence that these are the same or even similar neural changes. So numerous are the dissimilarities between the behavioral aspects of habituation and sensitization on the one hand and classical conditioning and maze learning on the other hand that to model the latter on the former would be like forcing a square peg into a round hole. Similarly, so much more elaborate are the neural networks and their responsiveness to stimulus patterns of associative learning than are the single synaptic interactions between sensory and motor cells responsive to single stimuli during habituation or sensitization that different physiological principles must apply for these diverse phenomena.

This not to say that ultimately, in retrospect, we might not discover overlap in cellular physiology among LTP, sensitization, conditioning, and maze learning. It is to say that the likelihood of finding such overlap *a priori* is low and even then in very limited, restricted contexts. And the usefulness of "simple" phenomena such as habituation as models for more complex phenomena such as maze learning is questionable at best. At worst, and more likely, the "simple" phenomena may simply obscure the more complex and necessitate additional controls rather than provide insight.

Implicit in our discussion thus far are really two kinds of models. One just mentioned concerns the use of "simple" phenomena as models for more complex phenomena. Another involves the use of one species' physiology as a model for another's.

Both types of models derive meaning not only insofar as they retain essential features of their object but also insofar as they approach uniqueness and acquire predictive capability. A model, whether it be a mathematical description or a close behavioral parallel between one species and another, approaches uniqueness as more and more constraints or conditions are satisfied.

The value of an interspecies model also depends on an implicit truth that ultimately has to be tested. This is that during the course of evolution, natural functions have been conserved. In the case of learning and memory, both the behavioral and the mechanistic levels must show such conservation for an interspecies model to have validity.

Although thus far few investigations of learning and memory have yielded a quantitatively descriptive model, there are many interspecies models. It is of interest to review point by point first what behavioral properties they show and then their underlying neural networks and cellular physiology. Finally, such a review might suggest some hypotheses concerning conservation of mechanisms.

3. FEATURES OF SPECIFIC MODELS

In Tables I and II are listed behavioral learning features for a number of the major experimental preparations used to date. No attempt has been made to be comprehensive;

TABLE I. Behavioral Features of Learning Models

Species	Type of learning	Temporal specificity	CS–UCS transfer	Duration
Aplysia	Sensitization	—	—	Weeks
	Conditioning	+	—	Days
Hermissenda	Pavlovian conditioning	+ (1.0 sec)	+	Weeks or longer
Rabbit	Pavlovian conditioning	+ (0.25 sec)	+	Months
Cat	Pavlovian conditioning	+	+	Months
Long-term potentiation	—	—	—	Days or weeks
Long-term depression	Vestibular–ocular adaptation	±	—	1–2 hr

only examples are included for which cellular mechanisms have been examined and/or proposed, and the best-understood example for each species was chosen. Superficial inspection reveals striking similarity for three of the models: *Hermissenda,* rabbit, and cat Pavlovian conditioning. All three models show temporal and stimulus specificity, CS–UCS transfer, extinction, a requirement for contingency, savings, and duration for weeks or longer. The single feature shared by these three models with *Aplysia* sensitization and LTP is duration of days or weeks. *Aplysia* conditioning appears to show both long duration and some temporal and stimulus specificity. Long-term depression shows some stimulus and temporal specificity but lasts only 1–2 hr, and none of the other properties have been demonstrated. Based then on behavioral criteria alone, we might expect *Hermissenda* associative learning to be a suitable model for rabbit and cat classical conditioning and possibly to depend on cellular mechanisms conserved during the course of evolution.

In Tables III and IV are listed characteristics of cellular mechanisms for the preparations of Tables I and II. These mechanisms have been arrived at, suggested, or speculated on (in varying degrees) after analysis of neuronal networks implicated in generating long-term behavioral and neural changes. Establishing causal relationships between neural and behavioral changes is, of course, not elementary. Extensive correlation between neural and behavioral change is a necessary but not sufficient condition. Such correlation has been possible for rabbit, cat, and *Hermissenda* classical conditioning but

TABLE II. Behavioral Features of Learning Models

Species	Type of learning	Stimulus specificity	Extinction	Contingency	Savings
Aplysia	Sensitization	—	—	—	—
	Conditioning	+	—	—	—
Hermissenda	Pavlovian conditioning	+	+	+	+
Rabbit	Pavlovian conditioning	+	+	+	+
Cat	Pavlovian conditioning	+	+	+	+
Long-term potentiation	—	—	—	—	—
Long-term depression	Vestibular–ocular adaptation	+	—	—	—

TABLE III. Cellular Features of Learning Models

Species	Type of learning	Postsynaptic	Initial depolarization	Initial elevation of Ca^{2+}	Initial K^+ current reduction
Aplysia	Sensitization	—	—	?	+
	Conditioning	—	—	?	?
Hermissenda	Pavlovian conditioning	+	+	+	+
Rabbit	Pavlovian conditioning	+	?	?	?
Cat	Pavlovian conditioning	+	?	?	?
Long-term potentiation	—	±	±	±	?
Long-term depression	Vestibulo–ocular adaptation	+	+	+	?

not yet for *Aplysia* conditioning. For *Hermissenda,* cat, and rabbit conditioning, it has also proven possible to produce the behavioral change by appropriate paired stimulation of the sensory pathways (and elements thereof) that mediate the CS and the UCS (see Chapters 4 and 14). For *Hermissenda* it was also possible by producing membrane changes of individual neurons to produce the learned behavioral change (Farley *et al.*, 1983). The intrinsic nature of conditioning-specific biophysical modifications was demonstrated for *Hermissenda* at a new level of localization: identified neuronal membranes after isolation of the individual neurons from all other elements of the nervous system. Remarkably, the magnitude of reduction of well-characterized K^+ currents (which regulate excitability of type B cells) predicts the magnitude of conditioning-specific changes of CS-elicited responses of postsynaptic neurons (including output motor neurons) within the visual pathway (Goh *et al.*, 1985). Furthermore, the degree of K^+ current reduction predicts as well the extent of training-induced acquisition of conditioned behavioral responses of intact animals.

In summary, considerable evidence has been gathered for biophysical changes having a causal role in the acquisition and retention of classically conditioned responses of *Hermissenda*. Although such a causal role has not yet been defined for rabbit and cat classical conditioning, some of the relevant criteria (e.g., correlation, synthesis by path-

TABLE IV. Cellular Features of Learning Models

Species	Type of learning	Persistent K^+ current reduction	Monoamine amplification	C-kinase implicated	Ca/CaM kinase implicated	Cyclic AMP kinase implicated
Aplysia	Sensitization	?	+	—	—	+
	Conditioning	?	?	—	—	?
Hermissenda	Pavlovian conditioning	+	?	+	+	?
Rabbit	Pavlovian conditioning	+	?	+	?	?
Cat	Pavlovian conditioning	+	?	?	+	?
Long-term potentiation	—	?	+	+	?	?
Long-term depression	Vestibulo–ocular adaptation	?	+	?	?	?

way stimulation) have been satisified. Because the *Hermissenda* classical conditioning provides an acceptable model for classical conditioning of the rabbit and cat, it would not be entirely unexpected were all three learning behaviors to share some common network and cellular mechanisms. Recently it has proven possible to obtain indirect measurements in rabbit hippocampus for one of the same K^+ currents reduced by *Hermissenda* classical conditioning. This K^+ current, the calcium-dependent K^+ current or $I_{Ca^{2+}-K^+}$, was measured as an "afterhyperpolarization" of CA1 neurons of hippocampal slices isolated from classically conditioned as well as control rabbits at least 1 day after the last training experience (see Chapter 10). As for *Hermissenda*, $I_{Ca^{2+}-K^+}$ was reduced for a certain number of CA1 neurons only in conditioned animals (Disterhoft *et al.*, 1986). This conditioning-specific reduction of $I_{Ca^{2+}-K^+}$ was present in varying degrees in approximately the same proportion of cells (50–60%) as showed (*in vivo*) increased excitability in response to CS presentations to living animals (Berger *et al.*, 1983; Thompson *et al.*, 1984).

Thus, modification of the same K^+ current in the same postsynaptic compartment is retained long after training of either the rabbit or *Hermissenda*. Increased postsynaptic excitability for days after training has only been observed for one other preparation: cat conditioning and sensitization (Woody *et al.*, 1984). It will, of course, be extremely interesting to identify the biophysical sequence of steps leading to these K^+ reductions for all three preparations. For *Hermissenda*, pairing and stimulus-specific depolarization accumulate during acquisition, and one is accompanied by elevation of intracellular Ca^{2+} (Alkon, 1980; Crow and Alkon, 1980; Connor and Alkon, 1984). Prolonged depolarization and Ca^{2+} elevation in turn result in and are followed by prolonged K^+ current reduction (cf. Table IV). There is some indication that these conditions (e.g., prolonged depolarization, Ca^{2+} elevation) also precede and help cause the long-term neural modifications of LTP and LTD (cf. Table III).

Some indication of conserved molecular mechanisms of memory storage has also recently emerged. In *Hermissenda*, phosphorylation of a 20,000-dalton protein was shown to undergo conditioning-specific modification measured hours after training (Neary *et al.*, 1981). This protein was shown with *in vivo* and *in vitro* assays to be a substrate for Ca^{2+}/calmodulin type II kinase as well as the Ca^{2+}/lipid-dependent kinase or C-kinase (Neary *et al.*, 1986; Naito and Alkon, 1986; Alkon and Naito, 1986). Depolarizing conditions that simulate electrophysiological changes during acquisition also modified 20,000-dalton protein phosphorylation. Finally, iontophoresis of either enzyme into the type B neuron after elevation of $Ca^{2+}{}_i$ (which also occurs during acquisition) produced the same reduction of the K^+ currents (I_A and $I_{Ca^{2+}-K^+}$) as occurred during classical conditioning of intact *Hermissenda* (Acosta-Urquidi *et al.*, 1984; Sakakibara *et al.*, 1986; Alkon *et al.*, 1986). These and other results, taken together, support the hypothesis that a synergistic activation of Ca^{2+}/calmodulin-dependent kinase and C-kinase is involved in the induction of long-term biophysical modifications during classical conditioning (Alkon *et al.*, 1986).

Routtenberg and his colleagues have found that induction of the neural modification, LTP, causes increased C-kinase activity within the membrane fraction of hippocampal slices 1 hr after electrical stimulation (Akers *et al.*, 1986). More recently, Bank *et al.* (1986) showed increased C-kinase activity in the membrane fraction of CA1 neurons from rabbit hippocampal slices isolated 24 hr after classical conditioning (but not control) procedures. Thus, persistent changes of C-kinase activity distribution were demonstrated for the same neurons for which evidence of conditioning-specific reduction of $I_{Ca^{2+}-K^+}$ had previously been obtained (Disterhoft *et al.*, 1986). In still another experimental context, this $I_{Ca^{2+}-K^+}$ of the CA1 neurons was shown to be reduced by C-kinase activation (Baraban *et al.*, 1985) just as it has been in *Hermissenda* (Alkon *et al.*, 1986; Farley and Auerbach, 1986). Finally, Woody *et al.*, (1984) obtained indirect measurements of K^+ current reduction *in vivo* from cat motor neurons in response to injection of Ca^{2+}/calmodulin-dependent kinase.

In summary, a Ca^{2+}/CaM- and C-kinase-mediated reduction of K^+ currents in postsynaptic compartments of *Hermissenda* has thus far been found to have considerable relevance to examples of vertebrate classical conditioning as well as LTP. This represents the first confirmation of the value of a "simple" associative learning model for elucidating conserved cellular mechanisms of learning of far more "complex" preparations.

REFERENCES

Acosta-Urquidi, J., Alkon, D. L., and Neary, J. T., 1984, Ca^{2+}-dependent protein kinase injection in a photoreceptor mimics biophysical effects of associative learning, *Science* **224:**1254–1257.

Akers, R. F., Lovinger, D. M., Colley, P. A., Linden, D. J., and Rottenberg, A., 1986, Translocation of protein kinase C activity may mediate hippocampal long-term potentiation, *Science* **231:**587–588.

Alkon, D. L., 1980, Membrane depolarization accumulates during acquisition of an associative behavioral change, *Science* **210:**1375–1376.

Alkon, D. L., 1984, Calcium-mediated reduction of ionic currents: A biophysical memory trace, *Science* **226:**1037–1045.

Alkon, D. L., and Naito, S., 1986, Biochemical mechanisms of memory storage, *J. Physiol. (Paris)* **81:**252–260.

Alkon, D. L., Kubota, M., Neary, J. T., Naito, S., Coulter, D., and Rasmussen, H., 1986, C-kinase activation prolongs Ca^{2+}-dependent inactivation of K^+ currents, *Biochem. Biophys. Res. Commun.* **134:**1254–1253.

Bank, B., Coulter, D., Kuzirian, A., Rasmussen, H., Alkon, D. L., and Chute, D. L., 1986, Effects of NMR conditioning on intracellular levels of protein kinase C, *Proc. Natl. Acad. Sci. U.S.A.* (in press).

Baraban, J. M., Snyder, S. H., and Alger, B. E., 1985, Protein kinase C regulates ionic conductance in hippocampal pyramidal neurons: Electrophysiological effects of phorbol esters, *Proc. Natl. Acad. Sci. U.S.A.* **82:**2538–2542.

Berger, T. W., Rinaldi, P. C., Weisz, D. J., and Thompson, R. F., 1983, Single-unit analysis of different hippocampal cell types during classical conditioning of the rabbit nictitating membrane response, *J. Neurophysiol.* **50:**1197–1219.

Connor, J., and Alkon, D. L., 1984, Light- and voltage-dependent increases of calcium ion concentration in molluscan photoreceptors, *J. Neurophysiol.* **51:**745–752.

Disterhoft, J. F., Coulter, D. A., and Alkon, D. L., 1986, Conditioning–specific membrane changes of rabbit hippocampal neurons measured *in vitro, Proc. Natl. Acad. Sci. U.S.A.* **83:**2733–2737.

Farley, J., and Auerbach, S., (1986), Protein kinase C activation induced conductance changes in *Hermissenda* photoreceptors like those seen in associative learning, *Nature* **319:**220–223.

Farley, J., Richards, W. G., Ling, L. J., Linman, E., and Alkon, D. L., 1983, Membrane changes in a single photoreceptor cause associative learning in *Hermissenda, Science* **221:**1201–1203.

Goh, Y., Lederhendler, I., and Alkon, D. L., 1985, Input and output changes of an identified neural pathway are correlated with associative learning in *Hermissenda, J. Neurosci.* **5:**536–543.

Kandel, E. R., and Schwartz, J. H., 1982, Molecular biology of learning: Modulation of transmitter release, *Science* **218:**433–443.

Naito, S., Bank, B., and Alkon, D. L., 1987, Transient and persistent depolarization-induced changes at protein phosphorylation in a molluscan nervous system, *J. Neurochem.* (in press).

Neary, J. T., Crow, T. J., and Alkon, D. L., 1981, Change in a specific phosphoprotein band following associative learning in *Hermissenda, Nature* **293:**658–660.

Neary, J. T., Naito, S., and Alkon, D. L., 1986, Ca^{2+}-activated phospholipid-dependent protein kinase (C-kinase) activity in the *Hermissenda* nervous system, *J. Neurochem.* **47:**1405–1411.

Sakakibara, M., Alkon, D. L., DeLorenzo, R., Goldenring, J. R., Neary, J. T., and Heldman, E., 1986, Modulation of calcium-mediated inactivation of ionic currents by Ca^{2+}/calmodulin-dependent protein kinase II, *Biophys. J.* **50:**319–327.

Thompson, R. F., Barchas, J. D., Clark, G. A., Donegan, N., Kettner, R. E., Lavond, D. G., Madden, J. IV, Mauk, M. E., and McCormick, D. A., 1984, Neuronal substrates of associative learning in the mammalian brain, in: *Primary Neural Substrates of Learning and Behavioral Change* (D. L. Alkon and J. Farley, eds.), Cambridge University Press, New York p. 71.

Woody, C. D., Alkon, D. L., and Hay, B., 1984, Depolarization-induced efects of Ca^{2+}-calmodulin-dependent protein kinase injection, *in vivo,* in single neurons of cat motor cortex, *Brain Res.* **321:**192–197.

II

Long-Term Potentiation

6

Long-Term Potentiation of Synaptic Transmission in the Hippocampus Obeys Hebb's Rule for Synaptic Modification

HOLGER WIGSTRÖM, BENGT GUSTAFSSON, and YAN-YOU HUANG

1. INTRODUCTION

Cellular mechanisms for associative memory are generally assumed to involve synaptic changes dependent on a conjunction between activity in different neuronal elements, and various principles for such conjunctive control of synaptic modification have been proposed (Brindley, 1969; von Baumgarten, 1970). In the model for neuronal memory put forward by Hebb (1949), the strengthening of synaptic connections was postulated to occur as a result of nearly simultaneous firing of the pre- and the postsynaptic cells. Long-term potentiation (LTP) in the hippocampus is a synaptic strengthening process that has been implicated in learning and memory (Bliss and Lømo, 1973; Teyler and Discenna, 1984), and recent studies have suggested that its induction is controlled by coincident pre- and postsynaptic activity in general agreement with Hebb's rule for synaptic modification (Wigström *et al.*, 1986; Kelso *et al.*, 1986).

According to a proposed model for the induction of LTP (Wigström and Gustafsson, 1985), this dependency of the induction process on the conjunction between pre- and postsynaptic activity is related to N-methyl-D-aspartate (NMDA) receptor channels, which coexist with the non-NMDA receptor channels in the spine membrane. The NMDA receptor channels are activated by simultaneous transmitter activation and postsynaptic depolarization, allowing for a local Ca^{2+} entry, which triggers the induction of LTP. That the postsynaptic condition is related to postsynaptic depolarization is also indicated by recent experimental findings (Gustafsson *et al.*, 1987; Kelso *et al.*, 1986). The question then arises to what extent this postsynaptic depolarization is critically related to firing of the postsynaptic cell, i.e., how close hippocampal LTP follows Hebb's principle. In the

HOLGER WIGSTRÖM, BENGT GUSTAFSSON, and YAN-YOU HUANG • Department of Physiology, University of Göteborg, S-400 33 Göteborg, Sweden. Y.-Y. H. is on sabbatical leave from Shanghai Institute of Physiology, Academia Sinica, People's Republic of China.

present chapter, some aspects of the LTP induced by conjunction between synaptic activation and intracellularly injected current pulses are taken up, with special emphasis on results pertinent to the above question. The study was carried out in the CA1 area of hippocampal slices from guinea pigs (cf. Skrede and Westgaard, 1971).

2. PROPERTIES OF CONJUNCTION-INDUCED LTP

As recently reported from our laboratory, hippocampal LTP can be produced by pairing single afferent volleys with intracellularly injected depolarizing current pulses (Wigström *et al.*, 1986). This result shows unequivocally the presence of a conjunctive control based on a requirement for concurrent pre- and postsynaptic activation. In the following we first consider the question of whether this conjunction-induced LTP is indeed identical with the LTP induced by tetanization. The timing requirements for the conjunctive stimulation, which can be analyzed in detail by the abovementioned experimental protocol, are also considered.

2.1. Relationship to Tetanus-Induced LTP

Several lines of evidence support the notion that the long-lasting increase in EPSP induced by pairing synaptic activation with depolarizing current pulses represents the same LTP process as that induced by tetanic activation (Gustafsson *et al.*, 1987). Thus, the conjunction-induced potentiation was specific for the pathway involved in the conjunctive stimulation, in agreement with the input specificity of tetanization-induced LTP. Moreover, it was blocked by the NMDA receptor antagonist 2-amino-5-phosphonovalerate (APV). A further demonstration of the equivalence between the conjunction-induced LTP and tetanus-induced LTP was obtained in experiments in which the induction of LTP by conjunction was followed by tetanic afferent activation, the latter producing little or no further LTP. A similar occlusion was also observed when the tetanization was delivered first and then followed by a conjunction test (see Fig. 1 for a demonstration of occlusion in both directions).

The finding that LTP can be induced without tetanic activation (see also Gustafsson and Wigström, 1986) demonstrates the difference between this process and purely presynaptic postactivation phenomena such as posttetanic potentiation. The tetanic afferent activation generally employed to produce LTP may then be viewed (primarily) as a means to obtain the necessary amount of postsynaptic activity.

2.2. The EPSP–Pulse Timing Requirements

The temporal requirements of the conjunctive stimulation were examined by using different intervals between the test EPSP and the conditioning pulse. The amount of peak potentiation induced by 50–60 pairing events was measured and normalized with respect to the amount of potentiation obtained with a short conditioning interval. As seen in Fig. 2A, potentiation appeared not only when the onset times of the EPSP and pulse were nearly simultaneous but also when the EPSP preceded the conditioning stimulation by up to 100–200 msec (forward conditioning). No potentiation was obtained when the EPSP immediately followed the depolarizing pulse. The general course of the timing curve shown in Fig. 2A is similar to the one obtained previously by extracellular methods, pairing single EPSPs with brief tetanic activation of a group of separate afferents (Gus-

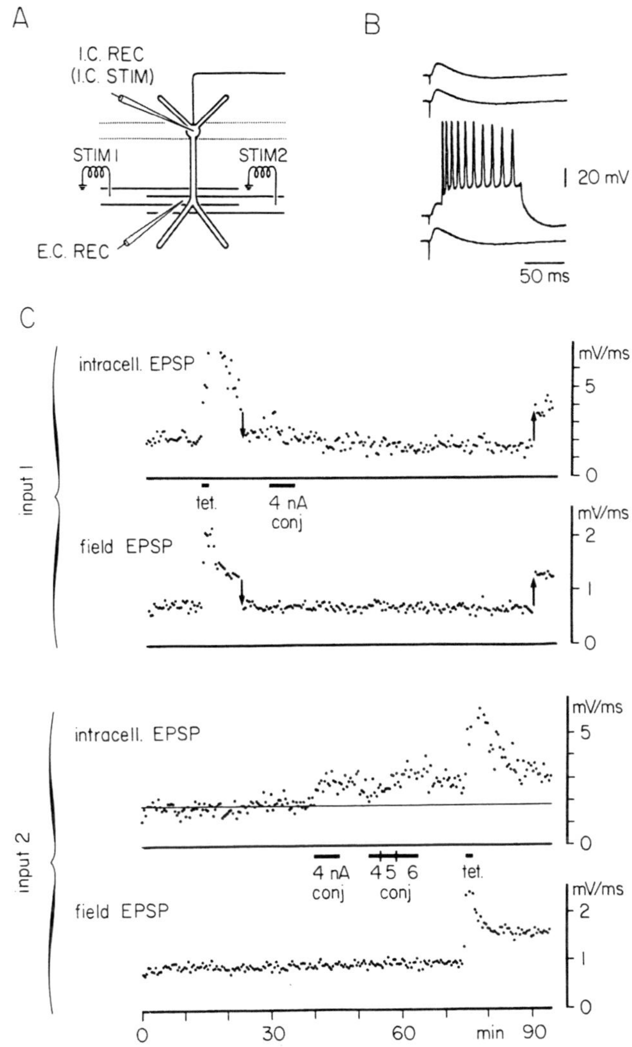

FIGURE 1. Equivalence between conjunction-induced and tetanus-induced LTP.(A) Schematic drawing showing arrangement of stimulating and recording electrodes in the hippocampal CA1 region. (B) Examples of intracellular potentials recorded in response to alternating activation of inputs 1 and 2 (STIM 1 and STIM 2) during testing (upper two records) and during conjunctive stimulation, i.e., pairing synaptic activation of one input with depolarizing current pulses injected through the recording electrode (lower two records). (C) Occlusion between potentiation evoked by homosynaptic tetanization and that by conjunction. Measurements of initial slopes of intracellular EPSPs and field EPSPs resulting from activation of STIM 1 (upper) and STIM 2 (lower) are shown for a series of responses. Input 1 was first homosynaptically tetanized with ten consecutive ten-impulse trains (bar). After the tetanization, and between the arrows, STIM 1 was lowered from 30 μA to 17 μA. During the time indicated by the bar, each EPSP in response to STIM 1 was followed by a 4-nA, 100-msec depolarizing current pulse. Input 2 was first paired with a depolarizing current pulse (bars) of indicated magnitude and was subsequently tetanized with ten consecutive ten-impulse trains. Experiment was performed with 0.1 mM picrotoxin in the bath. Adapted from Gustafsson *et al.* (1987).

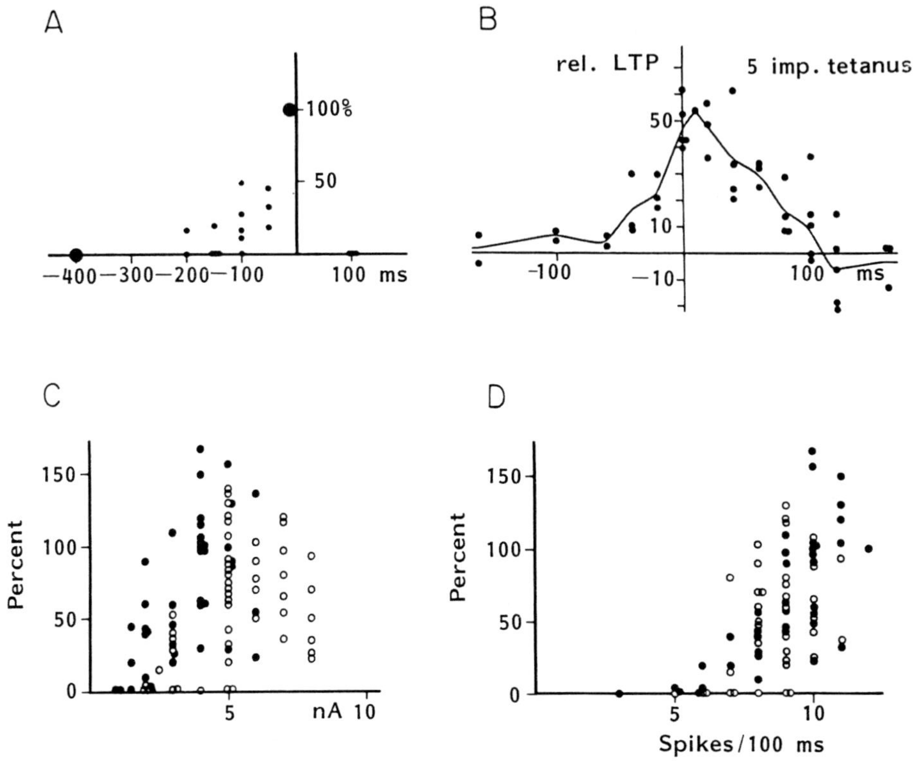

FIGURE 2. Amount of conjunction-induced LTP versus timing and magnitude of conditioning stimulation. (A) Time dependence of the conditioning effect of a current pulse. Zero time represents coincidence between the onset of EPSP and current pulse. Negative values (to the left) indicate the time by which the EPSP onset preceded current onset. The large filled circle at −400 msec signifies that no observable EPSP changes were found when this interval was used. The values at the other intervals (smaller filled circles) represent the maximal EPSP increase during the conjunction precedure, expressed relative to that found using a short interval (large filled circle at 100%). (B) Time dependence of the effect of a conditioning tetanus (five impulses, 50 Hz) to a separate population of afferents. Zero time represents coincidence between the single test volley and the first volley in the train (direction of time axis as in A). Potentiation of the field EPSP was measured 5 min after a sequence of five pairing events and is expressed as percentage of a standard potentiation (which measured about 30%). (C) Amount of conjunction-induced potentiation versus current injected. The maximal amount of EPSP increase reached during the conjunction procedure is plotted against the magnitude of the current pulse. Values obtained in normal solution are shown as open circles, and those obtained in the presence of picrotoxin (0.1 mM) as filled circles. (D) Same as in C, but the abscissa now represents the number of spikes appearing per 100-msec current injection. Adapted from Gustafsson *et al.* (1987) (A, C, D) and Gustafsson and Wigström (1986)(B).

tafsson and Wigström, 1986; see Fig. 2B). The somewhat narrower timing in the latter case may reflect methodological differences (see legend). It can be noted in particular that potentiation could be obtained at intervals at which there was no overlap between the EPSP and the conditioning pulse (forward conditioning intervals greater than 50 msec). However, as discussed by Wigström and Gustafsson (1987), an EPSP–pulse overlap may also exist in a sense at these long intervals, considering the "silent EPSP component" mediated by NMDA receptor channels.

3. WHAT IS THE NATURE OF THE POSTSYNAPTIC CONDITION?

In the following we address some questions that are related to Hebb's formulation of firing as the principal postsynaptic factor controlling synaptic modification. We ask whether cell firing is a necessary requirement for the production of LTP under different experimental conditions. Finally, we consider whether the observed relationships between cell firing and LTP induction are merely correlations or whether the postsynaptic spikes as such contribute to the postsynaptic control of LTP by means of the depolarization generated by them.

3.1. The Role of Cell Firing in Single-Volley Conjunction

The threshold current for producing LTP ranged from 1–2 nA to 5 nA in some cases, with a tendency to larger values in slices with intact inhibition (normal solution) and to smaller values in slices in which inhibition was blocked by treatment with picrotoxin. As seen in Fig. 2C, the potentiation–current relationship was quite scattered. A clearer relationship appeared when the amount of potentiation was plotted as a function of the number of spikes that occurred during a 100-msec depolarizing pulse (Fig. 2D). The relationship was essentially similar with and without picrotoxin in the bath. As shown in the diagram, LTP induction required firing of more than five to six spikes per 100 msec, corresponding to an average firing frequency during the pulse of 50–60 Hz. In order to obtain a large amount of potentiation, frequencies up to or exceeding 100 Hz were often needed.

3.2. Long-Term Potentiation Produced in the Absence of Firing in the Postsynaptic Cell

In a proposed model for the induction of LTP, the decisive factor is the opening of NMDA receptor channels by concurrent transmitter binding and postsynaptic depolarization (Wigström and Gustafsson, 1985). With this type of mechanism, postsynaptic firing would not be a requirement for LTP under all conditions, for instance, if large depolarization is achieved during an artificial blockade of spike generation. The above type of experiment was thus carried out also with the intracellular electrode filled with 0.1 mM QX-314 (Gustafsson *et al.*, 1987), thereby blocking sodium-dependent action potentials. Although the normal spike-generating mechanism was inactivated in this case, LTP was produced to the same extent by pairing single EPSPs with intracellular depolarizing current pulses. Moreover, LTP was induced in the case when synaptic stimuli were delivered during a period of constant suprathreshold depolarization leading to inactivation of the spike-generating mechanism (no treatment with QX-314). These experiments indicate that firing is not an absolute requirement for the induction of LTP.

3.3. The Role of Cell Firing during Brief Afferent Tetani

Long-term potentiation can thus be produced in the absence of postsynaptic firing under certain experimental conditions, such as during blockade of spike generation by QX-314. On the other hand, when single EPSPs were paired with intracellular depolarization that could elicit firing, LTP was only obtained when considerable firing was evoked, and the amount of LTP was correlated with the firing frequency. These results then suggest that although LTP induction is not directly dependent on the existence of sodium-mediated spikes, the postsynaptic depolarization must be such that a considerable spike activity is present. This need for postsynaptic spike activity appeared also to be true when current pulses were paired with brief trains of afferent activation instead of with single volleys, i.e., under more natural stimulating conditions, at least when using rather weak synaptic stimulation. An example from one such experiment is shown in Fig. 3A. The tetanus consisted of a five-impulse, 50-Hz train repeated five times at a rate of 0.1 Hz (picrotoxin-treated slice). It can be seen that LTP did not occur when the tetanus was paired with a hyperpolarizing current pulse (cf. Malinov and Miller, 1986) or evoked at resting level but appeared when the train was paired with depolarizing current pulses (see also Kelso *et al.*, 1986). Comparison with the records in Fig. 3A shows that this conditioning effect of membrane polarization was well correlated to the presence of spike activity during the tetanus, no LTP occurring until spikes appeared. However, a detailed quantitative relationship between postsynaptic firing frequency and amount of LTP (cf. Fig. 2D), including the magnitude of a possible threshold firing frequency, remains to be determined. It would also be important to carry out the above experiment with afferent firing frequencies other than 50 Hz and using different strengths of synaptic activation.

3.4. The Functional Significance of Spike Depolarization

Thus, in an operational sense, LTP might be considered to be related to cell firing in agreement with Hebb's postulate. However, this does not necessarily imply a direct causal relationship between the spikes and the production of LTP. Nevertheless, a more direct connection between postsynaptic firing and LTP induction might hold a functional advantage since it would lead to a reliable coupling between the two processes under varying conditions and thus secure operation according to Hebb's principle. Such a connection could be achieved if the depolarization generated by the spikes themselves was able to control LTP. In order to test this possibility, spikes were produced by trains (50–100 Hz) of brief (1-msec) depolarizing pulses. It was shown that LTP could be produced when weak trains of synaptic activation, by themselves not causing LTP (see above), were paired with such spike trains (picrotoxin-treated slices). Since the brief current pulses are likely to contribute relatively little to the interspike depolarization, the results indicate that the spikes contribute significantly to the conditioning effect on the induction of LTP.

To eliminate further any direct contribution of the injected current to cell depolarization, other experiments were performed in which the depolarizing current pulses were balanced by a superimposed hyperpolarizing pulse (300 msec). The current strengths were adjusted so that the mean value of the injected current (during the tetanus) was zero. An experiment of this type is shown in Fig. 3B. As can be seen, the initial tetanization (no current injection) did not produce LTP. During the next tetanic period, the cell was stimulated by nine brief depolarizing current pulses (five times), each evoking one spike, superimposed on a longer hyperpolarizing pulse. Despite the fact that the mean injected current was zero, a substantial LTP developed, indicating that the depolarization produced by the action potentials was able to activate the LTP induction mechanism. To demonstrate that LTP induction did not merely result from repetition of the stimulation, tetanization alone was again applied and did not elicit LTP, whereas subsequent conditioning by five times nine spikes (zero net current) was again effective. Figure 3B also illustrates the fact that spike depolarization is not solely responsible for the postsynaptic control of LTP

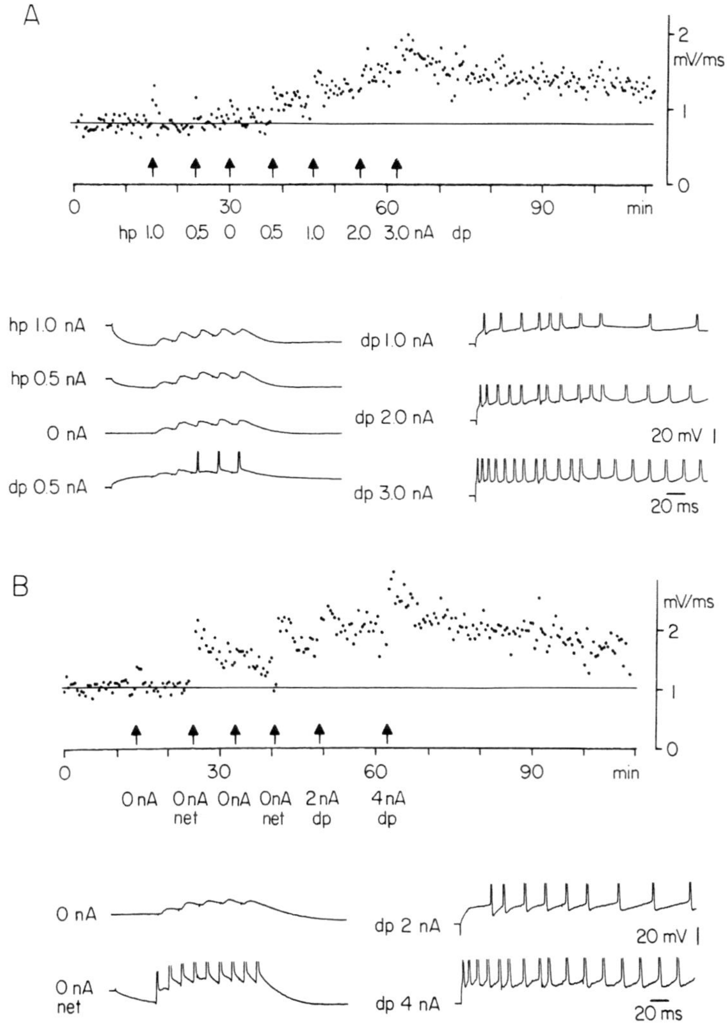

FIGURE 3. Relationship between cell firing and LTP induction. (A) Effect of polarizing current on spike discharge and induction of LTP. Upper diagram: measurements of initial slopes of intracellular EPSPs in response to successive test stimuli. Below: sample recordings of intracellular responses to brief tetanic activation paired with intracellularly injected hyperpolarizing or depolarizing current pulses. Arrows in upper diagram indicate the time for application of conjunctive stimulation. Each conjunction episode consisted of five pairings of a weak tetanus (five impulses, 50 Hz) with conditioning polarizing pulses (300 msec), (B) Conditioning effect of postsynaptic spikes. EPSP measurements and sample recordings as in A. During the conjunction episodes marked "0 nA net", firing was induced by a train of brief depolarizing current pulses (6 nA, 1.2 msec, 80 Hz) superimposed on a 300-msec hyperpolarizing pulse (0.6 nA). The current ratio of 10 : 1 in combination with a 10% "duty cycle" implies that the mean injected current (during tetanization) was zero. It can be seen that LTP is induced by conditioning with 0 nA net current (nine spikes produced) but not in the absence of current injection (0 nA; no spikes). Data from unpublished experiments.

but that the depolarization level between the spikes is likely to contribute. This can be appreciated from the fact that long depolarizing current pulses (300 msec, 2 and 4 nA) were able to produce a larger potentiation than the "balanced" conditioning current pulses despite the fact that the numbers of spikes occurring during the tetani were smaller (about five and eight, respectively, compared to nine).

4. CONCLUDING DISCUSSION: RELATIONSHIP TO HEBB'S MEMORY MODEL

In a general sense, the concept of a Hebb synapse refers to a synapse that is strengthened as a result of nearly simultaneous activity of the interconnected cells. As shown in several recent studies, hippocampal LTP fulfills this criterion (Wigström *et al.*, 1986; Kelso *et al.*, 1986; see also Malinov and Miller, 1986). This is in contrast to certain other forms of cellular memory, such as the presynaptic facilitation found in *Aplysia* (Hawkins *et al.*, 1983). The important postsynaptic factor appears to be membrane depolarization, as indicated by experiments demonstrating LTP induction by pairing synaptic activation and postsynaptic depolarization under conditions of pharmacologically blocked cell firing (Gustafsson *et al.*, 1987; Kelso *et al.*, 1986). These findings support the model proposed by Wigström and Gustafsson (1985) for LTP induction via a local Ca^{2+} influx controlled by a combination of agonist binding and postsynaptic depolarization. Furthermore, the observed timing requirements (see Fig. 2) are in general agreement with those predicted by this model (Gustafsson and Wigström, 1986).

Under conditions more resembling a natural situation—in the sense that brief trains of presynaptic activation were used—firing seemed to play a significant role not merely as an index of postsynaptic depolarization but by directly contributing to this depolarization. This is in agreement with the strict definition of a Hebb synapse, which requires that the strengthening should be related to simultaneous pre- and postsynaptic firing of action potentials. The relationship to firing, that is, to the actual input and output, is essential for the use of cells as building blocks in "cell assemblies," diffuse networks with connections formed according to the above rule (Hebb, 1949; see also Wigström, 1975).

In conclusion, hippocampal LTP seems to fulfill the basic requirements for a Hebb-type mechanism. However, a more complete characterization of the conditions that control LTP induction may require certain extensions of this basic principle. The possibility of a certain threshold firing frequency of the postsynaptic neuron was already pointed out (so far, this threshold has been clearly demonstrated only in the case with single afferent volleys). Another possible deviation is related to the fact that, in the original formulation by Hebb, the presynaptic cell was considered actually to take part in firing the postsynaptic cell. Although this may be true with LTP under certain conditions, it is not always the case. The present results demonstrate that EPSPs can be potentiated when the depolarizing pulse is positioned 100–200 msec after the EPSP, that is, at a time when the EPSP cannot influence cell firing. Furthermore, one might consider the possibility that activation of inhibitory circuits would counteract the global cooperativity effect by restricting the intracellular spread of depolarization. Such a behavior is indicated by the difficulty of inducing LTP by the "extracellular conjunction procedure" in slices with intact inhibition (Gustafsson and Wigström, 1986). Hence, it seems possible that in the presence of inhibition, hippocampal cells operate in a somewhat different manner than revealed here, for instance, according to the extended Hebb model used by Wigström (1973,1974).

Finally, the experiments by Bloch and Laroche (1985) demonstrating that LTP can be influenced by a delayed stimulation of the reticular formation may suggest a further extension of the number of controlling factors.

ACKNOWLEDGMENTS. This work was supported by the Swedish Medical Research Council (projects 05954 and 05180), the Swedish Natural Science Research Council (project 4018), Magnus Bergvalls Stiftelse, Stiftelsen Lars Hiertas Minne, and Anna Ahrenbergs Fond. Y.-Y. Huang was supported by the China Education Ministry.

REFERENCES

Bliss, T. V. P., and Lømo, T., 1973, Long-lasting potentiation of synaptic transmission in the dentate area of the anaesthetized rabbit following stimulation of the perforant path, *J. Physiol. (Lond.)* **232**:331–356.

Bloch, V., and Laroche, S., 1985, Enhancement of long-term potentiation in the rat dentate gyrus by posttrial stimulation of the reticular formation, *J. Physiol. (Lond.)* **360**:215–231.

Brindley, G. S., 1969, Nerve net models of plausible size that perform many simple learning tasks, *Proc. R. Soc. Lond. [Biol]* **174**:173–191.

Gustafsson, B., and Wigström, H., 1986, Hippocampal long-lasting potentiation produced by pairing single volleys and brief conditioning tetani evoked in separate afferents, *J. Neurosci.* **6**:1575–1582.

Gustafsson, B., Wigström, H., Abraham, W. C., and Huang, Y.-Y., 1987, Long-term potentiation in the hippocampus using depolarizing current pulses as the conditioning stimulus to single volley synaptic potentials, *J. Neurosci.* **7**:774–780.

Hawkins, R. D., Abrams, T. W., Carew, T. J., and Kandel, E. R., 1983, A cellular mechanism of classical conditioning in *Aplysia:* Activity-dependent amplification of presynaptic facilitation, *Science* **219**:400–405.

Hebb, D. O., 1949, *The Organization of Behavior,* John Wiley & Sons, New York.

Kelso, S. R., Ganong, A. H., and Brown, T. H., 1986, Hebbian synapses in hippocampus, *Proc. Natl. Acad. Sci. U.S.A.* **83**:5326–5330.

Malinov, R., and Miller, J. P., 1986, Postsynaptic hyperpolarization during conditioning reversibly blocks induction of long-term potentiation, *Nature* **320**:529–530.

Skrede, K. K., and Westgaard, R. H., 1971, The transverse hippocampal slice: A well defined cortical structure maintained *in vitro, Brain Res.* **35**:589–593.

Teyler, T. J., and Discenna, P., 1984, Long-term potentiation as a candidate mnemonic device, *Brain Res. Rev.* **7**:15–28.

von Baumgarten, R. J., 1970, Plasticity in the nervous system at the unitary level, in: *The Neurosciences: Second Study Program* (F. O. Schmitt, ed.), Rockefeller University Press, New York, pp. 260–271.

Wigström, H., 1973, A neuron model with learning capability and its relation to mechanisms of association, *Kybernetik* **12**:204–215.

Wigström, H., 1974, A model of a neural network with recurrent inhibition, *Kybernetik* **16**:103–112.

Wigström, H., 1975, Associative recall and formation of stable modes of activity in neural network models, *J. Neurosci. Res.* **1**:287–313.

Wigström, H., and Gustafsson, B., 1985, On long-lasting potentiation in the hippocampus: A proposed mechanism for its dependence on coincident pre- and postsynaptic activity, *Acta Physiol. Scand.* **123**:519–522.

Wigström, H., and Gustafsson, B., 1987, Presynaptic and postsynaptic interactions in the control of hippocampal long-term potentiation, in: *Long-Term Potentiation: From Biophysics to Behavior* (P. W. Landfield and S. Deadwyler, eds.), Alan R. Liss, New York, pp. 71–105.

Wigström, H., Gustafsson, B., Huang, Y.-Y., and Abraham, W. C., 1986, Hippocampal long-term potentiation is induced by pairing single afferent volleys with intracelluarly injected depolarizing current pulses, *Acta Physiol. Scand.* **126**:317–319.

7

Chloride-Mediated Feedforward Inhibition Is Not Involved in Long-Term Potentiation

B. P. C. MELCHERS, W. J. WADMAN, L. J. ZIJP, AND F. H. LOPES DA SILVA

1. INTRODUCTION

Long-term potentiation (LTP) can be defined as a semipermanent increase in synaptic efficacy that can be induced by tetanic stimulation of afferents. The cellular mechanisms underlying LTP are still unclear. Both presynaptic (Dolphin *et al.*, 1982; Lynch *et al.*, 1985) and postsynaptic (Lynch and Baudry, 1984) changes correlated with LTP have been reported.

In a number of studies the possible role of inhibitory processes in LTP has been investigated. However, the results are not conclusive: changes in IPSPs (Misgeld and Klee, 1985) and direct effects of inhibitory neurons (Buszaki and Eidelberg, 1982) have been reported during LTP. On the other hand, Haas and Rose (1982), using a paired-pulse facilitation paradigm, report that changes in inhibitory processes do not contribute to a great extent in the establishment of LTP.

Thus far no consideration has been given to possible changes in the distribution of current sources and sinks along the neuronal membranes during LTP. Morphological evidence (Desmond and Levy, 1983) indicates that such changes may occur during LTP. In order to test whether spatial aspects are of importance in LTP, we recorded field potentials evoked by perforant path stimulation at several sites along the soma and apical dendritic tree of the granular cells in the dentate gyrus of the *in vitro* hippocampal slice both before and after tetanization. On these recordings current source density (CSD) analysis was performed (Nicholson and Freeman, 1975). Secondly, we used this technique to reveal a possible involvement of Cl^--mediated inhibition in LTP by comparing the changes in CSD induced by tetanization with those induced by modulation of Cl^- currents, either by reducing the Cl^- concentration in the Ringer or by application of picrotoxin.

B. P. C. MELCHERS, W. J. WADMAN, L. J. ZIJP, and F. H. LOPES DA SILVA • Department of General Zoology, University of Amsterdam, 1098 SM Amsterdam, The Netherlands.

2. METHODS

2.1. Slice Preparation and Maintenance

Male Wistar rats (180–220 g) were decapitated after ether anesthesia. The brain was quickly removed and transferred into ice-cold (0–4°C), freshly oxygenated Ringer solution (124 mM NaCl, 5 mM KCl, 2 mM $CaCl_2$, 2 mM $MgSO_4$, 1.25 mM NaH_2PO_4, 26 mM $NaHCO_3$, 10 mM glucose). The hippocampus was prepared free and cut into slices of 500 μm with razor blades. The slices were transferred to the measuring chamber, which was constantly perfused with freshly oxygenated (95% O_2, 5% CO_2) Ringer solution kept at a temperature of 34–35°C. Low-Cl^- Ringer ($[Cl^-] = 30$ mM) was prepared by equimolar substitution of part of the NaCl by Na-isethionate. When used, picrotoxin was added to the Ringer in a final concentration of 20 μM. To prevent epileptiform discharges in the picrotoxin and in some of the low-Cl^- experiments, the amounts of $CaCl_2$ and $MgSO_4$ in the Ringer solution were doubled (cf. Wigstrom and Gustafsson, 1983).

2.2. Stimulation and Recording of Evoked Field Potentials

After 1 hr of incubation, recording and stimulation electrodes were placed under visual guidance. Stimulation electrodes consisted of two trimel-isolated stainless steel wires with a diameter of 50 μm with straight-cut tips. For stimulation, bipolar, biphasic current pulses of 200 μsec duration and a current intensity of 50–130 μA were used. The interval between successive test stimuli was 4 sec. Tetani consisted of a train of these pulses with a frequency of 50 Hz and a duration of 2 sec. The amplitude of each pulse was typically 50–60% of the stimulus intensity that elicited a maximal field EPSP. In general, two tetani were given with an interval of 5 min.

Recordings were made by means of glass microelectrodes filled with 3 M NaCl (tip resistance 10–15 MΩ). Evoked field potentials (EFPs) were sampled (4 kHz) and averaged by means of a Motorola Exorset 30 microprocessor. In order to record the field potentials at different positions along the cells, the recording electrode was fixed to a microprocessor-driven stepper-motor micromanipulator.

In general the following experimental scheme was used: after positioning of the recording and stimulation electrodes, four test stimuli were given, and the EFPs were sampled and averaged, and the recording electrode was moved to the next recording site. This was repeated until the field potentials at all selected recording sites were recorded. Usually five such EFP profiles were successively obtained both during the control period and after (1) tetanization, (2) lowering the Cl^- concentration, or (3) application of picrotoxin and were averaged afterwards. The second-order spatial derivative of the EFPs was calculated as a function of the position along the dendritic tree. The value obtained can, under certain restrictions, be interpreted as the current source density (CSD) along the dendritic tree of the neurons (Nicholson and Freeman, 1975).

3. RESULTS AND DISCUSSION

3.1. Field Potentials

In Fig. 1A an example of an EFP profile elicited by perforant path stimulation and recorded at seven consecutive recording sites (separated 50 μm) along the dendritic tree

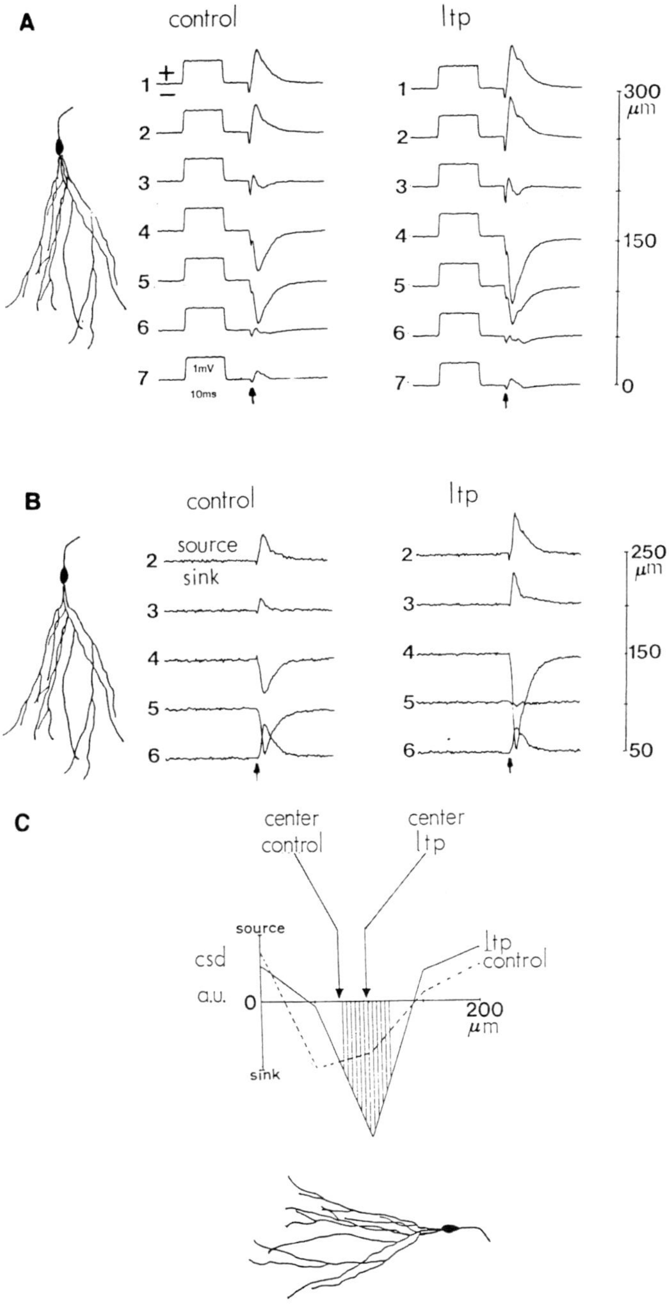

FIGURE 1. (A) An EFP profile elicited by perforant path stimulation and obtained at different recording sites along the dendritic tree of the granular cells. Pretetanus (control) and posttetanus (LTP) profiles of A are shown. (B) The CSDs corresponding to the EFP profiles of A are shown. (C) Plot of the CSD at 2.5 msec after the test stimulus as a function of the position along the cells during the control period (broken line) and after LTP induction. The center of the sink was defined as the middle of the region in which the sink was larger than a certain value. This value was chosen in such a way that this region covered 75% of the total surface of the sink as indicated in the figure.

of the granular cells is shown both before and after tetanization. In the stratum moleculare, an initially negative-going potential, representing the extracellular field EPSP, was encountered. This potential reverses polarity in the stratum granulosum. Tetanization resulted in a change in amplitude of the EFPs. However, as can be seen from Fig. 1A, this change in amplitude was highly dependent on the recording site. At recording site 4 a clear increase in EFP was found, whereas at recording site 5 the extracellular EPSP even decreases. This result indicates that spatial aspects may indeed play a role in LTP.

3.2. CSD Analysis

In order to interpret the change in EFP profiles in terms of transmembrane currents, we used CSD analysis. In Fig. 1B the CSDs corresponding to the EFP profiles discussed above are shown. In the stratum moleculare (recording sites 4 and 5), an extracellular sink was found, flanked by sources. From Fig. 1B the nonuniform distribution of LTP can be clearly seen: the sink at recording site 5 decreases after tetanization, whereas the more proximally located sink increases. Such a change in CSD can be caused by changes in extracellular conductivity. However, measurements of extracellular conductivity before and after tetanization gave no reason to expect that such changes are of importance in this respect (W. J. Wadman, unpublished observations).

To quantify the changes after tetanization, we calculated the CSD both at a time point in the rising phase of the EPSP, before the appearance of a population spike, and at a timepoint after the population spike(s). The estimates of membrane current density thus obtained at *ca.* 2.5 msec and *ca.* 7.5 msec after the test stimulus were described by two parameters. For the representation of the amplitude of the sink, the total surface of the sink was calculated; the position of the sink along the dendritic tree of the granular cells was defined as the point lying in the middle of the area that covers 75% of the total surface of the sink (as explained in Fig. 1C) and is called here the "center of the sink." In Fig. 1C the CSD at 2.5 msec after the test stimulus corresponding to the pre- and posttetanus CSDs shown in Fig. 1B is given.

At 2.5 msec after the test stimulus, a significant positive correlation was found between the change in amplitude of the sink after the induction of LTP and the change in the position of the sink (Spearman rank correlation coefficient $r_s = 0.40$, $P < 0.05$, $n = 18$), which means that slices in which the sink was enhanced tend to show a change in the position of the sink in the direction of the cell soma. No significant correlation was found between the two parameters at 7.5 msec after the test stimulus.

To obtain a general average in which the changes are put in evidence in a clear way, the slices showing an enlargement of the sink of more than 10% at *ca.* 2.5 msec after the test stimulus ($n = 8$) were selected. The CSDs obtained at *ca.* 2.5 and *ca.* 7.5 msec after the test stimulus were first normalized with respect to the pretetanus amplitude of the sink, and the centers of the sinks were aligned. An average with standard deviation was calculated. As can be seen from Fig. 2A, in which the average CSD at *ca.* 2.5 msec after the test stimulus is shown both before and after the tetanization, the tendency of the sinks to shift in the direction of the cell soma during LTP is clearly expressed. Significant changes between pretetanus and posttetanus CSD at 2.5 msec after the test stimulus were found only in a relatively small area located proximally to the main pretetanus dendritic sink, as indicated in Fig. 2A. Although no significant changes were observed in the average CSD of the potentiated slices at *ca.* 7.5 msec after the test stimulus, a tendency of the sink to shift in the direction of the cell soma could still be seen.

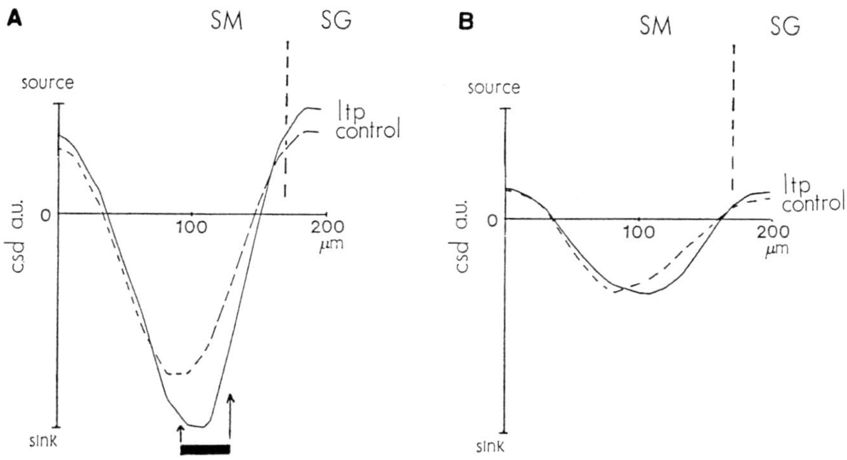

FIGURE 2. General average of the CSD at *ca*. 2.5 msec (A, $n = 8$) and *ca*. 7.5 msec (B, $n = 5$) after the test stimulus of the slices showing an enlargement of the sink of more than 10% in the early rising phase of the EPSP. (In three slices the recordings did not cover the complete late dendritic sink; therefore, the surfaces of the sinks of these slices could not be normalized.) The area in which significant differences between control (broken line) and posttetanus CSD (LTP) were found is indicated by the bar under the figure ($P < 0.05$, Student's t-test, two tailed). No significant changes were encountered in the CSD at 7.5 msec after the test stimulus. To give an indication of the position of the CSD along the neurons, the stratum moleculare (SM) and stratum granulosum (SG) are indicated in the figure.

It is not yet clear how these changes in CSD can be explained. One of the possible explanations is a localized decrease in Cl^--mediated (feedforward) inhibition during LTP; this Cl^- current should be situated in the dendrites to account for the increase in the dendritic sink. Dendritic feedforward inhibition has been shown in hippocampal pyramidal cells (Alger and Nicoll, 1982) and has also been implied in the dentate gyrus (Buszaki and Czeh, 1981). Therefore, we have tested the influence of a modulation of Cl^- currents on the CSD along the soma and dendritic membranes of the granular cells.

3.3. Effect of Modulation of Cl^- Currents

In order to test this, EFP profiles were obtained both before and during perfusion with low-Cl^- Ringer ($[Cl^-] = 30$ mM). This concentration of Cl^- has been shown to suppress or even reverse early IPSPs in CA3 pyramidal cells (Knowles *et al.*, 1984). This is reflected in the clear increase in the population spike amplitude that was encountered during perfusion with low-Cl^- Ringer. The CSDs of the EFP profiles were calculated at *ca*. 2.5 msec and 7.5 msec after the test stimulus. General averages of these CSDs were made as described above. Perfusion with low-Cl^- Ringer, in contrast to LTP induction, gave rise to no significant changes in CSD at 2.5 msec after the test stimulus (Fig. 3A) both with respect to the amplitude and the position of the sink. Also, picrotoxin, a potent Cl^- channel blocker (Akaike *et al.*, 1985), added in a concentration of 20 μM to the perfusion Ringer did not change the CSD obtained in the early rising phase of the EPSP, in a similar way as was found to accompany LTP.

An interesting observation during perfusion with low-Cl^- Ringer was a clear increase in both the amplitude of the dendritic sink and the amplitude of the source in the stratum granulosum at 7.5 msec after the test stimulus (Fig. 3B), which was also found in the

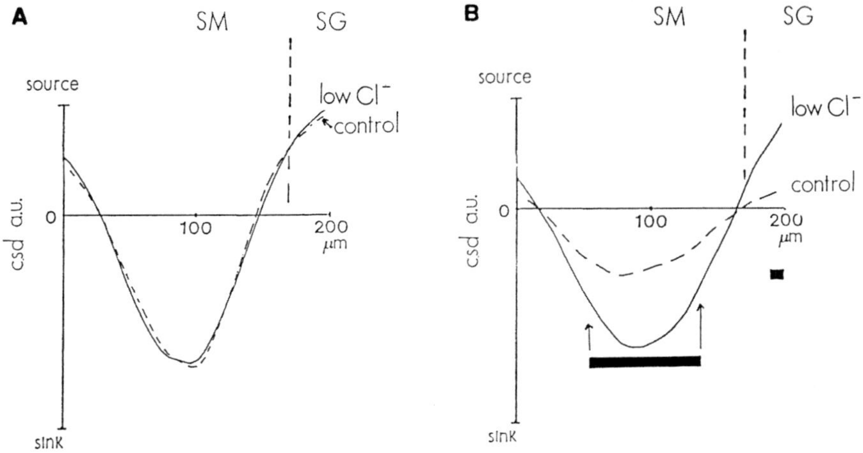

FIGURE 3. General averages (n = 16) of the CSD at *ca.* 2.5 (A) and *ca.* 7.5 msec (B) after the test stimulus before (control, broken line) and during perfusion with low-Cl^- RINGER (low Cl^-). No significant changes were observed between control and low Cl^- at 2.5 msec after the test stimulus; the bars indicate the regions in which significant (Student's t test, $P < 0.05$, two-tailed) changes were found in the CSD at *ca.* 7.5 msec after the test stimulus. SM, stratum moleculare; SG, stratum granulosum.

picrotoxin experiments. Apparently an important Cl^- current located in the dendritic tree is present at that instant in time. An inward flow of Cl^- does not seem to be of importance in the early rising phase of the EPSP, and therefore changes in Cl^--mediated inhibition cannot explain the changes in CSD during LTP.

To summarize, LTP is accompanied by changes in the distribution of current sinks and sources along the cells, which are manifested by an enhancement of the dendritic sink and a change in its position in the direction of the cell soma. Thus, the changes responsible for LTP in the dentate gyrus of the hippocampal slice are rather strictly localized on the dendritic tree of the neurons. The nature of the changes and the underlying mechanisms are unclear as yet. However, changes in Cl^--mediated inhibition seem not to play a role in the effects.

ACKNOWLEDGMENTS. This work was subsidized in part by Medigon grant 900-548-072 of the Dutch organization for the Advancement of Pure Research (ZWO).

REFERENCES

Akaike, N., Hattori, K., Oomura, Y., and Carpenter, D. O., 1985, Bicuculline and picrotoxin block γ-aminobutyric acid gated Cl^- conductance by different mechanisms, *Experientia* **41:**70–72.

Alger, B. E., and Nicoll, R. A., 1982, Feedforward dendritic inhibition in rat hippocampal pyramidal cells studied *in vitro*, *J. Physiol* (*Lond.*) **328:**105–123.

Buszaki, G., and Czeh, G., 1981, Commissural and perforant path interactions in the rat hippocampus, *Exp. Brain Res.* **43:**429–439.

Buszaki, G., and Eidelberg, E., 1982, Direct afferent excitation and long-term potentiation of hippocampal interneurons, *J. Neurophysiol.* **48:**597–607.

Desmond, N. L., and Levy, W. B., 1983, Synaptic correlates of associative potentiation/depression: An ultrastructural study in the hippocampus, *Brain Res.* **265:**21–30.

Dolphin, A. C., Errington, M. L., and Bliss, T. V. P., 1982, Long-term potentiation of the perforant path *in vivo* is associated with increased glutamate release, *Nature* **297:**496–498.

Haas, H. L., and Rose, G., 1982, Long-term potentiation of the excitatory synaptic transmission in the rat hippocampus: The role of inhibitory processes, *J. Physiol. (Lond.)* **329:**541–552.

Knowles, W. D., Schneiderman, J. H., Wheal, H. V., Stafstrom, C. E., and Schwartzkroin, P. A., 1984, Hyperpolarizing potentials in guinea pig hippocampal CA3 neurons, *Cell Mol. Neurobiol.* **4:**207–230.

Lynch, G., and Baudry, M., 1984, The biochemistry of memory: A new and specific hypothesis, *Science* **224:**1057–1063.

Lynch, M. A., Errington, M. L., and Bliss, T. V. P., 1985, Long-term potentiation of synaptic transmission in the dentate gyrus: Increased release of [^{14}C]glutamate without increase in receptor binding. *Neurosci. Lett.* **62:**123–129.

Misgeld, U., and Klee, M. R., 1985, Long-term potentiation and inhibition in CA3 neurons, *Exp. Brain Res. [Suppl.]* **9:**325–333.

Nicholson, C., and Freeman, J. A., 1975, Theory of current source-density analysis and determination of conductivity tensor for anuran cerebellum, *J. Neurophysiol.* **38:**356–368.

Wigstrom, H., and Gustafsson, B., 1983, Large long lasting potentiation in the dentate gyrus *in vitro* during blockade of inhibition, *Brain Res.* **275:**153–158.

8

β-Adrenergic Mechanisms in Long-Term Potentiation and Norepinephrine-Induced Long-Lasting Potentiation

JOHN M. SARVEY

1. INTRODUCTION

A brief train of high-frequency, repetitive electrical stimulation to the perforant path results in long-term potentiation (LTP) of the response to low-frequency stimulation of that pathway. In the dentate gyrus, LTP is manifested as an increased amplitude of the extracellularly recorded synchronous action potentials (population spike) of granule cells, a decrease in the latency of the population spike, and a steepening of the initial slope of the extracellularly recorded excitatory postsynaptic potential (EPSP) in response to stimulation of the perforant path (Bliss and Lømo, 1973; Bliss and Gardner-Medwin, 1973). Long-term potentiation has also been demonstrated in hippocampal fields CA1 and CA3 (Schwartzkroin and Wester, 1975; Alger and Teyler, 1976). In intracellular recordings, LTP is manifested primarily as an increased EPSP amplitude (Andersen *et al.*, 1977, 1980; Yamamoto and Chujo, 1978; Misgeld *et al.*, 1979; Wigström and Gustafsson, 1985; Barrionuevo *et al.*, 1986).

This phenomenon, which may represent the neural substrate for learning and memory (Swanson *et al.*, 1982; Morris and Baker, 1984; Lynch and Baudry, 1984), has sparked a great deal of interest. Although research efforts have been mounting steadily, its mechanism remains unknown. There is an absolute requirement for extracellular calcium and therefore, apparently, intact synaptic transmission (Dunwiddie *et al.*, 1978; Dunwiddie and Lynch, 1979). A postsynaptic requirement for calcium is suggested by the demonstration that injection of the calcium chelator EGTA into the postsynaptic neuron through an intracellular microelectrode also prevents the development of LTP (Lynch *et al.*, 1983). Phosphorylation of synaptic (Bär *et al.*, 1980; Chapter 46) and nonsynaptic (Browning *et al.*, 1981; Hoch *et al.*, 1984) proteins appears to be correlated with the appearance of LTP and may be involved in its production and/or maintenance. Protein synthesis has

JOHN M. SARVEY • Department of Pharmacology, Uniformed Services University of the Health Sciences, Bethesda, Maryland 20814-4799.

also been implicated in LTP (Duffy *et al.*, 1981; Stanton and Sarvey, 1984). Despite evidence for involvement of calcium, protein phosphorylation, and protein synthesis in LTP, no unifying hypothesis that integrates this information has emerged. It is not even clear whether the mechanism of LTP is presynaptic, postsynaptic, or a combination of both.

Although little is known about neurotransmitter actions in neuronal plasticity, interest has begun to focus on possible involvement of monoamine transmitters in LTP. Depletion of norepinephrine (NE) or serotonin has been shown to inhibit production of LTP in the dentate gyrus of anesthetized rat (Bliss *et al.*, 1983). The inhibition seen by that group could be accounted for purely by decreased potentiation of the EPSP, since the population spike still became enhanced relative to the EPSP after high-frequency trains (HFTs) of repetitive stimulation in NE-depleted animals (Bliss *et al.*, 1983). In contrast, Robinson and Racine (1985), recording from anesthetized or freely moving rats, reported increased potentiation of the EPSP and decreased potentiation of the population spike in reserpine-treated rats. Since both groups recorded both EPSPs and population spikes with a single electrode located in the granule cell body layer or dentate hilus, their contradictory findings might be attributable to inaccuracies in the EPSP measurements. In fact, NE, released during stimulation of the locus coeruleus (Dahl and Winson, 1985) or exogenously applied (Winson and Dahl, 1985), can reduce dendritically recorded EPSPs while not affecting EPSPs recorded at the granule cell soma in anesthetized rats. During repetitive stimulation of locus coeruleus, population spike amplitude increased in 69% of sites tested; but the potentiation did not persist (Dahl and Winson, 1985). In contrast, the population spike was actually depressed during iontophoresis but became potentiated for more than 20 min in 38% of the sites tested (Winson and Dahl, 1985).

Injection of the excitatory amino acid glutamate into the locus coeruleus of anesthetized rats induced a potentiation of the perforant-path-elicited population spike and variable effects on the EPSP recorded in the granule cell body layer. This potentiation, which was sensitive to propranolol, lasted more than 20 min in 37% of the animals tested (Harley and Milway, 1986). Neuman and Harley (1983) also saw no effect of iontophoretically applied NE on the EPSP recorded at the soma *in vivo* but reported a long-lasting potentiation (LLP) of the population spike in 39% of sites that potentiated in response to NE (30% of all sites exposed to NE). Recording at the granule cell body layer in hippocampal slices, Lacaille and Harley (1985) reported an LLP of both EPSP and population spike following bath application of NE. Similarly, dopamine has been shown to produce an LLP in field CA1 of rat hippocampal slices *in vitro* (Gribkoff and Ashe, 1984; Ashe, 1984).

In my laboratory, we have sought to investigate the regional specificity of the requirement for NE in LTP as well as the pharmacology and mechanism of action of NE in LTP and LLP. We chose the hippocampal slice preparation because of its stability and amenability to pharmacological manipulations. Slices (400 μm thick) from adult Sprague–Dawley rats were placed in an "interface" chamber (Schwartzkroin, 1975) at 35°C and perfused from beneath with a modified Krebs–Ringer buffer (Stanton and Sarvey, 1984, 1985a–c). Electrically evoked potentials were recorded in field CA1 and dentate as shown in Fig. 1A. In some experiments, extracellular EPSPs were also recorded by an electrode (not shown) placed in the dendrites where the maximal negative EPSP could be recorded (i.e., corresponding to the level of axons activated by the stimulating electrode). Responses were allowed to stabilize before data were taken. Then, at least three sets of stimulus duration–population spike amplitude curves were taken over a 15- to 30-min period to establish a base line before any drugs were added or a high-frequency train (HFT) of

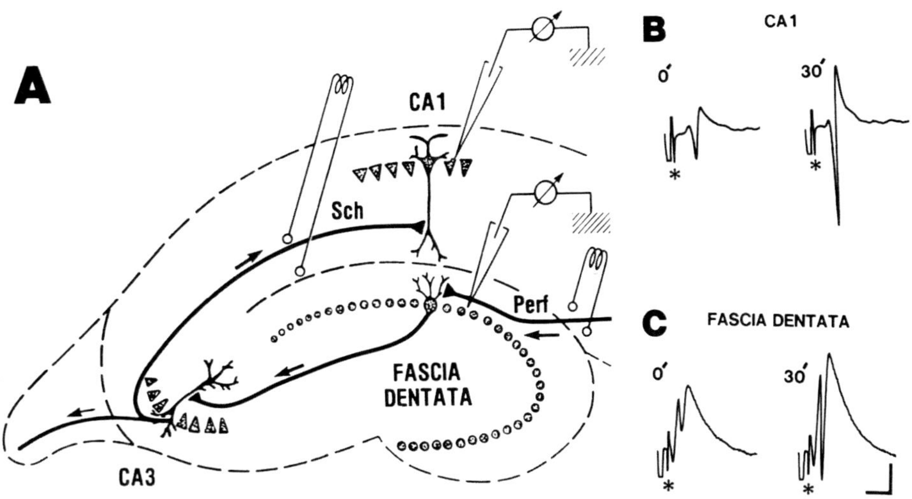

FIGURE 1. (A) The hippocampal slice preparation showing both recording sites (dentate and CA1) and both stimulus sites [perforant path (Perf) and Schaffer collaterals/commissural fibers (Sch)]. Only one set of recording and stimulus sites was employed in a given slice. (B) Control LTP in field CA1. The population spike was recorded in CA1 just prior to (0′) and 30 min after (30′) repetitive stimulation of the Schaffer collaterals/commissural fibers (Sch; 20 Hz for 10 sec). Asterisks denote stimulus artifact. (C) Control LTP in the dentate. The population spike was recorded in the dentate, and repetitive stimulation was applied to the perforant path (100 Hz for 2 sec). Calibration for B and C: 1 mV, 5 msec. From Stanton and Sarvey (1985a), reprinted with permission of the Society for Neuroscience.

repetitive stimulation was applied. The LTP was assessed 30 min after the HFT; NE-induced LLP (NELLP) was assessed 30 min after beginning washout of NE with drug-free physiological solution. In slices not given drugs or an HFT, responses remained stable for over 1 hr.

2. NOREPINEPHRINE DEPLETION

Figure 1 shows an example of LTP in field CA1 and the dentate gyrus (Stanton and Sarvey, 1985b). As shown in Fig. 2, LTP was significantly reduced in the dentate of slices from animals whose hippocampus had been depleted of norepinephrine (to 17% of untreated hippocampi). In contrast, in field CA1 of slices from depleted animals, LTP was not significantly affected (Stanton and Sarvey, 1985b). Thus, the effect of depletion of NE appears to be restricted to the dentate gyrus.

If the β-adrenergic antagonist propranolol (20 μM) or the β_1 antagonist metoprolol (20 μM) was added to the buffer perfusing the slice before, and left in during, the HFT, LTP in the dentate was significantly reduced (Stanton and Sarvey, 1985b). These concentrations of propranolol and metoprolol had no effect on the base-line orthodromic population spike amplitude or on the antidromic population spike elicited by stimulation of the mossy fibers in the dentate hilus. These data suggest that NE, acting through a β-adrenergic receptor, is required for LTP in the dentate.

Further evidence for a β-adrenergic mechanism is the finding that a 1 μM concentration of the adenylate cyclase stimulant forskolin restored LTP in dentate of slices from

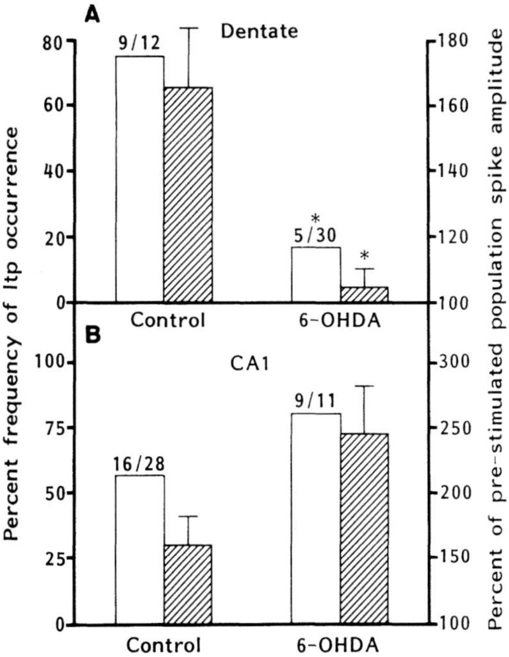

FIGURE 2. (A) Percentage of occurrence of LTP (open bars) and percentage of increase in population spike amplitude (mean ± S.E.M., hatched bars) in the dentate 30 min after repetitive stimulation of the perforant path. Above each open bar is the number of slices showing LTP per total number of slices tested. Long-term potentiation was defined as an increase in population spike amplitude greater than 2 S.D. over mean control amplitudes 30 min after repetitive stimulation. Depletion of NE by 6-OHDA virtually eliminated both the frequency of occurrence (*,χ^2, $P < 0.05$) and the increased amplitude (*, Student's *t*-test,$P < 0.05$) of LTP. (B) In contrast, depletion of NE by 6-OHDA did not impair LTP in field CA1 when repetitive stimulation was applied to the Schaffer collaterals. From Stanton and Sarvey (1985a), reprinted with permission of the Society for Neuroscience.

depleted rats (Stanton and Sarvey, 1985b). This concentration of forskolin had no effect on the population spike (Stanton and Sarvey, 1985a,c) and reportedly (Seamon *et al.*, 1981) has mainly a "priming" effect on adenylate cyclase.

Experiments were also performed with two recording electrodes, one in the cell body layer and one in the dendritic layer corresponding to the maximally activated excitatory synaptic input from the perforant path (Stanton and Sarvey, 1987). Normally, during LTP, the slope of the initial phase of the EPSP became steeper, that is, was potentiated, to 203 ± 39% ($n = 10$) of pre-HFT. In slices from NE-depleted rats, however, the slope was not significantly different from pre-HFT (117 ± 21%, $n = 7$; Stanton and Sarvey, 1987). Thus, depletion of NE inhibits LTP not only of the population spike but also of the EPSP.

Experiments in which 3′,5′-cyclic adenosine monophosphate (cAMP) levels in dentate gyrus were measured before and 30 min after an HFT further strengthened the case for a role of β-adrenergic receptors in LTP. In dentate of slices from untreated rats, cAMP levels more than doubled 1 min after an HFT to the perforant path but had declined to pre-HFT levels 30 min after the HFT. In contrast, in slices from NE-depleted rats, there was no significant increase in cAMP levels (Stanton and Sarvey, 1985b). These data suggest that an increase in cAMP levels in the dentate gyrus is required to initiate, but not to maintain, LTP.

Therefore, we can conclude that NE, acting through a β-adrenergic receptor linked to adenylate cyclase, appears to be required for LTP in the dentate, but not field CA1, of the hippocampus. Depletion of NE decreases LTP of both the population EPSP and the population spike.

3. NOREPINEPHRINE-INDUCED LONG-LASTING POTENTIATION

Exposure of slices to NE in the physiological solution bathing the slices resulted in a concentration-dependent potentiation of the perforant-path-evoked population spike in the dentate gyrus (Stanton and Sarvey, 1985c). This potentiation persisted for at least several hours after washout of NE following a 30-min exposure to NE. The concentration required to produce 50% of the maximal NE-induced long-lasting potentiation (NELLP) was 20 μM.

In field CA1, in contrast, bath application of NE produced no effect or a slight decrease in the population spike. After washout of NE, the population spike returned to the pre-NE amplitude (Stanton and Sarvey, 1985c). Thus, the effect of bath-applied NE, like that of NE depletion, is restricted to the dentate.

Several lines of evidence suggest that β-adrenergic receptors mediate NELLP. The β antagonists propranolol and metoprolol almost completely prevented the development of NELLP (Stanton and Sarvey, 1985c). Bath application of forskolin in a concentration (1 μM) that "primes" adenylate cyclase without producing any effect on the field potential shifted the log dose–response curve for NELLP to the left (Stanton and Sarvey, 1985c). Furthermore, as shown in Fig. 3, the β agonist isoproterenol (1 μM) could induce an LLP just like that induced by NE. Finally, NE increased cAMP levels more than threefold within 1 min; the level remained threefold higher 30 min after beginning washout of NE (Stanton and Sarvey, 1985b). By comparison with LTP in NE-depleted slices, it seems likely that only the initial increase in cAMP is required for NELLP, although this has not been established.

Figure 3 also shows that isoproterenol induced both an acute and a long-lasting potentiation of the EPSP recorded extracellularly in the dendrites. A 30-min exposure to 50 μM NE increased the absolute value of the maximum initial slope of the EPSP to 136

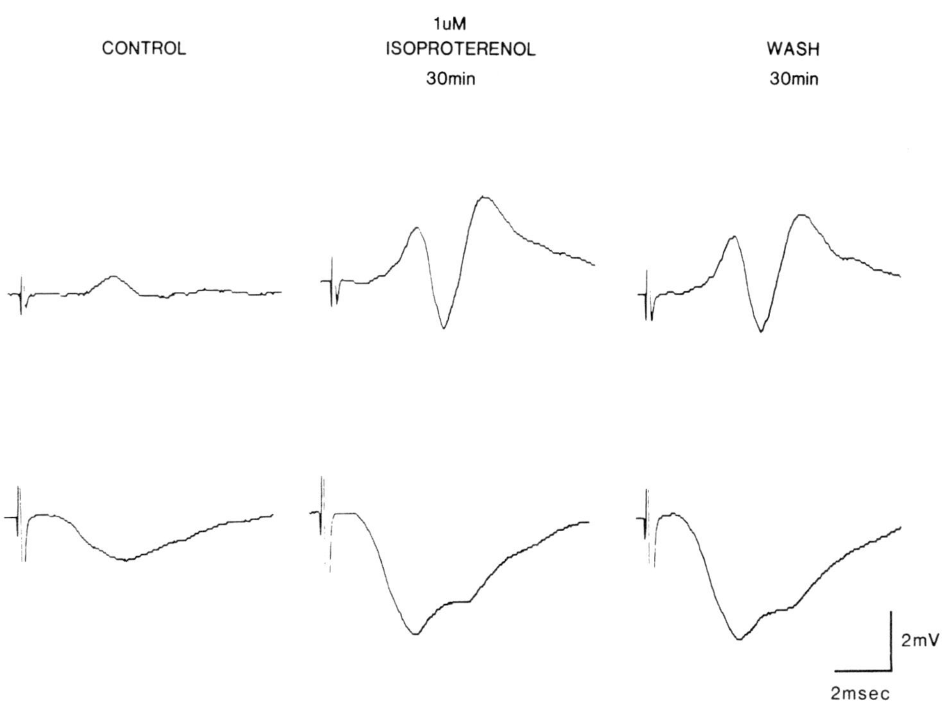

FIGURE 3. Long-lasting potentiation induced by isoproterenol in dentate gyrus. Control: response to stimulation of perforant path recorded in the granule cell body layer (upper trace) and molecular layer (i.e., dendrites; lower trace). Isoproterenol: 30-min exposure to 1 μM isoproterenol markedly potentiated both the population spike (upper trace) and the dendritically recorded EPSP (lower). Wash: the potentiation persisted through a 30-min wash with drug-free physiological solution.

± 9.7% ($n = 6$) of the pre-NE slope, and after a 30-min washout of NE, it was increased to 151 ± 11.9% of the pre-NE slope (Stanton and Sarvey, 1987). Thus, it appears that NE, acting through β-receptors, induces an LLP of the EPSP as well as the population spike.

4. PROTEIN SYNTHESIS

Several studies have implicated protein synthesis in LTP. During LTP, Duffy *et al.* (1981) found an increase in newly synthesized extracellular protein from hippocampal slices preloaded with radiolabeled valine. However, it is not clear whether the increase is required for production or maintenance of LTP. Repetitive stimulation of the perforant path in mice *in vivo* is followed by swelling of spines on the middle and distal thirds of the dendrites of dentate granule cells (Fifkova and Van Harreveld, 1977). Fifkova *et al.* (1982) demonstrated that intraperitoneal injection of anisomycin, an inhibitor of protein synthesis, 15 min before delivery of an HFT to the perforant path inhibited spine swelling 4 min but not 90 min after the HFT. It would be interesting to know what occurs in the intervening time. At 4 min, a number of changes apparently unrelated to LTP can occur (Misgeld *et al.*, 1979; McNaughton, 1982; Scharfman and Sarvey, 1985a,b; Stanton and Sarvey, 1984). Since electrophysiological recordings were not made by Fifkova *et al.* (1982), it is not clear whether LTP was expressed 90 min after the HFT in their preparation. Using rats with electrodes chronically implanted in dentate and perforant path, Krug *et al.* (1984) administered anisomycin 15 min before a series of HFTs and found that it had no effect on LTP for approximately the first 3 hr but then reversed LTP.

In order to test whether synthesis of proteins is required for LTP, we examined the effect of the inhibitors of protein synthesis emetine, cycloheximide, puromycin, and anisomycin on establishment of LTP in field CA1 of rat hippocampal slices (Stanton and Sarvey, 1984). In slices exposed to any of the four inhibitors in concentrations that blocked more than 85% of protein synthesis (measured as [^{3}H]valine incorporation into trichloroacetic-acid-precipitable macromolecules), the population spike measured 30 min after an HFT was not significantly larger than the pre-HFT population spike. A number of control experiments, including demonstration that emetine did not affect intracellularly recorded synaptic or membrane biophysical characteristics, suggest that these agents were inhibiting LTP by a specific effect on protein synthesis (Stanton and Sarvey, 1984). Furthermore, whereas a 30-min pre-HFT incubation with emetine completely blocked LTP measured 30 min post-HFT, addition of emetine at the same time as the HFT was not able to block LTP (Stanton and Sarvey, 1984).

The similarities between LTP and NELLP described above led us to test the effect of bath-applied emetine (15 μM) on NELLP. Perfusion of slices with emetine starting 30 min before perfusion with NE produced no effect on the acute potentiation in the presence of 50 μM NE but markedly decreased the LLP that is normally evident 30 min after beginning washout of NE (Stanton and Sarvey, 1985c). Thus, it appears that, as in LTP, establishment of NELLP requires either synthesis of new protein(s) or ongoing synthesis of some rapidly turned over protein(s).

5. REPETITIVE FIRING

Production of LTP appears to require firing of action potentials in the postsynaptic neurons during the HFT, as blockade of the population spike during the HFT prevents the production of LTP in fields CA1 and CA3 and the dentate gyrus (Scharfman and Sarvey, 1985a). As shown in Fig. 4, application of a discrete amount of GABA from a micropipette to the cell body layer can reversibly block the population spike without affecting the dendritically recorded EPSP. If this amount of GABA is applied coincident with the HFT so that no action potentials can fire during the HFT, LTP does not develop. The specificity of the effect is further underscored by the failure of the GABA injection to block short-term potentiation, which appears during the first few minutes after HFT (Fig. 4). Blockade of the population spike during the HFT with tetrodotoxin or pentobarbital also prevents production of LTP (Scharfman and Sarvey, 1985a). Thus, prevention of firing of action potentials during the HFT appears to be sufficient to prevent the production of LTP. Similar results were obtained by two groups who prevented action potential firing by voltage clamping (Kelso *et al.*, 1986) or current clamping (Malinow

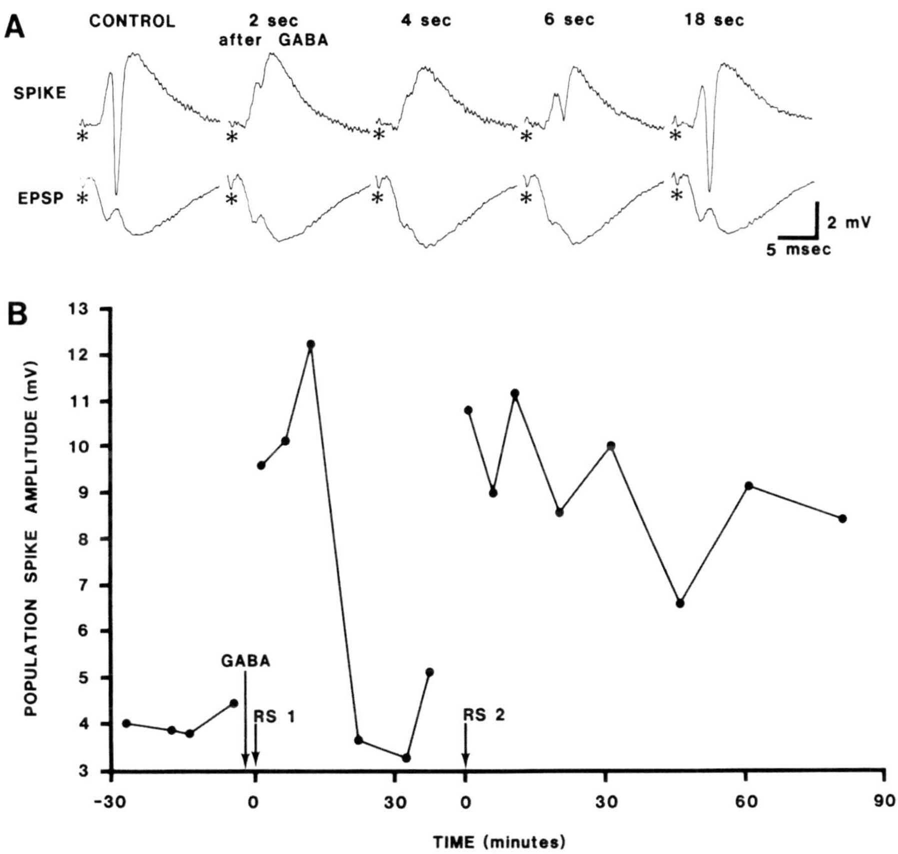

FIGURE 4. (A) Responses evoked by single test stimuli (asterisks; 35 μA, 190 μsec) to the Schaffer collateral/commissural axons before (control) and at various times after pressure ejection of GABA (15 psi, 200 msec). Note that the population spike was blocked for 4–6 sec and completely recovered in 18 sec, whereas the EPSP was unaffected. Neither the maximum d*V*/d*t* of the initial phase of the EPSP, amplitude at a fixed latency, nor peak amplitude of the EPSP changed. A lower stimulus intensity (35 μA, 170 μsec), which produced a small population spike, was used for repetitive stimulation. Each trace represents one evoked response. (B) GABA was applied (GABA; long arrow; 15 psi, 200 msec) immediately prior to repetitive stimulation (RS 1; 100 Hz for 2 sec; short arrow). Short-term potentiation (STP) occurred, but long-term potentiation (LTP) did not. A second, identical repetitive stimulation (RS 2; 100 Hz for 2 sec; short arrow), which was not preceded by a GABA pulse, was given 37 min later; LTP as well as STP was produced. From Scharfman and Sarvey (1985a) with permission.

and Miller, 1986) the postsynaptic neuron at a hyperpolarized potential. In contrast, Wigström *et al.* (1982) were unable to block LTP by current clamping neurons at a hyperpolarized potential, perhaps because of poor control of membrane potential (see their Fig. 1, for example) or because of heterosynaptic interactions at high stimulus frequencies between two overlapping synaptic fields. Action potentials may serve to depolarize the neurons to allow activation of the N-methyl-D-aspartate (NMDA) subclass of excitatory amino acid receptor (Watkins and Evans, 1981; Foster and Fagg, 1984; Nowak *et al.*, 1984; Mayer *et al.*, 1984), without which hippocampal LTP apparently cannot occur (Collingridge *et al.*, 1983; Harris *et al.*, 1984; Wigström and Gustafsson, 1984; Chapter 6), although the mossy fiber projection to field CA3 in guinea pig already appears to be an exception to this rule (Harris and Cotman, 1986). Of course, presynaptic NMDA receptors may also be involved (Collingridge, 1985).

6. CONCLUSIONS

In conclusion, LTP could be produced by several converging or parallel mechanisms. Norepinephrine, acting through a β-receptor-mediated increase in cAMP, can by itself induce an LLP. However, during normal induction of LTP, the second messengers calcium and protein kinase C (PKC) may (1) lead to phosphorylation of the same substrates as cAMP, (2) act in parallel with cAMP to produce similar effects (i.e., potentiation), or (3) enhance the activity of adenylate cyclase.

The site of action of NE could be presynaptic, postsynaptic, or both. In a postsynaptic model, NE, exogenously applied or released during an HFT, acts on postsynaptic β receptors to increase synthesis of cAMP, which then activates cAMP-dependent protein kinase. The kinase could then phosphorylate a number of substrates, some of which, such as membrane ion channels, may be involved in the establishment of potentiation. Additionally, cAMP-dependent protein kinase may translocate to the nucleus, where it could initiate transcription of specific mRNA(s) required for maintenance of potentiation. The protein(s) synthesized by this mRNA could be an ion channel or transmitter receptor, for example. There is evidence for cAMP-dependent transcription in a number of eukaryotic systems (Murdoch *et al.*, 1982; Eiden *et al.*, 1984; Montminy *et al.*, 1986).

A synergistic action of cAMP with calcium, granted access to a critical site near the NMDA receptor by activation of that receptor (Dingledine, 1983, 1986; MacDermott *et al.*, 1986), could enhance or serve as a trigger for the potentiation induced by NE. A similar synergy between calcium–calmodulin and cAMP at a serotonin receptor has been postulated for long-term sensitization of the gill-withdrawal reflex in *Aplysia* (Abrams, 1985). There is evidence in brain that certain excitatory amino acid receptors may be associated with inositol phospholipid turnover (Nicoletti *et al.*, 1986a), although it is still unclear whether these receptors are the electrophysiologically identified NMDA receptors (Watkins and Evans, 1981; Foster and Fagg, 1984; Nicoletti *et al.*, 1986b). The polyphosphoinositide hydrolysis product diacylglycerol activates PKC, which has been suggested to act in a synergistic or synarchic fashion with cAMP (Rasmussen, 1986a,b; Chapter 1). Perhaps activation of cAMP and PKC together can decrease the threshold for potentiation or enhance the ensuing potentiation.

Alternatively, both cAMP and PKC may act at different sites in LTP: cAMP at a postsynaptic site and PKC at a presynaptic site. Worley *et al.* (1986a) have shown that binding of [^{3}H]forskolin, which is specific for cAMP, is localized to granule cell dendrites and axons, whereas [^{3}H]phorbol-12,13-dibutyrate, which binds to PKC (Worley *et al.*, 1986b), binds preferentially to perforant path axons. As mentioned earlier, it is not clear whether the NMDA receptors critical for LTP are located pre- or postsynaptically.

Of course, it should be borne in mind that, in dentate, NE is required for LTP (Stanton and Sarvey, 1985a) and sufficient for NELLP (Stanton and Sarvey, 1985c). The relative necessity for activation of PKC during LTP has not been established, although activation of PKC by phorbol esters appears to be sufficient to induce an LLP in field CA1 (Malenka *et al.*, 1986) and can enhance LTP in dentate (Routtenberg *et al.*, 1986).

Additionally, or as another alternative, NE could be acting presynaptically, again via a β-mediated increase in cAMP, to enhance release of excitatory transmitter. A presynaptic site of action is not ruled out by Worley and co-workers' (1986a) findings, which show a relative, not an absolute, abundance of postsynaptic binding sites. Evidence for an acute presynaptic effect comes from Lynch and Bliss (1986), who found that NE, apparently acting via β receptors, enhances potassium-induced, calcium-dependent release of [^{14}C]glutamate from dentate gyrus but not CA1/CA3. It would be interesting to examine the long-lasting effects of NE on glutamate release.

In summary, NE, acting presynaptically, postsynaptically, or both, is required for initiation of LTP in the dentate gyrus but not field CA1. Similarly, NE by itself can initiate an LLP in dentate but not field CA1. The action of NE appears to be mediated by β receptors linked to adenylate cyclase, which generates cAMP. Concomitant pre- and postsynaptic activity is required for production of LTP; presumably, postsynaptic depolarization permits activation of NMDA receptors. The identity and function of macromolecules phosphorylated by cAMP-dependent processes, the relative contributions of calcium–calmodulin- and PKC-dependent phosphorylation, and the role of newly synthesized or rapidly turned over proteins in LTP are only beginning to be understood. The recent development of a monoclonal antibody that binds with high specificity to a 42-kDA cell-surface antigen in hippocampus (Moskal *et al.*, 1985) and blocks LTP with minimal effect on normal synaptic transmission (Stanton *et al.*, 1985) may provide a molecular tool to begin to determine the significance of protein synthesis and posttranslational processing in LTP.

ACKNOWLEDGMENT. This work was supported by USUHS Grant No. CO7514. The opinions or assertions contained herein are the private ones of the author and are not to be construed as official or reflecting the views of the DoD or the USUHS. The experiments reported herein were conducted according to the principles set forth in the "Guide for care and Use of Laboratory Animals," Institute of Animal Resources, National Research Council, DHEW Pub. No. (NIH) 74-23.

REFERENCES

Abrams, T. W., 1985, Activity-dependent presynaptic facilitation: An associative mechanism in *Aplysia*, *Cell. Mol. Neurobiol.* **5:**123–145.

Alger, B. E., and Teyler, T. J., 1976, Long-term and short-term plasticity in the CA1, CA3 and dentate regions of the rat hippocampal slice, *Brain Res.* **110:**463–480.

Andersen, P., Sundberg, S. H., Sveen, O., and Wigström, H., 1977, Specific long-lasting potentiation of synaptic transmission in hippocampal slices, *Nature* **266:**736–737.

Andersen, P., Sundberg, S. H., Sveen, O., Swann, J. W., and Wigström, H., 1980, Possible mechanism for long-term potentiation of synaptic transmission in hippocampal slices from guinea-pig, *J. Physiol.* (*Lond.*) **302:**463–482.

Ashe, J. H., 1984, A possible enabling and enhancing function for catecholamines in neuronal plasticity, in: *Memory Systems of the Brain* (N. M. Weinberger, J. L. McGaugh, and G. Lynch, eds.), Guilford, New York, pp. 107–119.

Bär, P. R., Schotman, P., Gispen, W. H., Tielen, A. M., and Lopes da Silva, F. H., 1980, Changes in synaptic membrane phosphorylation after tetanic stimulation in the dentate area of the rat hippocampal slice, *Brain Res.* **198:**478–484.

Barrionuevo, G., Kelso, S. R., Johnston, D., and Brown, T. H., 1986, Conductance mechanism responsible for long-term potentiation in monosynaptic and isolated excitatory synaptic inputs to hippocampus, *J. Neurophysiol.* **55:**540–550.

Bliss, T. V. P., and Gardner-Medwin, A. R., 1973, Long-lasting potentiation of synaptic transmission in the dentate area of unanaesthetized rabbit following stimulation of the perforant path, *J. Physiol.* (*Lond.*) **232:**357–374.

Bliss, T. V. P., and Lømo, T., 1973, Long-lasting potentiation of synaptic transmission in the dentate area of the anaesthetized rabbit following stimulation of the perforant path, *J. Physiol.* (*Lond.*) **232:**331–356.

Bliss, T. V. P., Goddard, G. V., and Riives, M., 1983, Reduction of long-term potentiation in the dentate gyrus of the rat following selective depletion of monoamines, *J. Physiol.* (*Lond.*) **334:**475–491.

Browning, M., Bennett, W. F., Kelly, P., and Lynch, G., 1981, Evidence that the 40,000 M_r phosphoprotein influenced by high frequency synaptic stimulation is the alpha subunit of pyruvate dehydrogenase, *Brain Res.* **218:**255–266.

Collingridge, G. L., 1985, Long term potentiation in the hippocampus: Mechanisms of initiation and modulation by neurotransmitters, *Trends Pharmacol. Sci.* **6:**407–411.

Collingridge, G. L., Kehl, S. J., and McLennan, H., 1983, Excitatory amino acids in synaptic transmission in the Schaffer collateral–commissural pathway of the rat hippocampus, *J. Physiol.* (*Lond.*) **334:**33–46.

Dahl, D., and Winson, J., 1985, Action of norepinephrine in the dentate gyrus. I. Stimulation of the locus coeruleus, *Exp. Brain Res.* **59:**491–496.

Dingledine, R., 1983, N-Methyl aspartate activates voltage-dependent calcium conductance in rat hippocampal pyramidal cells, *J. Physiol.* (*Lond.*) **343:**385–405.

Dingledine, R., 1986, NMDA receptors: What do they do? *Trends Neurosci.* **9:**47–49.

Duffy, C., Teyler, T. J., and Shashoua, V. E., 1981, Long-term potentiation in the hippocampal slice: Evidence for stimulated secretion of newly synthesized proteins, *Science* **212:**1148–1151.

Dunwiddie, T. V., and Lynch, G., 1979, The relationship between extracellular calcium concentrations and the induction of hippocampal long-term potentiation, *Brain Res.* **169:**103–110.

Dunwiddie, T., Madison, D., and Lynch, G., 1978, Synaptic transmission is required for initiation of long-term potentiation, *Brain Res.* **150:**413–417.

Eiden, L. E., Girand, P., Affolter, H. U., Herbert, E., and Hotchkiss, A. J., 1984, Alternative modes of enkephalin biosynthesis regulation by reserpine and cyclic AMP in cultured chromaffin cells, *Proc. Natl. Acad. Sci. U.S.A.* **81:**3949–3953.

Fifkova, E., and Van Harreveld, A., 1977, Long-lasting morphological changes in dendritic spines of dentate granular cells following stimulation of the entorhinal area, *J. Neurocytol.* **6:**211–230.

Fifkova, E., Anderson, C. L., Young S. J., and Van Harreveld, A., 1982, Effect of anisomycin on stimulation-induced changes in dendritic spines of the dentate granule cells, *J. Neurocytol.* **11:**183–210.

Foster, A. C., and Fagg, G. E., 1984, Acidic amino acid binding sites in mammalian neuronal membranes: Their characteristics and relationship to synaptic receptors, *Brain Res. Rev.* **7:**103–164.

Gribkoff, V. K., and Ashe, J. H., 1984, Modulation by dopamine of population responses and cell membrane properties of hippocampal CA1 neurons in vitro, *Brain Res.* **292:**327–338.

Harley, C. W., and Milway, J. S., 1986, Glutamate ejection in the locus coeruleus enhances the perforant path-evoked population spike in the dentate gyrus, *Exp. Brain Res.* **63:**143–150.

Harris, E. W., and Cotman, C. W., 1986, Long-term potentiation of guinea pig mossy fiber responses is not blocked by N-methyl D-aspartate antagonists, *Neurosci. Lett.* **70:**132–137.

Harris, E. W., Ganong, A. H., and Cotman, C. W., 1984, Long-term potentiation in the hippocampus involves activation of N-methyl-D-aspartate receptors, *Brain Res.* **323:**132–137.

Hoch, D. B., Dingledine, R. J., and Wilson, J. E., 1984, Long-term potentiation in the hippocampal slice: Possible involvement of pyruvate dehydrogenase, *Brain Res.* **302:**125–134.

Kelso, S. R., Ganong, A. H., and Brown, T. H., 1986, Hebbian synapses in hippocampus, *Proc. Natl. Acad. Sci. U.S.A.* **83:**5326–5330.

Krug, M., Lössner, B., and Ott, T., 1984, Anisomycin blocks the late phase of long-term potentiation in the dentate gyrus of freely moving rats, *Brain Res. Bull.* **13:**39–42.

Lacaille, J.-C., and Harley, C. W., 1985, The action of norepinephrine in the dentate gyrus: Beta-mediated facilitation of evoked potentials *in vitro*, *Brain Res.* **358:**210–220.

Lynch, G. S., and Baudry, M., 1984, The biochemistry of memory: A new and specific hypothesis, *Science* **224:**1057–1063.

Lynch, G., Larson, J., Kelso, S., Barrionuevo, G., and Schottler, F., 1983, Intracellular injections of EGTA block induction of hippocampal long-term potentiation, *Nature* **305:**719–721.

Lynch, M. A., and Bliss, T. V. P., 1986, Noradrenaline modulates the release of [^{14}C]glutamate from dentate but not from CA1/CA3 slices of rat hippocampus, *Neuropharmacology* **25**:493–498.

MacDermott, A. B., Mayer, M. L., Westbrook, G. L., Smith, S. J., and Barker, J.L., 1986, NMDA-receptor activation increases cytoplasmic calcium concentration in cultured spinal cord neurones, *Nature* **321**:519–522.

Malenka, R. C., Madison, D. V., and Nicoll, R. A., 1986, Potentiation of synaptic transmission in the hippocampus by phorbol esters, *Nature* **321**:175–177.

Malinow, R., and Miller, J. P., 1986, Postsynaptic hyperpolarization during conditioning reversibly blocks induction of long-term potentiation, *Nature* **320**:529–530.

Mayer, M. L., Westbrook, G. L., and Guthrie, P. B., 1984, Voltage-dependent block by Mg^{2+} of NMDA responses in spinal cord neurones, *Nature* **309**:261–263.

McNaughton, B. L., 1982, Long-term enhancement and short-term potentiation in rat fascia dentata act through different mechanisms, *J. Physiol.* (*Lond.*) **324**:249–262.

Misgeld, U., Sarvey, J. M., and Klee, M. R., 1979, Heterosynaptic postactivation potentiation in hippocampal CA 3 neurons: Long-term changes of the postsynaptic potentials, *Exp. Brain Res.* **37**:217–229.

Montminy, M. R., Low, J. M., Tapia-Arancibia, L., Reichlin, S., Mandel, G., and Goodman, R, H., 1986, Cyclic AMP regulates somatostatin mRNA accumulation in primary diencephalic cultures and in transfected fibroblast cells, *J. Neurosci.* **6**:1171–1176.

Morris, R., and Baker, M., 1984, Does long-term potentiation/synaptic enhancement have anything to do with learning or memory? in: *Neurobiology of Learning and Memory* (G. Lynch, J. L., McGaugh, and N. M. Weinberger, eds.), Guilford, New York, pp. 521–535.

Moskal, J. R., Schaffner, A. E., and Koller, K. J., 1985, The identification and partial characterization of a cell surface protein that modulates REM in neonatal rats and long term synapse plasticity in hippocampal slice preparations, *Soc. Neurosci. Abstr.* **11**:838.

Murdoch, G. H., Rosenfeld, M. G., and Evans, R. M., 1982, Eukaryotic transcriptional regulation and chromatin associated protein phosphorylation by cyclic AMP, *Science* **218**:1315–1317.

Neuman, R. S., and Harley, C. W., 1983, Long-lasting potentiation of the dentate gyrus population spike by norepinephrine, *Brain Res.* **273**:162–165.

Nicoletti, F., Meek, J. L., Iadorola, M. J., Chuang, D. M., Roth, B. L., and Costa, E., 1986a, Coupling of inositol phospholipid metabolism with excitatory amino acid recognition sites in rat hippocampus, *J. Neurochem.* **46**:40–46.

Nicoletti, F., Wroblewski, J. T., Novelli, A., Alho, H., Guidotti, A., and Costa, E., 1986b, The activation of inositol phospholipid metabolism as a signal-transducing system for excitatory amino acids in primary cultures of cerebellar granule cells, *J. Neurosci.* **6**:1905–1911.

Nowak, L., Bregestovski, P., Ascher, P., Hebet, A., and Prochiantz, A., 1984, Magnesium gates glutamate-activated channels in mouse central neurones, *Nature* **307**:462–465.

Rasmussen, H., 1986a, The calcium messenger system (first of two parts), *N. Engl. J. Med.* **314**:1094–1101.

Rasmussen, H., 1986b, The calcium messenger system (second of two parts), *N. Engl. J. Med.* **314**:1164–1170.

Robinson, B. G., and Racine, R. J., 1985, Long-term potentiation in the dentate gyrus: Effects of noradrenaline depletion in the awake rat, *Brain Res.* **325**:71–78.

Routtenberg, A., Colley, P., Linden, D., Lovinger, D., Murakami, K., and Sheu, F.-S., 1986, Phorbol ester promotes growth of synaptic plasticity, *Brain Res.* **378**:374–378.

Scharfman, H. E., and Sarvey, J. M., 1985a, Postsynaptic firing during repetitive stimulation is required for long-term potentiation in hippocampus, *Brain Res.* **331**:267–274.

Scharfman, H. E., and Sarvey, J. M., 1985b, γ-Aminobutyrate sensitivity does not change during long-term potentiation in rat hippocampal slices, *Neuroscience* **15**:695–702.

Schwartzkroin, P. A., 1975, Characteristics of CA1 neurons recorded intracellularly in the hippocampal *in vitro* slice preparation, *Brain Res.* **84**:424–436.

Schwartzkroin, P. A., and Wester, K., 1975, Long-lasting facilitation of the synaptic potential following tetanization in the *in vitro* hippocampal slice, *Brain Res.* **89**:107–119.

Seamon, K. B., Padgett, W., and Daly, J. W., 1981, Forskolin: Unique diterpine activator of adenylate cyclase in membranes and in intact cells, *Proc. Natl. Acad. Sci. U.S.A.* **78**:3363–3367.

Stanton, P. K., and Sarvey, J. M., 1984, Blockade of long-term potentiation in rat hippocampal CA1 region by inhibitors of protein synthesis, *J. Neurosci.* **4**:3080–3088.

Stanton, P. K., and Sarvey, J. M., 1985a, Depletion of norepinephrine, but not serotonin, reduces long-term potentiation in the dentate gyrus of rat hippocampal slices, *J. Neurosci.* **5**:2169–2176.

Stanton, P. K., and Sarvey, J. M., 1985b, The effect of high-frequency electrical stimulation and norepinephrine on cyclic AMP levels in normal versus norepinephrine-depleted rat hippocampal slices, *Brain Res.* **358**:343–348.

Stanton, P. K., and Sarvey, J. M., 1985c, Blockade of norepinephrine-induced long-lasting potentiation in the hippocampal dentate gyrus by an inhibitor of protein synthesis, *Brain Res.* **361**:276–283.

Stanton, P. K., and Sarvey, J. M., 1987, Norepinephrine regulates long-term potentiation of both the population spike and dendritic EPSP in hippocampal dentate gyrus, *Brain. Res. Bull.* **18**:115–119.

Stanton, P. K., Sarvey, J. M., and Moskal, J. R., 1985, A monoclonal antibody (MAb) to a cell-surface antigen which inhibits both production and maintenance of long-term potentiation (LTP) in rat hippocampal slice, *Soc. Neurosci. Abstr.* **11**:838.

Swanson, L. W., Teyler, T. J., and Thompson, R. F., 1982, Hippocampal long-term potentiation: Mechanisms and implications for memory, *Neurosci. Res. Prog. Bull.* **20**:613–769.

Watkins, J. C., and Evans, R. H., 1981, Excitatory amino acid transmitters, *Annu. Rev. Pharmacol. Toxicol.* **21**:165–204.

Wigström, H., and Gustafsson, B., 1984, A possible correlate of the postsynaptic condition for long-lasting potentiation in the guinea pig hippocampus *in vitro*, *Neurosci. Lett.* **44**:327–332.

Wigström, H., and Gustafsson, B., 1985, Facilitation of hippocampal long-lasting potentiation by GABA antagonists, *Acta Physiol. Scand.* **125**:159–172.

Wigström, H., McNaughton, B. L., and Barnes, C. A., 1982, Long-term enhancement in hippocampus is not regulated by postsynaptic membrane potential, *Brain Res.* **233**:195–199.

Winson, J., and Dahl, D., 1985, Action of norepinephrine in the dentate gyrus. II. Iontophoretic studies, *Exp. Brain Res.* **59**:497–506.

Worley, P. F., Baraban, J. M., De Sousa, E. B., and Snyder, S. H., 1986a, Mapping second messenger systems in the brain: Differential locations of adenylate cyclase and protein kinase C, *Proc. Natl. Acad. Sci. U.S.A.* **83**:4053–4057.

Worley, P. F., Baraban, J. M., and Snyder, S. H., 1986b, Heterogeneous localization of protein kinase C in rat brain: Autoradiographic analysis of phorbol ester receptor binding, *J. Neurosci.* **6**:199–207.

Yamamoto, C., and Chujo, T., 1978, Long-term potentiation in thin hippocampal sections studied by intracellular and extracellular recordings, *Exp. Neurol.* **58**:242–250.

9

Mechanisms of Noradrenergic Modulation of Dentate Gyrus Long-Term Plasticity

P. K. STANTON and U. HEINEMANN

1. INTRODUCTION

Since D. O. Hebb (1949) inaugurated the search for cellular mechanisms underlying brain function and behavioral plasticity, many strategies and model systems have been employed. One fruitful strategy has been the search for specific chemical transmitters able to modulate firing patterns of neurons in specific pathways for long periods of time. A promising model system has arisen from the discovery that brief, high-frequency stimulation of afferent pathways in the hippocampus leads to long-lasting enhancements of neuronal excitability whose persistence approach that of conditioned behavior (Bliss and Gardner-Medwin, 1973; Douglas and Goddard, 1975). The enhancement of evoked potentials after one such tetanus has been termed long-term potentiation (LTP; Bliss and Lømo, 1973; Schwartzkroin and Wester, 1975; Alger and Teyler, 1976), and repeated application of such stimulation yields the seizure state known as kindled epilepsy (Goddard *et al.*, 1969). In both cases, the location of such long-lasting plasticity in a brain structure implicated in memory processes (Milner, 1972; Berger, 1984), its production by brief (a few seconds) stimulation within the physiological range (10–400 Hz), and the long duration of the changes (months *in vivo*) all led to extreme interest in their underlying mechanisms (Swanson *et al.*, 1982).

It seems possible that long-lasting changes in excitability induced by repetitive stimulation may mimic the requirement for repetition observed in experience-dependent memory, whereas neuromodulators that modulate the strength and duration of such changes may correspond to the capability for associative learning. Such a potential modulatory neurotransmitter is the monoamine norepinephrine (NE), which has been strongly implicated both in memory processes (Stein *et al.*, 1975; Crow and Wendlandt, 1976; Mason and Iversen, 1977; Everitt *et al.*, 1983; Zornetzer, 1984) and hippocampal plasticity (McIntyre and Edson, 1982; Bliss *et al.*, 1983; Stanton and Sarvey, 1985a). Here we

---

P. K. STANTON • Department of Biophysics, John Hopkins University, Baltimore, Maryland 21218. U. HEINEMANN • Institute for Normal and Pathological Physiology, University of Cologne, 5000 Cologne 41, Federal Republic of Germany. This chapter is dedicated to the memory of Gary L. Stanton.

consider the cellular mechanisms of action of NE on hippocampal neurons and how these may relate to noradrenergic modulation of hippocampal LTP and behavioral plasticity.

2. NOREPINEPHRINE AND EXPRESSION OF LONG-LASTING HIPPOCAMPAL PLASTICITY

Evidence for a functional role for NE in hippocampal plasticity began with studies showing that prior depletion of NE markedly impairs perforant path LTP in the dentate gyrus both *in vivo* (Bliss *et al.*, 1983) and in isolated hippocampal slices (Stanton and Sarvey, 1985a). The slice work indicated a hippocampal site of action as well as specificity within the hippocampus, since LTP in field CA1 was unaffected. Furthermore, it was shown that β_1-receptor stimulation of cAMP is the most likely mediator of these effects (Stanton and Sarvey, 1985a–c).

Of equal interest was the finding that a brief application of NE in the absence of high-frequency electrical stimulation also produced a long-lasting potentiation of perforant path synaptic transmission in the dentate gyrus (Neuman and Harley, 1983; Stanton and Sarvey, 1985c). This potentiation showed similar area specificity and also appeared to be β_1-receptor mediated. An example of NE-induced long-lasting potentiation in the dentate gyrus is illustrated in Fig. 1.

Although noradrenergic modulation of long-term hippocampal plasticity is indicated, the mechanisms of action of NE in controlling excitability and information throughput in the hippocampus remain to be worked out. Doing so promises both to shed insight on normal hippocampal function and to suggest pharmacological interventions that may be able to alleviate memory deficits.

3. NOREPINEPHRINE ENHANCEMENT OF STIMULUS-EVOKED CHANGES IN EXTRACELLULAR CALCIUM AND POTASSIUM CONCENTRATION IN THE DENTATE GYRUS

In view of the role indicated for NE in LTP elicited by high-frequency stimulation, we employed ion-selective microelectrodes to measure the decreases in extracellular Ca^{2+} concentration ($\Delta[Ca^{2+}]_o$) and increases in extracellular K^+ concentration ($\Delta[K^+]_o$) as-

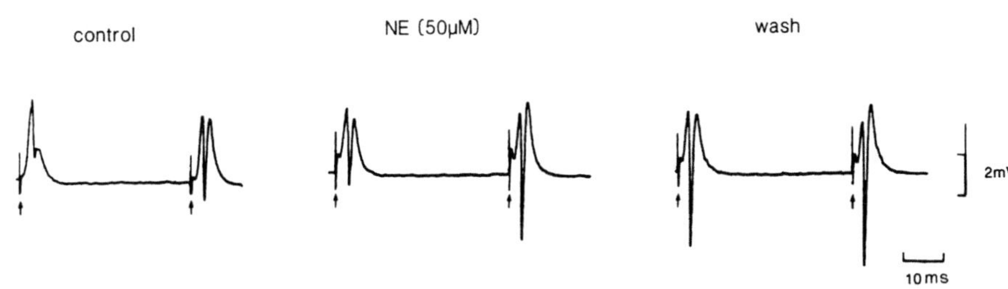

FIGURE 1. Norepinephrine (NE)-induced long-lasting potentiation of perforant-path-evoked population responses recorded in the dentate gyrus of hippocampal slices. Evoked field potentials are shown immediately prior to bath application of NE (control), after a 15-min NE application [NE (50 μM)], and after subsequent 30-min drug-free wash (wash). This long-lasting potentiation typically lasts for many hours. (Arrows denote single stimulus artifacts.)

sociated with such a tetanus. Hippocampal slices and ion-selective/reference electrodes were prepared by standard methods as described previously (Heinemann *et al.*, 1977; Stanton and Sarvey, 1985a). Orthodromic and antidromic field potentials, $\Delta[Ca^{2+}]_o$, and $\Delta[K^+]_o$ produced by high-frequency stimulation (20 Hz for 10 sec, 20–200 μA) were recorded in the cell body (stratum granulosum) and dendritic (stratum moleculare) layers of the dentate gyrus or in the CA1 pyramidal cell body (stratum pyramidale) and dendritic (stratum radiatum) layers.

Bath application of NE (15 min, 50 μM, Fig. 2) markedly potentiated both $\Delta[Ca^{2+}]_o$ and $\Delta[K^+]_o$ in the dentate gyrus granule cell layer evoked by high-frequency perforant path stimulation (Stanton and Heinemann, 1986). This potentiation was associated with an enhancement of slow negative field potentials accompanying ionic changes during the tetanus (Fig. 2A, f.p.). The observed changes were largely or completely reversible in

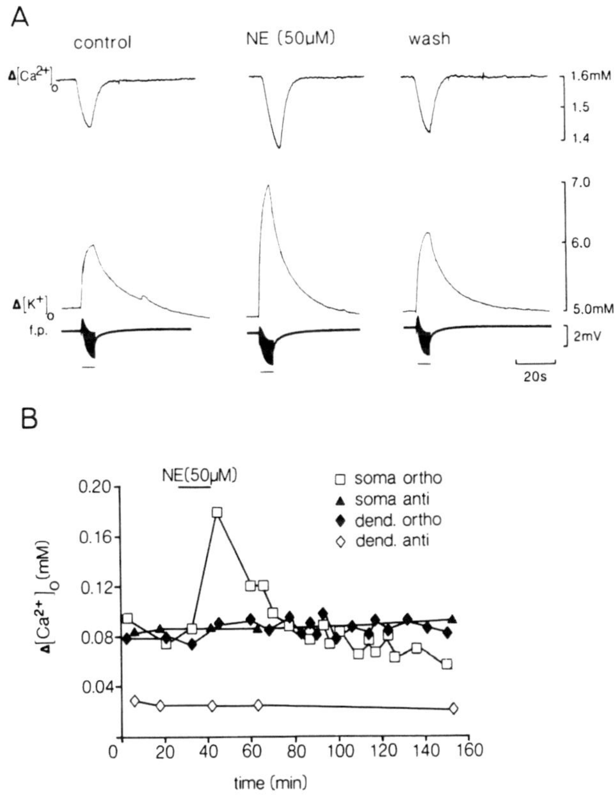

FIGURE 2. Norepinephrine reversibly enhances changes in calcium concentration ($\Delta[Ca^{2+}]_o$), potassium concentration ($\Delta[K^+]_o$), and slow negative field potentials recorded in the dentate granule cell layer (stratum granulosum) and elicited by high-frequency perforant path stimulation (20 Hz for 10 sec). (A) $\Delta[Ca^{2+}]_o$, $\Delta[K^+]_o$, and slow negative field potential shifts before (control), at the end of a 15-min bath application of NE [NE (50μM)], and after a subsequent 30-min drug-free wash (wash). Horizontal bars indicate stimulation. (B) Norepinephrine enhancement of $\Delta[Ca^{2+}]_o$ was specific for orthodromically (perforant path) induced responses recorded in the granule cell layer (open squares, stratum granulosum). In contrast, NE did not affect $\Delta[Ca^{2+}]_o$ produced by antidromic mossy fiber stimulation (closed triangles) or by either ortho- (closed diamonds) or antidromic (open diamonds) stimulation when recording 200 μm away in the dendritic layer (stratum moleculare). In this experiment, NE (50 μM) produced a 200% of control increase in somatic orthodromic $\Delta[Ca^{2+}]_o$, which completely reversed after a 30-min drug-free wash.

a subsequent 30-min drug-free wash (Fig. 2A, wash). Average NE-induced increases were $\Delta[Ca^{2+}]_o$ 173.3 ± 16.1% of control pre-NE base line (mean ± S.E.M., $P < 0.05$, $n = 18$); $\Delta[K^+]_o$ 147.4 ± 18.9% ($P < 0.05$, $n = 9$).

Interestingly, there is a high degree of specificity to the noradrenergic enhancement of these ionic fluxes. In contrast to the effect of NE on orthodromically evoked ionic fluxes in the granule cell layer, NE did not alter stimulus-evoked $\Delta[Ca^{2+}]_o$ recorded simultaneously 200 μm away in the dendritic layer where perforant path axons synapse (Fig. 2B; Stanton and Heinemann, 1986). Furthermore, antidromic mossy-fiber-evoked $\Delta[Ca^{2+}]_o$ was not enhanced in either the dentate cell body or dendritic layers at a time when orthodromic $\Delta[Ca^{2+}]_o$ was 200% percent of pre-NE base line.

In view of the previous observation of a lack of effect of NE depletion on production of LTP in field CA1 (Stanton and Sarvey, 1985a), we compared NE effects on dentate ionic fluxes with those in field CA1. Consistent with the earlier findings, NE (15 min, 50 μM) did not significantly alter Schaffer collateral stimulus-evoked $\Delta[Ca^{2+}]_o$ in the CA1 pyramidal cell or apical dendritic layers (Stanton and Heinemann, 1986).

Similarly, since previous studies (Stanton and Sarvey, 1985b,c) have suggested β-receptor involvement in NE-induced long-lasting potentiation in the dentate, we tested the β antagonist propranolol for its ability to impair noradrenergic stimulation of $\Delta[Ca^{2+}]_o$. Propranolol (1 μM) markedly antagonized these noradrenergic effects as well in slices that were able to exhibit enhancement in the absence of propranolol.

To summarize, NE appears to enhance preferentially $\Delta[Ca^{2+}]_o$ and $\Delta[K^+]_o$ produced by high-frequency synaptic stimulation in the dentate gyrus granule cell layer. Like NE-induced long-lasting potentiation, this effect is β-receptor mediated but, unlike the long-term effects of NE, is reversible in nature, suggesting a trigger role for these effects.

4. A POSSIBLE ROLE FOR NMDA RECEPTORS IN NORADRENERGIC MODULATION OF HIPPOCAMPAL PLASTICITY

Although it was clear that NE modulates dentate granule cell excitability, ionic fluxes, and LTP, it remained to search for possible mechanisms of alteration in efficacy of excitatory and/or inhibitory inputs. In this regard, the most likely candidate receptor to be involved was N-methyl-D-aspartate (NMDA). This is because, although there is little evidence for an NMDA receptor contribution to normal synaptic transmission in the dentate gyrus, recent work has suggested an important role of activation of NMDA receptors in long-term increases in hippocampal excitability. It has been shown that micromolar concentrations of the specific NMDA receptor antagonist D-2-amino-5-phosphonovaleric acid (APV) can block production of LTP in hippocampal slices (Collingridge *et al.*, 1983; Wigstrom and Gustafsson, 1984; Harris *et al.*, 1984). Additionally, APV abolishes burst activity observed when Mg^{2+}-free bathing solution is applied to relieve the Mg^{2+}-dependent blockade of NMDA receptors (Herron *et al.*, 1985; Walther *et al.*, 1986).

Therefore, we tested the effect of NE on $\Delta[Ca^{2+}]_o$ and $\Delta[K^+]_o$ elicited by iontophoretic application in the granule cell layer of either NMDA or the excitatory amino acid quisquilate (QUIS), which activates a different subpopulation of receptors. Figure 3 illustrates such an experiment in which NE (15 min, 50 μM) produced a marked enhancement of NMDA-evoked $\Delta[Ca^{2+}]_o$ and a smaller increase in $\Delta[K^+]_o$ (Fig. 3, NMDA). In contrast, NE did not elicit such marked increases in these fluxes when

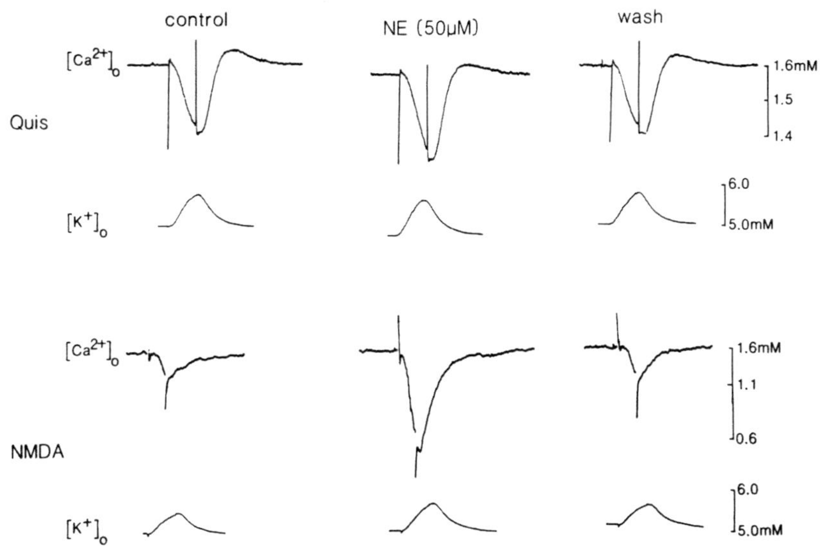

FIGURE 3. Norepinephrine reversibly enhances $\Delta[Ca^{2+}]_o$ and $\Delta[K^+]_o$ produced in the granule cell layer by the excitatory amino acid N-methyl-D-aspartate (NMDA) but not those elicited by the excitatory amino acid quisquilate (QUIS), which activates a different subpopulation of receptors. Responses are shown just before (control), after a 15-min application of NE [NE (50 μM)], and after a subsequent 30-min drug-free wash (wash). The vertical artifacts in $\Delta[Ca^{2+}]_o$ signals are the on and off artifacts produced by a 15-sec (QUIS) or 10-sec (NMDA) iontophoretic pulse (15 nA).

produced by application of QUIS (Fig. 3, QUIS). Interestingly, the largest enhancement of $\Delta[Ca^{2+}]_o$ by NE was seen for orthodromic stimulus-evoked fluxes (169 ± 22% of base line, $n = 6$), slightly smaller increases in NMDA responses (144 ± 12.7%, $n = 6$), and only very small enhancements for QUIS (116.4 ± 9.6%, $n = 4$).

Since these results pointed to an involvement of NMDA receptor activation in noradrenergic mechanisms, we tested the NMDA antagonist APV for its ability to impair noradrenergic enhancements in $\Delta[Ca^{2+}]_o$ in the dentate granule cell layer. APV (30 μM) was able completely to block NE-induced enhancement of $\Delta[Ca^{2+}]_o$ elicited by high-frequency perforant path stimulation. In fact, in a few experiments APV alone reduced the high-frequency-evoked $\Delta[Ca^{2+}]_o$ when high stimulus intensities were used, indicating activation of NMDA receptors during the tetanus. In slices in which APV blocked NE-induced increases in $\Delta[Ca^{2+}]_o$, a similar NE application after washout of APV elicited the usual enhancement.

From these data, we conclude that NE exerts a priming effect that can increase NMDA receptor activation, depolarization, $\Delta[Ca^{2+}]_o$, and $\Delta[K^+]_o$ during repetitive stimulation, thereby enhancing the amplitude of the resulting long-term synaptic changes.

5. INTRACELLULAR CORRELATES OF NORADRENERGIC EFFECTS

It becomes more and more clear that the actions of NE in the central nervous system, and in particular in the hippocampus, are quite complex. Early studies characterized NE simply as an inhibitory neurotransmitter (i.e., Segal and Bloom, 1974). Later studies in field CA1 have indicated that NE has both β-receptor-mediated excitatory and α-receptor-mediated inhibitory actions (Mueller *et al.*, 1981). To date, the most potent excitatory action of NE to be demonstrated intracellularly is the reduced accommodation of action potential firing in CA1 pyramidal cells during an intracellularly applied depolarizing current step of long duration (>100 msec; Madison and Nicoll, 1982; Haas and Konnerth, 1983). This effect is accompanied by a reduction in the accompanying long-lasting afterhyperpolarization usually attributed in large part to activation of a Ca^{2+}-dependent K^+ current ($G_{K(Ca)}$) and is probably mediated by β_1-receptor activation. This has led to the hypothesis that NE may preferentially enhance fast repetitive firing of the type associated with induction of LTP and to the idea that the signal-to-noise ratio of hippocampal inputs would thereby be improved (Langmoen *et al.*, 1981; Segal, 1982).

However, there is relatively little data concerning intracellular correlates to noradrenergic modulation of both direct and synaptic excitation in the dentate gyrus, the

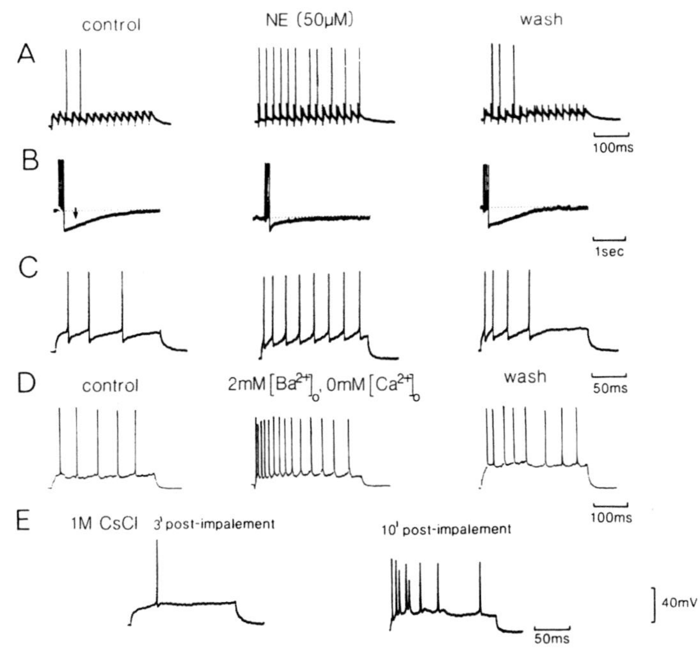

FIGURE 4. Intracellular correlates of noradrenergic actions in dentate granule cells. (A) Norepinephrine blocks accommodation of spike firing during a brief high-frequency perforant path synaptic stimulation (50 Hz for 300 msec). Intracellular action potential firing of a dentate granule cell in response to synaptic activation is shown before (control), after a 15-min NE application [NE (50 μM)], and after a subsequent 15-min drug-free wash (wash). RMP = −73 mV; R_N = 24 MΩ. (B) In the same cell, NE also markedly reduced the late afterhyperpolarizing $G_{K(Ca)}$ (arrow, cell depolarized by constant current to −55 mV to activate fully the $G_{K(Ca)}$). (C) Also in the same cell, NE blocked accommodation of spike firing during a direct depolarizing current step (150 msec; 1.5 nA). (D) In another granule cell, replacement of Ca^{2+} by Ba^{2+} in the extracellular medium mimicked the effects of NE in blocking accommodation of spike firing during a depolarizing current step (300 msec; 2 nA; RMP = −72 mV; R_N= 20 MΩ). (E) Similarly, impalement of another granule cell with an electrode filled with 1 M CsC1 also mimicked the action of NE in blockade of accommodation during a current step (150 msec; 0.2 nA; RMP = −73 mV; R_N = 60 MΩ).

hippocampal area where NE appears to play the largest role in regulating long-term plasticity. Therefore, we employed intracellular recordings from dentate granule cells to assess effects of NE and other treatments that can block $G_{K(Ca)}$ on direct and synaptically evoked firing.

Bath application of NE (10–50 μM) quite potently reduced both $G_{K(Ca)}$ (Fig. 4B, arrow) and accommodation of repetitive firing (Fig. 4C) elicited in dentate granule cells by a direct depolarizing current step, consistent with data from CA1 pyramidal neurons. Further, accommodation of granule cell spike firing was also observed in response to very brief high-frequency perforant path stimulation, and this synaptic accommodation was also markedly reduced by NE (Fig. 4A).

Experiments replacing extracellular Ca^{2+} with Ba^{2+} (2 mM), an ion that passes through calcium channels but does not activate $G_{K(Ca)}$ (Hotson *et al.*, 1979), were suggestive of the importance of this current in regulating granule cell excitation. After 15 min in $[Ba^{2+}]_o$, marked increases in granule cell excitability were observed. Extracellular evoked potentials in the granule cell layer exhibited pronounced multiple epileptiform spikes. Intracellularly, accommodation of spike firing (Fig. 4D) and the associated $G_{K(Ca)}$ were markedly reduced, and synaptic accommodation was also blocked. After 20 min, the cell was strongly depolarized (from −72 to −50 mV) with a large increase in input resistance and began firing spontaneous bursts and plateau spikes similar to those observed by others (Fricke and Prince, 1984). Furthermore, granule cells in Ba^{2+} that were then treated with NE showed no further blockade of accommodation or increase in excitability.

Granule cell intracellular recordings using CsCl (1 mM)-filled electrodes to block K^+ currents also suggest an important role for K^+ currents in granule cell excitability. Although Cs^+ also blocks other K^+ currents, it is very potent in blocking $G_{K(Ca)}$ (Puil and Werman, 1981). Figure 4E illustrates the Cs^+-induced blockade of accommodation, which was accompanied by reductions in $G_{K(Ca)}$, marked depolarization (from −73 to −57 mV), and spontaneous firing. Similar to the experiments with Ba^{2+}, NE had little or no further effects in cells impaled with Cs^+ electrodes, supporting the conclusion that a major action of NE to enhance excitability in granule cells is blockade of $G_{K(Ca)}$.

6. OVERVIEW

Perhaps the clearest message from studies of noradrenergic regulation of neuronal plasticity is the capacity for sensitive control offered by the complex spatial and temporal summation of a number of different noradrenergic actions at different sites. An important effect of NE is in permitting many more spikes to fire with prolonged stimuli. However, there is a dichotomy of noradrenergic importance to LTP in the dentate versus CA1, although accommodation is markedly reduced in both areas. The explanation may lie in the relative importance of the $G_{K(Ca)}$ to repetitive firing in the two areas. We observed in intracellular recordings from CA1 pyramidal neurons that, rather than rapidly accommodating, there is usually a frequency facilitation followed much later in the tetanus by accommodation of firing (Lux and Heinemann, 1982; P. K. Stanton and U. Heinemann, unpublished results). Although a strong controlling role of $G_{K(Ca)}$ in granule cell firing may be the whole story, it is also quite possible that granule cell electrotonic properties, a relative preponderance of β versus α receptors, differing spatial distribution of NE receptors, and/or differing distributions of other transmitter receptors (such as NMDA) may all be involved.

Indeed, a most significant finding is the interaction of NMDA and NE receptors in long-term plasticity. It is now clear that NMDA receptors are important to expression of LTP. An ability of NE to regulate NMDA receptor activation is shown to play an important role in the firing patterns of granule cells during high-frequency stimulation of a type that elicits LTP. Blockade of a strong $G_{K(Ca)}$ in granule cells may more strongly depolarize them during repetitive stimulation, which could, in turn, relieve the voltage-dependent blockade of NMDA receptors and facilitate their activation. Since these NMDA receptors are normally much more quiescent in the dentate than CA1, this represents a large gating capacity in the dentate, which can be turned on by NE.

Of equal interest is the question of noradrenergic mechanisms in more pathological examples of hippocampal long-term plasticity such as the kindled epilepsy. It is already known that depletion of NE accelerates hippocampal kindling (McIntyre and Edson, 1982). Furthermore, recent work in our laboratory has shown that dormant NMDA receptors become functional after kindling via either the Schaffer collaterals (Wadman *et al.*, 1985) or commissures (Mody and Heinemann, 1987) and that APV antagonizes kindled seizures (Peterson *et al.*, 1984). The expected modulatory connection is made complete by data we have indicating that the spreading depression that can be stimulus evoked in the dentate when extracellular [Mg^{2+}] is reduced has a lowered threshold and prolonged duration in the presence of NE (Stanton *et al.*, 1987).

However, this gives rise to the contradiction that NE is excitatory in the dentate gyrus, yet depletion of NE accelerates production of the kindled epilepsy. In relation to this problem, we have recently found that, in combined entorhinal cortex–hippocampal slices, NE has complementary properties. In the dentate gyrus, lowering extracellular Mg^{2+} markedly increases the amplitude of evoked potentials and leads to tetanically evoked spreading depression (SD) (Mody *et al.*, 1987). Even more interestingly, one observes in entorhinal cortex spontaneous epileptiform activity, spontaneous SDs, and SDs in response to a single stimulus (Walther *et al.*, 1986; Stanton *et al.*, 1987). Whereas NE enhances the stimulus-evoked spreading depression in the dentate, it eliminates both the spontaneous and evoked epileptiform activity and spreading depression in layer V of the entorhinal cortex (Stanton *et al.*, 1987). This may explain why depletion of NE facilitates kindling.

With these complementary actions, NE can dampen inputs from cortex to the dentate while increasing granule cell excitability. By simultaneously suppressing unwanted noise in the input and boosting the gain of the dentate filter, NE may well prove to improve the signal-to-noise ratio even more effectively than first suspected. If so, the vistas for pharmacological intervention in a number of disorders in neuronal plasticity are vast. The presence of NE has a strong associative effect on the magnitude of hippocampal plasticity and may thereby allow this modulator to exert an important enabling function in the association of synchronized inputs during learning. We have a transmitter with perhaps more specific pharmacological tools than any other, which may both enhance the throughput of important patterned signals (as in memory storage and retrieval) and regulate paroxysmal epileptiform activity in complex ways.

ACKNOWLEDGMENTS. This work was supported by DFG grant SFB220-B3 to U.H. and an Alexander von Humboldt Foundation fellowship to P.K.S. We thank Drs. B. Hamon, R. S. G. Jones, and I. Mody for helpful discussion, and B. Muffler for technical assistance.

REFERENCES

Alger, B. E., and Teyler, T. J., 1976, Long-term and short-term plasticity in the CA1, CA3, and dentate regions of the hippocampal slice, *Brain Res.* **110:**463–480.

Berger, T. W., 1984, Long-term potentiation of hippocampal synaptic transmission affects rate of behavioral learning, *Science* **224:**627–630.

Bliss, T. V. P., and Gardner-Medwin, A. R., 1973, Long-lasting potentiation of synaptic transmission in the dentate area of the unanaesthetized rabbit following stimulation of the perforant path, *J. Physiol.* (*Lond.*) **232:**357–374.

Bliss, T. V. P., and Lømo, T., 1973, Long-lasting potentiation of synaptic transmission in the dentate area of the anaesthetized rabbit following stimulation of the perforant path, *J. Physiol.* (*Lond.*) **232:**331–356.

Bliss, T. V. P., Goddard, G. V., and Riives, M., 1983, Reduction of long-term potentiation in the dentate gyrus of the rat following selective depletion of monoamines, *J. Physiol.* (*Lond.*) **334:**475–491.

Collingridge, G. L., Kehl, S. J., and McLennan, H., 1983, Excitatory amino acids in synaptic transmission in the Schaffer collateral–commissural pathway of the rat hippocampus, *J. Physiol.* (*Lond.*) **334:**33–46.

Crow, T. J., and Wendlandt, S., 1976, Impaired acquisition of a passive avoidance response after lesions induced in the locus coeruleus by 6-OH-dopamine, *Nature (Lond.)* **259:**42–44.

Douglas, R. M., and Goddard, G. V., 1975, Long-term potentiation of the perforant path–granule cell synapse in the rat hippocampus, *Brain Res.* **86:**205–215.

Everitt, B. J., Robbins, T. W., Gaskin, M., and Fray, P. J., 1983, The effects of lesions to ascending noradrenergic neurons on discrimination learning and performance in the rat, *Neuroscience* **10:**397–410.

Fricke, R. A., and Prince, D. A., 1984, Electrophysiology of dentate gyrus granule cells, *J. Neurophysiol.* **51:**195–209.

Goddard, G. V., McIntyre, D. C., and Leech, C. K., 1969, A permanent change in brain function resulting from daily electrical stimulation, *Exp. Neurol.* **25:**295–300.

Haas, H. L., and Konnerth, A., 1983, Histamine and noradrenaline decrease calcium-activated potassium conductance in hippocampal pyramidal cells, *Nature (Lond.)* **302:**432–434.

Harris, E. W., Ganong, A. H. and Cotman, C. W., 1984, Long-term potentiation in the hippocampus involves activation of N-methyl-D-aspartate receptors, *Brain Res.* **323:**132–137.

Hebb, D. O., 1949, *The Organization of Behavior*, John Wiley & Sons, New York.

Heinemann, U., Lux, H. D., and Gutnick, M. J., 1977, Extracellular free calcium and potassium during paroxysmal activity in the cerebral cortex of the cat, *Exp. Brain Res.* **27:**237–243.

Herron, C. E., Lester, R. A. J., Coan, E. J., and Collingridge, G. L., 1985, Intracellular demonstration of an N-methyl-D-aspartate receptor mediated component of synaptic transmission in the rat hippocampus, *Neurosci. Lett.* **60:**19–23.

Hotson, J. R., Prince, D. A., and Schwartzkroin, P. A., 1979, Anomalous rectification in hippocampal neurons, *J. Neurophysiol.* **42:**889–895.

Langmoen, I. A., Segal, M., and Andersen, P., 1981, Mechanisms of norepinephrine actions on hippocampal pyramidal cells *in vitro*, *Brain Res.* **208:**349–362.

Lux, H. D., and Heinemann, U., 1982, Consequences of calcium-electrogenesis for the generation of paroxysmal depolarization shift, in: *Epilepsy and Motor System* (E. J. Speckmann and H. Elger, eds.), Urban and Schwarzenberg, Munich, pp. 101–119.

Madison, D. V., and Nicoll, R. A., 1982, Noradrenaline blocks accommodation of pyramidal cell discharge in the hippocampus, *Nature (Lond.)* **299:**636–638.

Mason, S. T., and Iversen, S. D., 1977, Effects of selective noradrenaline loss on behavioral inhibition in the rat, *J. Comp. Physiol. Psychol.* **91:**165–173.

McIntyre, D. C., and Edson, N., 1982, Effect of norepinephrine depletion on dorsal hippocampus kindling in rats, *Exp. Neurol.* **77:**700–704.

Milner, B., 1972, Disorders of learning and memory after temporal lobe lesions in man, *Clin. Neurosurg.* **19:**421–446.

Mody, I., and Heinemann, U., 1987, NMDA receptors of dentate gyrus granule cells participate in synaptic transmission following kindling, *Nature (Lond.)* **326:**701–704.

Mody, I., Lambert, J. D. C., and Heinemann, U., 1987, Low extracellular magnesium induces epileptiform activity and spreading depression in rat hippocampal slices, *J. Neurophysiol.* **57:**869–888.

Mueller A. L., Hoffer, B. J., and Dunwiddie, T. V., 1981, Noradrenergic responses in rat hippocampus: Evidence for mediation by α and β receptors in the in vitro slice, *Brain Res.* **214:**113–126.

Neuman, R. S., and Harley, C. W., 1983, Long-lasting potentiation of the dentate gyrus population spike by norepinephrine, *Brain Res.* **273:**162–165.

Peterson, D. W., Collins, J. F., and Bradford, H. F., 1984, Anticonvulsant action of amino acid antagonists against kindled hippocampal seizures, *Brain Res.* **311**:176–180.

Puil, E., and Werman, R., 1981, Internal cesium ions block various K conductances in spinal motoneurons, *Can. J. Physiol. Pharmacol.* **59**:1280–1284.

Schwartzkroin, P. A., and Wester, K., 1975, Long-lasting facilitation of a synaptic potential following tetanization in the *in vitro* hippocampal slice, *Brain Res.* **89**:107–119.

Segal, M., 1982, Norepinephrine modulates reactivity of hippocampal cells to chemical stimulation *in vitro*, *Exp. Neurol.* **77**:86–93.

Segal, M., and Bloom, F. E., 1974, The action of norepinephrine in the rat hippocampus. I. Iontophoretic studies, *Brain Res.* **77**:79–97.

Stanton, P. K., and Heinemann, U., 1986, Norepinephrine enhances stimulus-evoked Ca^{2+} and K^{+} concentration changes in dentate granule cell layer, *Neurosci. Lett.* **67**:233–238.

Stanton, P. K., Jones, R. S. G., Mody, I., and Heinemann, U., 1987, Epileptiform activity induced by lowering extracellular [Mg^{2+}] in combined hippocampal–entorhinal cortex slices: Modulation by receptors for norepinephrine and N-methyl-D-aspartate, *Epilepsy Res.* **1**:53–62.

Stanton, P. K., and Sarvey, J. M., 1985a, Depletion of norepinephrine, but not serotonin, reduces long-term potentiation in the dentate of rat hippocampal slices, *J. Neurosci.* **5**:2169–2176.

Stanton, P. K., and Sarvey, J. M., 1985b, The effect of high-frequency electrical stimulation and norepinephrine on cyclic AMP levels in normal versus norepinephrine-depleted rat hippocampal slices, *Brain Res.* **358**:343–348.

Stanton, P. K., and Sarvey, J. M., 1985c, Blockade of norepinephrine-induced long-lasting potentiation in the hippocampal dentate gyrus by an inhibitor of protein synthesis, *Brain Res.* **361**:276–283.

Stein, L., Beluzzi, J. D., and Wise, C. D., 1975, Memory enhancement by central administration of norepinephrine, *Brain Res.* **84**:329–335.

Swanson, L. W., Teyler, T. J., and Thompson, R. F., 1982, Hippocampal long-term potentiation: Mechanisms and implications for memory, *Neurosci. Res. Prog. Bull.* **20**:612–769.

Wadman, W. J., Heinemann, U., Konnerth, A., and Neuhaus, S., 1985, Hippocampal slices of kindled rats reveal calcium involvement in epileptogenesis, *Exp. Brain Res.* **57**:404–407.

Walther, H., Lambert, J. D. C., Jones, R. S. G., Heinemann, U., and Hamon, B., 1986, Epileptiform activity in combined slices of the hippocampus, subiculum and entorhinal cortex during perfusion of low magnesium medium, *Neurosci. Lett.* **69**:156–161.

Wigstrom, H., and Gustafsson, B., 1984, A possible correlate of the postsynaptic condition for long-lasting potentiation in the guinea pig hippocampus *in vitro*, *Neurosci. Lett.* **44**:327–332.

Zornetzer, S. F., 1984, Brain substrates of senescent memory decline, in: *Neuropsychology of Memory*, Guilford, New York, pp. 588–600.

III

Conditioning

10

Conditioning-Specific Biophysical Alterations in Rabbit Hippocampus

JOHN F. DISTERHOFT, DOUGLAS A. COULTER, and DANIEL L. ALKON

1. INTRODUCTION

A series of studies was initiated to explore the possibility that conditioning-specific biophysical alterations could be demonstrated in *in vitro* brain slices prepared from conditioned rabbits. Hippocampus was chosen as the region to begin our brain slice studies for several reasons. First, hippocampus in humans is generally considered to be importantly involved in acquisition of cognitive learning tasks (probably for transferring information from immediate to long-term memory) because of profound learning deficits observed in patients with bilateral hippocampectomy (Scoville and Milner, 1957; Squire, 1982). Second, systematic unit recording studies in freely moving rats showed a profound involvement of hippocampus and its subregions in a tone-signaled food acquisition conditioned response (Disterhoft and Segal, 1978; Olds *et al.*, 1972; Segal, 1973). Berger, Thompson, and their collaborators showed in a series of studies during nictitating membrane (NM) conditioning in rabbits that hippocampal neurons showed learned alterations early in training (before CR acquisition), that hippocampal CA1 and CA3 neuron PST histograms to CS onset exhibit "models" of the NM conditioned response, and that there is a high degree of correlation between hippocampal unit firing and the acquisition of and characteristics of the NM conditioned response (Berger et al., 1980; Berger and Thompson, 1978; Thompson *et al.*, 1980). Recently, they have shown in a single-neuron study that it is the CA1 and CA3 neurons that model the NM conditioned response (Berger *et al.*, 1983). Third, the hippocampal slice has been one of the most widely studied *in vitro* brain slice preparations because of its regular lamellar organization and its viability *in vitro* (Langmoen and Andersen, 1981; Schwartzkroin, 1981). Thus, there exists a large body of *in vitro* biophysical data in addition to the *in vivo* recording data. In fact, much

JOHN F. DISTERHOFT • Department of Cell Biology and Anatomy, Northwestern University Medical School, Chicago, Illinois 60611; and Laboratory of Biophysics, National Institute of Neurological and Communicative Disorders and Stroke, National Institutes of Health at the Marine Biological Laboratory, Woods Hole, Massachusetts 02543. DOUGLAS A. COULTER • Laboratory of Biophysics, National Institute of Neurological and Communicative Disorders and Stroke, National Institutes of Health at the Marine Biological Laboratory, Woods Hole, Massachusetts 02543. DANIEL L. ALKON • Section on Neural Systems, Laboratory of Biophysics, IRP, National Institute of Neurological and Communicative Disorders and Stroke, National Institutes of Health at the Marine Biological Laboratory, Woods Hole, Massachusetts 02543.

of the *in vitro* data are from the CA1 and CA3 pyramidal cells, the population that shows the neural modeling of the NM conditioned response.

The NM/eye retraction preparation in rabbit was chosen for the conditioning paradigm because it is a well-controlled behavioral paradigm and has been used extensively as a model system for studying learning in mammals (Disterhoft *et al.*, 1977; Thompson, 1976). Studies have been done on the auditory CS pathway (Kraus and Disterhoft, 1982) and on cerebellar and brainstem CR circuitry (Disterhoft *et al.*, 1985; McCormick and Thompson, 1984) as well as the hippocampus.

We asked two questions in our studies. First, can we demonstrate learned alterations in biophysical parameters in the slice? That is, can we train an animal, remove a part of the brain known to be altered by associative learning, and show that effects of the learning are retained *in vitro* (even though neurons in the slice are denied their normal afferent and efferent connections)? If effects of associative conditioning are demonstrable in the slice, then we can exploit the advantages of the slice technique to analyze them (e.g., intracellular recording is easier, and penetrations are more stable; the external milieu may be manipulated to isolate currents; intracellular injections of substances such as HRP are relatively easy; single-electrode voltage clamping can be done in addition to current-clamp studies).

Our second question in these studies was to determine if common mechanisms may underlie learning in mammalian and invertebrate systems. Alkon and his collaborators have been investigating a biophysical sequence of alterations that occur in the type B photoreceptor of *Hermissenda* and underlie associative learning in this mollusc (Alkon, 1982, 1984). Very briefly, they have found an excitability increase in the type B cell that was caused by a reduction in an early, rapidly inactivating outward K^+ current, I_A, and in a Ca^{2+}-dependent outward K^+ current, I_C. They are the first to demonstrate, in an invertebrate preparation, ionic alterations intrinsic to a specific cell (medial type B photoreceptor) that underlies associative learning. Similar currents to those Alkon and his group have been studying have been demonstrated in mammalian hippocampal neurons (Gustafsson *et al.*, 1982; Brown and Griffith, 1983; Lancaster and Adams, 1986; Zbicz and Weight, 1985). We wanted to see if comparable ionic changes exist in mammalian neurons after classical conditioning.

2. METHODS

In brief, young adult male albino rabbits were well trained in the NM paradigm. The rabbits received three 80-trial conditioning sessions, at which time they exhibited CRs on 80–100% of trials (Disterhoft *et al.*, 1977). A short-delay paradigm in which a 400-msec monaural white noise or a 2.8-KHz free-field tone CS overlapped and coterminated with a 150-msec periorbital shock US was used. Pseudoconditioned rabbits, which received unpaired presentations of the CS and US, and naive rabbits served as the two control groups. The rabbits were sacrificed 24 hr after the final training period. Slices of the hippocampus were prepared and maintained in a surface (Kelso *et al.*, 1983) or submerged (Zbicz and Weight, 1985) recording chamber at 30–32° using standard procedures (Kelso *et al.*, 1983; Teyler, 1980). Intracellular recordings were done with a high-input-impedance bridge amplifier. The microelectrodes used were filled with 3 M KCl and had impedances of 70–90 MΩ. Data were taken by hand from a storage oscilloscope on line or from tape recordings. Cells included in the analysis had spike amplitudes of 60 mV or more, resting membrane potentials of 50 mV or greater, and input resistances of 20MΩ or higher.

3. RESULTS

3.1. Afterhyperpolarization

Slices were made from conditioned, pseudoconditioned, or naive rabbits. We then impaled CA1 pyramidal cells and studied their membrane properties. The alteration that is of major interest concerns the afterhyperpolarization (AHP) responses. Action potentials of CA1 neurons are followed by an AHP, which has been attributed to a Ca^{2+}-mediated K^+ current (Brown and Griffith, 1983; Madison and Nicoll, 1984; Wong and Prince, 1981). In our first study of about 75 cells, we found a conditioning-specific decrease in the AHP following injection of sufficient depolarizing current to cause one or two spikes to occur (Disterhoft *et al.*, 1986). Figure 1 shows examples of single AHP responses measured from three different neurons. We measured the slow, not the fast, AHP. The arrows in Fig. 1 demonstrate the points where we made the measurements after the depolarizing current pulse was terminated. It is the slow AHP that is attributed to the Ca^{2+}-dependent K^+ current (Hotson and Prince, 1980; Lancaster and Wheal, 1984). An important feature of the AHP decrease should be stressed: it was present in the conditioned group of cells in the absence of any alteration in spike height, resting potential, or input resistance from the other groups of cells (Table I).

Another interesting point is that during conditioning there was the separation of about half the cells into a group with very low one-spike AHP (Fig. 2). The remaining cells appeared to overlap with the naive and pseudoconditioned groups. Berger *et al.* (1983) showed that 62% of pyramidal cells increased their firing rate before US onset after conditioning. Figure 2 shows an interesting convergence between the *in vivo* and *in vitro* data: about 50% of pyramidal cells from conditioned rabbits showed AHP responses < 0.5 mV. If we assume that a cell with a markedly reduced AHP is a "learned" cell, the convergence of the two types of data is quite striking.

The evidence for AHP being caused by a Ca^{2+}-mediated K^+ current includes its reduction by EGTA iontophoretic injections, by Ca^{2+} channel blockers such as Mn^{2+} or Co^{2+}, and by substitution of Ba^{2+} for Ca^{2+} in the external perfusion medium (Brown

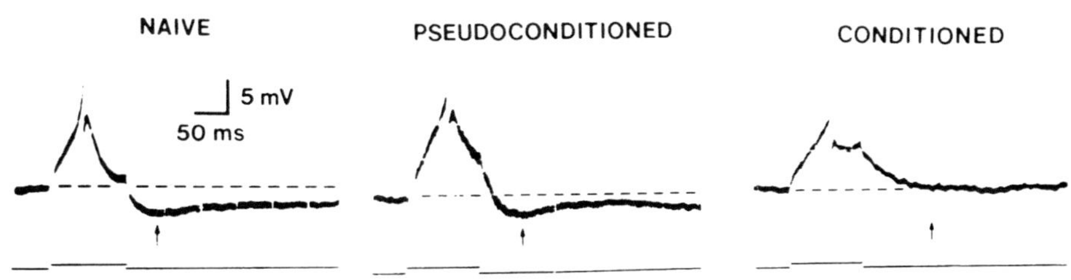

FIGURE 1. Afterhyperpolarization after one spike in three separate cells from a naive, a pseudoconditioned, and a conditioned rabbit. A depolarizing 100-msec current pulse just adequate to elicit one spike was injected into each cell. The arrows under each individual voltage trace point to where the AHP amplitude measurements were made. The AHP was considerably reduced in the conditioned as compared to the naive and pseudoconditioned neurons (Disterhoft *et al.*, 1986).

TABLE I. Measured Parameters and Statistics[a]

Parameter	Group[b] C	PC	N	Statistics Comparison	P^c
AHP (mV)					
One spike	−0.98 ± 0.80 (21)	−1.7 ± 0.83 (23)	−2.0 ± 0.79 (26)	C vs. PC C vs. N	<0.004 <0.0001
Two spikes	−1.89 ± 1.47 (19)	−2.26 ± 1.27 (18)	−2.82 ± 0.92 (24)	C vs. N	<0.02
Spike height (mV)	80.3 ± 11.8 (23)	80.8 ± 12.7 (24)	77.5 ± 11.5 (26)	—	—
Resting potential (mV)	−65.0 ± 6.1 (15)	−67.2 ± 7.5 (17)	66.0 ± 6.5 (21)	—	—
Input resistance (mΩ)	59.3 ± 22.6 (26)	61.2 ± 16.2 (23)	60.6 ± 17.6 (27)	—	—

[a] From Disterhoft *et al.* (1986).
[b] All values are expressed as mean ± S.D.; *n* is given in parentheses for each value.
[c] Two-tailed significance level.

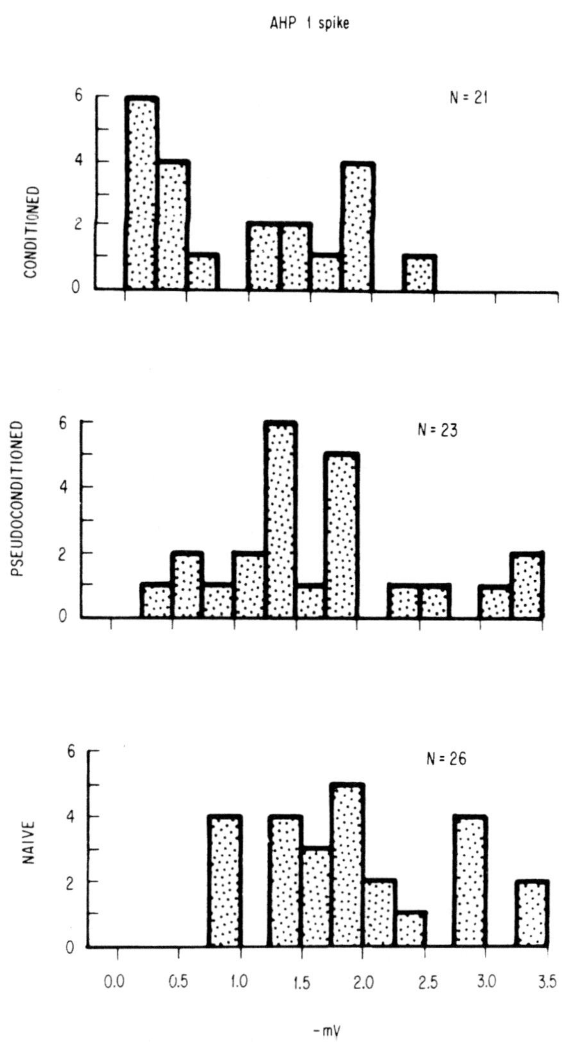

FIGURE 2. One-spike AHP responses for neurons in the conditioned, pseudoconditioned, and naive groups. It appears that the AHP amplitude distributions largely overlap except for those equal to or less than −0.5 mV in the conditioned group (Disterhoft *et al.*, 1986).

and Griffith, 1983; Hotson and Prince, 1980; Madison and Nicoll, 1984; Wong and Prince, 1981). When we perfused our slices with the Ca^{2+} channel blockers Co^{2+} (2.3 mM) or Cd^{2+} (100–200 μM), we reliably reduced or eliminated the AHP (−1.67 mV versus −0.67 mV after Co^{2+}, Cd^{2+}; $P < 0.02$). In addition, we found that EGTA injection and perfusion with 1–5 μM carbachol reduced AHPs.

Functionally, this AHP is presumed to shut the hippocampal pyramidal cells off during bursts. A reduction in AHP after conditioning would make the cells more excitable. As mentioned above, *in vivo* recording shows that hippocampal pyramidal cells are more responsive to a CS after conditioning. A reduction in the AHP could certainly cause, or contribute to, this increased excitability.

The data described thus far are from the original study we did, which included about 75 criterion cells. We have done an entirely separate replication and extension of the work (Coulter *et al.*, 1985). This study was done on 180 criterion CA1 pyramidal neurons. We have shown again that the AHP amplitude is reduced by conditioning in CA1 pyramidal neurons (as compared to pseudoconditioned and naive cell groups). In this study, we have also demonstrated the AHP reduction after enough current was injected to elicit one, two, three, or four spikes.

Figure 3 shows the kind of differences in AHP size we observed after four spikes were elicited. Note that the AHP was smallest in the conditioned group ($\bar{X} = -2.43$ mV) and approximately the same in the pseudoconditioned ($\bar{X} = -5.71$ mV) and naive ($\bar{X} = -4.52$ mV) groups, as we had previously observed after one spike. These conditioned AHP values were significantly reduced from the other two groups ($P < 0.01$ and $P < 0.05$, respectively), again in the absence of alterations in input resistance, resting potential, or spike height (input resistance in the pseudoconditioned group was slightly increased). Note also that at this slower time base (500 msec as compared to 50 msec in Fig. 1), it is obvious that the AHP duration, as well as the amplitude, is decreased. The AHP mean duration after four spikes was 1.32 sec for the conditioned group, 2.7 sec for the pseudoconditioned group, and 2.74 sec for the naive group. These differences were statistically significant. The time constant of AHP decay was also markedly reduced in conditioned as compared to the two control groups. Finally, the conditioning effect became more marked as the number of spikes in the burst became larger and was largest after four spikes in our sample.

3.2. Blind Measurements

In both of the brain slice studies we have completed thus far, a sufficient sample of cells has been gathered "blind" (without experimenter knowledge of the behavioral experience of the animal) to insure that the conditioning-specific ionic alterations we have observed are not attributable to experimenter bias. In the first study (Disterhoft *et al.*, 1986a), we carried out two blinding procedures. First, a representative sample of seven cells was chosen from each group. Photographic records of taped AHP measurements were analyzed by a technician not familiar with the experimental protocol. This analysis yielded mean AHP values in which the conditioned group was significantly lower than both pseudoconditioned and naive groups. In addition, a separate subpopulation of five neurons from each group was gathered in double-blind fashion. The one-spike AHP was significantly smaller in the conditioned than the naive cell group; the conditioned group was also smaller than the pseudoconditioned group, although not significantly so.

In the second study (Coulter *et al.*, 1985), we gathered all of the conditioned ($N = 45$) and pseudoconditioned ($N = 24$) and a small number of the naive ($N = 5$)

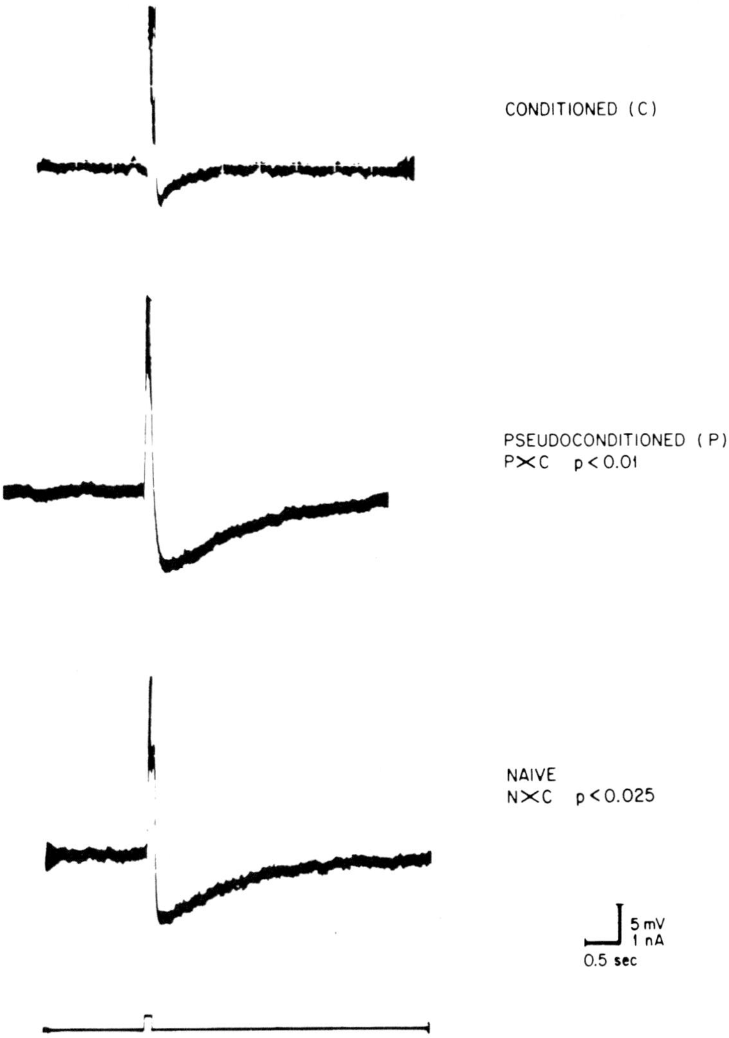

FIGURE 3. The AHP is reduced in both amplitude and duration after conditioning. A 100-msec depolarizing pulse adequate to elicit four spikes was injected into each cell. Note that the time base (500 msec) is slower in this figure than in Fig. 1 (50 msec) (Coulter *et al.*, 1985).

neurons in a blind fashion. The AHP measured after from one to four spikes was significantly smaller in the conditioned than in the pseudoconditioned and naive groups. In both studies, the blind subsets gathered were entirely consistent with the remaining data. Finally, as described below, the acquisition study is being carried out in a blind fashion.

3.3. Functional Consequence of AHP Reduction

The presumed function of the AHP in CA1 pyramidal cells is to act as one brake on neuronal firing during action potential bursts. In an attempt to explore the functional consequences *in vivo* of an AHP alteration recorded in the *in vitro* brain slice, we presented CA1 neurons with a series of depolarizing current pulses at varying intervals (Coulter *et al.*, 1985). We found that cells from conditioned animals were significantly more excitable

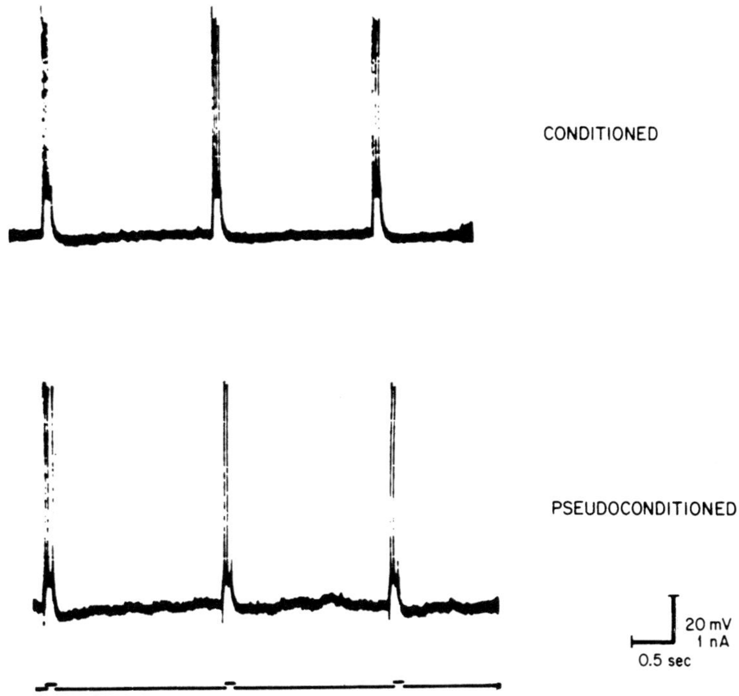

FIGURE 4. Functional consequence of AHP reduction. Repetitive pulse activation of conditioned cells (top trace) as compared to pseudoconditioned cell (lower trace). A depolarizing pulse just adequate to elicit four spikes was given repetitively at decreasing interpulse intervals. The conditioned cell sustained the four-spike response, whereas it dropped from four to two spikes in the pseudoconditioned cell. Calibration: 20 mV, 0.5 sec, 1 nA. The conditioned cells showed sustained more spikes than pseudoconditioned and naive cells at all intervals less than 2 sec (Coulter *et al.*, 1985).

in response to repetitive activation than cells from the pseudoconditioned rabbits (Fig. 4). That is, at all interpulse intervals less than 2 sec, a current pulse that initially elicited four spikes caused significantly fewer spikes in successive presentations in control neurons. The reduction of number of spikes elicited in pseudoconditioned neurons was associated with a buildup of the AHP and a reduction in excitability across the pulse train.

3.4. Acquisition

Some of our recent experiments concern determining when, during the course of conditioned response acquisition in the simple delay paradigm, the decrease in AHP size occurs (Disterhoft *et al.*, 1986). In all of the work described thus far, we have trained the rabbits for three 80-trial sessions to insure that the animals are very well trained. In the acquisition studies, we have been training the rabbits for two 80-trial sessions. Some rabbits are very well trained at this point, but others have shown almost no CRs.

As is illustrated in Fig. 5, the 17 cells from the "low"-CR group ($\bar{X}$ = 2.5% CRs) showed four-spike AHPs similar in amplitude to the naive and pseudoconditioned groups we have studied previously. The 18 cells from the "high"-CR group ($\bar{X}$ = 79% CRs) showed four-spike AHP amplitude values that tended toward the previous conditioned groups, and they were clearly statistically different from the "low"-CR group (−2.9

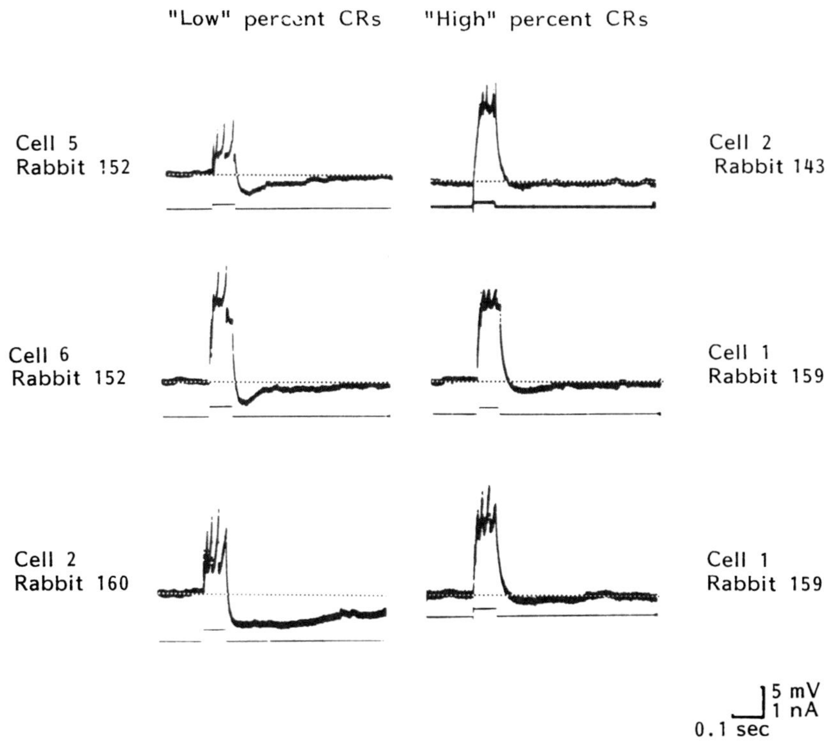

FIGURE 5. Acquisition of the reduced AHP. The AHPs after four spikes are shown for three individual CA1 neurons in the "low-CR" group as compared to three CA1 neurons in the "high-CR" group. All rabbits in this study received two 80-trial training sessions. Hippocampal neuron AHPs from those animals that had acquired the conditioned response tended to be reduced in amplitude and duration, as illustrated here (Disterhoft *et al.*, 1986).

versus −4.0 mV, $P < 0.05$). This same difference was seen in the average AHP response amplitudes after one to three spikes. As mentioned above, all cells in this study are being gathered blind; i.e., the person doing the recording does not know the number of conditioned responses that the rabbit had exhibited during training.

The important things about the acquisition study are (1) that it links the decreased AHP after conditioning more closely to behavioral performance and (2) it is further evidence that the decreased AHP is not some strange aberration of the conditioning paradigm. All animals in the acquisition study have received 160 CS–US pairings. Only those that learned, as evidenced by the number of CRs they showed, had CA1 pyramidal cells with reduced AHPs. This links the AHP reduction closely to conditioned response acquisition and the actual performance of the conditioned response. The acquisition group serves as another type of "control" in addition to the pseudoconditioned and naive groups.

3.5. Decreased Ca^{2+} Entry as Well as AHP Reduction

In another series of ongoing experiments (Coulter *et al.*, 1986), we have been examining the question of whether AHP reduction after conditioning results from direct effects on the calcium-dependent potassium current ($I_{Ca^{2+}K^{+}}$) or indirectly from effects on the voltage-dependent calcium current ($I_{Ca^{2+}}$). At this point, we have examined 12 conditioned and 13 control cells. When slices were maintained in standard medium, CA1 neurons from conditioned rabbits had reduced AHPs and shorter mean spike durations (suggesting a reduced calcium contribution to the spike). When a bathing medium containing 0.5 μM TTX and 5 mM TEA was used to examine calcium spikes, we found that conditioned neurons had a higher threshold to elicit calcium spikes ($\bar{X}$ = 1.29 nA) than did controls ($\bar{X}$ = 0.62 nA, $P < 0.02$). However, after a calcium spike had been elicited, the AHP was still reduced in conditioned neurons ($\bar{X}$ = 5.13 mV) as compared to controls ($\bar{X}$ = 8.84 mV, $P < 0.05$). This AHP reduction was obvious even though the calcium spike amplitude and duration were similar.

4. DISCUSSION

We are the first to demonstrate an ionic mechanism *in vitro* that underlies associative learning in mammalian brain. The AHP response in hippocampal pyramidal cells is primarily caused by a calcium-mediated potassium current. Functionally, this AHP is presumed to shut these neurons off during bursts (Hotson and Prince, 1980). A reduction in the AHP could certainly cause, or contribute to, the increased excitability (responsivity to the CS) seen in hippocampal pyramidal neurons after conditioning (Berger and Thompson, 1978). The AHP has also been demonstrated to be altered by aging and alcohol consumption, conditions that alter CA1 pyramidal cell excitability (Durand and Carlen, 1984; Landfield and Pitler, 1984).

A second important fact is the very strong indication that the ionic alteration is postsynaptic and intrinsic to the CA1 pyramidal cell. The AHP reductions are present even when the CA1 pyramidal cells are separated from the rest of the conditioned reflex arc. Thus, they are not dependent on afferent neuronal drive or a circulating neurohormone of some sort for their expression. At this point we have no evidence for altered EPSP or IPSP size onto the pyramidal cells after conditioning (but will examine this point more fully in future experiments). The increased glutamate binding found in hippocampus after blink conditioning (Mamounnas *et al.*, 1984) suggests that we will be able to detect altered excitatory synaptic drive after learning. The only alterations we have seen thus far are demonstrated by intracellular current injection. These alterations are present even in the presence of TTX (abolishing sodium spike-dependent synaptic transmission), further confirming that the membrane changes are entirely postsynaptic.

4.1. Potassium Currents Are Altered

Our major finding in the slice experiments was that the AHP response was reduced in magnitude and duration. The AHP in CA1 and CA3 pyramidal cells has been shown to be primarily caused by a Ca^{2+}-mediated K^+ conductance (Brown and Griffith, 1983; Hotson and Prince, 1980; Lancaster and Adams, 1986). One very successful analysis of the cellular substrates of associative learning in invertebrates has been carried out by Alkon and associates in *Hermissenda* (Alkon, 1982, 1984). They have found reductions in at least two K^+ currents causing an excitability increase in the type B photoreceptor sufficient to cause the visual avoidance response that is learned. The two reduced K^+ currents in *Hermissenda* are an early, rapidly inactivating outward K^+ current, I_A, and a Ca^{2+}-dependent outward K^+ current, I_C. Our experiments have shown that AHP, attributed to a Ca^{2+}-dependent K^+ current, is reduced in hippocampal pyramidal cells of rabbit after learning as well. We have not yet examined I_A but plan to do so. Potassium currents have also been demonstrated to decrease in other organisms during plastic modification (Byrne, 1985; Farley and Alkon, 1985; Kandel and Schwartz, 1982; Woody, 1986).

The reductions in potassium currents we have seen are postsynaptic events. How might such events mediate learned behaviors such as blink conditioning? We have studied the AHP alterations with intracellular recording and current injection in the pyramidal cell soma for technical reasons. But the cellular compartment in which the AHP is generated has not been defined—it could well be dendritic. We may be recording a somatic reflection of a current that is produced in a portion of the dendritic tree; it is not necessary that the entire dendritic tree of the cell be altered. Such a selective reduction would insure that a cell responds with an excitatory response to the tone CS^+ afferent drive and much less or not at all to other afferents after training. Spatially localized reductions of K^+ currents would assign uniqueness to the excitability increase, insuring that the cell did not fire indiscriminately to any stimulus. Spatial coding of learned information storage at the cellular level seems a quite reasonable hypothesis (Alkon, 1984; Woody, 1982). The fact that both nonassociative (Andersen *et al.*, 1980) and associative LTPs (Kelso and Brown, 1986) are specific to the portion of the dendritic tree that has been potentiated provides support for such a hypothesis.

4.2. Why Do So Many Hippocampal Cells Change during Learning?

We have described a possible mechanism by which AHP changes may achieve informational specificity at the cellular level; i.e., these changes may be localized spatially within the dendritic tree of CA1 pyramidal neurons. But another facet of the specificity issue regards the percentage of the total population of cells that are altered during learning. We have observed AHP reductions in almost 50% of the cells we have studied in hippocampal slices from conditioned rabbits (Fig. 2). One might ask what role such a widespread change in (at least CA1 region of) hippocampus may have in conditioning? A rabbit, like any mammal, learns many things during its lifetime. Do (or can) such ubiquitous changes occur in hippocampus during each learning event?

Our response is twofold, empirical and theoretical. Empirically, the unit recording data are quite clear in demonstrating that a large percentage of hippocampal neurons show firing rate changes to conditioned stimulus onset in both rats (in a appetitive food retrieval task) (Disterhoft and Segal, 1978; Olds *et al.*, 1972; Segal, 1973) and rabbits (in the aversive eye blink paradigm) (Berger *et al.*, 1980; Thompson *et al.*, 1980) in simple classical conditioning tasks. And, as we mentioned earlier, Berger *et al.* (1983) have recently shown that more than 60% of hippocampal pyramidal neurons show increased firing to CS onset after blink conditioning. In fact, this is one reason we chose hippocampal slices for a biophysical study—we wanted a region where a large number of cells changed during learning to facilitate determining the cellular mechanism of that change. It should be stressed that the rats and rabbits are placed in a somewhat artificial situation, which should highlight the importance of these simple associative tasks. The animals are living singly in simple laboratory cages in an environment that does not place the task demands or provide them with the environmental stimuli that a "natural" environment would. Considered within this context, being trained in an associative eye blink or food retrieval paradigm is probably the most important cognitive event occurring in the life history of these animals during the time we are studying them. So the hippocampus is free to be fully engaged in processing what may appear to be a relatively simple learning task.

Such massive hippocampal involvement may well not occur in a more normal situation. It would be interesting to train animals in several different associative learning tasks simultaneously and determine whether individual neurons are differentially involved in them. Alternatively, different portions of the dendritic tree could show uniquely enhanced excitability to the stimuli mediating each task, and as a result most pyramidal neurons could be involved in most tasks. Of course, the cells making up the involved cell population would have to be unique for each learning task in order for the across-fiber output volley of action potentials to have a unique signature for each learning situation.

Most theories of hippocampal function assume that its role is to transfer information from immediate to long-term memory (Squire, 1982). The fact that H.M., after a bilateral hippocampectomy, can recall information stored prior to his surgery but not afterward suggests that long-term memories are not stored in hippocampus *per se* (Scoville and Milner, 1957). Hippocampus is generally thought to be relatively plastic and neurons there easily changed during learning, consistent with its role as an intermediate storage site (Olds, 1969; Teyler and Di Scenna, 1985). So it may not be surprising that a large percentage of pyramidal neurons show altered excitability during the acquisition of a particularly important associative learning task. After returning to their basal state, these neurons would presumably be similarly altered in future learning situations.

Both empirical and theoretical considerations lead us to conclude that biophysical alterations occurring in a large percentage of hippocampal neurons during eye blink conditioning can serve a functionally meaningful role in the acquisition of this associative learning task. We would expect that neocortical regions postulated to serve a role as long-term storage sites (Mishkin *et al.*, 1984) would be obliged to exhibit considerably more cellular specificity as regards involvement during conditioning in order to identify uniquely the many learning tasks that occur during the life-span.

4.3. Modulation by Acetylcholine and/or Norepinephrine

Acetylcholine (ACh) is an important neurotransmitter in hippocampus. Scopolamine, a cholinergic blocker, retards NM acquisition only when the hippocampus is intact (Solomon *et al.*, 1983). The major cholinergic input to hippocampus is from medial septum. Medial septal lesions, like ACh blockade, retard acquisition of the NM conditioned response (Berry and Thompson, 1979). Medial septal neurons respond to both the tone CS and air-puff US with a "sensorylike" responsiveness that declines across conditioning in a negatively correlated manner with NM response acquisition (Berger and Thompson, 1978). These characteristics suggest that medial septum may play an arousal role in reference to hippocampus during NM conditioning. The important point in reference to our biophysical findings is that ACh application reduces the AHP in hippocampal pyramidal neurons (Benardo and Prince, 1982; Cole and Nicoll, 1984). This AHP reduction produced by acetylcholine does not cause a generalized excitability increase in hippocampal pyramidal cells but appears to increase the signal-to-noise ratio of specific afferent inputs (Krnjevic and Ropert, 1982).

Norepinephrine (NE) also serves as a transmitter in hippocampus (from locus coeruleus). In a striking parallel with ACh, NE also serves to reduce the AHP without necessarily altering membrane potential or input resistance (Haas and Konnerth, 1983; Madison and Nicoll, 1982, 1986). Locus coeruleus lesions, like hippocampal lesions, slow extinction of the NM conditioned response (McCormick and Thompson, 1982). The extinction of that portion of the conditioned response with which hippocampal unit firing is particularly correlated is most slowed. Cohen has demonstrated that visual system plasticity (in lateral geniculate nucleus neurons) during heart rate classical conditioning in pigeons can be eliminated by locus coeruleus lesions. Conversely, this plasticity can be caused by locus coeruleus stimulation (Cohen, 1985). Locus coeruleus neurons recorded in the alert animal respond to environmental stimuli (such as tones) in a fashion that suggests that their function is to enhance the processing of important external stimuli (Aston-Jones and Bloom, 1981).

The functional consequences of both ACh and NE are not to increase hippocampal pyramidal neuron firing directly but rather to enhance neuronal firing to other more specific synaptic inputs (Krnjevic and Ropert, 1982; Woodward *et al.*, 1979). We postulate that during NM conditioning, the auditory CS input is likely being projected to CA1 hippocampus via an entorhinal cortex–dentate–CA3–CA1 pathway. Tone CS onset, and especially periorbital shock US onset, may cause relatively long-lasting excitability increases via medial septum or locus coeruleus input or both. Neurons in both structures appear to serve an arousal function for interesting environmental stimuli. The fact that both ACh and NE effects last for seconds to minutes puts them in a time frame appropriate for mediating synaptically driven excitability increases in CA1 pyramidal neurons. By reducing the AHP, both of these modulating neurotransmitters have the capacity to strengthen informationally specific inputs across training trials in the absence of large generalized increases in firing rate or membrane potential. This type of neuromodulatory process seems likely to be involved in laying down the postsynaptic ionic alterations localized to hippocampus that we have demonstrated.

4.4. Role of Hippocampus in Blink Conditioning

It is obvious that hippocampus is not the only region that is changing in rabbit brain during NM conditioning. For example, recent work has emphasized the role of the cerebellum in this task (Lincoln *et al.*, 1982; McCormick and Thompson, 1984; Yeo *et al.*, 1985). We are not proposing that we have characterized *the* cellular ionic basis of learning in rabbit brain. The conditioned reflex arc subserving this task is likely to be distributed (Cohen, 1985; Squire, 1986; Thompson *et al.*, 1980; Woody, 1982). Other brain regions may utilize a reduction in AHP to increase cellular excitability during learning. It is likely that other mechanisms are involved as well. But we have demonstrated one mechanism that is operative in hippocampus.

It is well known that hippocampal lesions do not eliminate the ability of rabbits to acquire eye blink conditioning in the short-delay paradigm such as we have used (Schmaltz and Theios, 1972; Solomon and Moore, 1975). Indeed, eye blink conditioning has been demonstrated in the hippocampectomized human (Weiskrantz and Warrington, 1979). The human data suggest that the hippocampus is probably involved in storing information regarding the cognitive aspects of the task (declarative memory) rather than its motor program (procedural memory or habit) (Mishkin *et al.*, 1984; Squire and Cohen, 1984). But it should be stressed that disruption of the intact hippocampus by cholinergic blockers (Solomon *et al.*, 1983) or by medial septal nucleus lesions (Berry and Thompson, 1979) does retard blink conditioning even in the simple delay paradigm. We have already mentioned the massive engagement of hippocampal neurons shown in the unit-recording studies (Berger *et al.*, 1983). Thus, the hippocampal lesion result may not be as straightforward as it seems. A role for the hippocampus even in short-delay conditioning should not be dismissed on the basis of the lesion experiments (see Thompson *et al.*, 1980; Woody, 1986 for a more complete discussion of this issue).

4.5. An Identified Storage Site Can Be Studied at the Cellular Level

Our *in vitro* studies have demonstrated conditioning-specific reductions in the AHP response in CA1 pyramidal neurons. These alterations are not dependent on afferent drive from the remainder of the conditioned reflex arc for their expression. Thus, the evidence is very strong that the hippocampus is one local storage site for alteration during associative learning. By using biophysical measures from hippocampal slices, we will be able to make significant progress in understanding how learned information is stored in identifiable neurons during an associative learning task.

6. SUMMARY

A series of studies have been done to analyze biophysical alterations in CA1 pyramidal neurons from hippocampal slices of nictitating membrane/eyeblinking-conditioned rabbits. We have shown conditioning-specific reductions in the magnitude and duration of the afterhyperpolarization recorded intracellularly in CA1 pyramidal neurons after current-elicited action potentials. The AHP is attributed to a calcium-mediated potassium current. The presumed function of the AHP *in vivo* is to slow or stop the firing of the pyramidal neurons during a burst of action potentials. The AHP reduction after conditioning would tend to make the neuron fire more readily to conditioned stimulus afferent drive, as has been observed with *in vivo* unit-recording studies in rats and rabbits. We have seen the AHP to be reduced in two separate replications of the *in vitro* brain slice study. We have also observed that the AHP is reduced in brain slices from rabbits that have acquired the behavioral conditioned response as compared to those from rabbits that have received the same amount of training but have not yet learned the response. Recent experiments suggest that the calcium spike threshold is raised, in addition to and separate from the calcium-mediated potassium current reduction, after conditioning.

We are the first to demonstrate an ionic mechanism *in vitro* that underlies associative learning in mammalian brain. The experiments described have demonstrated conditioning-specific AHP reductions in CA1 pyramidal neurons. Since they are present in hippocampal brain slices, these alterations are not dependent on afferent drive from the remainder of the conditioned reflex arc for their expression. Therefore, our work provides strong evidence that the hippocampus is one local storage site for ionic alterations during associative learning. At this point, our experiments have concerned alterations that are sorted at postsynaptic sites. Possible alterations in other currents and/or in synaptically mediated events remain to be investigated.

ACKNOWLEDGMENTS. This research was partially supported by research grants from the National Science Foundation (BNS-8302488 and BNS-8607670) and by a National Institute of Neurological and Communicative Disorder and Stroke IPA Fellowship to J.F.D.

REFERENCES

Alkon, D. L., 1982, A biophysical basis for molluscan associative learning, in: *Conditioning, Representation of Involved Neural Functions* (C. D. Woody, ed.), Plenum Press, New York, pp. 147–170.

Alkon, D. L., 1984, Calcium-mediated reduction of ionic currents. A biophysical memory trace, *Science* **226:**1037–1045.

Andersen, P., Sundberg, S. H., Sveen, O., Swann, J. W., and Wigstrom, H., 1980, Possible mechanisms for long-lasting potentiation of synaptic transmission in hippocampal slices from guinea-pigs, *J. Physiol.* (*Lond.*) **302:**463–482.

Aston-Jones, G., and Bloom F. E., 1981, Norepinephrine-containing locus coeruleus neurons in behaving rats exhibit pronounced responses to non-noxious environmental stimuli, *J. Neurosci.* **1:**887–900.

Benardo, L. S., and Prince, D. A., 1982, Ionic mechanisms of cholinergic excitation in mammalian hippocampal pyramidal neurons, *Brain Res.* **249:**333–344.

Berger, T. W., and Thompson, R. F., 1978, Neuronal plasticity in the limbic system during classical conditioning of the rabbit nictitating membrane response. I. The hippocampus, *Brain Res.* **145:**323–346.

Berger, T. W., Clark, G. A., and Thompson, R. F., 1980, Neuronal plasticity recorded from limbic system brain structures during classical conditioning, *Physiol. Psychol.* **8:**155–167.

Berger, T. W., Rinaldi, P. C., Weiss, D. J., and Thompson, R. F., 1983, Single-unit analysis of different hippocampal cell types during classical conditioning of rabbit nictitating membrane response, *J. Neurophysiol.* **50:**1197–1219.

Berry, S. D., and Thompson, R. F., 1979, Medial septal lesions retard classical conditioning of the nictitating membrane response in rabbits, *Science* **205:**209–211.

Brown, D. A., and Griffith, W. H., 1983, Calcium activated outward current in voltage clamped hippocampal neurones of the guinea-pig, *J. Physiol.* (*Lond.*) **337:**287–301.

Byrne, J. H., 1985, Neural and molecular mechanisms underlying information storage in *Aplysia:* Implications for learning and memory, *Trends Neurosci.* **8:**478–482.

Cohen, D. H., 1985, Some organizational principles of a vertebrate conditioning pathway: Is memory a distributed property? in: *Memory Systems of the Brain* (N. M. Weinberger, J. L. McGaugh, and G. Lynch, eds.), Guilford, New York, pp. 27–48.

Cole, A. E., and Nicoll, R. A., 1984, Characterization of slow cholinergic postsynaptic potential recorded *in vitro* from rat hippocampal pyramidal cells, *J. Physiol.* (*Lond.*) **352:**173–188.

Coulter, D. A., Kubota, M., Morre, J. W., Disterhoft, J. F., and Alkon, D. L., 1985, Conditioning-specific reduction of CA1 afterhyperpolarization amplitude and duration in rabbit hippocampal slices, *Soc. Neurosci. Abstr.* **11:**891.

Coulter, D. A., Disterhoft, J. F., and Alkon, D. L., 1986, Decreased Ca^{2+} entry and K^{+} conductance contribute to AHP reductions with conditioning in rabbit hippocampus, *Soc. Neurosci. Abstr.* **12:**181.

Disterhoft, J. F., and Segal, M., 1978, Neuron activity in the rat hippocampus and motor cortex during discrimination reversal, *Brain Res. Bull.* **3:**583–588.

Disterhoft, J. F., Kwan, H. H., and Lo, W. D., 1977, Nictitating membrane conditioning to tone in the immobilized albino rabbit, *Brain Res.* **137:**127–143.

Disterhoft, J. F., Quinn, K. J., Weiss, C., and Shipley, M. T., 1985, Accessory abducens nucleus and conditioned eye retraction/nictitating membrane extension in rabbit, *J. Neurosci.* **5:**941–950.

Disterhoft, J. F., Coulter, D. A., and Alkon, D. L., 1986a, Conditioning-specific membrane changes of rabbit hippocampal neurons measured *in vitro, Proc. Nat. Acad. Sci. U.S.A.* **83:**2733–2737.

Disterhoft, J. F., Golden, D., Read, H., Coulter, D. A., and Alkon, D. L., 1986b, AHP reductions in rabbit hippocampal neurons during conditioning require acquisition of the learned response, *Soc. Neurosci. Abstr.* **12:**180.

Durand, D., and Carlen, P. L., 1984, Ethanol in low doses augments calcium-mediated mechanisms measured in hippocampal neurons, *Science* **224:**1359–1361.

Farley, J., and Alkon, D. L., 1985, Cellular mechanisms of learning, memory and information storage, *Annu. Rev. Psychol.* **36:**419–494.

Gustafsson, B., and Wigstrom, H., 1981, Evidence for two types of afterhyperpolarization in CA1 pyramidal cells in the hippocampus, *Brain Res.* **206:**462–468.

Gustafsson, B., Galvan, M., Grafe, P., and Wigstrom, H., 1982, A transient outward current in mammalian central neurons blocked by 4-aminopyridine, *Nature* **299:**252–254.

Haas, H. L., and Konnerth, A., 1983, Histamine and noradrenaline decrease calcium-activated potassium conductance in hippocampal pyramidal cells, *Nature* **302:**432–434.

Hotson, J. R., and Prince, D. A., 1980, A calcium-activated hyperpolarization follows repetitive firing in hippocampal neurons, *J. Neurophysiol.* **43:**409–419.

Kandel, E. R., and Schwartz, J. H., 1982, Molecular biology of learning: Modulation of transmitter release, *Science* **216:**433–443.

Kelso, S. R., and Brown, T. H., 1986, Differential conditioning of associative synaptic enhancement in hippocampal brain slices, *Science* **232:**85–87.

Kelso, S. R., Nelson, D. A., Silva, N. L., and Boulant, J. A., 1983, A slice chamber for intracellular and extracellular recording during continuous perfusion, *Brain Res. Bull.* **10:**853–857.

Kraus, N., and Disterhoft, J. F., 1981, Response plasticity of single neurons in rabbit auditory association cortex during tone-signalled learning, *Brain Res.* **246:**205–215.

Krnjevic, K., and Ropert, N., 1982, Electrophysiological and pharmacological characteristics of facilitation of hippocampal population spikes by stimulation of the medial septum, *Neuroscience* **7:**2165–2183.

Lancaster, B., and Adams, P. R., 1986, Calcium-dependent current generating the afterhyperpolarization of hippocampal neurons, *J. Neurophysiol.* **55:**1268–1282.

Lancaster, B., and Wheal, H. V., 1984, The synaptically evoked later hyperpolarization in hippocampal CA1 pyramidal cells in resistant to extracellular EGTA, *Neuroscience* **12:**267–275.

Landfield, P. W., and Pitler, T. A., 1984, Prolonged Ca^{2+}-dependent afterhyperpolarization in hippocampal neurons of aged rats, *Science* **226:**1089–1092.

Langmoen, I. A., and Andersen, P., 1981, The hippocampal slice *in vitro,* in: *Electro-Physiology of Isolated Mammalian CNS Preparations* (I. A. Kerkut and H. V. Wheal, eds.), Academic Press, New York, pp. 51–105.

Lincoln, J. S., McCormick, D. A., and Thompson, R. F., 1982, Ipsilateral cerebellar lesions prevent learning of the classically conditioned nictitating membrane/eyelid response, *Brain Res.* **242:**190–193.

Madison, D. V., and Nicoll, B. A., 1982, Noradrenaline blocks accommodation of pyramidal cell discharge in the hippocampus, *Nature* **299:**636–638.

Madison, D. V., and Nicoll, R. A., 1984, Control of the repetitive discharge of rat CA1 pyramidal neurons *in vitro, J. Physiol. (Lond.)* **354:**319–332.

Madison, D. V., and Nicoll, R. A., 1986, Actions of noradrenaline recorded intracellularly in rat hippocampal CA1 pyramidal neurons *in vitro, J. Physiol. (Lond.)* **372:**221–244.

Mamounnas, L. A., Thompson, R. F., Lynch, G., and Baudry, M., 1984, Classical conditioning of the rabbit eyelid response increases glutamate receptor binding in hippocampal synaptic membranes, *Proc. Natl. Acad. Sci. U.S.A.* **81:**2548–2552.

McCormick, D. A., and Thompson, R. F., 1982, Locus coeruleus lesions and resistance to extinction of a classically conditioned response: Involvement of the neocortex and hippocampus, *Brain Res.* **245:**239–249.

McCormick, D. A., and Thompson, R. F., 1984, Cerebellum: Essential involvement in the classically conditioned eyelid response, *Science* **223:**296–299.

Mishkin, M., Malamut, B., and Bachevalier, J., 1984, Memories and habits: Two neural systems, in: *Neurobiology of Learning and Memory* (G. Lynch, J. L. McGaugh, and N. M. Weinberger, eds.), Guilford, New York, pp. 65–77.

Nicoll, R. A., and Alger, B. E., 1981, Synaptic excitation may activate a calcium-dependent potassium conductance in hippocampal pyramidal cells, *Science* **212:**957–959.

Olds, J., 1969, The central nervous system and the reinforcement of behavior, *Am. Psychol.* **24:**114–132.

Olds, J., Disterhoft, J. F., Segal, M., Kornblith, C. L., and Hirsh, R., 1972, Learning centers of rat brain mapped by measuring latencies of conditioned unit responses, *J. Neurophysiol.* **35:**202–219.

Schmaltz, L. W., and Theios, J., 1972, Acquisition and extinction of a classically conditioned response in hippocampectomized rabbits (*Oryctalagus cuniculus*), *J. Comp. Physiol. Psychol.* **79:**328–333.

Schwartzkroin, P. A., 1981, To slice or not to slice, in: *Electrophysiology of Isolated Mammalian CNS Preparations* (G. A. Kerkut and H. V. Wheal, eds.), Academic Press, New York, pp. 15–49.

Scoville, W. B., and Milner, B., 1957, Loss of recent memory after bilateral hippocampal lesions, *J. Neurol. Neurosurg. Psychiatry* **20:**11–21.

Segal, M., 1973, Flow of conditioned responses in the limbic telecephalic system of the rat, *J. Neurophysiol.* **36:**840–854.

Solomon, P. R., and Moore, J. W., 1975, Latent inhibition and stimulus generalization of the classically conditioned nictitating membrane response in rabbits (*Oryctolagus cuniculus*) following dorsal hippocampal ablations, *J. Comp. Physiol. Psychol.* **89:**1192–1203.

Solomon, P. R., Solomon, S. D., Vander Schaaf, E., and Perry, H. E., 1983, Altered activity in the hippocampus is more detrimental to classical conditioning than removing the structure, *Science* **220:**329–331.

Squire, L. R., 1982, The neuropsychology of human memory, *Annu. Rev. Neurosci.* **5:**241–273.

Squire, L. R., 1986, Mechanisms of memory, *Science* **232:**1612–1619.

Squire, L. R., and Cohen, N. J., 1984, Human memory and amnesia, in: *Neurobiology of Learning and Memory* (G. Lynch, J. L. McGaugh, and N. M. Weinberger, eds.), Guilford, New York, pp. 3–64.

Teyler, T. J., 1980, Brain slice preparation: Hippocampus, *Brain Res. Bull.* **5:**391–403.

Teyler, T. J., and Di Scenna, P., 1985, The role of hippocampus in memory: A hypothesis, *Neurosci. Biobehav. Rev.* **9**:377–389.

Thompson, R. F., 1976, The search for the engram, *Am. Psychol.* **31**:209–227.

Thompson, R. F., Berger, T. W., Berry, S. D., Hoehler, R. K., Kettner, R. E., and Weisz, D. J., 1980, Hippocampal substrate of classical conditioning, *Physiol. Psychol.* **8**:262–279.

Weiskrantz, L., and Warrington, E. G., 1979, Conditioning in amnesic patients, *Neuropsychologia* **17**:187–194.

Wong, R. K. S., and Prince, D. A., 1981, Afterpotential generation in hippocampal pyramidal cells, *J. Neurophysiol.* **45**:86–97.

Woodward, D. A., Moises, H. C., Waterhouse, B. D., Hoffer, B. J., and Freedman, R., 1979, Modulating actions of norepinephrine in the central nervous system, *Fed. Prod.* **38**:2109–2116.

Woody, C. D., 1982, *Memory, Learning, and Higher Function, A Cellular View*, Springer-Verlag, New York.

Woody, C. D., 1986, Understanding the cellular basis of memory and learning, *Annu. Rev. Psychol.* **37**:433–493.

Yeo, C. H., Hardiman, M. J., and Glickstein, M., 1985, Classical conditioning of the nictitating membrane response of the rabbit. II. Lesions of the cerebellar cortex, *Exp. Brain Res.* **60**:99–113.

Zbicz, K. L., and Weight, F. F., 1985, Transient voltage and calcium-dependent outward currents in hippocampal CA3 pyramidal neurons, *J. Neurophysiol.* **53**:1038–1058.

11

Serotonin and Aversive Conditioning in Adult and Juvenile Snails

PAVEL M. BALABAN and
IGOR S. ZAKHAROV

1. INTRODUCTION

An important role of serotonin (5-HT) in sensitization and conditioning was disclosed by experiments in which noxious stimulation or noxious reinforcement was used in *Aplysia* (Kandel and Schwartz, 1982; Klein *et al.*, 1982). The present chapter describes the results of investigations of the role of 5-HT in sensitization and aversive conditioning in adult and juvenile land snails, *Helix lucorum*.

Conventional electrophysiological techniques for intracellular recording of identified neurons were used. Neurophysiological experiments were carried out in (1) a semi-intact preparation (Maximova and Balaban, 1984), (2) a lip–central nervous system (CNS) preparation, or (3) an isolated CNS preparation. Suction electrodes were used for electrical stimulation of nerves. In behavioral experiments, the snail's shell was glued to a holder, and the foot of the animal was on a plastic ball floating in water. This experimental setup allowed the animal to move freely while repeated tactile stimuli were delivered to the same place on the skin. The amplitude of the behavioral reactions was recorded by a photocell or using a video monitor. Tactile stimuli of different intensities were applied by means of calibrated von Frey hairs. Injections of 5,7-dihydroxytroptamine (5,7-DHT, Sigma, St. Louis, MO) dissolved 20 mg/kg of body weight, in a saline solution with 0.1% ascrobic acid were made into the hemocoelom of the snails 5–15 days before the experiments.

2. ROLE OF 5-HT IN SENSITIZATION AND AVERSIVE CONDITIONING IN ADULT SNAILS

The role of 5-HT in behavioral and neuronal plasticity was investigated by manipulating 5-HT levels either (1) directly, by increasing the concentration of 5-HT in the

---

PAVEL M. BALABAN and IGOR S. ZAKHAROV • Laboratory of Conditioned Reflexes and Physiology of Emotions, Institute of Higher Nervous Activity and Neurophysiology, USSR Academy of Sciences, Moscow, USSR.

medium bathing the CNS, or (2) indirectly, by selectively destroying serotonergic nerve terminals using 5,7-DHT injections into intact animals.

In the neurophysiological experiments, it was established that increases in 5-HT concentrations of up to 5×10^{-5} M elicited a sensitization of the synaptic responses evoked by nerve stimulations or tactile stimulations of the skin in the command neurons triggering the avoidance reactions (for a detailed description of command neuron functional identification, see Balaban, 1979, 1983). Usually, repeated presentations of noxious stimuli elicited sensitization, consisting of a change in the avoidance reaction amplitude in the behavioral experiments and enhancement of the synaptic responses in command neurons controlling avoidance behavior after the second or third stimulus in a series. In 5,7-DHT-injected animals, evidence for sensitization was absent in both behavioral and neurophysiological experiments.

These results suggest that sensitization of avoidance behavior in *Helix* is mediated by 5-HT and that the most probable site of 5-HT action is the synaptic contact between sensory and command neurons.

It was shown earlier (Maximova and Balaban, 1984) that after five to 15 paired presentations of a drop of carrot juice on the lip (lip–CNS preparation) and a strong pallial nerve stimulation, postsynaptic potentials and action potentials, which were absent prior to training, could be recorded in command neurons controlling avoidance behavior in response to a drop of carrot juice. A novel spike response may be considered equivalent to the avoidance behavior of the animal since action potentials in command neurons controlling this form of behavior always trigger avoidance reactions (Balaban, 1979, 1983). Unpaired presentation of the same stimuli had no effect on the responsiveness of command neurons.

In the next series of experiments, electrical reinforcement was replaced by a short-term increase in the 5-HT concentration in the bath (up to 5×10^{-5} M). It was established that only paired presentations of juice and 5-HT led to significant changes in command neuron responsiveness. These changes are seen as novel action potentials in response to food in neurons triggering the avoidance behavior. Thus, the transmitter responsible for sensitization is satisfactory as a reinforcing factor. The necessity of this transmitter for aversive conditioning was investigated in animals treated with the selective neurotoxin 5,7-DHT.

After five paired presentations of food and electrical stimulation, the number of feeding responses to food decreased from 90% to 15% in a group of control, intact snails (sham-injected). This decrease was characteristic of aversive learning in untreated animals. In both the experimental group, injected previously with 5,7-DHT, and the sham-injected pseudoconditioned group, the number of feeding responses did not change with pairing. These results suggest an essential role for serotonergic neural systems in aversive conditioning. It is important to note that in 5,7-DHT-treated snails it was possible to obtain conditioned responses with food reinforcement; i.e., in treated animals only conditioning to noxious stimuli was impaired. Injection of 5,7-DHT after acquisition of an aversive conditioned response did not impair expression of the conditioned response. This suggests that 5-HT is necessary only at the stage of acquisition of the conditioned response.

A comparison of the role of 5-HT in sensitization and aversive conditioning suggests that the presence of sensitization during the first stages of aversive conditioning is not a coincidence but that the same mechanism involving 5-HT underlies both processes.

3. SENSITIZATION AND AVERSIVE CONDITIONING IN JUVENILE SNAILS

In behavioral experiments, it was found that the characteristic increase in response amplitude (sensitization) to the second or third stimulus in a series seen in adult snails was absent in juvenile snails under 1 month old (Zakharov and Balaban, 1983). Only a gradual decrease in response amplitude was noted in snails of this age. In addition, no restoration of the response amplitude was seen after a strong extra stimulus in juvenile snails. These differences in the behavior of adult and juvenile snails are probably not attributable to the small size of the animal at this age (4–5 mm), since snails 5–6 months old are not much bigger (6–7 mm), and sensitization is clearly present. The absence of sensitization in identified command neurons during repeated nerve stimulation as well as after an extra stimulus was seen in neurophysiological experiments using juvenile snails under 1 month old.

With the same methods for conditioned reflex elaboration as were used for adult snails, the aversive conditioning paradigm was applied to a group of newborn snails and to a group of 4 to 5-month-old snails. Before training sessions, all animals exhibited feeding responses in 90–100% of trials. After nine to 20 pairings of the food with the noxious stimulus, the hungry snails 4–5 months of age refused to take food in 90% of test trials, whereas the snails under 1 month old ate food in 90–100% of test trials after 20–24 paired trials. Pseudoconditioned snails exhibited feeding responses in 95–100% of the test trials. Thus, in the early stages of postnatal development, sensitization is absent in snail behavior, as is the capacity to acquire aversively conditioned responses. The coincidence of these gaps in the behavioral repertoire of juvenile snails and in 5,7-DHT-treated adult animals allows us to suggest that it is the difference in 5-HT levels in juvenile, as compared to adult, snails that explains the absence of aversive conditioning.

4. HISTOCHEMICAL INVESTIGATION OF 5-HT IN HELIX CNS DURING DEVELOPMENT

For localization of 5-HT-containing neurons, the CNS of the snail was examined with the glyoxylic fluorescence histochemical technique modified for cryostat-sectioned nervous tissue (Lindvall and Björklund, 1974; de la Torre and Surgeon, 1976). Serially sectioned (40 μm) whole-animal preparations from juvenile snails and CNS preparations from adults were examined with a fluorescence microscope using a 405-nm light for excitation.

The presence and localization of biogenic amines were studied in serial sections from the CNS of six adult (4–6 years) animals, eight animals 7 days old, and nine snails 4 months old. In accord with previously reported results (Sakharov, 1974), 5-HT-containing neurons were found in cerebral, pedal, visceral, and right parietal ganglia of adult snails. However, in the CNS of newborn animals, 5-HT-containing neurons were not seen. It was thought that the cells in the newborn animals might be too small to be revealed by this technique, but in the CNS of the 4-month-old snails, 5-HT-containing neurons were readily seen in cerebral and pedal ganglia, and the CNS at this age was only slightly larger than that in the newborns. Absence of 5-HT-containing neurons in parietal and visceral ganglia in 4-month-old snails points to the possibility of a heterochronological development of transmitter-producing neuronal systems. It is assumed that the age dependence of behavioral and neuronal reactions can be caused by some delay in the development of serotonergic systems.

Thus, in the postnatal period, the serotonin content is very low in the nervous system of the snail. This result suggests that the absence of sensitization, as well as the snails' inability to be aversively conditioned is related to 5-HT levels.

The present results suggest that 5-HT is a necessary factor for the acquisition of aversive conditioning in molluscs. It is important to note that this conclusion is valid only for reflexes to stimuli evoking avoidance, escape, or defensive behavioral responses.

REFERENCES

Balaban, P. M., 1979, A system of command neurons in snail's escape behavior, *Acta Neurobiol. Exp.* **39:**97–107.

Balaban, P. M., 1983, Postsynaptic mechanism of withdrawal reflex sensitization in the snail, *J. Neurobiol.* **14:**365–375.

de la Torre, J. C., and Surgeon, J. W., 1976, A methodological approach to rapid and sensitive monoamine histofluorescence using a modified glyoxylic acid technique: The SPG method, *Histochemistry* **49:**81–93.

Kandel, E. R., and Schwartz, J. H., 1982, Molecular biology of learning, *Science* **218:**433–443.

Klein, M., Camardo, J., and Kandel, E. R., 1982, Serotonin modulates a specific potassium current in the sensory neurones that show presynaptic facilitation in *Aplysia, Proc. Natl. Acad. Sci. U.S.A.* **79:**5713–5717.

Lindvall, O., and Björklund, A., 1974, The glyoxylic acid fluorescence histochemical method: A detailed account of the methodology for the visualization of central catecholamine neurons, *Histochemistry* **39:**97–127.

Maximova, O. A., and Balaban, P. M., 1984, Neuronal correlates of aversive learning in command neurons for avoidance behavior of *Helix lucorum L., Brain Res.* **292:**139–149.

Sakharov, D. A., 1974, *Geneology of Neurons,* Nauka, Moscow.

Zakharov, I. S., and Balaban, P. M., 1983, Changes in defensive reflexes of *Helix lucorum* in ontogeny, *Neurosci Behav. Physiol.* **13:**248–251.

12

Catecholaminergic and Opioid Mechanisms in Conditioned Food Intake Behavior of the Monkey Amygdala

YUTAKA OOMURA, YASUHIKO NAKANO, LÁSZLÓ LÉNÁRD, HITOO NISHINO, and SHUJI AOU

1. INTRODUCTION

Amygdalar neurons respond to complex visual stimuli and activity changes at the sight of food (Ono *et al.*, 1983; Sanghera *et al.*, 1979). The amygdala (AM) is among the richest of the brain areas in opioid receptors and endogenous ligands, but the functional role of its opioid system in feeding behavior is not yet understood.

The AM receives noradrenergic (NA) inputs from the locus coeruleus and medullary NA and dopaminergic (DA) inputs from the ventral tegmental area. It has been reported that reward-related behavior can be modulated by catecholamines (CA) (Aou *et al.*, 1983; Beninger, 1983; Heffner *et al.*, 1981). Destruction of NA and DA terminals in the AM produces hyperphagia and hypophagia, respectively (Lénárd and Hahn, 1982), and injection of a β-adrenoceptor antagonist immediately after training decreases retention of passive avoidance conditioning (Gallagher *et al.*, 1981). Norepinephrine and DA should thus influence AM neurons in the regulation of feeding and other reward-related behavior, but studies have been behavioral and anatomic.

The central nucleus of the AM is connected to the lateral hypothalamus (LHA) and the nucleus tractus solitarii (NTS), both of which contain glucose-sensitive (GS) neurons identified by activity decrease in response to electrophoretic glucose. Their close involvement in the regulation of feeding has been shown in monkeys (Aou *et al.*, 1984). Close relationships among monkey LHA, GS, and morphine-sensitive cells during operant feeding are reported (Oomura *et al.*, 1986), but no such chemosensitive neurons have been reported in the AM. This is a report of CA, glucose, and morphine functions in the AM during reward-related behavior.

YUTAKA OOMURA • Department of Physiology, Faculty of Medicine, Kyushu University, Fukuoka 812, Japan; and Department of Biological Control Systems, National Institute for Physiological Sciences, Okazaki 444, Japan. YASUHIKO NAKANO, LÁSZLÓ LÉNÁRD, HITOO NISHINO, and SHUJI AOU • Department of Biological Control Systems, National Institute for Physiological Sciences, Okazaki 444, Japan.

2. MATERIAL AND METHODS

Male *Macaca fuscata* monkeys (4–6 kg) were used. A monkey was seated in a primate chair facing a panel that had a cue lamp, a bar, and a reward box. Experimental procedures have been published previously (Aou *et al.*, 1983). The monkeys were trained to a high-fixed-ratio schedule consisting of (1) a cue light to signal the start of bar pressing, (2) a high-fixed-ratio bar-press task (FR 20), (3) presentation of a short cue tone followed by a food reward, and (4) ingestion of the food (reward). A small ball of bread was the standard reward, but more palatable food (raisin or small piece of banana) was sometimes used. Task-related neuronal activity was evaluated in five consecutive trials with the same reward except during extinction. Electrophoresis and extracellular single-neuron recording were through a conventional multibarrel pipette. A neuron was considered responsive if the firing rate changed at least 20% from its basal level and was dose dependent on electrophoresis application current. A response was considered to be blocked by an antagonist if its magnitude was attenuated at least 50%. Pipette proximity to the cell was checked by glutamate, and nonspecific current effect by Na^+ and Cl^-. The centromedial and basolateral AM are shown in Fig. 1B.

3. RESULTS

3.1. Task-Related Activity Changes

Of 292 cells investigated with CA during the task, 68% responded. Of these, 36% discriminated the cue light and tone. Cells that responded to the cue light did not respond to a similar red light at another place on the panel or to a green light in the same position as the usual red light. Cue-tone-responsive cells did not respond to a random cue tones independent of the presentation of food. The response latency to the cue light was 158.3 ± 75.9 msec (mean ± S.D., n = 62), and to the cue tone 171.1 ± 80.5 msec (n = 19). More neurons responded during bar press (64%) and/or reward (35%). Of these, 24% also responded to the cue light or tone. Of the 127 cells that responded during bar press, 51 increased and 76 decreased activity. In the reward period, 41 cells increased and 29 decreased activity. Reward-related response changes often continued after the monkey swallowed the food. In trials with more palatable food, neuronal responses were significantly enhanced during bar press (59%) and/or reward (60%). In trials with no reward, 73% of the neuronal responses during the bar press period diminished significantly, and 33% also decreased during the intertrial intervals. The animal usually stopped responding after five or six trials.

Activity changed in at least one phase of the task in 75% of the cells in the centromedial AM and in 61% in the basolateral AM. Thus, more centromedial cells responded ($P < 0.01$, χ^2 test). There were more bar-press-responsive cells (decrease or increase) in the centromedial AM (49%) than in the basolateral AM (38%, $P < 0.05$, χ^2 test) (Fig. 1Aa), and more centromedial cells (15%) than basolateral cells (5%) decreased firing in the reward period ($P < 0.01$, χ^2 test). The proportions of cells responding to cue light or cue tone and the response latencies were similar in the two parts.

Of 199 cells tested with NA, 46% responded: 44% decreased and 2% increased. Of 177 cells tested with DA, 23% responded: 16% increased and 7% decreased. Both NA and DA were tested on 163 cells; 11% responded to both and 39% to either (29% to NA, 10% to DA). Figure 1Ab shows no regional differences in CA sensitivity.

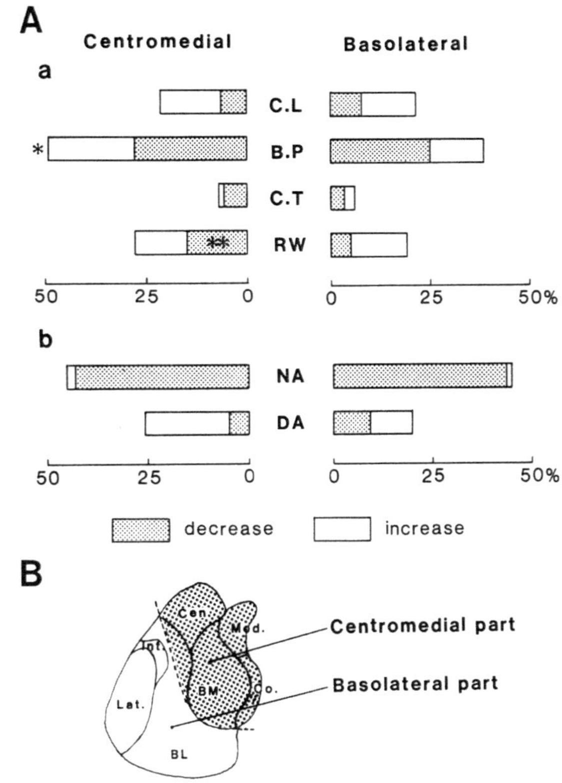

FIGURE 1. Distribution of cell populations (A) and tentative anatomic division of amygdala (B). (A) Differences of the task-responsive cells (a) and catecholamine (CA)-sensitive cells (b) between centromedial (CM) and basolateral (BL) parts of the amygdala. Dotted columns, percentage of cells that decrease firing in each period of feeding task or to CA; open columns, those that increase firing. C.L., cue light; B.P., bar press; C.T., cue tone; RW, reward period; NA, norepinephrine; DA, dopamine. $^*P < 0.05$; $^{**}P < 0.01$ (χ^2 test). Number of cells: (a) 144 cells in CM and 148 in BL; (b) NA, 102 cells in CM and 97 in BL; DA, 85 cells in CM and 92 in BL, coronal section of left amygdala 20 mm anterior from ear bar. Dotted area, centromedial AM tentatively defined in the present experiment to include cortical, medial, central, and basomedial nuclei, anterior amygdaloid area, and dorsal apex of the basolateral nucleus; open area, basolateral AM, remaining area. Med, medial; Co, cortical; Cen, central; BM, basomedial nucleus; int, intercalated cell population; BL, basolateral; Lat, lateral nucleus. AM coordinates: *A*, 18–22; *L*, 9–14; *V*, +2 to −7 (Kusama and Mabuchi, 1970; Nakano *et al.*, 1987).

3.2. Catecholamine Sensitivity and Task-Related Activity

Norepinephrine or DA sensitivity and task-related responses were studied in 205 cells. Of 189 cells tested with NA, 73% NA-sensitive and 68% -insensitive cells responded during the task. Of 171 cells tested with DA, 89% DA-sensitive and 63% DA-insensitive cells responded. Task-related responses occurred more often in DA-sensitive than in DA-insensitive cells ($P < 0.01$, χ^2 test). In the reward period, NA-sensitive cells decreased activity more frequently than NA-insensitive cells ($P < 0.01$, χ^2 test), and DA-sensitive cells increased in firing more frequently than DA-insensitive cells ($P < 0.05$, χ^2 test).

Reward-related responses and CA sensitivity correlated only in the centromedial part of the AM, and DA sensitivity and task-related responses correlated only in the basolateral AM ($P < 0.05$, χ^2 test). To investigate CA effects in the centromedial AM in the feeding task, an α- or β-adrenoceptor antagonist was applied continuously during several trials. In five cells examined with sotalol (SOT), a β-adrenoceptor blocker, three in the central and two in the basomedial nucleus, the reward-related responses of four were attenuated at least 50% without affecting other responses during the task. Figure 2A shows an example of the SOT effect. Phenoxybenzamine (PBZ) did not affect reward-related activity changes in seven cells (Fig. 2B). In four cells that increased activity during the reward period, spiroperidol (SPP), a DA antagonist, was applied continuously during the task to three cells in the central nucleus and one in the medial. As shown in Fig. 3, activity increase during the reward period was attenuated at least 50% in three of the four. The facilitatory effect of glutamate remained intact (Fig. 3Bb,c).

Of 12 cells that did not change activity during the reward period, two that were

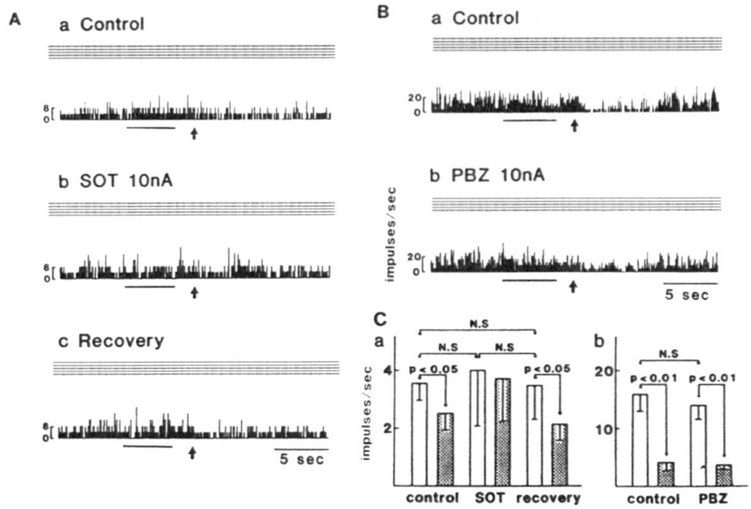

FIGURE 2. Effect of sotalol (SOT) and phenoxybenzamine (PBZ) on reward-related activity decrease. A recorded in basomedial and B in central nucleus (different cell). (A) a, control, decrease in activity in reward period of feeding task. b, 1 min after continuous application of SOT, attenuation of decrease of firing. c, recovery, 20 min after end of SOT application. (B) a, control, decreased activity in reward period. b, 1 min after continuous application of PBZ, no change in activity pattern. Each histogram is the sum of five trials. (C): Statistical analysis of firing rate in each series of trials (Student's *t* test). a corresponds to A, and b to B. Open column, intertrial interval (mean ± S.D.); dotted columns, reward period. Arrows connected to significant level (Student's *t* test). N.D., not significant (Nakano *et al.*, 1987).

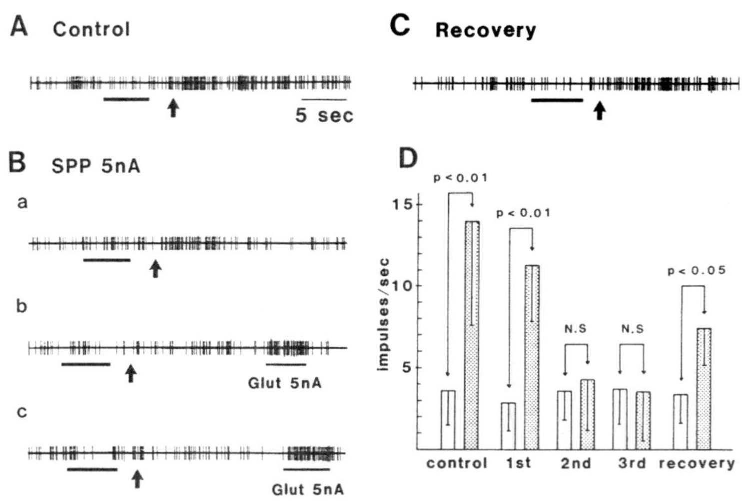

FIGURE 3. Effect of spiperone (SPP) on reward-related response in central nucleus. (A) Control, activity increase during reward period. (B) Continuous application of SPP; a, increase in activity remained in first trial 1 min after beginning of application; b, attenuation of increase in activity in second trial 2.5 min after application; c, disappearance of increase in third trial 4 min after application; c, disappearance of increase in third trial 4 min after application; glutamate effect intact. (C) Recovery 10 min after SPP. (D) Statistical analysis of firing rate in each trial (Student's *t* test). Open column, firing rate calculated for 5 sec in intertrial interval (mean ± S.D.); dotted column, reward period (Nakano *et al.*, 1987).

tested decreased activity significantly in the reward period during SPP application and increased it significantly during SOT application (Fig. 4A). To investigate the effects of NA antagonists, PBZ and SOT were applied continuously to neurons. Phenoxybenzamine did not affect the original firing pattern (Fig. 4D); SOT decreased the firing rate in the intertrial interval but significantly increased activity during the reward period (Fig. 4E,F). Although the firing rate during the intertrial interval still remained low 20 min after termination of SOT application, the increased response during the reward period completely disappeared. Although excitation by glutamate still remained intact (not shown), the decrease in activity in the intertrial interval during and after SOT application may reflect a weak anesthetic effect. During all trials shown in Fig. 4, the availability of food reward did not change. Neither the latency to the first bar press nor the duration of bar pressing changed significantly in any trial. This indicates that modulation of the reward-related neuronal activity was not caused by a change in reward value. The results suggest that neuronal activity during the reward period is simultaneously modulated oppositely by NA and DA.

Of 108 cells tested with glucose, 12% responded by decreasing activity. Of 167 cells tested with morphine, 32% responded (31% decreased, 1% increased). Of ten GS cells also tested with morphine, 70% decreased firing. Of 73 glucose-insensitive (GIS) cells tested with morphine, 32% decreased activity. Thus, more GS than GIS cells were morphine sensitive ($P < 0.05$, χ^2 test).

All GS cells and 68% of the GIS cells changed activity during the task. During the task, 77% of the morphine-sensitive cells and 65% of the insensitive cells responded. There were more task-related responses among GS cells than among GIS cells ($P < 0.05$, χ^2 test). More GS than GIS cells responded in the bar-press period ($P < 0.05$, χ^2 test).

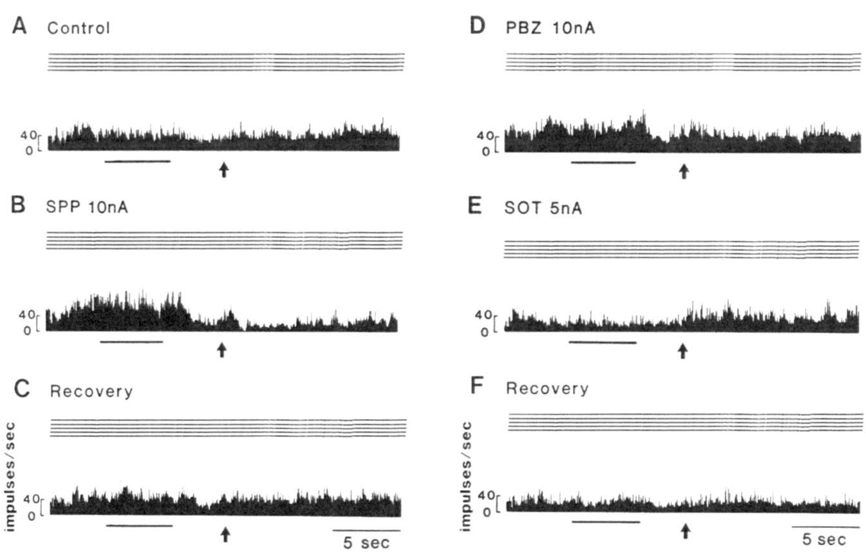

FIGURE 4. Effect of catecholamine antagonists on reward-related neuronal activity. (A–F) Same neuron in central nucleus. (A) Control, no change during reward period. (B) Decreased activity in reward period 1 min after start of application of spiperon (SSP). (C) Recovery of activity in reward period 5 min after stopping SPP. (D) No effect on firing pattern 1 min after starting application of phenoxybenzamine (PBZ). (E) Firing rate in intertrial interval decreased 1 min after starting application of sotalol (SOT); activity increased in reward period. (F) Recovery, 20 min after stopping SOT (Nakano *et al.*, 1987).

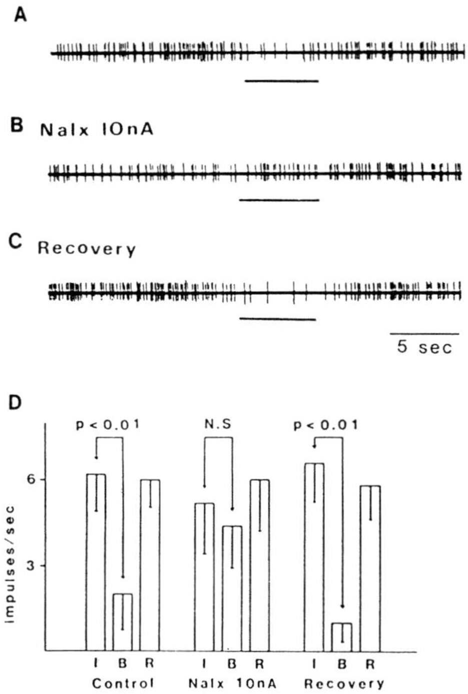

FIGURE 5. Naloxone (Nalx) effect and distribution of glucose- and morphine-sensitive cells. Centromedial AM neuron. Activity decrease in bar-press period blocked by Nalx. Recorded in central nucleus. (A) Control, activity decrease during bar press. (B) 2 min after start of Nalx application, attenuation of decreased response during bar press. (C) Recovery, 5 min after stopping Nalx. (D) Each histogram mean ± S.D. firing rate for 5 sec during intertrial interval (I), bar press (B), and reward period (R). Arrows connect significantly different levels (Student's *t* test). N.S., not significant.

The GS cells decreased activity more (56%) in the bar-press period than GIS cells (22%) ($P < 0.05$, χ^2 test). Since most GS cells were in the centromedial part of the AM, relationships between glucose sensitivity and task related responses were not analyzed separately in the centromedial and basolateral AM.

In the centromedial and basolateral AM, morphine and GS sensitivity were similar. In the centromedial AM, 50% of the morphine-sensitive cells and 24% of the insensitive cells decreased firing during bar press. Thus, more morphine-sensitive than morphine-insensitive cells in the centromedial AM responded in the bar press period ($P < 0.05$, χ^2 test). In two of four trials, naloxone reversibly attenuated decreased responses during the bar-press period (Fig. 5).

4. DISCUSSION

4.1. Task-Related Neuronal Activity

Tonic activity changes were evident in the bar-press (64%) and/or reward periods (35%) during feeding-related behavior (total 68%). Activity changes in both the bar-press and reward periods depended on reward value; the bar-press period was shorter when more palatable food was available. Neuronal responses were attenuated or disappeared, and latency to the first bar press was significantly longer, in extinction trials. The firing rate during the intertrial interval also changed if the reward changed. These results suggest involvement of some amygdalar neurons in evaluation of reward value. In extinction, activity changes in 27% of the cells were not attenuated during bar press until behavior responses to cue light on disappeared, so the AM is involved in the execution of simple motor activity.

Phasic responses to cue light were observed in 31% and to cue tone in 10% of the responsive cells. It has been reported that 10–20% of the amygdalar cells respond to visual stimuli in various task paradigms (Ono *et al.*, 1983; Sanghera *et al.*, 1979). Our results agree with these findings, but the changes we observed were predominantly in the bar-press and reward periods. This suggests that although the AM may be involved in food-related sensations, it is more active during food acquisition (bar press) and consumption (reward period).

4.2. Regional Differences in Task-Related Neuronal Activity and Its Catecholaminergic Involvement

We found some significant differences between the centromedial and basolateral AM during the task (Fig. 1) but no reciprocal activity changes. The AM, especially the basolateral part, receives inputs from the temporal cortex. This pathway has been considered to be involved in visual information processing. The centromedial AM, especially the central nucleus, receives inputs from the hypothalamus and lower brainstem. These are closely involved in regulation of internal conditions. There are also intimate intra-amygdaloid connections between different AM nuclei. It is thus possible that both external sensory and endogenous inputs are integrated in the same AM cell. This might explain why centromedial and basolateral AM cells in the monkey did not respond reciprocally during the task.

Catecholamine involvement in task-related activity changes was significantly different in the centromedial and basolateral AM. Correlation between reward-related activity changes and CA sensitivity appeared only in the centromedial AM, and that between activity changes during the task and DA sensitivity was seen only in the basolateral part.

4.3. Significance of Catecholamines in Task-Related Responses

The most interesting finding in the present experiment was the functional correlation between task-related neuronal activity and CA sensitivity. It seems that NA and DA inputs in the centromedial division innervate different cell groups and modulate neuronal activity during the reward period in different ways. The reward-related activity increase of the DA-sensitive cells was blocked by a DA antagonist, whereas the reward-related decrease of the NA-sensitive cells was blocked by a β-adrenoceptor but not by an α-adrenoceptor antagonist. In all cases in which no activity change was observed during the reward period, when there was no chemical stimulation, the β-adrenoceptor and DA antagonists induced opposite activity changes. These results suggest that NA and DA inputs converging on a single cell could modulate opposite neuronal responses and may explain the absence of reward-related activity changes in more than half of the CA-sensitive cells. In the basolateral AM, activity of DA-sensitive cells changed during the task more often than that of insensitive cells. This suggests that DA inputs in this division generally influence neuronal activity in the entire sequence of the feeding task.

Catecholamine systems have been reported to be involved in reward-related behavior and associative learning (Aou *et al.*, 1983; Heffner *et al.*, 1981; Leibowitz, 1980). Injection of a β-adrenoceptor antagonist into the rat AM dose-dependently decreased retention of passive avoidance conditioning (Gallagher *et al.*, 1981), and neurochemical lesion of the NA system increased resistance to extinction in several task paradigms (Ellis, 1984; Mason and Fibiger, 1978). The DA system may be implicated in the acquisition and maintenance of rat lever pressing for food (Wise and Schwarz, 1981), and dopamine metabolism in the amygdala has been reported to increase during fixed-ratio (FR 5) lever pressing for a reward (Heffner *et al.*, 1981). In addition, the AM, especially the basolateral nucleus, projects to the premotor cortex in the monkey. Although the exact function of this area is not yet clear, it is believed to be involved in sensory-oriented motor function (Halsband and Passingham, 1982). This and the present results suggest some characteristics of the amygdalar CA involvement in reward-related behavior. It may be that NA and DA inputs in the centromedial part of the AM are related to learning or memory that associates reward with reinforcement.

The AM, especially the central nucleus, has intimate interconnections with the LHA and the NTS. In the present experiment, 77% of the GS cells were in the centromedial AM, and 86% were in the central nucleus. This suggests that amygdalar GS cells might be in a chemosensitive neuronal circuit involved in integration of endogenous chemical information. Both GS and morphine-sensitive cells decreased firing during bar press more often than insensitive cells. Significantly more GS than GIS cells were morphine sensitive. Relationships between morphine sensitivity and bar-press-related neuronal responses were predominant in the centromedial AM, and GS cells were also concentrated in the same part. It is possible that GS cells are a subpopulation of the centromedial morphine-sensitive cells and are involved in the control of food acquisition (bar press).

In the LHA, GS cells are usually morphine sensitive and respond more often in the bar-press and reward periods with firing decrease, but only the reward-related activity decrease is attenuated by naloxone (Oomura *et al.*, 1986). We found that centromedial morphine-sensitive cells responded more often in the bar-press period, and this was attenuated by naloxone (Nakano *et al.*, 1986). Although opioid-containing perikarya may project to the LHA from the paraventricular, dorsomedial, ventromedial hypothalamic nuclei, periaqueductal gray, and central nucleus of the AM in the rat, this is not yet clear in the monkey. The AM is one of the richest of the monkey brain areas in opioid-containing cells, and the LHA and AM differ in the localization of opioid receptors (Goodman *et al.*, 1980). In contrast to the LHA, which is rich in μ receptors, the AM has relatively many δ receptors. These anatomic differences may explain functional differences between the LHA and AM in opioid sensitivity. The GS cells respond more during bar press or reward than the cue light or tone, although this difference is not as great as that for GIS cells. This suggests that GS cells are related to integration of endogenous information, whereas GIS cells might process external sensory information. A similar observation was made in the LHA using the same feeding task (Oomura *et al.*, 1986). It is likely that both GS and GIS cells in the AM correlate with those in the LHA and cooperate to execute hunger-motivated behavior.

5. SUMMARY AND CONCLUSIONS

Norepinephrine (NA), dopamine (DA), and morphine and their antagonists were studied in the amygdala of alert monkeys in a feeding task: (1) cue light to signal bar pressing, (2) high-fixed-ratio bar pressing (fixed ratio 20), (3) cue tone at last bar press to signal presentation of food, and (4) ingestion, reward.

Most cells (68%) responded during the task. Changes during bar press (64%) and/or reward (35%) were tonic, and those to the cue light (31%) and cue tone (10%) were phasic.

Norepinephrine elicited firing decreases (97%) and increases (3%). Dopamine elicited increases (70%) and decreases (30%). Inhibition by NA was blocked by α- and/or β-adrenoceptor antagonists.

Norepinephrine-sensitive cells decreased, and DA-sensitive cells increased, activity in the reward period more often than insensitive cells. Antagonists of β-adrenoceptors or of DA attenuated responses in the reward period, but an β-adrenoceptor antagonist did not affect reward-related responses.

Centromedial AM cells responded to tsk events significantly more than basolateral cells. More centromedial than basolateral cells responded (decrease) during bar press and reward, but equal numbers in both parts responded to sensory cues. More DA-sensitive cells responded in the basolateral AM, but reward responses and NA or DA sensitivity were related only in the centromedial AM.

Glucose-sensitive and morphine-sensitive cells, which were mostly in the centromedial AM, decreased firing during bar press more often than insensitive cells. Naloxone attenuated this activity decrease.

Norepinephrine and DA inputs to the centromedial AM may be related to reward-reinforcement association; DA inputs to the basolateral and glucose and morphine inputs to the centromedial AM may be involved in operant behavior.

Since cue light and tone responses were specific to task-related events, and bar press and reward-related responses extinguished along with behavior, and all responses were affected by catecholamines or opioids, AM responses were concluded to be conditioned.

REFERENCES

Aou, S., Oomura, Y., Nishino, H., Inokuchi, A., and Mizuno, Y., 1983, Influence of catecholamines on reward-related neuronal activity in monkey orbitofrontal cortex, *Brain Res.* **267:**165–170.

Aou, S., Oomura, Y., Lénárd, L., Nishino, H., Inokuchi, A., Minami, T. and Misaki, H., 1984, Behavioral significance of monkey hypothalamic glucose-sensitive neurons, *Brain Res.* **302:**69–74.

Beninger, R. J., 1983, The role of dopamine in locomotor activity and learning, *Brain Res. Rev.* **6:**173–196.

Ellis, M. E., 1984, Manipulation of the amygdala noradrenergic system impairs extinction of passive avoidance, *Brain Res.* **324:**129–133.

Gallagher, M., Kapp, B. S., Pascoe, J. P., and Rapp, P. R., 1981, A neuropharmacology of amygdala systems which contribute to learning and memory, in: *The Amygdaloid Complex* (Y. Ben-Ari, ed.), Elsevier, Amsterdam, pp. 343–354.

Goodman, R. R., Snyder, S. H., Kuhar, M. J., and Young, W. S., III, 1980, Differentiation of delta and mu opiate receptor localization by light microscopic autoradiography, *Proc. Natl. Acad. Sci. U.S.A.* **77:**6239–6243.

Halsband, V., and Passingham, R., 1982, The role of premotor and parietal cortex in the direction of action, *Brain Res.* **240:**368–372.

Heffner, T. G., Luttinger, D., Hartman, J. A., and Seiden, L. S., 1981, Regional changes in brain catecholamine turnover in the rat during performance of fixed ratio and variable interval schedules of reinforcement, *Brain Res.* **214:**215–218.

Kusama, T., and Mabuchi, M., 1970, *Stereotoxic Atlas of the Brain of Macaca fuscata*, University of Tokyo Press, Tokyo.

Leibowitz, S. F., 1980, Neurochemical systems of the hypothalamus: Control of feeding and drinking behavior and water–electrolyte excretion, in: *Handbook of the Hypothalamus*, Volume 3, Part A (P. J. Morgane and J. Panksepp, eds.), Marcel Dekker, New York, pp. 299–437.

Lénárd, L., and Hahn, Z., 1982, Amygdalar noradrenergic and dopaminergic mechanisms in the regulation of hunger and thirst-motivated behavior, *Brain Res.* **233:**115–132.

Mason, S. T., and Fibiger, H. C., 1978, 6-OHDA lesion of the dorsal noradrenergic bundle alters extinction of passive avoidance, *Brain Res.* **152:**209–214.

Nakano, Y., Lénárd, L., Oomura, Y., Nishino, H., Aou, S., and Yamamoto, T., 1987, Functional involvement of catecholamines in reward related neuronal activity of the monkey amygdala, *J. Neurophysiol.* **57:**72–91.

Nakano, Y., Oomura, Y., Lénárd, L., Nishino, H., Aou, S., Yamamoto, T., and Aoyagi, K., 1986, Feeding related activity of glucose and morphine sensitive neurons in the monkey amygdala, *Brain Res.* **399:**167–172.

Ono, T., Fukuda, M., Nishino, H., Sasaki, K., and Muramoto, K., 1983, Amygdaloid neuronal responses to complex visual stimuli in an operant feeding situation in the monkey, *Brain Res. Bull.* **11**:515–518.
Oomura, Y., Nishino, H., Aou, S., and Lénárd, L., 1986, Opiate mechanism in reward-related neuronal responses during operant feeding behavior of the monkey, *Brain Res.* **365**:335–339.
Sanghera, M. K., Rolls, E. T., and Roper-Hall, A., 1979, Visual responses of neurons in the dorsolateral amygdala of the alert monkey, *Exp. Neurol.* **63**:610–626.
Wise, R. A., and Schwartz, H. V., 1981, Pimozide attenuates acquisition of lever-pressing for food in rats, *Pharmacol. Biochem. Behav.* **15**:655–656.

The Neural Circuitry Subserving Aversive Conditioning of Contact Placing in Cats

The Necessity of the Internal Pallidum and Dispensability of the Cerebellum

VAHE E. AMASSIAN and CHRISTIAN WERTENBAKER

1. INTRODUCTION

1.1. Neural Circuitry Subserving Contact Placing: Limitations of the Sensorimotor Control Model

Before considering aversive conditioning of contact placing (CP), it is useful to review the circuitry subserving this behavior. Previous studies from our laboratory utilizing lesions, cold blocks, and recordings from individual somatosensory and motor control neurons led to the proposal that CP in the adult cat is dynamically subserved by at least two major thalamocortical projection systems, the ventralis posterior to sensorimotor cortex and the cerebellum to ventralis lateralis and anterior to motor cortex (Amassian *et al.*, 1972a,b). More recently, the rubral projection system was shown also to contribute dynamically to CP (Batson and Amassian, 1986), the major contributing input deriving from the deep cerebellar nuclei (Amassian and Batson, 1983). In the neonatal kitten, the sensorimotor cortex initially does not contribute to CP, but during the second week, the pyramidal tract (PT) begins contributing tonically to CP, the adult pattern of control of CP subsequently emerging after 6–7 weeks (Amassian and Ross, 1978). The claim that full hindlimb placing can be elicited in the spinal kitten (Forssberg *et al.*, 1974) has been refuted in several laboratories (Amassian *et al.*, 1977; Bradley *et al.*, 1983; Bregman and Goldberger, 1982), the lower brainstem being essential, especially the vestibular complex (Amassian *et al.*, 1977).

A major difficulty in explaining CP solely in terms of the specific thalamocortical and rubral systems so far studied (Fig. 1) in relation to CP is the lack of any obvious

VAHE E. AMASSIAN • Department of Physiology, State University of New York Health Science Center at Brooklyn, Brooklyn, New York 11203. CHRISTIAN WERTENBAKER • Departments of Ophthalmology and Neurology, Albert Einstein College of Medicine of Yeshiva University, Bronx, New York 10461.

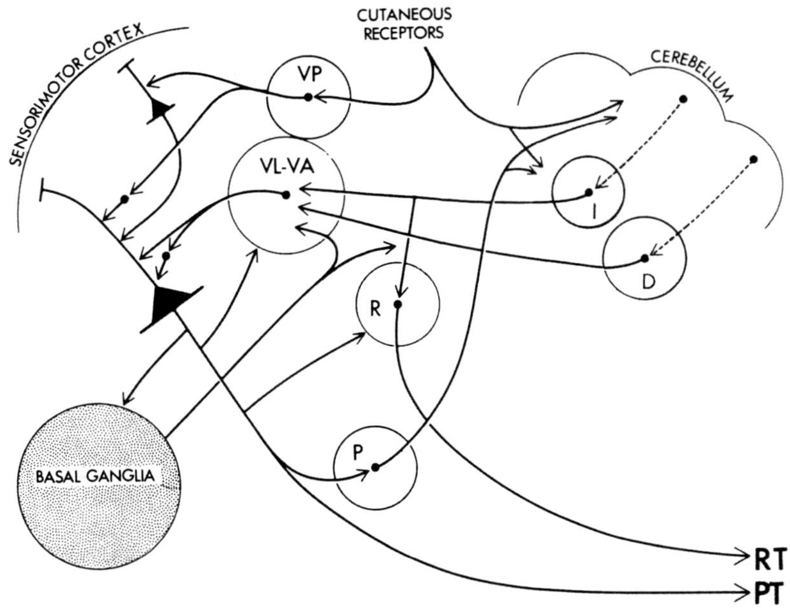

FIGURE 1. Diagram of some of the neural circuitry related to CP in the cat. Abbreviations: D, nucleus dentatus; I, n interpositus; P, pontine n; PT, pyramidal tract; R, red n; RT, rubrospinal tract; VL–VA, n ventralis lateralis and anterior; VP, n ventralis posterior. Although a possible dynamic role of basal ganglia (stippled) in normal CP has not yet been investigated, they are now known (Section 3.5) to participate in water conditioning of CP. Diagram is modified from Fig. 9 in Amassian *et al.* (1972b), which gives references for the synaptology.

explanation for the marked effects of such behavioral variables as degree of arousal or emotionality of the cat. Thus, a drowsy cat usually fails to place (Amassian *et al.*, 1972b); a frightened cat may merely ventroflex the paw at the wrist without appreciably lifting it or may display a marked hypermetria that is reduced during successive trials. Other cats struggle on repeated testing, requiring their abandonment as subjects (Wertenbaker *et al.*, 1973). Such observations suggest that at least limbic and reticular activities modulate the sensorimotor control circuitry so far shown to subserve.

1.2. Properties of Aversive Conditioning of Contact Placing

The account that follows summarizes our earlier findings (Wertenbaker *et al.*, 1973). Prompted by the commonplace observation that cats are reluctant to walk in puddles, we found that CP could readily be inhibited by substituting a container of water for the solid surface on which the forepaw normally lands. Passively resting the paw on a solid surface for a long period, e.g., 20 sec, "disinhibited" the paw, and subsequent testing resulted in placing on the solid surface. The inhibition is confined to the conditioned paw, even to the conditioned aspect of the paw; e.g., inhibition produced by conditioning of the ulnar aspect of the forepaw was absent when the radial aspect was tested. Water inhibition was often accompanied by a flexed posture of the forelimb between trials. Several patterns of inhibited CP were observed accompanied by changes in the biceps electromyogram. The patterns of inhibition are described in Section 3.1.

Water inhibition of CP has memory functions with at least two, quite different, time relations. Thus, if within a water-conditioning session a pause of either 10 sec or 60 sec was introduced after four consecutive failures had occurred, on resuming the wet-side testing the inhibition was usually found to survive a 10-sec but not a 60-sec pause. A second memory function with much longer time relations was identified by finding that the number of trials required for inhibition fell when water conditioning sessions were repeated at intervals from 30 min to a day.

Given a complex behavioral function such as water inhibition of CP, the possible effect of a lesion is more easily identified if the functional processes underlying the inhibition are first differentiated. These include (1) transmission of the sensory input representing the water stimulus, (2) interpretation of this input as aversive, (3) a very-short-term memory function operating within a training session, which links contact and/or placing with subsequent immersion of the paw, (4) either inhibition at some point in the known dynamic circuitry of CP (cf. Fig. 1) or substitution of an alternative type of motor response, and (5) a longer-term memory function manifested by savings between training sessions, e.g., those given on different days.

1.3. Effects of Lesions on Associative Conditioning

In our paradigm, the cat must make placing movements before being aversively stimulated by water; thus, water inhibition should be classified as a passive avoidance type of associative learning. The literature on the effects of lesions on passive avoidance is too extensive to be reviewed in detail here (see Gray, 1982). In many experimental conditions, bilateral lesions of such portions of the limbic system as septum, hippocampus, and amygdala impair passive avoidance. After septal lesions, important influences on the degree of impairment include increased age of the rat (Bengelloun *et al.*, 1977), increased punishment (Bengelloun *et al.*, 1976), and early postoperative testing (Miczek *et al.*, 1972). After hippocampal lesions, the passive avoidance deficit was only found when the animal had to travel to reach the site of punishment. The deficit could be overcome by increasing the punishment or by increasing the intertrial interval from 60 sec to 60 min (Cogan and Reeves, 1979). Black *et al.* (1977) relate the deficit seen after hippocampal lesions to a deficit in locating the site of punishment.

Bilateral amygdala lesions have yielded results that are conflicting. Ursin (1965) found both impaired passive and active avoidance in cats. Pellegrino (1968) demonstrated impaired passive avoidance in rats; Werka *et al.* (1978) differentiated central nuclear lesions, which impaired active avoidance (possibly by reducing fear), from basolateral lesions, which impaired passive avoidance. By contrast, Grossman *et al.* (1975) found facilitation of active and impairment of passive avoidance.

Since our effective lesions were all unilateral, reports of deficits after such lesions are of especial interest. Green and Schwartzbaum (1968) found that unilateral septal lesions facilitated active avoidance behavior. A spatial alternation task in the cat was greatly impaired by a unilateral entorhinal cortex lesion, but only if the animals were tested too soon (3 days) after the lesion to permit reinnervation (Steward, 1982). An effect with unilateral lesions might have been obtained in other studies if no time had been allowed for possible reinnervation after a lesion.

The effects of unilateral lesions on certain classical conditioning paradigms are well substantiated. Lincoln *et al.* (1982) demonstrated that ipsilateral cerebellar lesions prevented acquisition of classical conditioning of the nictitating membrane and eyelid closure response in the rabbit. McCormick *et al.* (1982) found that sectioning the brachium conjunctivum abolished the learned response ipsilaterally without altering the unconditioned response to an air puff. Later analysis (Thompson *et al.*, 1984) included unit recording and disclosed that nuclei (n) interpositus (I) and dentatus (D) were the cerebellar regions important in acquisition and retention of the learned response. Curiously, cerebellar cortical lesions did not alter the conditioning. Subsequently, Yeo *et al.* (1985) found that localized cerebellar cortical lesions were effective. However, the marked reduction in rubral activity following cooling as lesions of n, I, and D (Amassian and Batson, 1983) suggests the possibility that a site of conditioning is in the red n. The deep cerebellar nuclei were also required for conditioning leg flexion. Woody *et al.* (1974) found that bilateral motor cortical lesions abolished the conditioned eye blink response; however, given the bilateral input resulting from the unconditioned stimulus, a tap to the glabella, unilateral lesions would not be expected to be effective.

Our study started with the incidental observation that unilateral lesions of the zona incerta and Forel's fields H_1 and H_2 (made to "deafferent" nVL–VA) impaired water inhibition of CP, whereas lesions in the cerebellar deep nuclei failed to do so (Wertenbaker *et al.*, 1972). In a preliminary study, unilateral lesions of nVL–VA were also found to impair water inhibition, implying that these nuclei subserved water inhibition by processing input from a structure other than the cerebellum (Amassian *et al.*, 1974). However, when CP recovers following a lesion of nI and nD, the circuitry for aversive conditioning might also be changed; i.e., a possible role of Purkinje-cell-mediated inhibition in water conditioning in the intact cat was not excluded. Therefore, this study includes the effects of lesions of cerebellar cortex, which would enhance rather than remove deep nuclear contribution to CP and could possibly abolish water conditioning. More significantly, the major input to nVL–VA subserving water inhibition is now identified as n entopeduncularis.

2. MATERIALS AND METHODS

2.1. Lesion Making

The data were derived from a total of 32 cats, the numbers with a given lesion being identified in the appropriate section. The cats were anesthetized with Na pentobarbital, usually 40 mg/kg i.p. Except for n entopeduncularis (nE), deeply placed nuclei were destroyed by passing radio-frequency current (Grass, Model LM3) at a number of positions through a stereotactically oriented glass-insulated silver electrode. The metal protruded from 1 to 3 mm beyond the glass depending on whether a small or a large lesion was desired.

Intermediate and lateral zone cerebellar cortical lesions were made by aspirating cerebellum dorsal and lateral to a stereotactically oriented stainless steel strip whose long axis lay at 16–17° to the horizontal. In cross section, the strip had a vertical limb that was placed just lateral to the vermis and a horizontal limb just dorsal to nI and D.

The proximity of nE to the internal capsule precluded making a lesion of this nucleus with RF current. However, kainic acid selectively destroys neuronal somata with minimal damage to axons of passage (McGeer and McGeer, 1981). In the first cat, nE was approached vertically, but in the remaining cats, it was approached at a 45° angle to avoid even minor damage to the internal capsule. The kainic acid (Sigma, 2.5 mg/ml) was injected at the rate of 0.1 μl/min from 30-gauge hypodermic tubing which protruded 0.2 mm beyond a stereotactically oriented 24-gauge guide tube. The accuracy of placement of the guide tube was first established by locating electrophysiologically the optic tract, which lies just ventral to nE. This was done by inserting a 125-μm tungsten wire insulated with Teflon® down to the tip, which just protruded beyond the guide tube. After recording a response to visual stimulation, the guide tube could be accurately placed more dorsally in nE, and 0.2 to 0.4 μl kainic acid was injected at each of two to four rostrocaudal locations. To avoid leaving a trail of kainic acid in the brain, the inner (30-gauge) tube was retracted prior to withdrawal of the guide tube and reprotruded after the guide tube had been positioned in a different portion of nE.

2.2. Testing for CP and Aversive Conditioning

Our procedures were similar to those previously described (Amassian *et al.*, 1972b). Briefly, the cat is held with head tilted backwards so it can see neither the apparatus nor its paws. The forelimb to be tested is allowed to hang free, the other limbs and the body of the cat being held by the experimenter. The cat is then advanced until the free forepaw gently contacts the front of the apparatus. Subsequently, a lifting withdrawal of the forelimb occurs, which is converted into a landing extension phase when the forepaw clears the top corner of the apparatus. Usually, the dorsal surface of the forepaw is contacted, leading to a forward movement of the paw during the landing phase.

A two-compartment placing apparatus was used in which one compartment had a solid top and the other was filled with water, usually at 19–22°C (Fig. 2, top). In the majority of experiments, the pattern of response to contact was noted verbally on a tape recorder (see Section 3.1). When permanent records were required, a placing apparatus was used in which the initial contact of the paw with the front of the apparatus was signaled either by the paw interrupting a light beam falling on photocells (Fig. 2) or by contact with piezoelectric disks. Landing on the solid or wet side was signaled by amplifying the small potentials generated by dissimilar electrode potentials. The transducer data were stored on multichannel analogue tape at 15 ips.

In a previous account of aversive conditioning of CP, the criterion for inhibition was four successive failures in the first 20 wet trials (Wertenbaker *et al.*, 1973). In this study, the proportion of successes in a minimum of 30 trials on both the solid and the wet sides was considered more useful. In general, the solid-side trials were completed for each forepaw in turn. Subsequently, hours or, more recently, minutes later, the forepaws were tested in the same order on the wet side. Following the wet testing, the paw was passively rested on a solid surface for 20 sec and then given at least 15 trials on the solid side to see whether disinhibition of aversive conditioning had occurred. The order in which the paws were tested on subsequent days was alternated except that in the nE lesion series, the intact side was usually tested first on each occasion. (This sequence was used to maximize the opportunity for aversive conditioning.)

In most of our earlier experiments, prior to making any lesion, both CP and water conditioning were tested. Typically, preoperative testing for water inhibition was carried out in a single session with either five to 23 wet trials earlier in the day or 30 wet trials 15–16 days preoperatively. Because savings have previously been demonstrated in CP inhibition Wertenbaker *et al.*, 1973), cats preoperatively tested are distinguished from naive cats in Section 3.

2.3. Terminal Electrophysiology

To measure directly the functional effects of a lesion, cats were terminally anesthetized with Na pentobarbital, and transmission through the lesioned nucleus was studied electrophysiologically. For example, the effect of a nVL–VA lesion or of a lesion in the zona incerta and Forel's fields H_1 and H_2, which "deafferents" nVL–VA, was measured

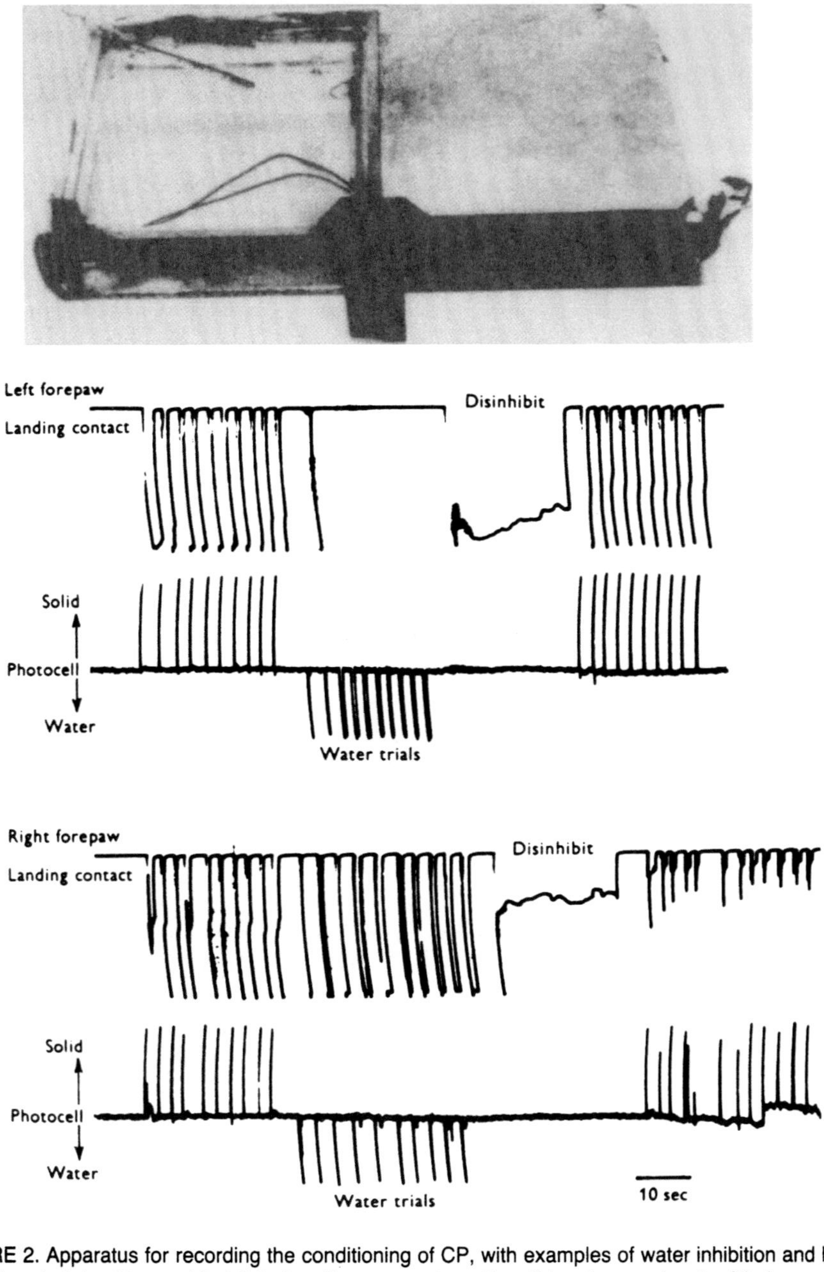

FIGURE 2. Apparatus for recording the conditioning of CP, with examples of water inhibition and lack thereof contralateral to an nVL–VA lesion. Top: apparatus with solid surface and water-filled compartments at right and left, respectively. Black box at front center houses light source. Contact of the paw with the front of the apparatus interrupts the light beam falling on photocells at each end, resulting at the solid and water sides in upward and downward deflections, respectively, in the lower of each pair of traces. Lifting withdrawal of the paw restores the photocell outputs. Landing on the solid or wet side is signaled on the upper of each pair of traces by amplifying the small potentials generated by dissimilar metals. Additional explanation in Section 3.1. (Figure from Fig. 1 of Amassian *et al.*, 1974, with permission).

by comparing the summed pericruciate responses on the intact and lesioned sides to stimulation of the contralateral brachium conjunctivum or nI. Adjoining bipolar 125-μm nichrome wires insulated down to the tip were used for such stimulation. Rectangular, 50–100 μsec duration pulses were used, the stimulus intensity being monitored with a current probe (Tektronix, Model P 6016).

2.4. Histology

The cats were perfused with 10% buffered formalin. Subsequently, the brain was sectioned at 40 to 75 μm with a freezing microtome and stained with cresyl violet to establish the site and size of the lesion.

3. RESULTS

3.1. Appearance of Aversive Conditioning

Examples of aversive conditioning (left forepaw) and absence of aversive conditioning (right forepaw) because of a prior VL–VA lesion are shown in Fig. 2, bottom. Contact of the left forepaw with the solid side is signaled by an upward deflection on the photocell trace. Lifting withdrawal of the paw restores the photocell output; subsequent landing on the solid is signaled by the deflection on the upper trace. Ten trials are shown on the solid side. The first contact on the wet side is followed by placing into water; placing failed to occur in the subsequent nine trials. The left forepaw was then rested on the solid side for 20 sec. Testing on the solid side then showed that CP had been restored; i.e., "disinhibition" had occurred. By contrast, during the same session, the right forepaw showed no inhibition, the paw placing on each occasion into water. In this example, a left-sided lesion of VL–VA had been made 57 days previously (Fig. 3). The sequence of testing shown in Fig. 1 was for the purposes of illustration; the sequence generally used is described in Section 2.2.

The transducers so far illustrated do not identify important differences observed in the pattern of water inhibition of CP and recorded verbally on tape (Wertenbaker *et al.*, 1973). Thus, after contact at least five patterns of inhibition may be distinguished: the forelimb may be immobile, show a vigorous ventroflexion at the wrist, flex at the elbow and shoulder joints with the forepaw hovering over the water, or sidestep as though searching for an alternative site to land, or the paw may move towards the experimenter's hand under the cat's chin. The most commonly encountered patterns are ventroflexion at the wrist, limb flexion with hovering over the water, and immobility. In any given session, the same forelimb may exhibit several patterns of inhibition.

After several wet trials, the forelimb often adopts a flexed posture between trials, with marked flexion at the elbow (angle <90°), and posterior flexion at the shoulder. However, we emphasize that water inhibition can occur regardless of whether the forelimb is flexed or extended prior to contact.

3.2. Lesions in nVL–VA

The data are derived from eight cats with nVL–VA lesions of differing sizes and sites; all cats showed at least a transient loss of water inhibition. Typically, the most effective lesions included anteroventral parts of nVL–VA. Figure 3 shows both water

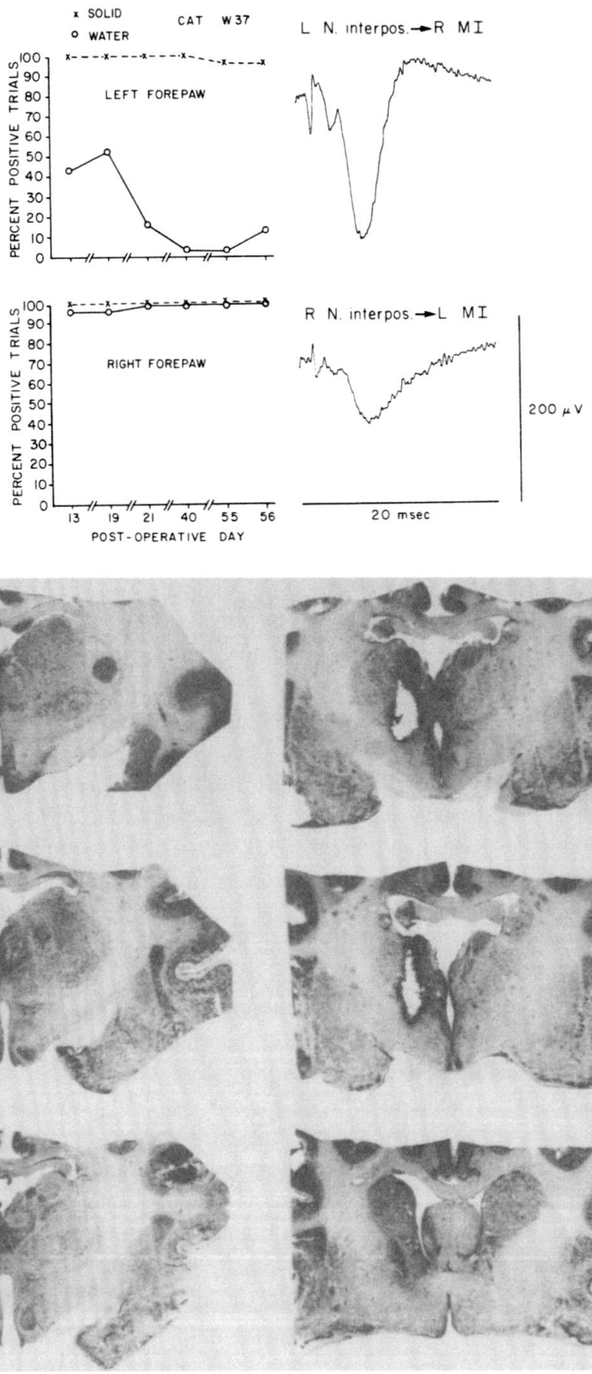

FIGURE 3. Percentage of positive trials during water conditioning of the control (left) forepaw and the forepaw contralateral to the nVL–VA lesion, with terminal electrophysiology and representative photomicrographs of lesion. Top right: 20 summed motor cortical responses to stimulation of contralateral nI at 4 mA.

inhibition and savings with repeated trials in the intact side (left forelimb). By contrast, after dorsal CP had recovered* contralateral to the left nVL–VA lesion (see Amassian *et al.*, 1972b), water inhibition could not be demonstrated. The lesion spared much of nVA dorsolaterally and VL posteriorly, probably accounting for the finding that the summed primary cortical response in the motor cortex to contralateral nI stimulation was as large as 43% of control value. This cat had not been preoperatively tested. Despite the lack of water inhibition, water was still aversive to the right forepaw as judged by its withdrawal after immersion. Remarkably, when dorsal CP has not completely recovered after the nVL–VA lesion, the percentage of positive trials on the wet side can exceed that on the solid side, possibly related to an increased arousal by the aversive water stimulation (Figs. 4 and 5).

Although Fig. 4 similarly shows the absence of water inhibition for most of the testing period following a large nVL–VA lesion, the site of the lesion is significantly more posterior than that shown in Fig. 3, perhaps accounting for the slight water inhibition seen 5 weeks postoperatively. This cat had been preoperatively tested. Transmission from brachium conjunctivum to pericruciate cortex was markedly reduced but not abolished by the lesion.

Smaller lesions of nVL–VA were made in two cats. In one of these, the deficit in water inhibition was transient and minor. In the other cat (Fig. 5), water inhibition of CP was initially absent. During the fourth postoperative week, the incidence of placing into water was diminished, implying that a degree of water inhibition was present. However, disinhibition was poor; i.e., a criterion of true water inhibition had not been met. Although the effect of the lesion on water inhibition was not the most prominent in the series, it abolished transmission from the brachium conjunctivum to motor cortex (see Section 3.4).

In this study, after repeated sessions of wet trials, a minority of the cats showed a remarkable diminution in the proportion of successful placings by the intact (left) paw on the solid side (Figs. 4 and especially 5). This effect evidences a long-term memory function and is not seen when relatively long periods intervene between wet trials (Fig. 3). The phenomenon is clearly unrelated to a lack of time allowed for disinhibition because the cat walks on the solid surface of its cage for one or more days between conditioning sessions. The small lesion in Fig. 5 did not prevent the reduction in successful placings on the solid side associated with repeated water conditionings. In another cat of this series, the incidence of placing on the solid side eventually fell with both the control and affected forelimb. In another cat not in this series, a low incidence of placing on the solid side acquired as a result of water conditioning did not increase at the expected time of recovery following a large nVL–VA lesion. Some cats display an antipathy to continued testing during a training session in which they are unable to prevent placing into the water, requiring "calming" to complete the testing. In addition, after repeated training sessions, cats may display a reluctance to be tested at all. Such findings suggest that the effect of nVL–VA lesions on water inhibition may be separable from that on a long-term memory function.

3.3. Lesions of Zona Incerta and Forel's Fields H_1 and H_2

Lesions caudal to nVL–VA but rostral to the red nucleus, if placed in the zona incerta and Forel's fields H_1 and H_2 cause a major interruption of cerebellothalamic input

* As we previously described (Amassian *et al.*, 1974), CP recovering from an nVL–VA lesion has a remarkably "jerky" quality (see Section 4.2).

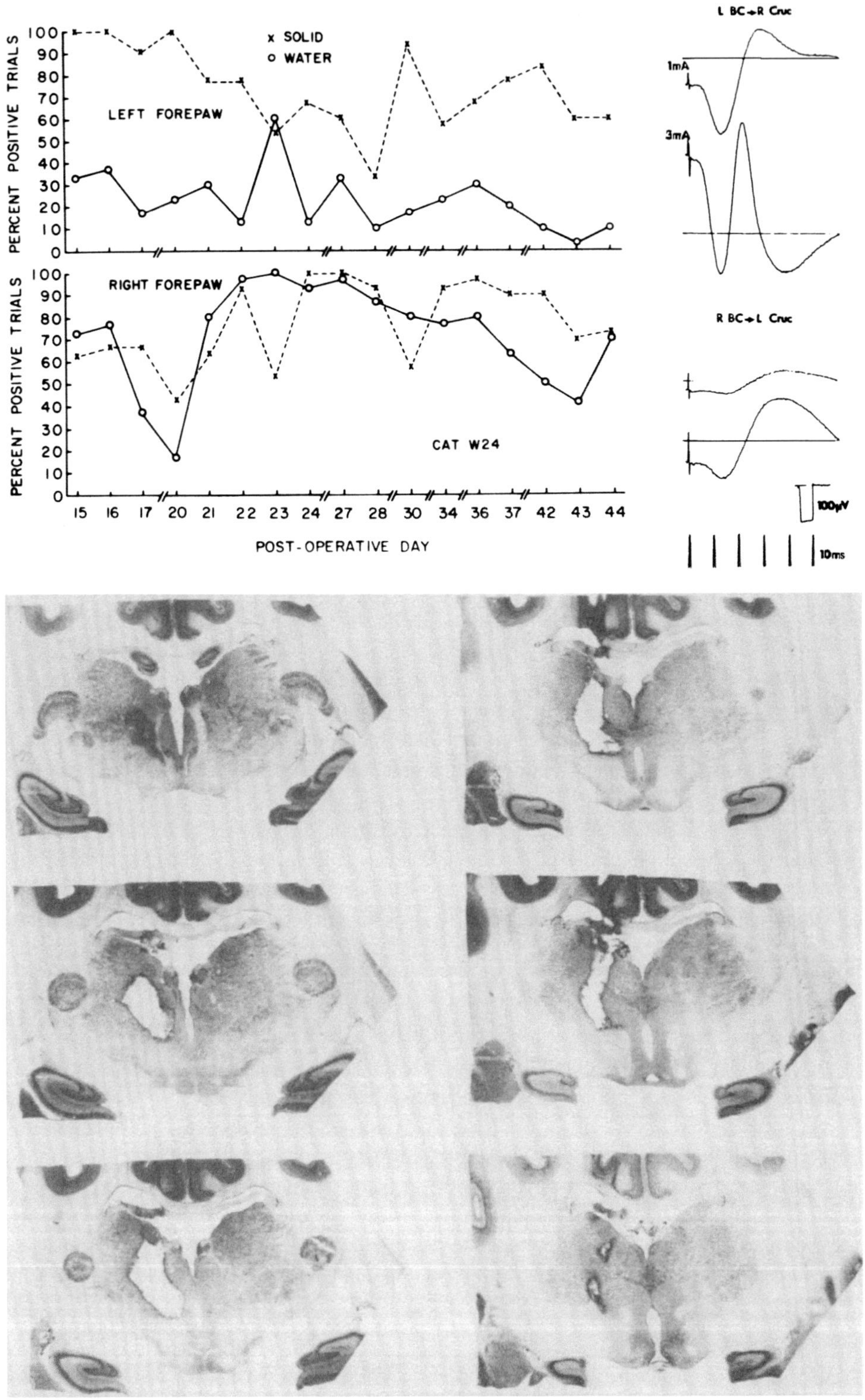

FIGURE 4. Percentage of positive CP trials during water conditioning of control (left) forepaw and the forepaw contralateral to nVL–VA lesion, with terminal electrophysiology and representative photomicrographs of the lesion. Top right: 20 summed motor cortical responses to stimulation of contralateral brachium conjunctivum at 1 mA and 3 mA on each side.

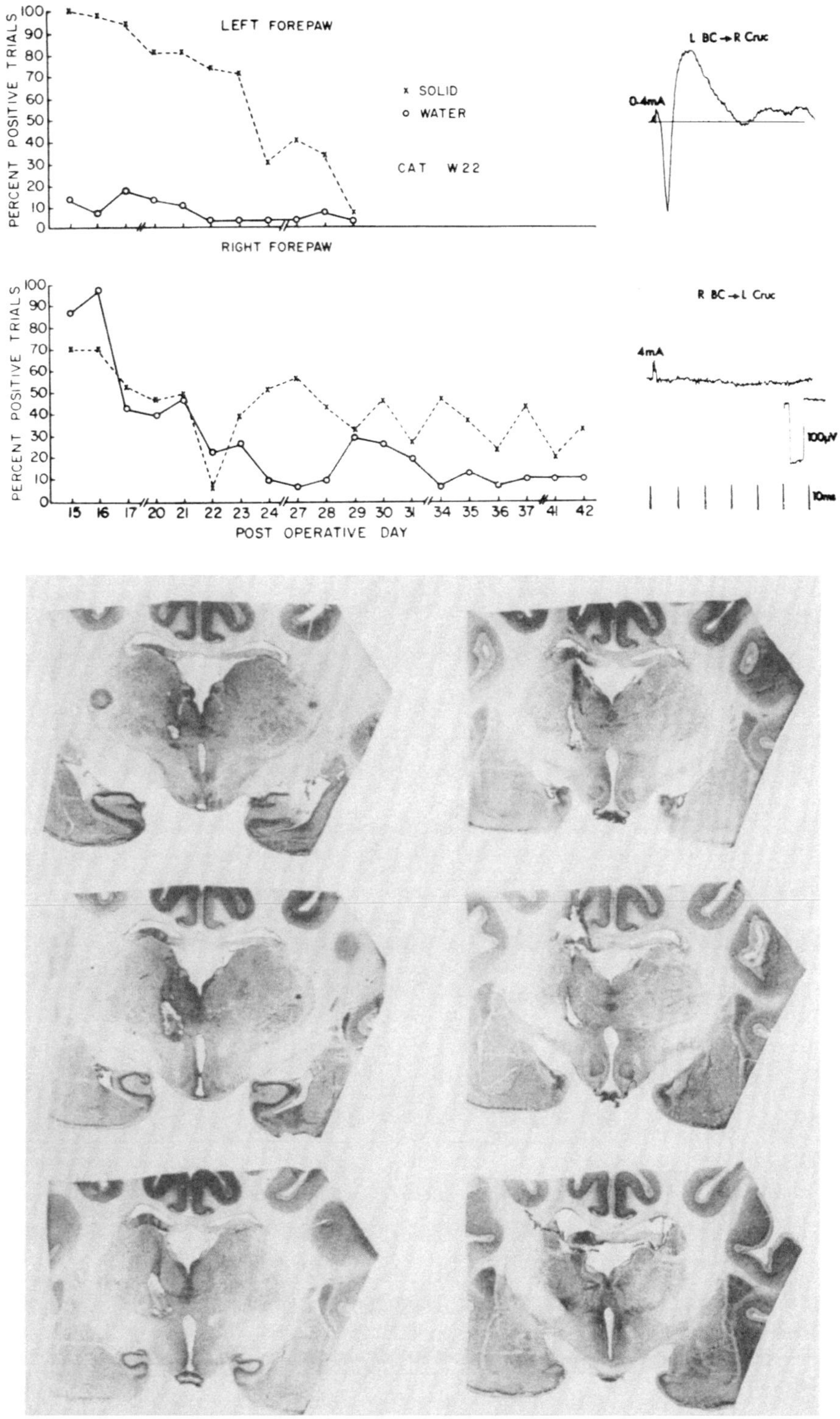

FIGURE 5. Percentage of positive CP trials during water conditioning of the control (left) forepaw and the forepaw contralateral to the nVL–VA lesion, with terminal electrophysiology and representative photomicrographs of the lesion. Top right: three summed motor cortical responses to stimulation of brachium conjunctivum contralateral to intact nVL–VA at 0.4 mA and contralateral to lesioned nVL–VA at 4.0 mA.

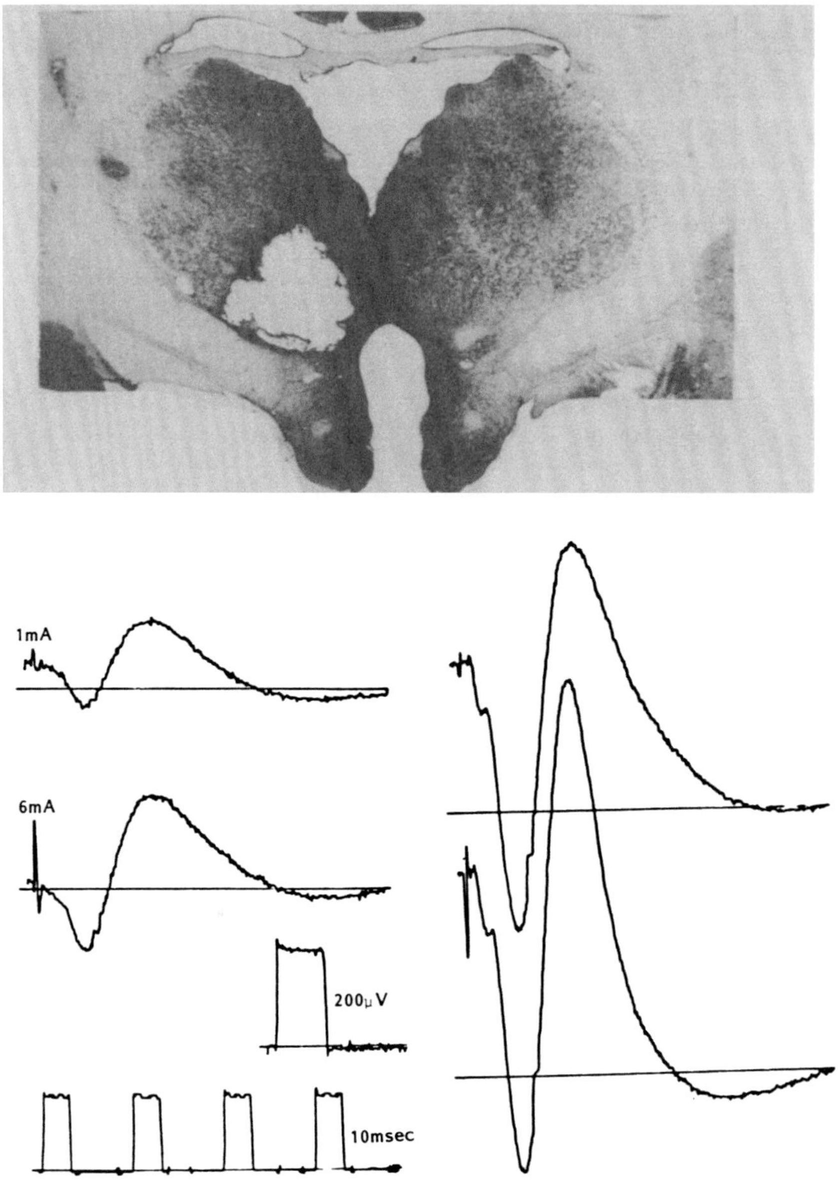

FIGURE 6. Photomicrograph of nVL–VA "deafferenting" lesion with terminal electrophysiology. Below: 20 summed motor cortical responses recorded on lesioned (left) and intact sides to stimulation of region of contralateral brachium conjunctivum and nl at 1 mA and 6 mA on each side. Lesion encroached on posterior nVL and VM.

(Amassian *et al.*, 1972). Such lesions could also interrupt pallidal input to nVL–VA, but this was not proved electrophysiologically either in the earlier or the present studies. The data are derived from five cats. Figure 6 shows a typical site of the nVL–VA "deafferenting" lesion and the resulting reduction in the summed pericruciate responses to cerebellar stimulation at two intensities. Water conditioning was initially tested 6 weeks postoperatively. On the control side, the left forepaw placed in 15/20 (75%) trials on the solid side but was inhibited on the wet side, placing in 9/24 (38%) trials. By contrast, the right forepaw placed in 16/27 (59%) trials on the dry and in 12/16 (75%) trials on the wet side; i.e., the incidence of placing was slightly increased on the wet side.

The effect of such lesions on water inhibition was the first observed, and our observations were generally fragmentary rather than systematic. Furthermore, the region of zona incerta and Forel's fields H_1 and H_2 is a "funnel" for many pathways, therefore implying that lesions had to be made separately in each of the major inputs to nVL–VA.

3.4. Lesions in the Cerebellum

The important differences in the effects of cerebellar cortical versus deep nuclear lesions demonstrated by Chambers and Sprague (1955a,b) required that the effects of such lesions on water conditioning be studied separately. The data are derived from eight cats. In four cats, nI was virtually destroyed combined with lesions of varying sizes in nD; in one other cat, nD alone was partially destroyed. The fastigial nucleus was destroyed in one cat contralateral to a previous lesion, mainly of lateral zone cerebellar cortex. Cerebellar cortical lesions were made in two other cats; the major lesion was in intermediate zone cortex with severe damage to lateral zone cortex. In none of the eight cats was water inhibition depressed.

Figure 7 illustrates the effects of the largest nI and D lesion in the series. Ipsilateral CP took unusually long to return, most likely because the lesion encroached on Dieter's nucleus (Amassian *et al.*, 1972b). To obtain CP, the dorsum of the paw had to be rubbed against the front of the apparatus in some of the conditioning sessions (indicated by arrows). Despite the need for such facilitation, water inhibition was as readily demonstrated as on the control side.

The lesion of nI and D in Fig. 8 is less extensive, sparing a small ventrolateral portion of nD but not encroaching on Deiter's nucleus. Postoperative recovery was swifter than in Fig. 7, but again water inhibition was as readily demonstrated as on the control side. Similarly, the partial lesion of nD did not affect water conditioning. Typically, lesions of nI and D result in the unsupported forelimb adopting a more extended posture than on the control side. Placing is hypermetric in both vertical and horizontal axes (Amassian *et al.*, 1972b).

The left-sided lesion of the fastigial nucleus was virtually complete (Fig. 9), but water conditioning of CP was retained.

A large lateral zone cerebellar cortical lesion was made initially in the cat of Fig. 9 and did not affect water conditioning of CP. Figure 10 illustrates the lack of effect of large lesions in both intermediate and lateral zone cortex on acquisition of water inhibition in two naive cats. Given the smaller volume of intermediate zone cortex, a greater proportion of intermediate than lateral zone was destroyed. Typically, lesions of intermediate and lateral zone cerebellar cortex are followed by a characteristic posture of the unsupported forelimb in which the elbow is flexed at 90° with the paw actively supported against gravity. The cat may spontaneously adopt this posture while standing on the other three limbs. Encroachment on the vermis (Fig. 10, top right) prevented such resting flexor hypertonus, but the difference in precontact posture did not affect water conditioning.

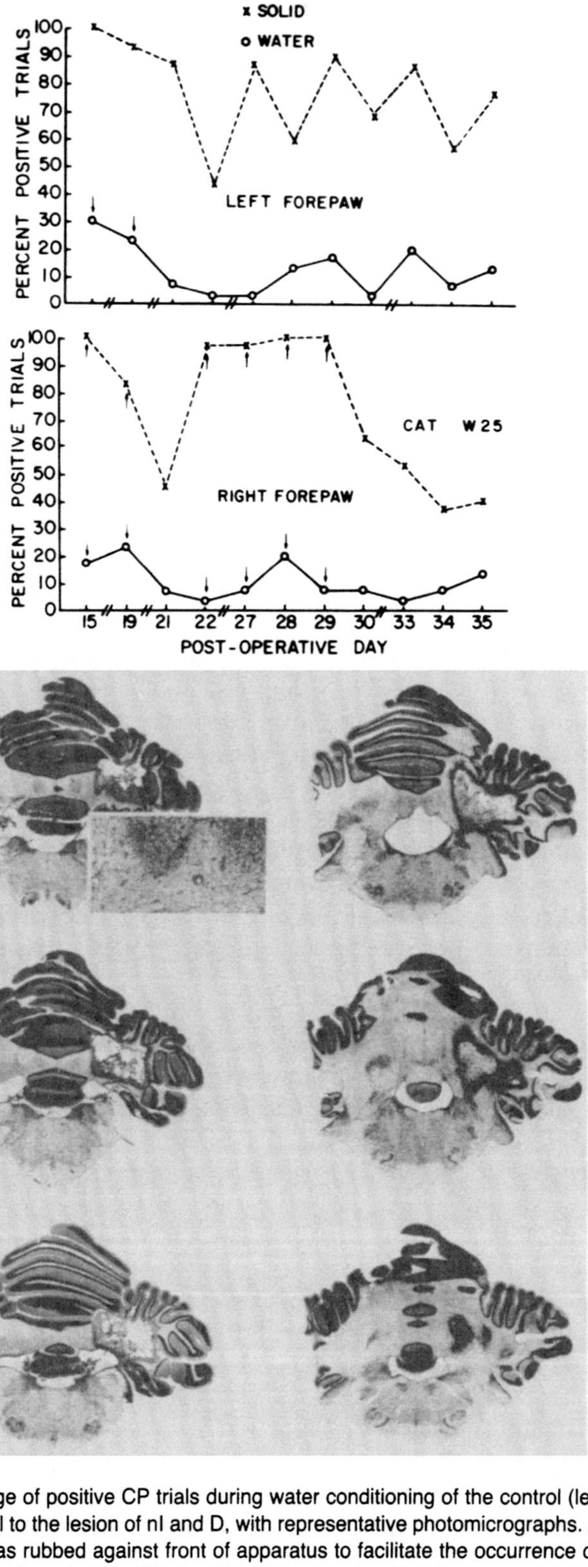

FIGURE 7. Percentage of positive CP trials during water conditioning of the control (left) forepaw and the forepaw ipsilateral to the lesion of nI and D, with representative photomicrographs. Arrows indicate trials in which paw was rubbed against front of apparatus to facilitate the occurrence of placing. Inset photomicrograph: ventral edge of lesion has scarring with loss of neurons.

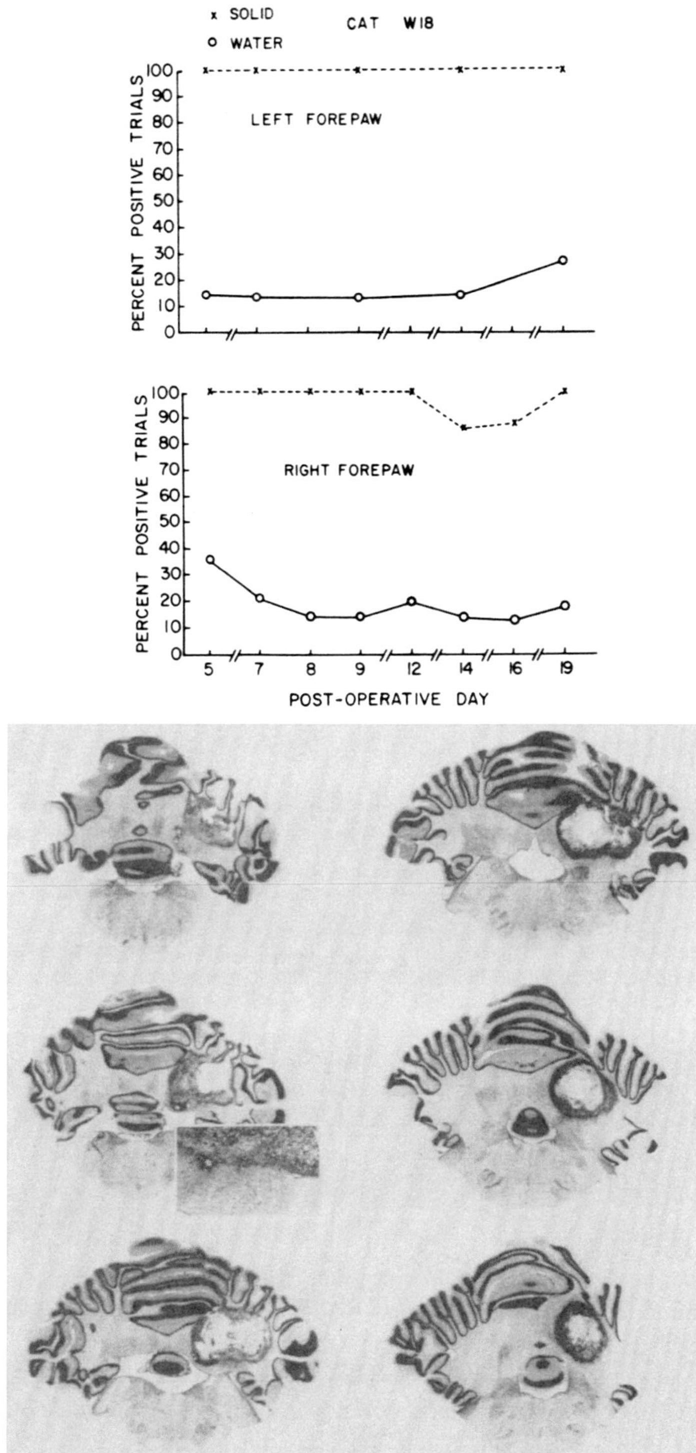

FIGURE 8. Percentage of positive CP trials during water conditioning of the control (left) forepaw and the forepaw ipsilateral to the lesion of nI and D, with representative photomicrographs. Inset photomicrograph: ventral edge of lesion has scarring with loss of neurons.

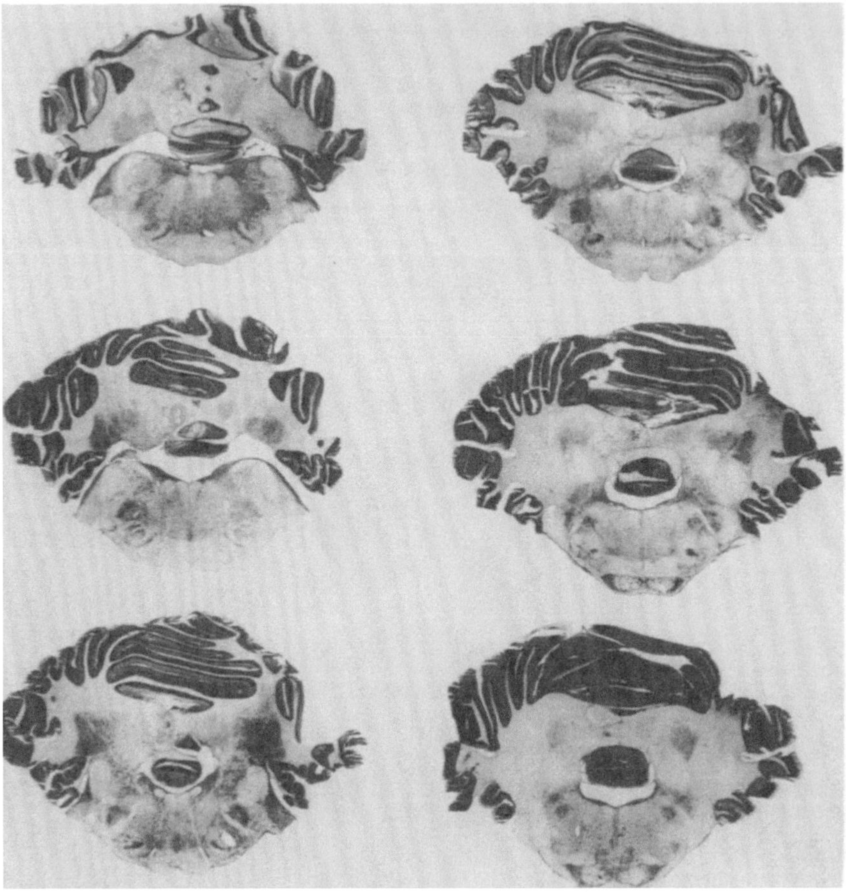

FIGURE 9. Representative photomicrographs of right, mainly lateral zone cortical lesion and left n fastigius nuclear lesion. Fastigial lesion made 2 months after cerebellar cortical lesion.

Contact placing was not only hypermetric but had a myoclonic, jerky quality as though contact delivered an electric shock to the paw. Despite the exaggerated response to contact, water inhibition occurred as readily as on the control side.

3.5. Lesions in Globus Pallidus and Nucleus Entopeduncularis

The data are derived from six cats, two with large but incomplete RF lesions of globus pallidus and the remaining four with kainic acid lesions of nE. In one each of the pallidal and nE lesions, the lesions extended into the nucleus centralis of the amygdala. The globus pallidus lesions failed to prevent a reduced incidence of placing on the wet side, but in both cats, disinhibition was usually impaired.

By contrast, nE lesions clearly prevented acquisition of water conditioning in all four naive cats. In two of the cats, the lesions were almost complete (e.g., Fig. 11); in the remaining pair, nE cells were spared rostrally and laterally, respectively, implying that destruction of the entire nucleus is not required to prevent water inhibition. Following the nE lesions, CP lacked the myoclonic quality seen after cerebellar cortical lesions (Section 3.4).

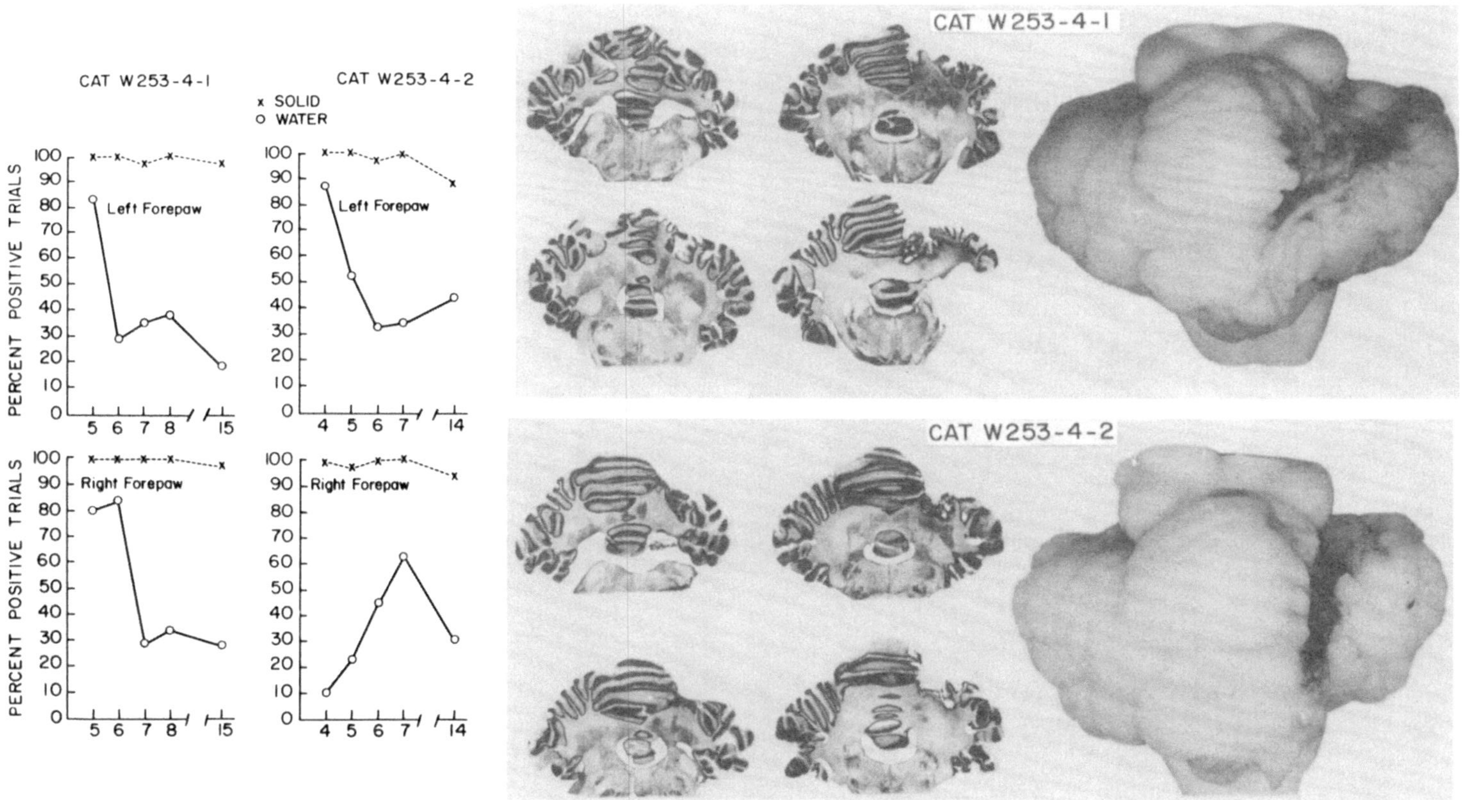

FIGURE 10. In two cats, percentage of positive CP trials during water conditioning of control (left) forepaw and the forepaw ipsilateral to the cerebellar cortical lesions, with photographs of cerebellum and representative photomicrographs. In cat W253-4-1, disinhibition of left forepaw not tested on eighth postoperative day.

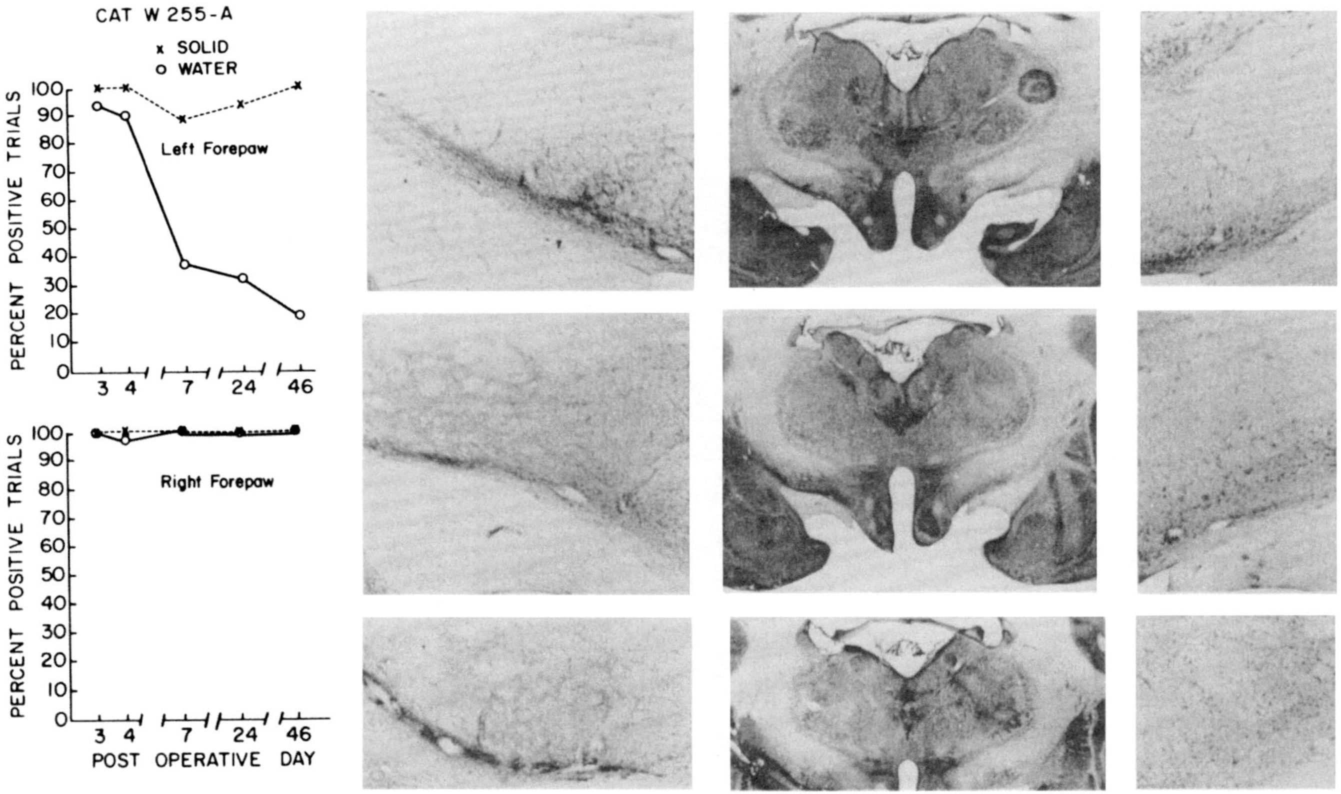

FIGURE 11. Percentage of positive CP trials during water conditioning of control (left) forepaw and the forepaw contralateral to the n entopeduncularis lesion, with representative photomicrographs. Lesion made with kainic acid after optic tract located electrophysiologically (Section 2.1). Regions of left and right nE shown at same higher magnifications.

3.6. Lesions of Other Thalamic Nuclei

The series included five cats in which unilateral RF lesions were made in several structures including n medialis dorsalis (nMD), n centrum medianum (nCM), n parafascicularis, the habenulopeduncular tract, and midline nuclei such as rhomboideus, periventricularis, and centralis medialis in various combinations. In three of these cats, all of the above structures were damaged, yet water inhibition of CP was demonstrated in two and transiently in the third. In a fourth cat, lesions were initially made in the right nCM and 4 months later in the left nDM. After each lesion, water inhibition of CP still occurred contralaterally, and water clearly continued to be an aversive stimulus. In the remaining cats, following the lesion of all the above nuclei except parafascicularis and the habenulopeduncular tract, water inhibition of CP was only transiently and feebly demonstrated. In summary, no consistent effects resulted from unilateral lesions of the above thalamic nuclei. Similarly, unilateral aspiration of prefrontal cortex failed to prevent water inhibition of CP contralaterally.

4. DISCUSSION

4.1. Role of Cerebellum

Our findings clearly implicate nVL–VA in water inhibition of CP. Which of the major inputs to these nuclei subserve this function? Although the Purkinje-cell-mediated inhibition of nI and D could have provided a plausible basis for water inhibition of CP, we have found no evidence that either ipsilateral intermediate or lateral zone cortex or the deep nuclei (nI, D, and fastigius) are required for acquisition or retention, respectively, of this type of passive avoidance. By contrast, ipsilateral nI is clearly part of the circuitry subserving associative conditioning of the nictitating membrane response and conditioned leg flexion in the rabbit (Thompson *et al.*, 1983). This discrepancy could reflect differences in the neural circuitry subserving types I and II conditioning. However, even within the category of associative (type I) conditioning, major differences in neural circuitry have been found; e.g., conditioning of the eyeblink in cats depends on the motor cortex (Woody *et al.*, 1974), whereas the conditioned eyeblink response in rabbits is preserved after decerebration (Norman *et al.*, 1977).

Because even the unconditioned response, CP, requires sensorimotor cortex, it is not surprising that the neural circuitry differs from that of the conditioned nictitating membrane response. It is apparent that a uniform inhibitory action by Purkinje cells could not account for all of the manifestations of water inhibition of CP. Although Purkinje cell activation, by inhibiting nI and D drive to the flexors, could lead to an extended, immobile forelimb following contact, other actions are needed to explain the more common patterns of water inhibition. Thus, a marked ventroflexion at the wrist, or flexion at the shoulder and elbow resulting in the paw hovering over the water, would require reduced Purkinje cell inhibition of nI and D drive to ventroflexors of the wrist and flexors of the elbow and shoulder, respectively. Sidestepping, or the paw moving towards the experimenter's hand under the chin, implies still more complex motor acts. Nevertheless, the cerebellar lesions were valuable in establishing that water inhibition is a direct consequence neither of the particular posture of the limb nor of a nonspecific, reduced agonist activation after contact. Water inhibition of CP where agonist output is exaggerated (hypermetric) after cerebellar cortical lesions attests to the robustness of the inhibitory mechanism.

A very-short-term "motor" memory function of cerebellum has previously been suggested in a preliminary study of the effects of lesions of cerebellar cortex or deep nuclei on the interrelations between forelimb joint angles during CP (Amassian *et al.*, 1984; Batson and Amassian, 1986). During normal, economical CP, the ratio of the angular velocities at the shoulder and elbow joints is conserved when the initial lifting (flexion) phase is replaced by the landing (extension) phase. After deep nuclear or cerebellar cortical lesions, the angular velocity of anterior flexion at the shoulder is increased relative to extension at the elbow, resulting in hypermetria in the *X* axis. Because the ratio of the angular velocities at shoulder and elbow joints may vary in successive trials, this particular type of "motor" memory operates only over the very short time period of CP, i.e., less than 1 sec. Theoretically, the cerebellum might similarly subserve an associative role between a contact and the landing in water shortly thereafter, but our data give no evidence of such a function. Given the lack of requirement for cerebellar circuitry in water conditioning in CP, it is not surprising that the depression of water conditioning by nVL–VA lesions is unrelated to the amount of interruption of the cerebellar projection to motor cortex.

4.2. Role of nE and nVL–VA

The effect of nE on nVL–VA neurons in the cat is described as predominantly inhibitory (Uno and Yoshida, 1975), precluding a simple measure of the effect of the nVL–VA lesion on transmission of nE influence on frontal cortex. However, the powerful depression of water conditioning by nE lesions suggests that the effect of the nVL–VA lesion is accounted for in part, if not wholly, by the effect of the nE influence on these thalamic nuclei. Such nE influence clearly can be exerted in the absence of nI and D input, implying that nVL–VA neurons are tonically driven by some other presynaptic source, which provides a background for nE inhibitory action. The possible interrelationships include the following. (1) The nVL–VA lesion may interrupt the inhibitory nE projections, "releasing" nVL–VA neurons that are facilitatory to motor cortex and thus accounting for the prominently "jerky" quality of CP recovering from such a lesion. Against this hypothesis are the lack of similar jerkiness after nE lesions and the fact that recovering CP tends to be jerky even after virtually complete nVL–VA lesions, implying a different mechanism of the jerky quality. (2) The nVL–VA lesion may damage two sets of neurons, the first set facilitatory to motor cortex, thus contributing to the lifting phase of CP (Amassian *et al.*, 1972b), and the second set either directly or indirectly inhibitory (Amassian *et al.*, 1974) to motor cortex. The two hypotheses might be distinguishable by recording from individual nE neurons during water conditioning, because they imply opposite changes in nE activity. However, as emphasized in Section 4.1, the motor patterns of water inhibition include some too complex to be caused simply by a general inhibition of CP agonists. Furthermore, disruption of an inhibitory process is a plausible but not the only possible explanation for the jerky quality of CP following a nVL–VA lesion. Sprouting of collaterals of intact excitatory presynaptic inputs to sensorimotor cortex, e.g., from nVP, would provide an alternative to release from inhibition, by analogy with the findings of Tsukahara (1978) in red nucleus.

Involvement of at least part of the basal ganglia in water inhibition suggests a similarity to the difficulty a parkinsonian patient has not only in initiating but also in altering or stopping complex movements. In the monkey, cooling the pallidum (Hore *et al.*, 1977; Hore and Vilis, 1980) and either cooling or making lesions in the internal pallidum (Trouche *et al.*, 1979) resulting in a slowing of learned movements and reduced accuracy in pointing, but the authors disagreed on whether the disability was reduced by permitting the animal visual information. However, their findings do not readily relate to the role of nE in water inhibition of CP; e.g., an inappropriately early cocontraction of antagonists occurring with cooling of the external pallidum (Hore and Vilis, 1980) could prevent CP from occurring, but this is the reverse of what is observed following an nE lesion.

4.3. Processes in Water Inhibition Potentially Modified by Lesions

In the Introduction (Section 1.2), the processes necessary for water inhibition and disinhibition of CP were outlined. Which of these processes are affected by the nVL–VA and nE lesions? Transmission of the sensory input resulting from water stimulation and its interpretations as aversive are probably minimally interfered with by the effective lesions. Although we have no quantitative measure of the aversive quality of the water, the affected forelimb was rapidly withdrawn after the initial landings in the water. Furthermore, during successive training sessions, cats tended to develop an antipathy to testing if unable to avoid placing into water. Indeed, given the bilateral projection of nociceptive and other inputs to the medially located thalamic nuclei, as evidenced both by electrophysiological recording (Albe Fessard and Kruger, 1962) and by the need for bilateral lesions to relieve unilateral pain (Tasker, 1984), it would be surprising if our unilateral lesions were effective through sensory or perceptual interference.

Impairment of the storage of a long-term memory aspect of water inhibition seems unlikely given the findings that, (1) following an nVL–VA lesion, repeated conditioning sessions may cause a progressive drop in incidence of CP on the solid side and (2) the low incidence of CP induced by such conditioning before making the lesion is not reversed by the lesion. The increasing antipathy shown by some cats to repeated blocks of trials in which they frequently place into water is probably a more general manifestation of the long-term memory process and is impaired by neither a nVL–VA nor a nE lesion. The retention of a conditioned inhibitory avoidance response in rats after decerebration rostral to the hypothalamus (Tomaz and Huston, 1986) implies a long-term memory storage site below the level of any of our effective lesions. Significantly, Fabre and Buser (1980) found that cats trained to press a moving lever for reward could still perform the task after bilateral nVL lesions. However, the cats could not learn this task if the lesions were made prior to the training. Such observations emphasized the role of nVL in acquisition of this particular motor skill rather than either in its execution or in its memory storage. Given the need for visual feedback in their experiments, a more elaborate paradigm including section of the corpus callosum and monocular occlusion would have been required to determine if a contralateral nVL lesion also resulted in the same deficit. Our findings with unilateral nVL–Va lesions indicate a lack of water inhibition of CP whether the cat was naive or had been preoperatively tested. Given the unlikelihood that long-term memory storage has been eliminated by a unilateral nVL–VA lesion, the loss of water inhibition in a preoperatively tested cat suggests disruption of either the short-term memory process (Wertenbaker *et al.*, 1973) or the motor control process subserving water inhibition.

The fact that many cats were given one conditioning session preoperatively and then readily showed water inhibition does not exclude the possible role of a short-term memory process in the postoperative training. In the naive cats, the inability to acquire water conditioning contralateral to the lesion in the first training session further indicates that the major deficit is either in the short-term memory storage or in the motor process subserving water inhibition. We have no evidence bearing on a possible short-term memory function. However, preliminary recording from individual nVL–VA neurons during water conditioning have shown that early postcontact discharges can be abolished in trials in which CP is inhibited but not in trials resulting in landing in water (Fig. 13 in Amassian, 1979). This finding is consistent with nVL–VA mediating the motor process subserving at least some of the patterns of water inhibition of CP.

5. SUMMARY

Contact placing in cats can readily be inhibited by substituting a container of water for the solid on which the forepaw normally lands. Passively resting the paw on a solid surface for 20 sec disinhibits the paw, and subsequent testing results in placing. Water inhibition of CP has short-term (<60 sec) and long-term memory functions. Unilateral radio frequency lesions of nVL–VA, especially the anteroventral portion, markedly impair the occurrence of water inhibition contralaterally in both naive and preoperatively tested animals. However, a gradual diminution of placing onto the solid side was not prevented by these lesions, implying that a long-term memory process had not been disrupted. Unilateral lesions of a major input to nVL–VA, the cerebellar deep nuclei, did not impair water conditioning in either forelimb. Lesions of intermediate and lateral zone cerebellar cortex were also ineffective. By contrast, unilateral kainic acid lesions of another major input to nVL–VA, n entopeduncularis, impaired acquisition of water inhibition contralaterally. The lesions effective in impairing water conditioning most likely did not disrupt either the afferent input representing the aversive stimulus, its interpretation as aversive, or the long-term storage mechanism. Preliminary evidence suggests that the effective lesions disrupt the motor substrate of inhibition but does not exclude disruption also of a short-term memory process.

REFERENCES

Albe-Fessard, D., and Kruger, L., 1962, Duality of unit discharges from cat centrum medianum in response to natural and electrical stimulation, *J. Neurophysiol.* **25:**3–20.

Amassian, V. E., 1979, The use of contact placing in analytical and synthetic studies of the higher sensorimotor control system, in: *Integration in the Nervous System* (H. Asanuma and V. J. Wilson, eds.), Igaku-Shoin Press, New York, pp. 279–304.

Amassian, V. E., and Batson, D., 1983, Effects on feline rubral units of cooling or lesions of cerebellar deep nuclei, *J. Physiol (Lond.)* **343:**72P.

Amassian, V. E., and Ross, R. J., 1978, Developing role of sensorimotor cortex and pyramidal tract neurons in contact placing in kittens, *J. Physiol. (Paris.)* **74:**165–184.

Amassian, V. E., Weiner, H., and Rosenblum, M., 1972a, Neural systems subserving the tactile placing reaction: A model for the study of higher level control of movement, *Brain Res.* **40:**171–178.

Amassian, V. E., Ross, R., Wertenbaker, C., and Weiner, H., 1972b, Cerebellothalamocortical interrelations in contact placing and other movements in cats, in: *Corticothalamic Projections and Sensorimotor Activities* (T. Frigyesi, E. Rinvik, and M. D. Yahr, eds.), Raven Press, New York, pp. 395–444.

Amassian, V. E., Reisine, H., and Wertenbaker, C., 1974, Neural pathways subserving plasticity of contact placing, *. Physiol. (Lond.)* **242:**67–69P.

Amassian, V. E., Ross, R., and Zipser, B., 1977, A role of the vestibular nuclei in contact placing by kittens, *J. Physiol. (Lond.)* **266:**97–98P.

Amassian, V. E., Batson, D., and Eberle, L., 1984, Cerebellar lesions change joint angle interrelations during cat limb trajectories, *J. Physiol. (Lond.)* **353:**40P.

Batson, D. E., and Amassian, V. E., 1986, A dynamic role of rubral neurons in contact placing in the adult cat, *J. Neurophysiol.* **56:**835–856.

Bengelloun, W. A., Burright, R. G., and Donovick, P. J., 1976, Nutritional experience and spacing of shock opportunities alter the effects of septal lesions on passive avoidance acquisition by male rats, *Physiol. Behav.* **16:**583–587.

Bengelloun, W. A., Burright, R. G., and Donovick, P. J., 1977, Septal lesions, cue availability and passive avoidance acquisition by hooded male rats of two ages, *Physiol. Behav.* **18:**1033–1037.

Black, A. H., Nadel, L., and O'Keefe, J., 1977, Hippocampal function in avoidance learning and punishment, *Psychol. Bull.* **84:**1107–1129.

Bradley, N. S., Smith, J. L., and Villablanca, J. R., 1983, Absence of hindlimb tactile placing in spinal cats and kittens, *Exp. Neurol.* **82:**73–88.

Bregman, B. S., and Goldberger, M. E., 1982, Anatomical plasticity and sparing of function after spinal cord damage in neonatal cats, *Science* **217:**553–554.

Cogan, D. C., and Reeves, J. L., 1979, Passive avoidance learning in hippocampectomized rats under different shock and intertrial interval conditions, *Physiol. Behav.* **22:**1115–1121.

Fabre, M., and Buser, P., 1980, Structures involved in acquisition and performance of visually guided movements in the cat, *Acta Neurobiol. Exp.* **40:**95–116.

Forssberg, H., Grillner, S., and Sjöström, A., 1974, Tactile placing reactions in chronic spinal kittens, *Acta Physiol. Scand.* **92:**114–120.

Gray, J. A., 1982, *The Neuropsychology of Anxiety,* Clarendon Press, Oxford, Oxford University Press, New York.

Green, R. H., and Schwartzbaum, J. S., 1968, Effects of unilateral septal lesions on avoidance behavior, discrimination reversal and hippocampal EEG, *J. Comp. Physiol. Psychol.* **65:**388–396.

Grossman, S. P., Grossman, L., and Walsh, L.,1975, Functional organization of the rat amygdala with respect to avoidance behavior, *J. Comp. Physiol. Psychol.* **88:**829–850.

Hore, J., and Vilis, T., 1980, Arm movement performance during reversible basal ganglia lesions in the monkey, *Exp. Brain Res.* **39:**217–228.

Hore, J., Meyer-Lohmann, J., and Brooks, V. B., 1977, Basal ganglia cooling disables learned arm movements of monkeys in the absence of visual guidance, *Science* **195:**584–586.

Lincoln, J. S., McCormick, D. A., and Thompson, R. F., 1982, Ipsilateral cerebellar lesions prevent learning of the classically conditioned nictitating membrane/eyelid response, *Brain Res.* **242:**190–193.

McCormack, D. A., Guyer, P. E., and Thompson, R. F., 1982, Superior cerebellar peduncle lesions selectively abolish the ipsilateral classically conditioned nictitating membrane/eyelid response of the rabbit, *Brain Res.* **244:**347–350.

McGeer, E. G., and McGeer, P. L., 1981, Neurotoxins as tools in neurobiology, *Int. Rev. Neurobiol.* **22:**173–204.

Miczek, K. A., Kelsey, J. E., and Grossman, S. P., 1972, Time course of effects of septal lesions on avoidance, response suppression and reactivity to shock, *J. Comp. Physiol. Psychol.* **79:**318–327.

Norman, R. J., Buchwald, J. S., and Villablanca, J. R., 1977, Classical conditioning with auditory discrimination of the eye blink in decerebrate cats, *Science* **196:**551–553.

Pellegrino, L., 1968, Amygdaloid lesions and behavioral inhibition in the rat, *J. Comp. Physiol. Psychol.* **65:**483–491.

Steward, O., 1982, Assessing the functional significance of lesion-induced neuronal plasticity, *Int. Rev. Neurobiol.* **23:**197–254.

Tasker, R. R., 1984, Stereotaxic surgery, in: *Textbook of Pain* (P. O. Wall, and R. Melzack, eds.), Churchill Livingstone, Edinburgh, pp. 639–655.

Thompson, R. F., Clark, G. A., Donegan, N. H., Lavond, D. G., Madden, J. IV, Mamounas, L. A., Mauk, M. D., and McCormick, D., 1984, Neuronal substrates of basic associative learning, in: *Neuropsychology of Memory* (L. Squire and N. Butters, eds.), Guilford Press, New York, pp. 424–442.

Tomaz, C., and Huston, J. P., 1986, Survival of a conditioned inhibitory avoidance response after decerebration, *Exp. Neurol.* **93:**188–194.

Trouche, E., Beaubaton, D., Amato, G., Legallet, E., and Zenatti, A., 1979, The role of the internal pallidal segment on the execution of a goal directed movement, *Brain Res.* **175:**362–365.

Tsukahara, N., 1978, Synaptic plasticity in the red nucleus neurons, *J. Physiol. (Paris)* **74:** 339–345.

Uno, M., and Yoshida, M., 1975, Monosynaptic inhibition of thalamic neurons produced by stimulation of the pallidal nucleus in cats, *Brain Res.* **99:**377–380.

Ursin, H., 1965, Effect of amygdaloid lesions on avoidance behavior and visual discrimination in cats, *Exp. Neurol.* **11**:298–317.

Werka, T., Skar, J., and Ursin, H., 1978, Exploration and avoidance in rats with lesions in amygdala and piriform cortex, *J. Comp. Physiol. Psychol.* **92**:672–681.

Wertenbaker, C. T., Ross, R. J., and Amassian, V. E., 1972, Aversive conditioning of contact placing in cats and its developmental aspects, *Neurosci. Abstr.* **2**:191.

Wertenbaker, C., Ross, R., and Amassian, V. E., 1973, Modification of contact placing by aversive conditioning,*Brain Behav. Evol.* **8**:304–320.

Woody, C. D., Yarowsky, P., Owens, J., Black-Cleworth, P., and Crow, T., 1974, Effect of lesions of cortical motor areas on acquisition of conditioned eye blink in the cat, *J. Neurophysiol.* **37**:385–394.

Yeo, C. H., Hardiman, M. J., and Glickstein, M., 1985, Classical conditioning of the nictitating membrane response of the rabbit: II lesions of the cerebellar cortex, *Exp. Brain Res.* **60**:99–113.

14

Essential Involvement of Mossy Fibers in Projecting the CS to the Cerebellum during Classical Conditioning

JOSEPH E. STEINMETZ, CHRISTINE G. LOGAN, and RICHARD F. THOMPSON

1. INTRODUCTION

A number of lesion and recording studies provide evidence that the cerebellum is essentially involved in classical conditioning of skeletal muscle responses (e.g., McCormick *et al.*, 1982a; McCormick and Thompson, 1984). In an attempt to delineate the essential neural circuitry involved in classical eyelid conditioning, we have adopted the following working hypothesis: (1) Information concerning the occurrence of the conditioned stimulus (CS) is projected to the cerebellum along mossy fibers that originate in the pontine nuclei. (2) Information concerning occurrence of the unconditioned stimulus (US) is projected to the cerebellum along climbing fibers that originate in the inferior olive. (3) Neural plasticity associated with acquisition and retention of the classically conditioned response occurs in regions of the cerebellum where CS mossy fibers and US climbing fibers converge.

In previous studies, we demonstrated complete abolition of classically conditioned responses (CRs) in rabbits after discrete electrolytic or chemical lesions were placed in the interpositus nucleus or its immediate efferents (i.e., superior cerebellar peduncle and red nucleus) (Clark *et al.*, 1984; Haley *et al.*, 1983; Lavond *et al.*, 1985; McCormick *et al.*, 1982b). In addition, multiple and extracellular single-unit recordings from cerebellar cortex and the interpositus nucleus have revealed behaviorally related patterns of neuronal discharge that precede the execution of the CR as well as other cells that discharge in response to presentations of the CS and US (Donegan *et al.*, 1985; Foy *et al.*, 1984; Foy and Thompson, 1986; McCormick and Thompson, 1984). In short, these lesion and recording studies provide evidence that the cerebellum is critically involved in the classical conditioning of skeletal muscle responses.

JOSEPH E. STEINMETZ • Department of Psychology, Indiana University, Bloomington, Indiana 47405. CHRISTINE G. LOGAN and RICHARD F. THOMPSON • Department of Psychology, University of Southern California, Los Angeles, California 90089.

We have also demonstrated a sufficient and necessary involvement of climbing fibers in projecting US information to the cerebellum. Discrete electrolytic lesions placed in the rostromedial portion of the dorsal accessory olive (RM-DAO) caused a gradual decrement in the number and amplitude of eyelid CRs in rabbits previously trained with a tone CS and airpuff US even though paired training was continued after the RM-DAO lesions (McCormick *et al.*, 1985; Steinmetz *et al.*, 1984). The rate of CR decrement in the RM-DAO-lesioned animals was comparable to the rate of extinction observed in nonlesioned animals given CS-alone presentations after paired acquisition training (i.e., similar to CS-alone extinction animals, the RM-DAO-lesioned animals appeared to be no longer receiving information concerning the occurrence of the airpuff US). In other studies, stimulation of the DAO has been used as a US in place of the peripheral airpuff (Mauk *et al.*, 1986; Mauk and Thompson, 1984; Steinmetz *et al.*, 1985b). We have observed a variety of discrete responses (e.g., eyeblinks, leg flexions, and head turns) when stimulation is delivered to the DAO. Furthermore, when rabbits are given paired tone CS and DAO-stimulation US presentations, CRs that are identical to the unconditioned response elicited by the DAO-stimulation US develop.

Lesions of the interpositus nucleus abolish both the conditioned and unconditioned responses in animals trained with a tone CS and DAO-stimulation US. This finding indicates not only that the interpositus nucleus forms a portion of the essential CR pathway but also that the unconditioned response produced by DAO stimulation is routed through the cerebellum. Together these studies demonstrate a sufficient (DAO stimulation study) and necessary (DAO lesion study) involvement of climbing fibers in projecting the US to the cerebellum during classical conditioning.

Recent evidence concerning the involvement of mossy fibers in projecting the CS to the cerebellum is presented here. In brief, the results of four studies are described. (1) Stimulation of the pontine nuclei or middle cerebellar peduncle can serve as a sufficient CS for classical conditioning. (2) Bilateral lesions of the middle cerebellar peduncle prevent acquisition and abolish retention of classically conditioned responses. (3) Recordings from lateral regions of the pontine nuclei reveal patterns of neuronal discharge that are related to the occurrence of acoustic CSs. (4) Bilateral lesions of lateral pontine nuclear regions selectively abolish CRs established with an acoustic CS.

2. MOSSY FIBER STIMULATION AS A CS

If, as we have hypothesized, the CS were projected to the cerebellum along mossy fibers, it seems likely that stimulation of mossy fibers as a CS would be sufficient for classical conditioning. To test this possibility, we have attempted to use direct electrical stimulation of mossy fibers arising from the pontine nuclei as a CS for classical eyelid conditioning (Steinmetz *et al.*, 1985a, 1987a,b). Rabbits were first chronically implanted with stimulating electrodes in the dorsolateral, lateral, or medial pontine nuclei or in the middle cerebellar peduncle. Subsequent paired presentations of a mossy fiber stimulation CS (a 350-msec train of 60–120 μA, 200 Hz stimulation) and a coterminating 100-msec airpuff US produced robust conditioned eyelid responses that were well established within a single training session (108 trials). The topography of the CRs observed with mossy fiber stimulation CSs was similar to the topography of CRs obtained when peripheral CSs were used (i.e., onset latencies, peak latencies, and response amplitudes were nearly identical). After acquisition training, extinction of the learned response was observed when either CS alone or explicitly unpaired presentations of the CS and US were given. Similarly, no conditioning was observed when unpaired training was given to a group of rabbits before paired acquisition training, although the unpaired presentations appeared to retard the rate of acquisition once paired training was instituted. Finally, lesions of the interpositus nucleus after training abolished conditioned eyelid responses established with the mossy fiber stimulation CS. These data indicate that stimulation of mossy fibers (the middle cerebellar peduncle) or their cell bodies (the pontine nuclei) as a CS is sufficient to produce classically conditioned responses. Furthermore, CRs produced with a mossy fiber stimulation CS appear to be behaviorally identical to CRs established with peripheral CSs.

3. LESIONS OF THE MIDDLE CEREBELLAR PEDUNCLE

In another attempt to demonstrate that the CS is projected to the cerebellum along mossy fibers, Paul Solomon and associates at Williams College and we have evaluated the effect that lesioning the middle cerebellar peduncle (MCP) has an acquisition and retention of the classically conditioned eyelid response (Lewis *et al.*, 1987; Solomon *et al.*, 1986). The MCP is the major source of mossy fiber projections to the cerebellum. In Solomon's study acoustic, visual, and tactile CSs were used to classically condition the nictitating membrane response. The rabbits were trained with one CS, electrolytic lesions were bilaterally placed in the MCPs, and the animals tested for retention of CRs established with the CS. In addition, the rabbits were tested for postlesion acquisition effects by pairing the second and third CSs with the US. Solomon's results indicate that complete bilateral lesions of the MCP abolished retention of conditioned responses established with the first CS and prevented acquisition of conditioned responses with the second and third CSs. Similarly, we have trained rabbits with both a tone and a pontine nuclei stimulation CS and bilaterally lesioned the MCP. The electrolytic MCP lesions permanently abolished the conditioned eyelid responses previously established with both the tone and stimulation CSs. These studies demonstrate that lesions of the MCP, a major source of mossy fiber projections to the cerebellum, prevent acquisition and abolish retention of conditioned responses when a variety of CSs are used. These data suggest that the mossy fibers within the MCP may provide the cerebellum with essential information concerning the occurrence of the CS during classical conditioning.

4. ACOUSTIC-RELATED NEURONAL ACTIVITY IN THE PONTINE NUCLEI

Because acoustic CSs are frequently used during classical conditioning, we have recently concentrated our efforts on delineating the CS pathway from the cochlea to the cerebellum. The results of the MCP lesion studies described above indicate that the acoustic CS may be relayed to the cerebellum by way of mossy fibers contained within the MCP. Since a majority of mossy fibers within the MCP originate in the pontine nuclei, it is likely that an acoustic CS may be relayed to the cerebellum through the pontine nuclei. To locate these potential relay sites, we have systematically recorded multiple- and single-unit activity from the pontine nuclei and surrounding regions in acute, anesthetized and awake, behaving animals (Logan *et al.*, 1986).

In acute, anesthetized rabbits, we have systematically mapped the entire extent of the pontine nuclei for acoustic-related responses. In brief, field potentials evoked by a click stimulus were recorded from localized areas within the pontine nuclei. These areas include the dorsolateral pontine nucleus (5–7 msec latency responses), caudal portions of the lateral pontine nucleus (3–5 msec latency responses), and a rostral portion of the paramedian pontine nucleus (4–5 msec latency responses). The observation of auditory-related units in the lateral and dorsolateral pontine nuclei is in agreement with previous recording and anatomic studies (Aitkin and Boyd, 1978; Boyd and Aitkin, 1976). Field potentials recorded in the dorsolateral and lateral nuclei were approximately the same amplitude, and paramedian nucleus potentials were somewhat smaller. Small electrolytic marking lesions were placed in the pontine recording sites, and potentials recorded at these sites could be distinguished from potentials recorded from surrounding brainstem auditory structures (e.g., the nucleus of the lateral lemniscus and the superior olive).

With a chronically implanted micromanipulator system, extracellular single units were recorded from the dorsolateral and lateral pontine nuclei of rabbits trained with a tone or white-noise CS and airpuff US. A number of acoustic-related units that demonstrated a variety of discharge patterns were isolated. Some units discharged only at the onset of the CS, whereas other units fired for the duration of the 350-msec period. A few cells inhibited during presentation of the acoustic CS. The same patterns of unit activity were also observed in untrained, awake animals given tone-alone or white-noise-alone presentations. In summary, recording data obtained from the anesthetized rabbits suggest three potential regions of the pontine nuclei that may relay information about the acoustic stimuli to the cerebellum. In addition, data from the awake, behaving rabbits demonstrate that the lateral pontine regions are activated by the acoustic CSs used during classical conditioning.

5. LESIONS OF THE PONTINE NUCLEI

If the pontine nuclear regions in which acoustic-related responses can be recorded are essential relays for projecting acoustic CSs to the cerebellum, lesions of these areas should produce a selective abolition of conditioned responses established with a tone CS. To test this possibility, we chronically implanted bilateral lesion electrodes into the dorsolateral, lateral, and paramedian nuclei of several rabbits (Steinmetz *et al.*, 1986, 1987c). The animals were then classically conditioned by pairing a tone CS and airpuff US and then trained with a light CS and the airpuff US. Bilateral electrolytic pontine nuclear lesions were then given, and paired training was reinstated. Our results indicate that bilateral lesions that include most of the lateral pontine nucleus and a caudal portion of the dorsolateral pontine nucleus selectively abolish CRs to the tone CS while leaving intact CRs to the light CS. Lesions that include only the lateral pontine nucleus severely impair the learned response to tone, but the CRs are reestablished with an additional 4–5 days of training. Lesions confined to the dorsolateral or paramedian nuclei have no effect on the learned response. These data suggest that the essential relay(s) to the cerebellum for acoustic CS information may reside in lateral regions of the pontine nuclei (i.e., the acoustic CS may be projected to the cerebellum by way of mossy fibers that originate in the lateral and/or dorsolateral pontine nuclei).

6. DISCUSSION

The four studies presented here provide strong evidence that CSs used for classical conditioning are projected to the cerebellum by way of mossy fibers that originate, at least for acoustic CSs, in the pontine nuclei. The stimulation study revealed that direct stimulation of mossy fibers (i.e., stimulation of the pontine nuclei or middle cerebellar peduncle) could serve as an effective CS for eyelid conditioning and produce CRs that are behaviorally identical to CRs created with peripheral stimuli such as tones and lights. Lesions of the middle cerebellar peduncle, the major source of mossy fiber projections to the cerebellum, abolish previously learned responses and prevent acquisition of new learned responses. Of importance is the fact that the MCP lesions were not specific to one sensory modality but rather affected conditioned responses established with acoustic, visual, tactile, and mossy fiber stimulation CSs. This observation suggests that all CSs may be routed to the cerebellum along mossy fibers contained within the MCP. Since acoustic CSs are most frequently used for classical conditioning, we have chosen to begin tracing potential acoustic CS pathways to the cerebellum. In addition, our stimulation and lesion studies suggest that the pontine nuclei may be the source of the acoustic CS projections. Neuronal recordings from the pontine nuclei of anesthetized and awake rabbits have revealed three regions that respond to auditory input: the dorsolateral, lateral, and paramedian nuclei. However, we have observed that bilateral lesions that include only two of these auditory regions, the dorsolateral and lateral nuclei, selectively abolish conditioned responses to an acoustic CS while leaving intact conditioned responses to a visual CS.

From the results of these studies, we propose that the essential CS pathway for an acoustic CS includes mossy fibers that originate in lateral regions of the pontine nuclei. Furthermore, we propose that the mossy fibers from the lateral pontine nuclear region terminate in areas of the cerebellum that receive climbing fiber input from a rostromedial region of the dorsal accessory olive (the airpuff US pathway). Finally, we propose that neuronal plasticity underlying acquisition of conditioned responses established with an acoustic CS occurs in loci within the cerebellum where mossy fibers from the lateral pontine nuclei and climbing fibers from the dorsal accessory olive converge.

REFERENCES

Aitkin, L., and Boyd, J., 1978, Acoustic input to the lateral pontine nuclei, *Hearing Res.* **1:**17–77.

Boyd, J., and Aitkin, L., 1976, Responses of single units in the pontine nuclei of the cat to acoustic stimulation, *Neurosci. Lett.* **3:**259–263.

Clark, G. C., McCormick, D. A., Lavond, D. G., and Thompson, R. F., 1984, Effects of lesions of cerebellar nuclei on conditioned behavioral and hippocampal neuronal responses, *Brain Res.* **291:**125–136.

Donegan, N. H., Foy, M. R., and Thompson, R. F., 1985, Neuronal responses of the rabbit cerebellar cortex during performance of the classically conditioned eyelid response, *Soc. Neurosci. Abstr.* **11:**835.

Foy, M. R., and Thompson, R. F., 1986, Single unit analysis of Purkinje cell discharge in classically conditioned and untrained rabbits, *Soc. Neurosci. Abstr.* **12:**518.

Foy, M. R., Steinmetz, J. E., and Thompson, R. F., 1984, Single unit analysis of the cerebellum during classically conditioned eyelid responses, *Soc. Neurosci. Abstr.* **10:**122.

Haley, D. A., Lavond, D. G., and Thompson, R. F., 1983, Effects of contralateral red nuclear lesions on retention of the classically conditioned nictitatng membrane/eyelid response, *Soc. Neurosci. Abstr.* **9:**643.

Lavond, D. G., Hembree, T. L., and Thompson, R. F., 1985, Effect of kainic acid lesions of the cerebellar interpositus nucleus on eyelid conditioning in the rabbit, *Brain Res.* **326:**179–182.

Lewis, J. L., LoTurco, J. J., and Solomon, P. R., 1987, Middle cerebellar peduncle lesions disrupt acquisition and retention of the rabbit's classically conditioned nictitating membrane response, *Behav. Neurosci.* **101:**151–157.

Logan, C. G., Steinmetz, J. E., and Thompson, R. F., 1986, Acoustic related responses recorded from the region of the pontine nuclei, *Soc. Neurosci. Abstr.* **12:**754.

Mauk, M. D., and Thompson, R. F., 1984, Classical conditioning using stimulation of the inferior olive as the unconditioned stimulus, *Soc. Neurosci. Abstr.* **10:**122.

Mauk, M. D., Steinmetz, J. E., and Thompson, R. F., 1986, Classical conditioning using stimulation of the inferior olive as the unconditioned stimulus, *Proc. Natl. Acad. Sci. U.S.A.* **83:**5349–5353.

McCormick, D. A., and Thompson, R. F., 1984, Neuronal responses of the rabbit cerebellum during acquisition and performance of a classically conditioned nictitating/eyelid response, *J. Neurosci.* **4:**2811–2822.

McCormick, D. A., Clark, G. C., Lavond, D. G., and Thompson, R. F., 1982a, Initial localization of the memory trace for a basic form of learning, *Proc. Natl. Acad. Sci. U.S.A.* **79:**2731–2735.

McCormick, D. A., Guyer, P. E., and Thompson, R. F., 1982b, Superior cerebellar peduncle lesions abolish the ipsilateral classically conditioned nictitating membrane/eyelid response of the rabbit, *Brain Res.* **245:**347–350.

McCormick, D. A., Steinmetz, J. E., and Thompson, R. F., 1985, Lesions of the inferior olivary complex cause extinction of the classically conditioned eyeblink response, *Brain Res.* **359:**120–130.

Solomon, P. R., Lewis, J. L., LoTurco, J. J., Steinmetz, J. E., and Thompson, R. F., 1986, The role of the middle cerebellar peduncle in acquisition and retention of the rabbit's classically conditioned nictitating membrane response, *Bull. Psychon. Soc.* **24:**75–78.

Steinmetz, J. E., McCormick, D. A., Baier, C. A., and Thompson, R. F., 1984, Involvement of the inferior olive in classical conditioning of the rabbit eyelid, *Soc. Neurosci. Abstr.* **10:**122.

Steinmetz, J. E., Lavond, D. G., and Thompson, R. F., 1985a, Classical conditioning of the rabbit eyelid response with mossy fiber stimulation as the CS, *Bull. Psychon. Soc.* **28:**245–248.

Steinmetz, J. E., Lavond, D. G., and Thompson, R. F., 1985b, Classical conditioning of skeletal muscle responses with mossy fiber stimulation CS and climbing fiber stimulation US, *Soc. Neurosci. Abstr.* **11:**982.

Steinmetz, J. E., Logan, C. G., Rosen, D. J., Lavond, D. G., and Thompson, R. F., 1986, Lesions in the pontine nuclear region selectively abolish classically conditioned eyelid responses in rabbits, *Soc. Neurosci. Abstr.* **12:**753.

Steinmetz, J. E., Rosen, D. J., Chapman, P. R., Lavond, D. G., and Thompson, R. F., 1987a, Classical conditioning of the rabbit eyelid response with a mossy fiber stimulation CS. I. Pontine nuclei and middle cerebellar peduncle stimulation, *Behav. Neurosci.* **100:**871–880.

Steinmetz, J. E., Rosen, D. J., Woodruff-Pak, D. S., Lavond, D. G., and Thompson, R. F., 1987b, Rapid transfer of training occurs when direct mossy fiber stimulation is used as a conditioned stimulus for classical eyelid conditioning, *Neurosci. Res.* **3:**606–616.

Steinmetz, J. E., Logan, C. G., Rosen, D. J., Thompson, J. K., Lavond, D. G., and Thompson, R. F., 1987c, Initial localization of the acoustic conditioned stimulus projection system to the cerebellum during classical eyelid conditioning, *Proc. Natl. Acad. Sci. U.S.A.* **84:**3531–3535.

15

Plasticity in Inferotemporal Cortex–Amygdala–Lateral Hypothalamus Axis during Operant Behavior of the Monkey

TAKETOSHI ONO, MASAJI FUKUDA, HISAO NISHIJO, and KIYOMI NAKAMURA

1. INTRODUCTION

Evidence indicates that the amygdala (AM) and lateral hypothalamic area (LHA) have intimate anatomic and functional interrelations (Oomura *et al.*, 1970; Krettec and Price, 1978; Amaral *et al.*, 1982; Kita and Oomura, 1982; Ono *et al.*, 1985a) and are deeply involved in discrimination, learning, and reward functions (Rolls *et al.*, 1976; Ono *et al.*, 1981, 1983). The AM receives highly processed information from the sensory association cortex including the inferotemporal cortex (ITCx) (Aggleton *et al.*, 1980; Turner *et al.*, 1980) and in turn projects to the hypothalamus, which is essential to motivation-related learning (Fukuda *et al.*, 1986; Nakamura and Ono, 1986; Ono *et al.*, 1986).

Lesions of the monkey ITCx or AM frequently induce the Klüver–Bucy syndrome in which the lesioned animal cannot discriminate between food and nonfood (Klüver and Bucy, 1939; Horel *et al.*, 1975; Aggleton and Passingham, 1981). It has been proposed that disconnection of visual information from the visual cortex to limbic centers such as the AM or to other cortices is one cause of the Klüver–Bucy syndrome (Geschwind, 1965). The primary aim of this study was to examine plasticity of LHA and AM neurons in operant learning behavior in monkeys. The secondary objective was to analyze dynamic interactions of the ITCx–AM–LHA axis in terms of a disconnection syndrome by means of reversible deficits produced by ITCx or AM cooling. In this task, the monkey had to discriminate between food, potable-associated objects, and nonfood according to visual and/or auditory cue signals.

TAKETOSHI ONO, MASAJI FUKUDA, HISAO NISHIJO, and KIYOMI NAKAMURA • Department of Physiology, Faculty of Medicine, Toyama Medical and Pharmaceutical University, Sugitani, Toyama 930-01, Japan.

2. MATERIAL AND METHODS

Eight *Macaca fuscata* monkeys (4–6 kg) were used. The experimental procedures have been described in detail elsewhere (Ono *et al.*, 1980, 1981; Fukuda *et al.*, 1986). Each monkey, restrained by a head holder in a chronic stereotactic apparatus developed in our laboratory, sat in a monkey chair facing a panel containing two shutters (Fig. 1). One shutter was a one-way mirror, S1, in front of a turntable, and the other was an opaque cover, S2, that prevented access to the right-hand bar. The monkey could not see the object on the turntable through the one-way mirror when the turntable side was dark. When the light (L) was turned on, the monkey could see an object, either food or nonfood, on the turntable but could not take the object because of the intervening shutter, S1. After a delay of at least 2 sec, the shutter S2 was opened automatically, and the animal could obtain the object or the indicated reward by pressing the bar a predetermined number of times (fixed ratio, FR 5–30). When the FR criterion was met, either shutter S1 opened automatically so that available food could be taken and eaten, or a drop of potable juice or water portended by some symbolic objects (column, cube, etc.) could be drunk from a small spout.

The feeding task was divided into four phases: control, visual (discrimination), bar press (operant responding), and ingestion (reward). Active avoidance and auditory discrimination tasks were added to the program during AM unit recording experiments. In avoidance tasks, one of two objects, a brown cylinder associated with electric shock or a roll of tape that was not associated with electric shock, was presented along with a

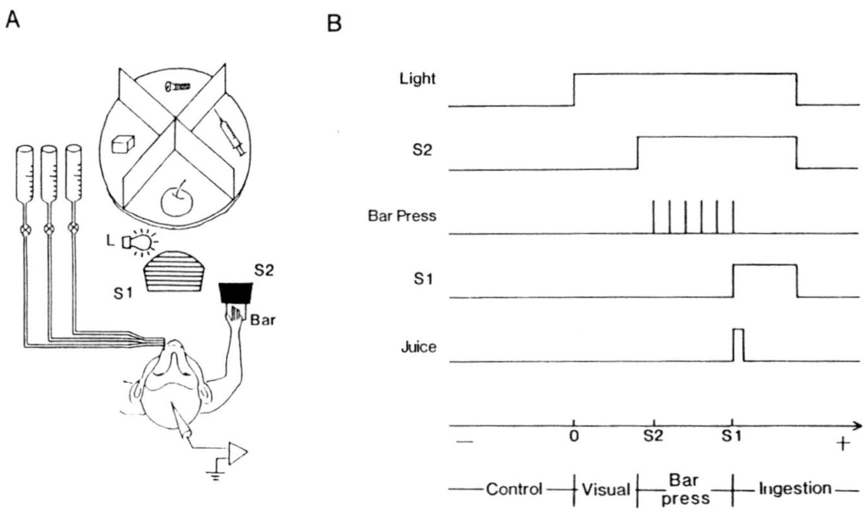

FIGURE 1. Schematic diagram of experimental arrangement (A) and timing diagrams (B). The monkey sat in front of a panel that had a bar and two shutters. One shutter was a one-way mirror, S1, concealing the turntable, and the other, S2, was opaque and concealed the bar. When the experimenter turned the light on, the monkey could see material on the turntable and discriminate it to be food or nonfood. This period is called the visual phase. After 2 sec or longer, S2 was opened. If the material on the turntable was food, the monkey had to press the bar a predetermined number of times (FR 5–30) to obtain it. This period is called the bar-press phase. After opening of S1 at the final bar press, the monkey could take the food and eat it. This is called the ingestion phase. The juice reward was delivered from the small spout at the monkey's mouth. Electrical shock was delivered between the ears (connections not shown) unless it was avoided.

1200-Hz tone. If the animal saw the brown cylinder and heard the 1200-Hz tone, it had to complete a FR schedule within a predetermined period (4–6 sec) to avoid mild electric shock applied to the ears. In auditory discrimination tasks, two sounds were synthesized in a microcomputer; one (buzzer noise) was associated with food (cookie or raisin), and the other (fundamental, 800 Hz) with juice. When the animal heard one of the sounds, it had to complete a FR schedule to obtain food or juice as in the visual feeding task, except that the food or object was not visible behind the one-way mirror.

3. LATERAL HYPOTHALAMIC AREA NEURON RESPONSES

Of 669 LHA neurons tested, 158 (23.6%) responded in one or more phases, 106 (15.8%) of those in the visual phase (Fig. 2). Of 80 visual-related neurons tested systematically, 33 (41.2%) responded selectively to the sight of food or nonfood objects associated with a juice reward but not to the sight of nonfood or to objects associated with aversive stimuli. These neurons responded only to visual cue signals related to reward, and the responses did not depend on the kind of reward. Figure 3 is an example of reward-related LHA neuron responses; it shows raster displays (Aa, Ba), histograms (Ab, Bb), responsiveness (C) of the same food-responsive neuron, and latency of the first bar press in each trial (D). The dots below each line of the raster (Aa) and below the EMG record (Ae) indicate individual bar presses. Responsiveness (R/Rc) is defined as the ratio of mean firing rate (R) during the visual phase between light-on (L) and the opening of S2 (S2) (in this case the 2 sec from light-on to S2 opening) to the mean pretrial control rate (Rc) for 2 sec. Latency of the first bar press was measured from the opening of S2. This neuron responded selectively at the sight of food (A, cookie), and not at the sight of nonfood (B, syringe).

In extinction tests, when the monkey could not obtain the reward even if it pressed

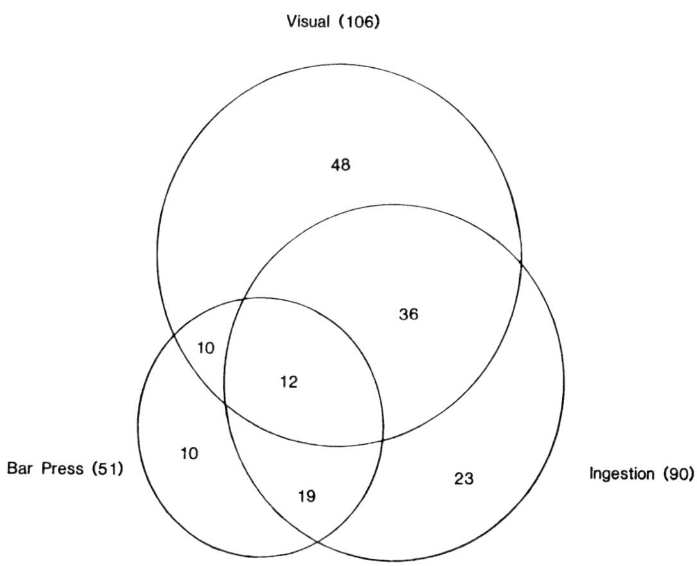

FIGURE 2. Responding neurons in the LHA sorted according to phase(s) in which responses occurred. Tested neurons, 669; responding neurons, 158.

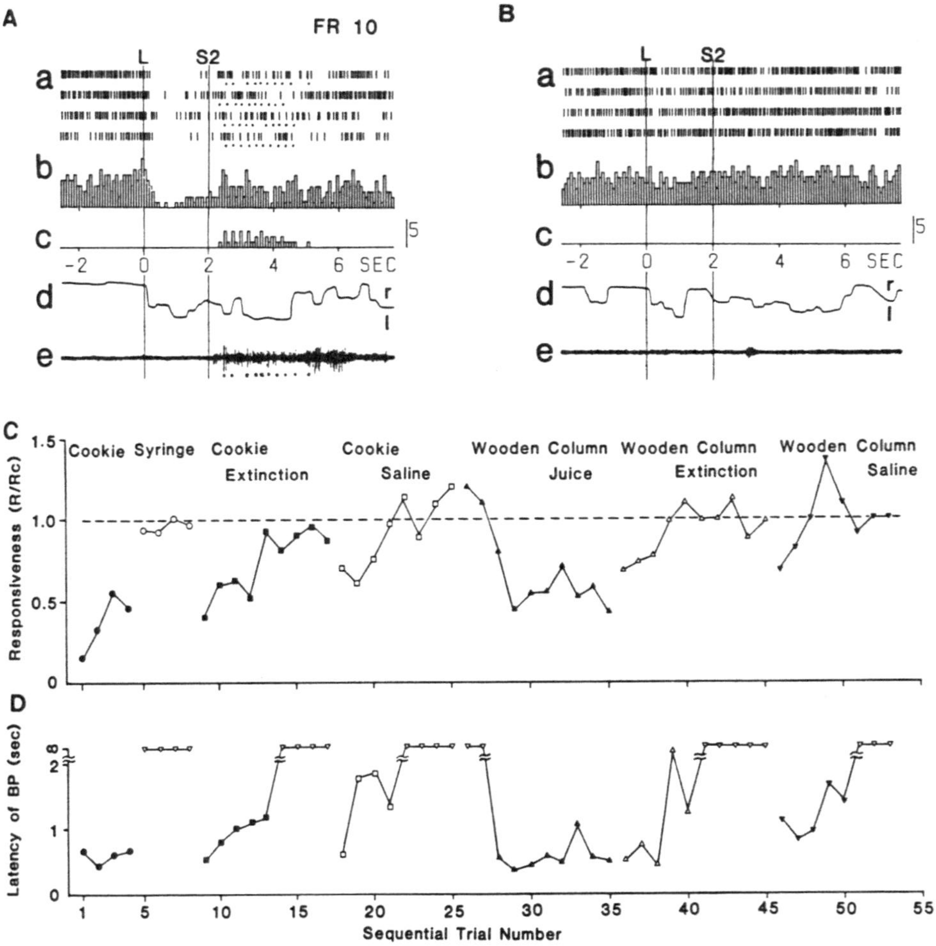

FIGURE 3. Responses of one food-related LHA neuron. (A) Response to sight of food (cookie). (B) Response to sight of nonfood (syringe). (a) Raster display; (b) Histogram (100-msec bins, 12.8 sec) of neuron activity in response to sight of food (cookie); (c) bar presses for four trials; (d) EOG; (e) and EMG in first trial. Dots under each raster line and EMG indicate individual bar presses. Time 0 and L, light-on; S2, shutter S2 in front of bar opened. (C) Responsiveness (R/Rc) during visual phase under various experimental conditions. R/Rc defined as ratio of mean firing rate (R) for 2 sec during visual phase to the mean pretrial control firing rate (Rc). Responsiveness of 1.0 (broken line) means no response change in visual phase. (D) Latency of first bar press from S2 opening. Neuron responded to sight of food or to nonfood associated with juice reward but not to sight of nonfood or food treated with aversive saline. Selective visual responses disappeared in extinction or reversal tests. Latency of first bar press depended on experimental conditions. Filled circle, food (cookie) trial; open circle, nonfood (syringe) trial; filled square, extinction test to food (cookie); open square, reversal test with food associated with delivery of aversive saline; upright filled triangle, wooden column associated with juice reward; upright open triangle, test of extinction to wooden column; inverted filled triangle, reversal test with wooden column associated with aversive saline; inverted open triangle, no bar press. Abscissa: sequential trial number.

the bar, the neuronal responses to the sight of food diminished in successive trials. An example of neuron activity during excitation is shown in Fig. 3C, third session (filled squares) and sixth session (upright open triangles). The inhibitory responses to the sight of food (cookie) became weaker in successive trials and disappeared in the fifth trials during the visual phase. After the sixth trial, the monkey stopped its procurement behavior even at the sight of a cookie. The visual response of the neuron to a juice-associated wooden column also disappeared in later trials when juice was not delivered (Fig. 3C, sixth session, upright open triangles). During extinction, bar-press latency gradually increased (D). In the association test the bar-press and neuronal responses were gradually facilitated as the animal learned to associate a wooden column with juice delivery (Fig. 3C and 3D, fifth session, upright filled triangles). In the first and second trials, when the monkey occasionally did not press the bar, the activity of the neuron did not change during the visual phase, but it did decrease at the sight of the wooden column when the monkey began to press the bar to get juice in the third and later trials. The neuron responded, and the latency to the first bar press diminished after the animal could be presumed to have learned to associate the wooden column with juice delivery.

During extinction or association learning, there was a negative correlation between the magnitude of neuronal responses to the sight of food and latency to the first bar press, which suggests that stronger reward-related visual responses might induce earlier bar presses. In satiation tests, the responses diminished in late trials when the animal did not ingest all of the food or the intermeal intervals became longer; but the neuron responded again, strongly, when another highly preferred food (cookie) was presented.

Of the 669 neurons recorded, 51 (7.6%) responded in the bar press phase; ten of these responded only during bar-press phase, 22 of the responses continued from the visual phase, and 29 occurred only during the bar-press phase or the bar-press and ingestion phases (Fig. 2). There was no correlation between neuronal responses and individual bar-press movement.

Of the 669 neurons tested, 90 (13.5%) responded in the ingestion phase; 23 in the ingestion phase only, 48 in both the visual and ingestion phases, and 31 in the bar-press and ingestion phases. The distribution and overlap of these responses can be seen in Fig. 2, which also indicates why the total exceeds 90. Some neurons habituated to water (not highly palatable) or were facilitated by juice (highly palatable) delivery, which suggests that these responses during the ingestion phase might be involved in higher integrated functions such as perception of reward or its learning.

Thus, LHA neurons showed complete, rapid reversibility in reversal and extinction tests and responded only when food or other reward was available.

4. AMYGDALA NEURON RESPONSES

Neuronal responsiveness to visual, auditory, somesthetic, and gustatory stimuli was investigated in the monkey AM in feeding, active avoidance, and auditory discrimination tasks related to reward or punishment. Of 585 AM neurons observed, 312 (53.3%) responded to one or more stimuli. Based on their responses, these neurons were categorized into five major groups: vision related, audition related, ingestion related, multimodal, and selective (Table I).

Forty (6.8%) and 26 (4.4%) neurons responded only to visual (vision-related) and auditory (audition-related) stimuli, respectively. These two groups were each further divided into two classes: vis-I, vis-II, aud-I, and aud-II. Types vis-I and aud-I responded

TABLE I. Classification and Number of Each Response Type in the AM

		Number of neurons[a]		
		E	I	Total (%)
Vision related	Vis-I	26	0	40 (6.8)
	Vis-II	14	0	
Audition related	Aud-I	8	0	26 (4.4)
	Aud-II	18	0	
Ingestion related	Gustation	27	0	
	Gustation + vision	11	0	41 (7.0)
	Gustation + audition	3	0	
Multimodal	Phasic	36	4	117 (20.0)
	Tonic	69	8	
Selective		14	0	14 (2.4)
Unclassified				74 (12.6)
No response				273 (46.8)
Total				585 (100)

[a] E, excitation; I, inhibition.

to familiar food and reward-associated sounds, respectively; types vis-II and aud-II did not. All four subtypes usually responded to unfamiliar stimuli and/or to some aversive nonfood stimuli but seldom responded to neutral familiar nonfood objects. The vis-I and aud-I neurons responded consistently to significant stimuli, whether rewarding or punishing, but did not respond to familiar neutral stimuli. These neurons usually habituated to affectively nonsignificant stimuli but responded again to those stimuli when they were associated with significant stimuli (electric shock, etc.).

There were 41 neurons (7.0%) that responded primarily during ingestion (ingestion related). Of these, 27 responded only in the ingestion phase, and 14 responded in the ingestion phase plus either the visual or auditory discrimination phase. Of the 41 neurons, five tended to habituate to certain gustatory stimuli in successive trials. Normal food responses of five out of 13 ingestion-related neurons tested were attenuated by saline or heavily salted food in sequential trials.

The responses of 117 neurons (20.0%) were multimodal. Of these, 40 responded only in the discrimination phase and regardless of the affective significance. The discrimination responses of 77 neurons continued into the following phases. Among the four modalities tested, visual stimuli were usually the most effective, and somesthetic the least. The most effective visual stimuli were unfamiliar objects, and the least effective were familiar, neutral, nonfood objects.

Responses of 14 neurons (2.4%) were highly selective to only one or another of certain familiar objects or sounds (selective). Of these, six were selective for one or another specific food item or an object associated with reward. In reversal tests, four neurons that were selective to certain food were tested by strongly salting the food, and their selective responses diminished or disappeared in a few sequential trials. After hand feeding of nonsalted food by the experimenter, responses of these neurons reappeared in nonsalted food trials. One neuron that was selective for the white cylinder, which was associated with juice, was tested by extinction. In 48 trials the behavioral (bar-press) and neuronal responses extinguished in correlation.

In the study reported here, AM neuronal responses to sensory stimuli were easily modified by changing various learning situations. The most prominent factor in modulating neuronal responses was the affective significance. These data, consistent with lesion studies (Jones and Mishkin, 1972; Spiegler and Mishkin, 1981), suggest that the AM is involved in recognizing the affective significance of a stimulus.

5. EFFECTS OF COOLING ITCx OR AM

The effects of ITCx cooling on AM neuron activity and of AM cooling on LHA neuron activity were investigated. Cooling probes were chronically implanted bilaterally over the dura of the anterior ITCx in one monkey and in the lateral part of the AM in two monkeys. The temperature in the ITCx ranged from 18 to 23°C, and in most of the regions surrounding the cooling probes in the AM, temperatures were below 20°C, which is sufficient to depress synaptic transmission (Jasper *et al.*, 1970).

Of 43 AM neurons tested 38 responded to visual stimulation before cooling the ITCx. The ITCx cooling depressed the spontaneous firing rates of 13 neurons and increased those of two. Of six neurons that were primarily excited in response to the sight of food (food responsive), two had their responses to the sight of food depressed by ITCx cooling (Table II). Of 11 neurons that were primarily excited in response to the sight of nonfood objects (nonfood responsive), cooling depressed the responses of four to the sight of nonfood, and two became responsive to the sight of both nonfood and food items (Table II). Of 21 neurons that responded to both food and nonfood (nondifferential), responses of three were enhanced, and seven were depressed. Thus, of 17 neurons that responded primarily to the sight of food or nonfood, eight became nondiscriminative during ITCx cooling. In contrast to the responses in the visual discrimination phase, responses in the ingestion phase were not affected by ITCx cooling. Figure 4 shows the neuronal activity of a food-responsive neuron before (a) and during ITCx cooling (b) and after rewarming (c). This neuron consistently responded at the sight of food (Aa) and responded at the sight of nonfood in the first trial, but the response disappeared in the second and later trials (Ba). About 3 min after the start of ITCx cooling, the visual responses to the sight of food disappeared in every one of the four trials (Ab). However, responses of this particular neuron during ingestion (Ab) and to the sight of nonfood were not changed by ITCx cooling (Bb). On rewarming (about 2 min after stopping cooling), the neuronal responses returned to the control values (Ac, Bc).

Of 55 LHA neurons tested, 50 responded during at least one phase of the task before cooling. Of these, 44 responded during the visual discrimination phase, eight during the bar-press phase, and 12 during ingestion. Amygdala cooling changed the spontaneous firing rates of 21 LHA neurons (12 increased; 9 decreased). The directions of the changes

TABLE II. Effects of ITCx Cooling on AM Activity[a]

ITCx cooling	Food responsive *n* (E, I)	Nonfood responsive *n* (E, I)	Nondifferential *n* (E, I)	No response *n* (N)	Total *n* (E, I, N)
Enhanced	0 (0, 0)	2 (2, 0)	3 (2, 1)	0 (0)	5 (4, 1, 0)
Decreased	2 (2, 0)	4 (4, 0)	7 (5, 2)	0 (0)	13 (11, 2, 0)
Unchanged	4 (4, 0)	5 (5, 0)	11 (8, 3)	5 (5)	25 (17, 3, 5)
Total	6 (6, 0)	11 (11, 0)	21 (15, 6)	5 (5)	43 (32, 6, 5)

[a] E, excitation; I, inhibition; N, no response.

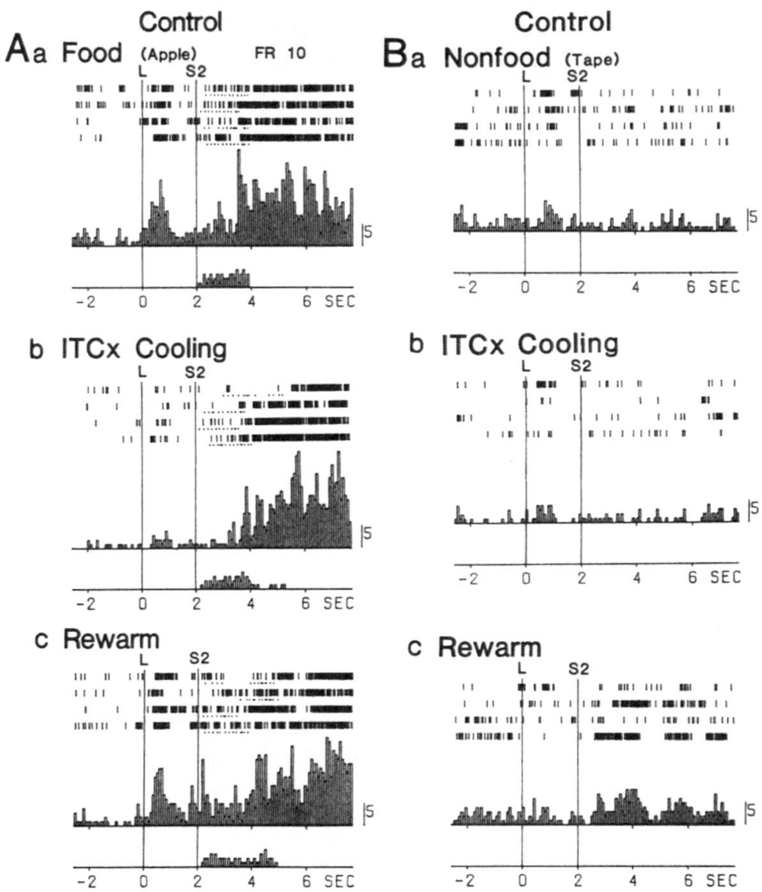

FIGURE 4. Effects of cooling bilateral ITCx on responses of food-related AM neuron. Raster display and histograms (80-msec bins, 10.24 sec) of neuron activity changes in response to (A) sight of food (apple) and (B) nonfood (tape) before (a) and during (b) ITCx cooling and after rewarming (c) and bar presses (lower histogram of each set) for four trials. Time L (0), light-on; S2, shutter S2 in front of bar opened. Dots under each raster line indicate individual bar presses. Neuron responded at sight of food and during ingestion phase in each of four trials before cooling. Response to sight of food selectively suppressed, but ingestion responses remained during functional deficit by ITCx cooling. After rewarming, visual responses returned to control level. Visual responses to sight of nonfood not changed by ITCx cooling except in first trial.

induced by AM cooling, increase or decrease, were significantly the same as the directions during responses in the task ($P < 0.01$). This is consistent with electrophysiological results that AM effects on LHA neurons are primarily inhibitory (Oomura *et al.*, 1970). Of 22 neurons that responded preferentially to the sight of food (food responsive), the responses of nine were weakened or depressed by AM cooling (Table III). In contrast to the effects of ITCx cooling on AM neuron responses in the ingestion phase, AM cooling depressed the responses of three of 12 neurons that responded in the ingestion phase

TABLE III. Effects of AM Cooling on LHA Activity[a]

AM cooling	Food responsive n (E, I)	Ingestion responsive n (E, I)	Nondifferential n (E, I)	No response n (N)	Total n (E, I, N)
Enhanced	0 (0, 0)	0 (0, 0)	0 (0, 0)	0 (0)	0 (0, 0, 0)
Decreased	9 (9, 0)	3 (1, 2)	2 (2, 0)	0 (0)	14 (12, 2, 0)
Unchanged	13 (9, 4)	3 (0, 3)	20 (9,11)	5 (5)	41 (18,18, 5)
Total	22 (18, 4)	6 (1, 5)	22 (11,11)	5 (5)	55 (30,20, 5)

[a] E, excitation; I, inhibition; N, no response.

(Table III, ingestion related). Cooling of the AM affected the responses of only two of 22 neurons that responded to both food and nonfood (Table III, nondifferential).

Results of the experiments described here suggest dynamic ITCx–AM–LHA interactions. The ITCx deficits mainly depressed visual responses and spontaneous firing in the AM, and AM deficits mainly depressed excitatory or inhibitory LHA responses related to visual and ingestion signals and depressed or increased spontaneous firing in the LHA.

6. GENERAL DISCUSSION

There is anatomic and physiological evidence of massive connections between the ITCx and the AM (Herzog and van Hoesen, 1976; Aggleton *et al.*, 1980; Turner *et al.*, 1980) and between the AM and the LHA (Oomura *et al.*, 1970; Krettec and Price, 1978; Price and Amaral, 1981; Amaral *et al.*, 1982; Kita and Oomura, 1982; Ono *et al.*, 1985a). It has been reported that ITCx neurons responded to physical properties of objects or graphic displays (Gross, 1979; Rolls *et al.*, 1977; Sato *et al.*, 1980). In this study, some selective-type neurons in the AM responded to both physical properties of objects (i.e., responses were selective to a specific object) and to affective significance (rewarding or nonrewarding). The LHA neurons responded to any rewarding object, and their responses to food were independent of the kind of food.

Taken together, these data suggest that visual information is sequentially processed from the ITCx to the LHA. The ITCx is involved in analysis of physical properties, the AM is involved in association of affective significance of certain stimuli that are processed in the ITCx and transfered into the AM, and information from various AM neurons that respond selectively to certain stimuli converge on LHA neurons, since LHA neurons responded nondifferentially to any kind of food and to reward-associated nonfoods. A previous study arrived at the same conclusion on the ground that latency to visual stimuli increased from the ITCx to the LHA (Rolls, 1981). This is consistent with our data that cooling of the ITCx and AM changed AM and LHA neuron responses, respectively.

The LHA and AM neuron activities were sharply modulated in reversal tests, which suggests that their responses depended on affective significance (rewarding or nonrewarding). In addition, their responses were correlated with the latency of behavioral responses. Recently, we reported rat LHA neurons that discriminated cue tone signals associated with reward or aversion (Ono and Nakamura, 1985; Ono *et al.*, 1985b, 1986; Nakamura and Ono, 1986). There was also a correlation between those neuronal responses and operant responses. These data suggest that the LHA and AM are involved in associative learning in which the animal must relate certain objects or stimuli to reward. In contrast to LHA and AM neurons, it has been reported that neuronal activity in the ITCx was not changed in reversal tests (Rolls *et al.*, 1977).

In summary, AM and LHA neuron responses changed in various learning situations, and their responses were highly correlated with behavioral responses. We suggest that the ITCx–AM–LHA axis is important in discriminating between food and nonfood, i.e., reward or no reward, and consequently in stimulus reinforcement.

ACKNOWLEDGMENTS. We thank Dr. A. Simpson, Showa University, for help in preparing this manuscript and Miss M. Furusaki and A. Hatayama for typing. This work was supported by the Japanese Ministry of Education, Science, and Culture Grants-in-Aid for Scientific Research 60440028 and 60216010.

REFERENCES

Aggleton, J. P., Burton, M. J., and Passingham, R. E., 1980, Cortical and subcortical afferents to the amygdala of the rhesus monkeys (*Macaca mulatta*), *Brain Res.* **190:**347–368.

Aggleton, J. P., and Passingham, R. E., 1981, Syndrome produced by lesions of the amygdala in monkeys (*Macaca mulatta*), *J. Comp. Physiol. Psychol.***95:**961–977.

Amaral, D. G., Veazey, R. B., and Cowan, W. M., 1982, Some observations on hypothalamo–amygdaloid connections in the monkey, *Brain Res.* **252:**13–27.

Fukuda, M., Ono, T., Nishino, H., and Sasaki, K., 1986, Visual responses related to food discrimination in monkey lateral hypothalamus during operant feeding behavior, *Brain Res.* **374:**249–259.

Geschwind, N., 1965, Disconnexion syndromes in animals and man, *Brain* **88:**237–294.

Gross, C. G., 1979, Activity of inferior temporal neurons in behaving monkeys, *Neuropsychologia* **17:**215–229.

Herzog, A. G., and van Hoesen, G. W., 1976, Temporal neocortical afferent connections to the amygdala in the rhesus monkey, *Brain Res.* **115:**57–69.

Horel, J. A., Keating, E. G., and Misantone, L. J., 1975, Partial Klüver–Bucy syndrome produced by destroying temporal neocortex or amygdala, *Brain Res.* **94:**347–359.

Jasper, H., Shacter, D. G., and Montplaisir, J., 1970, The effects of local cooling upon spontaneous and evoked electrical activity of cerebral cortex, *Can. J. Physiol. Pharmacol.* **48:**640–652.

Jones, B., and Mishkin, M., 1972, Limbic lesions and the problem of stimulus–reinforcement associations, *Exp. Neurol.* **36:**362–377.

Kita, H., and Oomura, Y., 1982, An HRP study of the afferent connections to rat lateral hypothalamic region, *Brain Res. Bull.* **8:**63–71.

Klüver, H., and Bucy, P. C., 1939, Preliminary analysis of functions of the temporal lobe in monkeys, *Arch. Neurol. Psychiatr.* **42:**979–1000.

Krettec, J. E., and Price, J. L., 1978, Amygdaloid projections to subcortical structures within the basal forebrain and brainstem in the rat and cat, *J. Comp. Neurol.* **178:**225–254.

Nakamura, K., and Ono, T., 1986, Lateral hypothalamus neuron involvement in integration of natural and artificial rewards and cue signals, *J. Neurophysiol.* **55:**163–181.

Ono, T., and Nakamura, K., 1985, Learning and integration of rewarding and aversive stimuli in the rat lateral hypothalamus, *Brain Res.* **346:**368–373.

Ono, T., Nishino, H., Sasaki, K., Fukuda, M., and Muramoto, K., 1980, Role of the lateral hypothalamus and the amygdala in feeding behavior, *Brain Res. Bull.* **5**(Suppl. 4)**:**143–149.

Ono, T., Nishino, H., Sasaki, K., Fukuda, M., and Muramoto, K., 1981, Monkey lateral hypothalamic neuron response to sight of food, and during bar press and ingestion, *Neurosci. Lett.* **21:**99–104.

Ono, T., Fukuda, M., Nishino, H., Sasaki, K., and Muramoto, K., 1983, Amygdaloid neuronal responses to complex visual stimuli in an operant feeding situation in the monkey, *Brain Res. Bull.* **11:**515–518.

Ono, T., Luiten, P. G. M., Nishijo, H., Fukuda, M., and Nishino, H., 1985a, Topographic organization of projections from the amygdala to the hypothalamus of the rat, *Neurosci. Res.* **2:**221–239.

Ono, T., Sasaki, K., Nakamura, K., and Norgren, R., 1985b, Integrated lateral hypothalamic neural responses to natural and artificial rewards and cue signals in the rat, *Brain Res.* **327:**303–306.

Ono, T., Nakamura, K., Nishijo, H., and Fukuda, M., 1986, Hypothalamic neuron involvement in integrating of reward, aversion, and cue signals, *J. Neurophysiol.* **56:**63–79.

Oomura, Y., Ono, T., and Ooyama, H., 1970, Inhibitory action of the amygdala on the lateral hypothalamic area in rats, *Nature* **228:**1108–1110.

Price, J. L., and Amaral, D. G., 1981, An autoradiographic study of the projections of the central nucleus of the monkey amygdala, *J. Neurosci.* **1:**1242–1259.

Rolls, E. T., 1981, Processing beyond the inferior temporal visual cortex related to feeding, memory, and striatal function, in: *Brain Mechanisms of Sensation* (Y. Katsuki, R. Norgren, and M. Sato, eds.), John Wiley & Sons, New York, pp. 241–269.

Rolls, E. T., Burton, M. J., and Mora, F., 1976, Hypothalamic neuronal responses associated with the sight of food, *Brain Res.* **111:**53–66.

Rolls, E. T., Judge, S. J., and Sanghera, M. K., 1977, Activity of neurones in the inferotemporal cortex of the alert monkey, *Brain Res.* **130:**229–238.

Sato, T., Kawamura, T., and Iwai, E., 1980, Responsiveness of inferotemporal single units to visual pattern stimuli in monkeys performing discrimination, *Exp. Brain Res.* **38:**313–319.

Spiegler, B. J., and Mishkin, M., 1981, Evidence for the sequential participation of inferior temporal cortex and amygdala in the acquisition of stimulus–reward associations, *Behav. Brain Res.* **3:**303–317.

Turner, B. H., Mishkin, M., and Knapp, M., 1980, Organization of the amygdalopetal projections from modality-specific cortical association areas in the monkey, *J. Comp. Neurol.* **191:**515–543.

16

Responses of Nucleus Basalis of Meynert Neurons in Behaving Monkeys

RUSSELL T. RICHARDSON, SUSAN J. MITCHELL, FRANK H. BAKER, and MAHLON R. DeLONG

1. INTRODUCTION

Cholinergic systems have long been implicated in memory processes since anticholinergic drugs disrupt performance on memory tasks in both human (Drachman, 1977; Mewaldt and Ghoneim, 1979) and nonhuman primates (Bartus, 1978; Ridley *et al.*, 1984) and other species (Buresova *et al.*, 1964; Deutsch, 1971; Squire *et al.*, 1971). The source of cholinergic afferents to cerebral cortex has recently been found to lie primarily in the nucleus basalis of Meynert (NBM) (Lehman *et al.*, 1980; Johnston *et al.*, 1981; Mesulam *et al.*, 1983; Pearson *et al.*, 1983). Moreover, both the number of neurons in the NBM (Whitehouse *et al.*, 1982; Arendt *et al.*, 1983; Rogers *et al.*, 1985) and the levels of cholinergic markers in cortex (Davies and Maloney, 1976; Perry *et al.*, 1977; White *et al.*, 1977) have consistently been found to be reduced in patients with Alzheimer's disease. These findings have led to the hypothesis that the NBM may play an important role in learning and memory.

Our laboratory has been studying the possible functions of the NBM by recording the activity of single NBM neurons in awake monkeys performing a series of behavioral tasks: a visuomotor tracking task, a delayed response task, and a go/no-go task. A consistent finding has been that a substantial proportion (approximately 50%) of all task-related NBM neurons have altered discharge rates following the water reward delivery in these tasks. In addition, NBM neurons appear to be particularly responsive to the presentation of stimuli that consistently precede the water reward delivery. These findings suggest that the NBM may be involved in mechanisms of positive reinforcement and may contribute to the neural basis of appetitive conditioning.

RUSSELL T. RICHARDSON, SUSAN J. MITCHELL, and FRANK H. BAKER • Department of Neurology, Johns Hopkins University, Baltimore, Maryland 21205. MAHLON R. DeLONG • Departments of Neuroscience and Neurology, Johns Hopkins University, Baltimore, Maryland 21205.

2. GENERAL PROCEDURES

Rhesus monkeys were trained on one of the three tasks. After the animals had become proficient in the task, they were surgically prepared for chronic extracellular recording procedures. Under sterile conditions, a stainless-steel cylinder with its axis directed at the NBM was affixed to the skull over a trephine hole. For recording sessions, a hydraulic microdrive apparatus was attached to the cylinder and was used to lower glass-coated platinum–iridium microelectrodes through the brain to the region of the NBM.

Neurons in the NBM were selected for study on the basis of three criteria: location, discharge pattern, and spike duration. First, the neuron had to be located ventral to the globus pallidus. The globus pallidus, putamen, and anterior commissure were all easily identified electrophysiologically, which made it possible to determine the anatomic location of a recorded neuron relative to these structures. Second, the selected neuron had to have a steady discharge pattern and a moderate discharge rate (approximately 20 Hz), which distinguishes NBM neurons from neighboring pallidal neurons (DeLong, 1971; Lamour *et al.*, 1986). Finally, the duration of the initial negative phase of the cell's action potential had to be greater than 180 μsec at 200-Hz to 10-kHz filtering. The action potential duration of pallidal neurons is generally less than 180 μsec (Mitchell *et al.*, 1987a). The location of each cell was subsequently confirmed histologically.

The activity of each selected neuron was recorded during a minimum of 30 trials of the behavioral task on which the monkey had been trained. Rasters and histograms of neuronal discharges were constructed for the various phases of the task. Subsequent data analysis was based on comparisons of a cell's discharge rate in the period preceding a distinct event with the discharge rate in the period following the event. The events used for the analysis were the appearance of target lights on the animal's viewing screen, the application of a torque to the handle, and the delivery of a small volume of water (see Fig. 1). A neuronal response was defined as a significant ($P < 0.01$) increase or decrease in firing rate in the 500-msec postevent period relative to the 500-msec preevent period using a paired t test. A one-way ANOVA ($P < 0.01$) was used to compare the changes in discharge rate associated with two different events.

3. VISUOMOTOR TRACKING TASK

Our initial study of NBM neuronal activity utilized a visuomotor tracking (VMT) task in which the monkey made a skilled elbow flexion or extension movement to a designated target (Richardson *et al.*, 1983; Mitchell *et al.*, 1987a). This task was used so that comparisons could be made between neuronal activity in the NBM and the neighboring globus pallidus. In addition, unpublished observations from a previous study (Georgopoulos *et al.*, 1983) suggested that NBM neurons had altered discharge rates during trained arm movements.

The VMT task is illustrated in Fig. 2. The monkey faced a display of two horizontal rows of 128 light-emitting diodes (LEDs) rather than the oscilloscope screen shown in Fig. 1 (for illustrations, the display used in the two subsequent tasks is adapted to the VMT task). Only three of the LEDs in the top row were used: one on the left, one on the right, and one in the center. The lights on the bottom row corresponded to the animal's arm position. By moving a handle attached to the forearm, the animal determined which of the bottom row LEDs was illuminated. A trial began when the center upper light turned

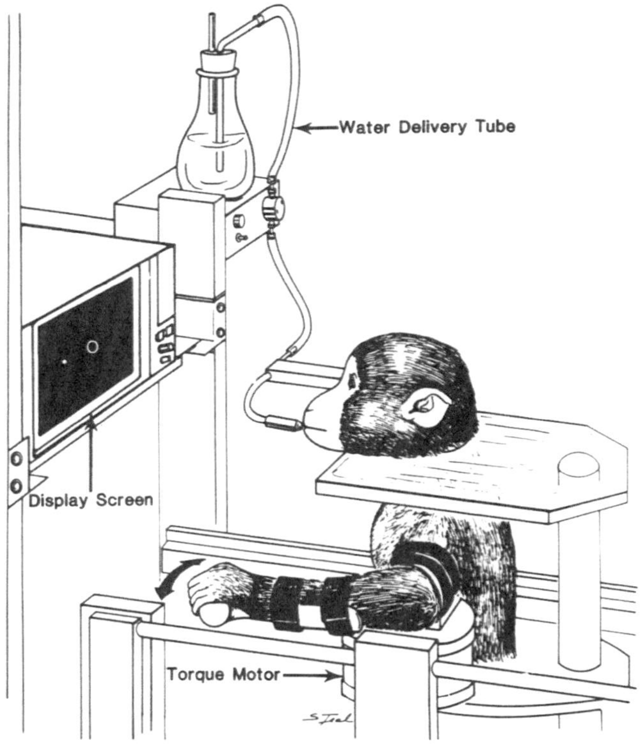

FIGURE 1. Behavioral testing apparatus. The monkey views the display screen on which a circle and a cursor are displayed. The monkey controls the position of the cursor by moving the handle strapped to his forearm. Torques can be applied to the handle via the attached torque motor. Correct positioning of the cursor according to the requirements of the behavioral task is rewarded with water delivered through a metal tube at the monkey's mouth.

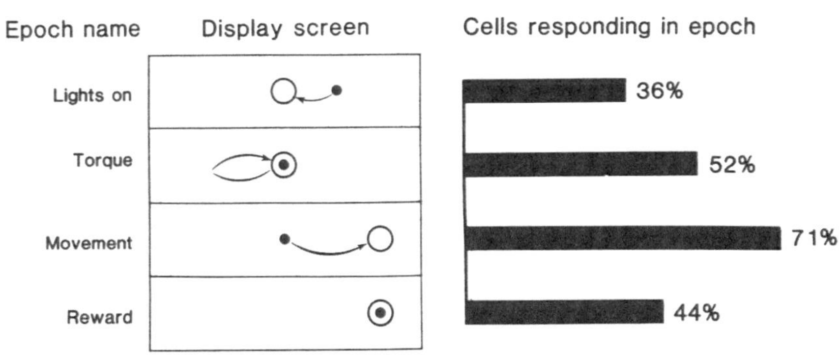

FIGURE 2. Diagram of the visuomotor tracking (VMT) task. The visual display seen by the monkey at the beginning of each epoch of a single trial is shown in the "Display screen" column. The arrows indicate movements of the cursor, which are controlled by the monkey. The direction of the torque and the movement are counterbalanced so that left and right torques and movements occur with equal frequency. The bar graph on the right shows the percentage of all task-related NBM neurons that had significant changes in discharge rate in the initial 500 msec of the specified epoch. Task-related neurons were those that responded in at least one epoch of the task.

on and the monkey moved its arm to the central position to illuminate an LED below the center light. After a variable hold time of 1 to 3 sec, a torque to the left or right was applied to the manipulandum on 75% of the trials. On 50% of all trials, the torque persisted throughout the remainder of the trial. On 25% of all trials, the torque lasted 20 msec and was followed 1 to 2 sec later by the delivery of a small volume of water into the monkey's mouth. On the remaining trials, no torque was applied. One to two seconds after the animal repositioned its arm in the center following a sustained torque application, the upper center LED was extinguished, and either the left or the right LED was illuminated according to a pseudorandom sequence. The animal then made a rapid and accurate arm flexion or extension to illuminate an LED below the illuminated upper row LED. After 1 to 2 sec, a small volume of water was delivered, the display lights were extinguished, and any torques were removed. The next trial began after a 1-sec intertrial interval. If the animal made an error at any time, the trial was terminated, and a new trial began after 1 sec.

In three hemispheres in two monkeys, 57 neurons were recorded in the main body of the NBM ventral to the globus pallidus, 44 of which (77%) had responses in at least one phase of the task. In this task and all subsequent tasks, roughly half of the neuronal responses in each epoch consisted of an increase in discharge rate, and half consisted of a decrease. Only a small percentage of the responses were biphasic judging from the neuronal discharge histograms.

Of the 44 task-related NBM neurons, 52% responded following the torque application, 71% responded following the signal to move, and 44% responded following the reward delivery (see Fig. 2). The majority of these responses were classified as nonspecific since only 15% of the torque epoch responses and 18% of the move signal responses were significantly greater for one of the two directions of torque or movement. Twelve NBM cells (27%) responded in each of the above epochs, i.e., following the torque, move signal, and reward. In contrast, neurons in the globus pallidus had changes in discharge rates that were more closely related to specific movement parameters. Over 80% of the task-related cells in the external and internal globus pallidus had differential responses depending on the direction of the torque, and over 65% had differential responses depending on the direction of the arm movement (Mitchell *et al.*, 1987b). These findings indicated that NBM neurons responded rather nonspecifically following a variety of events in the VMT task.

An interesting dichotomy was observed, however, in the responses of NBM cells to the application of a sustained torque and a brief torque (perturbation). Of the 33 neurons that had significant changes in discharge rate following the torque or perturbation onset, 45% had responses that were significantly different following one stimulus compared to the other. In nine cells the response to the perturbation was larger than the response to the torque, and in three cells the torque response was larger. In the remaining three neurons, the perturbation and torque responses were of different polarities; that is, one was an increase in discharge rate and the other was a decrease. An example of an NBM neuron with an enhanced perturbation response is shown in Fig. 3. Thus, several NBM cells had significantly larger responses following the perturbation than following the torque, although the perturbation was of a shorter duration and produced less arm displacement than the sustained torque. However, the perturbation was consistently followed by a water reward, whereas the torque was followed by subsequent phases of the task. This finding provided one of the first suggestions that the activity of NBM neurons may be influenced by the association of a stimulus with an upcoming reward delivery.

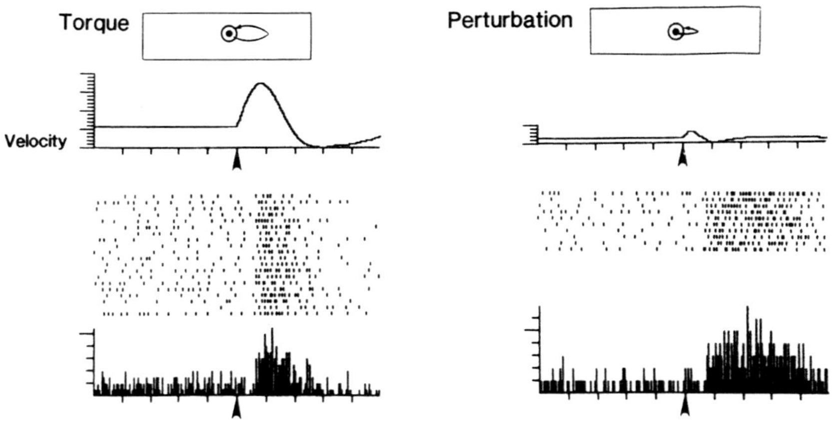

FIGURE 3. Neuronal responses to torque and perturbation in the VMT task. The top box illustrates the display screen as in Fig. 2. The velocity trace is averaged from all trials on which the specified stimulus occurred at the time indicated by the arrowhead. Each mark on the vertical axis represents 5.5°/sec. Neuronal discharges before and after the event are represented by dots on the raster, and each horizontal row represents one trial. Below the raster display is the histogram centered on the designated stimulus. Bin width is 5 msec, and each mark on the horizontal axis represents 100 msec. Divisions on the vertical axis of the histogram represent discharge rates of 20 Hz per bin. Twenty trials with rightward torques and ten trials with rightward perturbations were presented during the recording of this neuron. Neuronal firing was much greater following the perturbation than the torque, although the perturbation displaced the animal's arm considerably less.

4. DELAYED RESPONSE TASK

Several lines of evidence have suggested that the NBM may be directly involved in the performance of a delayed response (DR) task. Anticholinergic drugs can disrupt performance on DR tasks in monkeys (Bartus and Johnson, 1976; Bartus, 1978), and lesions of the basal forebrain regions containing cholinergic neurons disrupt DR performance in rodents (Dunnett, 1985; Hepler *et al.*, 1985). The brain region that has been most strongly implicated in DR performance in primates is the dorsolateral prefrontal cortex (DLPC) (Rosvold and Szwarcbart, 1964; Fuster and Alexander, 1970; Goldman and Rosvold, 1970; Rosenkilde, 1979), which receives cholinergic afferents directly from the NBM (Mesulam *et al.*, 1983; Pearson *et al.*, 1983). Many neurons in the DLPC have changes in discharge rate in the delay period during which the animal must remember a spatial location or an intended movement to a location (Fuster, 1973; Niki and Watanabe, 1976; Kubota and Funahashi, 1982). Furthermore, DLPC neurons often respond differentially in the delay period depending on the specific location or movement being remembered. We hypothesized that the activity of NBM neurons may also reflect a remembered location or movement and may thereby contribute to the differential delay period responses of DLPC neurons. To test this hypothesis, we incorporated a delay period into the VMT task described above to create a delayed-response memory task (Richardson and DeLong, 1984, 1987).

A modified display was used for the DR task, with an oscilloscope replacing the two rows of LEDs as shown in Fig. 1. Target locations (left, right, and center) were provided by 1-cm-diameter circles, and the monkey's arm position was designated by a

1-mm dot. Elbow flexion and extension movements moved the dot horizontally on the display screen (see Fig. 4). The initial epochs of the DR task trials were similar to those of the VMT task. However, the first arm movement to the side circle was not rewarded but served as a cue in the DR task. This "cuing" phase was followed by a return of the target circle to the center position. The monkey recentered the cursor and remained motionless for a delay period of 4 to 6 sec. In the ensuing "choice" phase, both left and right circles appeared, and the monkey was required to move to the previously cued target. Correct choices were followed 1 to 2 sec later by the delivery of 0.01 ml of water. One second later, the display lights and any torques were turned off.

A total of 183 NBM neurons were recorded in two hemispheres of one monkey performing the DR task. Of this sample, 74% had significant responses in at least one phase of the task. Of these task-related cells, 14% had significant changes in firing rate in the delay period. However, none of these 19 neurons had differential responses reflecting the position of the preceding cue. A differential response was defined as a significant difference ($P < 0.01$, ANOVA) between the magnitude of the neuronal response on flexion and extension trials. Neurons that did have delay period responses were found throughout the NBM rather than being grouped in the more anterior regions, which appear to contain the largest number of cells projecting to DLPC (Mesulam *et al.*, 1983; Pearson *et al.*, 1983). It therefore appears unlikely that the differential delay period responses of DLPC neurons can be accounted for by input from the NBM.

Cells in the NBM were much more responsive in other phases of the DR task, with 51% responding in the torque epoch, 64% responding in the choice epoch, and 67% in the reward epoch. These proportions are comparable with the results of the VMT task for similar epochs. For example, 71% of the task-related NBM cells responded during the arm movement in the VMT task compared to 64% in the choice phase of the DR

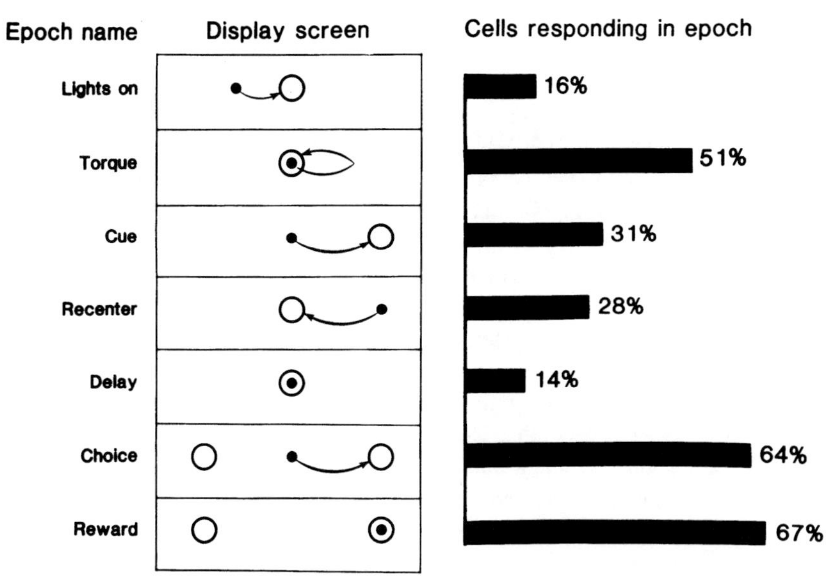

FIGURE 4. Diagram of the delayed response (DR) task. Conventions are the same as in Fig. 2. Note that the same movement was required in the cue and choice epochs, yet far more neurons responded in the choice epoch.

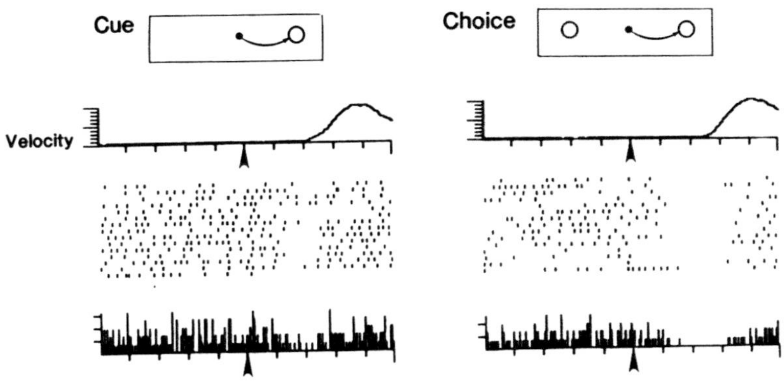

FIGURE 5. Neuronal responses in the cue and choice epochs of the DR task. Conventions are the same as in Fig. 3. Trials are aligned on the appearance of the side circle(s), which instructed the animal to move the cursor to the circle on the right. Note that the decrease in discharge rate is stronger in the choice epoch than in the cue epoch.

task. In contrast, only 31% of the NBM cells responded in the cue epoch of the DR task, although the arm movements made in the cue and choice epochs were almost identical.

A further analysis was made of the 89 neurons that had responses in the cue or choice epochs. Forty-four cells had responses only in the choice epoch, whereas five cells had responses only in the cue epoch. Moreover, of the remaining 40 cells with responses in both the cue epoch and the choice epoch, 16 cells (40%) had significantly larger choice responses than cue responses (paired t test, $P < 0.01$). Thus, 60 of 89 cells (67%) had more pronounced changes in discharge rate in the choice epoch relative to the cue epoch. An NBM neuron with an enhanced response in the choice phase is shown in Fig. 5. These results suggest that the enhanced NBM neuronal responses were caused by some feature of the choice phase other than the movement itself. The salient feature of the choice phase may be that it immediately precedes the reward delivery. This possibility is consistent with the differential NBM neuronal responses in the VMT task in which enhanced responses following the perturbation appeared to be related to the upcoming reward delivery.

5. GO/NO-GO TASK

In order to examine further the role of movement and reward contingencies in the responses of NBM neurons, we have begun to study NBM activity during a go/no-go (GNG) task (Fig. 6) (Richardson and DeLong, 1986). The GNG task is similar to the DR task, but the animal does not move the cursor towards the left or right circle when it appears in the cuing phase. The monkey must keep the cursor in the center position throughout the cuing phase and the following 4- to 6-sec delay period. In the choice phase, either the left or the right circle appears according to a pseudorandom sequence. The animal is given a water reward for moving the cursor to the circle when it appears on the same side as the previously cued circle (go trials), or for not moving from the center position when the circle appears on the opposite side of the cued circle (no-go trials).

There are two key features of the GNG task. First, it provides an assessment of the

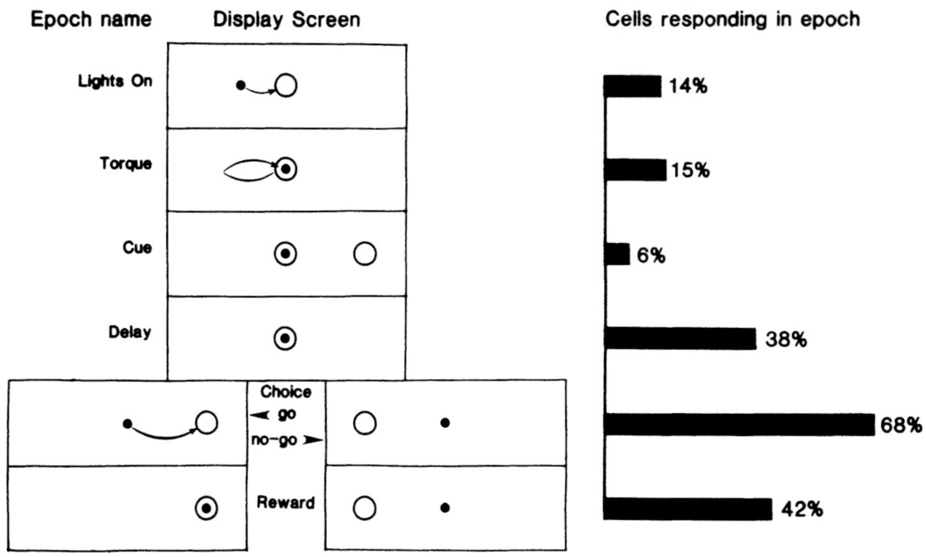

FIGURE 6. Diagram of the go/no-go (GNG) task. Conventions are the same as in Fig. 2. The highest percentage of neurons responded in the choice epoch, with the lowest percentage responding in the cue epoch. As in the DR task, no NBM neurons had differential responses in the delay period.

effect of movement *per se* on NBM activity by comparing neuronal responses in the go and no-go conditions. Second, it provides an assessment of neuronal responses to the same visual stimulus (i.e., the left and right side circles) under two different conditions (cuing and choice). For example, the neuronal response to the left circle presented as a cuing stimulus can be compared to the response to the left circle as a no-go signal.

Of 65 task-related NBM neurons recorded thus far, 44 (68%) responded in the choice epoch. The change in discharge rate in the choice epoch on no-go trials was not significantly different ($P < 0.01$, ANOVA) from that on go trials in 89% of all cells with choice epoch responses. This finding indicates that NBM neurons are highly responsive in the choice phase of the GNG task irrespective of whether the animal makes an arm movement or not. Hence, changes in NBM activity during movements do not appear to be related to the movements *per se*.

In contrast to the 68% of task-related neurons that responded in the choice epoch, only 6% responded in the cue epoch. Comparisons were made between the responses of 31 cells that had significant changes in discharge rate in the cue epoch or the no-go condition of the choice epoch, since no arm movements occurred in either case. Of these neurons, 87% had responses in the no-go condition but not in the cuing condition, whereas 10% had responses in the cuing condition but not in the no-go condition. Hence, NBM neurons showed a strong tendency to be more responsive to the presentation of a side circle in the choice phase of the task than in the cuing phase. An example of this response pattern is shown in Fig. 7.

During the course of the GNG task, 19 NBM neurons were tested for responses to the water delivery when it occurred during the 3-sec intertrial interval. Although the responses of these cells were not recorded in the intertrial interval and therefore could not be analyzed statistically, there were clear changes in discharge rate observed in at least nine (47%) of these neurons. Hence, it was not necessary for the animal to be actively performing the task for neuronal responses following the water delivery to occur.

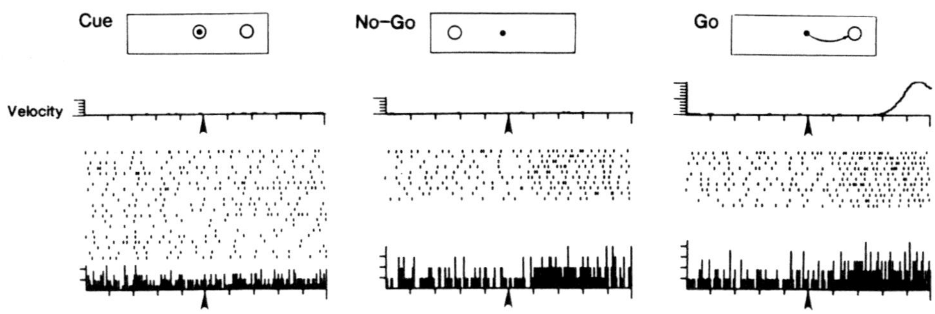

FIGURE 7. Neuronal responses in the GNG task. Conventions are the same as in Fig. 3. No change occurs in the neuron's discharge rate when the circle appears on the right side of the display screen in the cue epoch. A significant increase in discharge rate occurs when the opposite circle appears in the no-go condition, and an equivalent increase occurs when the cued circle appears in the go condition. Comparable results were obtained when the cued circle was on the left.

6. ACTIVITY IN THE NBM RELATED TO REWARD

The above studies indicate that the responses of most NBM neurons are not determined by specific sensory stimuli or motor acts. Rather, NBM neuronal responses appear to be strongly influenced by the conditions in which stimuli or movements occur. In the GNG task, the appearance of a side circle in the choice no-go condition produced a more robust response in NBM cells than did the same stimulus in the cuing condition. Similarly, in the DR task, an arm movement in the choice condition was accompanied by a larger NBM response than essentially the same movement in the cuing condition. Furthermore, the actual movements that occur in the choice condition do not appear to be essential for the enhanced neuronal responses, since equivalent responses occur in the go and no-go conditions of the GNG task. Thus, specific sensory or motor events do not appear to be responsible for the robust changes in discharge rate of NBM neurons in the choice phases of these tasks.

One possible explanation for the enhanced NBM responses in the choice condition is that the actions of the animal depend on remembered information. In the DR task, the choice movement was to a remembered location, whereas the cuing movement was to a specified target. Likewise, the go or no-go decisions in the GNG task depended on memory for the previously cued location on the display screen. However, this hypothesis cannot account for the high percentage (71%) of NBM responses associated with the movement in the VMT task, which was not dependent on remembered information.

A feature that is common to the choice phases of the DR and GNG tasks and the arm movement phase of the VMT task is that each of these phases immediately precedes the reward phase in which water is delivered to the animal. That is, in each task, a consistently large number of NBM neuronal responses occurred in the epoch that preceded the reward delivery. It appears likely that this close association with the reward delivery may account for the enhancement of NBM responses in the choice conditions of the DR and GNG tasks as well as the movement phase of the VMT task. Hence, NBM neurons may be particularly sensitive to any event (such as a visual stimulus or limb movement) that consistently precedes a reward.

Additional support for this hypothesis was obtained in the VMT task. A number of NBM neurons had significantly larger changes in discharge rate with the perturbation than with the torque. Again, an obvious difference between the torque and the perturbation that may account for the enhanced NBM neuronal responses is that the perturbation directly precedes the reward delivery, whereas the torque does not.

Rolls and associates have also found a relationship between neuronal activity in the region of the NBM and appetitive stimuli in awake, behaving primates (Rolls *et al.*, 1976, 1979). They have described a population of neurons in the lateral hypothalamus and substantia innominata (which contains much of the NBM) that respond selectively to visual stimuli that are associated with food or juice. These neurons developed differential responses to appetitive and aversive stimuli as the animal learned to discriminate behaviorally between the stimuli, i.e., to accept the appetitive stimulus (fruit juice) and reject the aversive stimulus (saline) (Mora *et al.*, 1976). Furthermore, these cells ceased to respond to a visual stimulus when it was no longer followed by a reward (Mora *et al.*, 1976) or when the animal became satiated (Burton *et al.*, 1976). More recently, they have found a small population of neurons within the NBM that respond differentially to visual stimuli that precede delivery of appetitive or aversive stimuli (Wilson *et al.*, 1984). The findings of Rolls's group are consistent with ours in that NBM neurons are found to be responsive to stimuli that immediately precede a reward delivery. Although Rolls and associates appear to find a much smaller percentage (typically less than 10%) of such neurons, this apparent discrepancy may be related to differences in behavioral paradigms and data analysis methods.

Thus, considerable evidence indicates that NBM neurons are particularly responsive to stimuli that precede food or water rewards. There is also substantial evidence that NBM neurons are highly responsive to the reward presentation itself. In an early study of NBM neuronal activity, DeLong (1971) recorded NBM neurons in monkeys trained to make a series of repeated arm movements to obtain a water reward. In that study, sampled NBM neurons had altered discharge rates primarily following the delivery of a water reward. In each of the three tasks described above, approximately half of the task-related NBM neurons had altered discharge rates following the reward delivery. Furthermore, during the GNG task, some NBM cells responded to water deliveries that were not contingent on the animal's behavior. However, in all of these studies, a distinctive click sound originated from the solenoid valve that dispensed the water reward. It was therefore not possible to determine whether the neuronal responses were associated with the auditory stimulus or the reward itself. Further studies are needed to address this issue.

7. POSSIBLE RELATIONS TO CONDITIONING

The close association between NBM neuronal activity and food or water rewards suggests that this structure may be related to the positive reinforcement properties of these rewards. We postulate that changes in the discharge rates of NBM neurons near the time of reward delivery may contribute to the development of appetitively conditioned responses. Consider the fact that the NBM projects to the cortical area that is involved in the conditioning of an eyeblink response in cats (Woody, 1982). A key component of the neural basis of this conditioned response appears to be an increase in the responsiveness of cortical neurons as a result of increased input resistance. A similar increase in input resistance can result from iontophoretic application of ACh to a cortical neuron that is being stimulated intracellularly (Woody *et al.*, 1978). This effect can persist for over an hour. Similarly, a brief application of ACh to a cortical neuron that is concurrently being stimulated by glutamate or afferent stimulation can also alter the cell's responsiveness for over an hour (Metherate *et al.*, 1985). Thus, ACh may induce long-lasting changes in a neuron's responsiveness if applied when the neuron is firing faster than its normal rate.

In an appetitive conditioning paradigm, cortical neurons may develop enhanced responses to afferent input as a result of the timing of ACh release onto that neuron. For example, consider a simple shaping paradigm in which a monkey is given a water reward each time it flexes its arm at the elbow. The movement is accompanied by an increase in the discharge rates of certain motor cortex neurons, and the subsequent reward is followed by an increase in the discharge rates of certain NBM neurons. The increase in NBM activity may increase ACh release onto numerous cortical neurons (Casamenti *et al.*, 1986), including the motor cortex cells that discharge more rapidly during the elbow flexion movement. The combination of increased ACh release and increased cell firing may occur close enough in time to enhance the responsiveness of the cortical neurons, making them more likely to discharge in the future. This would increase the probability that the specific arm movement would occur again, hence shaping that specific movement. This simple example could be expanded to apply to a number of appetitive conditioning paradigms and could also account for stimulus and response specificity. In each case, the crucial feature would be the coupling of increased ACh release with increased cortical cell firing to enhance the cell's responsiveness.

Despite the speculation that the NBM may be involved in positive reinforcement, other possible functions of this structure must not be overlooked. For example, it has been suggested that cholinergic neurons of the basal forebrain form the most anterior extent of the reticular activating system (Das and Kreutzberg, 1968; Das, 1971) and is involved in the desynchronization of cortical EEG activity (Shute and Lewis, 1967). Lesions of this anatomic region in rodents have produced significant alterations in cortical EEG patterns that are consistent with this hypothesis (LoConte *et al.*, 1982; Stewart *et al.*, 1984). Our studies have also provided indications that the role of the NBM may not be restricted to positive reinforcement. A large number of NBM neurons have been found to respond to the application of a torque or perturbation with latencies as brief as 40–50 msec (Mitchell *et al.*, 1987a). As mentioned earlier, NBM responses to a perturbation appear to be modified by its association with an upcoming reward, but the significance of the torque responses, which are unassociated with rewards, remain obscure. The significance of the short latency of these responses is also unclear. Hence, the NBM does not appear to be solely related to aspects of positive reinforcement.

In summary, the activity of NBM neurons has been recorded in awake, behaving monkeys engaged in three different tasks. Significant increases or decreases in discharge rates were most commonly associated with events (visual stimuli or limb movements) that immediately preceded a reward delivery or with the reward delivery itself. Identical events that occurred earlier in a trial and did not precede a reward delivery elicited fewer and smaller responses in NBM neurons. These results have led to speculation that NBM neuronal activity may be related to positive reinforcement and may facilitate the neural changes that underlie appetitively conditioned responses.

ACKNOWLEDGMENTS. The authors wish to acknowledge the assistance of Denise Wienecke and Melissa Zebley in the preparation of the manuscript and the valuable comments of Dr. Garrett Alexander on the content of the manuscript. Support for these studies are provided by the National Institutes of Health (NS 15417, NS 20471, and NS 07269).

REFERENCES

Arendt, T., Bigl, V., and Tennstedt, A., 1983, Loss of neurons in the nucleus basalis of Meynert in Alzheimer's disease, paralysis agitans and Korsakoff's disease, *Acta Neuropathol.* **61:**101–108.

Bartus, R. T., 1978, Evidence for a direct cholinergic involvement in the scopolamine-induced amnesia in monkeys: Effects of concurrent administration of physostigmine and methylphenidate with scopolamine, *Pharmacol. Biochem. Behav.* **9:**833–836.

Bartus, R. T., and Johnson, H. R., 1976, Short-term memory in the rhesus monkey: Disruption from the anti-cholinergic scopolamine, *Pharmacol. Biochem. Behav.* **5:**39–46.

Buresova, O., Bures, J., and Bohdanecky, Z., 1964, Effect of atropine on learning, extinction, retention, and retrieval in rats, *Psychopharmacology* **5:**255–263.

Burton, M. J., Rolls, E. T., and Mora, F., 1976, Effects of hunger on the responses of neurons in the lateral hypothalamus to the sight and taste of food, *Exp. Neurol.* **51:**668–677.

Casamenti, F., Deffenu, G., Abbamondi, A. L., and Pepeu, G., 1986, Changes in cortical acetylcholine output induced by modulation of the nucleus basalis, *Brain Res. Bull.* **16:**689–695.

Das, G. D., 1971, Projections of the interstitial nerve cells surrounding the globus pallidus: A study of retrograde changes following cortical ablations in rabbits, *Z. Anat. Entwickl. Gesch.* **133:**135–160.

Das, G. D., and Kreutzberg, G. W., 1968, Evaluation of interstitial nerve cells in the central nervous system: A correlative study using acetylcholinesterase and Golgi techniques, *Ergeb. Anat. Entwickl. Gesch.* **41:**1–58.

Davies, P., and Maloney, A. J. F., 1976, Selective loss of central cholinergic neurons in Alzheimer's disease, *Lancet* **2:**1403.

DeLong, M. R., 1971, Activity of pallidal neurons during movement, *J. Neurophysiol.* **34:**414–427.

Deutsch, J. A., 1971, The cholinergic synapse and the site of memory, *Science* **174:**788–794.

Drachman, D. A., 1977, Memory and cognitive function in man: Does the cholinergic system have a specific role, *Neurology (Minneap.)* **27:**783–790.

Dunnett, S. B., 1985, Comparative effects of cholinergic drugs and lesions of nucleus basalis of fimbria–fornix on delayed matching in rats, *Psychopharmacology* **87:**357–363.

Fuster, J. M., 1973, Unit activity in prefrontal cortex during delayed-response performance: Neuronal correlates of transient memory, *J. Neurophysiol.* **36:**61–78.

Fuster, J. M., and Alexander, G. E., 1970, Delayed response deficit by cryogenic depression of frontal cortex, *Brain Res.* **20:**85–90.

Georgopoulos, A. P., DeLong, M. R., and Crutcher, M. D., 1983, Relations between parameters of step-tracking movements and single cell discharge in the globus pallidus and subthalamic nucleus of the behaving monkey, *J. Neurosci.* **3:**1586–1598.

Goldman, P. S., and Rosvold, H. E., 1970, Localization of function within the dorsolateral prefrontal cortex of the rhesus monkey, *Exp. Neurol.* **27:**291–304.

Hepler, D. J., Olton, D. S., Wenk, G. L., and Coyle, J. T., 1985, Lesions in nucleus basalis magnocellularis and medial septal area of rats produce qualitatively similar memory impairments, *J. Neurosci.* **5:**866–873.

Johnston, M. V., McKinney, M., and Coyle, J. T., 1981, Neocortical cholinergic innervation: A description of extrinsic and intrinsic components in the rat, *Exp. Brain Res.* **43:**159–172.

Kubota, K., and Funahashi, S., 1982, Direction-specific activities of dorsolateral prefrontal and motor cortex pyramidal tract neurons during visual tracking, *J. Neurophysiol.* **47:**372–376.

Lamour, Y., Dutar, P., Rascol, O., and Jobert, A., 1986, Basal forebrain neurons projecting to the rat frontoparietal cortex: Electrophysiological and pharmacological properties, *Brain Res.* **362:**122–131.

Lehman, J., Nagy, J. I., Armadja, S., and Fibiger, H. C., 1980, The nucleus basalis magnocellularis: The origin of a cholinergic projection to the neocortex of the rat, *Neuroscience* **5:**1161–1174.

LoConte, G., Bartolini, L., Casamenti, F., Marconcini-Pepeu, I., and Pepeu, G., 1982, Lesions of cholinergic forebrain nuclei: Changes in avoidance behavior and scopolamine actions, *Pharmacol. Biochem. Behav.* **17:**933–937.

Mesulam, M., Mufson, E. J., Levey, A. I., and Wainer, B. H., 1983, Cholinergic innervation of cortex by the basal forebrain: Cytochemistry and cortical connections of the septal area, diagonal band nuclei, nucleus basalis (substantia innominata), and hypothalamus in the rhesus monkey, *J. Comp. Neurol.* **214:**170–197.

Metherate, R., Tremblay, N., and Dykes, R. W., 1985, Changes in neuronal function produced in cat primary somatosensory cortex by the iontophoretic application of acetylcholine, *Soc. Neurosci. Abstr.* **11:**753.

Mewaldt, S. P., and Ghoneim, M. M., 1979, The effects and interactions of scopolamine, physostigmine, and methamphetamine on human memory, *Pharmacol. Biochem. Behav.* **10**:205–210.
Mitchell, S. J., Richardson, R. T., Baker, F. H., and DeLong, M. R., 1987a, The primate nucleus basalis of Meynert: Neuronal activity related to a visuomotor tracking task, *Exp. Brain Res.* (in press).
Mitchell, S. J., Richardson, R. T., Baker, F. H., and DeLong, M. R., 1987b, The primate globus pallidus: Neuronal activity related to direction of movement, *Exp. Brain Res.* (in press).
Mora, F., Rolls, E. T., and Burton, M. J., 1976, Modulation during learning of the responses of neurons in the lateral hypothalamus to the sight of food, *Exp. Neurol.* **53**:508–519.
Niki, H., and Watanabe, M., 1976, Prefrontal unit activity and delayed response. Relation to cue location versus direction of response, *Brain Res.* **105**:79–88.
Pearson, R. C. A., Gatter, K. C., Brodal, P., and Powell, T. P. S., 1983, The projection of the basal nucleus of Meynert upon the neocortex in the monkey, *Brain Res.* **259**:132–136.
Perry, E., Perry, R., Blessed, G., and Tomlinson, B., 1977, Necropsy evidence of central cholinergic deficits in senile dementia, *Lancet* **1**:189.
Richardson, R. T., and DeLong, M. R., 1984, Activity of nucleus basalis of Meynert neurons during a delayed response task, *Soc. Neurosci. Abstr.* **10**:128.
Richardson, R. T., and DeLong, M. R., 1986, Differential responses of nucleus basalis of Meynert neurons in a go/no-go task in monkey (*Macaca mulatta*), *Soc. Neurosci. Abstr.* **12**:356.
Richardson, R. T., and DeLong, M. R., 1986, Nucleus basalis of Meynert neuronal activity during a delayed response task in monkey, *Brain Res.* **399**:364–368.
Richardson, R. T., Mitchell, S. J., Baker, F. H., and DeLong, M. R., 1983, Activity of neurons in the macaque nucleus basalis of Meynert in a visuomotor tracking task, *Soc. Neurosci. Abstr.* **9**:951.
Ridley, R. M., Bowes, P. M., Baker, H. F., and Crow, T. J., 1984, An involvement of acetylcholine in object discrimination learning and memory in the marmoset, *Neuropsychologia* **22**:252–263.
Rogers, J. D., Brogan, D., and Mirra, S. S., 1985, The nucleus basalis of Meynert in neurological disease: A quantitative morphological study, *Ann. Neurol.* **17**:163–170.
Rolls, E. T., Burton, M. J., and Mora, F., 1976, Hypothalamic neuronal responses associated with the sight of food, *Brain Res.* **111**:53–66.
Rolls, E. T., Sanghera, M. K., and Roper-Hall, A., 1979, The latency of activation of neurones in the lateral hypothalamus and substantia innominata during feeding in the monkey, *Brain Res.* **164**:121–135.
Rosenkilde, C. E., 1979, Functional heterogeneity of the prefrontal cortex in the monkey: A review, *Behav. Neurol. Biol.* **25**:301–345.
Rosvold, H. E., and Szwarcbart, M. K., 1964, Neural structures involved in delayed-response performance, in: *The Frontal Granular Cortex and Behavior* (K. Akert, ed.), McGraw-Hill, New York, pp. 1–15.
Shute, C. C. D., and Lewis, P. R., 1967, The ascending cholinergic reticular system: Neocortical, olfactory and subcortical projections, *Brain* **90**:497–519.
Squire, L. R., Glick, S. D., and Goldfarb, J., 1971, Relearning at different times after training as affected by centrally and peripherally acting cholinergic drugs in the mouse, *J. Comp. Physiol. Psychol.* **74**:41–45.
Stewart, D. J., Macfabe, D. F., and Vanderwolf, C. H., 1984, Cholinergic activation of the electrocorticogram: Role of the substantia innominata and effects of atropine and quinuclidinyl benzilate, *Brain Res.* **322**:219–232.
White, P., Goodhardt, M. J., and Keet, J. P., 1977, Neocortical cholinergic neurons in elderly people, *Lancet* **1**:668–671.
Whitehouse, P. J., Price, D. L., Struble, R. G., Clark, A. W., Coyle, J. T., and DeLong, M. R., 1982, Alzheimer's disease and senile dementia: Loss of neurons in the basal forebrain, *Science* **215**:1237–1239.
Wilson, F. A., Rolls, E. T., Yaxley, S., Thorpe, S. J., Williams, G. V., and Simpson, S. J., 1984, Responses of neurons in the basal forebrain of the behaving monkey, *Soc. Neurosci. Abstr.* **10**:128.
Woody, C. D., 1982, Acquisition of conditioned facial reflexes in the cat: Cortical control of different facial movements, *Fed. Proc.* **41**(6):2160–2168.
Woody, C. D., Swartz, B. E., and Gruen, E., 1978, Effects of acetylcholine and cyclic GMP on input resistance of cortical neurons in awake cats, *Brain Res.* **158**:373–395.

17

Neurophysiological Investigations of Tonic Mechanisms of Conditioning

BORIS I. KOTLYAR and NATALYA O. TIMOFEEVA

1. INTRODUCTION

Tonic phenomena accompanying acquisition and performance of conditioned reflexes (CRs) have been noted repeatedly during neurophysiological investigations of temporal connections at different stages of their development. Manifestations of tonic phenomena in conditioning have been described physiologically as first- and second-order reflexes preceding the CR and adjusting brain structures for appropriate performance (Kupalov, 1963). Such adjustments may serve (1) to create a state of readiness for CR performance, i.e., as tonic reflexes developed during switching of the meaning of the conditioned stimulus (CS) signal (Asratyan, 1963), (2) as preparatory reflexes arising on the basis of interconnections among experimental conditions, motivation, and the CS (Konorski, 1967), or (3) as physiological processes supporting evaluation of the key stimulus and decision making, i.e., anticipatory integration, during which evaluation of environmental information, motivation, memory, and emotions takes place (Anokhin, 1968).

In recent years, there has been growing interest in identifying tonic mechanisms involved in associative processes. These mechanisms underlie the tonic nature of acquisition and performance of CRs and involve all types of tonic reflexes mentioned. Without identification of the fundamental physiological mechanisms, there can be no adequate explanation for such properties of conditioning as (1) the latent period and learning acquisition dynamics, (2) the sizable CR latency, which is much greater than the time needed for spreading of excitation through a simple reflex arc, or (3) the absence of conditioned reaction in response to some presentations of the conditioned stimulus in spite of a high background motivational level.

In the present chapter we report results of neurophysiological investigations that have allowed us to identify tonic mechanisms of conditioning and demonstrate their widespread significance in various types of associative phenomena.

BORIS I. KOTLYAR and NATALYA O. TIMOFEEVA • Department of Higher Nervous Activity, Lomonosov Moscow State University, Moscow, USSR.

2. TONIC ACTIVITY OF NEURONS DURING CR ACQUISITION

The activity of 72 neurons in dorsal hippocampus was recorded extracellularly at different stages of avoidance CR acquisition in unrestrained adult male rabbits (Kotlyar *et al.*, 1981). After an isolated CS (400-Hz tone), electrocutaneous stimulation of the ear was performed. Reinforcement was abolished when the animal completed a shaking movement of the ear during the isolated action (4 sec) of the signal. Changes in tonic neuronal activity (mean firing frequency during interstimulus intervals) were noted after one to five paired trials. At the first stage of CR acquisition (when only 50–60% of CRs occurred), the tonic firing frequency increased significantly in 75% of recorded units with each paired trial. In 12.5% of units, a decrease in firing frequency was noted. At this stage the changes in phasic neuronal responses to the CS, acquiring signal meaning, were noted, as was a correlation between the neuronal activity pattern and the appearance of motor responses. Pseudoconditioning procedures did not lead to such changes.

After a level of 50–60% correctly performed conditioned movements had been achieved, changes in tonic activity were over, and during the next stage of CR acquisition (70–100% of conditioning performance), the frequency of tonic neuronal activity fluctuated near the base-line mean without a tendency to increase or decrease. For most neurons (83.3%) investigated at this stage, a significant difference was established between tonic firing frequency preceding CR performance and the firing frequency in the absence of performance. During the consolidation stage of CR acquisition, tonic neuronal activity appeared to be a precise indicator of the absence or delay of conditioning. Extinction of the CR was accompanied by a significant change of tonic neuronal activity, which could be manifested as either an increase or a decrease (in different neurons) in the mean firing frequency. After full extinction of the CR, no changes in tonic activity were noted between presentations of nonreinforced stimuli. In every unit investigated during restoration of avoidance CRs, the tonic firing frequency underwent changes that were opposite to the changes in activity noted in the same unit during extinction.

Similar results were obtained in hippocampal neurons of rabbits during operant food conditioning during acquisition of specialization of conditioned feeding (Kotlyar and Timofeeva, 1983). If, during an isolated CS (5 sec), the rabbit performed a complex movement, it received a food reinforcement. Most hippocampal neurons investigated showed significant changes in reactivity to the CS as well as in the rate of tonic activity, depending on the stage of CR acquisition.

3. TONIC NEURONAL ACTIVITY DURING CONDITIONING WITH CHANGEABLE REINFORCEMENT AND SWITCHING OF CRs

The dynamics of EEG activity recorded during modifications of the animal's behavior allow one to trace patterns of the brain's tonic state development using the level of brain activation to characterize "energetic" aspects of this process.

Qualitative specifics of this state can be disclosed by investigating unit activity during acquisition of different types of CRs. Activity of hippocampal neurons was investigated during conditioning with changeable reinforcement. The aim of these experiments was the identification within single neurons of changes accompanying any type of CR and specific changes attributable only to a specific type of conditioning. After elaboration of feeding and defensive operant CRs, 90.3% of neurons manifested significant changes in tonic activity as well as in responses to a CS (sound, light) that acquired new meaning after pairing. These data demonstrate that different types of conditioning occur at different levels of tonic base-line rates of neuronal activity.

In a specially designed series of experiments, unit activity was investigated in different brain structures during conditioned switching of operant and classical CRs with changeable reinforcement (Kotlyar *et al.*, 1985; Timofeeva *et al.*, 1984). The term "switching of CRs" refers to the experimental procedure in which one conditioned stimulus possesses two or more signal meanings, each of which can be manifested as an independent CR. According to this procedure, switching of CRs was signaled by an additional stimulus, single food or ear-shock presentations in our experiments. Analysis of unit activity revealed regularities in changes of tonic activity during switching of operant CRs. Comparison of neuronal activity dynamics with the behavior of the animals demonstrated a correlation among the level of acquisition of each reflex that underwent switching, the number of switchings, and patterns of changes in unit activity.

Switching of operant (food and defensive) reflexes underwent several stages. When defensive CRs were not acquired, when both CRs were impaired, or when only a defensive reflex was present, switching of the CS meaning led to distinctive changes in tonic unit activity. This process manifested itself in significant changes of firing frequency (during the interstimulus period) that depended on the numbers of conditioned stimulus and changed reinforcement pairings, e.g., electric skin stimulation or food. In averages made after 15–20 paired trials, a new level of tonic activity was observed that differed significantly from the rate characteristic of the previous reflex. The moment at which the new level could be detected coincided with the first conditioned responses of the animal that were appropriate for the new conditions as well as with new reactions of recorded units to the conditioned stimulus. At the stage of switching, characterized by 50–60% or more correct CRs, in both experimental paradigms (food or defensive reinforcement), significant changes of tonic activity were found in 81.8% of 37 investigated hippocampal neurons. It was noted that the rate of neuronal firing changed suddenly immediately after presentation of the switched CS. Associative plastic changes in neuronal responses to the CS as well as significant differences in responses to the same CS after switching were maximally expressed at this stage.

Frequencies of tonic firing in units that exhibited plastic changes of response to the CS and were recorded during switching differed significantly when tonic activity was followed by the correct conditioned reaction and when the CR was absent in defensive reflexes as well as in food reflexes. In addition, a correlation was established during interstimulus periods between significant changes of tonic unit activity depending on switching of CS meaning and significant changes in responses of the same neuron to the CS in either situation. Changes in responses to the CS were observed in 92% of recorded neurons after switching, as determined by changes in tonic activity. During repeated switching, it was found that each switching procedure was accompanied by a specific pattern of tonic activity in hippocampal neurons that was characteristic of the respective CS (Timofeeva *et al.*, 1984). In classical conditioning, reflex switching from food CRs to defensive CRs produced changes in tonic activity and phasic responses of 52 units recorded from hippocampus and visual and parietal association cortex. These changes in activity were similar to those occurring during operant conditioning (Kotlyar and Timofeeva, 1983; Kotlyar *et al.*, 1985). The number of units that exhibited plastic changes in tonic activity was 67.3% in hippocampus, 44.9% in parietal cortex, and 58.5% in visual cortex (Table I). In the remainder of the recorded cells, the switching of CRs elicited nonsignificant changes in firing frequency, but there was a tendency for a new level of tonic base-line activity to be present.

TABLE I. Frequencies of Neuronal Tonic Activity in Archicortex and Neocortex and Their Differences while Switching from Food to Defensive CR

Neuron	Food CR (impulses/sec)	Avoidance CR (impulses/sec)	Reliability of differences
Hippocampus			
1	2.9	2.0	$P < 0.05$
2	2.1	0.97	$P < 0.01$
3	6.9	1.4	$P < 0.001$
4	3.0	6.2	$P < 0.01$
5	7.3	5.5	$P < 0.05$
6	3.2	1.6	$P < 0.05$
7	10.8	6.5	$P < 0.05$
8	7.8	2.9	$P < 0.01$
9	10.8	8.4	$P < 0.05$
10	4.3	0.7	$P < 0.01$
11	11.8	5.3	$P < 0.01$
12	7.5	11.4	$P < 0.01$
13	4.4	0.5	$P < 0.01$
14	15.3	10.7	$P < 0.01$
15	3.6	10.6	$P < 0.01$
Parietal cortex			
1	4.2	1.7	$P < 0.01$
2	6.8	1.6	$P < 0.01$
3	10.5	6.1	$P < 0.05$
4	2.3	6.9	$P < 0.01$
5	8.0	3.9	$P < 0.05$
6	2.3	4.2	$P < 0.01$
7	11.4	5.7	$P < 0.01$
8	4.8	1.1	$P < 0.001$
9	8.9	5.4	$P < 0.01$
10	14.2	25.2	$P < 0.05$
11	12.1	3.5	$P < 0.01$
12	3.8	1.3	$P < 0.005$
13	3.7	2.2	$P < 0.01$
14	2.3	4.2	$P < 0.01$
15	5.6	2.6	$P < 0.01$
Visual cortex			
1	5.8	0.6	$P < 0.01$
2	10.6	4.1	$P < 0.01$
3	21.6	10.2	$P < 0.01$
4	10.7	3.1	$P < 0.001$
5	14.0	7.7	$P < 0.01$
6	5.2	1.5	$P < 0.01$
7	5.7	9.2	$P < 0.01$
8	5.1	1.3	$P < 0.01$
9	11.5	9.7	$P < 0.05$
10	6.7	3.4	$P < 0.05$
11	6.8	14.7	$P < 0.01$
12	14.3	24.0	$P < 0.01$
13	15.8	3.8	$P < 0.01$
14	15.1	11.3	$P < 0.01$
15	2.0	7.8	$P < 0.01$

As in the case of operant conditioning, repeated switching of classical CRs led to reappearance of tonic activity level and phasic responses characteristic of the CS for neurons in all structures investigated. Analysis of tonic firing level contributions to the phasic response changes revealed that in 73.1% of hippocampal units, 73.9% of units of the parietal cortex, and 75% of units of the visual cortex, the changes in response evoked by switching were caused by changes in levels of tonic activity.

Conditioned responses of the animals as well as emergence of conditioned neuronal responses, therefore, coincide with certain levels of tonic activity. The present data suggest that this aspect of tonic activity changes, observed in all investigated brain structures, characterizes a process involved in the formation of temporal connections and switching of CRs as well. Levels of tonic activity reflect levels of excitability in the nervous system necessary for a given reflex performance.

Investigation of neurophysiological mechanisms of classical and operant CR switching has allowed us to evaluate the contribution of several factors that affect tonic neuronal activity. Properties of the CS, the type of motivation underlying CR acquisition, and the properties of conditioned effector responses were studied. Analysis of tonic neuronal activity during switching of CRs revealed that motivation is a major factor that determines the level of tonic base-line activity of neurons of the brain (Kotlyar and Timofeeva, 1982).

The experimental environment was a constant factor during classical CR switching, whereas reinforcement was a variable factor. A single presentation of food or an electric stimulus as a switching factor signified a transition to a new conditioned situation, changed the motivational status of the animal, and created a new state of brain neurons characteristic of the current motivation and forthcoming reinforcement. This transition was manifested as a new level of tonic firing and a new phasic response to the CS observed in recorded neurons. These results imply a leading role of motivational excitation underlying a given reflex in formation of tonic neuronal activity and neuronal response to the CS. Created in this way, the specific level of tonic neuronal activity determines the choice of the conditioned behavior program.

4. CONCLUSION

These neurophysiological investigations have disclosed tonic mechanisms of conditioning considered to be a necessary neurophysiological base for acquisition and performance of temporal connections of different complexity.

Investigations of neuronal activity in the brain demonstrate that in addition to the associative changes caused by stimulus pairing, changes in tonic activity regularly occur that are maximally expressed during elaboration and extinction of CRs, during changes in conditioning with changeable reinforcement, and during switching of CRs.

In this connection, it is evident that the latent period of learning depends on the formation of specific tonic states within brain structures, i.e., a tonic state of readiness, which forms the background from which a specific CR can develop. If the necessary level of tonic activity is not attained, the CR does not manifest itself.

Reproduction of tonic neuronal activity levels characteristic of a given type of temporal connection formed during replacement of CRs elaborated to a different CS using changeable reinforcement as well as during repeated switching of operant and classical reflexes provides evidence that various levels of tonic neuronal activity formed during learning are of great importance for all types of conditioning.

Thus, tonic mechanisms of conditioning are manifested as changes in tonic neuronal activity in different brain structures during various associative phenomena. Associative changes in tonic neuronal activity precede development of other forms of conditioned reactions and are necessary for elaboration of any temporal connection. Conditioning can manifest itself with production of a CR only when tonic neuronal activity achieves the level specific for its realization. Different types of associative activity correlate strongly with different levels of tonic activity in the majority of cortical neurons studied. Dominating motivation is the main factor for determining the tonic activity level characteristic of a given CR.

REFERENCES

Anokhin, P. K., 1968, *Biology and Neurophysiology of the Conditioned Reflex*, Meditsina, Moscow.

Asratyan, E. A., 1963, Tonic conditioned reflexes as a form of integral activity, *Zh. Vyssh. Nervn. Deyat.* **13:**781–788.

Konarski, J., 1967, *Integrative Activity of the Brain*, University of Chicago Press, Chicago.

Kotlyar, B. I., and Timofeeva, N. O., 1982, Electrical activity of hippocampus during different forms of motivated behaviour, in: *Recent Developments of Neurobiology in Hungary*, Vol. 10 (K. Lissak, ed.), Akademiai Kiado, Budapest, pp. 269–280.

Kotlyar, B. I., and Timofeeva, N. O., 1983, Tonic component of conditioning and its functional role, *Zh. Vyssh. Nervn. Deyat.* **33:**1059–1066.

Kotlyar, B. I., Timofeeva, N. O., and Semikopnaya, I. I., 1981, Dynamics of hippocampal neuron activity during formation of avoidance conditioning, *Zh. Vyssh. Nervn. Deyat.* **31:**521–530.

Kotlyar, B. I., Timofeeva, N. O., and Popovich, L. D., 1985, Phenomenon of conditioned reflex switching: Neurophysiological mechanisms, *Vestnic Moscow Univ.* [*Biol.*]. **16:**3–16.

Kupalov, P. S., 1963, Reflectory theory and its perspectives, in: *Philosophy Aspects of Physiology of Higher Nervous Activity and Psychology*, Publishing House of the Soviet Academy of Sciences, Moscow, pp. 106–155.

Timofeeva, N. O., Kotlyar, B. I., and Popovich, L. D., 1984, Analysis of the neuronal mechanism of conditioning reflex switching, *Neurosci. Behav. Physiol.* **14:**146–152.

18

Hippocampal Units during Single-Alternation Conditioning in the White Rat

TOSHITSUGU HIRANO

1. INTRODUCTION

The response of animals to environmental stimuli is influenced not only by the physical properties of the stimulus but also by its acquired behavioral significance through learning. It is also true that even physically similar stimuli could exert different influences on the animals, depending on the occasion, once they have acquired meaning as stimuli. The experimental design of temporal single-alternation (SA) paradigm is such that the task requires the animal to process internal cues based on aftereffects of a prior stimulus, whether it was a reinforced (R) or a nonreinforced (N) trial.

In a recent study using classical conditioning of the rabbit nictitating membrane response, Hoehler and Thompson (1979) found that an SA schedule produced clear discrimination of the differential significance of R and N trial stimulus aftereffects in hippocampal unit activity in which two consequences were recognized: (1) pre-CS background unit activity levels that were lower on R trials than on N trials and (2) levels of unit activity in the CS periods that were higher on R trials than on N trials. These results were interpreted as supporting the notion of a critical role for the hippocampus in working memory. To examine the generality of their findings, we conducted experiments to study hippocampal function in a rewarding type of conditioning. It was also of interest to use rats as subjects, since this species is known to exhibit a well-developed patterning behavior (Capaldi, 1967). Thus, if hippocampal function were related to the memory process, we might expect to find differential unit activity developing in the hippocampus over the course of appetitive conditioning with the SA paradigm.

2. SINGLE ALTERNATION WITH REWARDING BRAIN STIMULATION

In a previous study (Hirano, 1984), we found that conditioning of an auditory stimulus reinforced by rewarding brain stimulation could be used as a paradigm for studying unit

TOSHITSUGU HIRANO • Department of Psychology, Faculty of Letters, Kyoto University, Sakyoku, Kyoto 606, Japan.

changes in different structures. We applied this conditioning procedure to a temporal SA paradigm following pseudoconditioning training (Hirano and Yamaguchi, 1985). All unit recordings were obtained from semimicroelectrodes chronically implanted in the dorsal hippocampus and related areas. Each animal's movement was recorded through an output of voltage by an open-ended wire, which was wrapped around a bundle of cables running from the rat's head. Bipolar stimulating electrodes were implanted in the lateral hypothalamus. Those animals that showed positive rewarding effects to hypothalamic stimulation in a self-stimulation test served as subjects.

Classical conditioning was run on three successive days. During the first session on the first day, pseudoconditioning was given, in which the CS (auditory stimulus, 2 KHz, 1 sec) and US (lateral hypothalamic stimulation) were randomly delivered with an average interstimulus interval (ITI) of 40 sec. During the second training session on day 1 and thereafter, a schedule of alternating R and N trials with CS–US onset intervals of 500 msec and ITI of 60 sec was followed. The animals were given 150 R and 150 N trials in each 6-hr session, with R trials regularly alternated with N trials.

Multiple unit activity and animal's movements were sampled for 480 msec from the onset of the CS. A poststimulus histogram for every session in terms of Z scores (Olds *et al.*, 1972) for successive intervals (480 msec in 24-msec bins) was plotted for each unit response, and the movement data are shown in Fig. 1. To compare characteristic features, if any, for unit locations, further analysis of the grouped data was obtained. For this purpose, the 480-msec period after the CS presentation was divided into four quarters of 120 msec each, and averaged groupings in terms of Z scores were calculated for each group. The summary analysis is shown in Fig. 2.

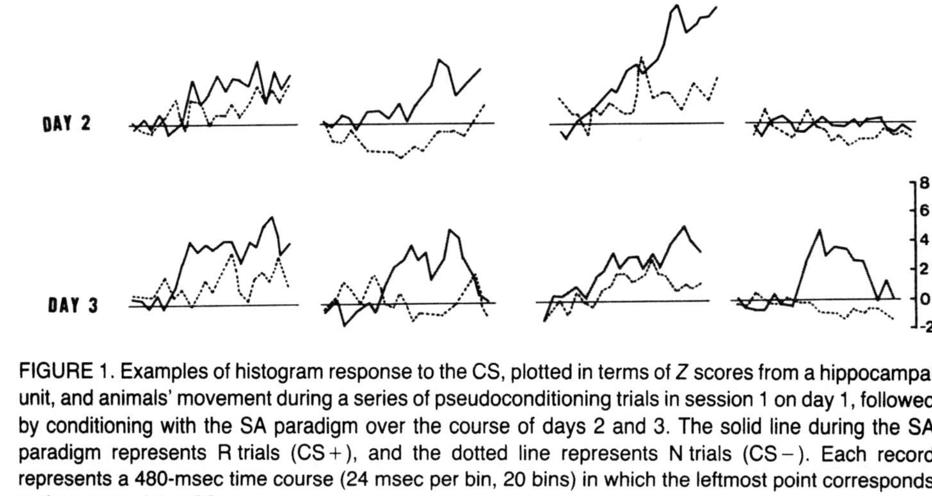

FIGURE 1. Examples of histogram response to the CS, plotted in terms of Z scores from a hippocampal unit, and animals' movement during a series of pseudoconditioning trials in session 1 on day 1, followed by conditioning with the SA paradigm over the course of days 2 and 3. The solid line during the SA paradigm represents R trials (CS+), and the dotted line represents N trials (CS−). Each record represents a 480-msec time course (24 msec per bin, 20 bins) in which the leftmost point corresponds to the onset of the CS.

The result of the experiment indicates a differential relationship of hippocampal unit activity to the behavioral change found in the session immediately following a shift from pseudoconditioning to the SA paradigm. Although there was no change in behavior after training with the SA paradigm on day 1, dentate and hippocampal unit activity significantly increased their firing rates to the CS. The increase in the unit response in these areas was, at an early stage, a generalized excitatory pattern for both N and R trials, indiscriminately. This seems to have reflected the function of the dentate–hippocampus in monitoring any change in the environmental stimulus contingencies. Following SA training, the animals acquired a differential movement response gradually over training by the third day of the experiment. The conditioned movement of the animals developed with an increased response to the paired stimulus (R trials) and an inhibitory response to the unpaired stimulus (N trials). In contrast to this inhibitory pattern in the movement data, hippocampal and dentate units acquired an enhanced response to the CS. This was found even on N trials, although to a lesser degree than on R trials. Thus, the differential unit responses were acquired through generalized excitation of the nonpaired stimulus at full strength in early phases and its gradual decrease across training.

The finding of higher levels of unit activity in the CS period on R trials than on N trials is consistent with a previous study of the rabbit nictitating membrane response (Hoehler and Thompson, 1979). However, when we examined the pre-CS level of R and N trials in the present experiment, no significant differences were detected. This failure to find lower pre-CS background unit activity levels prior to the R trials than prior to N trials contradicts hypotheses that relate hippocampal function to working memory exclusively. The temporal SA paradigm used in the present experiment includes in itself regularity of the stimulus event, since R trials occurred at 120-sec intervals regardless of the animal's behavior. To examine further hippocampal activity to the CS and the pre-CS background activity level, an experiment with varied intervals was next performed.

3. SINGLE ALTERNATION WITH OPERANT CONDITIONING

Since it was found that the increase in hippocampal unit activity to R trials might be influenced partly by the temporal regularity of the fixed ITI in the SA paradigm itself, it was of interest to determine whether there was a similar differential enhancement when the temporal regularity of the fixed ITI was excluded from the paradigm. We were also interested in determining precisely the behavioral learning stage that correlated with the unit response. Thus, we designed an SA schedule with operant conditioning of variable intervals, reinforced by bar pressing, that could provide a behavioral latency as a measure for the stage of learning.

During preliminary training, shaping of animals to a lever press for pellets was conducted on a CRF schedule. Over the course of preliminary training, a retractable lever was introduced for magazine training. All animals were trained to press the lever when it was inserted into a chamber before go/no-go training was initiated. The SA go/no-go training was started on the day following completion of the preliminary training. Each animal received 240 R trials and 240 N trials per day for 3 consecutive days. Each trial consisted of presentation of a tone of 1 sec (CS) followed by a 10-sec presentation of a retractable lever 0.5 sec after the onset of the CS. Each trial was followed by a variable interval (50 to 70 sec with 5-sec steps) during which the lever was retracted. On the odd-numbered trials, the animals were reinforced with a pellet for pressing, and the lever was retracted. On the even-numbered trials, the rats were not reinforced. The latency from

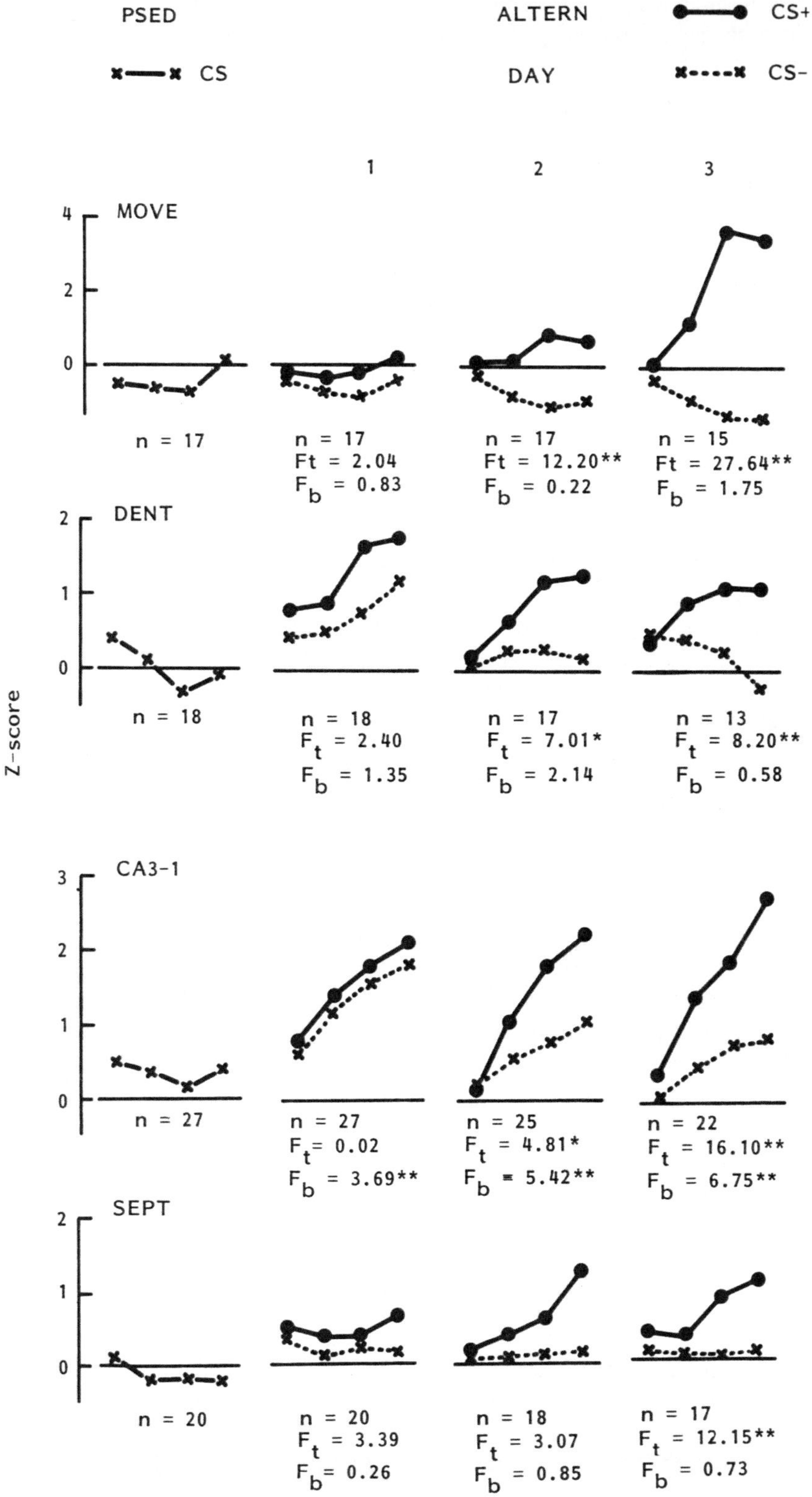
PSED
CS
ALTERN
DAY
CS+
CS-
1
2
3
Z-score
MOVE
4
2
0
n = 17
n = 17
Ft = 2.04
F_b = 0.83
n = 17
Ft = 12.20**
F_b = 0.22
n = 15
Ft = 27.64**
F_b = 1.75
DENT
2
1
0
n = 18
n = 18
F_t = 2.40
F_b = 1.35
n = 17
F_t = 7.01*
F_b = 2.14
n = 13
F_t = 8.20**
F_b = 0.58
CA3-1
3
2
1
0
n = 27
n = 27
F_t = 0.02
F_b = 3.69**
n = 25
F_t = 4.81*
F_b = 5.42**
n = 22
F_t = 16.10**
F_b = 6.75**
SEPT
2
1
0
n = 20
n = 20
F_t = 3.39
F_b = 0.26
n = 18
F_t = 3.07
F_b = 0.85
n = 17
F_t = 12.15**
F_b = 0.73

the onset of the CS to the lever press was recorded. Prior to the CS and during the CS presentation, hippocampal unit responses and animals' movements were recorded.

The basic paradigm is illustrated in Fig. 3 with data showing the behavioral latency at the top, animals' movements in the middle, and hippocampal unit activity at the bottom, each obtained on day 1 (left side) and day 3 (right side), respectively. The latency data are plotted across trials separately for R trials (upper portion of the panel) and N trials (lower portion of the panel). It should be noted that the animal responded to the lever indiscriminately on day 1 whether on R trials or N trials. It is natural to observe such short-latency responses for both types of trials in this period, since the animal was trained with a CRF schedule in the preceding sessions during preliminary training. The movement data, which were grouped for 240 R trials and 240 N trials separately, also exhibited nondifferential movement to the CS presentation on day 1, which corresponded well with the latency data. During this period, in which no behavioral differentiation appeared, hippocampal unit activity increased indiscriminately to the CS as shown at the bottom portion of Fig. 3.

When SA training was continued on day 3, however, the animal clearly developed differential conditioning to the CS as indicated by the latency as well as the movement data. Short-latency responses to R trials were maintained as on day 1, but longer-latency responses were frequently interposed between N trials. The movement data suggested that differentiation to the CS was mainly a result of suppression of the behavioral response to N trials, whereas excitatory responses to R trials were preserved as before. In accordance with such behavioral differentiation, hippocampal unit activity showed differential increases to R trials while acquiring suppressed activity to N trials. In agreement with a previous study (Hoehler and Thompson, 1979) and reports of our experiments of classical conditioning with rewarding brain stimulation, the enhancements of hippocampal unit activity were relative to R trials even when a variable ITI was introduced into the SA paradigm.

One of the aims of the present experiments was to examine any different levels of hippocampal unit activity during the pre-CS period in the SA paradigm with appetitive operant conditioning. In this respect, there were no detectable differences in base-level activity between R trials and N trials prior to the CS presentation, as shown in Fig. 3. Although these experiments are still at a preliminary stage, results have so far failed to demonstrate any differences in hippocampal unit activity in the pre-CS background period between paired and nonpaired trials.

4. DISCUSSION

In agreement with a previous study (Hoehler and Thompson, 1979), the present experiments with appetitive conditioning and an SA paradigm in rats confirm a greater conditioned increase in unit activity on paired CS–US trials regardless of the type of

FIGURE 2. Averaged response pattern for the various areas shown during four experimental sessions. Each pattern is composed of four successive 120-msec periods of the CS–US interval. The values under each averaged curve are the results of a two-way analysis of variance. An asterisk represents a significant difference at the 0.05 level; two asterisks represent a significant difference at the 0.01 level. The solid line during the SA paradigm represents R trials (CS+), and the dotted line represents N trials (CS−). Abbreviations: Fb, difference between blocks (successive 120-msec periods); Ft, difference between treatments (R trials versus N trials); PSED, pseudoconditioning; ALTERN, conditioning with the SA paradigm (Hirano and Yamaguchi, 1985).

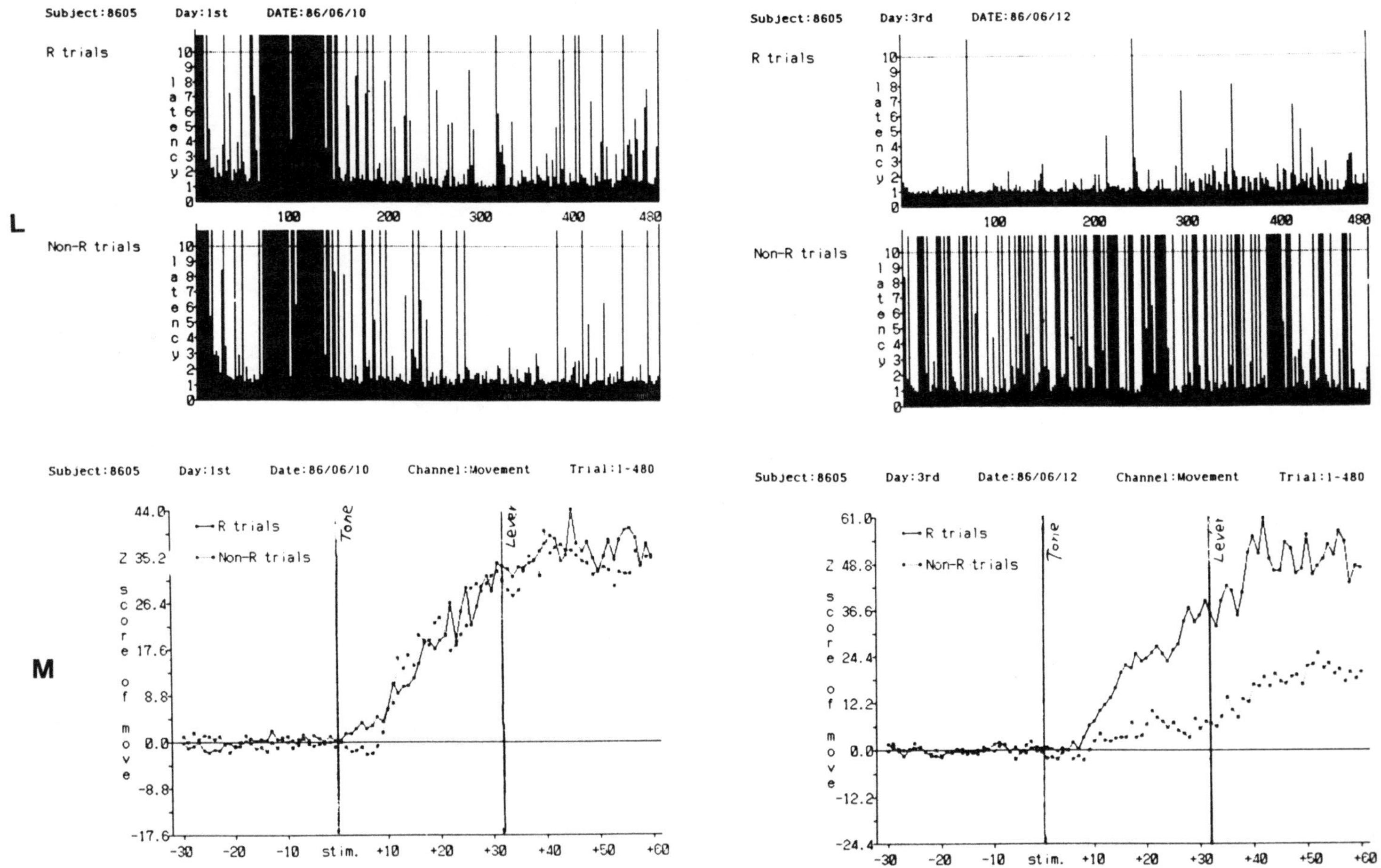
L
Subject:8605
Day:1st
DATE:86/06/10
R trials
Non-R trials
latency
Subject:8605
Day:3rd
DATE:86/06/12
R trials
Non-R trials
latency
M
Subject:8605
Day:1st
Date:86/06/10
Channel:Movement
Trial:1-480
R trials
Non-R trials
Z score of move
Tone
Lever
stim.
Subject:8605
Day:3rd
Date:86/06/12
Channel:Movement
Trial:1-480
R trials
Non-R trials
Z score of move
Tone
Lever
stim.

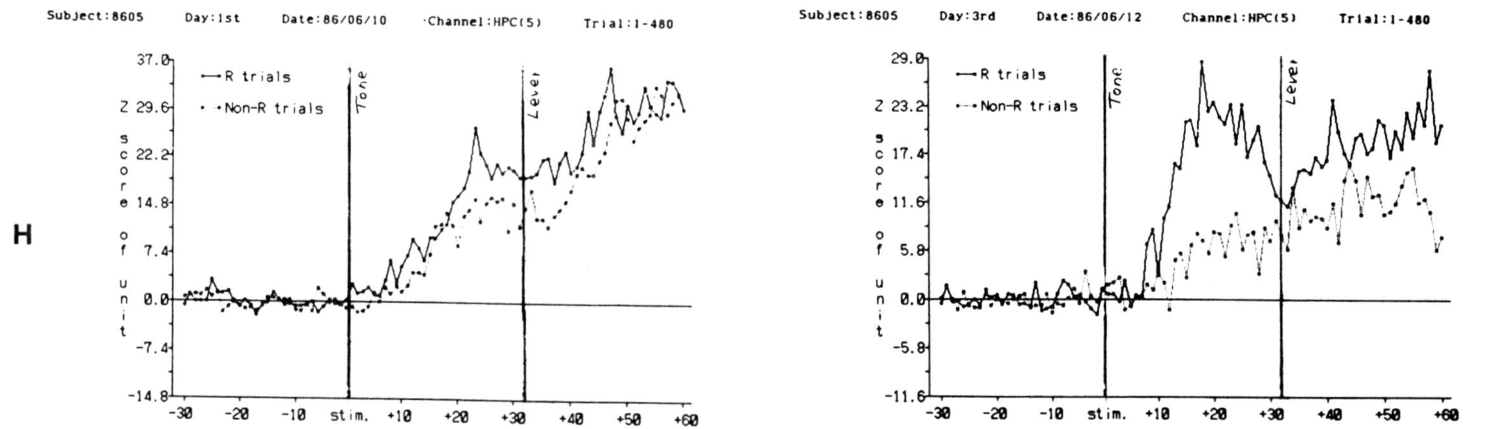

FIGURE 3. Records of one animal (8605) during operant conditioning with the SA paradigm. Behavioral latency (L) to bar press at the top, animals' movement (M) at the middle, and hippocampal unit activity (H) at the bottom, each obtained on day 1 (the left panel) and day 3 (the right panel). The latency records represent required latency from the onset of the CS to the lever pressing across trials. The upper panel shows the record for R trials, and the lower panel for N trials. The data of movement and unit activity illustrate averaged histograms over 240 R trials (solid line) and 240 N trials (dotted line) across intervals of 480 msec (16 msec per bins, 30 bins) prior to the CS and successive 960-msec periods after the onset of CS presentations. The vertical line (middle portion) represents the CS onset, and the second line to the right represents presentation of the bar to the rat.

training. Hippocampal unit activity showed a differential response to CS presentation depending on whether it was classical or operant conditioning. Such differential activity was recognized during the SA paradigm with variable intervals as well as with fixed intervals. This was expected, since rats trained with appetitive conditioning should exhibit a more well-developed patterned behavior than rabbits trained with nictitating membrane conditioning. In fact, development of differential unit activity in the hippocampus seems to be increased by R trials and reduced by N trials over the course of training. This suggests a critical role for the hippocampus in evaluating the new meaning of the stimulus based on the sequential regularity that is inherent in the SA paradigm. The acquired suppression of activity in N trials over the course of conditioning would be related to the hippocampal function of suppressing an irrelevant stimulus as evidenced by other lesion studies (Solomon and Moore, 1975). If so, one might also suppose that animals in the present experiments might be able to develop a sequential patterned behavior in relation to reference memory based on regular alternating sequences of stimulus events. Considering the failure in the present experiments to obtain lowered pre-CS levels of unit activity prior to paired CS–US trials, hippocampal function might be conceived as evaluating acquired meaning of the stimulus based on the interaction of working and reference memory.

5. SUMMARY

To evaluate hippocampal function on working memory, two experiments involving appetitive conditioning in rats with a temporal single-alternation (SA) paradigm were conducted. One used classical conditioning of a tone (CS) reinforced by rewarding brain stimulation, and the other used operant conditioning of a lever press reinforced by a pellet. In the classical conditioning experiments, animals that showed positive self-stimulation to the lateral hypothalamus were trained on an SA schedule with 60-sec intervals in which reinforced trials (R; CS–US) were regularly alternated with nonreinforced trials (N; CS alone). Unit activity from the hippocampus and the animals' movements were recorded during the period of the auditory CS. In the operant conditioning experiments, alternating go/no-go training with variable intervals was given, in which presentation of a tone (CS) was followed by presentation of a retractable lever. Animals were reinforced with a pellet for pressing during R trials. Unit activity, the animals' movement, and the latency to the lever press were recorded.

In both of these experiments, hippocampal unit activity increased in the CS periods on R trials relative to N trials when a differential behavioral response also appeared. The hippocampal activity showed a generalized excitatory response for both R and N trials in the first session and gradually developed a differential response over the course of training with an increase in R trials and a decrease in N trials. In contrast to this differentiation, no detectable differences in the background (pre-CS) levels of hippocampal unit activity, which might be expected to support the working memory hypothesis of hippocampal function, were found between R and N trials. The results are discussed in terms of the interaction between working memory and reference memory.

ACKNOWLEDGMENTS. This study was supported by grants in aid for scientific research from the Ministry of Education, Japan. I acknowledge with pleasure the collaboration of Eisuke Akase, who conducted the operant conditioning experiment discussed in this report.

REFERENCES

Capaldi, E. J., 1967, A sequential hypothesis of instrumental learning, in: *The Psychology of Learning and Motivation*, Volume 1 (K. W. Spence and J. T. Spence, eds.), Academic Press, New York, pp. 67–156.

Hirano, T., 1984, Unit activity of the septo-hippocampal system in classical conditioning with rewarding brain stimulation, *Brain Res.* **295:**41–49.

Hirano, T., and Yamaguchi, M., 1985, Hippocampal unit response during temporal single alternation of classical conditioning with rewarding brain stimulation in the rat, *Physiol. Psychol.* **13**(1):7–14.

Hoehler, F. K., and Thompson, R. F., 1979, The effect of temporal single alternation on learned increases in hippocampal unit activity in classical conditioning of the rabbit nictitating membrane response, *Physiol. Psychol.* **7:**345–351.

Olds, J., Disterhoft, J. F., Segal, M., Kornblith, C. L., and Hirsch, R., 1972, Learning centers of rat brain mapped by measuring latencies of conditioned unit responses, *J. Neurophysiol.* **35:**202–219.

Solomon, P. R., and Moore, J. W., 1975, Latent inhibition and stimulus generalization of the classically conditioned nictitating membrane response in rabbits (*Oryctolagus cuniculus*) following dorsal hippocampal ablation, *J. Comp. Physiol. Psychol.* **89:**1192–1203.

19

Conditioning and Habituation of the Arousal Response

A Historical Perspective

HERBERT H. JASPER

1. INTRODUCTION

In order to provide some historical perspective on the remarkable advances made during recent years in our understanding of microphysiological, biophysical, and possible molecular mechanisms involved in single nerve cells, synapses, and neuronal circuits during habituation and conditioning, I would like to review briefly some early studies of Pavlovian conditioning of the occipital α rhythm in man, begun about 50 years ago. I would like then to review the highlights of implanted electrode studies of habituation of electrical and behavioral arousal responses in normal, freely moving cats carried out with Seth Sharpless over 30 years ago.

Finally, I would like to mention briefly a contribution made to the Tokyo International Congress of Physiological Sciences in 1965 suggesting possible parallels between plasticity in developmental neurobiology and learning in a paper entitled "Mechanisms for the selection and preservation of acquired stimulus response patterns" (Jasper, 1965).

2. CONDITIONING THE OCCIPITAL α RHYTHM IN MAN

In the early days of the discovery and development of electroencephalography in the 1930s, it was noted (often accidentally) by several investigators that the occipital α rhythm, which was regularly blocked by visual attention, could readily be blocked by an auditory stimulus as the CS in a conditioning paradigm with photic stimulation as the UCS and α blocking as the CR (Durup and Fessard, 1935; Loomis *et al.*, 1936; Jasper and Cruikshank, 1937; Cruikshank, 1937; Walter, 1938; Travis and Egan, 1938).

HERBERT H. JASPER • Center for Research in Neurological Sciences, Department of Physiology, University of Montreal, Montreal, Quebec, Canada H3C 3J7; The Montreal Neurological Institute, McGill University, Montreal, Quebec, Canada H32 1E7.

2.1. Simple CR

Systematic studies of occipital α conditioning in selected human subjects (with a good regular α rhythm) was then undertaken with an enthusiastic student by the name of Charles Shagass from the Psychology Department at McGill University. We were able to confirm that a simple CR was easily established by only 10–15 paired trials with sound preceding (0.5–1.0 sec) and overlapping the light stimulus (Jasper and Shagass, 1941a,b). The CR was rather unstable, becoming extinguished after only five to ten repetitions of the sound without light reenforcement. The latency of the CR was about the same as that of the UCR (about 0.5 sec). An example of simple conditioning of the α-blocking response to sound is shown in Fig. 1 after only nine paired presentations of sound–light. Differential frequency-specific habituation was shown after 70 trials.

2.2. Cyclic Conditioned α-Blocking Responses

Cruikshank, in our original EEG laboratories at the Bradley Hospital of Brown University in Providence, Rhode Island, investigated the latency of the α-blocking response to various intensities of light stimulus and noted that a preparatory sound signal, given repeatedly at equal intervals before a light stimulus, established in a short while what appeared to be a CR to the constant-time-interval, sound–light, so that there seemed to be a negative latency to the light stimulus, the α block preceding the light stimulus, i.e., an anticipatory response. Grey Walter also noted this phenomenon under similar conditions and spoke of it as "time conditioning" (Walter, 1938).

Cyclic conditioning of the α-blocking response to a regularly repeated light stimulus at 10-sec intervals was then studied systematically with Shagass in the Montreal labo-

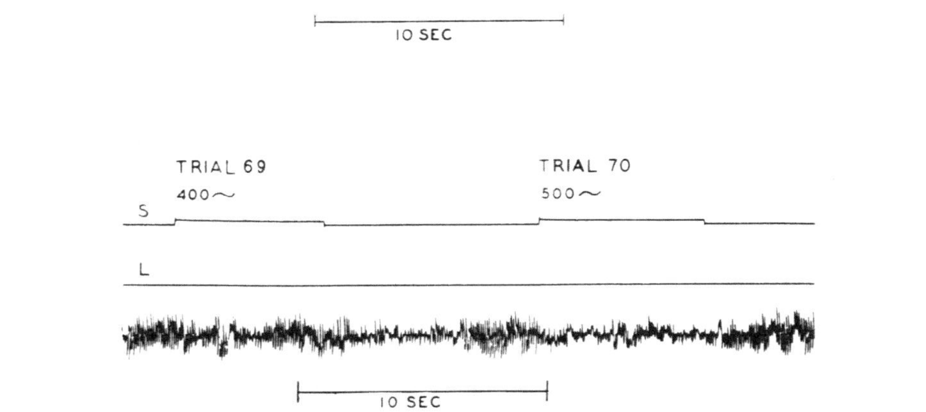

FIGURE 1. Simple-blocking CR to a 500–Hz sound stimulus after nine paired presentations of sound–light (above). Differential CR to a 500-Hz tone following 70 trials of paired sound–light interspersed with 400-Hz tones not paired with light stimulus (below).

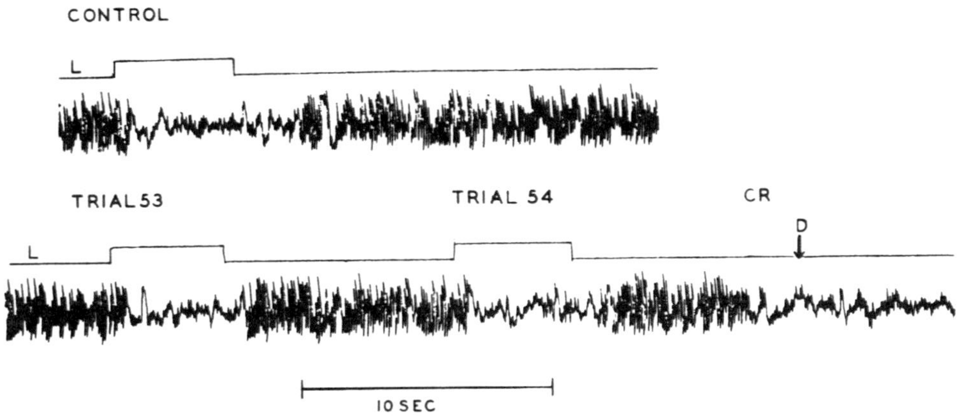

FIGURE 2. Cyclic "time" CR following 54 regular repetitions of visual stimulation alone at 14-sec intervals. Note anticipatory response at a 12-sec interval with omission of the light stimulus following trial 54.

ratories just before and during the onset of World War II (Jasper and Shagass, 1941a). After 10–20 regular repetitions of the light, the latency of the blocking response began to decrease, to become negative as an anticipatory response. Then it was possible to omit the light stimulus. A prolonged α-blocking response occurred to the time interval (about 1 sec shorter in anticipation) without an external stimulus, as shown in Fig. 2.

The cyclic CR usually lasted longer than the UCR, possibly because of some anxiety caused by the omission of an expected stimulus in a regular series. In one subject, after cyclic conditioning, a CR occurred lasting 90 sec, during which seven periodic light stimuli were given at 10-sec intervals before the α waves returned between stimuli.

2.3. Delayed α CR

A delayed α-blocking CR to sound for periods up to 30 sec could be readily established in 20 to 40 trials depending on the length of the delay. After the delayed CR was thoroughly established, there was no initial response to the CS. The α waves continued during the entire delay period and then suddenly were blocked shortly before the expected light stimulus, for example, after a 27-sec delay instead of the 30-sec delay used in conditioning trials (Fig. 3).

Delayed CRs were remarkably accurate in delay time even though anticipating the conditioned interval by about 1–3 sec. In fact, it was found that conscious time judgments of the delay interval were much less accurate and were usually somewhat longer than the conditioned delay interval (Jasper and Shagass, 1941b).

2.4. Trace CR

Trace CRs were established in about 20 trials by first presenting a 5-sec (700-Hz) sound and then allowing 9–10 sec delay interval to elapse before the light stimulus. Accuracy of the trace interval was less than for the delayed response interval. Trace intervals for CR tended to be shorter than the conditioning trials, sometimes with α blocking occurring as early as 5 sec delay for a conditioned 10-sec trace interval. An example is shown in Fig. 4.

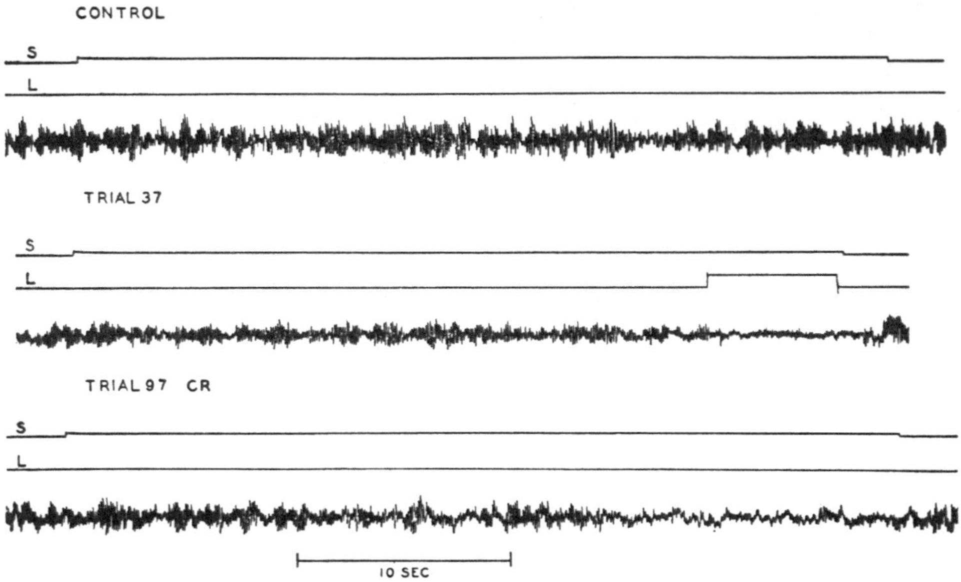

FIGURE 3. Delayed CRT following 97 sound–light pairs with light delayed for 30 sec after onset of sound, which was continued to overlap the light stimulus at end of delay period. Note anticipatory block with delay of about 27 sec following 97th trial.

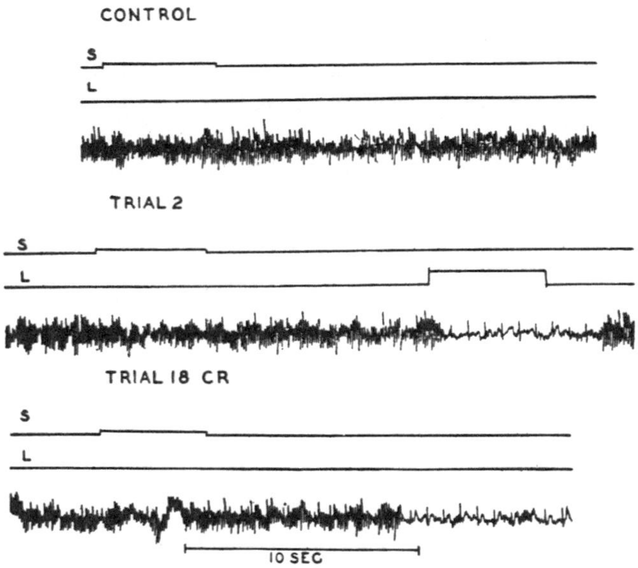

FIGURE 4. Trace CR after 18 paired trials with a delay of about 14 sec between sound and light without overlap (5-sec sound). Note anticipatory block 13 sec following the sound in the 18th trial, with no light stimulus.

2.5. Differential CR

A differential CR to a specific tonal frequency required many more trials than a simple CR. A simple CR was first established to a 500-Hz tone. Then a 400-Hz tone was introduced with the light stimulus. For a CR to occur only to the 500-Hz tone, with no response to a 400- or 475-Hz tone, required 70 to 80 trials (Fig. 1). Differential conditioning involves simultaneous habituation or inhibition of responses to sound frequencies not paired with the visual stimulus combined with positive reenforcement of the paired tone–light sequence. This is an important part of many conditioning paradigms in vertebrate preparations, as shown in our microelectrode studies of cortical cells during conditioned avoidance responses in the monkey, reviewed at the Asilomar Symposium in 1981 (Jasper, 1981; Jasper *et al.*, 1960).

2.6. Differential Delayed CR

This CR was the same as the simple differential CR above except for a delay of about 10 sec between the sound and light during conditioning. Such conditioning required over 100 trials before a delayed differential CR could be reliably produced to a 700-Hz tone with absence of response to a 500-Hz tone after a 10-sec delay period, as illustrated in Fig. 5.

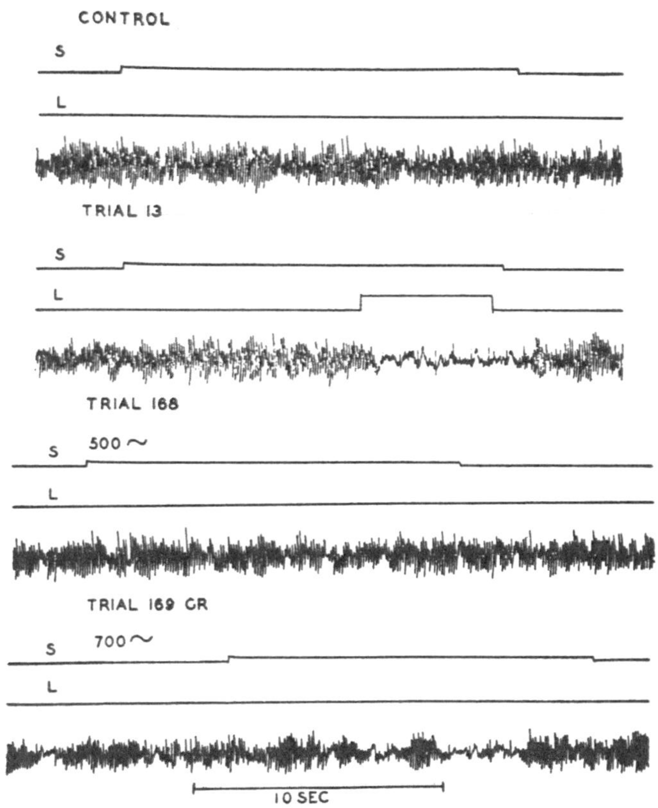

FIGURE 5. Differential delayed CR pairing 700–Hz tone with light after delay of 9.5 sec interspersed with a 500-Hz tone presented without the light stimulus. Note absence of delayed response to 500-Hz tone in 168th trial, with delayed response (8.5 sec) to 700-Hz conditioned tone in the 169th trial.

2.7. Backward CR

A backward α-blocking CR to sound could be established even though the light preceded the sound by 1–2 sec, overlapping with the sound stimulus. This required over 100 paired presentations of the two stimuli. When the light preceded but did not overlap the sound, it was not possible to elicit a CR even after about 400 paired trials.

3. HABITUATION OF THE EEG AND BEHAVIORAL AROUSAL REACTION

In the mid-1930s Margaret Rheinberger joined our EEG laboratories at the Bradley Hospital of Brown University to help develop the implanted electrode technique for recording the electrical activity directly from many areas of cortex bilaterally in unanesthetized, freely moving cats. She had acquired her animal neurosurgical technique while working in John Fulton's Department of Physiology at Yale. We soon found that the pattern of electrical activity from all cortical areas showed striking changes in relation to behavioral states of sleep or waking and arousal or attention and in response to both external and internal (autonomic) stimulation, as seen in the "cat box effect." This was called the "activation pattern" long before Moruzzi and Magoun described the ascending reticular activating system, which controls states of sleep and waking in experimental animals (in 1949). We observed that this "activation pattern" with behavioral arousal soon became habituated to repetition of the same stimulus, returning abruptly with a novel stimulus (Rheinberger and Jasper, 1937).

It was 20 years later, at the Montreal Neurological Institute, following the notable

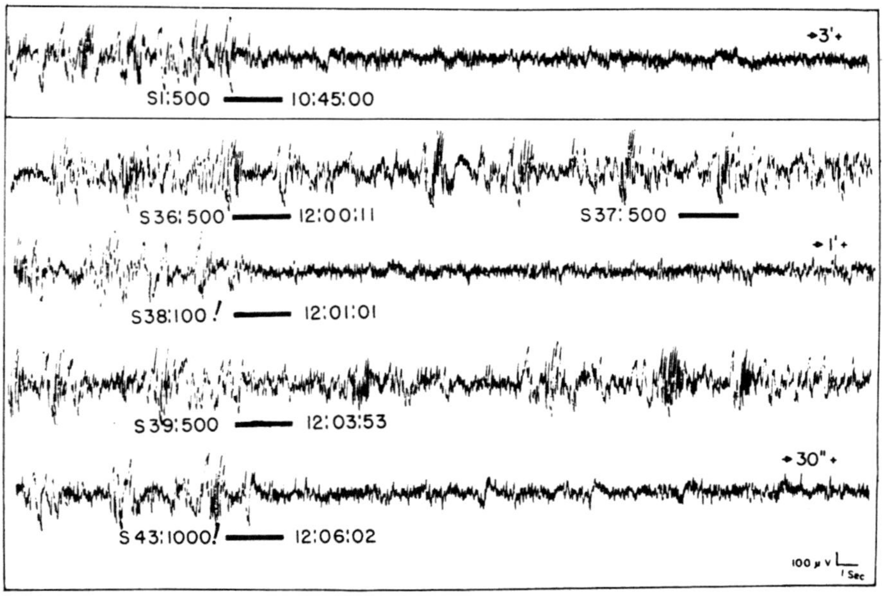

FIGURE 6. Differential habituation of arousal response to a 500-Hz tone without affecting arousal to 100-Hz or 1000-Hz tones in 38th and 43rd trials.

Laurentian Symposium on "Brain Mechanisms and Consciousness" that I was joined by Seth Sharpless, a talented graduate student from Donald Hebb's Department of Psychology at McGill. We undertook a more comprehensive and detailed study of habituation of the EEG and behavioral arousal response in cats using an improved implanted electrode technique similar to that used with Margaret Rheinberger 20 years before (Sharpless and Jasper, 1956). Auditory stimuli (tones or clicks) were repeatedly administered to sleeping cats until they were no longer awakened or aroused while we recorded electrical activity from sigmoid, suprasylvian, and ectosylvian cortical areas and from subcortical regions in the diencephalon and midbrain. Activation and sleep patterns of electrical activity were similar in cortical and subcortical structures.

Habituation to repetition of a pure tone during sleep occurred after 20–40 trials. Frequency-specific habituation required more trials, especially for tones of closely similar frequency (e.g., 450 and 500 Hz). Stimulus specificity was remarkable even when stimuli were administered only when animals were deeply asleep. A change in the intensity or duration of a given tone, a change in the pattern of a series of identical tones (ascending or descending), or the administration of light tactile stimulus (puff or air) would cause arousal in an animal that had become completely habituated to a specific auditory stimulus (Figs. 6 and 7).

Frequency-specific habituation could be established following excision of all of the primary auditory and adjacent cortex bilaterally, as shown in Fig. 8. Habituation to specific patterns was not possible after cortical excision.

By using a repetitive click stimulus, it was possible to record auditory evoked potentials directly from primary auditory cortex during habituation. There was no attenuation (even some increase) in cortical auditory evoked potentials to click stimuli to which the animals had become completely habituated for the behavioral or EEG arousal response (see Fig.9).

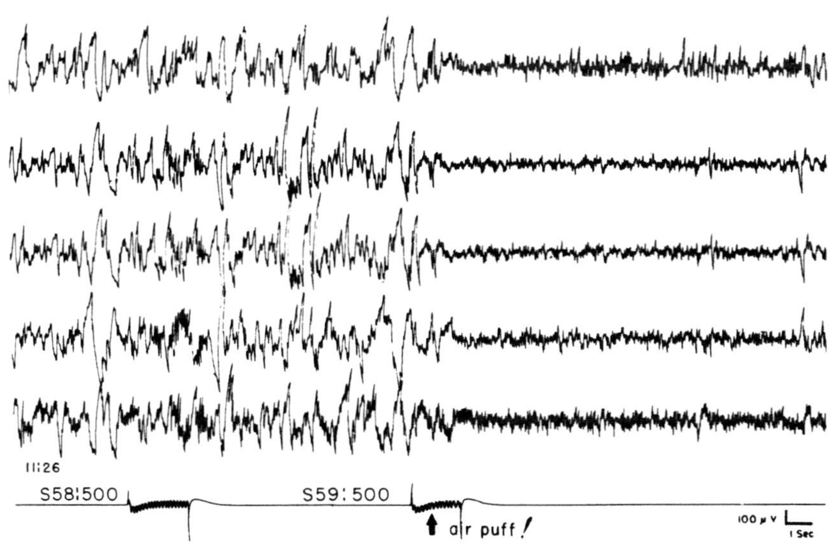

FIGURE 7. Habituation of arousal response to 500-Hz tone after 58th trial, with arousal by puff of air in 59th trial, showing lack of general inhibition in the habituation process even during sleep.

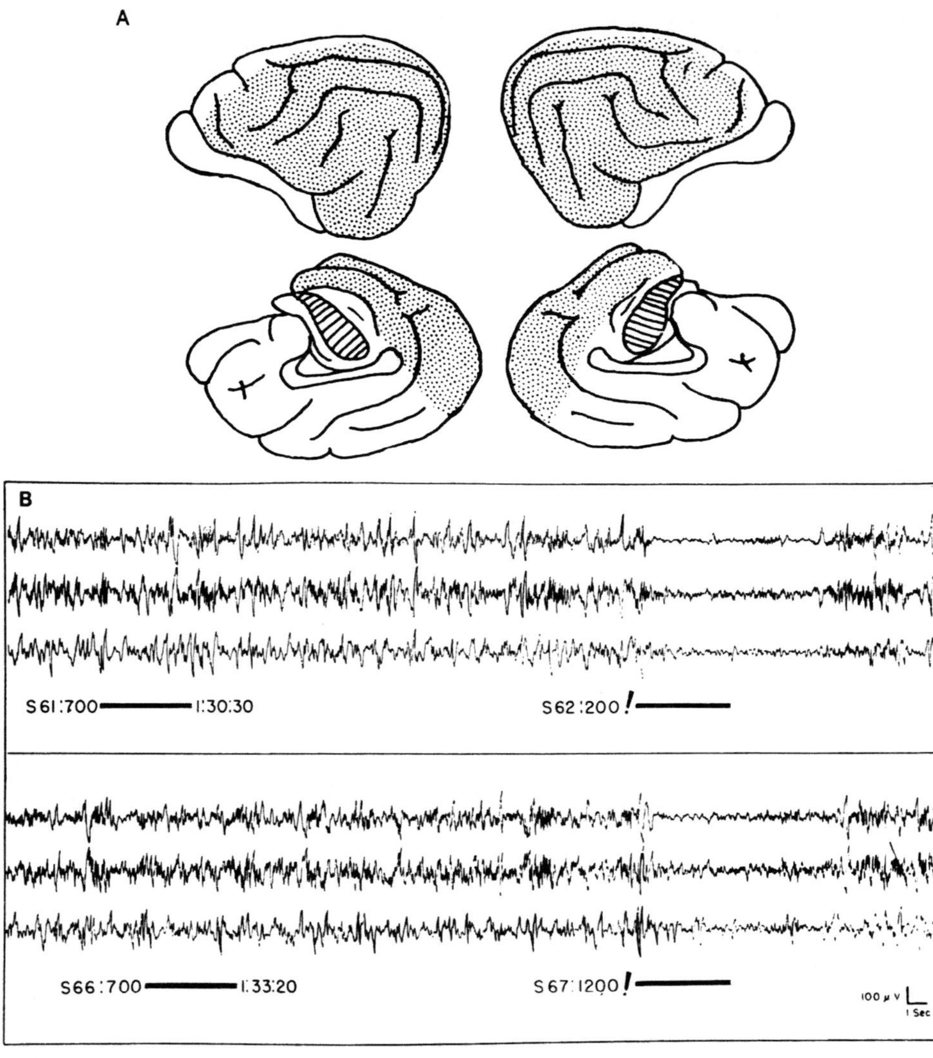

FIGURE 8. Differential frequency-specific habituation to 700-Hz tone after over 60 trials in cat with large neocortical excisions as shown above (A). A short lasting arousal still occurred to 200-Hz and 1200-Hz tones. Frequency-specific habituation was abolished by lesions of the branchium of the inferior colliculus bilaterally in other animals.

4. SIGNIFICANCE OF CONDITIONING AND HABITUATION OF THE AROUSAL REACTION FOR STUDIES OF CELLULAR AND MOLECULAR MECHANISMS OF CONDITIONING

The fact that many forms of Pavlovian conditioning can be precisely reproduced by using the occipital α rhythm of the human EEG as a conditioned response to specific tonal frequencies of sound, including differential, delayed, and trace conditioning, conditioning with delayed responses of 10–30 sec, and cyclic conditioning to the time interval (10 sec) of regular repetition of the light stimulus alone suggests that the physiological mechanisms of EEG activation may play an important role in conditioning mechanisms

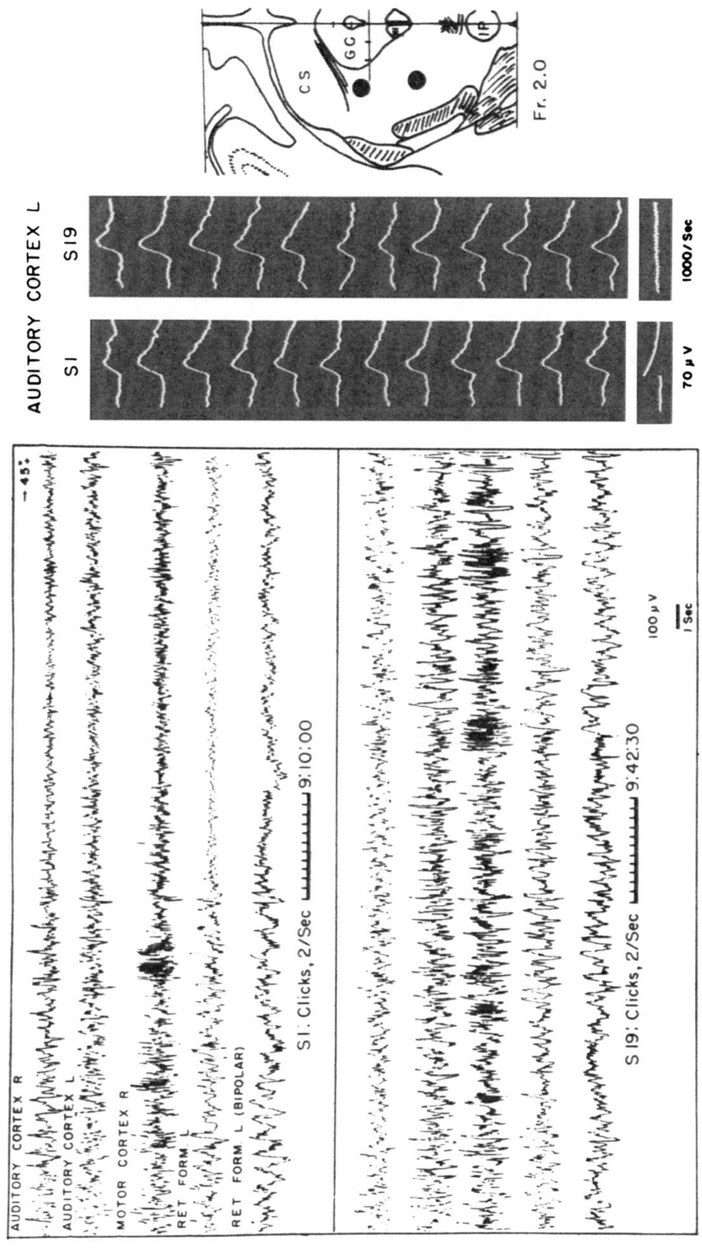

FIGURE 9. Cortical auditory I evoked potentials to 2/sec clicks following habituation of arousal response to this stimulus after 19 repetitions. Note absence of attenuation of cortical evoked response in spite of complete lack of the arousal response to the same stimulus.

in higher vertebrates, including man. Behaviorally, this EEG response is related to visual attention or "arousal," known to have an important cholinergic component to its neurochemical mediation (Jasper, 1969). The inhibition of the response during delayed and trace conditioning and the establishment of a response to time intervals alone, without an external CS, present some interesting and important problems for future studies of the cellular and circuit mechanisms of conditioning in vertebrates.

Habituation of the EEG and behavioral arousal response in cats to a specific frequency of auditory stimulation without attentuation of auditory cortical evoked potentials to the habituated stimulus implies subcortical mechanisms other than those of the specific auditory sensory pathways. Habituation to a pattern of tones, with each tone remaining constant, is established with greater difficulty and is not possible after large cortical excisions. This implies that the cerebral cortex is essential for pattern discrimination even in the sleeping animal.

There are important lessons to be learned in reviewing some of these very old experiments on the electrical activity of the brain during conditioning in higher vertebrates, which may require mechanisms different in kind or location from those found in simple invertebrate preparations such as *Aplysia* or *Hermissenda*. In vertebrates, specialized neuronal systems in hippocampus, cerebellum, basal forebrain, hypothalamus, and brainstem seem to provide parts of the mechanisms for special aspects of a complex conditioned response or for the selective inhibition of irrelevant responses in habituation. It is believed that inhibitory mechanisms will play a larger role in our studies of cellular and molecular mechanisms of conditioning in the future. Differential frequency- or modality-specific conditioning requires simultaneous inhibitory and facilitatory mechanisms in the conditioning process or perhaps simultaneous habituation and positive reinforcement.

It was at another satellite symposium on learning in conjunction with the International Congress of Physiological Sciences in Tokyo in 1965 (Jasper, 1965) that the suggestion was made that we might look to developmental neurobiology for mechanisms governing the selection and preservation of only a small percentage of the "exuberant" growth of nerve cells and synapses by some sort of Darwinian selection process, with atrophy of all others. It was suggested also that physiological and neurochemical mechanisms involved in alerting and attention, such as acetylcholine, might be important in the selection and preservation of only a few of the many synaptic events continually taking place without their preservation in long-term memory.

We had shown previously that EEG activation or arousal pattern could be reproduced by the local application of an anticholinesterase and acetylcholine to the cortical surface. It had long been known that muscarinic blocking agents such as atropine and scopolamine would block the EEG activation pattern during behavioral arousal. We had measured the liberation of acetylcholine from the cortical surface and found the rate of liberation to be increased three- to fivefold during the physiological arousal of an unanesthetized animal and with the arousal pattern induced by stimulation of the brainstem reticular activating system (Sie *et al.*, 1965; Celesia and Jasper, 1966; Jasper, 1969).

We then undertook to test the effect of scopolamine on learning of tactile pattern discrimination in the monkey with George Murray. Murray's work has taken on greater interest since the demonstration of the importance of acetylcholine in Alzheimer's disease. Direct evidence for the importance of cholinergic mechanisms, and cGMP as a second intracellular messenger, has recently been obtained by the elegant intracellular microiontophoretic studies of Charles Woody and associates in pyramidal cortical cells in the cat (Woody *et al.*, 1978).

REFERENCES

Celesia, G. G., and Jasper, H. H., 1966, Acetylcholine released from the cerebral cortex in relation to state of activation, *Neurology (Minneap.)* **16:**1053–1064.

Cruikshank, R. M., 1937, Human occipital brain potentials as affected by intensity–duration variables of visual stimulation, *J. Exp. Psychol.* **21:**625–641.

Durup, G., and Fessard, A., 1935, L'Électroencéphalogramme de l'homme, *Ann. Psychol.* **36:**1–32.

Jasper, H. H., 1969, Neurochemical mediators of specific and non-specific cortical activation, in: *Attention in Neurophysiology* (C. R. Evans and T. B. Mulholland, eds.), Butterworth, London, pp. 377–395.

Jasper, H. H., 1965, Mechanisms for the selection and preservation of acquired stimulus–response patterns, *Excerpta Med. Int. Cong. Ser.* **87:**641–644.

Jasper, H., 1981, Reflections on early studies of cortical unit activity during conditioning in the monkey, in: *Conditioning: Representation of Involved Neural Functions*, Volume 26 (C. D. Woody, ed.), Plenum Press, New York, pp. 319–332.

Jasper, H.H., and Cruikshank, R. M., 1937, Electroencephalography II, Visual stimulation and the after image as affecting the occipital alpha rhythm, *J. Gen. Psychol.* **17:**29–48.

Jasper, H. H., and Shagass, C., 1941a, Conditioning the occipital alpha rhythm in man, *J. Exp. Psychol.* **28:**373–388.

Jasper, H., and Shagass, C., 1941b, Conscious time judgments related to conditioned time intervals and voluntary control of the alpha rhythm, *J. Exp. Psychol.* **28:**503–508.

Jasper, H. H., Ricci, G., and Doane, B., 1960, Microelectrode analysis of cortical cell discharge during avoidance conditioning in the monkey, *J. Electroencephalogr. Clin. Neurophysiol. [Suppl.]* **13:**137–155.

Loomis, A. L., Harvey, E. N., and Hobart, G. A., 1936, Electrical potentials of the human brain, *J. Exp. Psychol.* **19:**249–279.

Murray, G., 1969, The role of cholinergic mechanisms in the learning of a tactile discrimination in the monkey, M.Sc. Thesis, Faculty of Medicine, University of Montreal, Montreal.

Rheinberger, M. B., and Jasper, H. H., 1937, The electrical activity of the cerebral cortex in the unanesthetized cat, *Am. J. Physiol.* **119:**186–196.

Sharpless, S., and Jasper, H., 1956, Habituation of the arousal reaction, *Brain* **79:**655–680.

Sie, G., Jasper, H. H., and Wolfe, L., 1965, Rate of ACh release from cortical surface in 'encephale' and 'cerveau isole' cat preparations in relation to arousal and epileptic activation of the ECoG, *Electroencephalogr. Clin. Neurophysiol.* **18:**206.

Travis, L. E., and Egan, J. P., 1938, Conditioning of the electrical response of the cortex, *J. Exp. Psychol.* **22:**524–531.

Walter, W. G., 1938, The technique and application of electroencephalography, *J. Neurol. Psychiatry* **1:**359–385.

Woody, C. D., Swartz, B. E., and Gruen, E., 1978, Effects of acetylcholine and cyclic GMP in awake cats, *Brain Res.* **158:**373–395.

20

A Preliminary Note on Spatial EEG Correlates of Olfactory Conditioning

KAMIL A. GRAJSKI

1. INTRODUCTION

1.1. Olfactory System Organization

The vertebrate olfactory epithelium (100–200 μm thick) covers roughly 100 cm^2 and contains 10^6–10^8 receptor cells. Each receptor cell has a broad response characteristic to odorants (Lancet, 1986). Odorant information is conveyed topographically to the main olfactory bulb (OB) via the primary olfactory nerve (PON) as spatial patterns of spike trains generated by the noninteracting receptor cells. The PON fibers converge onto 10^3 glomeruli in the OB glomerular layer. Within each glomerulus (80–150 μm in diameter), 10^3–10^5 PON fibers synapse with the apical dendrites of 10^2–10^3 mitral and tufted cells. The mitral and tufted cells form reciprocal synapses with 10^3–10^4 deep-lying granule cells. Mitral–granule cell interactions are mediated by a sigmoidal gain function (Eeckman and Freeman, 1986). Under synaptic driving, the nonspiking granule cells generate electric dipoles with common instantaneous orientation perpendicular to the bulbar surface. These field potentials form the predominant component of the EEG activity recorded at the bulbar surface (Freeman, 1975). Mitral cell axons converge to form the lateral olfactory tract (LOT), which projects topographically to the anterior olfactory nucleus but nontopographically to the prepiriform cortex. These structures project back to the bulb in a similar fashion.

1.2. A Model of Olfactory Discrimination

The spatial distribution of PON fibers over the OB surface is not uniform. Input fibers converge into roughly 2000 discrete glomeruli. Each glomerulus defines a functional unit or paleocortical "column" in analogy to neocortical columns (Edelman and Mountcastle, 1978). We model a bulbar column as the complex of cells driven synaptically under one glomerulus. Lateral excitatory interactions support coactive cell populations across columns. Under a noradrenergic-dependent process, coactive synapses are strength-

KAMIL A. GRAJSKI • Graduate Group in Biophysics, University of California at Berkeley, Berkeley, California 94720. *Present address:* Coleman and Epstein Memorial Laboratories, University of California, San Francisco, San Francisco, California 94143.

ened, thus forming a neural assembly (Gray *et al.*, 1986). This assembly provides a basis for generalization over receptor input, odor discrimination, and behavioral responding. The model has been presented in detail elsewhere (Freeman and Skarda, 1986; Baird, 1986a,b).

1.3. Spatial EEG Recording

We test our model by sampling collective neural dynamics of the olfactory bulb. We use an eight by eight electrode array (4.0 by 4.0 mm) implanted directly on the lateral surface of the bulb in rabbits. Volume conduction smooths and attenuates (tenfold per 0.5 cycle/mm) high spatial frequencies of EEG at the surface. With a center–center electrode distance of 0.5 mm, each electrode measures the activity of a neighborhood of four to eight "columns." Data consist of spatial patterns of EEG activity from 10–20% of total bulbar surface area. The spatial EEG data transform the rabbits' olfactory discrimination problem into our statistical pattern recognition problem.

2. STATISTICAL METHODOLOGY

2.1. Quantifying Behavioral Responding

Behavioral responding is quantified through statistical analysis of digitized pneumograph and electromyograph (EMG) traces over standard 6-sec trials (3 sec control, 3 sec test). Sniffing is detected by a significant change in test respiratory rate relative to the control rate (Davis and Freeman, 1982). Sniff latency is given by the time of the first significantly different cycle; magnitude by their number. The EMG signals are passed through a window discriminator to detect occurrence and latency of jaw movements during licking. Figure 1 depicts the relationship between bulbar EEG on one channel selected from the array of 64 and respiration.

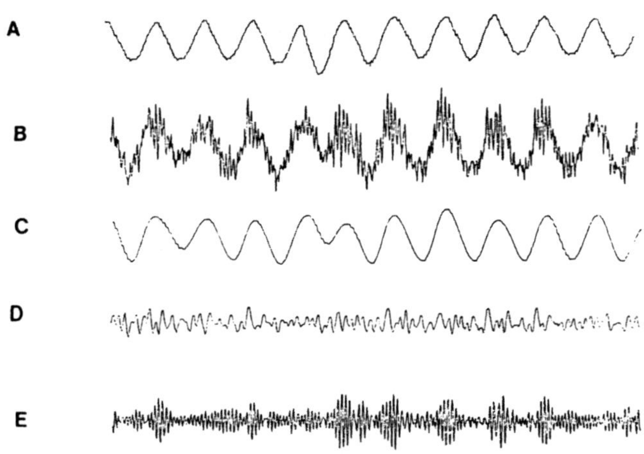

FIGURE 1. Temporal relationship between bulbar EEG and respiration. (A) Pneumograph trace. (B) Raw EEG from one channel selected at random from an array of 64 electrodes. (C) Low-pass digital filter (15-Hz cutoff) applied to B. (D) Band-pass filter applied to B (15–55 Hz band). (E) Band-pass filter applied to B (55–95 Hz band). (Time is 76 msec; voltage is 100 μV.)

2.2. Spatial EEG Data Analysis

We focus on the 40 to 80-Hz bursting EEG riding on the "slow-wave" accompanying respiration. Within each 75 to 100-msec-long burst, individual channel series are smoothed, detrended, and set to zero mean. Figure 2 shows such a burst sampled on 64 channels. For the ensemble average time series, a Fourier transform yields a peak frequency and phase. These are used as initial guesses in a nonlinear fit of an amplitude- and frequency-modulated cosine wave to the ensemble average. The modulations are set as linear functions about the burst center time. This procedure is repeated recursively for the next

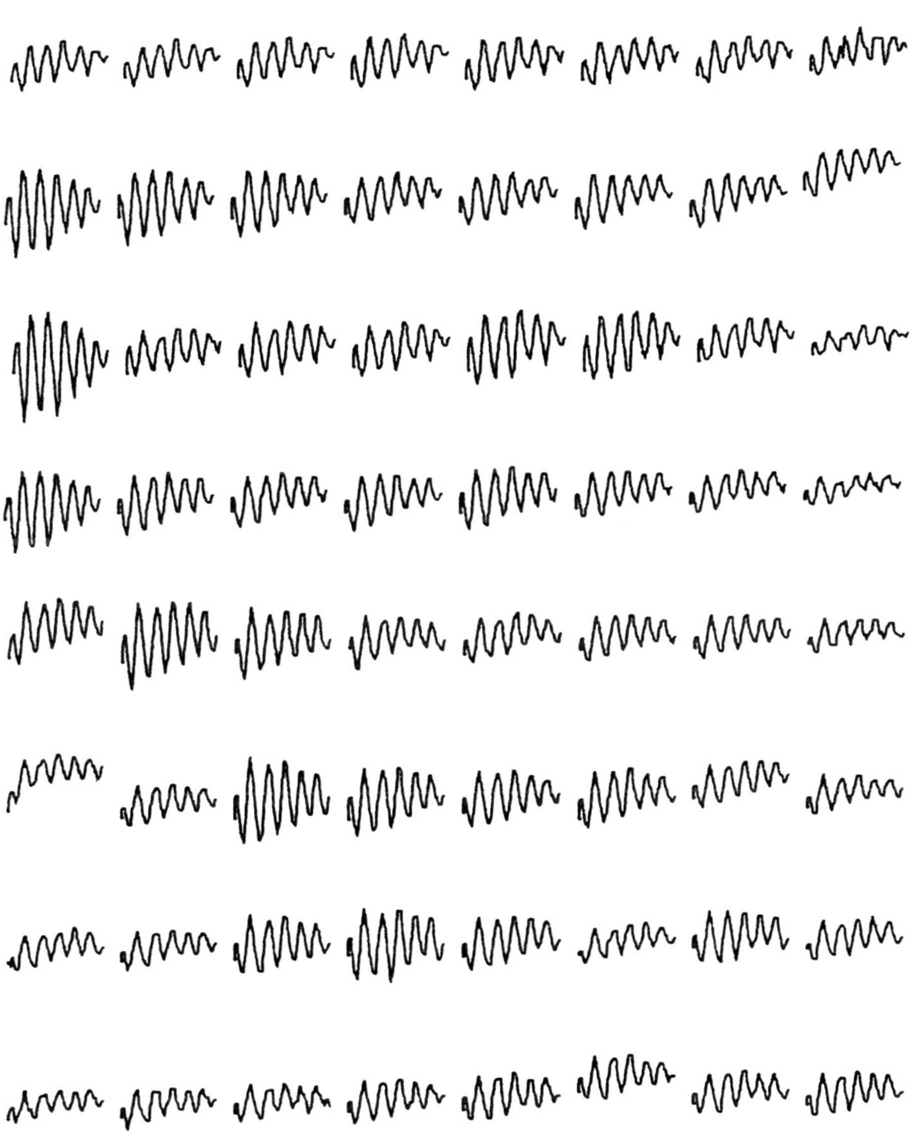

FIGURE 2. A typical spatial sample of bursting EEG recorded from the rabbit olfactory bulb. The burst has been smoothed with a triangular window. (Time is 76 msec; voltage is 100 μV.)

four residuals. Thus, the ensemble average time series is decomposed into five components. The sum of these five components is fitted by linear regression to each of the 64 channels to yield five eight-by-eight amplitude patterns. A spatial filter and deconvolution are applied to each amplitude pattern to compensate for volume conduction. Goodness of fit is measured by the percentage of total spectral energy captured by each component. The component with highest percentage, generally the first, is termed the dominant component. Spatial amplitude patterns of the dominant component form the basis for statistical pattern analysis.

2.3. Statistical Pattern Recognition of EEG Spatial Data

Classification is by both parametric and nonparametric multivariate techniques. Emphasis is placed on deriving lower- and upper-bound performance estimates. To obtain these estimates, data sets are repeatedly (10–20 times) divided randomly into "learning" and "test" subsamples. Classifiers are constructed using the "learning" sample. The upper-bound estimate is the percentage of "learning" cases correctly classified; lower-bound estimate is the percentage of "test" cases correctly classified. Parametric methods evaluated include linear and quadratic discriminant analysis, linear stepwise discriminant analysis, and factor analysis. Nonparametric methods include tree-structured methods (Grajski *et al.*, 1986). Most recently we have employed a newly developed nonlinear generalized and stepwise discriminant analysis method based on optimal scaling and iterative smoothing techniques (Breiman and Ihaka, 1984).

3. CORRELATION OF EEG WITH BEHAVIOR

3.1. Electroencephalographic Burst Analysis: Overview

Analysis of 10^5 bursts from 30 rabbits provides evidence for the following. First, the burst is a unitary event involving the entire bulb. During a burst, local regions oscillate at different amplitudes but at the same common frequency (see Fig. 2). No reproducible phase relationship occurs across bursts (Freeman and Baird, 1986). Second, across bursts, there is a common amplitude or "signature" pattern unique to each animal. For robust classification, this pattern is "removed" by a combination of spatial filtering and normalization by channel to zero mean and unit standard deviation (Freeman and Grajski, 1987; Grajski *et al.*, 1986). Third, bursts fall into two classes: orderly and disorderly. The dominant components of orderly bursts contains a sharp spectral peak in the 55 to 85-Hz range with 50% of total spectral energy, less than 25% frequency modulation, and a stable spatial pattern (within-group Fisher Z-transformed correlation is 0.92). Disorderly bursts have a broad distribution of spectral energy across components, frequency modulation in excess of 25%, and an unstable spatial pattern (within-group Fisher Z-transformed correlation is 0.12).

For burst amplitude pattern classification, the orderly bursts are selected. The disorderly bursts, however, are not merely statistical outliers. It has been suggested that they manifest a failure of convergence to a stable spatial mode of oscillation in the bulb (Freeman and Skarda, 1986; Baird, 1986a,b).

TABLE I. Novel Odor Burst Classification (4-Way)[a]

Subject	Air	Odor A	Odor B	Odor C	N
1	69	14	24	20	221
2	62	16	5	22	362
3	71	47	24	26	497
4	90	0	0	29	152
5	87	46	30	15	167
6	80	29	0	1	525
Mean	76.5	25.3	13.8	18.83	
S.E.	4.5	7.7	5.6	4.1	

[a] Lower-bound percentage correct estimated by "jackknifing" in stepwise linear discrimination. Only lower-bound estimates are shown. Upper-bound values average 10% higher. Number of variables used for best discrimination averaged between three and seven.

3.2. Behavioral Experiments

3.2.1. Novel Odors

Our model predicts that no stable novel odor-specific patterns should exist in the bulbar EEG signal. Six rabbits were presented with ten trials each of three novel odors and ten air trials, randomly interspersed at 1-to 2-min intervals. The odors were *n*-butanol, benzaldehyde, and ethyl acetate diluted in water (1 : 1000); the final concentration at the nose cone was estimated to be 1 : 10,000 for all odors. On average, the rabbits sniffed on nine of ten odors and on three of ten air trials. Up to 24 bursts per 6-sec trial were selected for analysis to yield a data base of 2700 bursts. If novel odor-specific patterns emerge in the bulbar EEG, it should be possible to classify each burst as belonging to an air class or to one of the three odor classes.

In this data base, commonality of burst wave form, presence of a signature pattern, and orderly and disorderly bursts are observed. Disorderly bursts occur on average for 25% of air and 33% of odor bursts. Table 1 lists the lower-bound percentage correct classification obtained with a linear stepwise discriminant analysis. Upper-bound estimates averaged 10% higher. Across subjects, four-way discrimination between air and three odor groups shows that air bursts are separable from odor bursts; the odor bursts do not form stable, separable classes. To confirm this, all odor bursts are pooled into one group and discriminated against air bursts. Across subjects, lower-bound percentage correct classification is 85% for air, 49.5% for pooled odor; random classification gives 50% correct for each class. These results, shown in Table II, confirm earlier findings with factor analysis that although there is differentiation of air from pooled odor bursts, at present resolution no stable odor-specific burst amplitude patterns are found (Grajski and Freeman, 1986).

3.2.2. Habituation of the Sniff Response

Our model predicts that the formation of assemblies is a gradual process and thus may be wholly or in part suppressed by habituation to novel odors. This process is simulated by a decrease in the synaptic coefficient of the lateral excitatory connections in the bulb. Simulated burst amplitudes are suppressed, and the base-line shift is enhanced.

TABLE II. Novel Odor Burst Classification (2-Way)[a]

Subject	Air	Odor
1	83	44
2	85	47
3	72	42
4	88	70
5	88	50
6	95	44
Mean	85.2	49.5
S.E.	3.1	4.3

[a] Percentage correct as estimated by linear step-wise discrimination between air and pooled odor bursts. Only lower-bound estimates are shown (see Table I).

We have begun to study habituation in array-implanted rabbits. We measure habituation by the rate of occurrence and magnitude (number of respiratory cycles) of a sniffing response. A standard session consists of 10 to 12 6-sec trials (3 sec control, 3 sec test) with an ITI near 2 min. Experiments from two groups of rabbits (total $N = 16$) provide evidence for short- (within session) and long-term (between sessions) and odor-specific habituation. Such evidence is summarized by Figs. 3 and 4 for a typical subject.

Figure 3 shows pneumograph traces recorded during the first, second, ninth, and tenth trials to amyl acetate (1 : 1000) during the first session. The first two traces show a short-latency protracted sniff response, which is diminished by the end of the session. Figure 4 shows traces from a fifth habituation session to amyl acetate, during which the odor specificity of habituation is tested. The first trace shows a weak sniffing response to amyl acetate (compare to first trace of Fig. 3). The second and fourth traces are the

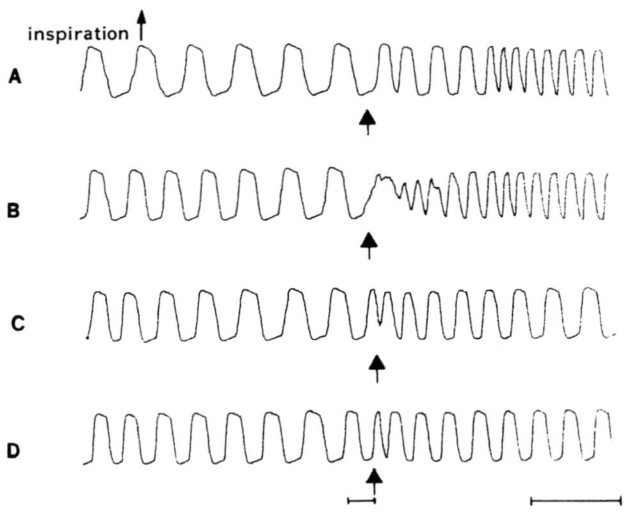

FIGURE 3. Pneumograph traces from standard 6-sec trials. All trials are from the first session of habituation to amyl acetate (1 : 1000): A, first trial; B, second trial. Note diminished sniff response by the ninth and tenth odor trial (C,D). Arrow denotes time of onset of odorant in the nose cone. (Short bar indicates 350 msec dead time between switching of solenoid and earliest odor arrival. Long bar, 1 sec.)

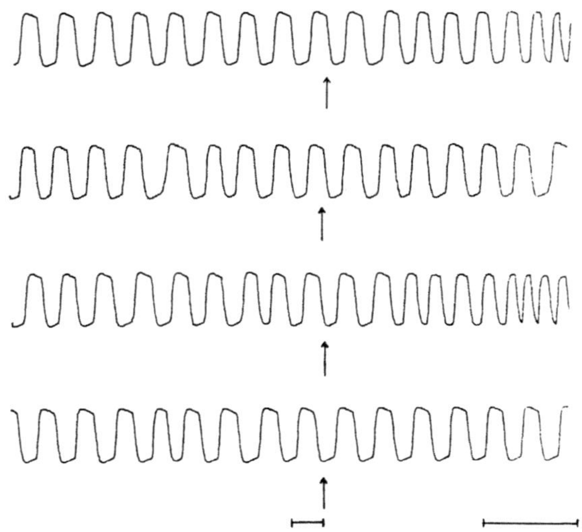

FIGURE 4. All traces are from fifth habituation session to amyl acetate in which odor-specific habituation was tested. Note diminished response on first (A) trial to amyl acetate, strong response to novel (C) odor (1 : 1000 butyric acid), and maintained diminished responding on amyl acetate trials preceding (B) and following (D) novel odor trial (see Fig. 5A). (Bars, see Fig. 3.)

trials immediately preceding and following a novel odor trial (butyric acid, 1 : 1000). The sniff response is of short latency, sustained duration, and increasing strength. Its vigor is less than that to the first trial of amyl acetate and may reflect nonspecific effects of adaptation to the test chamber previously observed (Viana di Prisco and Freeman, 1986). Concurrently recorded EEGs are being studied for short- and long-term spatial pattern changes and shifts in frequency spectra in both burst and extended EEG segments; results are presented elsewhere (Grajski and Freeman, 1987).

Odor-specific changes in sniff probability and magnitude with habituation are being analyzed as one of a set of appropriate controls in conditioning of the sniff response (Freeman *et al.*, 1983; Gormezano *et al.*, 1983). Parametric changes in habituation stimulus and training are also under analysis (Thompson and Spencer, 1966; Grajski and Freeman, 1987; Gray *et al.*, 1986).

3.2.3. Differential Conditioning

Our model further predicts that with learning, odor-specific patterns will emerge and stabilize in the spatial patterns of bursting EEG. Five mildly thirsty rabbits were appetitively conditioned to respond with an anticipatory licking to a CS + odor; a CS − odor served as a discriminative control. Three CS +, CS − combinations were studied, each such stage spanning six sessions (stage I: A + ,B −; stage II: C + ,B −; stage III: C + ,A −). Using a standard 6-sec trial, air, CS +, and CS − odor bursts were recorded concurrently with behavioral responding. Figure 5 shows the average jaw movement conditioned response expressed in percentage of trials for each session (Viana di Prisco and Freeman, 1986).

For classification, bursts from the last three sessions within each stage are pooled. Typical results are those from stage I sessions 4 through 6. Only the dominant components of orderly bursts from trials with "correct" responding are included. Bursts are classified into air, CS +, and CS − groups. Linear stepwise discriminant analysis percentage lower-

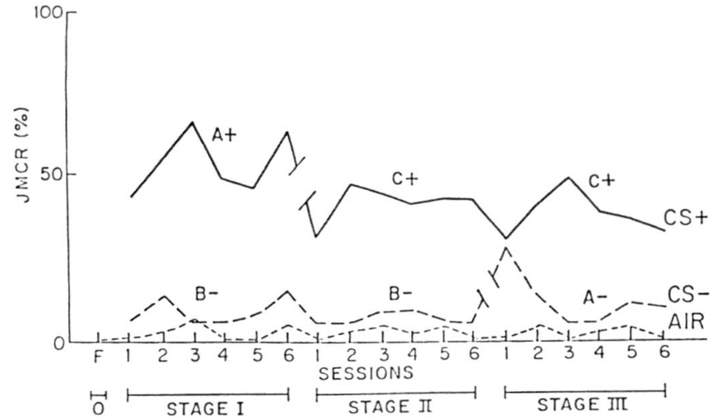

FIGURE 5. Curves for average jaw movement conditioned response expressed in percentage of odor(−), odor(+), and blank (air) trials. In stage I, odor A is reinforced, and odor B is not; in stage II, novel odor C is reinforced, A is discontinued, and B is not reinforced; in stage III, odor C is reinforced, and A is reintroduced but not reinforced. F is familiarization, stage 0.

bound correct classification by class listed in Table III. Averaging over the four subjects that showed discriminative responding, 72.0% of air, 73.3% of CS+, and 60.5% of CS− bursts are correctly classified. Upper-bound values averaged 10% higher. A second method is the nonlinear generalization of discriminant analysis, which determines variable transformations optimal for separating groups. Table IV lists upper and lower overall correct classification for each subject. Figure 6 plots bursts for one subject in reduced class space. The coordinates are similar to those obtained with linear discriminant analysis, but the transformations that place data on them are nonlinear. Thus, with two additional independent methods, the finding of odor-specific EEG spatial patterns with differential conditioning to odors is confirmed (Freeman and Grajski, 1987; Grajski *et al.*, 1986).

4. DISCUSSION

The olfactory bulb is, to first order, a massively parallel interconnected network of identical units, with globally synchronized oscillatory dynamics. A generalization of memory and computational properties in such systems (Grossberg, 1976; Hinton and Anderson, 1981; Hopfield, 1982; Kohonen, 1984) provides a model mechanism for olfactory discrimination (Freeman, 1979; Baird, 1986a,b). Through spatial analysis of

TABLE III. Percentage Correct Classification of Conditioned Odor Bursts (3-Way)[a]

Subject	Air	CS+	CS−	Cases	Vars.
1	62	72	48	110	3
2	80	76	67	118	6
3	88	88	70	123	6
4	58	57	57	122	6
5	86	70	38	118	6
Mean (1–4)	72.0	73.3	60.5		
S.E.	7.2	6.4	5.0		

[a] Percentage correct as estimated by linear stepwise discrimination among air, CS+, and CS− bursts. Only lower-bound estimates are shown. Number of variables used for best discrimination also shown.

TABLE IV. Percentage Correct Classification of Conditioned Odor Burst (3-Way)[a]

Subject	Upper	Lower	S.E.
1	100	76	.05
2	100	91	.03
3	97	73	.04
4	98	73	.06
5	97	40	.10
Mean (1–4)	98.75	78.25	.05

[a] Nonlinear discrimination among air, CS +, and CS − bursts. Overall percentage correct classification upper- and lower-bound estimates are shown. The standard error of ten "learning" and "test" set pairs used to determine lower-bound estimates are also shown.

EEG, we have sought experimental evidence for the operations of such a system in the olfactory bulb of rabbits. Two experiments show that (1) stable odor-specific patterns do not emerge for novel odors and (2) under associative learning, stable, coexisting, odor-specific burst amplitude patterns emerge with learning, persist for weeks, and reorganize with new learning. A complete analysis of EEG recorded concurrently with short- and long-term habituation to odors and odor-specific habituation is presented elsewhere (Grajski and Freeman, 1987).

The present findings are a necessary step in a program to define how collective neural dynamics underlie vertebrate olfactory discrimination. Further studies are needed on the relationship between EEG and unit activity. Further studies are also needed on bulbar projection sites such as the anterior olfactory nucleus and prepiriform cortex. These structures share bulbar structural and dynamical properties, perform spatiotemporal integration on their input, feed back to one another, and undergo changes in EEG activity patterns with learning (Bressler, 1982). Lastly, mathematical analysis is needed to define more rigorously neural dynamics in model systems and state-space reconstructions from experimental data.

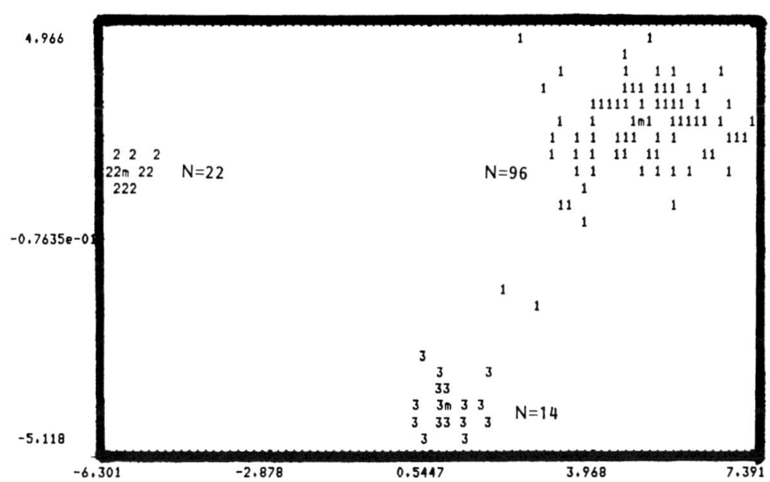

FIGURE 6. Results of nonlinear discriminant analysis applied to a representative set of data from one subject undergoing differential odor conditioning. "1" denotes air bursts, "2" CS− bursts, "3" CS+ bursts. Upper and lower overall percentage correct classification are 98.5, 73.0 ± 0.04, respectively.

ACKNOWLEDGMENTS. The support and encouragement of W. J. Freeman are gratefully acknowledged. This work was supported by NIH Systems and Integrative Biology Training Grant 5-T32-GM07379.

REFERENCES

Baird, B., 1986, Nonlinear dynamics of pattern formation and pattern recognition in rabbit olfactory bulb, *Physica D* **22:**150–175.

Baird, B., 1986b, Bifurcation analysis of oscillating network model of pattern recognition in the rabbit olfactory bulb, in: *Neural Networks for Computing* (J. S. Denker, ed.), American Institute of Physics Conference Proceedings, American Institute of Physics, New York, Volume 151. pp. 29–34.

Breiman, L, and Ihaka, R., 1984, *Discriminant Analysis via Scaling and ACE*, Technical Report No. 40, UC Berkeley Department of Statistics, Berkeley.

Bressler, S. L., 1982, Response of olfactory bulb and cortex to conditioned odor stimulation, *Soc. Neurosci. Abstr.* **8:**314.

Davis, W., and Freeman, W. J., 1982, On-line detection of respiratory events applied to behavioral conditioning in rabbits, *IEEE Trans. Biomed. Eng.* **29:**453–456.

Edelman, G. M., and Mountcastle, V.B., 1978, *The Mindful Brain: Cortical Organization and the Group-Selective Theory*. MIT Press, Cambridge, Massachusetts.

Eeckman, F., and Freeman, W. J., 1986, The sigmoid nonlinearity in neural computation: An experimental approach, in: *Neural Networks for Computing* (J. S. Denker, ed.), American Institute of Physics Conference Proceedings Volume 151, American Institute of Physics, New York, pp. 135–139.

Freeman, W. J., 1975, *Mass Action in the Nervous System*, Academic Press, New York.

Freeman, W. J., 1979, Nonlinear dynamics of paleocortex manifested in olfactory EEG, *Biol Cybernet.* **35:**21–37.

Freeman, W. J., and Baird, B., 1987, Correlation of olfactory EEG with behavior: Spatial analysis. *Behav. Neurosci.* **101:**393–408.

Freeman, W. J., and Grajski, K. A., 1987, Correlation of olfactory EEG with behavior: Factor analysis. *Behav Neurosci.* (in press).

Freeman, W. J., and Skarda, C., 1986, Spatial EEG patterns, non-linear dynamics and perception: The neo-Sherringtonian view, *Brain Res. Rev.* **10:**147–175.

Freeman, W. J., Viana di Prisco, G., Davis, G. W., and Whitney, T. M., 1983, Conditioning of relative frequency of sniffing by rabbits to odors, *J. Comp. Psychol.* **97**(1):12–23.

Gormezano, I., Kehoe, E. J., and Marshall, B. S., 1983, Twenty years of classical conditioning research with the rabbit, *Prog. Psychobiol. Psychol.* **10:**197–275.

Grajski, K. A., and Freeman, W. J., 1987, Spatial EEG correlates of non-associative and associative olfactory learning in rabbits (submitted for publication).

Grajski, K. A., and Freeman, W. J., 1986, A dynamic model of olfactory discrimination, in: *Neural Networks for Computing* (J. S. Deuker, ed.), American Institute of Physics Conference Proceedings, Volume 151, American Institute of Physics, New York, pp. 188–193.

Grajski, K. A., Breiman, L., Viana di Prisco, G., and Freeman, W. J., 1986, Classification of EEG spatial patterns with a tree structured methodology: CART, *IEEE Trans. Biomed. Eng.* **33:**(12)**:**1076–1086.

Gray, C. M., Freeman, W. J., and Skinner, J. E., 1986, Chemical dependencies of learning in the rabbit olfactory bulb: Acquisition of the transient spatial pattern change depends on norepinephrine, *Behav. Neurosci.* **100:**585–596.

Grossberg, S.,1976, Adaptive classification and universal coding, *Biol. Cybernet.* **23:**187.

Hinton, G., and Anderson, J., 1981, *Parallel Models of Associative Memory*, Lawrence Erlbaum Associates, Hillsdale, NJ.

Hopfield, J. J., 1982, Neural networks and physical systems with emergent collective computational abilities, *Proc. Natl. Acad. Sci U.S.A.* **79:**2554–2557.

Kohonen, T., 1984, *Self-Organization and Associative Memory*, Springer-Verlag, Berlin.

Lancet, D., 1986, Vertebrate olfactory reception, *Ann. Rev. Neurosci.* **9:**329–355.

Thompson, R. F., and Spencer, W. A., 1966, Habituation: A model phenomenon for the study of neuronal substrates of behavior, *Psychol. Rev.* **73**(1):16–43.

Viana di Prisco, G., and Freeman, W. J., 1986, Odor-related bulbar EEG spatial pattern analysis during appetitive conditioning in rabbits, *Behav. Neurosci.* **99:**964–978.

21

A Theoretical Neuronal Learning Mechanism That Predicts the Basic Categories of Classical Conditioning Phenomena

A. HARRY KLOPF

1. THE NEURONAL MODEL AND LEARNING MECHANISM

It is suggested that the neuronal model proposed by Hebb (1949) be modified in the following ways to make it consistent with the animal learning phenomena it is intended to explain: (1) instead of correlating pre- and postsynaptic levels of activity, changes in pre- and postsynaptic levels of activity should be correlated; (2) instead of correlating approximately simultaneous pre- and postsynaptic signals, earlier presynaptic signals should be correlated with later postsynaptic signals. More precisely and consistent with the first modification, earlier changes in presynaptic signals should be correlated with later changes in postsynaptic signals. Thus, sequentiality replaces simultaneity in the model. (3) A change in the efficacy of a synapse should be proportional to the current efficacy of the synapse, accounting for the initial positive acceleration in the classic S-shaped acquisition curves observed in animal learning.

The resulting neuronal model is an extension of the Sutton–Barto (1981) model, which, in turn, may be seen as a temporally refined extension of the Rescorla–Wagner (1972) model. The model to be proposed is termed a drive–reinforcement model because it suggests that nervous system activity can be understood in terms of two classes of signals: drives, which are defined to be signal levels, and reinforcers, which are defined to be changes in signal levels. Mathematically, the learning mechanism of the drive reinforcement model may be characterized as follows:

$$\Delta w_i(t) = \Delta y(t) \sum_{j=1}^{\tau} c_j | w_i(t - j)| \Delta x_i(t - j) \qquad (1)$$

where $w_i(t)$ is the weight or efficacy of synapse i at discrete time, t; $\Delta w_i(t)$ is the change in efficacy of synapse i at time, t, and is equal to $w_i(t + 1) - w_i(t)$; τ is the longest

A. HARRY KLOPF • Avionics Laboratory, Air Force Wright Aeronautical Laboratories, Wright-Patterson Air Force Base, Ohio 45433.

interstimulus interval over which conditioning is effective; c_j is a learning rate constant, which is proportional to the efficacy of conditioning when the interstimulus interval is j; $x_i(t)$ is a measure of the frequency of action potentials at synapse i; $y(t)$ is a measure of the frequency of firing of the neuron; $\Delta x_i(t) = x_i(t) - x_i(t-1)$; and $\Delta y(t) = y(t) - y(t-1)$.

Two refinements of the above equation further improve the model's ability to predict animal learning phenomena. The first refinement involves allowing only positive changes in x to contribute to a change in a synaptic weight, w. Negative changes in x are set equal to zero for the purpose of calculating changes in synaptic efficacy. The second refinement involves assigning a new role to negative changes in x. A negative change in x serves to reduce the effective magnitude of an earlier positive change in x if the negative change in x comes between a positive change in x and a change in y. This way of utilizing negative changes in x is termed the trace conditioning mechanism. When the trace conditioning mechanism is added to the drive–reinforcement model, the effective change in presynaptic signal level at time t, $[\Delta x_i(t - \mathrm{j})]_{e(t)}$ may be computed as follows:

$$[\Delta x_i(t - j)]_{e(t)} = \min \{x_i(t - j), x_i(t - j - 1), \ldots, x_i(t - T - 1)\} - x_i(t - j - 1) \quad (2)$$

where $0 \leqq [\Delta x_i(t - j)]_{e(t)} \leqq \Delta x_i(t - \mathrm{j})$, $j > T$, and T is the maximal trace interval (measured in time steps) that occur without a negative change in presynaptic signal level causing a diminution in the rate of conditioning. Utilizing equation 2, equation 1 then becomes:

$$\Delta w_i(t) = \Delta y(t) \sum_{\mathrm{j}=l}^{T} c_{\mathrm{j}} |w_i(t - j)| [\Delta x_i(t - j)]_{e(t)} \quad (3)$$

If $j \leqq T$, then $[\Delta x_{\mathrm{i}}((t - j)]_{e(t)}$ is equal to $\Delta x_i(t - j)$. To consider an example of how the trace-conditioning mechanism works, if T equals 2, then a negative change in presynaptic signal level may occur two time steps or less prior to the change in the postsynaptic signal level without the rate of conditioning being diminished.

A lower bound is set on the absolute values of the synaptic weights, w_i. The bound is near but not equal to zero because synaptic weights appear as factors on the right side of equation 2. It can be seen that the learning mechanism would cease to yield changes in synaptic efficacy for any synapse whose efficacy reached zero; i.e., $\Delta w_{\mathrm{i}}(t)$ would henceforth always equal zero. A nonzero lower bound on the efficacy of synapses models the notion that a synapse must have some effect on the postsynaptic neuron in order for the postsynaptic learning mechanism to be triggered.

Computer simulations of the drive–reinforcement neuronal model are discussed below. In these stimulations, each input to the neuron is made available via both an excitatory and an inhibitory synapse so that the neuronal learning mechanism has, for each input, both an excitatory and an inhibitory weight available for modification. In the classical conditioning experiments that were simulated, the weights associated with synapses carrying unconditioned stimuli were assumed to be fixed (nonplastic), and the remaining synaptic weights were assumed to be variable (plastic).

2. PREDICTIONS OF THE NEURONAL MODEL

It can be demonstrated that the drive–reinforcement neuronal model accounts for the basic animal learning phenomena that have been observed. This includes, within the category of classical or Pavlovian conditioning, delay conditioning, trace conditioning, simultaneous conditioning, conditioned stimulus duration and amplitude effects, unconditioned stimulus amplitude effects, interstimulus interval effects, second- and higher-order conditioning, conditioned inhibition, habituation and extinction, reacquisition effects, backward conditioning, blocking, overshadowing, and serial compound conditioning.

Examples of conditioning phenomena predicted by the drive–reinforcement neuronal model are shown in Fig. 1. In the figure, comparisons are made of the predictions of three models: Hebb's (1949) learning mechanism, the Sutton–Barto (1981) learning mechanism, and the drive–reinforcement learning mechanism. For all three models, the following neuronal input–output relationship is assumed:

$$y(t) = \sum_{i=1}^{n} w_i(t)\, x_i(t) - \theta \tag{4}$$

where $y(t)$ is a measure of the postsynaptic frequency of firing at discrete time, t; n is the number of synapses impinging on the neuron; $w_i(t)$ is the efficacy of synapse i; $x_i(t)$ is a measure of the frequency of action potentials at synapse i; and θ is the neuronal threshold. The synaptic efficacy, w_i, can be positive or negative, corresponding to excitatory or inhibitory synapses, respectively. Also, $y(t)$ is bounded such that $y(t)$ is greater than or equal to zero and less than or equal to the maximal output frequency, $y'(t)$, of the neuron.

The learning mechanism for the drive–reinforcement neuronal model has already been specified in equation 3. The Hebbian learning mechanism may be specified as follows:

$$\Delta w_i(t) = c x_i(t) y(t) \tag{5}$$

where c is the learning rate constant and the other symbols are defined as above. The Sutton–Barto learning mechanism is specified by the following equations:

$$\Delta w_i(t) = c\bar{x}_i(t)\,[y(t) - \bar{y}(t)] \tag{6}$$

$$\bar{x}_i(t) = \alpha\, \bar{x}_i(t-1) + x_i(t-1) \tag{7}$$

where $\bar{y}(t) = y(t-1)$ for the theoretical curves shown in Fig. 1, α is a positive constant, and the other symbols are defined as above.

In Fig. 1, predicted acquisition (learning) curves are compared for the Hebbian, Sutton–Barto, and drive–reinforcement learning mechanisms. A delay-conditioning paradigm is utilized; this is a classical (Pavlovian) conditioning paradigm in which the onset of the conditioned stimulus (CS) precedes the onset of the unconditioned stimulus (UCS) and the offset of the CS occurs at or after the onset of the UCS. The CS and UCS were presented once in each trial, and the values of the synaptic weights at the end of each trial are shown in the graphs. (Data points are not shown on these graphs because they fall exactly on the computed theoretical curves.)

In Fig. 1, a comparison of the predictions of the three models for delay conditioning is made for each of three cases: CS offset at the time of UCS onset in the case of CS_1, CS offset at the time of UCS offset in the case of CS_2, and CS offset after UCS offset in the case of CS_3. Thus, predicted effects of CS duration are examined for each model. Experimentally, it is known that conditioned excitation (corresponding to positive synaptic weights) is observed in all three cases, with the acquisition curve positively accelerating initially and negatively accelerating subsequently (Spence, 1956; Kamin, 1965). It can

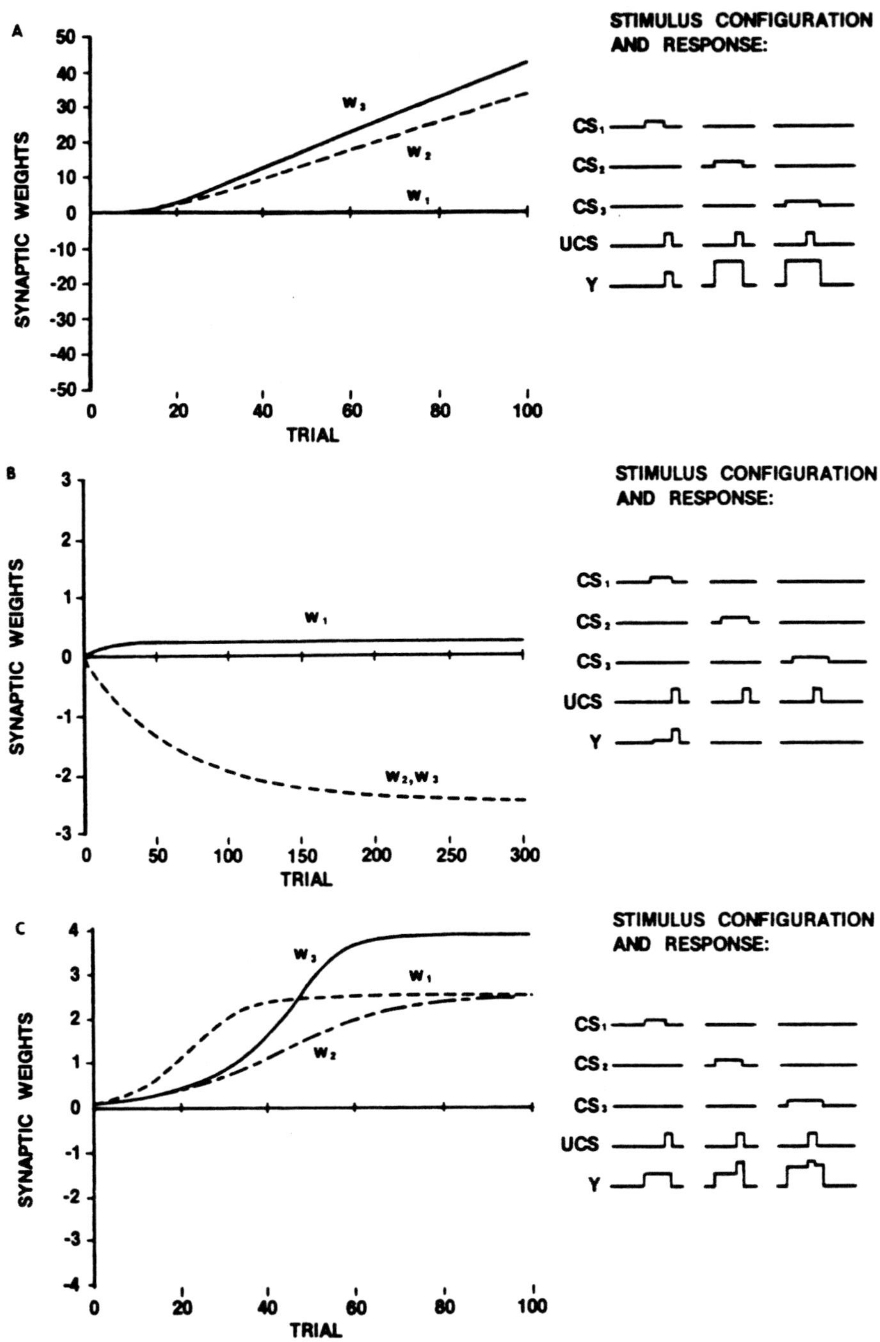

FIGURE 1. A comparison of predicted acquisition (learning) curves for classical (Pavlovian) conditioning for the (A) Hebbian, (B) Sutton–Barto, and (C) drive–reinforcement models of single-neuron function. See text for additional explanation and discussion.

be seen in Fig. 1c that the drive–reinforcement model's predictions are consistent with the experimental evidence in all three cases. The Hebbian model predicts conditioned excitation for two of the three cases (CS_2 and CS_3) and no conditioning for the third case (CS_1), as can be seen in Fig. 1a. Furthermore, the predicted acquisition curves for the Hebbian model are essentially linear. The Sutton–Barto model predicts conditioned excitation for one case (CS_1) and conditioned inhibition for the other two cases (CS_2 and CS_3), with acquisition curves that are negatively accelerated, as seen in Fig. 1b.

The theoretical acquisition curves shown in Fig. 1 were obtained by means of computer simulations of each neuronal model. The CSs and UCSs are represented in the neuronal models as presynaptic signal levels, x_i. The CSs and UCSs have a base-line level of zero. Y is equal to $y(t)$ for the last trial shown on the graph. Parameter values utilized in the computer simulations of the three neuronal models are as follows: $0 \leqq y(t) \leqq 1.0$, $\theta = 0.0$, $n = 4$, UCS (fixed) weight $= 1.0$, CS amplitudes $= 0.2$, UCS amplitudes $= 0.5$, CS onsets at $t = 10$, CS_1 offset at $t = 13$, CS_2 offset at $t = 14$, CS_3 offset at $t = 15$, UCS onsets at $t = 13$, UCS offsets at $t = 14$; in a,$c = 0.5$, initial plastic weight values $= 0.0$; in b,$c = 0.5$, $\alpha = 0.9$, initial plastic weight values $= 0.0$; in c, initial values of excitatory plastic weights and lower bound $= 0.1$, initial values of inhibitory weights and lower bound $= 0.0$, $\tau = 5$, $c_1 = 5.0$, $c_2 = 3.0$, $c_3 = 1.5$, $c_4 = 0.75$, $c_5 = 0.25$. In c, the relative values for c_j are consistent with the assumption that a time step represents (nominally) 0.5 sec. For the trace conditioning mechanism, $T = 0$.

3. CONCLUSIONS

The accuracy of the drive–reinforcement model's predictions, as shown in Fig. 1c, is typical of its predictions for the other categories of classical conditioning phenomena noted earlier. This provides reason to believe that the drive–reinforcement learning mechanism specified by equation 1 may model the learning mechanism employed by single living neurons. The model also suggests a possible theoretical linkage between single-neuron behavior and whole-animal behavior.

ACKNOWLEDGMENTS. This research was supported by the Life Sciences Directorate of the Air Force Office of Scientific Research under Task 2312 R1. Jim Morgan is acknowledged for the software he wrote for the single-neuron simulator employed in this research.

REFERENCES

Hebb, D. O., 1949, *The Organization of Behavior,* John Wiley & Sons, New York.

Kamin, L. J., 1965, Temporal and intensity characteristics of the conditioned stimulus, in: *Classical Conditioning: A Symposium* (W. F. Prokasy, ed.), Appleton-Century-Crofts, New York, pp. 118–147.

Rescorla, R. A., and Wagner, A. R., 1972, A theory of Pavlovian conditioning: Variations in the effectiveness of reinforcement and non-reinforcement, in: *Classical Conditioning II: Current Research and Theory* (A. H. Black and W. F. Prokasy, eds.), Appleton-Century-Crofts, New York, pp. 64–69.

Spence, K. W., 1956, *Behavior Theory and Conditioning,* Yale University Press, New Haven.

Sutton, R. S., and Barto, A. G., 1981, Toward a modern theory of adaptive networks: Expectation and prediction, *Psychol. Rev.* **88:**135–170.

IV

Anatomy and Cable Properties

22

Dendritic Spine Synapses, Excitable Spine Clusters, and Plasticity

WILFRID RALL and IDAN SEGEV

1. INTRODUCTION

These computations focus on dendritic spines and the possibility that some spine heads possess excitable nerve membrane. It will be shown that such spines can provide synaptic amplification and, what is more important, that interactions between such spines can result in a local chain reaction involving clusters of such spines. Whether a cluster fires depends (with nonlinear sensitivity) on changes in synaptic excitation and inhibition and on changes in spine stem resistance (a possible locus for plasticity related to conditioning and/or learning) or other spine parameters. Several computed examples indicate the rich repertoire of logical operations that could be implemented by excitable spine clusters.

Before modeling synaptic input to dendritic spines, it is necessary to know some of the earlier results and insights obtained from the application of cable theory to branched neurons and the subsequent modeling of different synaptic input distributions over the extensive dendritic surface.

2. PASSIVE CABLE PROPERTIES OF DENDRITES

2.1. Models of Dendritic Neurons

The assumptions and the equations used to model dendritic trees have been carefully presented and discussed in earlier publications; only a few points are summarized here. In one early paper (Rall, 1962), it was shown how the partial differential equation for a passive nerve cable can represent an entire dendritic tree and how this can be generalized from cylindrical to tapered branches and trees; this paper also showed how to incorporate synaptic conductance input into the mathematical model and presented several computed examples of nonuniform synaptic input distributions. Then another publication (Rall, 1964) showed how the same results can be obtained with compartmental modeling of

WILFRID RALL and IDAN SEGEV • Mathematical Research Branch, National Institute of Diabetes and Digestive and Kidney Diseases, National Institutes of Health, Bethesda, Maryland 20892. *Present address of I.S.:* Department of Neuroscience, Institute of Life Sciences, The Hebrew University, Jerusalem, Israel.

dendritic trees. This chapter also pointed out that such compartmental models are not restricted to the assumption of uniform membrane properties or to the family of dendritic trees that transform to an equivalent cylinder or to an equivalent taper and, consequently, that such compartmental models can be used to represent any arbitrary amount of nonuniformity in branching pattern, in membrane properties, and in synaptic input that one chooses to specify. The power of this generality has been overlooked by those who erroneously consider dendritic models to be valid only for equivalent cylinders with uniform membrane. Quite recently, this compartmental approach has been applied to detailed dendritic anatomy represented as thousands of compartments (Bunow *et al.*, 1985; Segev *et al.*, 1985; S. J. Redman and J. D. Clements, personal communication).

2.2. Effects of Dendritic Synaptic Input Location

Physiologically significant theoretical predictions and insights were obtained over 20 years ago by means of computations with a simple ten-compartment model. One computation predicted different shapes for the voltage transient (EPSP) expected at the neuron soma when identical brief synaptic inputs are delivered to different dendritic locations (see Fig. 6 of Rall, 1964; also Rall, 1967). These EPSP shapes and the resulting shape index loci (Rall *et al.*, 1967) represent theoretical predictions that have been experimentally confirmed in many laboratories (see Jack *et al.*, 1975; Redman, 1976; Rall, 1977; see also very recent detailed experimental tests by Redman and Walmsley, 1983; Henneman *et al.*, 1984).

2.3. Distinguishing Spatiotemporal Patterns of Synaptic Input

Another early computation demonstrated the possible functional importance of different spatiotemporal patterns of synaptic input to the dendrites (Fig. 7 of Rall, 1964). Using the same ten-compartment model to represent different dendritic locations along the cable distance from the neuron soma to the dendritic terminals (see upper half of Fig. 1), it was possible to compute and contrast the effects of two different spatiotemporal input patterns. The sequence A–B–C–D represents proximal dendritic input first, followed in time steps by inputs at successively more distal dendritic locations (this input pattern is represented schematically by the input locations marked on the four sets of ten compartments shown at left in the figure). The resulting computed voltage transient (EPSP at the soma) shows a rapid rise to an early peak, and this is then followed by a prolonged plateau; this plateau could be used to bias the soma at slightly subthreshold levels, poised for impulse discharge by small additional input.

The contrasting sequence D–C–B–A represents distal dendritic input first, followed by inputs at successively more proximal dendritic locations (this input pattern is indicated by the four sets of ten compartments shown at right in the figure). The resulting computed EPSP (at the soma) shows a delayed rise, but it then builds to a higher peak. This peak could exceed the threshold for impulse discharge and would thus provide a means of distinguishing between these two spatiotemporal sequences of dendritic synaptic input.

The intermediate dotted transient in this figure shows the computed result when the spatiotemporal pattern is eliminated by distributing the same total quantity of synaptic input uniformly over all four locations for the full-time duration of the previous patterns. Clearly, spatiotemporal pattern of input makes a difference. With suitable synaptic input distributions, this effect could be used to detect movement and to distinguish between different directions of movement. It may be noted that this theoretical prediction provided

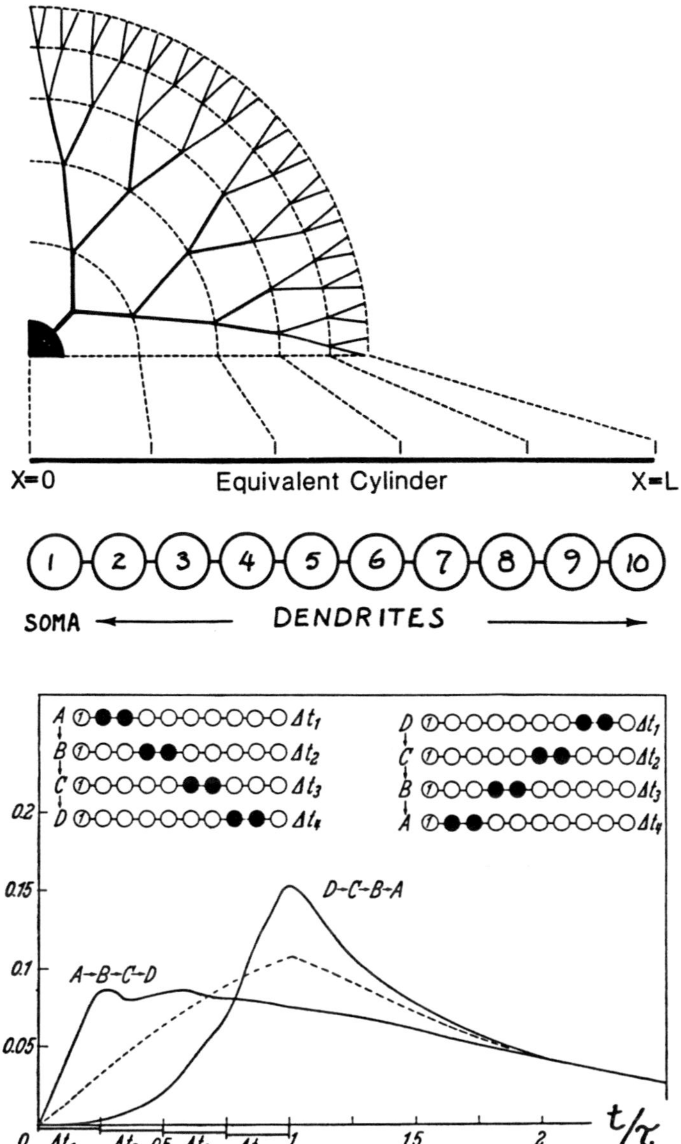

FIGURE 1. Effects of spatiotemporal input pattern. Upper diagram depicts an idealized dendritic tree that can be mathematically transformed into an equivalent cylinder (Rall, 1962) and into an approximately equivalent chain of ten compartments (Rall, 1964), where compartment 1 represents the neuron soma and compartments 2 through 10 correspond to progressive cable distance from the soma to the dendritic terminals. Middle diagrams show two sets of four ten-compartment chains. The set at left represents the spatiotemporal input pattern A–B–C–D to indicate the time sequence of the four input locations shown; this sequence is proximal first and then successively more distal. The set at right represents a contrasting spatiotemporal input pattern, D–C–B–A, to indicate a distal-to-proximal time sequence of the four input locations shown. The solid curves at bottom show computed EPSPs (i.e., normalized voltage transients at the soma) for the two contrasting input patterns, as labeled; the dotted curve shows the computed result when the same amount of input is smeared uniformly over the four locations and four time intervals (for further details, see text and Rall, 1964).

the basis for an interpretation of "asymmetric" firing patterns in cochlear neurons by Erulkar *et al.* (1968).

Different computed examples provided insight into the conditions for either linear or nonlinear combinations of the effects of synaptic inputs delivered to different dendritic locations for both excitatory and inhibitory synaptic inputs (Figs. 8 and 9 of Rall, 1964). Further discussion and references can be found elsewhere (Rall, 1964, 1967, 1977; Jack *et al.*, 1975; Redman, 1976; Segev and Parnas, 1983; Rall and Segev, 1985, 1987).

2.4. Effects of Input to a Single Dendritic Branch

Figure 2 illustrates an idealized dendritic neuron consisting of six equal dendritic trees. Current is injected at a single branch terminal designated I; this input branch is distinguished from its sibling branch (S) and its first- and second-cousin branches (C-1 and C-2). The resulting steady voltage distribution in the various branches of the input tree (shown in this figure) was computed from the general solution of this problem (Rall and Rinzel, 1973).

One noteworthy feature of these results is the contrasting decrement of voltage in the input branch and its sibling branch. Both branches have the same length and diameter in this idealized tree. However, the input branch is open to a large current flow into its parent branch; this permits a large flow of current along its cytoplasmic core, resulting

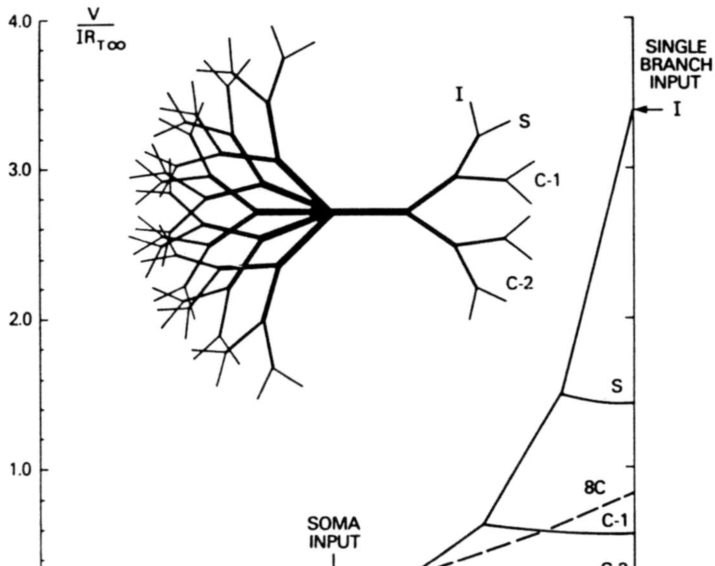

FIGURE 2. Diagram of idealized neuron model composed of six dendritic trees, and plot of steady-state voltage values for three different cases: current input at the soma (curve with short dashes), current input to a single distal branch terminal (input branch designated I; continuous curve), and the case of input divided equally among eight closely related branch terminals (I, S, C-1, and C-2; curve with long dashes). Modified from Rall and Rinzel (1973), which can be consulted for the mathematical statement and solution of this problem. The voltage scale is expressed relative to the product of the injected current and a reference input resistance (i.e., that of the dendritic trunk cylinder extended to semiinfinite length).

in a steep voltage decrement along the length of the input branch. In contrast, the sealed terminal of the sibling branch allows zero current to flow out of that end; also, little current flows across the high resistance of the cylindrical branch membrane; with so little current, this branch is almost isopotential. This contrast also applies to dendritic spines, with interesting functional consequences (see Section 4.7).

Another feature of these results is the contrast in input resistance values when the distal input location is compared with a central input location (at the soma). In this figure, the dashed curve shows the lower voltage values obtained when the same amount of current is injected at the soma as that previously injected at the distal branch. In this example, the distal input resistance is 16 times larger than the somatic input resistance, and still larger factors can result from additional orders of branching (Rall and Rinzel, 1973). This contrast in input resistance and its effect on voltage amplitude (i.e., local synaptic depolarization) are very important to the attainment of threshold conditions in excitable dendritic spines located on distal dendritic branches.

3. DENDRITIC SPINES

The existence of dendritic spines has been known for 100 years, since the classical studies of Ramón y Cajal; however, the demonstration of synaptic contacts on spines was accomplished much later by means of electron microscopic observations (Gray, 1959). For readers with limited knowledge of neuroanatomy, it may help to review the following facts: most neurons exhibit extensive dendritic branching; a single neuron has a dendritic surface area that is 20 to 100 times larger than that of the neuron soma; several important cell types have their dendrites studded with dendritic spines, numbering, say, 20,000 per neuron; such neurons also receive many synaptic contacts, numbering, say, 21,000 per neuron (in the case of neocortical pyramidal cells it has been reported that 95% of the synaptic contacts are on spines, Colonnier, 1968). Thus, there is a need to understand the functional possibilities for synaptic contacts on dendritic spines. Electrophysiological techniques have not yet been applied directly to dendritic spines, and it thus seems particularly appropriate to use computations to explore the functional possibilities of synaptic inputs to such spines.

3.1. Spine Stem Resistance

The variety in spine size and shape was demonstrated by Jones and Powell (1969) and by Peters and Kaiserman-Abramof (1970), as shown here in Fig. 3. Also included in this figure are our estimates of the electrical resistance to current flow inside the spine stem between the spine head and the spine base. These estimates depend on spine stem geometry and on the value of the intracellular resistivity, neither of which is known accurately; also, membranous inclusions within the spine stem cytoplasm may significantly increase the spine stem resistance (Wilson *et al.*, 1983; Miller *et al.*, 1985; Rall and Segev, 1987).

It was recognized by Chang (1952) that the high electrical resistance of the thin spine stem could be expected to attenuate the effect on the postsynaptic neuron of a synaptic input to a spine head. Because he expected significant attenuation, Chang concluded that summation of many synaptic inputs would be needed for effective excitation of such neurons.

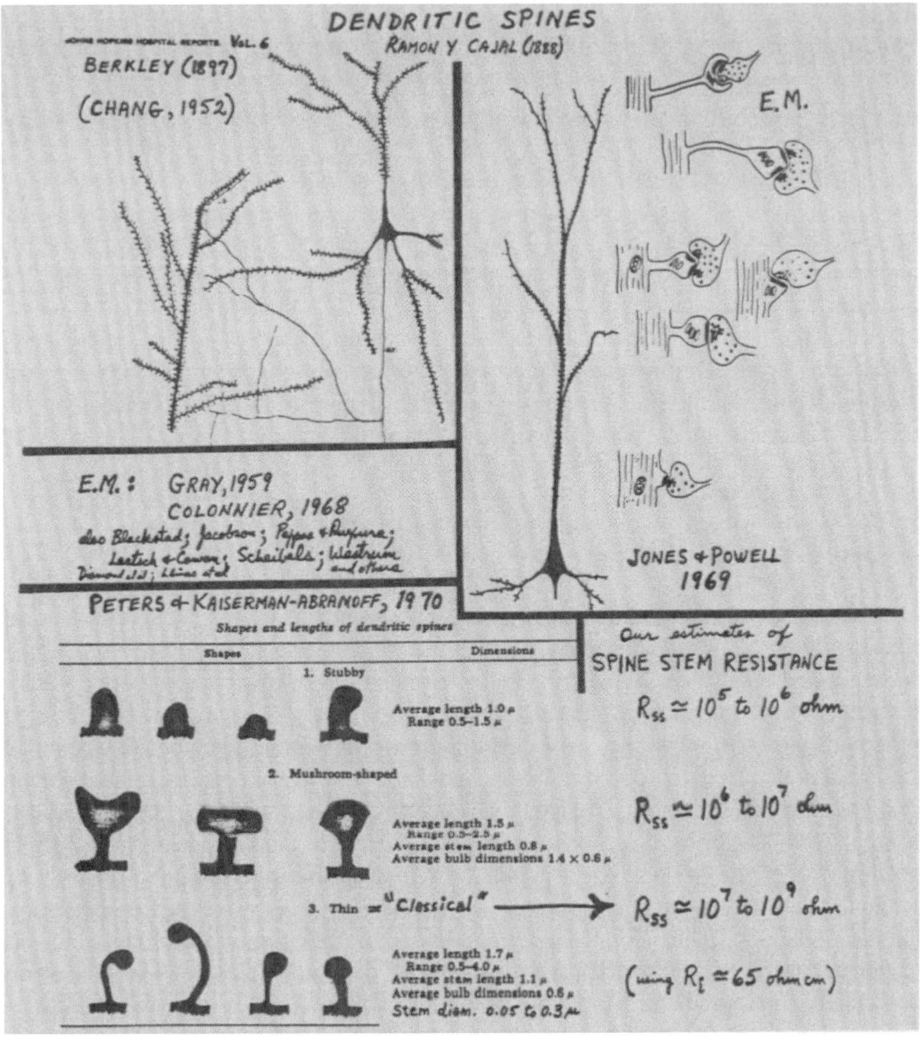

FIGURE 3. Neuroanatomic montage prepared as slide in 1971 to introduce dendritic spines. Upper left shows dendritic branches covered with dendritic spines, from an 1897 study of cortical neurons by Berkley, following Ramón y Cajal. Upper right shows diagrammatic pyramidal cell together with enlarged drawings of spines and synapses, based on electron microscopic observations, modified from Jones and Powell (1969). Lower left shows variety of dendritic spine shapes and sizes, modified from Peters and Kaiserman-Abramof (1970), and lower right gives our estimates (made in 1971) of the ranges of spine stem resistance values corresponding to the spines at left.

3.2. Changing Synaptic Efficacy by Changing Spine Stem Resistance

The idea that changes in spine stem resistance values would change the relative weighting (synaptic efficacy) of many different synapses was introduced by Rall and Rinzel (1971a,b). At that time, they computed both steady-state and transient responses for synaptic conductance input to a dendritic spine; this spine had only passive membrane, but a range of values was assumed for the amount of synaptic input and the values of spine stem resistance and other parameters (Rall, 1970, 1974, 1978; Rinzel, 1982; see also Diamond *et al.*, 1970; Jack *et al.*, 1975).

For steady-state conditions, a simple Ohm's law argument can be used to explain the effect of spine stem resistance. This is illustrated by Fig. 4; the spine stem current is given by the intracellular voltage difference from spine head to spine base divided by the spine stem resistance (this is true for both steady states and transients because more sophisticated analysis shows that negligible current crosses the membrane of the spine stem). This spine stem current times the input resistance (of the neuron at the branch input point where the spine is attached) gives the steady-state voltage at this branch input point (relative to the resting intracellular reference potential).

Using physical intuition (or the algebra summarized in Fig. 4), one can see, for example, that the voltage at the spine base will equal exactly half that at the spine head for the special case in which the spine stem resistance equals the branch input resistance; in other words, half of the total voltage drop occurs along the branches of the whole neuron (as illustrated in Fig. 2).

Both the graph and the equation in Fig. 4 show how the voltage ratio (spine base/spine head) depends on the ratio of spine stem resistance to branch input resistance; this illustrates the idea of an "operating range" for changes in synaptic efficacy determined by changes in spine stem resistance (relative to branch input resistance). A qualitatively similar "operating range" was found for transient responses to brief synaptic input, provided that the synaptic conductance input was sufficient for a large depolarization of the spine head. It may be noted that this did depend on nonlinearity in the spine head (approach to voltage saturation for depolarization caused by conductance input); for very small inputs (that would be of no physiological interest), these nonlinear effects are negligible, as was later also recognized by others (Koch and Poggio, 1983; Kawato and Tsukahara, 1984; Turner, 1984; Wilson, 1984).

In view of this "operating range" for changing spine stem resistance, it was suggested that evolution could have sacrificed maximal synaptic power in exchange for adjustability of relative synaptic weights. It was also pointed out that changes in the relative synaptic weights of large numbers of such synapses might contribute to neural plasticity underlying learning and memory (Rall and Rinzel, 1971a,b; Rall, 1974, 1978; Rinzel, 1982). This suggestion that the spine stem might be an important morphological locus for changes in synaptic weights did have an impact on anatomic studies; review of that literature is provided by Coss and Perkel (1985). Also noteworthy are the serial reconstructions of electron micrographs of dendritic branches, spines, and synapses recently reported by Harris *et al.* (1985; K. M. Harris, personal communication).

4. EXCITABLE SPINES

4.1. Spines with Excitable Spine Head Membrane

The possibility that the membrane of dendritic spine heads might be excitable was assumed by Diamond *et al.* (1970). Although this possibility was considered in various informal discussions at that time, it is noteworthy that Julian Jack analyzed this possibility carefully and published an early and astute discussion (Jack *et al.*, 1975); he pointed out that for an optimal range of parameter values, an action potential at the spine head could result in amplification of the synaptic effect.

Not until 1983, to the best of our knowledge, did anyone carry out transient computations for the response of an excitable spine head to brief synaptic input; then two

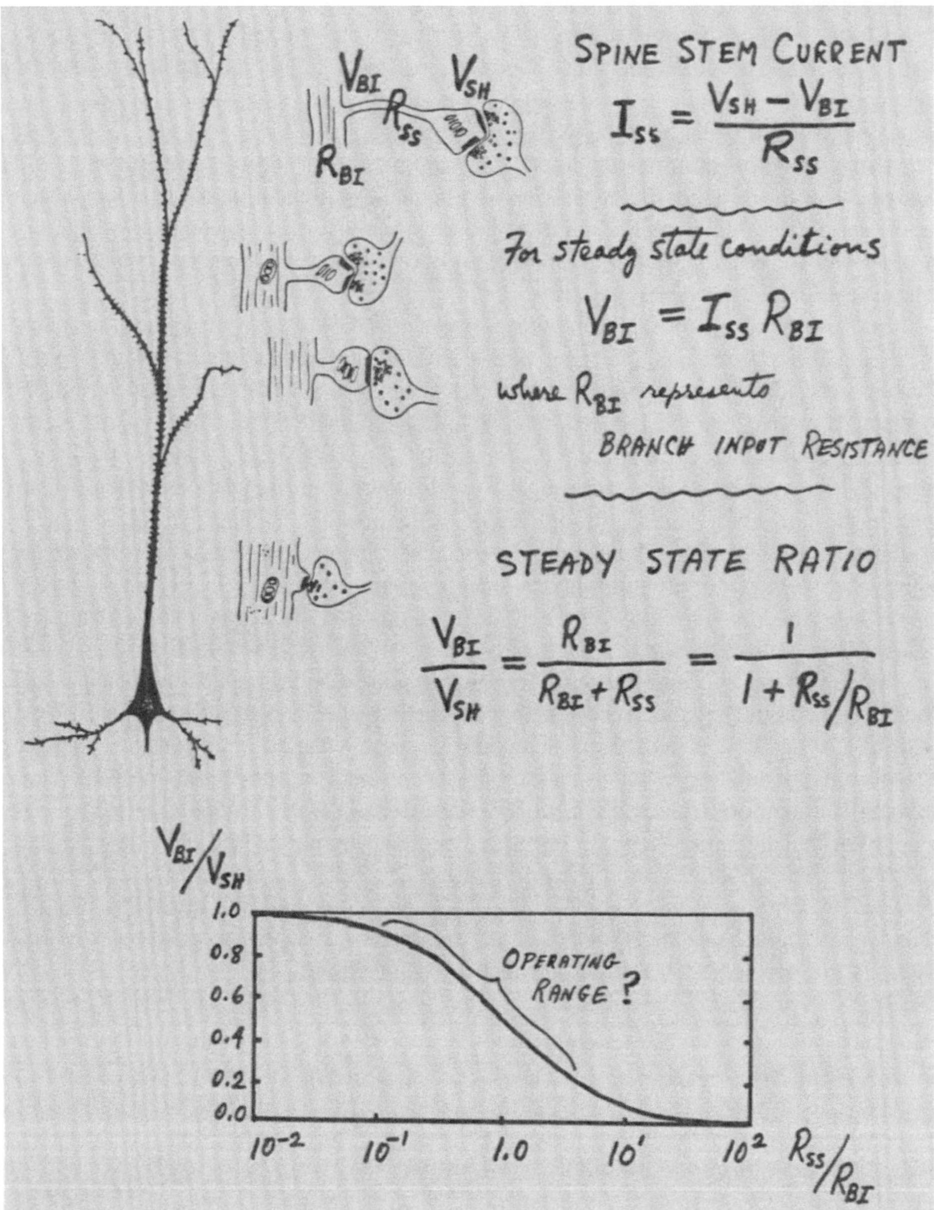

FIGURE 4. Diagram summarizing steady-state implications of Ohm's law for the currents, voltages, and resistances of one dendritic spine and the branch to which it is attached, also prepared as slide in 1971. The symbols I_{SS} and R_{SS} represent spine stem current and resistance, respectively; the voltages V_{SH} and V_{BI} are for the spine head and the spine base, which is also the branch input point, having branch input resistance designated R_{BI}. The plot at bottom displays the steady-state ratio dependence defined by the equation above; it suggests an "operating range" for adjustments of synaptic efficacy by changes of R_{SS} relative to R_{BI}.

independent research groups reported preliminary results at a symposium of the Society for Neuroscience. To acknowledge this coincidence, we arranged to submit paired short papers, first to *Nature* and then to *Science* (only to be told that these results and insights were of insufficient interest to a wide audience); paired papers were finally published in *Brain Research* (Miller *et al.*, 1985; Perkel and Perkel, 1985). Since then various functional implications of excitable dendritic spines have been explored in collaborative discussions (with John Miller, John Rinzel, and Gordon Shepherd). Miller has focused more on the conditions under which excitable dendritic spine interactions could generate bursts of spikes (Malinow and Miller, 1984). Shepherd has focused more on the possibility of saltatory propagation in distal dendrites, from one excitable spine head to another (Shepherd *et al.*, 1985). Our computations have focused on chain reaction effects in clusters of excitable spines (Rall and Segev, 1987).

4.2. Nonlinear Dependence on Spine Stem Resistance: Excitable Spines

Figure 5 illustrates computed results for a single dendritic spine located on a dendrite having an input resistance of 262 MΩ. A brief synaptic conductance input is delivered to the spine head, and the resulting voltage transients are shown for the spine head (upper left) and for the spine base (lower left, with different amplitude scale); the solid lines are for excitable spine head membrane, whereas the dashed lines are corresponding controls for passive membrane. The left side of this figure shows results for only two values of spine stem resistance: 630 MΩ (a) and 1000 MΩ (b). When the two spine head action potentials at upper left are compared, the shorter latency and greater amplitude of b indicate a secure spike, where the delay and the smaller amplitude of a indicate an insecure spike that barely succeeded under near-threshold conditions. The reason case b is secure is that the larger spine stem resistance results in a steeper and larger depolarization of the spine head in response to the same synaptic input; this is shown best by the dashed curves, which correspond to passive spine head membrane.

The right side of Fig. 5 summarizes results computed for 45 different values of spine stem resistance; a very strong nonlinearity is apparent. For spine stem resistance values less than 400 MΩ (for this set of parameter values), the response of the excitable spine head membrane differs negligibly from the passive controls; when this resistance is increased from 400 to 600 MΩ, some nonlinear deviation from the passive controls can be seen at the spine head; this is even more apparent at the spine base (see lower right). The greatest nonlinearity is over the range from 600 to 700 MΩ; this clearly corresponds to conditions just below and just above threshold for generation of an action potential in the spine head membrane.

4.3. Optimal Range for Maximal Amplification

Because the voltage delivered to the spine base (see lower right of Fig. 5) becomes smaller as spine stem resistance values increase above 700 MΩ, it follows that maximal amplification of synaptic efficacy occurs for an optimal range of values near threshold; in this case, a maximal amplification factor of about 6 occurs over an optimal range of about 630 to 670 MΩ. An intuitive explanation of the reduced amplification computed for the larger spine stem resistance values can be achieved by noting that the areas under the two action potentials, a and b at upper left, are approximately the same and that it is this time integral of spine head voltage (less spine base voltage) that drives the spine stem current and delivers charge to the neuron (at the spine base); thus, with approximately

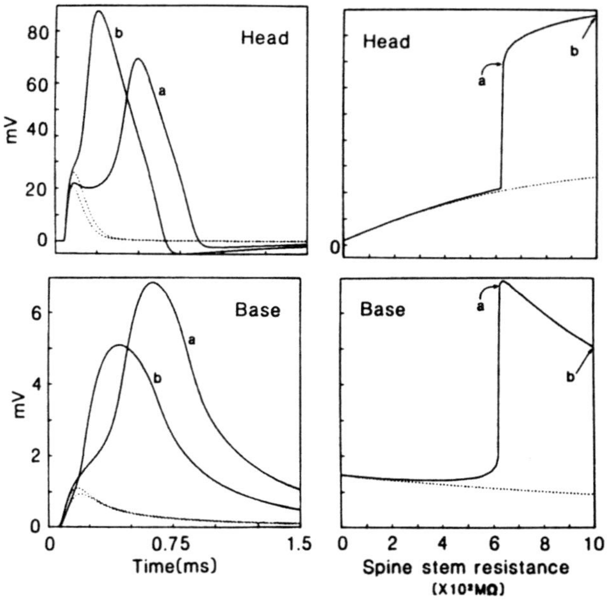

FIGURE 5. Dependence on spine stem resistance values, computed for depolarizing voltages at spine head and spine base in response to brief synaptic conductance input to the spine head, shown for excitable spine head membrane (continuous curves) and for passive spine head membrane (dotted curves). See text for description and discussion. Computations assumed a spine head area of 1.5 μm^2, of which one-eighth was assigned to the synaptic contact area and seven-eighths was either passive membrane (with parameters given below) or excitable membrane with Hodgkin and Huxley (1952) kinetics adjusted to ten times the channel density for squid axon at 22° using the computer program described by Parnas and Segev (1979); synaptic excitatory conductance had a peak value of 0.37 nS, with a reversal potential of 100 mV, and a time course proportional to $t \cdot \exp(-t/p)$, where the peak time p = 0.035 msec (with a 1.4-msec passive membrane time constant, this corresponds to a value of 40 for the usual α parameter); the dendrite was simplified to a 0.63-μm-diameter cylinder extending one length constant in both directions (with sealed ends) and with parameters (R_m = 1400 Ω cm^2, C_m = 1.0 μF/cm^2, R_i = 70 Ω cm) implying an input resistance of 262 MΩ,

equal voltage drive, Ohm's law implies that the current and charge delivered to the neuron must decrease as the spine stem resistance is increased. Clearly, changes in other parameters that shift threshold conditions will change the optimal range for this parameter.

4.4. Distal Branch Arbors and Spine Clusters

The linear density of dendritic spines on distal dendritic branches has been reported as about two spines per micrometer of length (Wilson *et al.*, 1983); higher values can result when corrections are made for spines hidden from view (Feldman and Peters, 1979), and the serial reconstructions by Harris *et al.* (1985; and K. M. Harris, personal communication) have yielded significantly higher densities for some neuron types. In any case, for distal branches of 25 μm length, it is quite conservative to allow 50 spines per branch. To simplify our computations, we idealized the distal dendritic arbors (Fig. 6) and assumed that every branch has exactly 50 spines and that exactly five of these possess excitable spine head membrane.

The inset in Fig. 6 shows more detail for distal branches, A and B, together with their parent branch, C. The symbolic notation is meant to indicate a particular case in

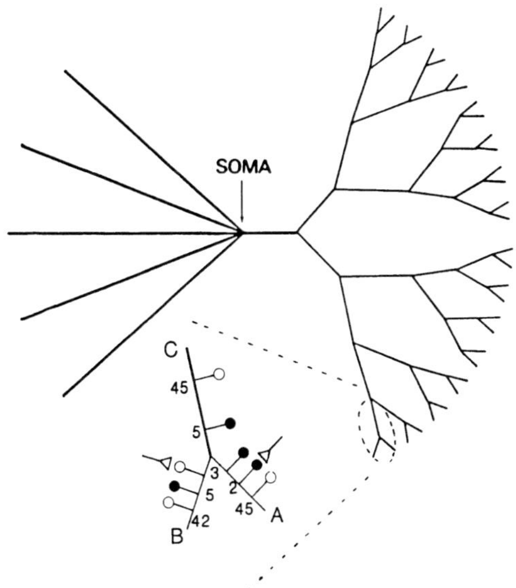

FIGURE 6. Diagram for focus on one distal dendritic arbor of an idealized neuron with six dendritic trees (of which five are represented by their equivalent cylinders; see Rall and Rinzel, 1973). Black spine heads are excitable, whereas those shown as open circles are passive; see text for further description and discussion of symbols and spine clusters. This idealized neuron was used for the computations summarized in Fig. 7. The branching is symmetrical and satisfies the 3/2 power diameter constraint for transformation to an equivalent cylinder (Rall, 1962, 1964, 1977); all branch lengths are set equal to 0.2 in dimensionless electrotonic length; here R_m was 2500 Ω cm², implying a 2.5-msec passive membrane time constant for the usual membrane capacity value; this R_m together with R_i of 70 Ω cm implies a 180-μm length constant for distal branches of 0.36 μm diameter; then the input resistance at the soma is 7.8 MΩ, and with five orders of branching, Table I of Rall and Rinzel (1973) shows that the distal branch input resistance must be about 50 times larger (here $L = 1.2$, $M = 5$, and $N = 6$). Computations for this model made use of the computer program SPICE; see Vladimirescu *et al.* (1980), Bunow *et al.* (1985), and Segev *et al.* (1985). Excitable spine head membrane used Hodgkin and Huxley (1952) kinetics adjusted to five times the channel density for squid axon at 22°.

which synaptic input is delivered to two of the excitable spines belonging to branch A and to three of the passive spines belonging to branch B but to none of the spines belonging to parent branch C. This particular case is one of the five different cases presented next in Fig. 7.

4.5. Processing of Different Synaptic Inputs

The use of symbols in Fig. 7 differs only slightly from Fig. 6. The left-hand column shows only those spines whose synapses are active in that particular trial; it is important to emphasize that the computation includes 45 passive spines and five excitable spines present on every branch for every trial. The middle column shows only those excitable spines that fired in response to that trial. The right-hand column shows only the resultant peak depolarization that reaches the neuron soma; some of the computed voltage values in the branches are mentioned in the following paragraphs.

Case 1 shows a synaptic input delivered to only one distal excitable spine; only that one spine fires, generating a local voltage peak of 8.5 mV (at the middle of branch A); this delivers 24 μV to the soma, double the amount that would result from the same

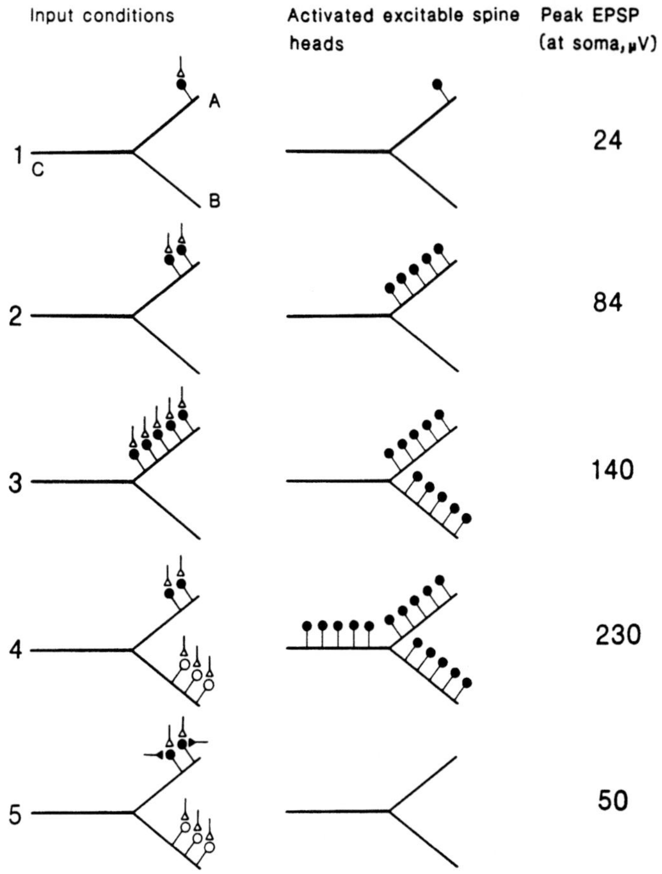

FIGURE 7. Excitable spine cluster firing contingencies; schematic summary for the five cases described and discussed in the text. Note that branches A, B, and C represent the same distal dentritic arbor shown enlarged in Fig. 6. The computations were based on the information summarized in the legend of that figure.

input to a distal passive spine. In this case, the excitable spine produces a small amplification of the response to the synaptic input, but there is no chain reaction between excitable spines. The same result would be expected for a single input to any one excitable spine on any of the terminal branches of this model.

Case 2 shows simultaneous synaptic inputs to two excitable spines on branch A; here the local depolarization of branch A is sufficient to reach the firing threshold of the other three excitable spines on branch A; this delivers an 84-μV peak depolarization at the soma. This result more than triples the result of case 1. It is not five times as great for two reasons: three spines fire with a slight delay relative to the first two; also, the driving potential for synaptic current is reduced by the depolarization of branch A, which peaks at 25.5 mV.

Several fascinating insights can be derived from this case. It is clear that the local depolarization is sufficient for the chain reaction to take place in branch A, but the 11.7-mV peak in branch B and the 8.6-mV peak in branch C are not quite sufficient to trigger chain reactions in the excitable spines of those branches. Note that the depolarization in branch B is larger than that in branch C; this was expected from an understanding of the asymmetric voltage attenuations in Fig. 2; this effect favors distally spreading chain reactions and limits central spread. Cases 3, 4, and 5 show examples of the different results obtainable by adding a small amount of synaptic input to that of case 2.

Case 3 shows simultaneous synaptic input to all five excitable spines belonging to branch A; here simultaneous firing of the five excitable spines of branch A produces sufficient local depolarization in branch B to fire (with slight delay) all five of its excitable spines; however, the local depolarization in parent branch C is not sufficient to fire its excitable spines (another example of the asymmetric decrement shown in Fig. 2). The result at the soma is less than twice that of case 2 and less than ten times that of case 1, but it is more than twice the control value for passive spines.

Case 4 shows a different combination of five simultaneous synaptic inputs with a significantly different result that can be attributed to more synchronous firing in branches A and B. The two inputs to branch A are the same as in case 2, but here three additional simultaneous inputs are delivered to any three of the 45 passive spines of branch B; this additional input produces enough additional local depolarization in branch B to fire its five excitable spines almost simultaneously with those of branch A; the resulting depolarization in parent branch C is more than that in case 3 and here proved sufficient to fire its five excitable spines. Thus, 15 spines fire, producing a 230-μV peak at the soma; this is nearly ten times that for case 1 and nearly three times that for case 2, the amplification factor is close to four relative to passive control.

Case 5 shows one of many possible examples of how effective and specific synaptic inhibitory input can be. Synaptic inhibition delivered to only one or two excitable spines can make the difference between success or failure in the firing of a large cluster of spines. The timing of inhibitory input relative to excitatory input can also be investigated along the lines previously explored (Rall, 1964; Jack *et al.*, 1975; Segev and Parnas, 1983).

Reviewing Fig. 7 suggests obvious additional cases, some of which we have already computed. If synchronous synaptic input is delivered to all five excitable spines of branch C, we find that all 15 excitable spines on these three branches do fire; the voltage peak in A and B occurs 0.78 msec later than that in C. In this case, the amount of depolarization that spreads into the sibling arbor is significant but not sufficient to fire those 15 excitable spines without some additional synaptic input to that arbor; with such an assist we would fire a cluster of 30 excitable spines in this pair of sibling arbors, but again, this could be blocked by very few strategically placed inhibitory synaptic inputs.

4.6. Brief Discussion of Excitable Spine Clusters

From the examples above it is clear that excitable spine clusters of different size can be fired and that success or failure to fire depends on a number of contingencies. This could provide a basis for logical processing of different synaptic input combinations and for the realization of various computations and integrative functions. All of this depends very nonlinearly on the various input parameters and system parameters, such as spine stem resistance. The spine stem resistance is still an attractive locus for changing synaptic weights (the effect is even more sensitive because of the steep nonlinearity displayed in Fig. 5); however, we do not suggest that this parameter represents the only one of importance. Also, it may be noted that the logical possibilities provided by the contingencies for the firing of various excitable spine clusters in distal dendritic arbors represent an updated version of an idea (about information processing in dendrites) that has been noted by several investigators over the years (Lorente de Nó and Condouris,

1959; Arshavskii *et al.*, 1965; Rall, 1970; Goldstein and Rall, 1974; R. FitzHugh, personal communication; Y. Y. Zeevi, personal communication).

4.7. Significance of Distal Branch Properties

Although this theoretical model can be made to work for different but interrelated ranges of parameter values, the importance of several insights noted with Fig. 2 can now be usefully reviewed. Without the large input resistance value found at distal dendritic locations, the local membrane depolarization produced by a few spine firings would be insufficient to fire other spines. The asymmetry of the voltage decrement has several interesting consequences: (1) the local voltage of the distal branch spreads with negligible decrement into the spine heads of the nearby spines that did not receive synaptic input; if this were not true, i.e., if there were a tenfold voltage decrement into these spine heads, as suggested by Diamond *et al.* (1970), these other spines would not reach threshold, and no chain reaction could occur; (2) without asymmetry in the arbors, the chain reaction would travel centrally, enveloping the entire neuron in an "all-or-nothing" response, which would destroy the richness made possible by distal cluster firings; (3) with the asymmetry, fractionation into distal clusters of different size occurs naturally. In addition to these points, large values of spine stem resistance (plus branch input resistance) are needed in order to reach threshold depolarization of the excitable spine head when it receives a reasonable amount of synaptic input. In other words, a rather special set of circumstances makes distal dendritic locations particularly well suited for excitable dendritic spines.

5. CONCLUSION

For all of the above reasons, we suggest that evolution has placed voltage-dependent ionic channels in the membrane of some distally located spine heads in sufficient number to make them excitable. Whether this suggestion is correct is not yet known; it is testable, in principle, by three different techniques, at least one of which can be expected to succeed in the next few years. These are (1) the use of antibodies to mark the locations of particular channels, (2) the use of a patch clamp or a suction electrode to record from individual spines, and (3) the use of voltage-sensitive dyes to record voltage transients and perhaps also voltage decrements in distal dendritic branches and dendritic spines.

ACKNOWLEDGMENT. Dr. Segev was a Fogarty Fellow at NIH. Some of the same results, figures, and insights have been presented at other symposia and may also appear in resulting publications.

REFERENCES

Arshavskii, Y. I., Berkinblit, M. B., Kovalev, S. A., Smolyaninov, V. V., and Chailakhyan, L. M., 1965, The role of dendrites in the functioning of nerve cells, *Dokl. Akad. Nauk SSSR* **163:**994–997.

Bunow, B., Segev, I., and Fleshman, J. W., 1985, Modeling the electrical behavior of anatomically complex neurons using a network analysis program: Excitable membrane, *Biol. Cybernet.* **53:**41–56.

Chang, H. T., 1952, Cortical neurons with particular reference to the apical dendrites, *Cold Spring Harbor Symp. Quant. Biol.* **17:**189–202.

Colonnier, M., 1968, Synaptic patterns on different cell types in the different laminae of the cat visual cortex. An electron microscope study, *Brain Res.*, **9**:268–287.

Coss, R. G., and Perkel, D. H., 1985, The function of dendritic spines: A review of theoretical issues, *Behav. Neural Biol.* **44**:151–185.

Diamond, J., Gray, E. G., and Yasargil, G. M., 1970, The function of the dendritic spines: An hypothesis, in: *Excitatory Synaptic Mechanisms* (P. Anderson and J. K. S. Jansen, eds.), Universitetsforlaget, Oslo, pp. 213–222.

Erulkar, S. D., Butler, R. A., and Gerstein, G. L., 1968, Excitation and inhibition in cochlear nucleus, II. Frequency-modulated tones, *J. Neurophysiol.* **31**:537–548.

Feldman, M. L., and Peters, A., 1979, A technique for estimating total spine numbers on Golgi-impregnated dendrites, *J. Comp. Neurol.* **118**:527–542.

Goldstein, S. S., and Rall, W., 1974, Changes of action potential shape and velocity for changing core conductor geometry, *Biophys. J.* **14**:731–757.

Gray, E. G., 1959, Axo-somatic and axo-dendritic synapses of the cerebral cortex: An electron microscopic study, *J. Anat.*, **93**:420–433.

Harris, K. M., Trogadis, J., and Stevens, J. K., 1985, Three dimensional structure of dendritic spines in the rat hippocampus (CA1) and cerebellum, *Abstract, Soc. Neurosci. Abstr.* **11**:306.

Henneman, E., Lüscher, H.-R., and Mathis, J., 1984, Simultaneously active and inactive synapses of single Ia fibres in cat spinal motoneurones, *J. Physiol. (Lond.)* **352**:147–161.

Hodgkin, A. L., and Huxley, A. F., 1952, A quantitative description of membrane current and its application to conduction and excitation in nerve, *J. Physiol. (Lond.)* **117**:500–544.

Jack, J. J. B., Noble, D., and Tsien, R. W. 1975, *Electric Current Flow in Excitable Cells,* Oxford University Press, London.

Jones, E. G., and Powell, T. P. S., 1969, Morphological variations in the dendritic spines of the neocortex, *J. Cell. Sci.* **5**:509–519.

Kawato, M., and Tsukahara, N., 1984, Electrical properties of dendritic spines with bulbous end terminals, *Biophys. J.* **46**:155–166.

Koch, C., and Poggio, T., 1983, A theoretical analysis of electrical properties of spines, *Proc. R. Soc. Lond. [Biol.]* **218**:455–477.

Lorente de Nó, R., and Condouris, G. A., 1959, Decremental conduction in peripheral nerve: Integration of stimuli in the neuron, *Proc. Natl. Acad. Sci. U.S.A.* **45**:592–617.

Malinow, R., and Miller, J. P., 1984, Interactions between active dendritic spines could generate bursts of spikes, *Soc. Neurosci. Abstr.* **10**:547.

Miller, J. P., Rall, W., and Rinzel, J., 1985, Synaptic amplification by active membrane in dendritic spines, *Brain Res.* **325**:325–330.

Parnas, I., and Segev, I., 1979, A mathematical model for conduction of action potentials along bifurcating axons, *J. Physiol. (Lond.)* **295**:323–343.

Perkel, D. H., and Perkel, D. J., 1985, Dendritic spines: Role of active membrane in modulating synaptic efficacy, *Brain Res.* **325**:331–335.

Peters, A., and Kaiserman-Abramof, I. R., 1970, The small pyramidal neuron of the rat cerebral cortex. The perikaryon, dendrites and spines, *Am. J. Anat.* **127**:321–356.

Rall, W., 1962, Theory of physiological properties of dendrites, *Ann. N.Y. Acad. Sci.* **96**:1071–1092.

Rall, W., 1964, Theoretical significance of dendritic trees for neuronal input–output relations, in: *Neural Theory and Modeling,* Stanford University Press, Stanford, pp. 73–97.

Rall, W., 1967, Distinguishing theoretical synaptic potentials computed for different soma–dendritic distributions of input, *J. Neurophysiol.* **30**:1138–1168.

Rall, W., 1970, Cable properties of dendrites and effect of synaptic location, in: *Excitatory Synaptic Mechanisms* (P. Andersen and J. K. S. Jansen, eds.), Universitetsforlaget, Oslo, pp. 175–187.

Rall, W., 1974, Dendritic spines, synaptic potency and neuronal plasticity, in: *Cellular Mechanisms Subserving Changes in Neuronal Activity* (C. D. Woody, K. A. Brown, T. J. Crow, and J. D. Knispel, eds.), Brain Information Service, Los Angeles, pp. 13–21.

Rall, W., 1977, Core conductor theory and cable properties of neurons, in: *Handbook of Physiology,* Vol. 1, Pt. 1, *The Nervous System, Cellular Biology of Neurons* (J. M. Brookhart, V. B. Mountcastle, and E. R. Kandel, eds.), American Physiological Society, Bethesda, pp. 39–97.

Rall, W., 1978, Dendritic spines and synaptic potency, in: *Studies in Neurophysiology* (R. Porter, ed.), Cambridge University Press, New York, pp. 203–209.

Rall, W., and Rinzel, J., 1971a, Dendritic spines and synaptic potency explored theoretically, *Proc. I.U.P.S.* **IX**:466.

Rall, W., and Rinzel, J., 1971b, Dendritic spine function and synaptic attenuation calculations, *Abstr. Soc. Neurosci.* **1**:64.

Rall, W., and Rinzel, J., 1973, Branch input resistance and steady attenuation for input to one branch of a dendritic neuron model, *Biophys. J.* **13**:648–688.

Rall, W., and Segev, I., 1985, Space clamp problems when voltage clamping branched neurons with intracellular microelectrodes, in: *Voltage and Patch Clamping with Microelectrodes* (T. G. Smith, Jr., H. Lecar, S. J. Redman, and P. Gage, eds.), American Physiological Society, Bethesda, pp. 191–215.

Rall, W., and Segev, I. 1987 Functional possibilities for synapses on dendrites and on dendritic spines, in: *Synaptic Function* (G. M. Edelman, W. E. Gall, and W. M. Cowan eds.), John Wiley & Sons, New York pp. 605–636.

Rall, W., Shepherd, G. M., Reese, T. .S, and Brightman, M. W., 1966, Dendrodendritic synaptic pathway in the olfactory bulb, *Exp. Neurol.* **14**:44–56.

Rall, W., Burke, R. E., Smith, T. G., Nelson, P. G., and Frank, K., 1967, Dendritic location of synapses and possible mechanisms for the monosynaptic EPSP in motoneurons, *J. Neurophysiol.* **30**:1169–1191.

Redman, S. J., 1976, A quantitative approach to the integrative function of dendrites, in: *International Review of Physiology: Neurophysiology II,* Vol. 10 (R. Porter, ed.), University Park Press, Baltimore, pp. 1–36.

Redman, S., and Walmsley, B., 1983, The time course of synaptic potentials evoked in cat spinal motoneurones at identified group Ia synapses, *J. Physiol. (Lond.)* **343**:117–133.

Rinzel, J., 1982, Neuronal plasticity (learning), in: *Some Mathematical Questions in Biology—Neurobiology,* Vol. 15, *Lectures on Mathematics in the Life Sciences* (R. M. Miura, ed.), American Mathematical Society, Providence, pp. 7–25.

Segev, I., and Parnas, I., 1983, Synaptic integration mechanisms. Theoretical and experimental investigation of temporal postsynaptic interaction between excitatory and inhibitory inputs, *Biophys. J.* **41**:41–50.

Segev, I., Fleshman, J. W., Miller, J. P., and Bunow, B., 1985, Modeling the electrical behavior of anatomically complex neurons using a network analysis program: Passive membrane, *Biol. Cybernet.* **53**:27–40.

Shepherd, G. M., and Brayton, R. K., 1979, Computer simulation of a dendrodendritic synaptic circuit for self- and lateral-inhibition in the olfactory bulb, *Brain Res.* **175**:377–382.

Shepherd, G. M., Brayton, R. K., Miller, J. P., Segev, I., Rinzel, J., and Rall, W., 1985, Signal enhancement in distal cortical dendrites by means of interactions between active dendritic spines, *Proc. Natl. Acad. Sci. U.S.A.* **82**:2192–2195.

Turner, D. A., 1984, Conductance transients on dendritic spines in a segmental cable model of hippocampal neurons, *Biophys. J.* **46**:85–96.

Vladimirescu, A., Newton, A. R., and Pederson, D. O., 1980, *SPICE Version 26.0 User's Guide,* EECS Department, University of California, Berkeley.

Wilson, C. J., 1984, Passive cable properties of dendritic spines and spiny neurons, *J. Neurosci.* **4**:281–297.

Wilson, C. J., Groves, P. M., Kitai, S. T., and Linder, I. C., 1983, Three dimensional structure of dendritic spines in the rat neostriatum, *J. Neurosci.* **3**:383–398.

23

Electrodiffusion Model of Electrical Conduction in Neuronal Processes

NING QIAN AND TERRENCE J. SEJNOWSKI

1. INTRODUCTION

The cable model of electrical conduction in neurons is central to our understanding of information processing in neurons. The conduction of action potentials in axons has been modeled as a nonlinear excitable cable (Hodgkin and Huxley, 1952), and the integration of postsynaptic signals in dendrites has been studied with analytic solutions to passive cables (Rall, 1977). Recently, several groups have examined the possibility of more complex signal processing in dendrites with complex morphologies and excitable membranes by numerical integration of the cable equations (Shepherd *et al.*, 1985; Koch *et al.*, 1983; Rall and Segev, 1985; Perkel and Perkel, 1985).

The cable equation is based on an electrical conductance model in which driving forces arising from ionic concentration differences across the membrane are represented by batteries in series with conductances. This model can be derived as an approximation to the Nernst–Planck equation for electrodiffusion. In this chapter we introduce an electrodiffusion model of electrical conduction in one dimension—along the longitudinal dimension of a thin process. In this preliminary report we determine conditions under which the electrical conductance model may not be valid. Complications such as cytoplasmic cisternae, membrane pumps, and ionic buffers will be considered in a later paper.

2. LIMITATIONS OF THE ELECTRICAL CONDUCTANCE MODEL

The membrane battery potentials in the electrical conductance model are usually obtained from the Nernst equation and are considered constants. This is a good approximation in the squid giant axon and large neurons but may introduce errors if the concentrations of some ions change significantly. This is more likely to occur in small processes and during synaptic events in small structures such as spines (Rall, 1978; Koch and Poggio, 1983).

A second limitation of the electrical conductance model is in the treatment of lon-

NING QIAN and TERRENCE J. SEJNOWSKI • Department of Biophysics, Johns Hopkins University, Baltimore, Maryland 21218.

gitudinal current spread within neurons. Only the potential gradient in the cytoplasm is considered, and not concentration gradients. This is usually a good assumption, but the concentration gradients can be large when spatial compartments are small and for some ions like Ca^{2+}, whose concentration can change dramatically under some circumstances.

One additional observation is that different ions may have different concentration-dependent cytoplasmic resistivities, but in the electrical conductance model only the total cytoplasmic resistivity is usually considered.

In the following sections we first derive a set of equations that govern the electrodiffusion of ions in thin cables and then present numerical solutions to these equations for an excitatory postsynaptic potential on a dendritic spine.

3. ELECTRODIFFUSION MODEL

The movement of ions in neurons is governed by the Nernst–Planck equation (Jack *et al.*, 1975):

$$\bar{J}_i = -D_i[\bar{\nabla} n_i + (n_i/\alpha_i)\,\bar{\nabla}\, V] \tag{1}$$

where V is the potential, $\bar{J}_i$ is the flux of ionic species i (number of particles per unit area), D_i is the diffusion constant, n_i is the concentration, and the constant α_i is defined as

$$\alpha_i = \alpha/z_i \tag{2}$$

with

$$\alpha = RT/F \tag{3}$$

where z_i is the charge per ion, R is the gas constant, F is the Faraday constant, and T is the temperature. The ionic concentrations and ionic currents must additionally satisfy the continuity equation:

$$\bar{\nabla} \cdot \bar{J}_i + \partial n_i/\partial t = 0 \tag{4}$$

The Nernst–Planck equation will be applied to a cylinder of diameter d. We assume that the longitudinal current and ionic concentrations are uniform across the transverse cross section of the cylinder and that the radial current is independent of angle around the axis of the cylinder. These assumptions reduce the problem of electrodiffusion to a one-dimensional problem along the axis of the cylinder. The constant-field approximation is made for the transverse currents passing through the cylinder (Goldman, 1943). The equations can be written in cylindrical coordinates and reduced to a single equation for the concentration as a function of the distance along the z axis of a cylinder:

$$\frac{\partial n_i}{\partial t} = D_i \frac{\partial^2 n_i}{\partial z^2} + \frac{D_i}{\alpha_i}\frac{\partial}{\partial z}\left(n_i \frac{\partial V}{\partial z}\right) - \frac{4P_iV}{\alpha_i d}\left[\frac{n_i^{out} - n_i\, e^{V/\alpha_i}}{1 - e^{V/\alpha_i}}\right] \tag{5}$$

where P_i is the permeability of the membrane and n_i^{out} is the concentration of ionic species i outside the membrane. The three terms on the right-hand side of this equation are, respectively, the contributions from pure diffusion, the potential gradient, and the membrane current. This equation must be supplemented by an additional constraint between the membrane potential and the ionic concentrations. We adopt the same capacitative model of the membrane used in the electrical conductance model:

$$V(t) = V(0) + \sum_i [n_i(t) - n_i(0)]\, z_i\, Fd/4c_m \tag{6}$$

where $V(0)$ is the initial voltage, $n_i(0)$ are the initial ionic concentrations, and the membrane has capacitance c_m per unit area.

If branches are allowed, then these equations must be solved on a tree rather than a line. At any jump in diameter, the continuity equation must be satisfied across the jump:

$$d_1^2 \left(\frac{\partial n_i}{\partial z} + \frac{n_i}{\alpha_i} \frac{\partial V}{\partial z} \right) \Bigg|_1 = d_2^2 \left(\frac{\partial n_i}{\partial z} + \frac{n_i}{\alpha_i} \frac{\partial V}{\partial z} \right) \Bigg|_2 \tag{7}$$

where the diameter of the process is d_1 on one side and d_2 on the other side. This implies that both the voltages and the ionic concentrations are continuous. However, by equation 6, the voltage will not be continuous at a diameter jump if the ionic concentrations are continuous; hence, the voltage at the jump is set to the average of the voltages on either side.

The coupled differential equations were solved by converting them to finite difference equations and solving them by an explicit method. The solutions at diameter jumps were obtained by solving the coupled nonlinear algrebraic equations derived from the matching conditions, equation 7. The calculation was performed for space and time steps of successively smaller size, and the values reported were ones for which further decrease to the step sizes made less than 2% difference to the solution.

4. RELATIONSHIP BETWEEN THE ELECTRODIFFUSION MODEL AND THE ELECTRICAL CONDUCTANCE MODEL

In large neurons, the internal and external ionic concentrations remain approximately constant during a transient excitation. The contribution of the diffusion of ions within the neuron then becomes negligible, and the longitudinal current is purely resistive. It can then be shown that

$$1/R_t = \sum_i (1/R_i) \tag{8}$$

$$1/R_i = (F_2/RT)\, D_i\, n_i\, z_i \tag{9}$$

where R_t is the total resistivity of the cytoplasm and R_i are the ionic resistivities for each species of ion. For the squid cytoplasm ($[K^+] = 400$ mM, $[Na^+] = 50$ mM) and for the D_i given in Table I, the estimated resistivities are $R_t = 29.7\ \Omega$ cm, $R_K = 33.4\ \Omega$ cm, and $R_{Na} = 267\ \Omega$ cm. There is a significant difference between the resistivities of the individual ionic species.

TABLE I. Parameters for Electrodiffusion Model in Figs. 2, 3, and 4

Symbol	Value	Parameter
D_K	1.96×10^{-5} cm^2/sec	Diffusion coefficient for K^+
D_{Na}	1.33×10^{-5} cm^2/sec	Diffusion coefficient for Na^+
P_K	3.64×10^{-6} cm/sec	Resting permeability of K^+
P_{Na}	6.07×10^{-8} cm/sec	Resting permeability of Na^+
$[K]_{in}$ (0)	140 mM	Initial internal K^+ concentration
$[Na]_{in}$ (0)	12 mM	Initial Na^+ concentration
$[K]_{out}$	4 mM	External K^+ concentration
$[Na]_{out}$	145 mM	External Na^+ concentration
P_M	6.07×10^{-3} cm/sec	Maximum Na^+ permeability of spine
t_p	0.25×10^{-3} sec	Time to reach peak permeability
T	20°C	Temperature
c_m	2 μF/cm^2	Membrane capacitance per unit area

5. ELECTRODIFFUSION MODEL OF A DENDRITIC SPINE

Many vertebrate and invertebrate neurons receive synaptic inputs on spines (Coss and Perkel, 1985). Because of the small size of dendritic spines, postsynaptic potentials can be accompanied by significant changes in the internal ionic concentrations. In this section we simulate an excitatory postsynaptic potential on a spine using the electrodiffusion model and compare the results with the conventional electrical conductance model.

The morphology of the dendritic spine used in the simulations is shown in Fig. 1. The synaptic input was modeled by a transient change in the Na^+ permeability of the membrane

$$P_{Na} = P_M \, (e \, t/t_p)^4 \, e^{-4t/t_p} \tag{10}$$

where P_M is the maximum Na^+ permeability and t_p is the time to reach peak (Kock and Poggio, 1983).

With the parameters of the model given in Table I, the resting potential was -78 mV. The total surface area of the spine head was 0.65 μm^2. The membrane potential

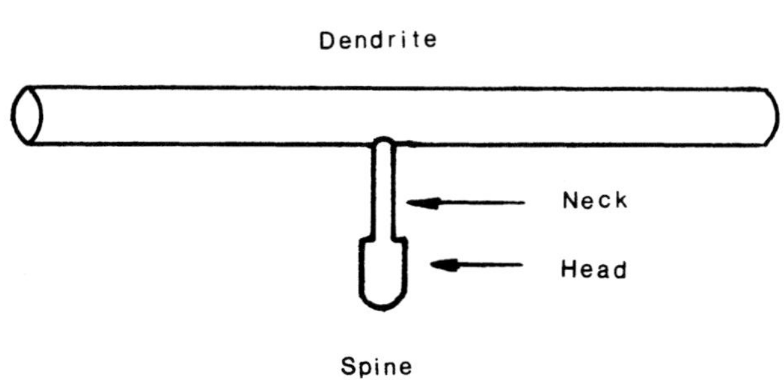

FIGURE 1. Geometry for the electrodiffusion model of a dendritic spine. The spine was in the center of a dendrite with a total length of 300 μm and a diameter of 1 μm; the spine neck was 1 μm long and 0.1 μm in diameter; the spine head was 0.69 μm long and 0.3 μm in diameter. Sample points in the dendrite were 10 μm apart, and the integration time step was 10^{-7} sec; in the spine head and neck the spacing was 0.173 μm and 0.167 μm, respectively, and the time steps were 10^{-9} sec. The model had a total of 41 sample points: 31 in the dendrite, six in the spine neck, and four in the spine head.

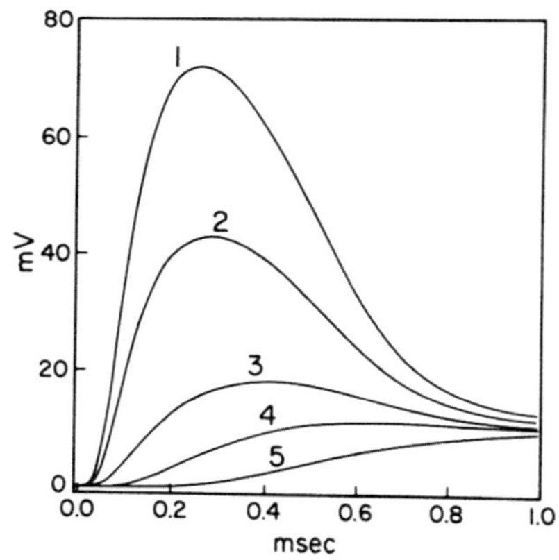

FIGURE 2. Excitatory postsynaptic potential modeled by electrodiffusion in a dendritic spine. The membrane potential relative to the resting potential is given as a function of time for the (1) middle of the spine head, (2) middle of the spine neck, (3) dendritic shaft at the base of spine, (4) dendrite 50 μm from spine, and (5) dendrite 150 μm from spine.

during the simulated excitatory postsynaptic potential is shown in Fig. 2, and the changes in the ionic concentrations of sodium and potassium are shown in Fig. 3. There is an increase in the sodium concentration inside the spine head of over threefold and a reduction in the concentration of potassium of 20%. In Fig. 4, the maximum response is shown as a function of the maximum sodium permeability during the excitatory postsynaptic potential.

The parameters given in Table II for the electrical conductance model were chosen so that the resting and equilibrium potentials and the resting currents of the electrodiffusion model closely matched those in the electrical conductance model. The transient change in the membrane conductance of Na^+ at the spine head during the synaptic input was modeled by

$$G_{Na} = G_M \, (e \; t/t_p)^4 \; e^{-4t/t_p} \tag{11}$$

where G_M is the maximum Na^+ conductance.

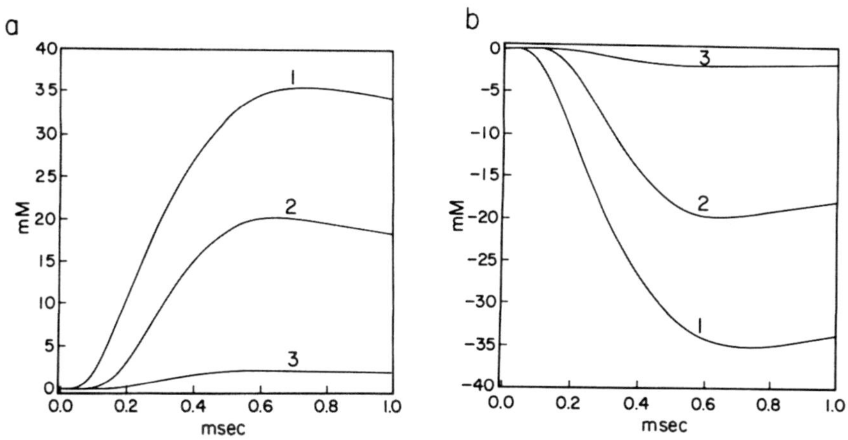

FIGURE 3. Ionic concentration changes for (a) Na^+ and (b) K^+ in a dendritic spine during an excitatory postsynaptic potential using the electrodiffusion model. Concentrations are given relative to the resting levels (see Table I) in the (1) middle of the spine head, (2) middle of the spine neck, and (3) dendritic shaft at the base of the spine.

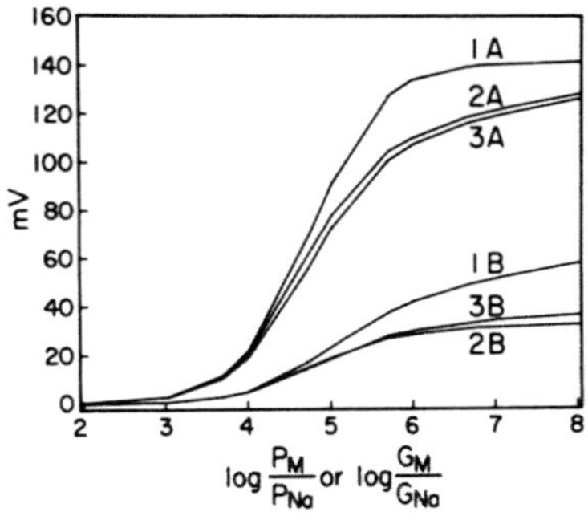

FIGURE 4. Maximum response during an excitatory postsynaptic potential for (1) the electrical conductance model and (2) the modified electrical conductance model as a function of the maximum conductance change and for (3) the electrodiffusion model as a function of the maximum permeability. The response functions for all three models are given at two locations: (A) the spine head and (B) the dendritic shaft at the base of the spine.

For small conductance changes, the two models predicted similar responses, as shown in Fig. 4. However, for large conductance changes, there were significant differences between the responses predicted by the electrical conductance model and the electrodiffusion model, especially at the base of the dendritic spine.

The saturation of the response in the electrical conductance model results from the approach of the membrane potential toward the sodium equilibrium potential. This saturation occurs at a lower membrane potential in the electrodiffusion model because of the increase in the internal sodium concentration and concomitant decrease of the sodium equilibrium potential.

The discrepancy between the two models can be reduced by using conductance changes in the electrical conductance model to match the membrane currents in the electrodiffusion model rather than the conductance changes given by equation 11. However, this procedure requires a complete solution of the electrodiffusion equations first. An alternative modification of the electrical conductance model is presented in the next section that is computationally less demanding.

6. MODIFICATIONS TO THE ELECTRICAL CONDUCTANCE MODEL

In the conventional electrical conductance model, the ionic concentrations inside the neuron are constant during changes in the membrane potential. This assumption can be relaxed by making several changes to the formalism:

1. Calculate the concentration of each ionic species in each compartment from the ionic currents flowing between compartments.

TABLE II. Parameters for Electrical Conductance Model in Fig. 4

Symbol	Value	Parameter
g_{iK}	5.56×10^{-3} S/cm	Cytoplasmic conductance of K^+
g_{iNa}	5.56×10^{-3} S/cm	Cytoplasmic conductance of Na^+
g_{mK}	2.31×10^{-4} S/cm^2	Resting membrane conductance of K^+
g_{mNa}	1.94×10^{-5} S/cm^2	Resting membrane conductance of Na^+
E_K	−89.8 mV	K^+ equilibrium potential
E_{Na}	62.9 mV	Na^+ equilibrium potential
G_M	1.26×10^{-8} S	Maximum sodium conductance of spine
t_p	0.25×10^{-3} sec	Time to reach peak conductance
c_m	2 $\mu F/cm^2$	Membrane capacitance per unit area

2. Compute the new equilibrium potentials during each time step and update the membrane batteries.
3. Replace the longitudinal resistance between compartments with parallel conductances in series with batteries and treat them in the same way as membrane conductances.

This modified electrical conductance model applied to the dendritic spine model in Fig. 1 gave qualitatively similar results for the changes in ionic concentrations compared with the electrodiffusion model. Without the above modifications, the predicted ionic concentration changes were markedly in error, in some cases having the wrong sign. The modified electrical conductance model also made predictions for the maximum responses that were qualitatively similar to those of the electrodiffusion model, with quantitative discrepancies of less than 10% over the entire range of conductance changes, as shown in Fig. 4.

7. DISCUSSION

In most circumstances, the electrical conductance model of electrical conduction in neurons gives accurate predictions for membrane potentials during transient electrical events. In this chapter we have developed an electrodiffusion model of electrical conduction for thin processes that reduces to the electrical conductance model for processes with large diameters.

This one-dimensional electrodiffusion model was used to study the changes in concentration of ions in dendritic spines during excitatory postsynaptic potentials. During a conductance change for Na^+ at the distal tip of a spine, the concentration of sodium can transiently increase by threefold, and the potassium concentration can decrease by 20%. Thus, significant errors can be made in estimating the membrane potential and concentration changes with the electrical conductance model if the effects of diffusion are not taken into account. We suggest a modification of the electrical conductance model to minimize these errors.

In a later paper we will extend the present model by including membrane pumps, buffers, and other ions, such as Ca^{2+}, that may also be important (Simon and Llinas, 1985; Fogelson and Zucker, 1985). For some problems it may be necessary to include spatially inhomogeneous diffusion within neurons, which would require the solution of the Nernst–Planck equation in three dimensions.

ACKNOWLEDGMENTS. We are grateful to Drs. Julian Jack, Wilfred Rall, and John Rinzel for helpful comments on earlier versions of this chapter.

REFERENCES

Coss, R. G., and Perkel, D. H., 1985, The function of dendritic spines, *Behav. Neural Biol.* **44:**151–185.

Fogelson, A. L., and Zucker, R. S., 1985, Presynaptic calcium diffusion from various arrays of single channels, *Biophys. J.* **48:**1003–1017.

Goldman, D. E., 1943, Potential, impedance and rectification in membranes, *J. Gen. Physiol.* **27:**37–60.

Hodgkin, A. L., and Huxley, A. F. 1952, Currents carried by sodium and potassium ions through the membrane of the giant axon of *Loligo, J. Physiol. (Lond.)* **116:**449–472.

Jack, J. J. B., Noble, D., and Tsien, R. W., 1975, *Electrical Current Flow in Excitable Cells,* Oxford University Press, Oxford.

Koch, C., and Poggio, T., 1983, A theoretical analysis of electrical properties of spines, *Proc. R. Soc. Lond. [Biol.]* **218:**455–477.

Koch, C., Poggio, T., and Torre, V., 1983, Nonlinear interaction in a dendritic tree: Location, timing, and role in information processing, *Proc. Natl. Acad. Sci. U.S.A.* **80:**2799–2802.

Perkel, D. H., and Perkel, D. J., 1985, Dendritic spines: Role of active membrane modulating synaptic efficacy, *Brain Res.* **325:**331–335.

Rall, W., 1977, Core conductor theory and cable properties of neurons, in: *Handbook of Physiology: The Nervous System* (E. R. Kandel, ed.), American Physiological Society, Bethesda, pp. 39–97.

Rall, W., 1978, Dendritic spines and synaptic potency, in: *Studies in Neurophysiology* (R. Porter, ed.), Cambridge University Press, Cambridge, pp. 203–209.

Rall, W., and Segev, I., 1987, Functional possibilities for synapses on dendrites and dendritic spines, in: *New Insights into Synaptic Function* (G. M. Edelman, W. F. Gall, and W. M. Cowan, eds.), John Wiley & Sons, New York (in press).

Shepherd, G. M., Brayton, R. K., Miller, J. P., Segev, I., Rinzel, J., and Rall, W., 1985, Signal enhancement in distal cortical dendrites by means of interactions between active dendritic spines, *Proc. Natl. Acad. Sci. U.S.A.* **82:**2192–2195.

Simon, S. M., and Llinas, R. R., 1985, Compartmentalization of the submembrane calcium activity during calcium influx and its significance in transmitter release, *Biophys. J.* **48:**485–498.

24

The Effectiveness of Individual Synaptic Inputs with Uniform and Nonuniform Patterns of Background Synaptic Activity

WILLIAM R. HOLMES and CHARLES D. WOODY

1. INTRODUCTION

The effectiveness of a synaptic input in changing the potential at the soma depends strongly on the resistance of the dendritic membrane (Barrett and Crill, 1974b; Rall, 1959), which, in turn, depends on the density and degree of opening of ionic channels in the membrane. Since ionic channels are presumed to be most highly concentrated near synapses, and since the distributions and types of synapses and their frequencies of activation may be highly nonuniform, the resistance of dendritic membrane can be expected to be nonuniform (Barrett, 1975; Eaton, 1980). Furthermore, synaptic activity, which is responsible for opening and closing ionic channels varies from moment to moment. Spatial and temporal variations in membrane resistance may cause the effectiveness of an individual synaptic input to be quite different at one moment than at another.

In this study, the effectiveness of single synaptic inputs in producing a change in potential at the soma is studied for particular cases of uniform and nonuniform background distributions of activated dendritic conductances in a model of a cortical pyramidal neuron. Since knowledge of the activated conductances and their distributions within cortical pyramidal cells remains limited, it was not known how accurately a particular distribution used in the simulations might mimic the true situation. Accordingly, several different distributions were tested. The simulations performed suggest that the effectiveness of a distal input could be markedly altered with different distributions of activated conductances and that it is important to determine possible distributions of activated synapses to understand neuronal function.

WILLIAM R. HOLMES • Mathematical Research Branch, National Institute of Diabetes and Digestive and Kidney Diseases, National Institutes of Health, Bethesda, Maryland 20892. CHARLES D. WOODY • Mental Retardation Research Center, Brain Research Institute, University of California at Los Angeles, Los Angeles, California 90024.

2. METHODS

2.1. Modeling Distributions of Synaptic Inputs in a Cortical Pyramidal Cell

Simulations were performed using a passive cable model that gave the transient potential response to individual synaptic conductance changes and distributions of ionic conductance changes (Holmes, 1986). The hundreds of discrete synaptically generated conductance changes of a particular type that might occur in a particular dendritic segment were approximated by an average constant conductance change distributed uniformly over the segment. By averaging inputs on each dendritic segment in this way, distributions of synaptic activity could be approximated. For a uniform distribution of synaptic activity, the particular average conductance values were the same for all dendritic segments (and thus R_m, the specific membrane resistance, was uniform). For nonuniform distributions of synaptic activity, different dendritic segments could have different average conductance values (and thus different R_m values).

Uniform and nonuniform distributions of synaptic activity were modeled in the layer V cortical pyramidal neuron pictured in Fig. 1. This cell was injected intracellularly with horseradish peroxidase (HRP) (Sakai *et al.*, 1978) and was serially reconstructed. Dimensions used in the simulations were obtained from a planar montage composed of photomicrographs taken at different, overlapping areas within several serial sections. Distributions of each type of conductance were assigned based on distance from the soma. Five regions were established (see Fig. 1): the soma, 0–40 μm from the soma, 40–100 μm from the soma, 100–200 μm from the soma, and beyond 200 μm from the soma. Thus, the conductance of a given type was assigned the same value over all dendrites at the same distance from the soma.

2.2. Types of Conductances Modeled

Three different types of conductances were used in the simulations described below. The first conductance, G_1 resulted from an excitatory input that operated by increasing conductance and had a reversal potential of 0 mV. The basis for this choice was reports of excitatory synapses in which glutamate or aspartate has been proposed to cause an increase in conductance to Na^+ or Ca^{2+} (Connors *et al.*, 1982; Flatman *et al.*, 1983; Stafstrom *et al.*, 1982, 1984, 1985). The second type of conductance modeled was caused by an excitatory input that operated by decreasing the magnitude of the conductance through (usually open) ionic channels that had a reversal potential of −85 mV. This conductance was modeled from reports of muscarinic synapses in which acetylcholine is the neurotransmitter. The conductance decreased at these synapses is thought to be a K^+ conductance. Since estimates for the reversal potential for this conductance range from −75 to −109 mV (Krnjevic *et al.*, 1971), −85 mV was chosen for use in the simulations. The third type of conductance change modeled was evoked by an inhibitory input in which the conductance was increased for an ion whose reversal potential was more negative than the resting potential. The reversal potential for these synapses was also chosen to be −85 mV. This choice of input was based on reports of synapses in which GABA or glycine has been proposed to increase the conductance to K^+ or Cl^-. Because the reversal potentials for the second and third types of conductances were chosen to be the same, differences between these two were indistinguishable in the model, and so they were considered together as one conductance, G_2. For a resting potential of −60 mV,

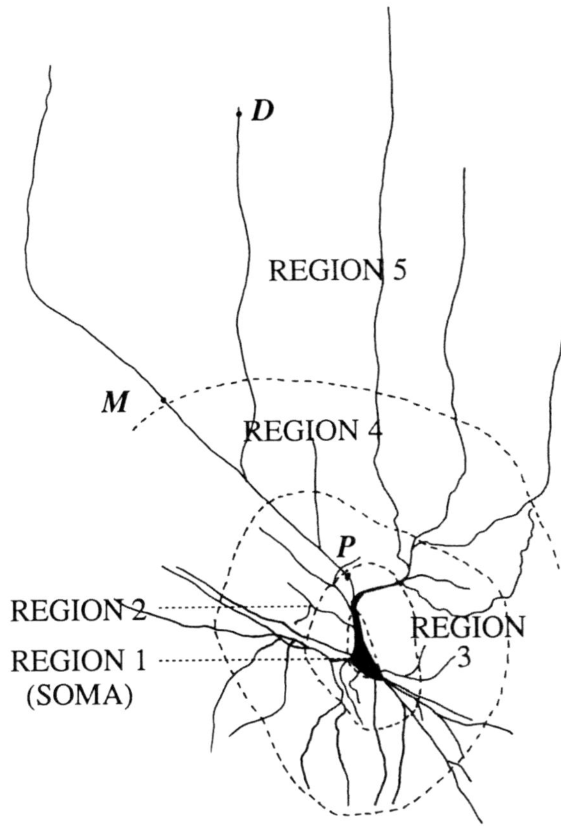

FIGURE 1. Planar projection of the reconstructed layer V cortical pyramidal cell used in the simulations. Distributions of synaptic conductances were modeled in five regions. Region 1 is soma. Region 2 comprises the first 40 μm of proximal dendrites. Region 3 encompasses all dendrites 40–100 μm from the soma. Dendrites at 100–200 μm are in region 4. Region 5 includes all dendrites beyond 200 μm from the soma. Single inputs were modeled at proximal (P), middendritic (M) and distal (D) locations as indicated.

over 70% of the total conductance was associated with G_2. The modeled conductances are passive conductances. No attempt was made to incorporate voltage dependencies.

Individual synaptic inputs were modeled at proximal (P), middendritic (M), and distal (D) locations as indicated in Fig. 1. The individual inputs were assumed to cause a 2-nS conductance change of type G_1 (reversal potential of 0 mV).

2.3. Synaptic Distributions Modeled

Five different distributions of background synaptic activity ("activation distributions") were modeled. These five were chosen so that soma resting potential, cell input resistance, total conductance G_1, and total conductance G_2 were approximately the same for all simulations. In the first distribution (D_1), both conductances G_1 and G_2 were distributed uniformly over all dendrites. The equivalent R_m for this distribution was 2000 $\Omega\text{-cm}^2$ throughout the dendritic tree. In the second and third distributions (D_2 and D_3), conductance G_1 (reversal potential of 0 mV) was distributed uniformly, and conductance

TABLE I. Background Synaptic Activity Distributions

Distribution	Conductance	
	G_1	G_2
D_1	Uniform all regions	Uniform all regions
D_2	Uniform all regions	Proximal regions
D_3	Uniform all regions	Distal regions
D_4	Proximal regions	Uniform all regions
D_5	Distal regions	Uniform all regions

G_2 (reversal potential of −85 mV) was distributed either only on the soma and on proximal dendritic membrane less than 40 μm from the soma (D_2) or primarily on distal dendritic membrane beyond 200 μm from the soma (D_3). In the final two distributions, conductance G_2 was uniformly distributed, and conductance G_1 was distributed either only on the soma and on proximal dendritic membrane less than 40 μm from the soma (D_4) or primarily on distal dendritic membrane beyond 200 μm from the soma (D_5). These distributions are summarized in Tables I and II.

3. RESULTS

3.1. "Resting" Membrane Potential

The membrane potentials at proximal (P), middendritic (M), and distal (D) locations as labeled in Fig. 1 are given in Table III for each of the five modeled distributions of background synaptic activity. When both conductances were uniformly distributed over the cell (D_1), the membrane potential was −59.5 mV at all locations in the cell. However, with nonuniform distributions of background synaptic activity, the membrane potential was found to be highly nonuniform. With conductance G_2 uniformly distributed and conductance G_1 found primarily on distal dendritic membrane (D_5), the membrane potential at the distal location (D) was found to be −32.8 mV. Conversely, when conductance G_1 was uniformly distributed and conductance G_2 was distributed primarily on distal dendritic membrane (D_3), the membrane potential at the distal location was −79.4 mV. In all cases the membrane potential at the soma was approximately −60 mV. Different distributions of background synaptic activity caused the membrane potential in different dendritic regions to be quite different from that seen at the soma.

TABLE II. Equivalent R_m Values (Ω-cm²) for Conductance Distributions D_1–D_5

		D_1	D_2	D_3	D_4	D_5
Region 1	(soma)	2000.0	754.7	555.6	1200.5	2564.1
Region 2	(0–40 μm)	2000.0	754.7	555.6	1200.5	2564.1
Region 3	(40–100 μm)	2000.0	8000.0	555.6	3003.0	2564.1
Region 4	(100–200 μm)	2000.0	8000.0	1282.1	3003.0	2564.1
Region 5	(<200 μm)	2000.0	8000.0	271.7	3003.0	719.4

TABLE III. Uniform and Nonuniform Distributions of Synaptic Activity[a]

Distribution	Input location		
	Proximal	Middendritic	Distal
A. D_1 (soma RP −59.5)			
Resting potential at site	−59.5	−59.5	−59.5
Electrotonic distance	0.05	0.57	1.47
Steady-state EPSP at soma	1.88	1.02	0.67
B. D_2 (soma RP −60.7)			
Resting potential at site	−60.2	−51.2	−47.4
Electrotonic distance	0.08	0.34	0.80
Steady-state EPSP at soma	1.91	1.14	0.85
C. D_3 (soma RP −59.9)			
Resting potential at site	−60.5	−72.1	−79.4
Electrotonic distance	0.03	0.57	3.09
Steady-state EPSP at soma	1.88	0.76	0.14
D. D_4 (soma RP −59.5)			
Resting potential at site	−59.9	−67.4	−70.7
Electrotonic distance	0.07	0.49	1.23
Steady-state EPSP at soma	1.90	1.27	0.96
E. D_5 (soma RP −60.5)			
Resting potential at site	−59.9	−45.2	−32.8
Electrotonic distance	0.04	0.50	2.04
Steady-state EPSP at soma	1.89	0.64	0.19

[a] The effectiveness of proximal, middendritic, and distal inputs is compared for five different distributions of synaptic conductances. The resting membrane potential in millivolts at the input site and the electrotonic distance from the soma to each site ($\Sigma\ l_i/\lambda_i$) is also given.

3.2. Electrotonic Distance

The electrotonic distance from the soma to synapses at the proximal (P), middendritic (M), and distal (D) locations was determined for each of the five "activation distributions" discussed above by summing the electrotonic lengths of each segment on the direct path from the soma to the synapse ($\Sigma l_i/\lambda_i$, where l_i is the length of segment i and λ_i is the space constant of segment i). Results are given in Table III. When both conductances were uniformly distributed, the distal synapse was found to be at an electrotonic distance of 1.47 from the soma. However, with different distributions of background synaptic activity, the electrotonic distance to this synapse was found to range from 0.8 to 3.1 (cf. distributions D_2 and D_3).

3.3. Effectiveness of Individual Synaptic Inputs

The steady-state soma potential change caused by a synaptic input resulting in a synaptic conductance change of 2 nS (reversal potential of 0 mV) was determined for inputs at three dendritic locations (P, M, and D in Fig. 1) for each of the five distributions of background synaptic activity.

With distribution D_2, conductance G_1 was distributed uniformly, but conductance G_2 was localized to the soma and most proximal dendrites. With this distribution, the driving force for the middendritic and distal inputs was reduced compared to the uniform case (D_1), but the electrotonic distance from the soma to these two input locations was also reduced. The reduction in electrotonic distance easily offset the reduction in driving force, and so the potentials at the soma caused by middendritic and distal synaptic inputs were larger than those seen with uniform conductances. In fact, the change in soma potential produced by the distal synapse increased by over 25%.

When conductance G_1 was uniformly distributed but G_2 was distributed primarily on distal dendrites (D_3), the membrane potentials at middendritic and distal locations were hyperpolarized relative to the uniform distribution case, making the driving force larger. However, the electrotonic distance from the soma doubled for the distal synapse. This increase in electrotonic distance dominated the increase in driving force, and the net result was that the potential change at the soma was reduced for the middendritic and distal synapses compared to the uniform case. The reduction for the distal synapse was over 75% in this simulation. Since G_2 comprises most of the total conductance in the cell, different distributions of G_2 can make the electrotonic distance to a distal synapse either small (D_2) or large (D_3). Since the soma resting potential was relatively close to the reversal potential of G_2, changes in driving force produced by nonuniform distributions of G_2 tended to be less important than changes in electrotonic distance.

In distributions D_4 and D_5, conductance G_2 was distributed uniformly, and G_1 was distributed either on the soma and first 40 μm of proximal dendrites (D_4) or distally beyond 200 μm from the soma (D_5). When G_1 was distributed proximally, the driving force was larger, and the electrotonic distance from the soma slightly smaller for the middendritic and distal synapses than it was in the uniform distribution case. Both of these factors increased the effectiveness of these synapses, particularly the distal one (over 40%). When G_1 was distributed distally, the opposite happened. The driving force was reduced, and the electrotonic distance from the soma was increased, compared to the uniform distribution case. These factors lessened the effectiveness of the middendritic and distal synapses. The reduction for the distal synapse was over 70%. Changes in the distribution of G_1 were more likely to affect the driving force at a synapse than the electrotonic distance because the reversal potential for G_1 was far from the soma resting potential and G_1 contributed far less to the total conductance than G_2.

4. DISCUSSION

One consequence of nonuniform background synaptic activity is nonuniform membrane potential in the cell. This finding is in agreement with experimental findings in cortical pyramidal neurons and other central neurons. Resting potentials in the −30 to −60 mV range have been reported for dendritic recordings of cortical pyramidal cells (Woody and Gruen, 1978; Woody *et al.*, 1984). A statistically significant difference in resting potentials between somatic and dendritic recordings has also been reported in thalamic cells (Jahnsen and Llinas, 1984a,b). Differences in soma and dendritic resting potentials in cerebellar Purkinje cells also have been found (Llinas and Sugimori, 1980a,b). The results reported here demonstrate how nonuniform distributions of background synaptic activity might produce such nonuniformities in the membrane potential. By producing a nonuniform driving force, nonuniform membrane potentials can have significant effects on the effectiveness of individual synaptic inputs.

Another consequence of nonuniform distributions of background synaptic activity is nonuniform membrane resistance. It has already been proposed that membrane resistance in nonuniform in motoneurons (Barrett and Crill, 1974a; Fleshman *et al.*, 1983), although the pattern of nonuniformity and the reasons for this nonuniformity have yet to be determined. The results reported here show that different distributions of background synaptic activity may produce highly nonuniform patterns of membrane resistance, which can have a significant influence on the effectiveness of distal synaptic inputs.

The effectiveness of a distal synaptic input in changing soma potential is a variable, dynamic property that depends on the distribution and types of background synaptic conductances activated at a given moment. Different distributions of activated synapses may significantly affect the effectiveness of a distal synaptic input at the soma by changing the membrane potential and hence the driving force at the synaptic site and by changing the electrotonic distance from the soma to the synaptic site. The geometry of the dendritic tree sets a constraint on the effectiveness of a distal synapse, but how remote that synapse is electrically from the soma depends on the distribution of activated conductances. The effectiveness of a proximal synapse in changing the soma potential is comparatively static. Nonuniform patterns of synaptic activity produce only small changes in the driving force and electrotonic distance from the soma for proximal synapses. However, different distributions of background synaptic activity may significantly alter the membrane potential at distal locations and the electrotonic distance to distal synapses and thus make distal synapses appear to be close to the soma or far away from the soma. Dendritic geometry sets the limits within which the "activation distribution" can make individual synapses highly effective or not effective at all.

Given the importance of the background synaptic activity in determining the resting membrane potential at different locations in a cell and the pattern of nonuniform membrane resistances, disturbances in background activity could distort the perception of the effectiveness of a distal synaptic input. In fact, electrophysiological properties of cells in which much of the tonic background synaptic activity has been eliminated may be quite different from those in which such activity is present, and this should be kept in mind by those working with experimental preparations.

REFERENCES

Barrett, J. N., 1975, Motoneuron dendrites: Role in synaptic integration, *Fed. Proc.* **34:**1398–1407.

Barrett, J. N., and Crill, W. E., 1974a, Specific membrane properties of cat motoneurones, *J. Physiol. (Lond.)* **239:**301–324.

Barrett, J. N., and Crill, W. E., 1974b, Influence of dendritic location and membrane properties on the effectiveness of synapses on cat motoneurones, *J. Physiol (Lond.)* **239:**325–345.

Connors, B. W., Gutnick, M. J., and Prince, D. A., 1982, Electrophysiological properties of neocortical neurons *in vitro, J. Neurophysiol.* **48:**1302–1320.

Eaton, D., 1980, How are the membrane properties of individual neurons related to information processing? in: *Information Processing in the Nervous System* (H. M. Pinsker and W. D. Willis, Jr., eds.), Raven Press, New York, p. 39–58.

Flatman, J. A., Schwindt, P. C., Crill, W. E., and Stafstrom, C. E., 1983, Multiple actions of N-methyl-D-aspartate on cat neocortical neurons *in vitro, Brain Res.* **266:**169–173.

Fleshman, J. W., Segev, I, Cullheim, S., and Burke, R. E., 1983, Matching electrophysiological with morphological measurements in cat alpha-motoneurons, *Soc. Neurosci. Abstr.* **9:**431.

Holmes, W. R., 1986, A continuous cable method for determining the transient potential in passive dendritic trees of known geometry, *Biol.Cybernet.* **55:**115–124.

Jahnsen, H., and Llinas, R., 1984a, Electrophysiological properties of guinea-pig thalamic neurons: An *in vitro* study, *J. Physiol.(Lond.)* **349:**205–226.

Jahnsen, H., and Llinas, R., 1986, Ionic basis for the electroresponsiveness and oscillatory properties of guinea-pig thalamic neurones, *in vitro, J. Physiol.(Lond.)* **349:**227–247.

Krnjevic, K. Pumain, R., and Renaud, L, 1971, The mechanism of excitation by acetylcholine in the cerebral cortex, *J. Physiol.(Lond.)* **215:**247–268.

Llinas, R., and Sugimori, M., 1980a, Electrophysiological properties of *in vitro* purkinje cell somata in mammalian cerebellar slices, *J. Physiol.(Lond.)* **305:**171–195.

Llinas, R., and Sugimori, M., 1980b, Electrophysiological properties of *in vitro* purkinje cell dendrites in mammalian cerebellar slices, *J. Physiol.(Lond.)* **305:**197–213.

Rall, W., 1959, Branching dendritic trees and motoneuron membrane resistivity, *Exp. Neurol.* **1:**491–527.

Sakai, M., Sakai, H., and Woody, C. D., 1978, Intracellular staining of cortical neurons by pressure microinjection of horseradish peroxidase and recovery by core biopsy, *Exp. Neurol.* **58:**138–144.

Stafstrom, C. E., Schwindt, P. C., Crill, W. E., and Flatman, J. A. 1982, Membrane currents in cat neocortical neurons, *in vitro, Soc. Neurosci. Abstr.* **8:**413.

Stafstrom, C. E., Schwindt, P. C., Crill, W. E., and Flatman, J. A., 1984, Properties of subthreshold response and action potential recorded in layer V neurons from cat sensorimotor cortex *in vitro, J. Neurophysiol.* **52:**244–263.

Stafstrom, C. E., Schwindt, P. C., Chubb, M. C., and Crill, W. E., 1985, Properties of persistent sodium conductance and calcium conductance of layer V neurons from cat sensorimotor cortex *in vitro, J. Neurophysiol.***53:**153–170.

Woody, C. D., and Gruen, E., 1978, Characterization of electrophysiological properties of intracellularly recorded neurons in the neocortex of awake cats: A comparison of the response to injected current in spike overshoot and undershoot neurons, *Brain Res.* **158:**343–357.

Woody, C. D., Gruen, E., and McCarley, K., 1984, Intradendritic recordings from neurons of the motor cortex of cats, *J. Neurophysiol.* **51:**925–938.

25

Passive and Active Properties of Motoneuron Dendrites

STEPHEN REDMAN and JOHN CLEMENTS

1. INTRODUCTION

The active and passive current–voltage characteristics of dendritic membrane must be determined if the integrative actions of central neurons are to be fully understood. Reliable measurements of these properties have been elusive, largely because the electrical transients generated by a somatically located microelectrode are dominated by the electrical properties of somatic and proximal dendritic membrane and hence give limited resolution of the properties of more distal dendritic membrane. Yet some progress has been made, mainly by combining electrophysiological measurements with intracellular staining of the same neuron (Lux *et al.*, 1970; Barrett and Crill, 1974; Turner and Schwartzkroin, 1983; Turner, 1984; Durand *et al.*, 1983). This chapter gives a brief report on results obtained by combining voltage clamp, current clamp, and intracellular HRP staining in the same motoneuron.

Excitatory postsynaptic potentials (EPSPs) evoked in dendrites will be larger and briefer than the same EPSPs recorded at the soma. The summed activity of dendritic EPSPs will have large voltage excursions occurring on a rapid time scale. With this in mind, it is important to determine the voltage range over which dendritic membrane remains passive (assuming some range exists), whether it is the same for all dendritic membrane, and what the passive membrane resistivity is in different regions of the dendrites. Beyond the voltage range for passive responses, the rapid fluctuations in dendritic membrane potential make it important to determine the kinetics of activation and inactivation of voltage-dependent conductances as well as the types of channels present and their distribution throughout the dendrites. One approach to answering some of these questions is to use dendritic EPSPs evoked at different and relatively localized regions of dendrites to generate dendritic depolarization. Changes in the amplitude and time course of these EPSPs at the soma after different substances known to block certain types of voltage-activated channels are applied intracellularly can give some insight into the types of channels present and their location. The results of experiments using this approach (Clements *et al.*, 1986) are discussed in this chapter.

STEPHEN REDMAN and JOHN CLEMENTS • Experimental Neurology Unit, John Curtin School of Medical Research, Australian National University, Canberra, A.C.T., Australia.

2. PASSIVE MEMBRANE PROPERTIES OF MOTONEURONS

2.1. A Two-Time-Constant Model of the Motoneuron

A widely used model of the passive electrical properties of neurons was developed by Rall (see Rall, 1977, for a review). This model assumes a uniform membrane resistivity (R_m) throughout the soma and dendrites and a uniform diameter for the cylindrical cable that represents the combined dendritic tree. Morphological reconstruction of motoneurons filled by intracellular injection of horseradish peroxidase or lucifer (Barrett and Crill, 1974; Egger and Egger, 1982; Ulfhake and Kellerth, 1981a,b, 1984) has revealed that the equivalent cylinder tapers towards its distal end, largely because terminal branches of dendrites end at variable electrical lengths from the soma. Attempts to match the response of the uniform-R_m, uniform-diameter cable model to the response measured in motoneurons after a brief current pulse was applied to the soma (Iansek and Redman, 1973) were largely unsuccessful. The most likely reason was that R_m for the somatic membrane was less than R_m for the dendritic membrane (see Jack, 1979, for discussion). Obviously, if procedures for measuring neuron model parameters are based on an inappropriate cable representations, incorrect values will result.

If electrical and geometric nonuniformities are to be usefully included in the motoneuron cable model, they must be resolvable from experimental measurements. We approached this problem by voltage clamping and current clamping each motoneuron and then injecting horseradish peroxidase (HRP). This procedure defined the morphology and provided two transients that could be used for parameter evaluation.

The results of one experiment are shown in Fig. 1. The reconstructed neuron is shown in Fig. 1A, the current response to a voltage step in Fig. 1C, and the voltage response to a brief current pulse in Fig. 1D. A compartmental model of the neuron was constructed by measuring soma dimensions, dendritic diameters, and lengths between branch points, similar to the approach described by many others (e.g., Rall, 1962; Johnston and Brown, 1983). Identical voltage steps and current pulses to those used experimentally were applied to the soma of the model. The neuron was assumed to have a membrane capacitance (C_m) of 1 μFcm^{-2}, and the cytoplasmic resistance (R_i) was assumed to be 70 Ωcm (Barrett and Crill, 1974). The membrane resistivity of the dendrites (R_{md}) was assumed to be uniform throughout the dendrites but different from the resistivity of the somatic membrane (R_{ms}). Values of R_{md} and R_{ms} were adjusted in the compartmental model until the best fit to both the experimental voltage and current transients was obtained, using a minimum sum of squared errors criterion. The transients that gave the best fits are shown in Fig. 1C,D overlying the experimental points and were obtained with R_{md} = 8.2 kΩcm^2 and R_{ms} = 0.26 kΩcm^2. This corresponds to a somatic shunt resistance (R_{sh}) of 2.2 MΩ.

The electrotonic length of each dendritic branch was calculated using R_{md} = 8.2 kΩcm^2, and the dendrites were then reduced to an equivalent dendritic cable using procedures developed by Rall (see Rall, 1977). Because all dendritic branches do not terminate at identical electrical lengths from the soma, the diameter of the equivalent dendritic cable progressively decreases, as shown in Fig. 1B. Once the equivalent cable is calculated in electrotonic length, it is a straightforward procedure to convert it to an equivalent diameter in physical distance. Both profiles show marked tapering, and the equivalent radius is halved at 1.1λ from the soma.

The final time constant of decay of an EPSP or the response to a current pulse is

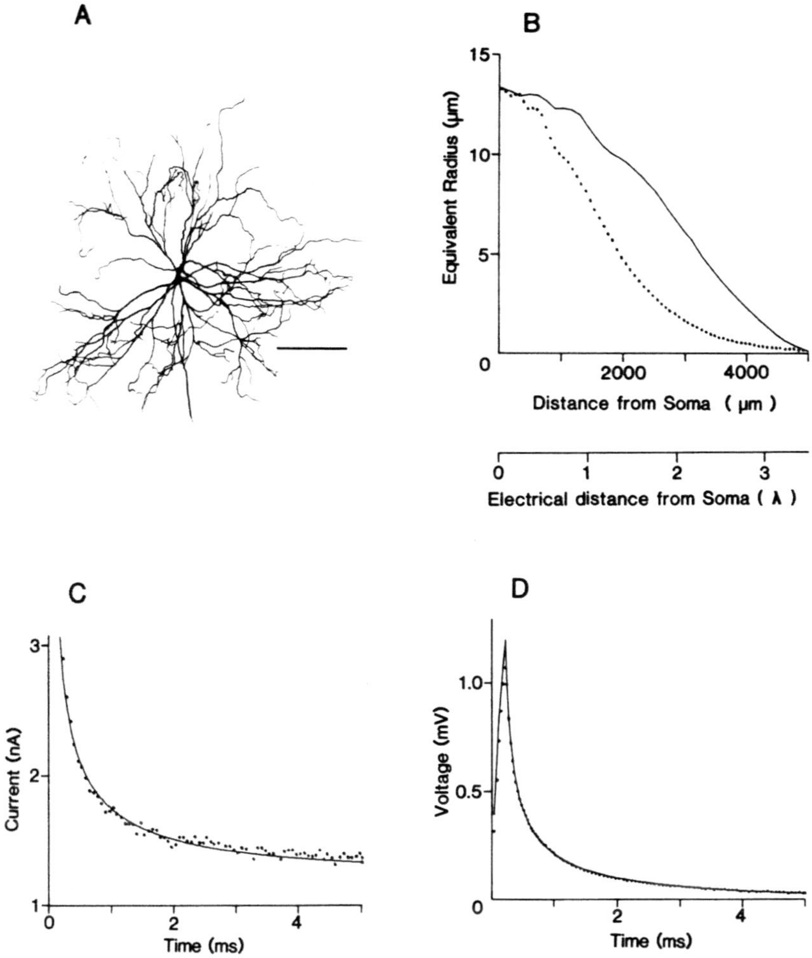

FIGURE 1. The reconstructed motoneuron is shown in A. (Calibration bar is 500 µm.) The current transient associated with a voltage step of 2 mV (hyperpolarization) is shown in C (dots). The voltage response to a 0.2-msec, 3-nA hyperpolarizing current is shown in D (dots). The transients generated by these voltage and current steps in a compartmental model of the reconstructed neuron in A for R_{md} = 8.2 kΩcm^2, R_{ms} = 0.26 kΩcm^2, R_i = 70 Ωcm, and C_m = 1 µFcm^{-2} are superimposed in C and D as continuous lines. These were the best fits that could be obtained by varying R_{md} and R_{ms} independently. The diameters and lengths of each dendritic branch, together with the R_i and R_{md} values, were used to convert the dendrites to an equivalent dendrite by combining branches at equal electrotonic distances from the soma. The dendritic profile in terms of electrotonic length is shown (as dots) in B. This profile can be converted into an equivalent profile in terms of physical distance from the soma (continuous line in B).

often measured and interpreted as the membrane time constant (τ_m). If it is assumed that R_m is uniform throughout the neuronal membrane, then $\tau_m = R_mC_m$. The final time constant measured for the current pulse response in Fig. 1D was 4.4 msec, which gives R_m = 4.4 kΩcm^2 (assuming C_m = 1 µFcm^{-2}), whereas R_{md} was calculated to be 8.2 kΩcm^2 from the full time course of both transients.

There are electrical and geometric uncertainties that can have a marked effect on the result of this analysis of membrane resistivity. One is the unknown amount of shrinkage that occurs in the neuron's dimensions as a result of histological processing. A linear expansion was incorporated into all diameters and lengths to compensate for shrinkage. Barrett and Crill (1974) estimated a volume shrinkage of up to 50%, which, if uniform in all directions, corresponds to a 20% shrinkage along any axis. Ulfhake and Kellerth (1981a) report up to 10% longitudinal shrinkage using a similar fixation procedure to that used in this study but made no allowance for shrinkage on the basis that it may not be uniform throughout the tissue. An inappropriate compensation can have a dramatic effect on the calculated R_{md}. For example, if no shrinkage had occurred in the neuron illustrated in Fig. 1B, the values for R_{md} and R_{ms} would be 14.3 and 0.2 kΩcm^2, respectively compared with 8.2 and a 0.26 kΩcm^2.

Another problematic measure is the value of R_i. This is difficult to measure in neurons, and estimates in the range 50–200 Ωcm have been reported for invertebrate axons (see Rall, 1977). The specific resistivity of a 100 mM KCl, 50 mM NaCl solution at 37°C in 1% agar is 43 Ωcm, and this must be close to the minimum R_i for neurons. Table I shows the values obtained for R_{md} and the somatic shunt resistance for the neuron analyzed in Fig. 1 when R_i was varied from 50 to 100 Ωcm. Both R_{md} and R_{sh} are very dependent on the value used for R_i. Acceptable fits were never obtained in any of the neurons analyzed (see below) for R_i = 50 Ωcm. For some neurons, including the neuron illustrated in Fig. 1, R_i = 100 Ωcm gave fits that were as good as those obtained with R_i = 70 Ωcm.

The results for six other neurons analyzed using the same procedures as those illustrated in Fig. 1 are presented in Table II. All but one neuron (cell 24) could be reliably fitted to the experimental transients using the two-time-constant model with R_i = 70 Ωcm. The input resistance (R_N) is dominated by the somatic shunt resistance, and the somatic shunt makes the apparent membrane resistivity of the soma much less than the dendritic membrane resistivity. The final time constant of decay of the current pulse response (τ_{cp}) generally underestimates the dendritic membrane resistivity. Notice that with the exception of cell 24, somatic shunts are required for all the neurons studied.

2.2. Other Cable Models

The two-time-constant model may be an oversimplification of the actual distribution of R_m, and it may be more correct to assume a gradual increase in membrane resistivity with distance from the soma (Fleshman *et al.*, 1983). In a compartmental model with the same geometry as the neuron analyzed in Fig. 1, we let R_{md} increase linearly with a defined gradient as a function of electrotonic distance from the soma. Transients were generated to identical voltage-clamp and current-clamp steps as those used experimentally,

TABLE I. Best-Fit Parameters to Pulse Transients[a]

R_i (Ωcm)	R_{md} (kΩcm^2)	r_{sh} (MΩ)	GOF index
50	4	3.3	4.4
70	8.2	2.2	2.3
100	24	2.0	2.0

[a] The best fits to both the voltage-clamp and current-pulse transients were calculated for R_i = 50, 70, and 100 Ωcm. The quality of the fit is expressed as a goodness-of-fit index (GOF), and it is the sum of squared errors between the fitted curve and the experimental transient, normalized by the same measure in the base-line region. The results for each of the two transients are added, and an ideal fit has a nominal GOF = 2. However, noise on the records adds statistical variability to this measure, and empirical testing indicates that GOF values between 1.4 and 2.6 can represent equally acceptable fits to the experiment data.

TABLE II. Analysis of Seven Motoneurons[a]

Model parameter	Cell number						
	10	11	14	22	24	113[b]	114[b]
τ_{cp} (msec)	3.6	5.1	6.5	6.5	8	4.3	3.7
R_{md} (kΩcm^2)	11	35	9	8	1.6	28	17
R_{ms} (kΩcm^2)	.14	.20	.42	.26	1.6	.22	.10
r_{sh} (MΩ)	1.8	2.0	3.1	2.2	∞	2.1	1.0
GOF index	2.4	2.6	2.1	2.3	4.0	2.6	2.3

[a] Cell 22 is illustrated in Fig. 1. τ_{cp} is the final time constant of decay of the response to a brief intracellular current pulse, and r_{sh} is the shunt resistance that must be placed in parallel with the soma to lower its specific membrane resistivity from R_{md} to R_{ms}.
[b] A single electrode voltage clamp was used in these neurons.

and the parameters of a two-time-constant model with the same geometry were sought that gave identical responses. A combination of R_{ms} and R_{md} could always be found such that the response generated by the two-time-constant model was indistinguishable from the response of a model in which R_{md} increased linearly with distance. For example, if $R_{ms} = 5$ kΩcm^2, and R_{md} increased linearly at 4 kΩcm$^2/\lambda$, this model generated voltage-clamp and current-clamp transients that were indistinguishable from those generated by a two-time-constant model in which $R_{ms} = 2.5$ kΩcm^2 and $R_{md} = 7.5$ kΩcm^2.

This lack of uniqueness was not present when the two-time-constant model was compared with the single-time-constant model ($R_{ms} = R_{md}$). The quality of the fit to the experimental transients was vastly inferior to that achieved with the two-time-constant neuron.

2.3. The Electrical Length of the Dendrites

The diameter profile of the equivalent dendritic cable at different electrotonic distances from the soma for the neuron analyzed in Fig. 1 is shown in Fig. 1B and together with another three motoneurons in Fig. 2. There is a region of approximately constant equivalent diameter 0.3–0.5λ, and then rapid tapering occurs to a negligible diameter at 2λ. The conventional definition of dendritic length for a uniform diameter cable that terminates abruptly does not apply here. Various definitions for electrotonic length could be used, such as the length at which the diameter is 0.71 of its initial value (i.e., the cross-sectional area is halved) or the length at which the initial diameter is halved. For either of these measures, the electrical length of all the equivalent dendrites in Fig. 2 is 1.2λ or less, indicating that the higher resistivity of the dendritic membrane in the two-time-constant model makes the dendrites more electrically compact than was previously found using the single-time-constant model.

2.4. The Somatic Shunt Resistance

The somatic shunt resistances were usually 2 to 3 mΩ, which makes the apparent resistivity of the somatic membrane very much less than R_{md}. The question arises as to whether all or part of this shunt should be attributed to the leak caused by microelectrode impalement (Jack, 1979). Gustaffson and Pinter (1984) estimate that up to 30% of the input conductance could be attributed to a somatic conductance shunt, and they argue that because this conductance shunt becomes smaller with more negative resting membrane

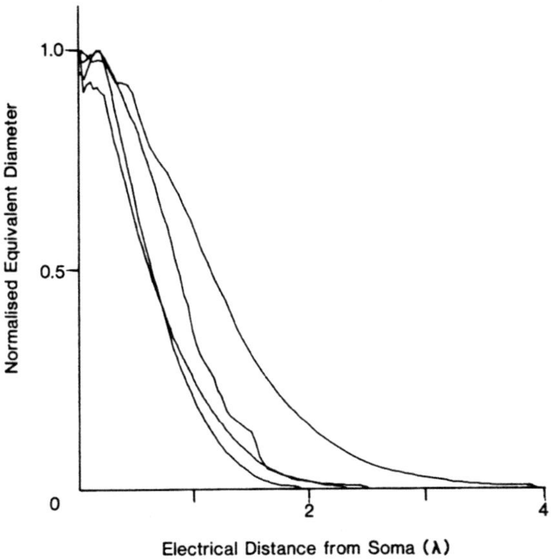

FIGURE 2. The R_{md} for each motoneuron has been used to calculate the electrotonic length of each dendritic branch, and these branches have been combined to form an equivalent dendrite. The diameter of each equivalent dendrite varies with electrotonic distance from the soma, mainly because terminal branches end at different electrotonic distances. The initial diameter has been normalized to allow a useful superposition of the four profiles. Neurons 10 and 11 were not included in this diagram because we suspected that the distal dendrites had not completely stained. Neuron 24 was excluded because the best R_{md} value was obtained with a very poor fit to the experimental transients.

potentials, it can be entirely associated with the electrode leak conductance. With the exception of cell 24, the shunt conductance varied from 60 to 90% of the input conductance, i.e., a much greater contribution than that estimated by Gustaffson and Pinter (1984). Double-barreled electrodes or large single-barreled electrodes with tip diameters around 2.5 μm were used in these experiments, and this may have caused a larger leak conductance. The resting potentials varied from −66 to −52 mV, compared with −83 to −65 mV found by Gustaffson and Pinter (1984).

The membrane potential cannot be used as a secure guide to the leak conductance. With the morphology of the neuron illustrated in Fig. 1, a somatic shunt of 2.2 MΩ with 0mV reversal potential, $R_m = 8.2$ kΩcm^2, and an unperturbed membrane potential of −70 mV, the compartmental model predicted a somatic membrane potential of −17 mV, whereas the recorded potential was −56 mV. If intracellular Ca^{2+} is increased by electrode damage, this could activate the Ca^{2+}-mediated K^+ conductance (Krnjević *et al.*, 1978), thereby hyperpolarizing the neuron and adding an additional conductance (Gustaffson and Pinter, 1984). For these reasons it is difficult to determine how much of the somatic shunt is caused by the microelectrode leak. Because the main objective in these experiments was to obtain a more accurate measure of R_{md}, the practical difficulty caused by an electrode leak is that as it becomes greater, the curve-fitting procedure becomes less sensitive to R_{md}.

2.5. Spread of EPSPs in Motoneuron Dendrites

Synaptic potentials evoked in motoneurons by impulses in single group Ia axons display a variety of time courses (Rall *et al.*, 1967). Some of this variation is caused by different membrane time constants for different motoneurons, but most of it can be attributed to different sites of termination of afferent fibers on dendrites of motoneurons (Jack *et al.*, 1971). Assuming passive dendritic membrane, the shape indices of an EPSP

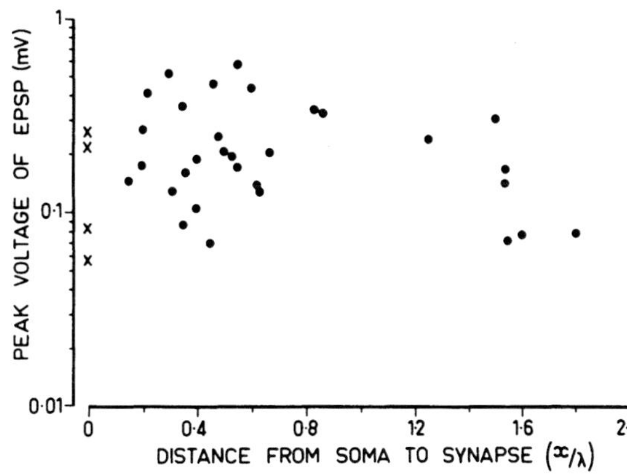

FIGURE 3. The average peak voltages of single-fiber group Ia EPSPs recorded in motoneurons have been pooled and plotted against the calculated electrotonic distance from the soma to the synapse at which each EPSP was generated. Crosses indicate somatic EPSPs. (From Iansek and Redman, 1973)

(rise time and half width) can be used to calculate the electrotonic distance from the soma to the region of termination. This calculation shows that EPSPs originate at various distances from the soma, with most connections within 0.6λ. The reliability of this procedure was recently confirmed (Redman and Walmsley, 1983a) by filling the afferent fiber used to evoke the EPSP with HRP as well as labeling the motoneuron in which the EPSP was recorded with HRP. The location of the reconstructed connection agreed with the predictions based on the shape indices of the EPSP.

Cable calculations of the spread of EPSPs in dendrites with passive membrane predict that the further from the soma an EPSP is generated, the smaller will be its amplitude at the soma, assuming the same current at all synapses. In Fig. 3, the peak amplitude of single-fiber group Ia EPSPs recorded at the soma is plotted against the calculated electrotonic distance to the synaptic site. Although considerable variation exists for different EPSPs originating at similar electrotonic distances, it is clear that the peak amplitude does not decrease as the synapse becomes more distal (Iansek and Redman, 1973). In fact, when the total charge delivered to the soma from the synaptic current is calculated (the area under the EPSP divided by the input resistance), distal synapses are shown to be more effective at depolarizing the soma than are proximal synapses. These observations can only be reconciled with passive dendritic membrane properties if a greater synaptic current is generated at distal synapses than at somatic synapses. A factor of about 10 is required. It is not possible to invoke a greater number of boutons in connections at distal synapses, as the morphology does not support this idea (Brown and Fyffe, 1981; Burke *et al.*, 1979). Furthermore, the EPSPs generated at a single bouton are similar for both distal and proximal connections (Edwards *et al.*, 1983).

Synaptic transmission at a single bouton occurs in an all-or-none manner (Jack *et al.*, 1981; Redman and Walmsley, 1983b). When transmitter is released, the amplitude of the postsynaptic response does not vary. One way this could happen is that the minimum quantity of transmitter released (one quantum) is sufficient to saturate all subsynaptic receptors (see also Rang, 1981). Boosting the synaptic current is then simply achieved by increasing the number of transmitter-activated channels in the subsynaptic membrane, assuming that the contents of one quantum can still open all channels. For the Ia synapse, it has been calculated that 100–250 channels would generate the measured current at a single somatic bouton, and 1000–2500 channels would generate the calculated current at a distal bouton (Finkel and Redman, 1983).

3. VOLTAGE-ACTIVATED CONDUCTANCES IN MOTONEURON DENDRITES

In vitro measurements of voltage-dependent currents in mammalian neurons have revealed the presence of calcium channels in dendrites (Llinás and Sugimori, 1980; Llinás and Yarom, 1981a,b). If these channels are activated by dendritic depolarization, the resulting inward current could support the spread of EPSPs. Depending on the density and distribution of these channels and of any other types of voltage-dependent channels

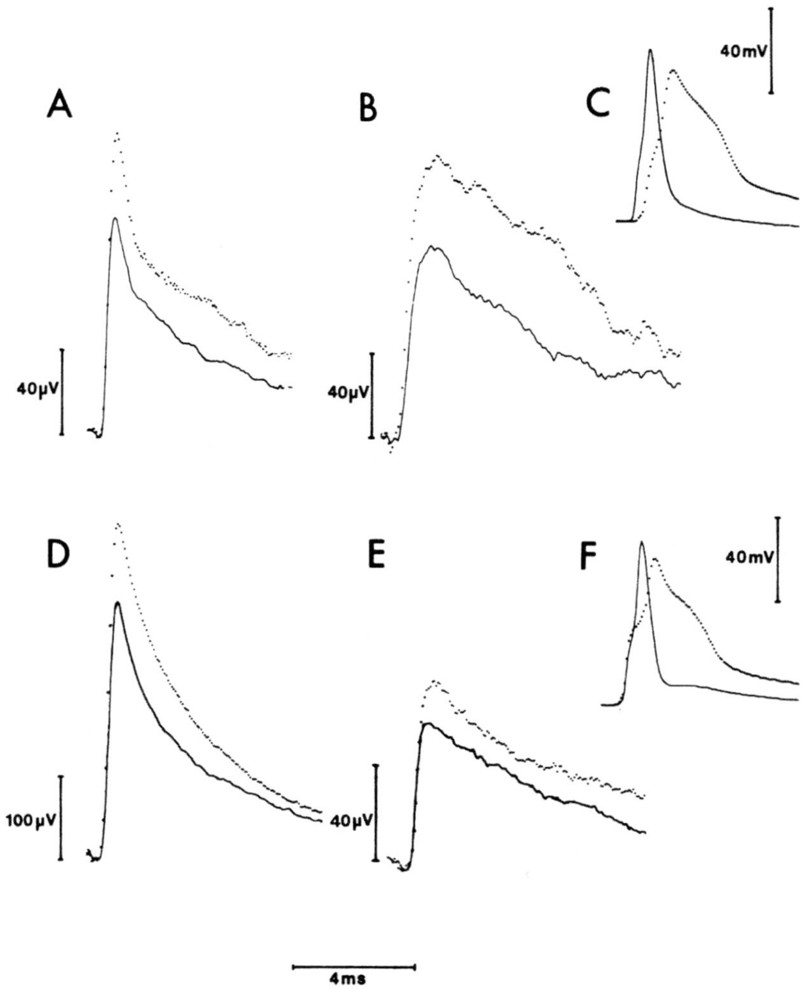

FIGURE 4. A and B are spike-triggered averages of single-fiber group Ia EPSPs, and C is an antidromic action potential, all recorded in the same motoneurons. The responses obtained after TEA injection are dotted. The EPSP in A has a time course indicative of a somatic synapse, whereas the EPSP in B has a slower time course and will have been generated at a dendritic synapse. The rise time and half width of the EPSP in A were unchanged after TEA, whereas both the rise time and half width of the EPSP in B were increased after TEA injection. D, E, and F were recorded in another motoneuron. D is a somatic EPSP averaged by spike triggering, and its time course was unaltered by TEA. E is a dendritic EPSP, and its rise time and half width were both prolonged by TEA. (From Clements *et al.*, 1986)

present, many different modes of dendritic propagation could be possible, ranging from full dendritic action potentials to passive spread.

Tetraethyl ammonium (TEA) ions applied internally block voltage-dependent potassium channels (Neher and Lux, 1972) and, in so doing, often reveal a small inward calcium current (Schwindt and Crill, 1980). If Ca^{2+} channels and voltage-dependent K^+ channels are present in motoneuron dendrites, then intracellular TEA combined with sufficient dendritic depolarization should generate a calcium current. Dendritic depolarization can be produced by stimulating group Ia axons that terminate at dendritic sites. Dendritic EPSPs spreading to the soma might be expected to become larger and more prolonged after internal perfusion with TEA. In contrast, EPSPs generated at somatic synapses, and rarely exceeding 0.5 mV, would not be expected to alter after TEA injection.

When this idea was tested (Clements *et al.*,1986), both dendritic and somatic EPSPs increased after internal perfusion with TEA, but only dendritic EPSPs became more prolonged (Fig. 4). It was shown that these changes were not caused by an increase in membrane time constant. The changes in dendritic EPSPs could be at least partially reversed by membrane hyperpolarization (Fig. 5), but the increased amplitude of somatic EPSPs was unaffected. This observation suggests that the prolonged time course of dendritic EPSPs and part of their amplitude increase are caused by a depolarization-activated inward current, presumably a calcium current. Results obtained when two EPSPs

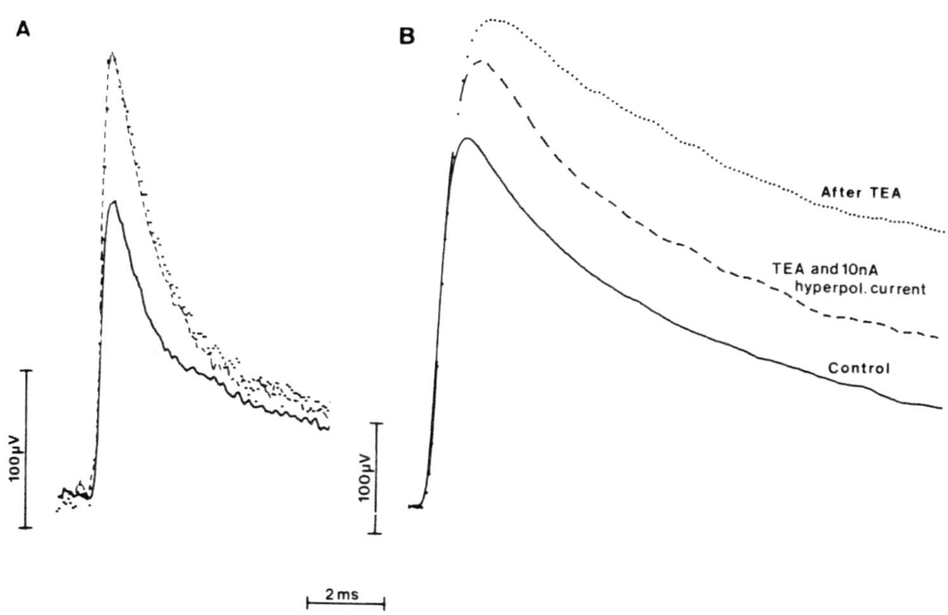

FIGURE 5. A and B are spike-triggered averages of EPSPs obtained in the same motoneuron. The continuous lines are the EPSPs before TEA perfusion. A has a time course indicating a somatic synapse, whereas B is a dendritic EPSP. The averages indicated by dots were obtained after sufficient TEA was injected to broaden the antidromic action potential, as shown in Fig. 4C,F. The averages indicated by dashed lines were obtained after TEA injection and during continuous hyperpolarization with 10 nA. The time course of the EPSP in A was unchanged by these procedures. The rise time/half width of the EPSP in B was 0.5/6 msec (control), 0.6/7.1 msec (TEA and current), and 0.88/11 msec (TEA alone). (From Clements *et al.*, 1986)

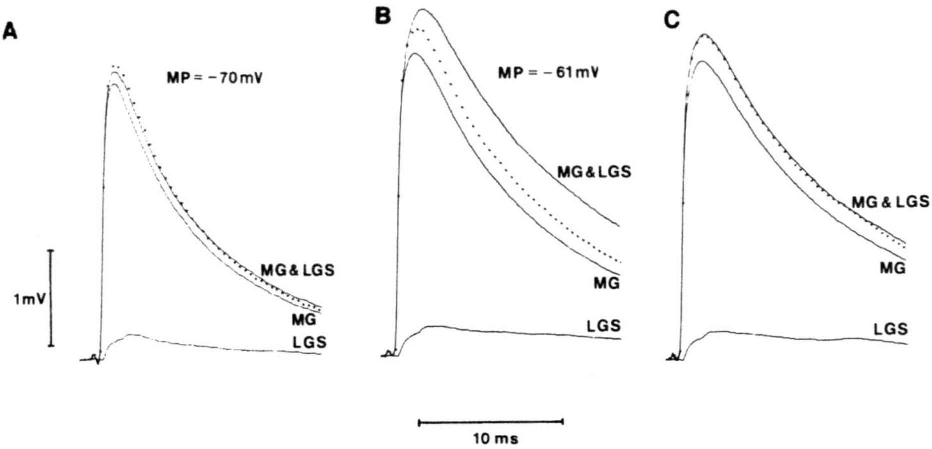

FIGURE 6. MG and LGS are composite EPSPs recorded in the same motoneuron by stimulating medial gastrocnemius and lateral gastrocnemius–soleus nerves, respectively. "MG & LGS" is a composite EPSP recorded in the same motoneuron when both nerves were stimulated simultaneously. The dots are the linear sum of the MG and the LGS EPSPs. A, before injection of TEA; B, after injection of TEA; C, after injection of TEA and 8nA hyperpolarizing current. (From Clements *et al.*, 1986)

were evoked separately and then simultaneously (Fig. 6A) also support the existence of a voltage-activated current. Linear summation of EPSPs was observed before TEA was injected (Fig. 6A). Greater-than-linear summation occurred in the presence of TEA, and this was reversed by membrane hyperpolarization (Fig. 6B,C).

The increase in somatic EPSPs and some of the increase in dendritic EPSPs can be attributed to an increase in reversal potential for the EPSP (Clements *et al.*, 1986). Either the ionic selectivity of the synaptic channel was altered by TEA to impede the outward movement of K^+ ions or ionophoresis of TEA significantly lowered the internal potassium concentration. This effect is secondary to the main observation that part of the increase in amplitude of dendritic EPSPs, the prolongation of dendritic EPSPs, and the sensitivity of these changes to hyperpolarizing current following TEA injection are caused by the presence of voltage-activated inward and outward currents in dendritic membrane. Most of the dendritic EPSPs examined were generated at 0.2 to 0.4λ (from the soma), which would usually correspond to either a first- or second-order dendritic branch. The voltage-dependent phenomena could occur either proximally or distally to the synaptic site, but the sensitivity of these phenomena to relatively small hyperpolarizing currents suggests that the proximal dendrites are most likely to be involved.

In normal motoneurons, two important observations suggest that voltage-dependent currents do not influence the amplitude and time course of dendritic EPSPs evoked at normal resting potential. One is that EPSPs invariably sum in a linear or less-than-linear manner (Burke, 1967). The other is that the synaptic location calculated using the time course of the EPSP at the soma and the cable properties of the neuron (which are assumed to be passive) coincides with the location actually observed when the same synapse is identified morphologically (Redman and Walmsley, 1983a). We suggest that under the experimental conditions in which single-fiber group Ia EPSPs have been studied in normal motoneurons, the depolarization accompanying dendritic EPSPs does not activate an inward current. Instead, the boosting mechanism needed to account for the size of dendritic EPSPs is more likely to be present at the synapse itself.

REFERENCES

Barrett, J. N., and Crill, W. E., 1974, Specific membrane properties of cat motoneurones, *J. Physiol. (Lond.)* **239:**301–324.

Brown, A. G., and Fyffe, R. E. W., 1981, Direct observations on the contacts made between 1a afferent fibers and α-motoneurones in the cat's lumbosacral cord, *J. Physiol. (Lond.)* **313:**121–140.

Burke, R. E., 1967, Composite nature of the monosynaptic excitatory postsynaptic potential, *J. Neurophysiol.* **30:**114–1137.

Burke, R. E., Walmsley, B., and Hodgson, J. A., 1979, HRP anatomy of group 1a afferent contacts on alpha motoneurones, *Brain Res.* **160:**347–352.

Clements, J. D., Nelson, P. G., and Redman, S. J., 1986, Intracellular tetra-ethyl ammonium ions enhance group 1a excitatory post-synaptic potentials evoked in cat motoneurones, *J. Physiol. (Lond.)* **377:**267–282.

Durand, D., Carlen, P. L., Gurevich, N., Ho, A., and Kunov, H., 1983, Electrotonic parameters of rat dentate granule cells measured using short current pulses and HRP staining, *J. Neurophysiol.* **50:**1080–1097.

Edwards, F. R., Jack, J. J. B., and Kullmann, D. M., 1983, The relationship between amplitude and time course of single fibre group 1a excitatory postsynaptic potentials in cat spinal motoneurones, *J. Physiol. (Lond.)* **345:**58p.

Egger, M. D., Egger, L. D., 1982, Quantitative morphological analysis of spinal motoneurones, *Brain Res.* **253:**19–30.

Finkel, A. S., and Redman, S. J., 1983, The synaptic current evoked in cat spinal motoneurones by impulses in single group 1a axons, *J. Physiol. (Lond.)* **342:**615–632.

Fleshman, J. W., Segev, I., Culheim, S., and Burke, R. E., 1983, Matching electrophysiological and morphological measurements in cat alpha motoneurones, *Soc. Neurosci. Abstr.* **9:**341.

Gustaffson, B., and Pinter, M. J., 1984, Relations among passive electrical properties of lumbar α-motoneurones of the cat, *J. Physiol. (Lond.)* **356:**401–431.

Iansek, R., and Redman, S. J., 1973, The amplitude, time course and charge of unitary excitatory post-synaptic potentials evoked in spinal motoneurone dendrites, *J. Physiol. (Lond.)* **234:**665–688.

Jack, J. J. B., 1979, An introduction to linear cable theory, in: *The Neurosciences* (F. O. Schmitt and F. G. Warden, eds.), MIT Press, Cambridge, pp. 423–437.

Jack, J. J. B., Miller, S., Porter, R., and Redman, S. J., 1971, The time course of minimal excitatory post-synaptic potentials evoked in spinal motoneurones by group 1a afferent fibres, *J. Physiol. (Lond.)* **215:**353–380.

Jack, J. J. B., Redman, S. J., and Wong, K., 1981, The components of synaptic potentials evoked in cat spinal motoneurones by impulses in single group 1a afferents, *J. Physiol. (Lond.)* **321:**65–96.

Johnston, D., and Brown, T. H. G., 1983, Interpretation of voltage-clamp measurements in hippocampal neurons, *J. Neurophysiol.* **50:**464–486.

Krnjević, K., Puil, E., and Werman, R., 1978, EGTA and motoneuronal after-potentials, *J. Physiol. (Lond.)* **275:**199–223.

Llinás, R., and Sugimori, M., 1980, Electrophysiological properties of *in vitro* Purkinje cell dendrites in mammalian cerebellar slices, *J. Physiol. (Lond.)* **305:**197–213.

Llinás, R., and Yarom, Y., 1981a, Electrophysiology of mammalian inferior olivary neurones *in vitro*. Different types of voltage-dependent ionic conductances, *J. Physiol. (Lond.)* **315:**549–568.

Llinás, R., and Yarom, Y., 1981b, Properties and distribution of ionic conductances generating electroresponsiveness of mammalian inferior olivary neurones *in vitro*, *J. Physiol. (Lond.)* **315:**569–584.

Lux, H. D., Schubert, P., and Kreutzberg, G. W., 1970, Direct matching of morphological and electrophysiological data in cat spinal motoneurones, in: *Excitatory Synaptic Mechanisms* (P. Andersen and J. K. S. Jansen, eds,), Universitetsforlaget, Oslo, pp. 189–198.

Neher, E., and Lux, H. D., 1972, Differential actions of TEA^+ on two K^+ current components of a molluscan neurone, *Pflugers Arch.* **336:**87–100.

Rall, W., 1962, Theory of physiological properties of dendrites, *Ann. N.Y. Acad. Sci.* **96:**1071–1092.

Rall, W., 1977, Core conductor theory and cable properties of neurons, in: *Handbook of Physiology, The Nervous System*, Section 1, Vol. 1 (E. R. Kandel, ed.), American Physiology Society, Bethesda, pp. 39–97.

Rall, W., Burke, R. E., Smith, T. G., Nelson, P. G., and Frank, K., 1967, Dendritic location of synapses and possible mechanisms for the monosynaptic e.p.s.p. in motoneurons, *J. Neurophysiol.* **30:**1169–1193.

Rang, H., 1981, The characteristics of synaptic currents and responses to acetylcholine of rat submandibular ganglion cells, *J. Physiol. (Lond.)* **311:**23–55.

Redman, S. J., and Walmsley, B., 1983a, The time course of synaptic potentials evoked in cat spinal motoneurones at identified group 1a synapses, *J. Physiol.* (*Lond.*) **343:**117–133.

Redman, S. J., and Walmsley, B., 1983b, Amplitude fluctuations in synaptic potentials evoked in cat spinal motoneurones at identified group 1a synapses, *J. Physiol.* (*Lond.*) **343:**135–145.

Schwindt, P. C., and Crill, W. E., 1980, Properties of a persistent inward current in normal & TEA-injected motoneurons, *J. Neurophysiol.* **43:**1700–1724.

Turner, D. A., 1984, Segmental cable evaluation of somatic transients in hippocampal neurons (CA1, CA3 and dentate), *Biophys. J.* **46:**73–84.

Turner, D. A., and Schwartzkroin, P. A., 1983, Electrical characteristics of dendrites and dendritic spines in intracellularly stained CA3 and dentate hippocampal neurones, *J. Neurosci.* **3:**2381–2394.

Ulfhake, B., and Kellerth, J.-O., 1981a, A quantitative light microscopic study of the dendrites of cat spinal α-motoneurones after intracellular staining with horseradish peroxidase, *J. Comp. Neurol.* **202:**571–583.

Ulfhake, B., and Kellerth, J.-O., 1981b, A quantitative morphological study of HRP-labelled cat α-motoneurones supplying different hindlimb muscles, *Brain Res.* **264:**1–19.

Ulfhake, B., and Kellerth, J.-O., 1984, Electrophysiological and morphological measurements in cat gastrocnemius and soleus α-motoneurones, *Brain Res.* **307:**167–179..

26

Some Conclusions Relevant to Plasticity Derived from Normal Anatomy

ALMUT SCHÜZ

1. INTRODUCTION

The cerebral cortex is supposed to be heavily involved in learning processes and has, therefore, been the object of many deprivation studies. However, even the study of the normal, not artifically perturbed brain during and after development may contribute to the question of anatomic traces of plasticity. The advantage of this alternative approach is that it is not necessary to expose animals to an artificial situation in which it may be difficult to distinguish between direct effects of learning and more indirect effects connected with the general condition of the animal. Here I summarize the results we have collected in recent years.

2. DO DENDRITIC SPINES INCREASE THE NUMBER OF SYNAPTIC SITES ON DENDRITES?

It is still an open question as to why some cells in the cerebral cortex have dendritic spines and others do not. Evidently, spines are specialized organelles for synapses. The great majority of synapses on spiny dendrites are located on spines, so few synapses would be left if all the spines were cut off. One wonders, therefore, how many synapses can be made by dendrites that have only a few spines or none at all. This is interesting not only in order to elucidate the role of spines but also because of the assumption that the two classes of cells—spiny and nonspiny ones—play very different roles in the cortical network, the spiny cells most probably being excitatory and the nonspiny ones inhibitory (e.g., Peters and Fairén, 1978; Somogyi and Cowey, 1981).

As a byproduct of a phosphotungstic acid study, we obtained an interesting stain that showed rather long pieces of nonspiny dendrites in electron microscopic preparations (Fig. 1). We used this opportunity for a study of the synaptic density on such dendrites.

ALMUT SCHÜZ • Max Planck Institute for Biological Cybernetics, 7400 Tübingen, Federal Republic of Germany.

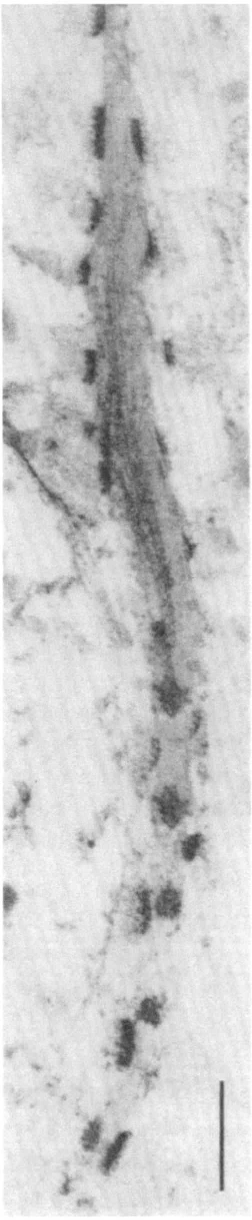

FIGURE 1. Electron micrograph of a nonspiny dendrite in the cerebral cortex of the mouse. It has been stained with phosphotungstic acid and shows how densely a dendrite can be covered with synaptic junctions. Bar, 1 μm.

The average density measured along 24 dendrites was 1.9 synapses per micrometer. As the sections were thinner than the dendrites, this value had to be corrected for synapses that were not contained in the sections. This was carried out in an approach similar to the Abercrombie method (Abercrombie, 1946), taking into account the size of the synapses and the thickness of the sections (A. Schüz and M. Dortenmann, in press). The real density then turned out to be 3.3 synapses per micrometer.

This result was compared to the density of the spines on pyramidal cells found in a previous Golgi study (Schüz, 1976). The average was 1.9 spines per micrometer, also corrected for the spines not seen in the section. The ratio between synapses on spines and those on dendritic shafts along spiny dendrites ranges between 2.1 : 1 and 4.9 : 1, as one can gather from studies on cat and mouse cortex (Vaughan and Peters, 1973; Peters and Feldman, 1977; White and Hersch, 1981). As most of the spines carry one synapse, one ends up with an average density between 2.3 and 2.8 synapses per micrometer of dendritic length, which is thus lower than on the nonspiny dendrites investigated.

This shows that dendritic spines are by no means necessary for the accommodation of a large number of synaptic contacts on a dendrite. They must, therefore, have a further function.

3. DOES THE FORMATION OF DENDRITIC SPINES AND SYNAPSES IN THE CEREBRAL CORTEX DEPEND ON LEARNING PROCESSES?

One function that has been attributed to the dendritic spines is their role in plasticity. The reduction of the spines in some deprivation experiments (Valverde, 1967; Fifková, 1968; Winkelmann *et al.*, 1976), their increase in enriched environment (Globus *et al.*, 1973; Schapiro and Vukovich, 1970) or after stimulation (Rutledge *et al.*, 1974), and their rapid increase soon after birth in many animals have led to this idea. The formation of a spine, or, more exactly, of the connection that is made by a spine, was supposed to be the result of a learning process.

In order to examine this hypothesis, I investigated the development of spines and synapses in the cerebral cortex of a precocial animal, the guinea pig, whose physical development, including that of the brain, is already quite advanced at birth but that has, of course, still the ability to learn. The formation of spines was investigated in Golgi preparations from about 3 weeks before birth to the adult stage (Schüz, 1981). It showed the following course (Fig. 2, solid lines): the first spines had already appeared about 18 days before birth; these then increased rapidly in number and, depending on the area, either reached or even exceeded their final density by term. An electron microscopic

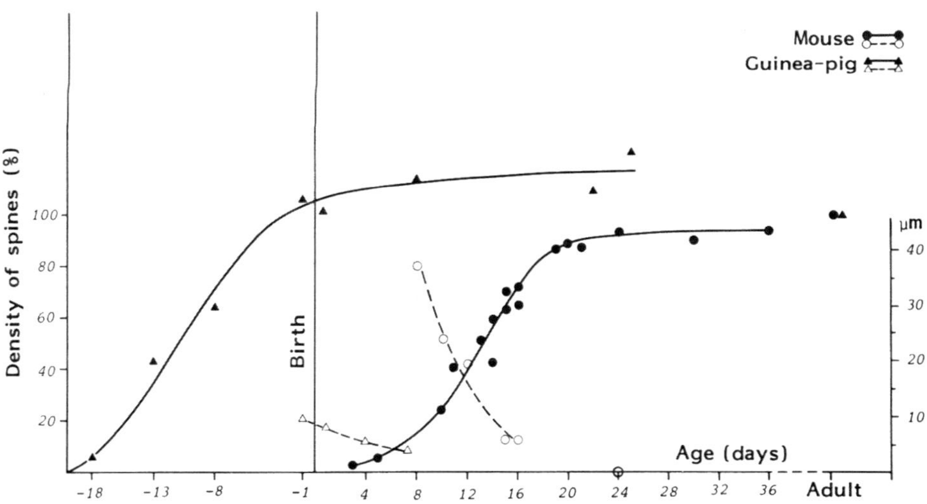

FIGURE 2. Solid lines: increase in the density of dendritic spines (left scale) with age in the cerebral cortex of mice and guinea pigs. The values for the mouse have been taken from Valverde (1971). Broken lines: decrease in the thickness of the external granular layer (right scale) in the cerebellum of the two species. (From Schüz and Hein, 1984)

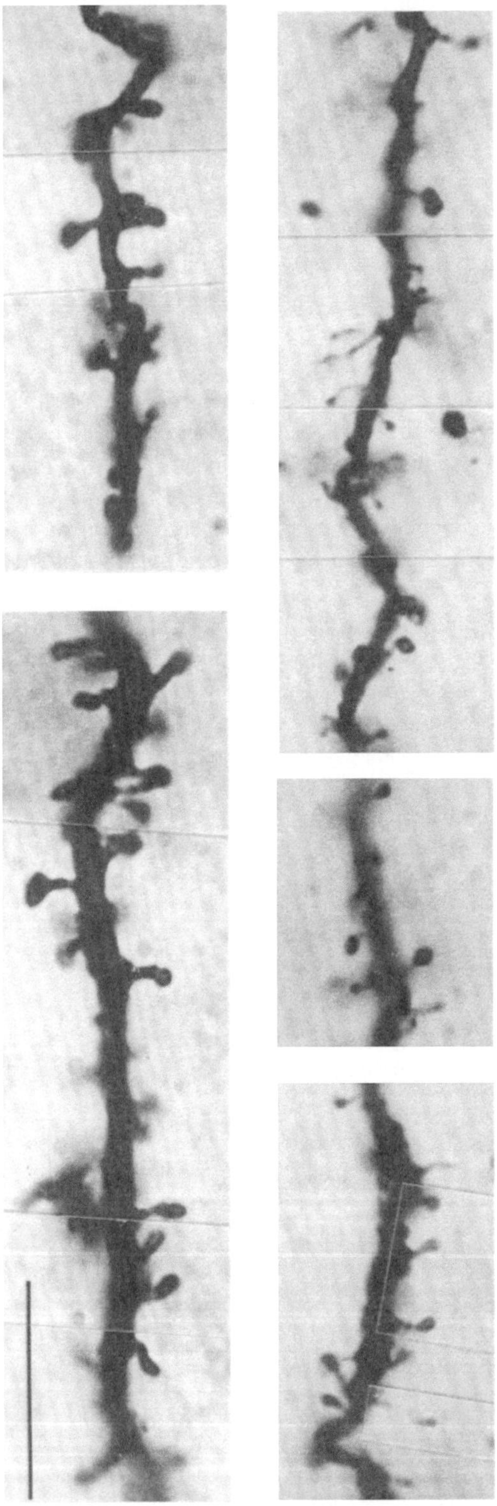

FIGURE 3. Examples that illustrate the difference in the appearance of the dendritic spines in adult (left) and newborn (right) guinea pigs. Golgi preparations of pyramidal cell dendrites in the cerebral cortex. Bar, 10 μm. Although thick and thin spines can be found in both age groups, one has the impression of a general tendency towards finer ones in the newborn animals.

investigation revealed that the synaptic density at birth did not differ significantly from the adult value either.

This shows that, at least in the guinea pig, the formation of spines and synapses in the cerebral cortex is not dependent on learning processes, which, of course, occur mainly after birth.

4. WHAT ANATOMIC CHANGES OCCUR DURING THE LEARNING PHASE IN THE GUINEA PIG CORTEX?

The answer to the previous question restricts the search for the mechanisms of plasticity to that for modifications of the already existing synapses or spines. Considering the structural changes that still take place after birth in the guinea pig cortex, the following may play a role: (1) an increase in the number of synaptic vesicles at individual synapses, (2) an increase in the postsynaptic thickening of type I synapses, or (3) a change in the dimensions of the dendritic spines.

The latter point was investigated in a quantitative Golgi study (Schüz, 1986). The length of the spines and of their necks, the thickness of their heads and necks, and their obliquity on the dendrites were measured. A qualitative impression of the differences between the spines of newborn and adult guinea pigs can be gained from Fig. 3.

No further changes in the average length of the necks were found after birth, but there was a significant increase in the thickness of both the necks (Fig. 4) and the heads.

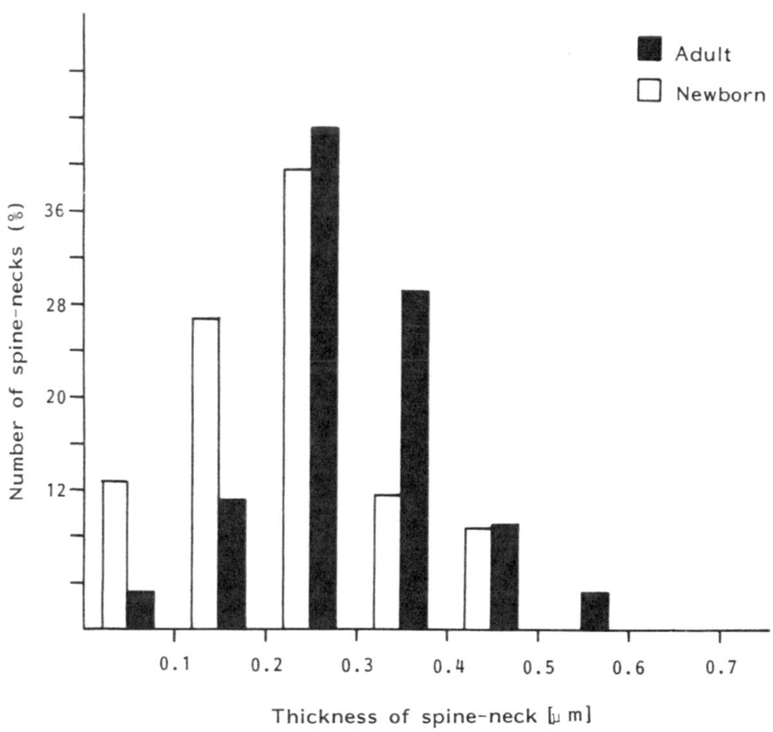

FIGURE 4. Frequency distribution of the thickness of the spine necks in newborn and adult guinea pigs.

The postnatal changes in the guinea pig cortex are thus consistent with the idea that the effectiveness of a synapse is controlled by a change in the form of the spine. It furthermore shows that, if this is true, it is the thickness rather than the length by which the effectiveness is controlled.

5. ARE THE LATE CHANGES IN CEREBELLAR ANATOMY NECESSARILY POSTNATAL AND THEREFORE POSSIBLY RELATED TO MOTOR LEARNING?

The cerebellum, too, is supposed to be involved in learning processes (Marr, 1968; Albus, 1971; Ito *et al.*, 1982). It is, therefore, also interesting in the case of the cerebellar cortex to compare its development in precocial and altricial species and to determine if it is somehow related to the event of birth, which entails new motor tasks, or if it is rather linked to an internal timetable. We chose the development of the granular layer because it is an aspect of cerebellar development that is easy to assess. The granular cells are formed at the surface of the cerebellum in the so-called external granular layer and subsequently migrate down through the molecular layer to form the internal granular layer. By a certain age, all the cells have migrated down, and the external granular layer has disappeared.

Figure 2 shows the decrease in the thickness of the external granular layer in mouse and guinea pig (broken lines). One can see that in this respect, too, the newborn guinea pig is much more mature than the newborn mouse, the external granular layer already being quite thin in the guinea pig.

The values have been drawn in the same diagram as those for the increase in dendritic spines in the cerebral cortex. Comparing the two species, one can see that the decrease in the external granular layer has no relationship to the event of birth but parallels the formation of dendritic spines in the cerebral cortex.

We did not investigate the formation of synapses between granular and Purkinje cells in the guinea pig. However, from the mouse, it is known (Larramendi, 1969) that synapse formation in the molecular layer parallels the migration of the granular cells through it. If one may draw an inference from the presence of spines on the Purkinje cells (Fig. 5) about the presence of synapses there (in analogy to the cerebral cortex), one may suppose that not only cell migration but also the connectivity of the cerebellar cortex is already very advanced in the newborn guinea pig.

The coincidence in the development of the cerebral and the cerebellar cortex reflects the high degree of coupling between the two parts of the brain (Glickstein and May, 1982). As we have seen, the development of both structures is, to a large extent, not related to external events but to an internal calendar. This suggests that the connectivity of both cortices is not the result of learning but precedes it. One gains the impression that both structures are intended to be ready at the moment at which the animal must be able to cope with the environment, which is, in the case of the guinea pig, immediately after birth and in the case of the mouse, at the time when it starts to become independent of its mother.

ACKNOWLEDGMENT. I thank Prof. V. Braitenberg for valuable discussions and for correction of this manuscript.

FIGURE 5. Golgi preparation of a Purkinje cell in the cerebellum of a newborn guinea pig. It shows that the dendrites are already densely covered with spines. Bar, 10 μm.

REFERENCES

Abercrombie, M., 1946, Estimation of nuclear population from microtome sections, *Anat. Rec.* **94:**239–247.

Albus, J. S., 1971, A theory of cerebellar function, *Math. Biosci.* **10:**25.

Fifková, E., 1968, Changes in the visual cortex of rats after unilateral deprivation, *Nature* **220:**379–380.

Glickstein, M., and May, J., 1982, Visual control of movement: The circuits which link visual to motor areas of the brain with special reference to the visual input to pons and cerebellum, in: *Sensory Physiology* (W. D. Neff, ed.), Academic Press, New York, pp. 103–145.

Globus, A., Rosenzweig, E., Bennett, L., and Diamond, M. C., 1973, Effects of differential experience on dendritic spine counts in rat cerebral cortex, *J. Comp. Physiol. Psychol.* **82:**175–181.

Ito, M., Sakurai, M., and Tongroach, P., 1982, Climbing fibre induced depression of both mossy fibre responsiveness and glutamate sensitivity of cerebellar Purkinje cells, *J. Physiol. (Lond.)* **324:**113–134.

Larramendi, L. M. H., 1969, Analysis of synaptogenesis in the cerebellum of the mouse, in: *Neurobiology of Cerebellar Evolution and Development* (R. Llinás, ed.), American Medical Association, Chicago, pp. 803–843.

Marr, D., 1968, A theory of cerebellar cortex, *J. Physiol. (Lond.)* **202:**437–470.

Peters, A., and Fairén, A., 1978, Smooth and sparsely-spined stellate cells in the visual cortex of the rat: A study using a combined Golgi–electron microscope technique, *J. Comp. Neurol.* **181:**129–172.

Peters, A., and Feldman, L., 1977, The projection of the lateral geniculate nucleus to area 17 of the rat cerebral cortex. IV. Termination upon spiny dendrites, *J. Neurocytol.* **6:**669–689.

Rutledge, L. T., Wright, C., and Duncan, J., 1974, Morphological changes in pyramidal cells of mammalian neocortex associated with increased use, *Exp. Neurol.* **44:**209–228.

Schapiro, S., and Vukovich, K. R., 1970, Early experience effects upon cortical dendrites: A proposed model for development, *Science* **167:**292–294.

Schüz, A., 1976, Pyramidal cells with different densities of dendritic spines in the cortex of the mouse, *Z. Naturforsch.* **31C:**319–323.

Schüz, A., 1981, Pränatale Reifung und postnatale Veränderungen im Cortex des Meerschweinchens: Mikroskopische Auswertung eines natürlichen Deprivationsexperimentes (English summary), *J. Hirnforsch.* **22:**93–127.

Schüz, A., 1986, Comparison between the dimensions of dendritic spines in the cerebral cortex of newborn and adult guinea pigs, *J. Comp. Neurol.* **224:**277–285.

Schüz, A., and Dortenmann, M., 1987, Synaptic density on non-spiny dendrites in the cerebral cortex of the house mouse. A phosphotungstic acid study, *J. Hirnforsch.* **28**:(in press).

Schüz, A., and Hein, F. M., 1984, Comparison between the developmental calendars of the cerebral and cerebellar cortices in a precocial and an altricial rodent, in: *Cerebellar Functions* (J. R. Bloedel, J. Dichgans, and W. Precht, eds.), Springer-Verlag, Berlin, Heidelberg, New York, pp. 318–321.

Somogyi, P., and Cowey, A., 1981, Combined Golgi and electron microscopic study on the synapses formed by double bouquet cells in the visual cortex of the cat and monkey, *J. Comp. Neurol.* **195**:547–566.

Valverde, F., 1967, Apical dendritic spines of the visual cortex and light deprivation in the mouse, *Exp. Brain Res.* **3**:337–352.

Valverde, F., 1971, Rate and extent of recovery from dark rearing in the visual cortex of the mouse, *Brain Res.* **33**:1–11.

Vaughan, D. W., and Peters, A., 1973, A three-dimensional study of layer I of the rat parietal cortex, *J. Comp. Neurol.* **149**:355–370.

White, E. L., and Hersch, S. M., 1981, Thalamocortical synapses of pyramidal cells which project from SmI to MsI cortex in the mouse, *J. Comp. Neurol.* **198**:167–181.

Winkelmann, E., Brauer, K., and Werner, L., 1976, Untersuchungen zu Spineveränderungen der Lamina-V-Pyramidenzellen im visuellen Kortex junger und subadulter Laborratten nach Dunkelaufzucht und Zerstörung des Corpus geniculatum laterale, pars dorsalis, *J. Hirnforsch.* **17**:495–506.

27

Anatomic Observations on Afferent Projections of Orbicularis Oculi and Retractor Bulbi Motoneuronal Cell Groups and Other Pathways Possibly Related to the Blink Reflex in the Cat

GERT HOLSTEGE, JOEP TAN, and JACQUELINE J. VAN HAM

1. INTRODUCTION

The orbicularis oculi muscle brings the eyelids together, gently as in spontaneous blinking and sleep or strongly in cases of painful tactile, auditory, or visual stimuli. The main function of the orbicularis oculi muscle is protection of the eye from injury and intense light. In addition, this muscle, together with other facial muscles, plays a role in expression of emotions (Ekman *et al.*, 1972). The orbicularis oculi motoneurons, according to retrograde degeneration (Papez, 1927; Courville, 1966) and HRP studies (Kume *et al.*, 1978), are grouped together in the dorsal part of the facial nucleus. This subgroup was called the intermediate facial subnucleus by Courville (1966) (Fig. 1).

The retractor bulbi (RB) muscle is an extraocular muscle divided into four slips, each of which inserts on the eyeball behind and beside the inferior and superior recti muscles. The four slips are thinner and shorter than the other extraocular muscles. The RB muscles are present in all vertebrates except fish, snake, chameleon, bat, and primates (Bolk *et al.*, 1938). The motoneurons innervating the retractor bulbi muscle are located in a small group in the brainstem at the level of the abducens nucleus just dorsal to the superior olivary complex (Grant *et al.*, 1979, 1981; Spencer *et al.*, 1980) (Fig. 2) an area also called the accessory abducens nucleus. The motoneurons in this cell group are often loosely arranged and number a total of only 80–120 (Grant *et al.*, 1979) or 66–79 (Spencer *et al.*, 1980). The functional role of the RB muscle is purely eye protection: it retracts the eyeball, forcing the intraorbital fat against the base of the nictitating membrane and causing the latter to sweep across the eyeball (Bach-Y-Rita, 1971). This event is

GERT HOLSTEGE, JOEP TAN, and JACQUELINE J. VAN HAM • Department of Anatomy II, Medical Faculty, Erasmus University Rotterdam, 3000 DR Rotterdam, The Netherlands.

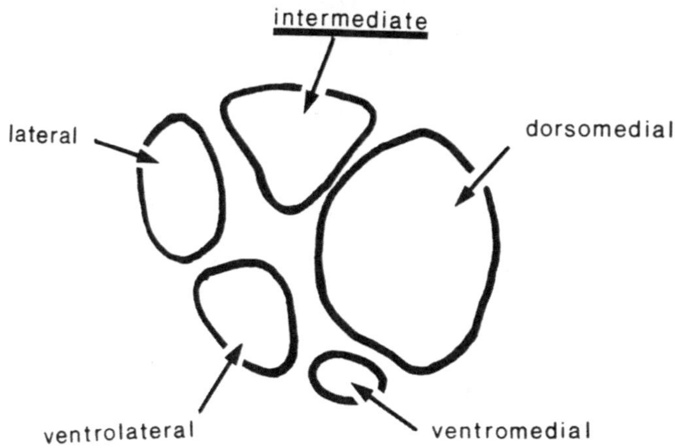

FIGURE 1. Schematic representation of the various subnuclei of the left facial nucleus. Motoneurons in the intermediate subnucleus innervate muscles around the eye; those in the lateral subnucleus innervate the upper part and those in the ventrolateral subnucleus the lower part of the muscles around the mouth. The motoneurons in the large dorsomedial facial subnucleus innervate ear muscles, and those in the ventromedial subnucleus innervate the platysma muscle.

also called the nictitating membrane response (NMR). The RB muscles always contract simultaneously with the orbicularis oculi muscle (McCormick *et al.*, 1982). In the remainder of this chapter we call this combined muscle action reflex blinking.

The blink reflex, which must be considered an eye-protective mechanism (Bach-Y-Rita, 1971), is extensively studied physiologically. The blink reflex in the cat consists of two components (Lindquist and Martensson, 1970) and has latencies of 9–12 msec (R_1) and 15–25 msec (R_2). R_1 is only ipsilateral; R_2 is bilateral in humans (Kugelberg, 1952) but ipsilateral in cats (Hiraoka and Shimamura, 1977). Most often, part of the blink reflex is investigated, for example, only the orbicularis oculi reflex (Hiraoka and Shimamura, 1977; Kimura *et al.*, 1969; Ongerboer de Visser and Kuypers, 1978; Shahani and Young, 1972; Shahani, 1973; Tamai *et al.*, 1986) or only the nictitating membrane response (Mis, 1977; Moore *et al.*, 1980; Rosenfield and Moore, 1983; Richardson and Thompson, 1984). This latter response is used in studying conditioned reflexes because

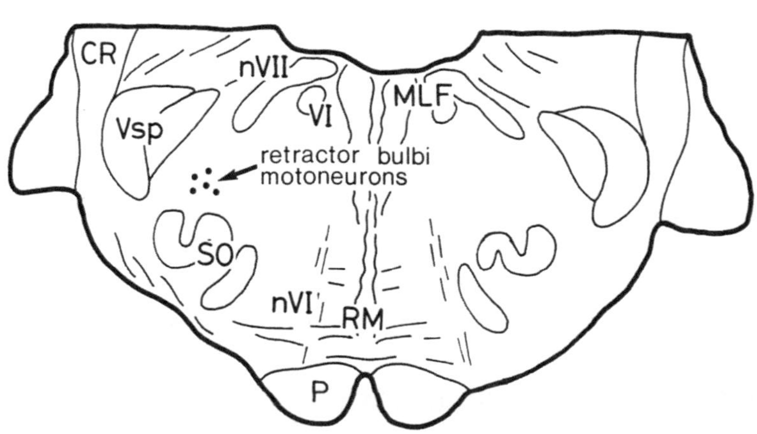

FIGURE 2. Schematic representation of the location of the RB motoneuronal cell group.

it appeared to provide the experimenter with a high degree of control over the sensory consequences of the unconditioned stimulus (Gormezano *et al.*, 1962).

As already mentioned, the blink reflex is a simultaneous action of the orbicularis oculi and RB muscles. Therefore, in this chapter we emphasize the connections of only those areas that project to both the orbicularis oculi and RB motoneuronal cell groups. For an account of the projections only to the orbicularis oculi motoneuronal cell group, we refer to Holstege *et al.* (1986a). On the other hand, we could not find any case of a projection to the RB nucleus without a simultaneous projection to the orbicularis oculi motoneuronal cell group.

2. AFFERENT PROJECTIONS TO BLINK MOTONEURONAL CELL GROUPS

Afferent connections to the orbicularis oculi and RB motoneuronal cell groups, which we call the blink motoneuronal cell groups, are known from the following brainstem structures: (1) the principal and spinal trigeminal nuclei and dorsolateral pontine tegmentum, (2) the ventral part of the lateral pontine tegmentum around the level of the motor trigeminal nucleus, (3) the medial tegmentum at levels of the hypoglossal nucleus, and (4) the red nucleus and the mesencephalic tegmentum just dorsal to it.

2.1. The Principal and Spinal Trigeminal Nuclei and Dorsolateral Pontine Tegmentum

Injections of HRP in the RB nucleus (Durand *et al.*, 1983) and in the lateral half of the facial nucleus (Takeuchi *et al.*, 1979; Panneton and Martin, 1983) resulted in HRP-labeled neurons in the ventral part of the trigeminal nuclei. According to our own observations (Holstege *et al.*, 1986a,b) with the autoradiographic tracing technique, the dorsolateral pontine tegmentum and the principal and rostral half of the spinal trigeminal nuclei distribute some labeled fibers to the RB motoneuronal cell group and to the intermediate facial subnucleus (Fig. 3), but these projections were much weaker than those observed from the areas described in Sections 2.2 and 2.3. The trigemino-blink motoneuronal projections probably play an important role in the R_1 component of the blink reflex because, according to Hiraoka and Shimamura (1977), this component is consistent with a disynaptic reflex arc with synapses in the trigeminal and facial nucleus.

2.2. The Ventral Part of the Lateral Pontine Tegmentum around the Level of the Motor Trigeminal Nucleus

According to autoradiographic tracing findings of Holstege *et al.* (1986a,b), this part of the pontine tegmentum strongly projects specifically to the blink motoneuronal cell groups ipsilaterally (Fig. 4). It must be stressed that this area lies outside the principal and/or spinal trigeminal nucleus, which indicates that it cannot play a role in the disynaptic R_1 component of the blink reflex.

2.3. The Medial Tegmentum at Levels of the Hypoglossal Nucleus

Bilateral projections from this part of the caudal brainstem tegmentum specifically to the blink motoneuronal cell groups have been described by Holstege *et al.* (1986a,b) (Fig. 5C,D). Interestingly, they also observed strong projections to a distinct part of the

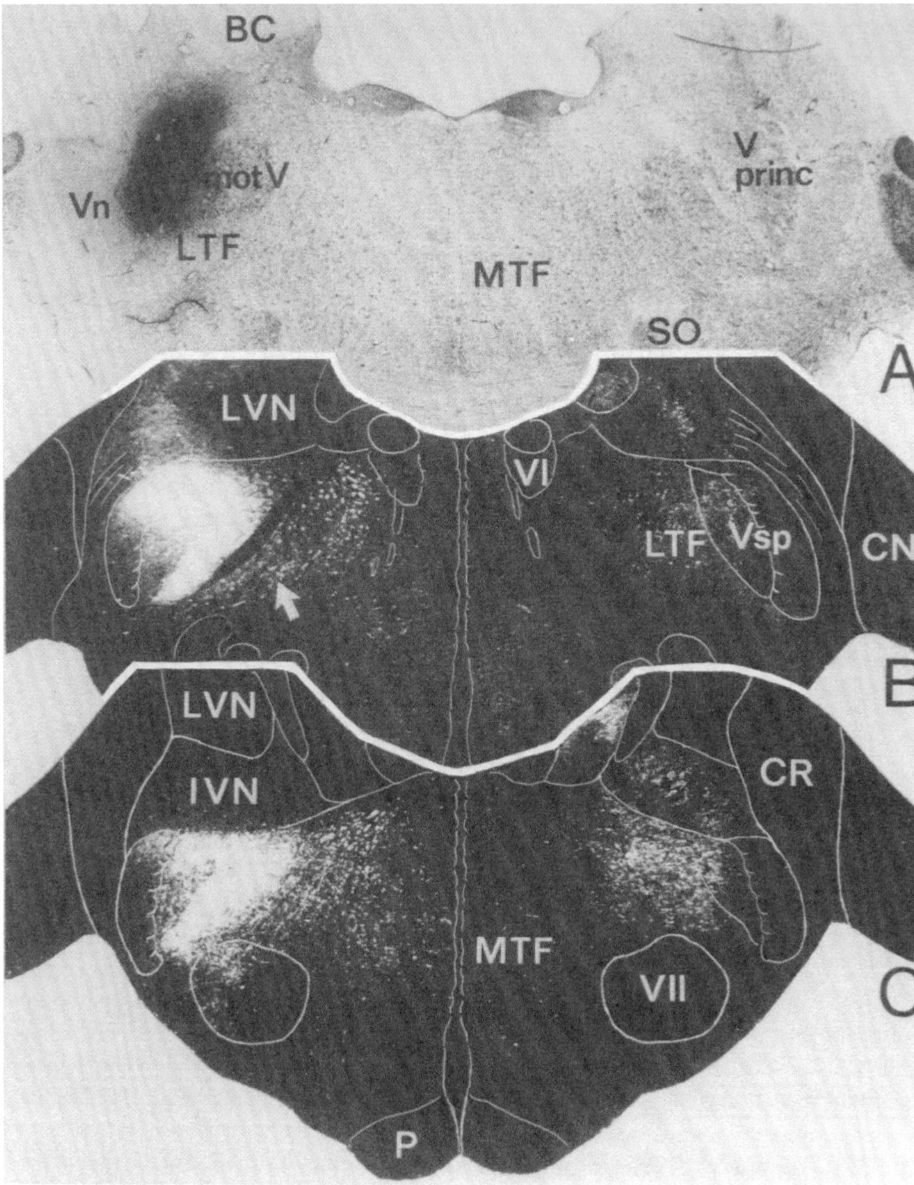

FIGURE 3. Bright-field (A) and dark-field (B and C) photomicrographs of case 1131 with an injection in the principal and rostral part of the spinal trigeminal nucleus. Note the dense ipsilateral projection to the intermediate facial subnucleus (C). Note further a limited projection to the RB motoneuronal cell group and surrounding lateral tegmentum (B).

pontine ventrolateral tegmental field, which may correspond to the area that projects strongly to the blink motoneuronal cell groups ipsilaterally as described in Section 2.2 (Fig. 5A,B; Fig. 6). It is important to note that almost all [^{3}H]leucine injection sites described in the study by Holstege *et al.* (1986a,b) were not confined to one side of the brainstem. Therefore, it was difficult for them to determine whether these projections were ipsi- or bilateral. A new case (1166) in which a unilateral injection was made in the medial tegmental field at the rostral pole of the hypoglossal nucleus showed these

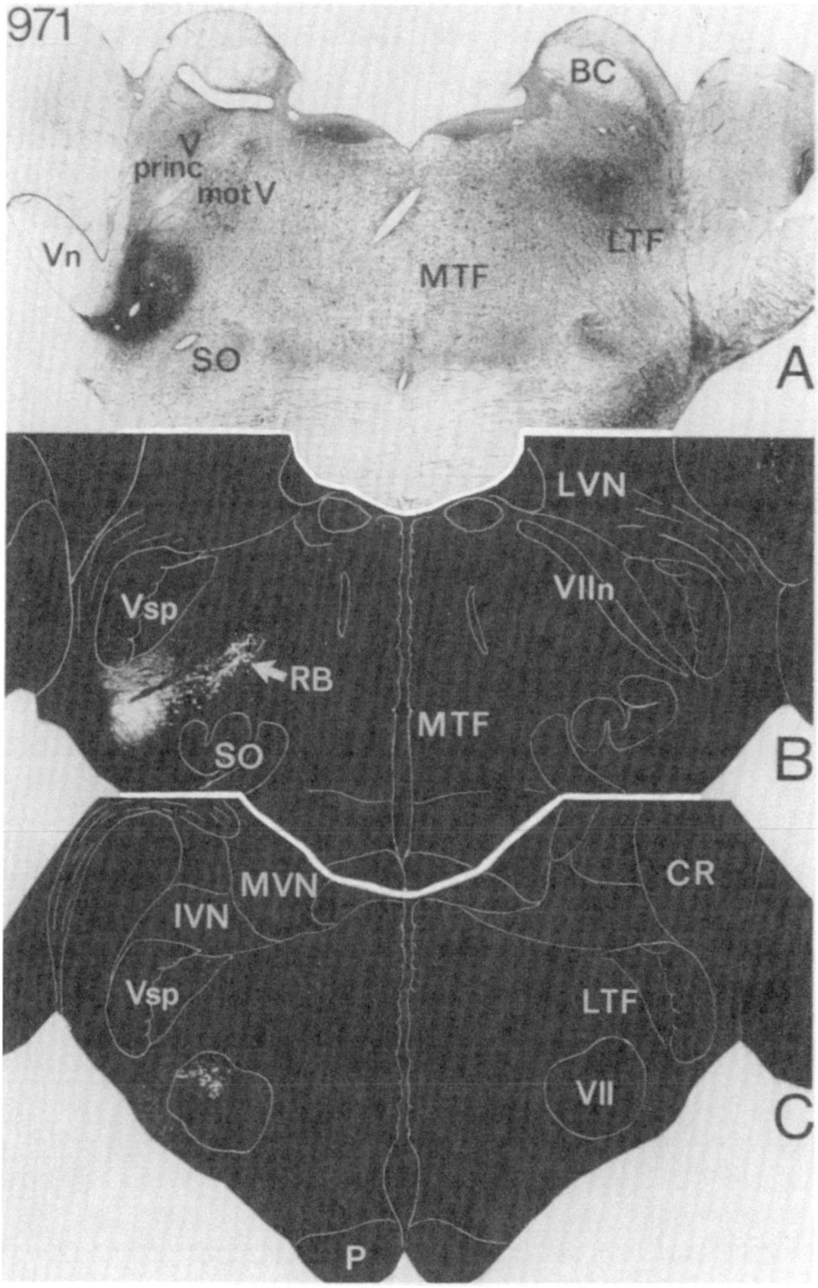

FIGURE 4. Bright-field (A) and dark-field (B and C) photomicrographs of case 971 with an injection in the caudal pontine ventrolateral tegmental field. Note the dense ipsilateral distribution of labeled fibers to the RB motoneuronal area (B) and the intermediate facial subnucleus (C). Note further the absence of labeled fibers in the spinal trigeminal nucleus, indicating that the injection site (A) did not involve the principal or spinal trigeminal nucleus.

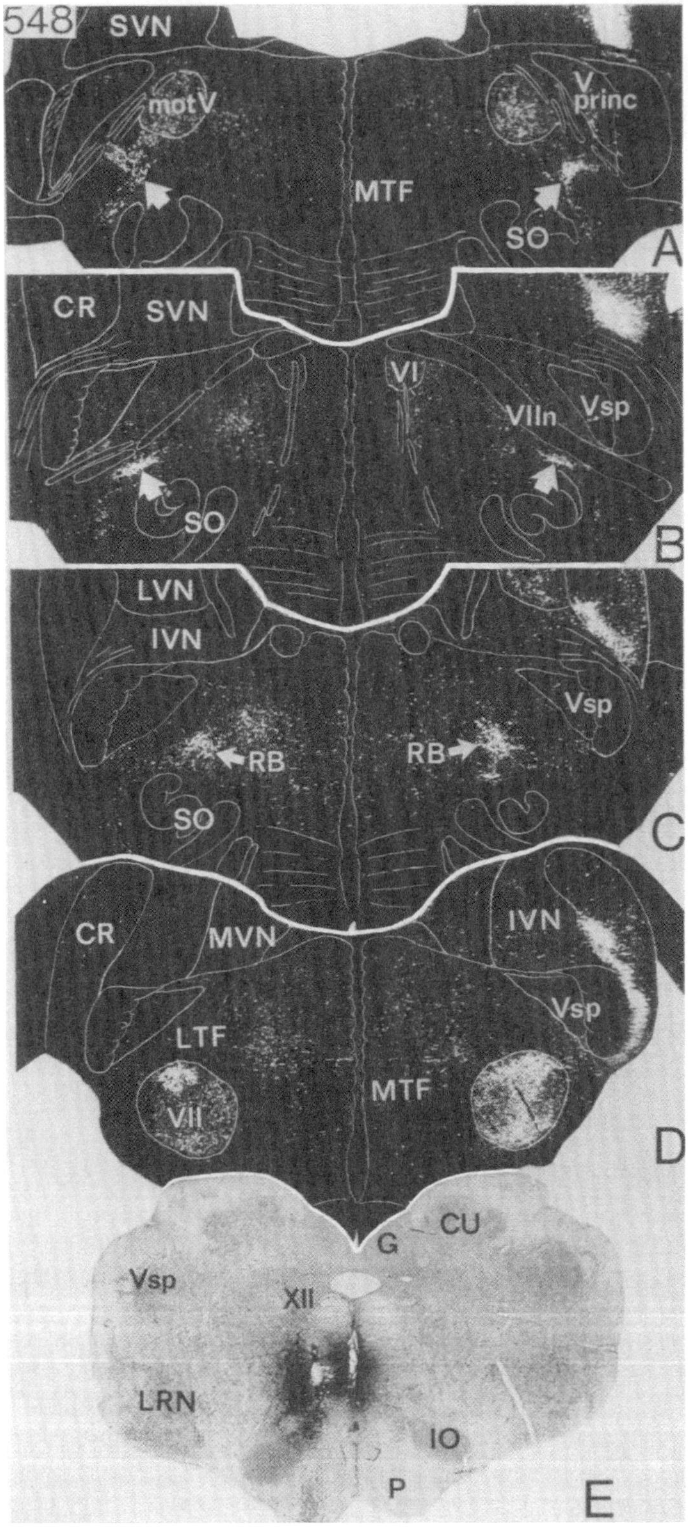

FIGURE 5. Dark-field (A, B, C, D) and bright-field (E) photomicrographs of case 548 with an injection in the medullary medial tegmentum at the level of the hypoglossal nucleus. Note the dense bilateral projection to the intermediate facial subnuclei (D), the RB motoneuronal cell group (C), and a cell group in the caudal pontine ventrolateral tegmentum (A and B).

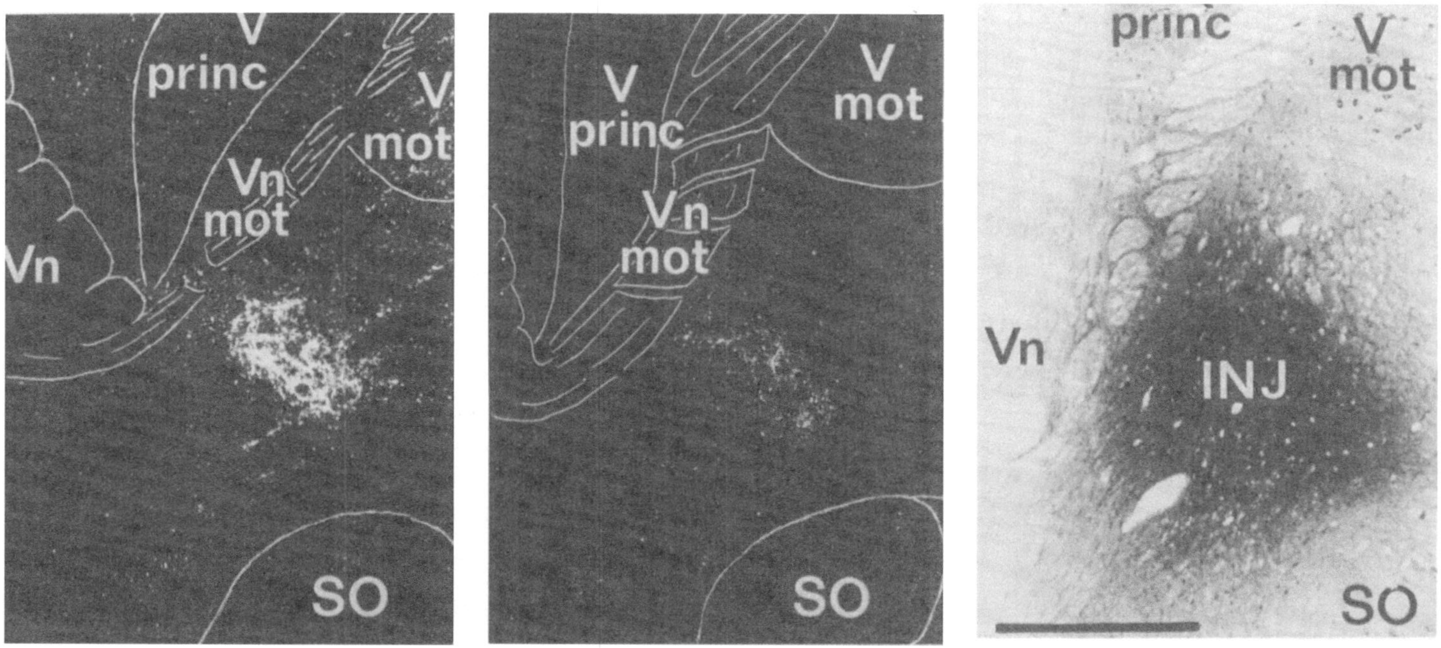

FIGURE 6. Left: Dark-field photomicrograph of case 1206 showing a dense projection to a cell group in the caudal pontine ventrolateral tegmentum. Middle: dark-field photomicrograph showing a projection to the caudal pontine ventrolateral tegmentum in case 1339 with an injection in the olivary pretectal nucleus and/or nucleus of the optic tract. Note that this projection is almost identical to the one observed in case 1206 (left). Right: Bright-field photomicrograph of case 1018 with an injection in the caudal pontine ventrolateral tregmentum. Note that the injection site involves the cell group receiving afferents from the caudal medullary medial tegmentum (cf. case 1206, left) and pretectum (case 1339, middle). Bar represents 1mm.

specific projections to the blink motoneuronal cell groups and ventrolateral pontine tegmentum, but only ipsilaterally. This indicates that at least part of the medullary projections are ipsilateral, although a strong bilateral component cannot be excluded. Like the ventrolateral pontine tegmentum described in Section 2.2, the medullary area is not located in the trigeminal nuclei and thus cannot be involved in the disynaptic organization of the R_1 blink reflex component.

This medullary area, together with the ventrolateral pontine area, is most probably involved in the R_2 blink reflex component for the following reasons. (1) The R_2 reflex component is not disynaptic but multisynaptic (Kugelberg, 1952; Lindquist and Martensson, 1970; Hiraoka and Shimamura, 1977; Ongerboer de Visser and Kuypers, 1978), and the response consists of several spikes (Berthier and Moore, 1983; Kugelberg, 1952). (2) The R_2 blink reflex component, according to Shahani and Young (1972), is responsible for actual closure of the eyelids. For such a motor performance, strong afferent connections are necessary. Holstege *et al.* (1986a,b) found such connections only from the two premotor areas described in Sections 2.2 and 2.3. (3) The medullary premotor area projects strongly to the pontine premotor area, indicating that both areas are involved in the same neuronal organization.

2.4. The Red Nucleus and the Mesencephalic Tegmentum Just Dorsal to It

Holstege *et al.* (1986a,b) also presented anatomic evidence for a contralateral projection to the blink motoneuronal cell groups directly derived from the dorsal part of the red nucleus and the area just dorsal to it. They emphasize, however, that these projections, which go by way of the rubrospinal tract, are much weaker than those observed from the pontine and medullary premotor areas.

In addition, according to Holstege *et al.* (1986b), from the contralateral rubrospinal tract fibers may be distributed to the pontine blink premotor area, although this projection is not specific but involves all parts of the pontine lateral tegmentum.

3. OTHER ANATOMIC PATHWAYS POSSIBLY RELATED TO THE BLINK REFLEX MECHANISM

As mentioned above, the pontine and medullary blink premotor areas cannot be involved in the R_1 blink reflex component. Assuming that they play a role in the R_2 blink reflex component, the question arises from which structures these two premotor areas receive their afferent connections. Because the afferent limb of both components of the blink reflex is the trigeminal nerve, the pontine and medullary blink premotor areas must receive trigeminal afferent input. Since both blink premotor areas do not receive direct projections from the trigeminal nuclei (Holstege *et al.*, 1986b), they must receive this input by way of other (relay) structures in the brainstem.

In this respect, it is interesting that the principal and spinal trigeminal nuclei project to the intermediate and deep layers of the superior colliculus (Huerta *et al.*, 1981; Holstege *et al.*, 1986b) (Fig. 7) and to the dorsal part of the red nucleus and dorsally adjoining area (Holstege *et al.*, 1986b) (Fig. 7). The intermediate and deep collicular layers project to the paramedian pontine reticular formation (PPRF) and upper medullary medial tegmental field (Altman and Carpenter, 1961; Martin, 1969; Harting *et al.*, 1973; Graham, 1977; Holstege and Collewijn, 1982). The PPRF and upper medullary medial tegmental field project to the medullary premotor area (Graybiel, 1977). Hiraoka and Shimamura

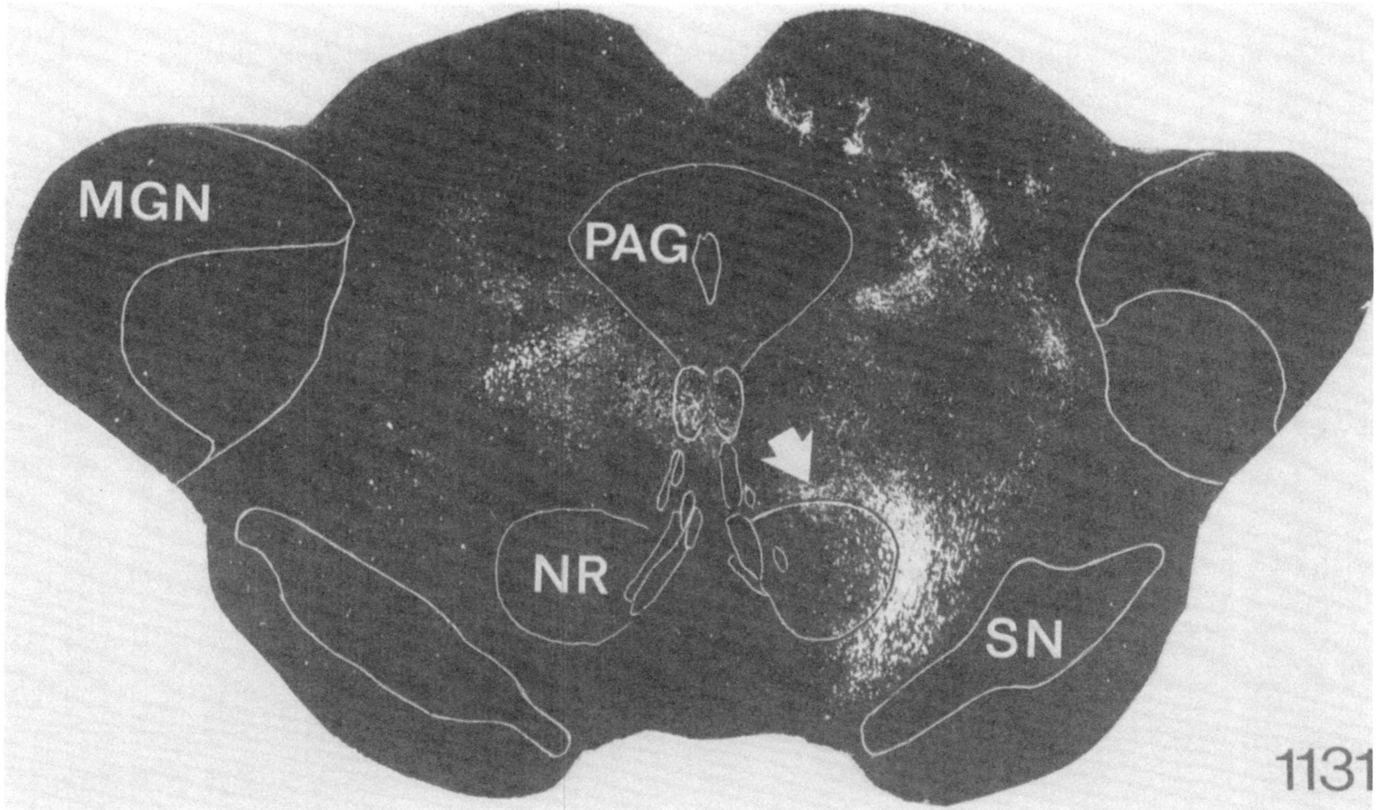

FIGURE 7. Dark-field photomicrograph of a section through the mesencephalon in case 1131 with an injection in the principal and the most rostral portion of the spinal trigeminal nucleus. Note the patchlike projections to the intermediate and deep collicular layers and the projection to the dorsal red nucleus and area just dorsal to it (arrow).

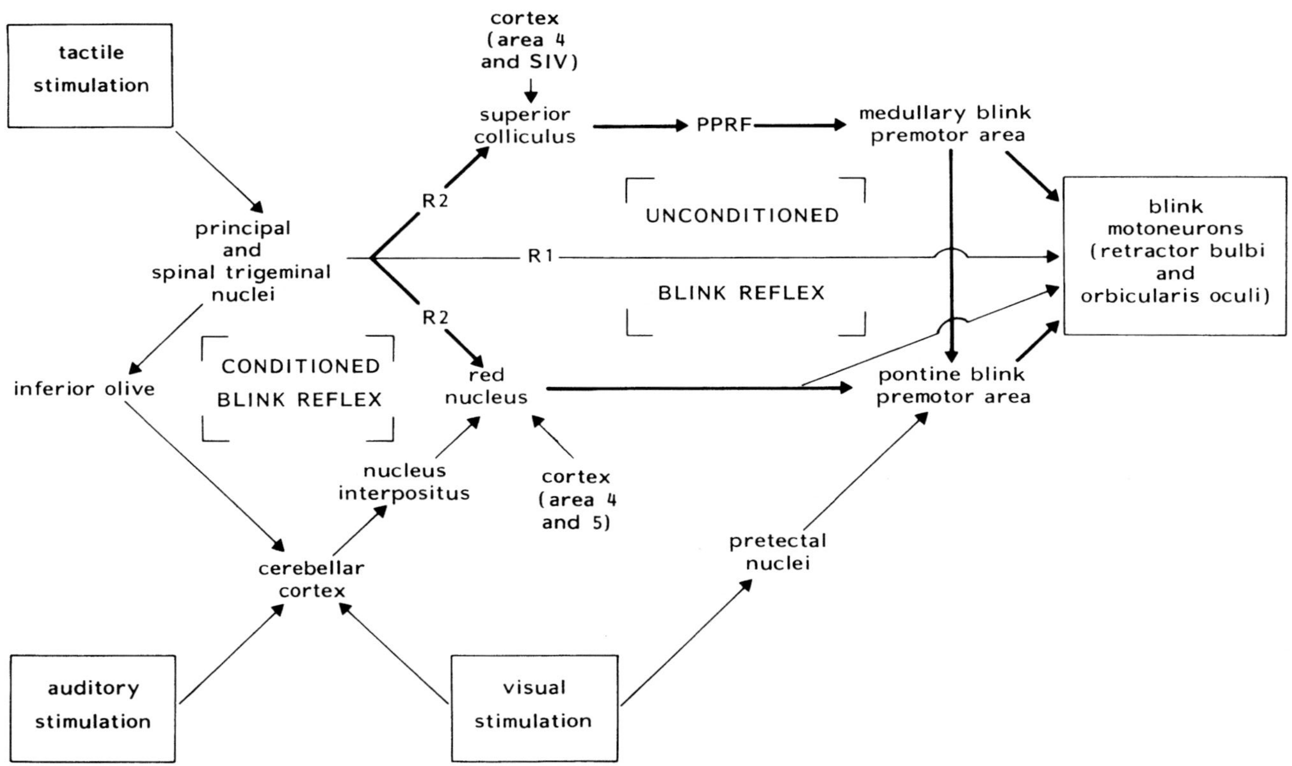

FIGURE 8. Schematic representation of the pathways possibly involved in the anatomic framework of the R_1 and R_2 blink reflex components.

(1977) recorded discharges in the PPRF and upper medullary medial tegmentum after corneal electrical stimulation. These discharges had latencies longer than 7.5 msec, which would be in keeping with involvement of these areas in the R_2 blink reflex component.

Thus, a possible neuronal circuit for the second blink reflex component is the pathway from the trigeminal nerve to the blink motoneuronal cell groups by way of (1) trigeminal nuclei, (2) intermediate and deep layers of the superior colliculus, (3) the PPRF and upper medullary medial tegmental field, and (4) the medullary blink premotor area. This pathway comprises five synapses (cf. Fig.8). Notably, important parts of this circuit are so-called oculomotor control systems, which would support the findings of several authors that blinking is accompanied by eye movements (Ginsborg, 1952; Esteban and Salinero, 1979; Evinger *et al.*, 1984; Collewijn *et al.*, 1985) and that saccadic eye movements tend to be accompanied by blinks (Zee *et al.*, 1983; Evinger *et al.*, 1984).

The dorsal red nucleus and dorsally adjoining area project directly, although not as strongly as the pontine and medullary blink premotor structures, to the blink motoneuronal cell groups and to the pontine blink premotor area. This latter structure may also receive afferents from the facial motor cortex (Kuypers, 1958) and from other parts of the lateral tegmental field of caudal pons and medulla (Holstege *et al.*, 1977). A very interesting finding was that the nucleus of the optic tract (NOT) and/or the olivary pretectal nucleus sends projections to the pontine blink premotor area (Holstege *et al.*, 1984, 1986a) (Fig. 6, middle). Holstege and Collewijn (1982) demonstrated a similar projection in the rabbit from the pretectum to the ventrolateral pontine reticular formation. This suggests that a pontine blink premotor area also exists in the rabbit. The pretectal nuclei receive retinal afferents (Scalia, 1972; Berman, 1977; Collewijn and Holstege, 1984) and thus could serve as relay structures in the neuronal organization of visually induced blink reflexes.

4. CORTEX

Bilateral lesions of the facial motor cortex in humans resulted in absence of voluntary eyelid closure but retention of normal reflex blinking to corneal stimulation and visual threat (Russell, 1980). Stimulation in cortical areas 4 and 5 and in the cortex of the anterior ectosylvian sulcus (S IV) in the cat can elicit blinks (Waters and Asanuma, 1983; Waters and Friedman, 1984). According to Waters and Friedman (1984), area 4 and S IV project to the superior colliculus in the same way as the trigeminal nuclei (Huerta *et al.*, 1981), whereas areas 4 and 5 project to the same area of the red nucleus as the trigeminal nuclei (cf. Fig. 7). These data further support the concept that the superior colliculus and the dorsal red nucleus with dorsally adjoining tegmentum play an important role in the neuronal organization of the blink reflex (Fig. 8).

5. CONDITIONED NICTITATING MEMBRANE RESPONSE

The classically conditioned nictitating membrane response (NMR) is a learned response. When a neutral stimulus (visual or auditory) is repeatedly followed by an air puff to the cornea, the animal will soon develop reflex blinking (i.e., NMR and closure of the eyelids) to the neutral stimulus alone. It has been shown that the pathway for this conditioned response is different from the unconditioned one, because in the rabbit lesioning either the rostromedial portion of the dorsal accessory inferior olive (McCormick *et al.*, 1985; Yeo *et al.*, 1986), lobule H VI of the cerebellar cortex (Yeo *et al.*, 1984a), the anterior portion of the interpositus nucleus (Yeo *et al.*, 1984b), or the medial portion of the red nucleus and adjoining areas (Rosenfield and Moore, 1983) abolishes the conditioned but not the unconditioned blink reflex. Cerebellar lobule H VI receives afferents from that part of the inferior olive that in turn receives trigeminal afferents (Miles and Wiesendanger, 1975). Furthermore, lobule H VI sends fibers, by way of the anterior interpositus nucleus (Yeo *et al.*, 1984b), to the dorsal part of the red nucleus and dorsally adjoining area (Kawamura *et al.*, 1982) (Fig. 8). At some point this separate pathway for the conditioned blink response must be connected with the neuronal circuit of the unconditioned response. The red nucleus, which receives afferents from the interpositus nucleus as well as from the trigeminal nuclei (cf. Fig. 7) and sends fibers to the pontine blink premotor area and to the blink motoneuronal cell groups, is a strong candidate for such a connection. On the other hand, a lesion in the red nucleus of the rabbit does not affect the unconditioned blink reflex (Rosenfield and Moore, 1983). In this respect, it must be recalled that another relay in the R_2 blink reflex component may exist, i.e., the intermediate and deep layers of the superior colliculus. The findings of Hiraoka and Shimamura (1977) support this assumption. They transected the brainstem rostral and caudal to the pontine and upper medullary levels. The caudal transections did not affect the blink reflex, but after mesencephalic transections the blink reflex disappeared immediately.

ACKNOWLEDGMENTS. The authors thank Mr. E. Dalm, Mr. A. M. Vreugdenhil, Mr. R. C. Boer, Mr. F. H. Klink, and Mrs. C. Bijker-Biemond for their technical help and Miss P. van Alphen for her help with the photography. They also thank Miss E. Klink for typing the manuscript.

Abbreviations used:

BC	Brachium conjunctivum	nVI	Nervus abducens
CN	Cochlear nucleus	nVII	Nervus facialis
CR	Corpus restiforme	P	Pyramidal tract
CU	Cuneate nucleus	PAG	Periaqueductal gray
G	Gracile nucleus	RB	Retractor bulbi nucleus
IO	Inferior olivary complex	RM	Nucleus raphe magnus
IVN	Inferior vestibular nucleus	SN	Substantia nigra
LRN	Lateral reticular nucleus	SO	Superior olivary complex
LTF	Lateral tegmental field	SVN	Superior vestibular nucleus
LVN	Lateral vestibular nucleus	Vn	Trigeminal nerve
MGN	Medial geniculate nucleus	Vprinc.	Principal trigeminal nucleus
MLF	Medial longitudinal fasciculus	Vn mot	Motor trigeminal nerve
motV	Motor trigeminal nucleus	Vsp.	Spinal trigeminal nerve
MTF	Medial tegmental field	VI	Nucleus abducens
MVN	Medial vestibular nucleus	VII	Facial nucleus
NR	Red nucleus	XII	Hypoglossal nucleus

REFERENCES

Altman, J., and Carpenter, M. B., 1961, Fiber projections of the superior colliculus in the cat, *J. Comp. Neurol.* **166:**157–177.

Bach-Y-Rita, P., 1971, Neurophysiology of eye movements, in: *The Control of the Eye Movements* (P. Bach-Y-Rita and C. L. Collins, eds.), Academic Press, New York, London, pp. 7–46.

Berman, N., 1977, Connections of the pretectum in the cat, *J. Comp. Neurol.* **174:**227–254.

Berthier, N. E., and Moore, J. W., 1983, The nictitating membrane response: An electrophysiological study of the abducens nerve and nucleus and the accessory abducens nucleus in rabbit, *Brain Res.* **258:**201–210.

Bolk, L., Groppert, E., Kallius, E., and Lubosch, W., 1938, *Handbuch der Vergleichenden Anatomie den Wirbeltiere*, Urban & Schwartzenberg, Berlin.

Collewijn, H., and Holstege, G., 1984, Effects of neonatal and late unilateral enucleation on optokinetic responses and optic nerve projections in the rabbit, *Exp. Brain Res.* **57:**138–150.

Collewijn, H., van der Steen, J., and Steinman, R. M., 1985, Human eye movements associated with blinks and prolonged eyelid closure, *J. Neurophysiol.* **54:**11–27.

Courville, J., 1966, The nucleus of the facial nerve; the relation between cellular groups and peripheral branches of the nerve, *Brain Res.* **1:**338–354.

Durand, J., Gogan, P., Gueritaud, J. P., Horcholle-Bossavit, G., and Tyc-Dumont, S., 1983, Morphological and electrophysiological properties of trigeminal neurones projecting to the accessory abducens nucleus of the cat, *Exp. Brain Res.* **53:**118–128.

Ekman, P., Friesen, W. V., and Ellsworth, P., 1972, *Emotion in the Human Face: Guidelines for Research and an Integration of Findings*, Pergamon Press, New York.

Esteban, A., and Salinero, E., 1979, Reciprocal reflex activity in ocular muscles: Implications in spontaneous blinking and Bell's phenomenon, *Eur. Neurol.* **18:**157–165.

Evinger, C., Shaw, M. D., Peck, C. K., Manning, K. A., and Baker, R., 1984, Blinking and associated eye movements in humans, guinea pigs and rabbits, *J. Neurophysiol.* **52:**323–339.

Ginsborg, B. L., 1952, Rotation of the eyes during involuntary blinking, *Nature* **169:**412–413.

Gormezano, I., Schneiderman, H., Deaux, E., and Fuentes, I., 1962, Nictitating membrane: Classical conditioning and extinction in the albino rabbit, *Science* **138:**33–34.

Graham, J., 1977, An autoradiographic study of the efferent connections of the superior colliculus in the cat, *J. Comp. Neurol.* **173:**629–654.

Grant, K., Gueritaud, J. P., Horcholle-Bossavit, G., and Tyc-Dumont, S., 1979, Anatomical and electrophysiological identification of motonèurones supplying the cat retractor bulbi muscle, *Exp. Brain Res.* **34:**541–550.

Grant, K., Guegan, M., and Bossavit, G., 1981, The anatomical relationship of the retractor bulbi and posterior digastric motoneurones to the abducens and facial nuclei in the cat, *Arch. Ital. Biol.* **119:**195–207.

Graybiel, A. M., 1977, Direct and indirect preoculomotor pathways of the brainstem: An autoradiographic study of the pontine reticular formation in the cat, *J. Comp. Neurol.* **175:**37–78.

Harting, J. K., Hall, W. C., Diamond, I. T., and Martin, G. F., 1973, Anterograde degeneration study of the superior colliculus in *Tupaia glis*: Evidence for a subdivision between superficial and deep layers, *J. Comp. Neurol.* **148:**361–386.

Hiraoka, M., and Shimamura, M., 1977, Neural mechanisms of the corneal blinking reflex in cats, *Brain Res.* **125:**265–275.

Holstege, G., and Collewijn, H., 1982, The efferent connections of the nucleus of the optic tract and the superior colliculus in the rabbit, *J. Comp. Neurol.* **209:**139–175.

Holstege, G., Kuypers, H. G. J. M., and Dekker, J. J., 1977, The organization of the bulbar fibre connections of the trigeminal, facial and hypoglossal motor nuclei. II. An autoradiographic tracing study in cat, *Brain* **100:**265–286.

Holstege, G., Tan J., van Ham, J. J., and Bos, A., 1984, Mesencephalic projections to the facial nucleus in the cat. An autoradiographical tracing study, *Brain Res.* **311:**7–22.

Holstege, G., van Ham, J. J., and Tan, J., 1986a, Afferent projections to the orbicularis oculi motoneuronal cell group. An autoradiographical tracing study in the cat, *Brain Res.* **374:**306–320.

Holstege, G., Tan, J., van Ham, J. J., and Graveland, G. A., 1986b, Anatomical observations on the afferent projections to the retractor bulbi motoneuronal cell group and other pathways possibly related to the blink reflex in cat, *Brain Res.* **374:**321–334.

Huerta, M. F., Frankfurter, A. J., and Harting, J. K., 1981, The trigemino-collicular projection in the cat: Patch-like endings within the intermediate gray, *Brain Res.* **211:**1–13.

Kawamura, S., Hattori, S., Higo, S., and Matsuyama, T., 1982, The cerebellar projections to the superior colliculus and pretectum in the cat: An autoradiographic and horseradish peroxidase study, *Neuroscience* **7:**1673–1689.

Kimura, J., Powers, J. M., and van Allen, M. W., 1969, Reflex response of orbicularis oculi muscle to supra orbital nerve stimulation, *Arch. Neurol.* **21:**193–199.

Kugelberg, E., 1952, Facial reflexes, *Brain* **75:**385–396.

Kume, M., Uemura, M., Matsuda, K., Matsushima, K., and Mizuno, N., 1978, Topographical representation of peripheral branches of the facial nerve within the facial nucleus: An HRP study in the cat, *Neurosci. Lett.* **8:**5–8.

Kuypers, H. G. J. M., 1958, An anatomical analysis of cortico-bulbar connections to the pons and lower brainstem in the cat, *J. Anat.* **92:**198–218.

Lindquist, C., and Martensson, A., 1970, Mechanisms involved in the cat's blink reflex, *Acta Physiol. Scand.* **80:**149–159.

Martin, G. F., 1969, Efferent tectal pathways of the opossum (*Didelphis virginiana*), *J. Comp. Neurol.* **135:**209–224.

McCormick, D. A., Lavond, D. A., and Thompson, R. F., 1982, Concomitant classical conditioning of the rabbit nictitating membrane and eyelid responses: Correlations and implications, *Physiol. Behav.* **28:**769–775.

McCormick, D. A., Steinmetz, J. E., and Thompson, R. F., 1985, Lesions of the inferior olivary complex cause extinction of the classically conditioned eyeblink response, *Brain Res.* **359:**120–130.

Miles, T. S., and Wiesendanger, M., 1975, Organization of climbing fibre projections to the cerebellar cortex from trigeminal cutaneous afferents and from the S I face area of the cerebral cortex in the cat, *J. Physiol.* (*Lond.*) **245:**409–424.

Mis, F. W., 1977, A midbrain–brainstem circuit for conditioned inhibition of the nictitating membrane response in the rabbit (*Oryctolagus cuniculus*), *J. Comp. Neurol.* **91:**975–988.

Moore, J. W., Yeo, C. J., Oakley, D. A., and Steele Russell, I., 1980, Conditioned inhibition of the nictitating membrane response in decorticate rabbits, *Behav. Brain Res.* **1:**397–409.

Ongerboer de Visser, B. W., and Kuypers, H. G. J. M., 1978, Late blink reflex changes in lateral medullary lesions. An electrophysiological and neuroanatomical study of Wallenberg's syndrome, *Brain* **101:**285–294.

Panneton, W. M., and Martin, G. F., 1983, Brainstem projections to the facial nucleus of the opossum. A study using axonal transport techniques, *Brain Res.* **267:**19–33.

Papez, J. W., 1927, Subdivisions of the facial nucleus, *J. Comp. Neurol.* **43:**159–191.

Richardson, R. T., and Thompson, R. F., 1984, Amygdaloid unit activity during classical conditioning of the nictitating membrane response in rabbit, *Physiol. Behav.* **32:**527–539.

Rosenfield, M. E., and Moore, J. W., 1983, Red nucleus lesions disrupt the classically conditioned nictitating membrane response in rabbits, *Behav. Brain Res.* **10:**393–398.

Russell, R. W. R., 1980, Supranuclear palsy of eyelid closure, *Brain* **103:**71–82.

Scalia, F., 1972, The termination of retinal axons in the pretectal region of mammals, *J. Comp. Neurol.* **145:**223–258.

Shahani, B. T., and Young, R. R., 1972, Human orbicularis oculi reflexes, *Neurology* (*Minneap.*) **22:**149–154.

Shahani, B. T., and Young, R. R., 1973, Blink reflexes in orbicularis oculi, in: *New Developments in Electromyography and Clinical Neurophysiology*, Vol. 3 (J. E. Desmedt, ed.), S. Karger, Basel, pp. 641–648.

Spencer, R. F., Baker, R., and McCrea, R. A., 1980, Localization and morphology of cat retractor bulbi motoneurons, *J. Neurophysiol.* **43:**754–770.

Takeuchi, Y., Nakano, K., Uemura, M., Matsuda, K., Matsushima, K., and Mizuno, N., 1979, Mesencephalic and pontine afferent fibre system to the facial nucleus in the rat. A study using the horseradish peroxidase and silver impregnation techniques, *Exp. Neurol.* **66:**330–342.

Tamai, Y., Iwamoto, M., and Tsujimoto, T., 1986, Pathway of the blink reflex in in the brainstem of the cat: Interneurons between the trigeminal nuclei and the facial nucleus, *Brain Res.* **380:**19–25.

Waters, R. S., and Asanuma, H., 1983, Movements of facial muscles following intra-cortical microstimulation (ICMS) along the lateral branch of the posterior bank of the ansate sulcus, area 5a and 5b, in the cat, *Exp. Brain Res.* **50:**459–463.

Waters, R. S., and Friedman, D. P., 1984, Brainstem projections from cortical motor output regions (CMORS) to facial muscles in the cat, *Soc. Neurosci. Abstr.* **10:**737.

Yeo, C. H., Hardiman,M. J., and Glickstein, M., 1984a, Discrete lesions of the cerebellar cortex abolish the classically conditioned nictitating membrane response of the rabbit, *Behav. Brain Res.* **13:**261–266.

Yeo, C. H., Hardiman, M. J., and Glickstein, M., 1984b, Cerebellar pathways in the conditioned nictitating membrane response, *Neurosci. Abstr.* **10:**793.

Yeo, C. H., Hardiman, M. J., and Glickstein, M., 1986, Classical conditioning of the nictitating membrane response of the rabbit. IV. Lesions of the inferior olive, *Exp. Brain Res.* **63:**81–92.

Zee, D. S., Chu, F. C., Leigh, R.J., Sarino, I. J., Schatz, N. J., Reingold, D. B., and Cogan, D., 1983, Blink–Saccade synkinesis, *Neurology* (*N.Y.*) **33:**1233–1236.

V

Storing and Retrieving Information

28

Mammalian Systems for Storing and Retrieving Information

JOHN C. ECCLES

1. LONG-TERM POTENTIATION IN THE HIPPOCAMPUS

In the first demonstration of long-term potentiation (LTP) by Bliss and Lømo (1973), the conditioning presynaptic stimulation was a brief tetanus (usually 15 Hz for 15 sec) of the perforating pathway to the fascia dentata (Fig. 1A) of the hippocampus, and the homosynaptic stimulation was by a single stimulus through the same electrode; i.e., it was a homosynaptic LTP. Figure 1B illustrates the extracellular potentials recorded either at the level of the synapses and hence largely a negative wave, a population EPSP, or at the level of the granule cell bodies and hence a favorable site for recording the cell discharges, the negative population spike. In Fig. 1C, four bursts of conditioning stimulations cause the population EPSP to increase to more than double and to remain large 10 hr after the last conditioning tetanus.

Most subsequent investigations on LTP continued with this same paradigm, and there was general agreement that LTP was clearly distinguishable from posttetanic potentiation (PTP), which has a much shorter duration of at most a few minutes and is clearly a presynaptic phenomenon. An important new development was the demonstration of heterosynaptic LTP (McNaughton *et al.*, 1978; Levy and Steward, 1979, 1983; Wigström and Gustafsson, 1983; Gustafsson and Wigström, 1986). With heterosynaptic LTP a weak presynaptic input that exhibited no homosynaptic LTP was effective for LTP production when it was coupled in the conditioning procedure with a different stronger presynaptic input.

In their investigations on heterosynaptic LTP, Gustafsson and Wigström (1986) had the advantage of virtually eliminating with picrotoxin the complications induced by superposition of the hyperpolarizing synaptic inhibition, as described by Wigström and Gustafsson (1983a, 1985). When this is done, a conditioning stimulation of only three to five bursts of three stimuli at 50 Hz repeated at 0.5-sec intervals is sufficient to evoke considerable LTP in CA_1 pyramidal cells. For example, in Fig. 2, stimuli 1 and 2 were applied to nonoverlapping bundles of Schäffer collaterals. A, B, and C show the experimental arrangements. In E, sizes of the extracellular ESPS responses evoked by input 1 or 2 every 5 sec are plotted (the testing stimulation), and the maximum slope (see scale

JOHN C. ECCLES • Contra CH6611, Switzerland.

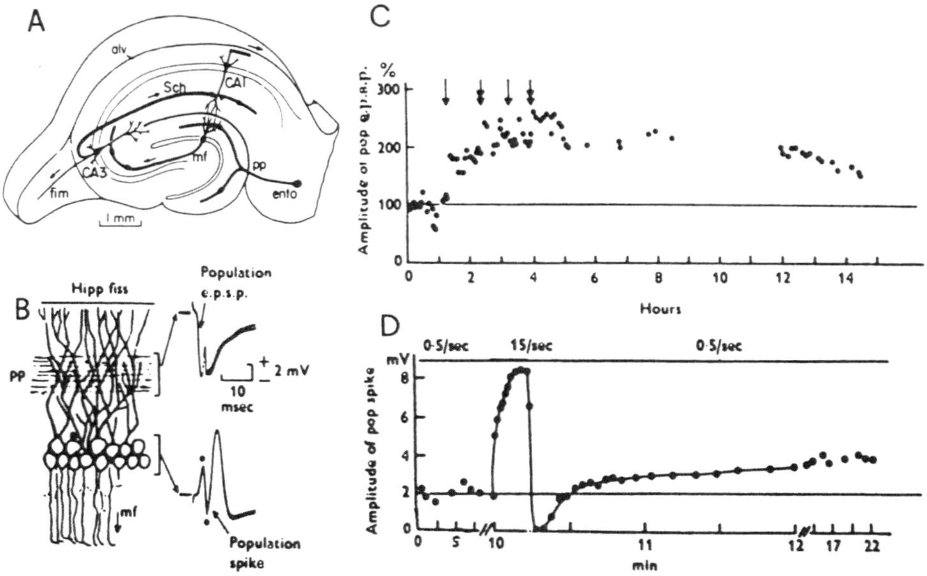

FIGURE 1. (A) Section of hippocampus. (B) Drawing of granule cells with their bodies, dendrites, and axons that form the mossy fibers (mf). The fibers of the perforating pathway (pp) are shown traversing the dendrites on which they make excitatory synapses. Recording the field potentials at the level of the pp synapses results in a large and prolonged negative potential, which is labeled population e.p.s.p. When the recording electrode is advanced to the level of the cell bodies, the sharp negative spike (population spike) signals the generation of impulses in the cell bodies. (C) The relative amplitudes of the population e.p.s.p. are plotted up to 10 hr after four conditioning trains of stimulation indicated by the arrows. At single arrows there was stimulation at 15/sec for 15 sec and at the double arrows 100/sec for 3 sec. The 100% line is drawn through the prestimulation responses at 0.5/sec, and after the conditioning tetanus the same low rate was resumed. (D) Time course of poststimulation potentiation as in C, but for the population spike. There was only a single conditioning tetanus of 15/sec for 15 sec (Bliss and Lømo, 1973).

at right) is shown as a dot, the successive dots making up the two horizontal lines during the initial control observations over more than 5 min. The timings of the three conditioning stimulation bursts (STIM 1) are shown by the arrows below. The homosynaptic potentiation of EPSP in the upper trace of E has an initial peak in each case that is probably largely attributable to PTP. It declines in 2 to 3 min to the steadily maintained LTP. With the third STIM 1, the additional LTP was quite small (cf. McNaughton, 1982). The actual EPSP records are shown in Fig. 2F at times a_1, b_1, c_1, and d_1 of E.

The input 2 EPSPs show no influence of the first two stimuli because the stimulations were applied with alternating timings in a spacing of 2.5 sec before and after STIM 1 as shown in the top row of C. However, when there was conjunction (second row of C), there was a considerable enduring heterosynaptic LTP of 15%, as may be seen in the superimposed traces of c_2 and d_2 of F. Gustafsson and Wigström (1986) report that heterosynaptic LTP always declines from an initial summit to reach the enduring level of LTP in several minutes, as in Fig. 2E (input 2).

In the conjunction of Fig. 2, STIM 2 was synchronized with the first stimulus of each STIM 1 burst (Fig. 2C). It is possible to test the effect of varying this conjunction interval on the size of the heterosynaptic LTP. In Fig. 3, the plotted points give the LTPs as percentages of the maximum homosynaptic LTP for a range of conjunction intervals relative to the start of the conditioning tetanus. In Fig. 3B, the time of the conditioning

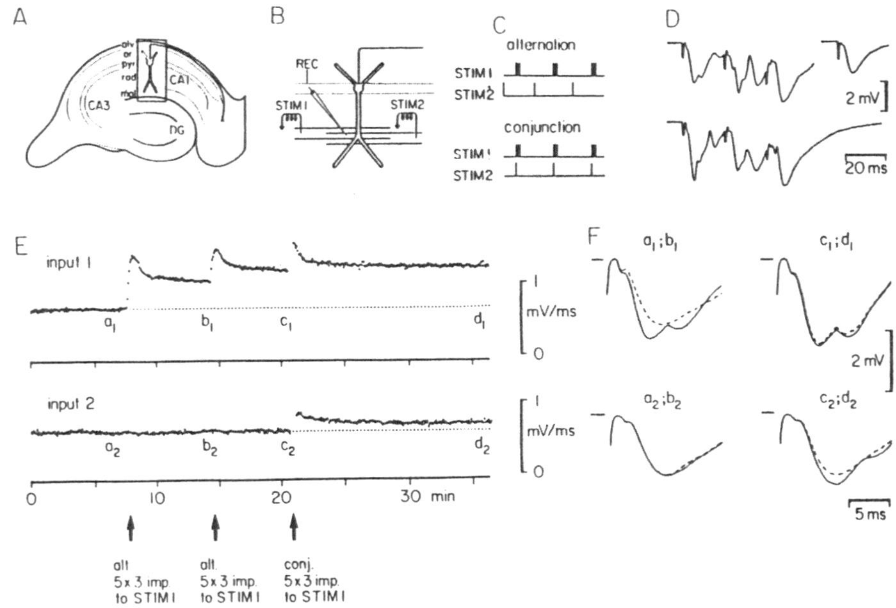

FIGURE 2. Long-lasting potentiation induced by combining single volleys and brief tetani to separate afferents. (A) Schematic drawing of the hippocampal slice with the studied CA_1 region boxed in. Abbreviations: alv, alveus; or, stratum oriens; pyr, stratum pyramidale; rad, stratum radiatum; mol, stratum moleculare. (B) Diagram showing the arrangement of stimulating (STIM) and recording (REC) electrodes. (C) Schematic drawing of the stimulus conditions during tetanization. In the alternation mode (upper half), the stimuli to STIM 1 and STIM 2 were kept 2.5 sec apart, whereas in the conjunction mode (lower half), the stimulus to STIM 2 coincided with the first stimulus to STIM 1. (D) Sample records of extracellular responses from the apical dendritic layer, stratum radiatum. Upper records show the responses to a three-impulse tetanus to STIM 1 and later to a single stimulus to STIM 2, whereas the lower record shows the response to a combined activation of STIM 1 and STIM 2 (conjuunction in C). (E) Measurements of the initial slopes of the field EPSPs resulting from activation of STIM 1 (imput 1) and STIM 2 (input 2) are shown for a series of test responses. Input 1 was subjected to a series of tetanic activations as indicated in the bottom of the graph. (F) Average records (n - 10) of the responses shown in E, taken at the indicated times. Upper records are reponses to STIM 1 (input 1), and lower records to STIM 2 (input 2) (Gustafsson and Wigstrom, 1986).

five-impulse tetanus is shown by the thick lower line. Figure 3B is remarkable in two respects: a considerable LTP is produced when the testing stimulation preceded the conditioning by up to 40 msec; the maximally induced heterosynaptic LTP was about half of the homosynaptic LTP and began to decline during the conditioning tetanus. In Fig. 3A, with a two-impulse tetanus there was a comparable heterosynaptic potentiation with conjunction in which the testing stimulation preceded the onset of the conditioning stimulation (see thick line below). In an earlier investigation, Levy and Steward (1983) had discovered this unexpected preconditioning interval for heterosynaptic LTP of dentate granule cells. The virtual elimination of inhibition may account for the preconditioning interval in Fig. 3 (40 msec) being longer than the 20 msec reported by Levy and Steward (1983).

Heterosynaptic LTP was observed between remote synaptic sites. In Fig. 4A, one stimulation was applied to the stratum oriens to activate synapses on the basal dendrites and another to the apical dendrites (Gustafsson and Wigström, 1986). With this remote conjunction, the heterosynaptic LTP had a mean value of 19%, which is about half of

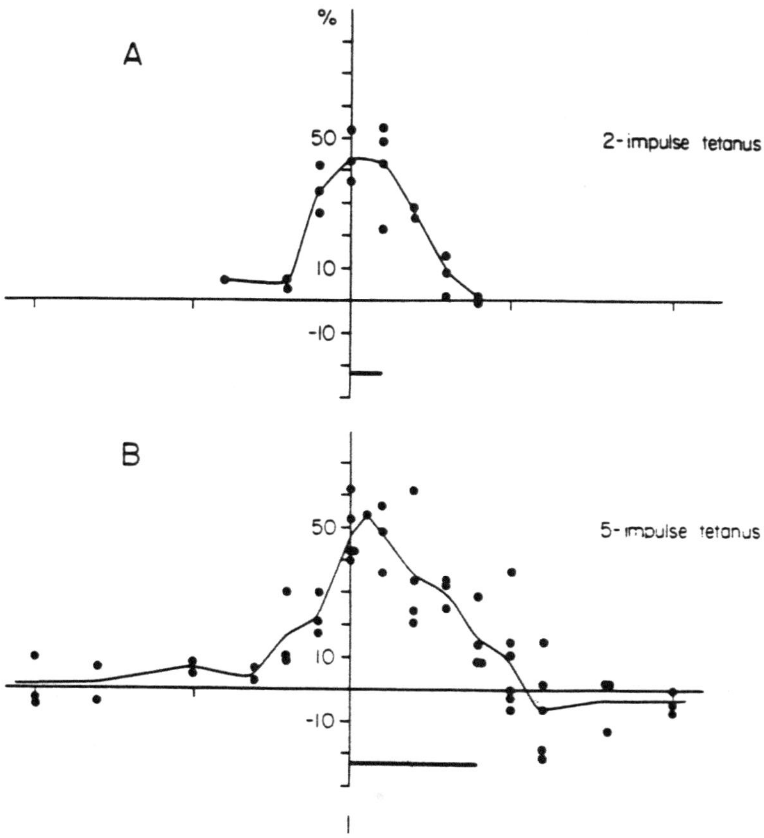

FIGURE 3. The time course of the effect of a conditioning tetanus. (A) Effect of a two-impulse tetanus. The origin of the time axis represents coincidence between the single test volley and the first volley in the train (100 msec scale). Negative values (to the left) on the same axis indicate that the test volley preceded the conditioning tetanus. The duration of the tetanus from the first to the last volley is shown below the graph as a thick line. In each slice, at most two (generally one) conditioning–test intervals were used, and the filled circles represent values obtained from each such trial. At each trial, the test volley was, after having reached a stable value (slope = v_1), combined with the conditioning train, and five such conditioning–test events were repeated at 0.2 Hz. The slope of the test field EPSPs was measured 5 min thereafter (slope = v_2). The test and conditioning inputs were subsequently twice tetanized together with a five-impulse tetanus, and the test EPSP was measured after 5 min (slope = v_3). The degree of potentiation given by the conditioning tetanus was then calculated as $[(v_2 - v_1)/(v_3 - v_1)] \times 100$, i.e., as the fraction of the total potentiation that was obtained. The line in the graph is drawn through the average value at each interval. (B) Same as in A but with a five-impulse conditioning tetanus (Gustafsson and Wigstrom, 1986).

the mean percentage observed for heterosynaptic LTP under the close apposition of Fig. 2B.

2. A HYPOTHESIS OF THE SYNAPTIC MECHANISMS CONCERNED IN LTP

For the purpose of this chapter, I chose to formulate a hypothesis of the synaptic factors directly concerned in LTP. The discoveries of the last 3 years have necessitated a radical revision of the hypothesis that attributed LTP to the widespread postsynaptic

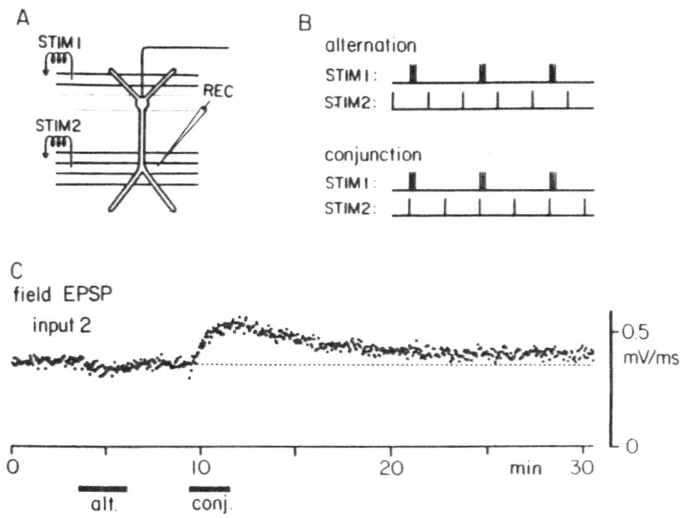

FIGURE 4. Conditioning effect of a stratum oriens input. (A) Schematic diagram of the arrangement of stimulating (STIM) and recording (REC) electrodes. STIM 1 was used as a conditioning stimulus, and the electrode was positioned in the stratum oriens. The test (STIM 2) and the recording electrodes were located in the middle of stratum radiatum. (B) Schematic drawing of the strimulus conditions during tetanization. During the conjunction tetanization, every second test volley coincided with the first volley of the conditioning tetanus, whereas in the alternation mode, the tetani occurred between the test volleys. (C) Measurements of the initial slope of the field EPSP resulting from stimulation of input 2 are shown for a series of responses. Two episodes of tetanization are indicated below the graph. During the alternation mode, there is a small decrease of the test response, which quickly recovers after the end of the conditioning tetanization. During the conjunction-type tetanization, there is a large increase in the test response, which then decays to a stable elevated level (Gustafsson and Wigstrom, 1986).

action of Ca^{2+} ions (Eccles, 1983). I now present a simplified version of the hypothesis recently published by Wigström and Gustafsson (1985). This hypothesis offers a coherent explanation of a wide range of experimental findings and provides several challenges to experimental testing. I develop my version in a series of seven stages together with the experimental evidence.

Stage 1. As illustrated diagrammatically in Fig. 5A, it is proposed that a spine synapse on dendrites of a granule cell or of a CA_1 pyramidal cell has a double postsynaptic receptor mechanism as defined by distinctive pharmacological and physiological properties (Watkins, 1984; and as formulated by Wigström and Gustafsson, 1985b). One component selectively combines with the amino acids glutamate and quisqualate (QQ). The other component also combines with glutamate, and in addition it selectively combines with N-methyl-D-aspartate (NMDA) and is selectively blocked by 2-amino-5-phosphonovalerate (APV) (Collingridge *et al.*, 1983a,b; Dingledine, 1983).

Stage 2. The glutamate liberated by a presynaptic impulse acts on the QQ receptors, opening up Na^+ and K^+ channels by an ionotropic action and resulting in the rapid depolarization of the EPSP. The glutamate also acts on the NMDA receptors so that in the picrotoxin-treated preparation a brief tetanus gives a small depolarization that continues for up to 60 msec and is identified by its great reduction by APV (Wigström and Gustafsson, 1984; Wigström *et al.*, 1985, 1986). If the postsynaptic membrane is strongly depolarized, Ca^{2+} channels are opened through the activated NMDA receptor zone, with a resulting large influx of Ca^{2+} ions (Dingledine, 1983; Mayer and Westbrook, 1985)

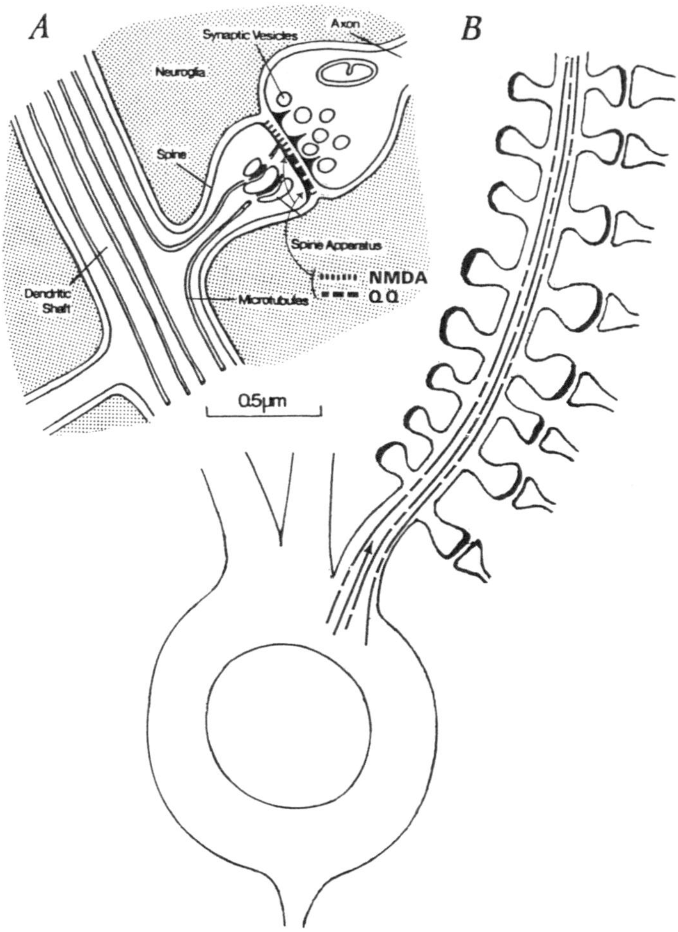

FIGURE 5. (A) Drawing of spine synapse of a dendrite of a hippocampal pyramidal cell. From the presynaptic membrane there are dense projections up to and between the synaptic vesicles. The postsynaptic density is shown with two distinct receptors for NMDA and QQ as described in the text and as labeled below the diagram (modified from Gray, 1982). (B) Drawing of a dentate granule cell showing dendrites in outline with synaptic spines drawn on both sides of one. Microtubules are drawn in the right dendrite by three interrupted lines.

and the subsequent LTP. The APV blockade of the NMDA receptors prevents this influx of Ca^{2+} and the LTP.

Stage 3. With the synaptic activation of any one spine synapse, the depolarization produced by its QQ receptors is inadequate to open the Ca^{2+} channels of the NMDA receptors. The combined depolarizing actions of bursts of impulses in many spine synapses is essential to give a sufficient Ca^{2+} influx to generate LTP. This is the phenomenon of cooperativity described by McNaughton (1982) and his associates. Even when the hyperpolarization of the complicating inhibitory action is largely depressed by picrotoxin, conditioning by at least five bursts of two stimuli (Fig. 3A) is necessary to produce sufficient depolarization to open the NMDA Ca^{2+} channels (Gustafsson and Wigström, 1986). The blocking of NMDA receptors by APV eliminates LTP (Collingridge *et al.*, 1983b; Wigström and Gustafsson, 1984) without appreciably reducing the EPSP.

Stage 4. If the opening of Ca^{2+} channels and the consequent production of LTP are dependent on the level of depolarization of the activated NMDA receptors, it would be expected that effective conditioning could be achieved by an intracellularly applied depolarizing current. This has been observed with single-volley EPSPs timed in relation to a current pulse of >5 nA for 100 msec (Gustafsson *et al.*, 1986). Thus, it is proposed that all cooperativity, both homosynaptic and heterosynaptic, is simply caused by the EPSPs generated by glutamate action on the QQ receptors. An apparent anomaly may be the failure of antidromic impulses to act as conditioning depolarizers for initiating an LTP. Presumably the depolarization of impulses is too brief to open effectively the Ca^{2+} channels of the NMDA receptors.

Stage 5. The preconditioning phenomenon illustrated in Fig. 3 can be attributed to the long duration (up to 60 msec) of the NMDA activation by a single presynaptic impulse (Wigström *et al.*, 1985, 1986). The enduring NMDA activation ensures that the later depolarization of the conditioning volleys is still able to open Ca^{2+} channels.

Stage 6. In contrast to the very effective electrotonic transmission of depolarization over the whole neuron, the Ca^{2+} input via the activated NMDA receptors must be restricted to that spine and effect there the postsynaptic changes giving the LTP, probably in the manner outlined by Lynch and Baudry (1984) and Eccles (1983). Only in this manner is it possible to account for the findings that with dentate granule cells or CA_1 pyramidal cells both homosynaptic and heterosynaptic LTPs are restricted to the activated synapses, as is shown, for example, in Fig. 2E, input 2 responses to the first and second conditioning stimuli (STIM 1). Presumably this restriction of the Ca^{2+} input to the spine is dependent in part on the spine apparatus that sequesters Ca^{2+} (Fifkova *et al.*, 1983). The key role of Ca^{2+} in inducing LTP is demonstrated by the prevention of LTP by intracellular injection of the calcium chelator EGTA (Lynch *et al.*, 1983).

Stage 7. The hypothesis of LTP generation leads to the concept that there are two basic principles, cooperativity and privacy. The *principle of cooperativity* arises from the global electrotonic transmission of depolarizations over the whole neuron. Normally these depolarizations are the EPSPs produced by glutamate action on QQ receptors. The *principle of privacy* results from the privacy of the Ca^{2+} inputs into a spine by the glutamate activation of its NMDA receptors when they are sufficiently depolarized by the global EPSP. The LTP results from the increased Ca^{2+} and is private to the QQ glutamate receptors of each spine. It is not known if the NMDA receptors become more effective in LTP.

It is important to recognize that the hypothesis of LTP is based largely on experiments on hippocampal slices that are usually only reliable for about 4 hr. At longer durations after the conditioning stimulation, particularly if there are repeated reinforcements (cf. Fig. 1C), slower-acting events may contribute to the LTP. Swelling of the spine and its stalk have been described (Fifkova and van Harreveld, 1977; Fifkova and Anderson, 1981), and this appears to be associated with synthesis of protein in the soma. In addition, with enduring synaptic potentiation for weeks (Bliss and Gardner Medwin, 1973; Barnes and McNaughton, 1980), one would expect flow of synthesized macromolecules such as membrane proteins and receptor proteins from the soma up the dendritic microtubules, as indicated by the arrow in Fig. 5B. Though the influx of Ca^{2+} through the NMDA receptors would be restricted to that spine, when Ca^{2+} is combined with calmodulin to form a second messenger system (cf. Eccles, 1983), it would be expected to travel as a signal to the synthesizing machinery in the soma. However, once the internal Ca^{2+} is transported from the spine of its origin, the explanation of the sharp restriction of LTP to the activated spines can no longer hold. Evidently we are far from a complete explanation of LTP. An additional unresolved problem is the assimilation of the evidence for a presynaptic component of the LTP (Bliss and Dolphin, 1982; Eccles, 1983). Nevertheless, we have a model for heterosynaptic LTP that can be utilized in developing a hypothesis for long-term memory in the cerebral cortex.

3. THE NEURAL PATHWAYS INVOLVED IN LAYING DOWN COGNITIVE MEMORIES

The hypothesis proposed here is developed from Kornhuber's theory (1973), which is illustrated in Fig. 6. The sensory association areas play a key role, first being on the input pathway to the limbic system and thence to the frontal cortex and second, in an intimate two-way relationship with the frontal cortex (write into LTM) that receives a "selection input" from the limbic system via the mediodorsal (MD) thalamus. It is to be noted that the hippocampus is given a dominant role in the limbic circuits. One circuit is the so-called Papez loop: hippocampus, subiculum, mammillary body, anterior thalamic nucleus, cingulate gyrus, parahippocampus, hippocampus. This circuit could function as a self-reexciting loop, providing a selective background excitation of the hippocampus. The other circuit is of special interest because it leads from the assocation cortices to the hippocampus via the cingulate gyrus and thence via the mediodorsal thalamus to the prefrontal cortex.

Figure 6 greatly simplifies the connections, particularly in the limbic system. For the present we may note the important operation of motivation in Fig. 6 and also the labeling of the limbic system as a selection unit. According to the present hypothesis, the hippocampal output does indeed participate in selection, but the actual operation in selection is done in the association cortex.

4. A MODEL FOR COGNITIVE MEMORY BUILT ON THE LTP

The proposed postsynaptic origin of LTP is of great significance with respect to the hypothesis developed by Marr (1970) that long-term memory in the cerebral cortex is encoded in synaptic potentiations that are set up by a conjunction process and that have an indefinitely long duration. Figure 7 represents the essential feature of this interaction together with the specificity of the synaptic potentiation as illustrated by the hypertrophied synapses for one horizontal fiber. In Fig. 7 there is conjunction between, on the one hand, the MD thalamic input by spiny stellate cells (Sst) to the cartridge synapses on the apical dendrites of pyramidal cells and, on the other hand, the crossing-over synapses made by horizontal fibers that come from the sensory inputs of Fig. 6 via the association (ASS), and commissural (COM) fibers and possibly Martinotti cell axons (MA). It was assumed by Marr (1970) that there was a selective potentiation of the crossing-over synapses on the pyramidal cell dendrites and that it happened only in those having impulses at an approximate conjunction in time. Hitherto, this conjecture has had no experimental support.

The situation may be transformed by the discovery of heterosynaptic LTP, as described in Section 1. It was originally proposed that the cartridge synapses effected LTP by causing a large dendritic depolarization and so opening the Ca^{2+} channels on the dendrites (Eccles, 1983). In view of the evidence presented in Sections 1 and 2 above, it can now be proposed that impulses in the cartridge fibers produce a large postsynaptic depolarization (cooperativity) that causes the opening of the NMDA Ca^{2+} channels for

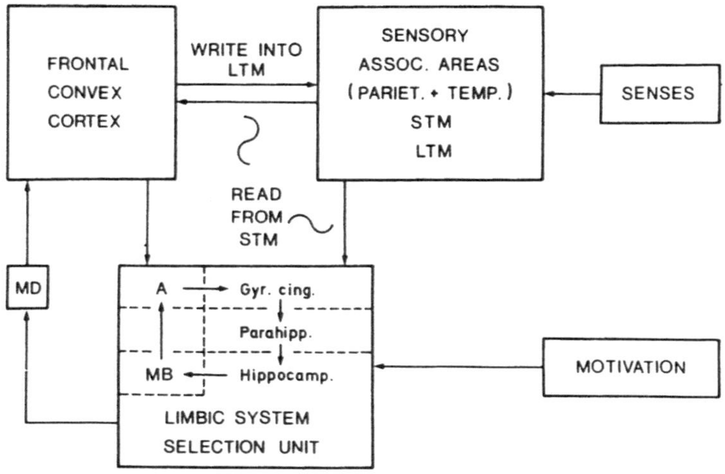

FIGURE 6. Scheme of anatomic structures involved in selection of information between short-term memory (STM) and long-term memory (LTM). MB, mammillary body; A, anterior thalamic nucleus; MD, mediodorsal thalamic nucleus (Kornhuber, 1973).

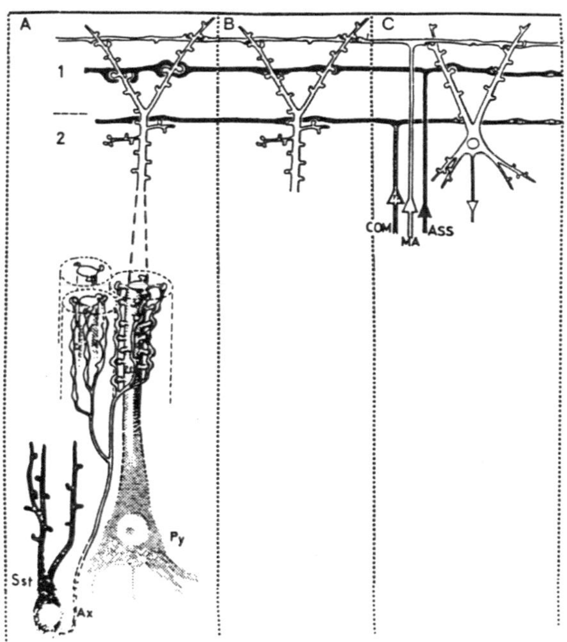

FIGURE 7. Simplified diagram of connectivites in the neocortex that is constructed in order to show pathways and synapses in the proposed theory of cerebray learning. The diagram shows three modules (A, B, C), which are vertical functional elements of the neocortex, each with about 4000 neurons. In lamina 1 and 2 there are horizontal fibers arising as bifurcating axons of commissural (COM) and association (ASS) fibers and also Martinotti axons (MA) from module C. The horizontal fibers make synapses with the apical dendrites of the stellate pyramidal cells in module C and of pyramidal cells in modules A and B. Deeper, a spiny stellate cell (Sst) is shown with axon (AX) making cartridge synapses with the shafts of apical dendrites of pyramidal cells (Py). Because of the conjunction hypertrophy, the association fiber from module C has enlarged synapses on the apical dendrites of the pyramidal cell in module A (modified from Szentagothai, 1970).

synapses of the horizontal fibers that are activated at about the same time. The increase in intracellular Ca^{2+} would be initially in the activated spines, just as in Fig. 5 (privacy). The selection is effected by the timing of the conjunction: horizontal fiber impulses before or during the depolarization generated by the cartridge synaptic activity. Selectivity would be sharpened by the cartridge synaptic activity. Selectivity would be sharpened by the requirement that there should be reinforcement by a succession of conjunctions.

The hypothesis of the postsynaptic origin of LTP based on increased intraspinous Ca^{2+} can thus be converted into a more developed variant of the Marr (1970) hypothesis. This is justified because Lee (1983) has found that slice preparations of several areas of the cerebral cortex exhibit an LTP matching that for the hippocampus. Thus, cerebral learning can be attributed to the calcium-induced LTP in the QQ receptors on the spine synapses of the pyramidal cells, particularly for synapses made by horizontal fibers in lamina I, as outlined in a recent commentary (Eccles, 1983). There is urgent need for the experimental testing of this comprehensive hypothesis. From the unitary effect illustrated in Fig. 7, a hypothesis has been developed of the manner in which cognitive memories can be stored and retrieved in the cerebral cortex (Eccles, 1981; Figs. 7 and 8).

The Papez loop illustrated in Fig. 6 would have the properties of a reverberatory circuit and so could be the basis of repetitive burst discharges from the hippocampus to the MD thalamus and thence, via the spiny stellate cells, to the cartridge synapses on the pyramidal cell dendrites (Fig. 7). It could thus resemble the repetitive burst discharges that are most effective in setting up LTPs in the hippocampus. So it can be envisaged that in the interaction illustrated in Fig. 7, the synapses of the horizontal fibers are normally weak but are potentiated by the conjunction. Repeated replays of the conjunctional interaction would serve to prolong the potentiation of the horizontal fiber synapses on the pyramidal cell (cf. Fig. 7), just as has been observed by Barnes and McNaughton (1980) and for the LTP of dentate granule cells; hence, there is a plausible explanation of memory consolidation in the cerebral cortex.

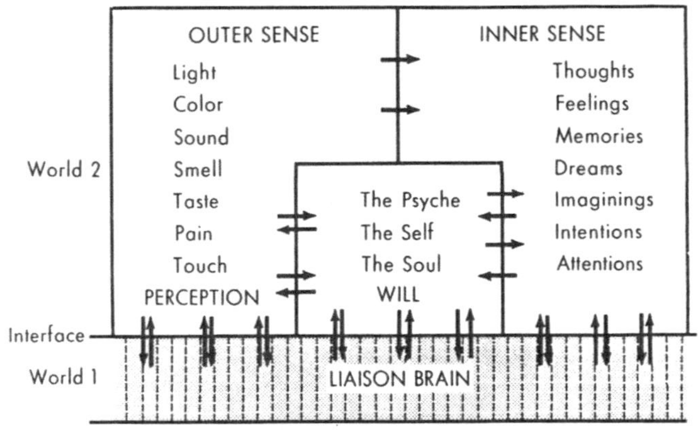

FIGURE 8. Information flow diagram for brain–mind interaction. The three components of World 2—outer sense, inner sense, and the psyche or self—are diagrammed with their connectivities. Also shown by reciprocal arrows are the lines of communication across the frontier between World 1 and World 2, that is, from the liaison brain to and from these World 2 components. The liaison brain has the columnar arrangement indicated. It must be imagined that the area of the liaison brain is enormous, with open modules probably numbering over a million and not just the 40 here depicted. Note that memories are located in the inner sense of World 2 as well as in the data banks of the modules of the liaison brain.

5. DURATIONS OF LTP AND COGNITIVE MEMORY

As we inquire into the applicability of the postsynaptic calcium model for providing an explanation of cerebral memory, the duration of the LTP is important. At this level of inquiry, we have to consider the question of long duration of the LTP, particularly after repeated reinforcement (Barnes and McNaughton, 1980).

In an extraordinary clinical case, the patient HM had a retrograde amnesia for events occurring 1 to 3 years before the hippocampectomy (Squire, 1982, 1983). There is a transiently experienced retrograde amnesia for a similar duration of events after bilateral electroshock therapy (Squire, 1982, 1983). The period of sensitivity to disruption correlates with observations on the normal course of forgetting. Thus, there seems to be a period of 1 to 3 years involved in the process of consolidation (Block, 1970) of a long-term memory so that it is no longer susceptible to loss in the process of forgetting or in the process of memory disruption by bilateral hippocampectomy or electroshock therapy. Squire (1983) suggests that "the medial temporal region [hippocampus?] directly brings about those changes in memory storage whereby memory becomes gradually resistant to disruption." This suggestion should be incorporated into the theory of hippocampally induced consolidation. Hitherto this role was assumed to be a relatively transient intervention as in Fig. 7, although later replays of the memory experience were assumed to add to the stability of the patterned synaptic hypertrophy. We have now to envisage that, in order to effect a "permanent" consolidation of a memory, the hippocampal input to the neocortex must be replayed much as in the initial experience in what we may name "recall episodes" for 1 to 3 years. Failure of this replay results in the ordinary process of forgetting. After 3 years, the memory codes in the cerebral cortex are much more securely established and apparently require no further adjuvenating hippocampal inputs; hence, they are not lost in the disruption of bilateral hippocampectomy or electroconvulsive therapy.

6. RECALL OF MEMORY

At one stage of consideration, the recall of a memory can be explained by the replay of neocortical circuits that were consolidated by the learning process. Thus, the replayed circuits would closely resemble those giving the original experience; hence, the remembered experience is recognized as genuine. But there are many problems overlooked in this simple story. No special difficulty would attend an explanation of a memory triggered by some related present experience or a sequence of memories in some train of thought. But we recognize that we can at will attempt to recall a memory. Evidently, we are now involved with the mind–brain problem as illustrated diagrammatically in Fig. 8. Even more challenging is the experience that we can assess the validity of a recalled memory, recognizing, for example, that the telephone number or ZIP code or name is almost correct, so that a new demand is made on the memory data banks for a deliverance that can be recognized as correct. Thus, we have to entertain the idea of two kinds of long-term memory: a data bank memory that is stored in the neocortex and a recognition memory that is in the conscious mind as located in the inner sense column of Fig. 8.

In retrieval of a memory, we have further to conjecture that the self-conscious mind is continuously searching to recover memories, e.g., words, phrases, sentences, ideas, events, pictures, melodies, by active scanning through the modular array and that, by its action on the preferred "open" modules, it tries to evoke the full neural patterned operation that it can read out as a memory. Largely this could be by a trial-and-error process. We are all familiar with the ease or difficulty of recall of one or another memory and with the strategies we discover in order to recover memories of names that for some unknown reason are refractory to recall. We can imagine that our self-conscious mind is under a continual challenge to recall the desired memory by discovering the appropriate entry into module operation that would, by development, give the appropriate patterned array of modules.

Penfield and Perot (1963) gave a most illuminating account of the experimental responses evoked in 53 patients by stimulation of the cerebral hemispheres during operations performed under local anesthesia. These responses differed from those produced by stimulation of the primary sensory areas, which were merely flashes of light or touches or paresthesia, in that the patients had experiences that resembled dreams, the so-called dreamy states. The stimulation acts as a mode of recall of past experiences. It can be suggested that the storage of these memories is likely to be in cerebral areas close to the effective stimulation sites. It is important to recognize, however, that the experiential recall is evoked from areas in the region of the disordered cerebral function that is displayed by the epileptic seizures. Conceivably, the effective sites are abnormal zones that are thereby able to act by association pathways to the much wider areas of the cerebral cortex that are the actual storage sites for memories.

The experiments of Roland and Friberg (1985) delineating the patchy activity of human cerebral activity during silent thinking are of particular importance in regard to the retrieval of memory. Essentially, the thinking procedures carried out by the subjects were the recovery of memories of a stereotyped kind: the progressive subtraction procedure of 3s from 50; the memory of the successive second words of a well known nonsense jingle of nine words in a ring; the sequential visual experiences in a specific route along well-known streets. The subjects were screened from all sensory inputs and were immobile except for breathing. Yet in this memory retrieval they activated, often strongly, specific zones of the cerebral cortex, particularly of the prefrontal lobes. If the silent memory was of a sequence of motor movements of a hand, the significant cerebral activity was bilateral in the supplementary motor area (SMA) (Roland *et al.*, 1980). So the mental events involved in the retrieval of a memory actually excite specific areas of the liaison brain as indicated in the communication flow diagram of Fig. 8, where memory is located in the World 2 of subjective experience in the inner sense compartment and is related to the modules of the liaison brain by reciprocal arrows that relate to the interiors of the modules.

It may be objected that this retrieval process proposed for memory entails the action of a nonmaterial mental event, the desired memory, on the neuronal activities of the liaison brain, which are in the World 1 of matter–energy. However, a recent development from proposals by the quantum physicist Margenau (1984) has led to a microsite hypothesis of mind–brain interaction that does not violate the conservation laws of physics (Eccles, 1986). Essentially, the hypothesis is based on the discovery that a presynaptic impulse effects, probabilistically, the exocytosis of a single vesicle of a synaptic bouton (Jack *et al.*, 1985; Korn *et al.*, 1981). This probability is always less than 1, the mean value being about 0.3 for boutons in diverse locations in the central nervous systems. The probability for a bouton can be varied up or down by physiological or pharmacological means, and it is proposed that mental events analogous to the probability field of quantum mechanics (Margenau, 1984) could also alter the probability of exocytosis from boutons (Eccles, 1986). In effecting an exocytosis from vesicles already in position in the presynaptic vesicular grid (Akert *et al.*, 1975), the mass involved would be no more than 10^{-18} g, which is a size well within the limits set by the Heisenberg uncertainty principle.

Hence, it can be claimed that as diagrammed in Fig. 8, the mental intention recalling a memory could influence the synaptic activation of a neuron in accord with quantum mechanical principles. The events at one microsite would be too small by many orders of magnitude, but one has to incorporate in the microsite hypothesis a global influence of the mental intention on the 10,000 spine synapses on one neuron and the grouping of pyramidal apical dendrites in bundles of 3–20 (Fleischhauer and Detzer, 1975; Feldman, 1984), as indicated for 3 in Fig. 7. In this way there can be several orders of magnitude of summation of microsites. Only in this way is it possible to account for the powerful action of all varieties of thinking on the cerebral cortex.

REFERENCES

Akert, K., Peper, K., and Sandri, C., 1975, Structural organization of motor end plate and central synapses, in: *Cholinergic Mechanisms* (P. G. Waser, ed.), Raven Press, New York, pp. 43–57.

Barnes, C. A., and McNaughton, B. L., 1980, Spatial memory and hippocampal synaptic plasticity in senescent and middle-aged rats, in: *Psychology of Aging* (D. Stein, ed.), Elsevier/North-Holland, Amsterdam, pp. 253–272.

Bliss, T. V. P., and Dolphin, A. C., 1982, What is the mechanism of long-term potentiation in the hippocampus? *Trends Neurosci.* **5**:289–290.

Bliss, T. V. P., and Gardner-Medwin, A. R., 1973, Long-lasting potentiation of synaptic transmission in the dentate area of the unanesthetized rabbit following stimulation of the perforant path, *J. Physiol.* (*Lond.*) **232**:357–374.

Bliss, T. V. P., and Lømo, T., 1973, Long-lasting potentiation of synaptic transmission in the dentate area of the anaesthetized rabbit following stimulation of the perforant path, *J. Physiol.* (*Lond.*) **232**:331–356.

Bloch, V., 1970, Facts and hypotheses concerning memory consolidation processes, *Brain Res.* **24**:561–575.

Collingridge, G. L., Kehl, S. J., and McLennan, H., 1983, Excitatory amino-acids in synaptic transmission in the Schaffer collateral–commissural pathway of the rat hippocampus. *J. Physiol.* (*Lond.*) **334**:33–46.

Dingledine, R., 1983, N-Methyl aspartate activates voltage-dependent calcium conductance in rat hippocampal pyramidal cells, *J. Physiol.* (*Lond.*) **343**:385–405.

Eccles, J. C., 1981, The modular operation of the cerebral neocortex considered as the material basis of mental events, *Neuroscience* **6**:1839–1856.

Eccles, J. C., 1983, Calcium in long-term potentiation as a model for memory, *Neuroscience* **10**:1071–1081.

Eccles, J. C., 1986, Do mental events cause neural events analogously to the probability fields of quantum mechanics? *Proc. R. Soc. Lond. [Biol.]* **227**:411–428.

Feldman, M. L., 1984, Morphology of the neocortical pyramidal neuron, in: *Cerebral Cortex,* Vol. 1 (A. Peters and E. G. Jones, eds.), Plenum Press, New York, pp. 123–200.

Fifková, E., and Anderson, C. L., 1981, Stimulation-induced changes in dimensions of stalks of dendritic spines in dentate molecular layer, *Exp. Neurol.* **74**:621–627.

Fifková, E., and van Harreveld, A., 1977, Long-lasting morphological changes in dendritic spines of dentate granular cells following stimulation of the entorhinal area, *J. Neurocytol.* **6**:211–230.

Fifková, E., Markham, J. A., and Delay, R. J., 1983, Calcium in the spine apparatus of dendritic spines in the dentate molecular layer, *Brain Res.* **266**:163–168.

Fleischhauer, K., and Detzer, K., 1975, Dendritic bundling in the cerebral cortex, in: *Advances in Neurology,* Vol. 12 (G. W. Kreutzberg, ed.), Raven Press, New York, pp. 71–77.

Gustafsson, B., and Wigström, H., 1987, Hippocampal long-lasting potentiation produced by pairing single volleys and brief conditioning tetani evoked in separate afferents, *J. Neurosci.* **6**:1575–1582.

Gustafsson, B., Wigström, H., Abraham, W. C., and Huang, Y.-Y., 1987, Long-term potentiation in the hippocampus using depolarizing current pulses as the conditioning stimulus to single volley synaptic potentials, *J. Neurosci.* (in press).

Jack, J. J. B., Redman, S. J., and Wong, K., 1981, The components of synaptic potentials evoked in cat spinal motoneurones by impulses in single group Ia afferents, *J. Physiol.* (*Lond.*) **321**:65–96.

Korn, H., Triller, A., Mallet, A., and Faber, D. S., 1981, Fluctuating responses at a central synapse: *n* of binomial fit predicts number of stained presynaptic boutons, *Science* **213**:898–901.

Kornhuber, H. H., 1973, Neural control of input into long-term memory: Limbic system and amnestic syndrome in man, in: *Memory and Transfer of Information* (H. P. Zippel, ed.), Plenum Press, New York, pp. 1–22.

Lee, K. S., 1983, Sustained modification of neuronal activity in the hippocampus and cerebral cortex, in: *Molecular, Cellular and Behavioural Neurobiology of the Hippocampus* (W. Seifert, ed.), Academic Press, New York, pp. 265–272.
Levy, W. B., and Steward, O., 1979, Synapses as asssociative elements in the hippocampal formation, *Brain Res.* **175**:233–245.
Levy, W. B., and Steward, O., 1983, Temporal contiguity requirements for long-term associative potentiation/depression in the hippocampus, *Neuroscience* **8**:791–797.
Lynch, G., and Baudry, M., 1984, The biochemical intermediates in memory formation: A new and specific hypothesis, *Science* **224**:1057–1063.
Lynch, G., Larson, J., Kelso, S., Barrionirevo, G., and Schottler, F., 1983, Intracellular injections of EGTA block induction of hippocampal long-term potentiation, *Nature* **305**:719–721.
Margenau, H., 1984, *The Miracle of Existence,* Ox Bow Press, Woodbridge, CT.
Marr, D., 1970, A theory for cerebral neocortex, *Proc. R. Soc. Lond. [Biol.]* **176**:161–234.
Mayer, M. L., and Westbrook, G. L., 1985, Divalent cation permeability of N-methyl-D-aspartate channels, *Soc. Neurosci. Abstr.* **11**:785.
McNaughton, B. L., 1982, Long-term synaptic enhancement and short-term potentiation in rat fascia dentata act through different mechanisms, *J. Physiol. (Lond.)* **324**:249–262.
McNaughton, B. L., Douglas, R. M., and Goddard, G. V., 1978, Synaptic enhancement in fascia dentata: Cooperativity among coactive afferents, *Brain Res.* **157**:277–293.
Penfield, W., and Perot, P., 1963, The brain's record of auditory and visual experience, *Brain* **86**:596–696.
Roland, P. E., and Friberg, L., 1985, Localization of cortical areas activated by thinking, *J. Neurophysiol.* **53**:1219–1243.
Roland, P. E., Larsen, B., Lassen, N. A., Skinhøj, E., 1980, Supplementary motor area and other cortical areas in organization of voluntary movements in man, *J. Neurophysiol.* **43**:118–136.
Squire, L. R., 1982, The neuropsychology of human memory, *Annu. Rev. Neurosci.* **5**:241–273.
Squire, L. R., 1983, The hippocampus and the neuropsychology of memory, in: *Molecular, Cellular and Behavioural Neurobiology of the Hippocampus* (W. Seifert, ed.), Academic Press, New York, pp. 491–507.
Szentágothai, J., 1970, Les circuits neuronaux de l'écorce cérébrale, *Bull. Acad. R. Med. Belg.* **7**:475–492.
Szentágothai, J., 1978, The neuron network of the cerebral cortex: A functional interpretation, *Proc. R. Soc. Lond. [Biol.]* **201**:219–248.
Watkins, J. C., 1984, Excitatory amino acids and central synaptic transmission, *Trends Pharmacol. Sci.* **5**:373–376.
Wigström, H., and Gustafsson, B., 1983a, Facilitated induction of hippocampal long-lasting potentiation during blockade of inhibition, *Nature* **301**:603–604.
Wigström, H., and Gustafsson, B., 1983b, Heterosynaptic modulation of homosynaptic long-lasting potentiation in the hippocampal slice, *Acta Physiol. Scand.* **119**:455–458.
Wigström, H., and Gustafsson, B., 1984, A possible correlate of the postsynaptic condition for long-lasting potentiation in the guinea pig hippocampus *in vitro, Neurosci. Lett.* **44**:327–333.
Wigström, H., and Gustafsson, B., 1985a, Facilitation of hippocampal long-lasting potentiation by GABA antagonists, *Acta Physiol. Scand.* **125**:159–172.
Wigström, H., and Gustafsson, B., 1985b, On long-asting potentiation in the hippocampus: A proposed mechanism for its dependence on coincident pre- and postsynaptic activity, *Acta Physiol. Scand.* **123**:519–522.
Wigström, H., Gustafsson, B., and Huang, Y.-Y., 1985, A synaptic potential following single volleys in the hippocampal CA_1 region possibly involved in the induction of long-lasting potentiation, *Acta Physiol. Scand.* **124**:475–478.
Wigström, H., Gustafsson, B., and Hunag, Y.-Y., 1986, Mode of action of excitatory amino acid receptor antagonists on hippocampal long-lasting potentiation, *Neuroscience* **17**:1105–1115.

29

Humanlike Characteristics of Visual Mnemonic System in Macaques

ROBERT W. DOTY, JAMES L. RINGO, and
JEFFREY D. LEWINE

1. INTRODUCTION

Although there is still significant controversy as to whether in man each memory endures indefinitely in unaltered form (Bahrich *et al.*, 1975; Hall and Loftus, 1984), any normal human being can unequivocally attest to a vast inventory of remembered material that is retained lifelong. The accuracy of this mnemonic store can seldom be assessed, but common experience suggests that the great majority of it is valid. In a few cases in which something approaching formal testing was achieved in, admittedly, possibly unique individuals (e.g., Luria, 1968; Hunter, 1977), trivial, unrehearsed material could be retrieved with astonishing accuracy after a lapse of 15 to 30 or more years (and see Neisser, 1982, for a collection of other examples). In tests of undergraduate college students, Standing (1973) was able to demonstrate, using 10,000 photographic slides that took several days to view, that a very large proportion of constantly changing visual scenes could still be recognized after a lapse of several days. Thus, the human mnemonic system can be characterized as being highly resistant to saturation (Nickerson, 1965) and capable of permanent retention and retrieval of information acquired in a single, brief exposure.

In seeking to understand the neural mechanisms subserving these remarkable phenomena, the degree of permanence of the mnemonic trace is clearly a critical factor (e.g., Doty, 1979). A very substantial effort, as evidenced in the present volume, is being devoted to analysis of synaptic alterations consequent to activation in "reduced" systems, i.e., invertebrates or truncated vertebrate preparations. There must remain some doubt, however, as to whether the cellular changes observed in these instances possess the permanence reflected in human memory, especially as the latter is acquired following momentary exposure. It is thus of some importance to determine if the qualities of human memory may also exist in certain animals and thereby be amenable to direct experimentation.

There is, of course, already substantial evidence that the "anatomy" of the mnemonic system in macaques bears striking similarity to that in man. In both species, loss or

ROBERT W. DOTY, JAMES L. RINGO, and JEFFREY D. LEWINE • Center for Brain Research, University of Rochester, Rochester, New York 14642.

inactivation of structures in the medial temporal lobe or anterior thalamus precipitates an amnestic condition in which there is a profound inability to store (or retrieve?) information concerning ongoing "events," at the same time leaving memory for "rules" or "skills" and the learning thereof essentially unperturbed (e.g., Bickford *et al.*, 1958; Milner *et al.*, 1968; Mishkin, 1978, 1982; Squire and Moore, 1979; Horel and Pytko, 1982; Mahut *et al.*, 1981; Speedie and Heilman, 1982; Zola-Morgan *et al.*, 1982; Halgren, 1984; Zola-Morgan and Squire, 1985). In addition, as summarized below, memory for visual images is demonstrably congruent in macaque and man, including presumptive evidence for permanence and lack of saturation (Overman and Doty, 1980; Sands and Wright, 1982; Owen and Butler, 1984; Ringo and Doty, 1985; Ringo *et al.*, 1987). The major problem in such endeavors is devising appropriate means of testing comparable performance in the two species.

2. METHOD

2.1. Subjects

The experiments have been performed with immature male *Macaca nemestrina* and undergraduate students of either sex. Several of the macaques had undergone transsphenoidal transection of the optic chiasm plus, in some instances, transection of one or the other of the forebrain commissures; but these intrusions are irrelevant to the present discussion save insofar as they might have diminished slightly the capabilities of these particular animals.

2.2. Tests

Three types of visual recognition tests were utilized, all involving the viewing of colored photographic slides for a few seconds. In each instance macaque or human subjects responded by pressing either the illuminated panel on which the image appeared or an immediately subjacent panel, depending on task. The same apparatus was used for both macaque and human subjects. The macaques were rewarded with fruit juice for a correct response and punished for errors by the sound of a raucous horn and a stream of air directed at their head, whereas for the human subjects correct and incorrect responses were signaled by auditory cues.

In the trial-unique delayed matching to sample (DMTS) (used only with the macaques), a "sample" image appeared on a central panel, which was pressed for a small reward. The image then disappeared and, after a delay of 2–15 sec, was projected again, together with a second, alternate image, on panels above or below the central panel. A reward of 1–2 ml of juice was procured by pressing the image that had been given as the "sample." A different pair of images was used for each trial, 40–60 trials per day.

For the running recognition task (RR), a series of colored photographic images, 100–280 per session, was rear-projected on a panel 12.5 × 10 cm (subtending roughly 30° for the macaques). The first time in the session that the image appeared, the correct response was to press the image, but a recurrence of an image previously presented in that session required pressing a subjacent blank panel. A large inventory of slides permitted a lapse of at least 2 weeks before any particular image was used again. There were two classes of images, ordinarily not intermixed in any given session. The common images were of scenes that included people, vehicles, buildings, or animals and photographs of tools, packages, toys, etc. The second class had two distinctive features: their right and left halves were equivalent (to allow for the split visual fields in the animals with transected optic chiasm), and they consisted of patterns, e.g., of fabric or wallpaper, devoid of identifiable objects and difficult for human subjects to describe or name.

The third task employed was the serial probe recognition test of Sternberg (1966, 1969, 1975), in which the subject must as rapidly as possible identify whether a given image is or is not a member of a previously specified target set. The critical feature of this test is that the reaction time for this decision is linearly related to the number of items in the target set, thus, apparently, reflecting the time taken to "search" the memory store for the designated items. For any given test there are one to six target items.

For the monkeys, both target and nontarget items are selected pseudorandomly from a group of 60 easily distinguishable images. Over the course of many training sessions with these images, the animal becomes familiar with the entire set, and, therefore, the distinction between target and nontarget items cannot be made on the basis of novelty. The same set of images is used for human subjects. The monkey is informed which items are the targets by presenting them individually three times concurrently with the presence of a blue light. A change in the color of light on the "start" panel signals to the monkey that test conditions prevail and that it must now distinguish target from nontarget images as they are presented individually. The animal holds its hand on the illuminated center panel to start a trial, thus assuring that responses from one trial to the next begin from a standard posture. Probe images are presented on a panel above the start panel. If it is a target image (positive probe item), that panel must be pressed, and if it is not a member of the target set (negative probe item), pressing the bottom panel is required. The amount of juice given for correct responses is steeply scaled inversely to the reaction time (after a window of 200 msec to forestall random responses), so that the faster the animal responds, the greater is its reward. The same procedure, slides, and equipment are used for the human subjects, except that they are more readily instructed as to the nature of the task.

3. RESULTS

3.1. When Is an Animal Trained?

With human subjects verbal instruction is sufficient to obtain performance immediately on the above tasks at very near the asymptotic level. Indeed, with the DMTS procedure at the delays used in the present experiments, human subjects almost never make an error. Similar performance can be attained by macaques, e.g., Fig. 1, where the best animal (RHD) made only eight errors in the last 1000 trials, i.e., 99.2% correct. However, as can be seen, this was achieved only after roughly 20,000 trials, during which continuous improvement was manifested. Although the prior transection of the optic chiasm in these animals could conceivably have contributed to the slow pace of improvement illustrated in Fig. 1, factors of attention and motivation are the more likely explanation, for an entirely similar rate of acquisition is displayed by intact macaques. In any event, the caution is that in what follows below some of the deficiencies in macaque versus human performance may arise simply because the animals had not yet achieved the level of accuracy of which they are capable.

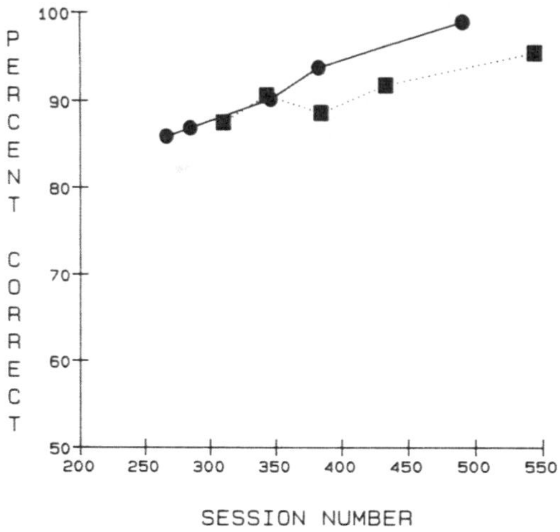

FIGURE 1. Continued improvement in performance of macaques RHD (circles) and HUD (squares) on the DMTS task even after thousands of trials (*ca.* 40–60 trials/session). Each point represents performance for 480–1300 trials combined across the relevant 12–30 sessions.

3.2. Degree of Permanence in Memory for Images

Using the DMTS task, Overman and Doty (1980) found that some macaques could correctly choose 90% of the sample images viewed 24 hr previously; and one animal attained 76% correct after a lapse of 4 days. These findings in animals that were only minimally trained for working at such long delays suggested that retention of visual images by macaques might approach that of man in its degree of permanence. The opportunity to test this possibility arose during the transfer of one macaque, TDY, from the DMTS to the RR task.

For RR the rule is that if an image has been perceived previously during a given session, it must be so identified by pressing a blank panel rather than the image itself. It was reasoned, however, that in the early stages of training TDY might misinterpret the rule to be that if the image has ever been seen before, i.e., in any session, whatever, then the blank panel should be pressed. This surmise, for the most part, turned out to be true, for TDY pressed this blank panel for about one-third of the images that, 6 or more months previously, it had seen during the DMTS training and that had not been viewed since that time (Ringo and Doty, 1985). When presented with entirely comparable material that it had never seen, TDY routinely hit the image itself, and this difference in responding to the "old" versus the "new" material had a high degree of statistical reliability ($P < 0.005$). It is particularly noteworthy that TDY "recognized" equally well images it had seen only 14–16 times for a total of 30 sec or less during the DMTS training as it did images with which it had had more extensive experience 9–18 months prior to the test. Thus, a rather brief exposure to these images had been sufficient to induce a memory enduring at least 6 months. Comparable observations have been made by Owen and Butler (1984), who found that macaques with transected fornix distinguished 180 objects, which they had seen on two to four occasions over a period of weeks, from objects that they had never seen before.

In the case of TDY, it was possible to make a direct comparison with a human subject. A medical student had participated in training the macaques on the DMTS procedure and also had not seen any of the slides for a period of 6 months. When tested in the same fashion and with the same material as TDY, he could correctly identify about two-thirds of the "old" material versus the "new." Of course, our instructions to this subject were far more specific than they could be to TDY, and, in addition, the student had had more extensive exposure to the material than had TDY.

3.3. Pictorial Memory in Man and Macaque When the Language Factor Is Reduced

Although some of the items in our usual inventory of photographic slides may have some inherent significance to the monkey, e.g., animals and fruit, most of this material is wholly beyond their ken, e.g., buildings, electronic components, and packages. Thus, for most of the material, the monkey's memory must be formed purely on the basis of the visual image, whereas human subjects have an added cue in their understanding of and, in most instances, ability to name or at least categorize the items displayed. In an attempt to place our human and macaque subjects more on a par with each other for evaluating their ability to remember visual images, we devised a series of photographic slides that the human subjects found very difficult to name or categorize. For the most part, these consisted of repeating colored patterns, either hand drawn or taken from a variety of designs. Despite this care, it is still likely that most human observers would have been able to classify at least the source of this material, i.e., fabric, weavings, drawings, etc.; but the primary impression is, nevertheless, simply visual and not one of identification in any linguistic sense.

Two macaques that had been highly overtrained on the RR task, although not with this type of material, were compared with six human subjects. For the test, 140 slides of this "nonnamable" material were presented *seriatim* in a given session, and 45 slides intervened between initial presentation of an item and its subsequent recurrence, when its recognition was required. The two macaques recognized 79% and 85%, respectively, under these conditions, whereas the six human subjects averaged 83% correct recognition (Ringo *et al.*, 1987). Although the human subjects made fewer errors of false recognition than did the monkeys, i.e., designating an item as a repeat when it was not, so that their overall performance exceeded that of the macaques, there was an overlap in the individual performances of the two species.

3.4. Adoption of Appropriate Strategy

In the RR task described above, when 20 or 45 images intervene before a repetition occurs for any given image, there are no repetitions in the early part of a session. Thus, the session is divided into three periods: an initial run of 20 presentations in which no repetition will occur; a second run during which repetitions may occur for images viewed 20 presentations earlier; and a final period in which repetitions may be occurring after either 20 or 45 intervening presentations. To maximize its reward, the monkey, in full accord with signal detection theory (Green and Swets, 1974), shifts its criterion for designating an image as a repetition according to response feedback. The optimum strategy followed is to relax the criterion from period 1 to 2 (Fig. 2A) and still further in passing from period 2 to 3 (Fig. 2B). Such shifts in criterion were also characteristic of the other macaque tested as well as of the six human subjects.

It is also of interest to note in Fig. 2, which is typical in this regard, that although the percentage correct is lower when more images intervene, there is no systematic diminution in performance as more and more items are committed to memory during the

A

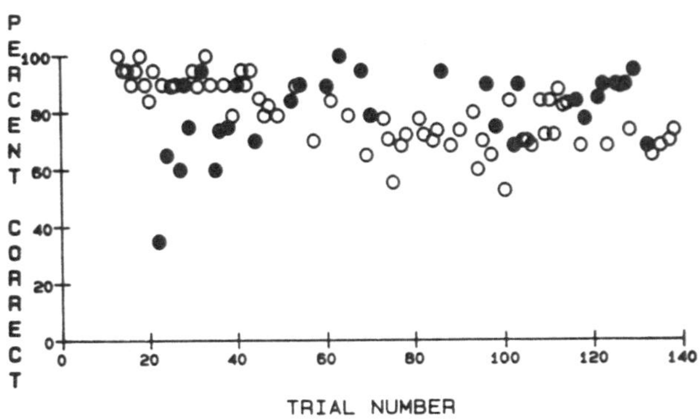

B

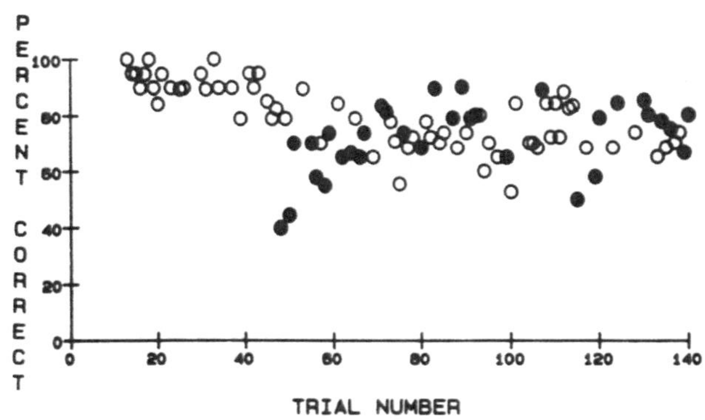

FIGURE 2. Change in strategy in accord with signal detection theory (Green and Swets, 1974); animal NEV working on the RR task. The number of images intervening prior to re-presentations was set at only two levels: 20 (filled circles in A) and 45 (filled circles in B). Thus, until 20 images have appeared, all images will be "new" (open circles), and the monkey's reward is highest if it pursues a corresponding strategy of responding to all images as "new." However, when images begin recurring after 20 trials, this strategy now produces errors when the image is a "repeat" (filled circles in A). The animal quickly takes account of this and changes its criterion, i.e., the percentage correct for the repeated images (filled circles) rises to the same level as in trials with "new" images (open circles). A similar shift is seen when repetitions also begin occurring after 45 intervening images (B). Also note that, as expected, performance is not as accurate with 45 images intervening as it is with 20 but that there is no deterioration in accuracy as the session progresses even though there is an ever-increasing number of images to be remembered. Each point represents 20 trials.

course of a session. In other words, there is no indication whatever for any degree of "saturation" of the system as the inventory of remembered items increases.

3.5. Sternberg Paradigm

Sands and Wright (1982) were able to train a macaque to perform a serial probe recognition task (Sternberg, 1969) and found the same linear relationship between reaction time and number of target items that characterizes such data from man. We have confirmed this finding, as illustrated in Fig. 3. An interesting feature of these data, again concordant with Sands and Wright (1982), is that the increment in reaction time per added target item is only some 8–12 msec for macaques, whereas in Fig. 3 and most human data (Sternberg, 1975), it is 20–30 msec for man.

4. DISCUSSION

It bears emphasis that all of the foregoing experiments involve memory for "events" as distinct from "rules" or "habits." That such distinction is highly relevant to endeavors

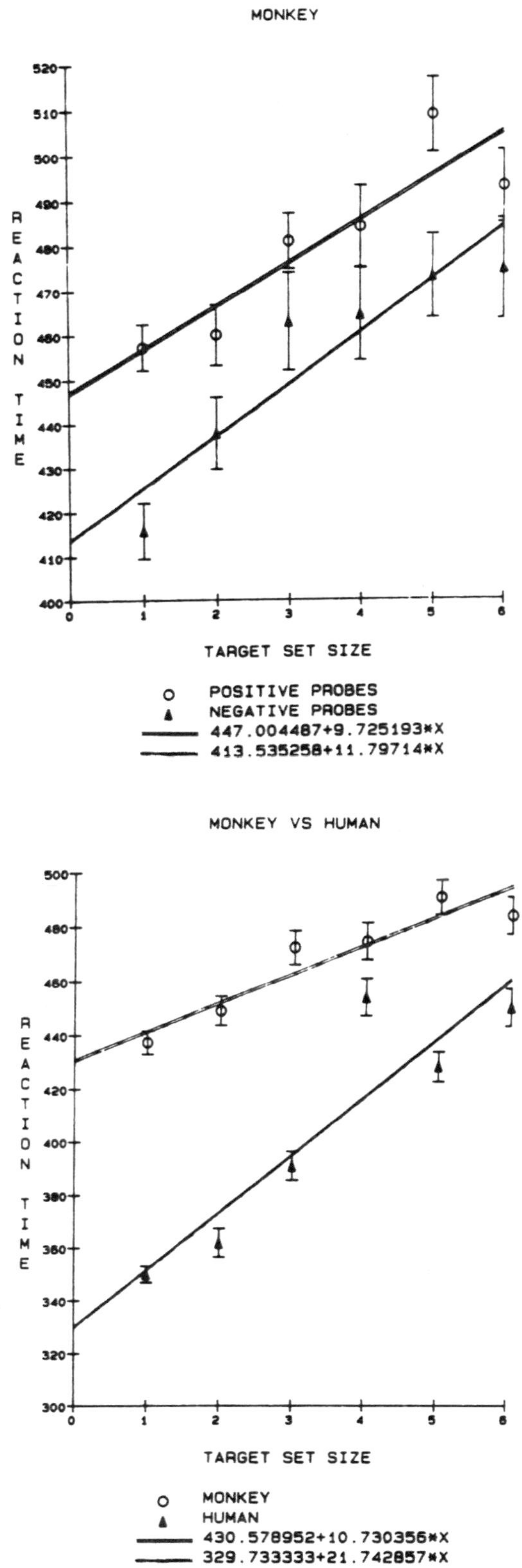

FIGURE 3. Comparison of macaque ARH with human subject on Sternberg serial probe recognition task. Note nearly identical slopes in A for responding to target items (positive probes) as to nontarget items (negative probes). The intercept for the human subject (B) is less than that for the macaque, probably because larger hand and arm size required a proportionately smaller angular displacement to make the response. The equations for the regression lines as a function of number of target items (x) account for 80% of the variance in the slope. In B, reaction times for positive and negative probe items have been combined. For A, each point represents the mean latency for 90–150 correct responses, performance never being <80% correct. In B, open circles (macaque) are for 180–300 correct responses, and filled circles (man) for 100–160.

to identify the neural substrate for memory in mammals is evidenced by a growing number of experiments in which diencephalic or medial temporal lesions disrupt the former while leaving the latter essentially intact (Olton *et al.*, 1979; Mahut *et al.*, 1981; Mishkin, 1982; Squire, 1982; Zola-Morgan *et al.*, 1982; Mishkin *et al.*, 1984; Thomas, 1984). In the past, the great majority of experimental observations on human memory have involved what is here termed "memory for events," i.e., specific, datable occurrences, whereas those studying memory in animals have commonly used the acquisition of various discriminatory or associative habits.

There thus has been little opportunity to make a valid comparison between human and animal mnemonic ability. True, there are conditioned taste aversion, one-trial avoidance learning, and the various forms of delayed alternation, including delayed matching to sample, in which a limited set of objects is used continually throughout a session. The first two, however, are actually "habits," although enduring for significant periods of time (the limits of which remain largely untested), and an almost life-threatening degree of motivation is required to establish them. Delayed alternation, or matching with highly repetitious samples, almost requires an impermanence of memory for the problem to be solved efficiently (Mishkin and Delacour, 1975; Overman and Doty, 1980; Thomas, 1984). Finally, in none of these is it feasible to employ a large variety of comparable stimuli.

Of course, it might be argued that the recognition displayed by TDY for images not encountered for 6 months represents some "habit" that was acquired during their use in the DMTS task. However, not only is this very unlikely, given the large number of different images to which the animal was exposed in the DMTS procedure and the brevity of exposure to many of the images subsequently recognized, but the response required is opposite in the two tasks. In DMTS, striking the image is rewarded, whereas in the RR task a blank panel is pressed for a recognized image and the image itself must not be touched.

We are led to the conclusion that there is a high degree of concordance between human and macaque mnenomic abilities in regard to visual recognition. Both show a capacity for errorless performance on the DMTS task (Section 3.1, above), a surprising permanence (Section 3.2), resistance to saturation by large numbers of such "events" (Sections 3.3. and 3.4), linearly increasing reaction time as a function of mnemonic load (Section 3.5), and optimization of response strategies (Fig. 2). Thus, macaque and man share not only a clear commonality in the anatomy and psychophysics of their visual systems (Scott and Powell, 1963; DeValois *et al.*, 1974; Cavonius and Robbins, 1973; Sarmiento, 1975; Boltz *et al.*, 1979; Cowey, 1979; Maguire *et al.*, 1980; Jacobs, 1981; Smith *et al.*, 1982; Golomb *et al.*, 1985) but also, it seems, a similar capacity to retain indefinitely massive amounts of information acquired via this system.

There is clearly a relationship between this mnemonic substrate and what Beritashvili (1972) studied as the phylogeny of "image-driven behavior" among vertebrates. Nevertheless, it seems premature to speculate on the phylogenetic origin of such prodigious mnemonic ability as manifested in our Old World primates. A possible clue to its significance for survival can be surmised, however, from the knowledge needed by *M. nemestrina* in its home territory on the Malay peninsula. Their food intake is roughly 25% of their body weight per day. To achieve this safely and efficiently, they must be able from one year to the next to distinguish the more than 100 often seasonal fruits, flowers, and leaves that are edible from those that are poisonous (Caldecott, 1986), a task requiring a precise and protracted memory, most appropriate if visual.

REFERENCES

Bahrick, H. P., Bahrick, P. O., and Wittlinger, R. P., 1975, Fifty years of memory for names and faces: A cross-sectional approach, *J. Exp. Psychol. Gen.* **104**:54–75.

Beritashvili, I. S., 1972, Phylogeny of memory development in vertebrates, in: *Brain and Human Behavior* (A. G. Karczmar and J. C. Eccles, eds.), Springer, New York, pp. 341–351.

Bickford, R. G., Mulder, D. W., Dodge, H. W., Svien, H. J., and Rome, H. P., 1958, Changes in memory function produced by electrical stimulation of the temporal lobe in man, in: *The Brain and Human Behavior. Association for Research in Nervous and Mental Disorders,* vol. 36, Williams & Wilkins, Baltimore, pp. 551–574.

Boltz, R. L., Harwerth, R. S., and Smith, E. L. III, 1979, Orientation anisotropy of visual stimuli in rhesus monkey; a behavioral study, *Science* **205**:511–513.

Caldecott, J. O., 1986, *An Ecological and Behavioural Study of the Pig-Tailed Macaque,* S. Karger, Basel.

Cavonius, C. R., and Robbins, D. O., 1973, Relationships between luminance and visual acuity in the rhesus monkey. *J. Physiol. (Lond.)* **232**:239–246.

Cowey, A., 1979, Cortical maps and visual perception. The Grindley Memorial Lecture. *Q. J. Exp. Psychol.* **31**:1–17.

DeValois, R. L., Morgan, H. C., Polson, M. C., Mead, W. R., and Hull, E. M., 1974, Psychophysical studies of monkey vision. I. Macaque luminosity and color vision tests, *Vision Res.,* **14**:53–67.

Doty, R. W., 1979, Neurons and memory: Some clues, in: *Brain Mechanisms in Memory and Learning: From the Single Neuron to Man* (M. A. B. Brazier, ed.), Raven Press, New York, pp. 53–63.

Golomb, B., Andersen, R. A., Nakayama, K., MacLeod, D. I. A., and Wong, A., 1985, Visual thresholds for shearing motion in monkey and man, *Vision Res.* **25**:813–820.

Green, D. M., and Swets, J. A., 1974, *Signal Detection Theory and Psychophysics,* Krieger, Huntington, NY.

Halgren, E., 1984, Human hippocampal and amygdala recording and stimulation: Evidence for a neural model of recent memory, in: *Neuropsychology of Memory* (L. R. Squire and N. Butters, eds.), Guilford Press, New York, pp. 165–182.

Hall, D. F., and Loftus, E. F., 1984, The fate of memory: Discoverable or doomed? in: *Neuropsychology of Memory* (L. R. Squire and N. Butters, eds.), Guilford Press, New York, pp. 25–32.

Horel, J. A., and Pytko, D. E., 1982, Behavioral effect of local cooling in temporal lobe of monkeys, *J. Neurophysiol.* **47**:11–22.

Hunter, I. M. L., 1977, An exceptional memory, *Br. J. Psychol.* **68**:155–164.

Jacobs, G. H., 1981, *Comparative Color Vision,* Academic Press, New York.

Luria, A. R., 1968, *The Mind of a Mnemonist. A Little Book about a Vast Memory* (translated by L. Solotaroff), Basic Books, New York.

Maguire, W. M., Meyer, G. E., and Baizer, J. S., 1980, The McCollough effect in rhesus monkey, *Invest. Ophthalmol.* **19**:321–324.

Mahut, H., Moss, M., and Zola-Morgan, S., 1981, Retention deficits after combined amygdalo-hippocampal and selective hippocampal resections in the monkey, *Neuropsychologia* **19**:201–225.

Milner, B., Corkin, S., and Teuber, H.-L., 1968, Further analysis of the hippocampal amnesic syndrome: 14-year follow-up study of H.M., *Neuropsychologia* **6**:215–234.

Mishkin, M., 1978, Memory in monkeys severely impaired by combined but not by separate removal of amygdala and hippocampus, *Nature* **273**:297–298.

Mishkin, M., 1982, A memory system in the monkey, *Phil. Trans. R. Soc. Lond. [Biol.]* **298**:85–95.

Mishkin, M., and Delacour, J., 1975, An analysis of short-term memory in the monkey, *J. Exp. Psychol. Anim. Behav. Proc.* **1**:326–334.

Mishkin, M., Malamut, M., and Bachevalier, J., 1984, Memories and habits: Two neural systems, in: *Neurobiology of Learning and Memory* (G. Lynch, J. L. McGaugh, and N. M. Weinberger, eds.), Guilford Press, New York, pp. 65–77.

Neisser, U., 1982, *Memory Observed. Remembering in Natural Contexts,* W. H. Freeman, San Francisco.

Nickerson, R. S., 1965, Short-term memory for complex meaningful visual configurations: A demonstration of capacity, *Can. J. Psychol.* **19**:155–160.

Olton, D. S., Becker, J. T., and Handelmann, G. E., 1979, Hippocampus, space, and memory, *Behav. Brain Sci.* **2**:313–365.

Overman, W. H., Jr., and Doty, R. W., 1980, Prolonged visual memory in macaques and man, *Neuroscience* **5**:1825–1831.

Owen, M. M., and Butler, S. R., 1984, Does amnesia after transection of the fornix in monkeys reflect abnormal sensitivity to proactive interference? *Behav. Brain Res.* **14**:183–192.

Ringo, J. L., and Doty, R. W., 1985, A macaque remembers pictures briefly viewed six months earlier, *Behav. Brain Res.* **18**:289–294.

Ringo, J. L., Lewine, J. D., and Doty, R. W., 1986, Comparable performance by man and macaque on memory for pictures, *Neuropsychologia* **24**:711–717.

Sands, S. F., and Wright, A. A., 1982, Monkey and human pictorial memory scanning, *Science* **216**:1333–1334.

Sarmiento, R. F., 1975, The stereoacuity of macaque monkey, *Vision Res.* **15**:493–498.

Scott, T. R., and Powell, D. A., 1963, Measurement of a visual motion after-effect in the rhesus monkey, *Science* **140**:57–59.

Smith, E. L., III, Harweth, R. S., Levi, D. M., and Boltz, R. L., 1982, Contrast increment thresholds of rhesus monkeys, *Vision Res.* **22**:1153–1161.

Speedie, L. J., and Heilman, K. M., 1982, Amnesic disturbance following infarction of the left dorsomedial nucleus of the thalamus, *Neuropsychologia* **20**:597–604.

Squire, L. R., 1982, The neuropsychology of human memory, *Annu. Rev. Neurosci.* **5**:241–273.

Squire, L. R., and Moore, R. Y., 1979, Dorsal thalamic lesion in a noted case of human memory dysfunction, *Ann. Neurol.* **6**:503–506.

Standing, L., 1973, Learning 10,000 pictures, *Q. J. Exp. Psychol.* **25**:207–222.

Sternberg, S., 1966, High-speed scanning in human memory, *Science* **153**:652–654.

Sternberg, S., 1969, Memory-scanning: Mental processes revealed by reaction-time experiments, *Am. Sci.* **57**:421–457.

Sternberg, S., 1975, Memory scanning: New findings and current controversies, *Q. J. Exp. Psychol.* **27**:1–32.

Thomas, G. J., 1984, Memory: Time binding in organisms, in: *Neuropsychology of Memory* (L. R. Squire and N. Butters, eds.), Guilford Press, New York, pp. 374–384.

Zola-Morgan, S., and Squire, L. R., 1985, Amnesia in monkeys after lesions of the mediodorsal nucleus of the thalamus, *Ann. Neurol.* **17**:558–564.

Zola-Morgan, S., Squire, L. R., and Mishkin, M., 1982, The neuroanatomy of amnesia: Amygdala–hippocampal versus temporal stem, *Science* **218**:1337–1339.

30

Neuronal Activity in the Inferomedial Temporal Cortex Compared with That in the Hippocampal Formation

Implications for Amnesia of Medial Temporal Lobe Origin

F. A. W. WILSON, M. W. BROWN, and I. P. RICHES

1. INTRODUCTION

Numerous studies have reported that bilateral lesions of the medial temporal lobes cause enduring anterograde amnesia in man (Scoville and Milner, 1957; Victor *et al.*, 1961; DeJong *et al.*, 1969; Van Buren and Borke, 1972). There are two major and as yet unsolved issues concerning the nature of this amnesia: the structures within the medial temporal lobes that are crucial for memory function and the precise processes that are impaired by the lesions. Published accounts of the temporal lobe damage accompanying the amnesia generally describe lesions of the hippocampus, adjacent temporal cortex, and the amygdala in varying degrees and combinations. Zola-Morgan *et al.* (1986) have reported a patient (case R.B.) with complete bilateral destruction of region CA1 of the hippocampus. This patient was amnesic, but differed from the patient H.M. (Scoville and Millner, 1957) in two major respects: first, the hippocampal lesions of H.M. were subtotal, sparing the posterior third of the hippocampus; second, H.M.'s impairment was much more severe than R.B. As noted by Squire (1987), damage to the medial temporal cortex exacerbates memory impairments. Thus, it is not clear what the specific contributions of the hippocampus, the amygdala, and the adjacent temporal cortex are to memory and related mechanisms.

The ability to perform visual recognition memory tasks appears to require the integrity of the medial temporal lobes. The performance of such tasks requires a judgment of the relative novelty/familiarity or recency of occurrence of the task stimuli. Amnesic patients typically have no conscious awareness of having previously seen faces, places, and the stimuli and tasks with which they have been tested (Jacoby and Witherspoon, 1982; Weiskrantz, 1982). Formal testing has shown impairments in some recognition memory tasks, although such patients do show evidence of residual learning and memory (Milner, 1970; Warrington, 1974; Warrington and Weiskrantz, 1979). Sidman, Stoddard, and

F. A. W. WILSON • Section of Neuroanatomy, School of Medicine, Yale University, New Haven, Connecticut 06510. M. W. BROWN and I. P. RICHES • Department of Anatomy, The Medical School, University of Bristol, Bristol BS8 1TD, England.

Mohre (1968) found that the patient H.M. was able to learn and perform a visual discrimination task but was unable to describe the task that he had been performing. They also found evidence of an impairment in a delayed matching-to-sample task when the sample and choices were separated by long delays. However, when the material to be learned was very unfamiliar, and using a forced-choice procedure, H.M. was able to select accurately the test material (Huppert and Piercy, 1978). Thus, the difficulty that amnesic patients have with recognition memory tasks may depend on the exact form of the test and the type of material used.

As a first step in examining the issues raised above, recordings have been made of neuronal activity in the inferomedial temporal cortex (IMC) and hippocampal formation of monkeys. The IMC includes the ento-, pro-, and perirhinal areas and areas Pro and a medial sector of TEl (Seltzer and Pandya, 1978) of the medial temporal cortex. The hippocampal formation comprises the hippocampus, dentate gyrus, and subicular cortex. Recordings have been made during the performance of delayed matching-to-sample (DMS) or recognition memory tasks in which stimuli varying in their novelty and familiarity were presented. These procedures are appropriate, for they resemble incidental and explicit recognition memory tasks in which amnesic patients are impaired. Furthermore, monkeys with damage to the hippocampus, amygdala, and IMC are severely impaired in delayed nonmatching-to-sample tasks, which require the judgment of relative novelty/familiarity (Mishkin, 1978; Zola-Morgan and Squire, 1985).

The initial recordings were made in the hippocampal formation. Although hippocampal units were found to be responsive during the performance of the DMS task (Brown, 1982), such units did not respond on the basis of the novelty/familiarity of the stimuli. Subsequent recordings made in the inferomedial temporal cortex found that neuronal activity in this cortex is influenced by the relative novelty/familiarity or the recency of presentation of visual stimuli.

2. METHODS AND MATERIALS

Monkeys were trained to perform a Konorski conditional delayed matching task in which two successively presented stimuli were displayed on a video monitor (Brown, 1982). Fruit juice was delivered following a right-panel press when both stimuli were the same and following a left-panel press when they differed. Two stimuli, most commonly both of the same shape but differing in size, were used in each block of 8–20 trials. The stimuli were then changed at the start of the next block. The animals were initially trained with a standard stimulus (white square). During the testing of each unit, this stimulus was presented in addition to a variety of colored, simple, and complex geometric shapes, varying in their familarity to the animals. The two successively presented stimuli (S_1 and S_2) were each displayed for 0.5 sec with an interstimulus interval of 0.5 sec. Presses resulting in juice had to occur between 0.5 and 5.0 sec after the offset of the S_2. A red cuing flash of 0.5-sec duration occurred 1.0 sec before the onset of the S_1.

Recordings of single units were made using a discriminator with two amplitude levels and a time window (Brown and Leendertz, 1979) and conventional amplification and filtering. Recordings were made during the delayed matching task and during presentation of objects and foods that were novel (not seen for 14 days or more) or highly familiar (seen several times daily). The objects and foods could be either viewed and obtained or merely viewed by the animal. Eye and hand movements were monitored and videorecorded. Recordings were made in the hippocampal formation and inferomedial temporal cortex of two *Macaca fascicularis* and five *Macaca mulatta*. The location of the recorded units was determined from x-ray photographs of the electrodes *in situ* and from the positions of marker lesions made at the sites of responsive units and subsequently identified on histological sections. The significance of changes in response was established using paired t tests or multivariate analyses of variance (MANOVA) as appropriate. Only statistically significant ($P < 0.05$) changes are reported.

3. RESULTS

3.1. The Inferomedial Temporal Cortex

One population of units (26 of 173 tested) recorded in the inferomedial temporal cortex of a *Macaca fascicularis* responded with maximal increases in firing rate to stimuli that were novel or had not been recently seen. The responses of these units declined significantly with repetition of a stimulus. Their locations are shown in Fig. 1. The units appeared to be located mainly in the deeper cortical layers, as indicated both by lesions made at the site of such units and by the proximity of the units to the white matter dorsal to the cortex, as judged during the recordings. The responses of one unit to the presentation of novel objects is shown in Fig. 2A. The responses of 21 of the 26 units declined with repeated presentations of objects. Ten units responded with decrements to repeated stimulus presentations during the performance of the DMS task (Fig. 2B). Five of 11 units tested during the presentation of objects and during performance of the DMS task showed significantly declining responses in both tasks.

The effect of distractors and/or the passage of time on the neuronal responses was examined. For seven of eight units tested, the decrements in response to repeated presentations of objects were maintained even after intervening presentations of other objects (Fig. 2A). The maximum delay tested was >100 sec, during which 15 other stimuli were presented. In the DMS task the response was no longer significantly reduced after intervention of one or more blocks of different stimuli (>40 sec) for nine of ten units.

Another group of units (45 of 80 visually responsive units) responded significantly differently to the highly familiar standard (square) stimulus compared to the other, novel or rarely seen (colored or nonsquare) stimuli used in the DMS task (Fig. 3). The responses of these units did not decline with repetition. A third group of eleven units was maximally activated by the presentation of novel objects and was significantly less responsive to familiar stimuli such as foods, highly familiar objects, and faces (Fig. 4). These units declined in response to initially novel stimuli over the few stimulus repetitions tested, and responded differently to highly familiar stimuli.

Occasionally units were tested for somatic and auditory responsiveness: two units located in the prorhinal cortex were responsive to auditory stimuli.

3.2. The Hippocampal Formation

In contrast to the effects of stimulus novelty/familiarity on the cortical units, such decremental neuronal responses have not been found for any of more than 650 units recorded in the hippocampus and subicular cortex, a proportion significantly less than that of the IMC (χ^2; $P < 0.001$). In three monkeys the activity of 395 units was recorded during the performance of a serial visual recognition memory task (Gaffan, 1977) employing objects varying in their novelty and familiarity. In this task monkeys obtained fruit juice for lick responses made during the presentation of a familiar stimulus; saline

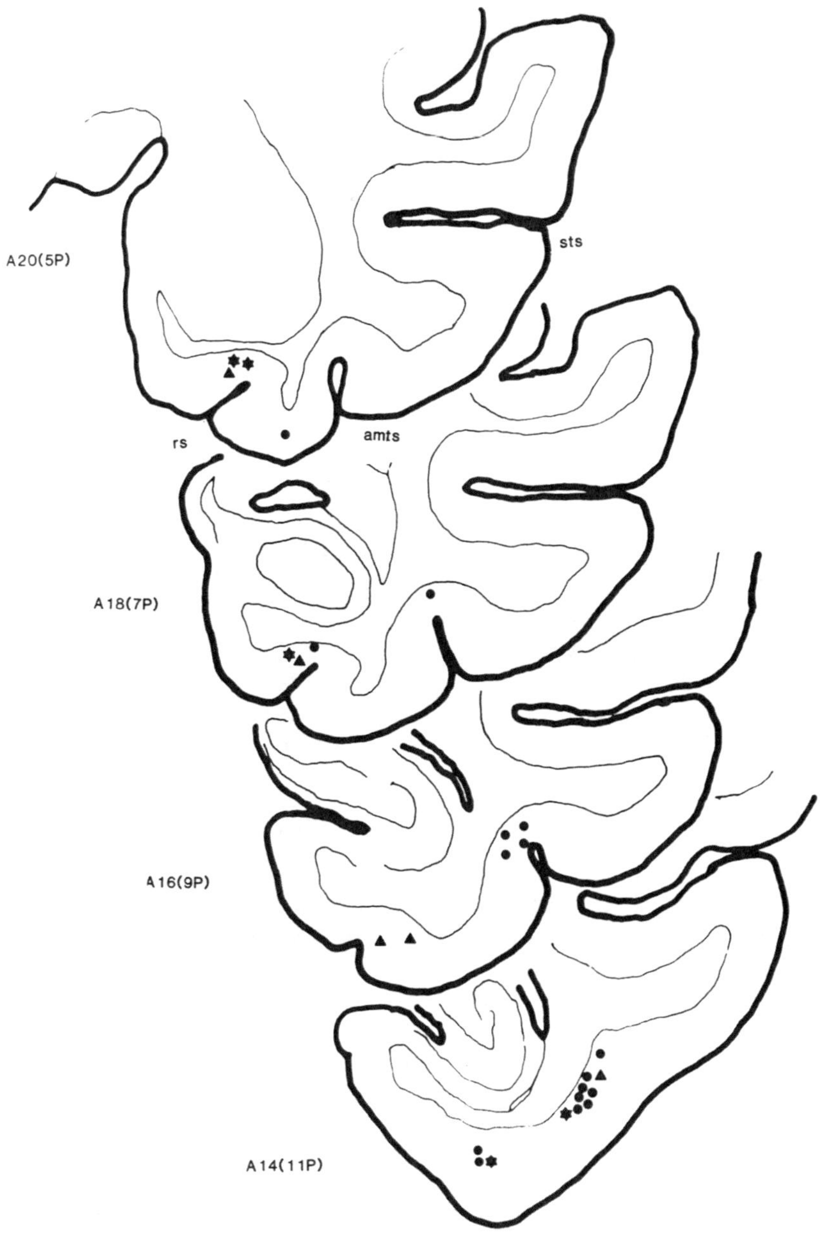

FIGURE 1. The positions of units with responses that declined to repeated presentations of objects (●) or following the introduction of new stimuli in the delayed matching task (▲) or in both situations (★) are shown on frontal sections at the indicated distances anterior (A) to the interaural line and posterior (P) to the sphenoid (Aggleton and Passingham, 1981). amts, anterior medial temporal sulcus; rs, rhinal sulcus; sts, superior temporal sulcus.

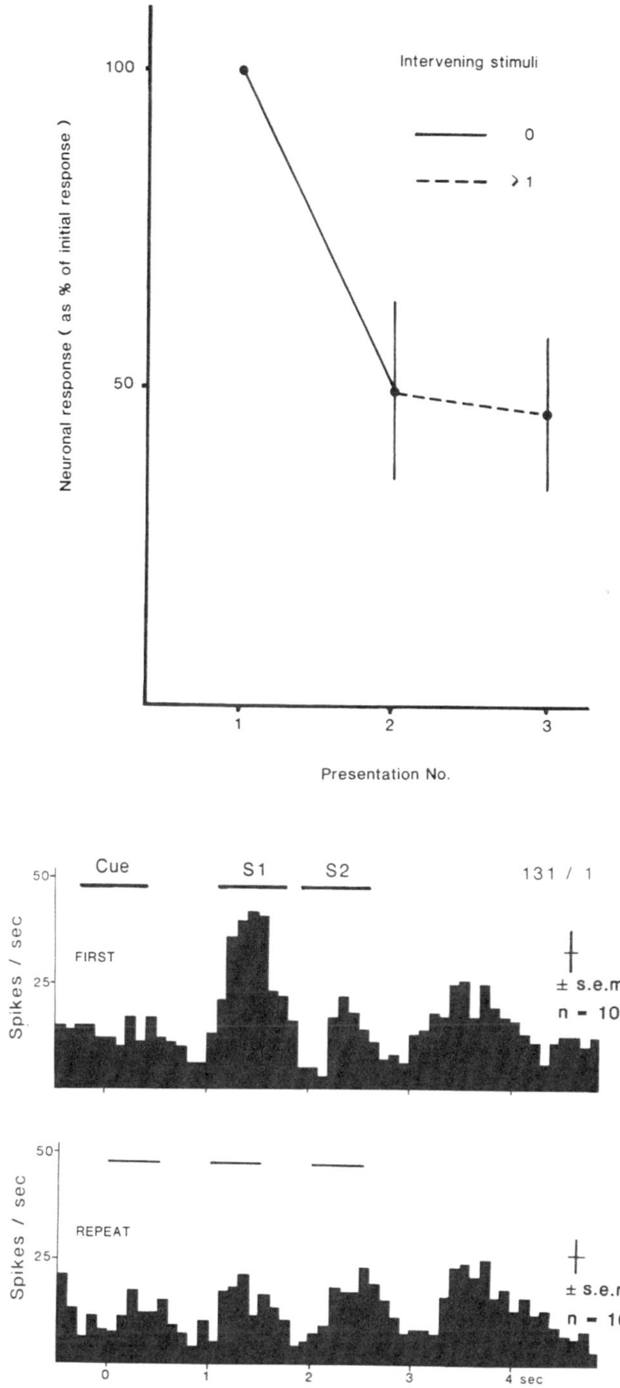

FIGURE 2. (A) The responses of unit 131/1 to repeated presentations of objects. The unit was located in prorhinal cortex. Each of nine objects was shown three times. There was no intervening presentation of other objects between first and second presentations but one or more between the second and third presentations. Response to initial presentations is taken as 100%. Note declining response to second presentation and lack of recovery to third presentation. Error bars represent S.E.M. (B) Activity of same unit during performance of delayed matching task. The upper histogram (FIRST) is the response averaged across trials during which ten stimuli that had not been seen recently were first presented. The lower histogram (REPEAT) is composed from the activity of the immediately following trials during which each of the ten stimuli occurred again. Cue, red cueing flash; S1, S2, the two stimuli to be matched. Bar to the right of each histogram represents two mean S.E.M.S. Note the decline in response to S1 stimulus on the repeat trials (MANOVA, $P < 0.001$). No other differences between the histograms were significant.

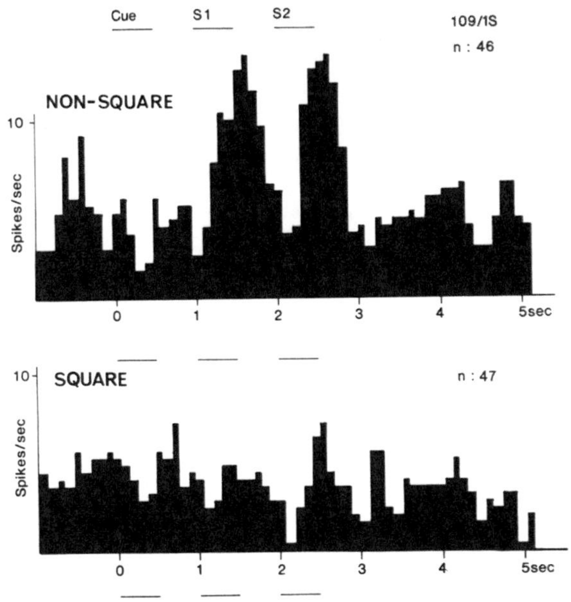

FIGURE 3. Greater (MANOVA, $P < 0.001$) response of stimulus-selective unit in area TE1 to unfamiliar non-square stimuli (upper histogram) than to highly familiar white squares (lower histogram). The S.E.M.s for bin 25 in the two histograms were ± 0.2 (non-square) and 1.0 (white square). Bin width, 100 msec. Conventions as in Fig. 2B.

was delivered for licks made during the presentation of novel stimuli. For none of these units were consistent differences found between trials on which novel stimuli were presented and those on which the stimuli reappeared as familiar. An additional >250 units tested with objects and in the DMS task for responsiveness to stimuli varying in their novelty/familiarity did not show consistent responses dependent on the novelty/familiarity of the stimuli. These neurons were recorded in the same experiments in which the decrementing units were recorded in the inferomedial temporal cortex.

Although no hippocampal units were found with responses that declined strongly on stimulus repetition, this was not because of a lack of responsiveness during the task. Hippocampal units were commonly active during the presentation of stimuli, during the behavioral response, or during both events. Hippocampal units were not active simply

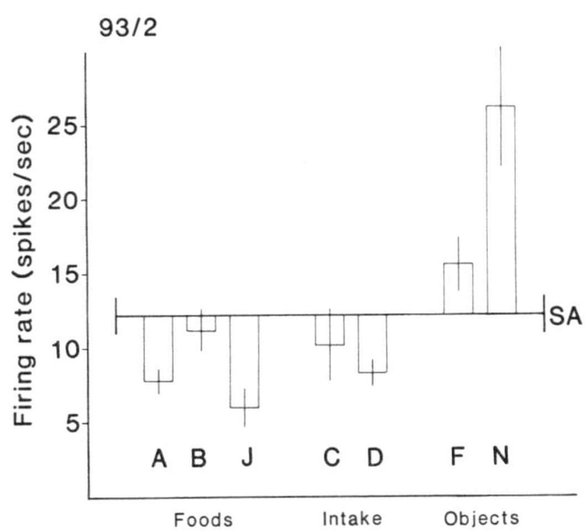

FIGURE 4. Mean responses of unit recorded in area TE1 to sight of pieces of apple (A), banana (B), and syringe containing fruit juice (J); during chewing (C) and drinking (D); and to sight of highly familiar (F) or novel (N) objects. Sight of novel objects was the only condition resulting in a significant (t test, $P < 0.05$) increase in activity over background. Horizontal line represents mean spontaneous activity (SA). Vertical bar represents $\pm$S.E.M. in each case.

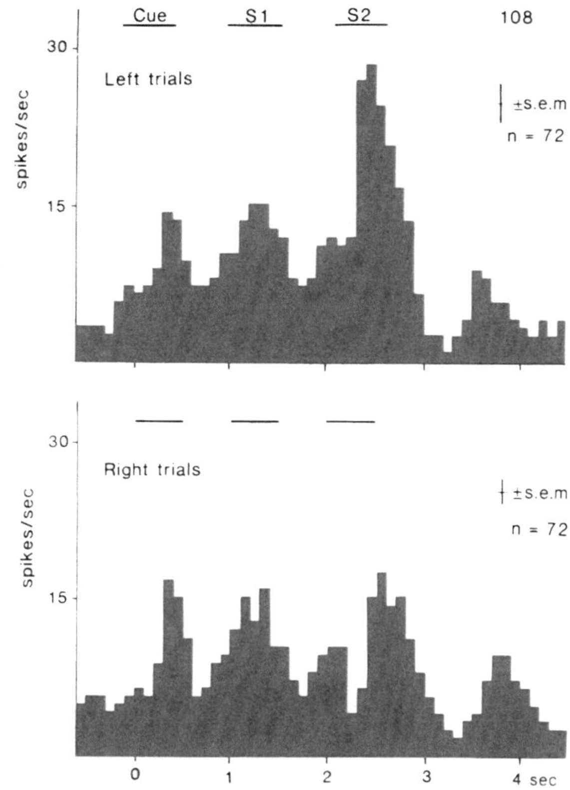

FIGURE 5. Activity of unit recorded in the hilar region of the dentate gyrus on left-press trials (nonmatch, upper trace) and right-press trials (match, lower trace). Conventions as in Fig. 2B. Note the greater (MANOVA, $P < 0.001$) response to the S2 stimulus on left- than on right-press trials.

during general behavioral activity but responded significantly (MANOVA, $P < 0.05$) differently on the left- compared to the right-press trials in the DMS task (Fig. 5). At the time of presentation of the S_2 stimulus, 102/303 (34%) units showed this effect. Fewer units (45/303 = 15%) responded significantly differently on left versus right trials at the time of the press. A further population (42/303 = 14%) of units responded specifically to the occurrence of a particular stimulus on a particular type of trial (match or nonmatch), so that the neuronal activity was not determined solely by the sensory stimulus or solely by the behavioral response. These latter responses demonstrate encoding of the occurrence of a particular sensory stimulus in a particular behavioral context (Brown, 1982).

4. DISCUSSION

Certain neurons recorded in the inferomedial temporal cortex responded to visual stimuli on the basis of their novelty, relative familiarity, or recency of presentation. This type of information concerning the previous occurrence or familiarity of stimuli is required for visual recognition memory. In contrast, similar responses based on memory for the previous occurrence of stimuli were not found for any of a large number of neurons recorded in the hippocampus and tested with the same techniques.

An objective of this discussion is to examine the hypothesis that the IMC, in conjunction with anatomically connected structures, provides a substrate for certain types of recognition memory. This hypothesis requires, for example, that the neuronal responses within such areas differentiate between novel and familiar stimuli in the presence of delays and distracting presentations of intervening stimuli, so paralleling the ability of monkeys to perform such recognition memory tasks (Mishkin and Delacour, 1975; Gaffan, 1977; Rolls *et al.*, 1982; Ringo and Doty, 1985; Zola-Morgan and Squire, 1985). It also requires that lesions of such areas disrupt performance of recognition memory tasks.

One group of IMC neurons responded maximally (under the conditions of these experiments) to the first presentation of stimuli that had not been seen recently, the responses declining with stimulus repetition. Some of these neurons showed evidence of memory for the previous occurrence of objects: the response to a previously seen stimulus was less than that to its initial presentation even when there had been intervening presentations of other stimuli. The maximum memory spans of these neurons have yet to be determined, and therefore it is not known whether they could be adequate as a general substrate for recognition memory. In contrast to their responses to objects, most of the units tested in the DMS task demonstrated only a short-lasting memory for the geometric stimuli used. This difference could be related to the properties of these two-dimensional stimuli, which, although discriminable, were not as distinctive as the three-dimensional objects: monkeys have difficulty in learning two-dimensional pattern discriminations in contrast to their learning of object discriminations (Zola-Morgan and Squire, 1984). Furthermore, the monkey was not required to remember from trial to trial the stimuli used in the DMS task, and, indeed, it may pay not to remember the stimuli, as proactive interference may make the task more difficult (Warrington and Weiskrantz, 1974). In summary, the decrementing neuronal responses show evidence of a memory for the previous occurrence of stimuli, but it remains to be demonstrated that they show memory over long time periods.

The decrementing neuronal responses are not accounted for by factors such as behavioral training, failure of attention to the stimuli, or the reinforcement value of the stimuli. Information about the novelty and familiarity of visual stimuli was incidental to the performance of the DMS task, indicating that the neuronal decrements were not a result of training and thus occur endogenously. Although the neuronal responses changed with stimulus repetition in the DMS task, the accuracy of task performance was unchanged. Thus, the decrements in the neuronal responses are not attributable to a failure to attend to the task stimuli or to changes in the behavioral responses (panel presses). Furthermore, because juice was available on all trials, the changes in neuronal responses cannot be attributed to variations in reinforcement. The response decrements can be distinguished from simple habituation (Thompson and Spencer, 1966): they occurred to stimuli being used to obtain reward in the DMS task and were maintained in the presence of distraction caused by intervening presentations of other stimuli.

A second group of cortical units responded maximally to novel objects as well as decrementing in response over the small number of stimulus repetitions tested. The response of these units to highly familiar objects, foods, and faces was markedly different from that to novel objects. If the basis of this difference is the relative novelty/familiarity of the stimuli, these units demonstrate evidence of a long-term memory since in these cases the highly familiar objects had not been seen recently (>30 min).

A third group of units responded differently to the highly familiar white square stimulus compared to novel colored or nonsquare stimuli. The basis of this difference in response could be either the relative familiarity of the stimuli or their physical appearance. Although the data do not permit a resolution of these alternatives, the fact that IMC units respond selectively to certain stimuli reflects access to information that would allow stimulus classification. Such classification is a prerequisite for the performance of recognition memory tasks.

The results outlined above suggest that the activity of units in the inferomedial temporal cortex may reflect their involvement in more than one memory mechanism.

4.1. Memory Deficits and the Inferomedial Temporal Cortex

The hypothesis that neuronal activity in the IMC reflects access to mnemonic information is consistent with putative memory functions of the medial temporal lobes. Combined bilateral removal of the hippocampus, amygdala, and inferomedial temporal cortex results in a severe impairment in the performance of delayed nonmatching-to-sample (DNMS) tasks in monkeys (Mishkin, 1978). However, sparing of the inferomedial temporal cortex with removal of the hippocampus and amygdala does not produce a severe impairment (Murray *et al.*, 1985). The implication is that integrity of the IMC is sufficient to support performance of the DNMS task. Cooling of the inferior temporal gyrus has been shown to impair severely the performance of a DMS task requiring recent or short-term memory (Horel and Pytko, 1982). Thus, it is likely that damage to the IMC is partially responsible for the memory impairment seen after temporal lobe damage.

There is further evidence that damage to the IMC is associated with memory dysfunction. Pathological changes have been observed in this cortex early in the course of dementia of the Alzheimer type (Brun and Englund, 1981; Pearson *et al.*, 1985). It is notable that memory impairments are an early symptom of Alzheimer's disease (Flicker *et al.*, 1985), and it is possible that the degeneration of this cortex may be associated with impairments of memory observed in the disease. The relationship between the disease and the inferomedial temporal cortex is strengthened by anatomic and electrophysiological studies. The IMC is one of the few cortical structures known to project to the basal forebrain (Mesulam and Mufson, 1984). Degeneration of the basal forebrain occurs in Alzheimer's disease (Whitehouse *et al.*, 1981). Recent electrophysiological studies of the basal forebrain have observed neuronal activity reflecting the novelty/familiarity of visual stimuli in monkeys performing recognition memory tasks (Rolls *et al.*, 1982; Wilson *et al.*, 1984). It is possible that neuronal activity in the basal forebrain could reflect the projections from the IMC and that pathology of the IMC could produce secondary degeneration in the basal forebrain. The evidence is therefore consistent with a mnemonic function for the inferomedial temporal cortex.

4.2. Fractionation of Recognition Memory and Implications for Its Neuronal Basis

Further experiments, including the recording of IMC units during the performance of a variety of carefully designed tasks, will be required before it is possible to determine the precise types of memory in which the area is involved. At present the evidence is strongest for an involvement in recognition memory. However, even in the case of recognition memory, it has become clear from studies of human memory that there are a number of related, though potentially separable types of information on which recognition may be based. For example, recognition may be based on (1) detection of novelty, i.e., that something has never occurred previously, (2) judgment of relative familiarity, i.e., how frequently or for how long a period something has been experienced, (3) assessment of the recency of presentation of a stimulus, i.e., whether a stimulus occurred a short or a long time ago (4) discrimination of context, i.e., when, where, how, and associated with what a previously presented stimulus occurred. (Such discrimination of context is particularly important when the target material of a recognition test is highly familiar, as is the case for humans for lists of everyday words.) Hence, under different conditions, judgments may be made, for instance, on the basis of trace strength or complex contextual discriminations or a range of available information.

These differing types of information are not uniformly impaired in human amnesics and hence must have differing neuronal substrates. For instance, discrimination of the frequency from the recency of occurrence is impaired in Korsakoff patients with presumed diencephalic lesions (Huppert and Piercy, 1978), but the ability of amnesics to detect novelty as opposed to the discrimination of relative familiarity is controversial (Biber *et al.*, 1981). Another phenomenon, priming (facilitation of tasks involving previously presented material without conscious awareness of prior presentation being necessary), has been found to be intact in amnesics of whatever etiology (Cermak *et al.*, 1985; Milner, 1970). Accordingly, it is unlikely that the responses of IMC units subserve priming. Conversely, the absence of hippocampal units responding on the basis of the previous occurrence of stimuli renders it unlikely that the hippocampus is important for this aspect of recognition memory.

4.3. The Inferomedial Temporal Cortex and Related Brain Areas

Structures other than the IMC also make contributions to memory. In man, amnesia has been reported after damage to the medial thalamus (Squire and Moore, 1979; Winocur *et al.*, 1984) and inferomedial prefrontal cortex (Wallesch *et al.*, 1983). Basal forebrain damage in conjunction with other cortical areas has also produced profound amnesia (Friedman and Allen, 1969; Gascon and Gilles, 1973; Damasio *et al.*, 1985). Impairments in object recognition tasks in monkeys follow damage to the medial thalamus (Aggleton and Mishkin, 1983), inferomedial prefrontal cortex (Mishkin and Bachevalier, 1983), and basal forebrain (Aigner *et al.*, 1984).

The IMC is anatomically connected with all of these medially located regions (Whitlock and Nauta, 1956; Jones and Powell, 1970; Aggleton *et al.*, 1986; Russchen *et al.*, 1985). The inferior medial regions of the temporal and prefrontal cortices have in common several anatomic features. First, they constitute the final destinations of sequential pathways originating in sensory and association cortices (Jones and Powell, 1970). Second, they exhibit similar cytoarchitectural organization (Mesulam and Mufson, 1982). Third, concentrations of AChE are higher in these regions than in other cortical areas (Mesulam *et al.*, 1984). Fourth, they constitute the only cortical regions known to project to the basal forebrain (Mesulam and Mufson, 1984). Projections to the medial thalamus, amygdala, and hippocampus have also been described (Van Hoesen and Pandya, 1975; Leichnetz and Astruc, 1976; Aggleton *et al.*, 1980, 1986). Thus, all these medially located regions appear to form an anatomically connected system in which each component part is known to make a contribution to memory function, though the individual contributions made by these structures are not yet known.

There is physiological evidence that the IMC influences subcortical structures implicated in memory function. Recordings of single-unit activity in monkeys performing recognition memory tasks have found neurons responding on the basis of the novelty and familarity of visual stimuli in the basal forebrain (Rolls *et al.*, 1982; Wilson *et al.*, 1984) and the amygdala (Wilson, 1985). Other studies have found similar units in medial regions of the head of the caudate nucleus (G. V. Williams, F. A. W. Wilson, and E. T. Rolls, unpublished observations), the tail of the caudate (Caan *et al.*, 1984), and, in the present experiments, in the ventral putamen. The IMC is known to project to these striatal regions (Van Hoesen *et al.*, 1981), which are also known to play a role in visual discrimination tasks (Divac *et al.*, 1967; Buerger *et al.*, 1974); their role in recognition memory is unknown. Since the units with decrementing responses were located mainly in the deeper cortical layers, it is possible that these units project to subcortical structures, which may explain why such neuronal responses were not observed in the hippocampus. The input to the hippocampus comes from layer III (Steward and Scoville, 1976).

The IMC receives inputs from the inferotemporal cortex (Van Hoesen and Pandya, 1975). This area does not share the anatomic features of the inferomedial regions but is of major importance for visual memory function (Gross, 1972; Mishkin, 1982) and projects to inferomedial cortical regions. Units recorded in the inferotemporal cortex show "habituation" (Desimone and Gross, 1979) but have very limited memory spans (Baylis and Rolls, 1983).

The IMC provides one route through which information may be transmitted to subcortical structures. More generally, information initially processed in the inferotemporal cortex is probably transmitted in parallel to a number of brain structures including the IMC (Horel, 1978; Mishkin, 1982). Such information might then be operated on in a specific way by each recipient structure, so providing substrates for the multiplicity of processes necessary for the performance of various recognition memory tasks. Each structure would thus perform one or more different, contributory functions. Possible examples of this division of function are described below for the hippocampus and amygdala.

4.4. The Hippocampus and Amygdala

The role of the hippocampus and amygdala in object recognition is not clear. There are no detailed reports of human amnesia following damage confined to the hippocampus; detailed clinical studies reporting amnesia following medial temporal damage have all reported significant involvement of the inferomedial temporal cortex (Scoville and Milner, 1957; Victor *et al.*, 1961; DeJong *et al.*, 1969; McLardy, 1970; Van Buren and Borke, 1972). The amnesia observed in these patients appears to be duplicated by the deficits found in object recognition tasks in monkeys after combined lesions of the hippocampus, amygdala, and IMC (Mishkin, 1978; Zola-Morgan and Squire, 1985). Lesions restricted to the amygdala do not produce such impairments (Mishkin, 1978; Zola-Morgan *et al.*, 1984). Some studies have found deficits after selective hippocampal lesions or fornix transections (Gaffan, 1974; Owen and Butler, 1981; Mahut *et al.*, 1982), but other studies have not (Mishkin, 1978; Murray and Mishkin, 1984). Thus, the experiments in monkeys have not settled the role of the hippocampus in object recognition.

In contrast to the negative evidence, numerous studies attest to the involvement of the hippocampus in a variety of tasks requiring memory for behavioral responses, e.g., delayed alternation (Pribram *et al.*, 1962), and for different forms of spatial tasks (Jones and Mishkin, 1972; O'Keefe and Nadel, 1978; Parkinson and Mishkin, 1982). The present results are consistent with these studies. Hippocampal units are reliably activated by visual stimuli and the behavioral responses they evoke. Large proportions of hippocampal units appear to encode information about the specific (left or right) responses made in the DMS task. Furthermore, hippocampal units are active in delayed alternation tasks in which a behavioral response is guided by the memory of the previous response rather than by exteroceptive stimuli (Wilson *et al.*, 1986). Previous studies in rat, rabbit, cat, and monkey have reported that hippocampal electrical activity is related to behavioral activity or to both sensory stimuli and behavioral responses (Adey *et al.*, 1960; Vanderwolf *et al.*, 1975; Ranck, 1973; Brown, 1982; Watanabe and Niki, 1985). In addition, studies in neurosurgical patients have reported choice- and movement-related activity (Halgren *et al.*, 1978; Arnolds *et al.*, 1980). Thus, the finding that hippocampal units are active in relation to behavioral responses in the monkey is consistent with findings for other species.

Recent work (Gaffan, 1985) has emphasized the importance of the hippocampus in learning to modify behavioral responses to sensory input. Hippocampal lesions impair discrimination tasks in which the reward associations of a stimulus are insufficient to guide behavior (for example, the present DMS task). The context-dependent activity described both by Brown (1982) and in the present results could be a reflection of such a function of the hippocampus in that neuronal activity is dependent on both a particular sensory input (e.g., a small square following a small square) and a particular behavioral response (press right panel). Note that the high percentage of units showing such task-dependent activity suggests that this activity may have been induced as a result of the animals' training. In the present experiments the activity of many hippocampal units recorded during the DMS task differed between right- and left-press trials. However, in the serial recognition task, no such differences in activity were found between lick and no-lick trials (go/no-go). These findings indicate a difference between hippocampal involvement in oral go/no-go and spatial conditional response tasks. In agreement with this suggestion, hippocampal lesions have been found to impair spatial but not go/no-go tasks (Mahut, 1972). Thus, if the hippocampus does prove to be important for linking newly acquired sensory memories to behavioral responses (Gaffan, 1985), lesions of this structure could produce deficits in performance of certain recognition tasks without a deficit in registering the previous occurrence of stimuli.

The functions of the hippocampus and amygdala can be dissociated, since their ablation results in impairments in different tasks. Damage to the amygdala does not impair performance of spatial tasks but does produce emotional changes (Weiskrantz, 1956). Similarly, the activity of units in the hippocampus and amygdala can be dissociated on the basis of their relationship to movement and of the types of stimuli that elicit responses. In studies of awake animals, amygdala units are responsive to stimuli such as faces, foods, and environmentally significant noises (Sawa and Delgado, 1963; O'Keefe and Bouma, 1969; Jacobs and McGinty, 1972; Leonard *et al.*, 1985; Wilson, 1985). Emotional content is a common element of these stimuli and may be important for eliciting neuronal activity in the amygdala, whereas neuronal activity related to movement has been observed in none of these studies. Conversely, hippocampal unit activity is related to behavioral responses but not to the presentation of foods and faces (present results).

Because the hippocampus and amygdala share certain common inputs (e.g., the IMC), it seems likely that these structures operate on specific aspects of similar sets of data, providing different contributions to the analysis of visual inputs and the responses made to these inputs. The impairments in recognition memory tasks found after large medial temporal lesions could be caused by the loss of several processes undertaken by the hippocampus, amygdala, and IMC. The types of processes affected by hippocampal damage have been discussed above. Amygdala damage could impair several contributory processes. For example, some amygdala neurons demonstrate limited memory spans for previously presented stimuli, thereby providing direct evidence of whether a stimulus has been seen recently (Wilson, 1985). As a further example, other units in the amygdala respond to specific stimuli that the monkeys have learned are reinforcing. One population responds maximally to primary reinforcers such as foods; another population responds differentially to positively and negatively reinforcing stimuli used in a visual discrimination task (Wilson and Rolls, 1985). When the DNMS task does not employ symmetrical reward of target and distractor items, the acquired reinforcement value of the target item could be used to solve the task if use of the obvious familiarity difference was impossible. Indeed, Spiegler and Mishkin (1981) have found that amygdalectomy impairs single-trial learning of object–reward associations. Finally, it is also possible that amygdala lesions contribute to recognition impairments by interrupting fibers of passage from the IMC.

4.5. Overview

The present experiments have provided evidence that neuronal activity in the IMC reflects memory for the previous occurrence of visual stimuli. This finding suggests that this cortex plays a role in recognition memory, in accord with lesion studies.

The IMC is anatomically connected to many cortical and subcortical structures also implicated in memory function. The pathways from the neocortex of the inferior temporal lobe enable visual information to be processed in parallel, though differing, ways by a variety of other structures. It has been emphasized that recognition memory may be based on more than one type of information. Further, it is clear that performance of recognition memory tasks requires the integrity of several different processes. Accordingly, the IMC may not be the only neural structure crucial for judging the previous occurrence of stimuli, and the need for such information to modify behavior must require the involvement of other areas. Thus, it is not surprising to find many brain regions involved in the performance of recognition memory tasks. These diverse structures probably have unique but additive functions. This suggestion is consistent with the various types of memory impairment found after lesions in different brain regions and is illustrated by the different types of neuronal activity in the hippocampus and amygdala.

The existence of such parallel processes provides an explanation for the functions remaining in amnesic patients after damage to restricted parts of the temporal regions. Correspondingly, the finding that combined lesions of limbic structures in man and monkey produce more severe deficits in memory tasks than do restricted lesions may be explained as a reduction in the number of processes available to solve a task.

ACKNOWLEDGMENTS. The contributions of C. M. de la Mahotiere, J. Leendertz, R. C. Chambers, and A. M. Somerset are gratefully acknowledged. F.A.W.W. is grateful for facilities provided by Dr. E. T. Rolls for part of this study. This research was supported by the Medical Research Council, U.K.

REFERENCES

Adey, W. R., Dunlop, C. W., and Hendrix, C. E., 1960, Hippocampal slow waves: Distribution and phase relationships in the course of approach learning, *Arch. Neurol.* **3**:74–90.

Aggleton, J. P., and Mishkin, M., 1983, Memory impairments following restricted medial thalamic lesions in monkeys, *Exp. Brain Res.* **52**:199–209.

Aggleton, J. P., and Passingham, R. E., 1981, Stereotaxic surgery under x-ray guidance in the rhesus monkey, with special reference to the amgydala, *Exp. Brain Res.* **44**:271–276.

Aggleton, J. P., Burton, M. J., and Passingham, R. E., 1980, Cortical and subcortical afferents to the amygdala of the rhesus monkey (*Macaca mulatta*), *Brain Res.* **190**:347–368.

Aggleton, J. P., Desimone, R., and Mishkin, M., 1986, The origin, course and termination of the hippocampothalamic projections in the macaque, *J. Comp. Neurol.* **243**:409–421.

Aigner, T., Mitchell, S., Aggleton, J., DeLong, M., Struble, R., Wenk G., Price, D., and Mishkin, M., 1984, Recognition deficit in monkeys following neurotoxic lesions of the basal forebrain, *Soc. Neurosci. Abstr.* **10:**116.11.

Arnolds, D. E. A. T., Lopes Da Silva, F. H., Aitink, J. W., Kamp, A., and Boeijinga, P., 1980, The spectral properties of hippocampal EEG related to behaviour in man, *EET Clin. Neurophysiol.* **50:**324–328.

Baylis, G. C., and Rolls, E. T., 1983, Responses of neurons in the inferior temporal visual cortex in short and long term memory tasks, *Soc. Neurosci. Abstr.* **9:**12.2.

Biber, C., Butters, N., Rosen, J., Gerstman, L., and Mattis, S., 1981, Encoding strategies and recognition of faces by alcoholic Korsakoff and other brain-damaged patients, *J. Clin. Neuropsychol.* **3:**315–350.

Brown, M. W., 1982, Effect of context on the response of single units recorded from the hippocampal region of behaviorally trained monkeys, in: *Neuronal Plasticity and Memory Formation* (C. Ajmone Marsan and H. Matthies, eds.), Raven Press, New York, pp. 557–573.

Brown, M. W., and Leendertz, J. A., 1979, A pulse-shape discriminator for action potential, *J. Physiol. (Lond.)* **298:**17–18P.

Brun, A., and Englund, E., 1981, Regional pattern of degeneration in Alzheimer's disease: Neuronal loss and histopathological grading, *Histopathology* **5:**549–564.

Buerger, A. A., Gross, C. G., and Rocha-Miranda, C. E., 1974, Effects of ventral putamen lesions on discrimination learning by monkeys, *J. Comp. Physiol. Psychol.* **86:**440–446.

Caan, W., Perrett, D. I., and Rolls, E. T., 1984, Responses of striatal neurons in the behaving monkey. 2. Visual processing in the caudal neostriatum, *Brain Res.* **290:**53–65.

Cermak, L. S., Talbot, N., Chandler, K., and Wolbarst, L. R., 1985, The perceptual priming phenomenon in amnesia, *Neuropsychologia* **23:**615–622.

Damasio, A. R., Graff-Radford, N. R., Eslinger, P. J., Damasio, H., and Kassel, N., 1985, Amnesia following basal forebrain lesions, *Arch. Neurol.* **42:**263–271.

DeJong, R. N., Itabashi, H. H., and Olson, J. R., 1969, Memory loss due to hippocampal lesions, *Arch. Neurol.* **20:**339–348.

Desimone, R., and Gross, C. G., 1979, Visual areas in the temporal cortex of the macaque, *Brain Res.* **178:**363–380.

Divac, I., Rosvold, H. E., and Szwarcbart, M. K, 1967, Behavioural effects of selective ablation of the caudate nucleus, *J. Comp. Physiol. Psychol.* **63:**184–190.

Flicker, C., Ferris, S. H., Crook, T., Bartus, R. T., and Reisberg, B., 1985, Cognitive function in normal aging and early dementia, in: *Senile Dementia of the Alzheimer Type* (J. Traber and W. H. Gispen, eds.), Springer-Verlag, Berlin, pp. 2–37.

Friedman, H. M., and Allen, N., 1969, Chronic effects of complete limbic lobe distruction in man, *Neurology (Minneap.)* **19:**679–689.

Gaffan, D., 1974, Recognition impaired and association intact in the memory of monkeys after transection of the fornix, *J. Comp. Physiol. Psychol.* **86:**1100–1109.

Gaffan, D., 1977, Monkeys' recognition memory for complex pictures and the effect of fornix transection, *Q. J. Exp. Psychol.* **29:**505–514.

Gaffan, D., 1985, Hippocampus: Memory, habit and voluntary movement, *Phil. Trans. R. Soc. Lond. [Biol.]* **308:**87–99.

Gaffan, D., Saunders, R. C., Gaffan, E. A., Harrison, S., Shields, C., and Owen, M. J., 1984, Effects of fornix transection upon associative memory in monkeys: Role of the hippocampus in learned action, *Q. J. Exp. Psychol.* **36B:**173–221.

Gascon, G. G., and Gilles, F., 1973, Limbic dementia, *J. Neurol. Neurosurg. Psychiatry* **36:**421–430.

Gross, C. G., 1972, Visual functions of inferotemporal cortex, in: *Handbook of Sensory Physiology,* Vol. VIII/3B (R. Jung, ed.), Springer-Verlag, Berlin, pp. 451–482.

Halgren, E., Babb, T. L., and Crandall, P. H., 1978, Activity of human hippocampal formation and amygdala neurons during memory testing, *Electroencephalogr. Clin. Neurophysiol.* **45:**585–601.

Horel, J. A., 1978, The neuroanatomy of amnesia. A critique of the hippocampal memory hypothesis, *Brain* **101:**403–445.

Horel, J. A., and Pytko, D. E., 1982, Behavioural effect of local cooling in temporal lobe of monkeys, *J. Neurophysiol.* **47:**11–22.

Huppert, F. A., and Piercy, M., 1978, The role of trace strength in recency and frequency judgements by amnesic and control subjects, *Q. J. Exp. Psychol.* **30:**347–354.

Jacobs, B. L., and McGinty, D. J., 1972, Participation of the amygdala in complex stimulus recognition and behavioural inhibition: Evidence from unit studies, *Brain Res.* **36:**431–436.

Jacoby, L. L., and Witherspoon, D., 1982, Remembering without awareness, *Can. J. Psychol.* **36:**300–324.

Jones, B., and Mishkin, M., 1972, Limbic lesions and the problem of stimulus–reinforcement associations, *Exp. Neurol.* **36**:362–377.
Jones, E. G., and Powell, T. P. S., 1970, An anatomical study of the converging sensory pathways within the cerebral cortex of the monkey, *Brain* **93**:793–820.
Leichnetz, G. R., and Astruc, J., 1976, The efferent projections of the medial prefrontal cortex in the squirrel monkey (*Saimiri Sciureus*), *Brain Res.* **109**:455–472.
Leonard, C. M., Rolls, E. T., Wilson, F. A. W., and Baylis, G. C., 1985, Neurons in the amygdala of the monkey with responses selective for faces, *Behav. Brain Res.* **15**:159–176.
Mahut, H., 1972, A selective spatial deficit in monkeys after transection of the fornix, *Neuropsychologia* **10**:65–74.
Mahut, H., Zola Morgan, S., and Moss, M., 1982, Hippocampal resections impair associative learning and recognition memory in the monkey, *J. Neurosci.* **2**:1214–1229.
McLardy, M., 1970, Memory function in hippocampal gyri but not in hippocampi, *Int. J. Neurol.* **1**:113-118.
Mesulam, M.-M., and Mufson, E. J., 1982, Insula of the Old World monkey. 1: Architectonics in the insulo-orbito-temporal component of the paralimbic brain, *J. Comp. Neurol.* **212**:1–22.
Mesulam, M.-M., and Mufson, E. J., 1984, Neural inputs into the nucleus basalis of the substantia innominata (Ch4) in the rhesus monkey, *Brain* **107**:257–274.
Mesulam, M.-M., Rosen, A. D., and Mufson, E. J., 1984, Regional variations in cortical cholinergic innervation: Chemoarchitectonics of acetylcholinesterase-containing fibres in the macaque brain, *Brain Res.* **311**:245–258.
Milner, B., 1970, Memory and the medial temporal regions of the brain, in: *Biology of Memory* (K. H. Pribram and D. E. Broadbent, eds.), Academic Press, New York, pp. 29–50.
Mishkin, M., 1978, Memory in monkeys severely impaired by combined but not separate removal of amygdala and hippocampus, *Nature* **273**:297–299.
Mishkin, M., 1982, A memory system in the monkey, *Phil. Trans. R. Soc. Lond. [Biol.]* **298**:85–95.
Mishkin, M., and Bachevalier, J., 1983, Object recognition impaired by ventromedial but not dorsolateral prefrontal cortical lesions in monkeys, *Soc. Neurosci. Abstr.* **9**:12.13.
Mishkin, M., and Delacour, J., 1975, An analysis of short-term visual memory in the monkey, *J. Exp. Psychol. Anim. Behav. Proc.* **1**:326–334.
Murray, E. A., and Mishkin, M., 1984, Severe tactual as well as visual memory deficits follow combined removal of the amygdala and hippocampus in monkeys, *J. Neurosci.* **4**:2565–2580.
Murray, E. A., Bachevalier, J., and Mishkin, M., 1985, Rhinal cortex: A third temporal-lobe component of the limbic memory system, *Soc. Neurosci. Abstr.* **11**:140.13.
O'Keefe, J., and Bouma, H., 1969, Complex sensory properties of certain amygdala units in the freely moving cat, *Exp. Neurol.* **23**:384–398.
O'Keefe, J., and Nadel, L., 1978, *The Hippocampus as a Cognitive Map,* Clarendon Press, Oxford.
Owen, M. J., and Butler, S. R., 1981, Amnesia after transection of the fornix in moneys: Long-term memory impaired, short-term memory intact, *Behav. Brain Res.* **3**:115–123.
Parkinson, J. K., and Mishkin, M., 1982, A selective mnemonic role for the hippocampus in monkeys: Memory for the location of objects, *Soc. Neurosci. Abstr.* **8**:11.7.
Pearson, R. C. A., Esiri, M. M., Hiorns, R. W., Wilcocks, G. K., and Powell, T. P. S., 1985, Anatomical correlates of the distribution of the pathological changes in the neocortex in Alzheimer disease, *Proc. Natl. Acad. Sci U.S.A.* **82**:4531–4534.
Pribram, K. H., Wilson, W. A., Jr., and Connors, J., 1962, Effects of lesions of the medial forebrain on alternation behaviour of rhesus monkeys, *Exp. Neurol.* **6**:36–47.
Ranck, J. B., Jr., 1973, Studies on single neurones in dorsal hippocampal formation and septum in unrestrained rats. Part 1. Behavioural correlates and firing repertoires, *Exp. Neurol.* **414**:461–531.
Ringo, J. L., and Doty, R. W., 1985, A macaque remembers pictures briefly viewed six months earlier, *Behav. Brain Res.* **18**:289–294.
Rolls, E. T., Perrett, D. I., Caan, A. W., and Wilson, F. A. W., 1982, Neuronal responses related to visual recognition, *Brain* **105**:611–646.
Russchen, F. T., Amaral, D. G., and Price, J. L., 1985, The afferent connections of the substantia innominata in the monkey, *Macaca fasicularis, J. Comp. Neurol.* **242**:1–27.
Sawa, M., and Delgado, J. M. R., 1963, Amygdala unitary activity in the unrestrained cat, *Electroencephalogr. Clin. Neurophysiol.* **15**:637–560.
Scoville, W. B., and Milner, B., 1957, Loss of recent memory after bilateral hippocampal lesions, *J. Neurol. Neurosurg. Psychiatry* **20**:11–22.
Seltzer, B., and Pandya, D. N., 1978, Afferent cortical connections and architectonics of the superior temporal sulcus and surrounding cortex in the rhesus monkey, *Brain Res.* **149**:1–24.

Sidman, M., Stoddard, L. T., and Mohr, J. P., 1968, Some additional quantitative observations of immediate memory in a patient with bilateral hippocampal lesions, *Neuropsychologia* **6:**245–254.

Spiegler, B. J., and Mishkin, M., 1981, Evidence for the sequential participation of inferior temporal cortex and amygdala in the acquisition of stimulus–reward associations, *Behav. Brain Res.* **3:**303–317.

Squire, L. R., and Moore, R. Y., 1979, Dorsal thalamic lesion in a noted case of chronic memory dysfuntion, *Ann. Neurol.* **6:**503–506.

Squire, L. R., 1987, *Memory and Death,* Oxford University Press, New York.

Steward, O., and Scoville, S. A., 1976, Cells of origin of entorhinal cortical afferents to the hippocampus and fascia dentata of the rat, *J. Comp. Neurol.* **169:**347–370.

Thompson, R. F., and Spencer, W. A., 1966, Habituation: A model phenomenon for the study of neuronal substrates of behaviour, *Psychol. Rev.* **173:**16–43.

Van Buren, J. M., and Borke, R. C., 1972, The mesial temporal substratum of memory: Anatomical studies in three individuals, *Brain* **95:**599–632.

Vanderwolf, C. H., Kramis, R., Gilespi, L. A., and Bland, B. H. 1975, Hippocampal rhythmic slow activity and neocortical low-voltage fast activity: Relations to behaviour, in: *The Hippocampus,* Vol. 2 (R. L. Isaacson and K. H. Pribram, eds., Plenum Press, New York, pp. 101–128.

Van Hoesen, G. W., and Pandya, D., 1975, Some connections of the entorhinal (area 28) and perirhinal cortices of the rhesus monkey. 1. Temporal lobe afferents, *Brain Res.* **95:**1–24.

Van Hoesen, G. W., Yeterian, E. H., and Lavizzo-Mourey, R., 1981, Widespread corticostriate projections from temporal cortex of the rhesus monkey, *J. Comp. Neurol.* **199:**205–219.

Victor, M., Angevine, J. B., Mancall, E. L., and Fisher, C. M., 1961, Memory loss with lesions of hippocampal formation, *Arch. Neurol.* **5:**244–263.

Wallesch, C. W., Kornhuber, H. H., Kollner, C., Haas, H. C., and Hufnagl, J. M., 1983, Language and cognitive deficits resulting from medial and dorsolateral frontal lobe lesions, *Arch. Psychiatr. Nervenkr.* **233:**279–296.

Warrington, E. K., 1974, Deficient recognition memory in organic amnesia, *Cortex* **10:**289–291.

Warrington, E. K., and Weiskrantz, L., 1974, The effect of prior learning on subsequent retention in amnesic patients, *Neuropsychologia* **12:**419–428.

Warrington, E. K., and Weiskrantz, L., 1979, Conditioning in amnesic patients, *Neuropsychologia* **17:**187–194.

Watanabe, T., and Niki, H., 1985, Hippocampal unit activity and delayed response in the monkey, *Brain Res.* **325:**241–254.

Weiskrantz, L., 1956, Behavioural changes associated with ablation of the amygdaloid complex in monkeys, *J. Comp. Physiol. Psychol.* **49:**381–391.

Weiskrantz, L., 1982, Comparative aspects of studies of amnesia, *Phil. Trans. R. Soc. Lond. [Biol.]* **298:**97–109.

Whitehouse, P. J., Price, A. W., Clark, A. W., Coyle, J. T., and DeLong, M. R., 1981, Alzheimer's disease: Evidence for a selective loss of cholinergic neurons in the nucleus basalis, *Ann. Neurol.* **10:**122–126.

Whitlock, D. G., and Nauta, W. J. H., 1956, Subcortical projections from the temporal neocortex in *Macaca mulatta, J. Comp. Neurol.* **106:**184–207.

Wilson, F. A. W., 1985, *Neuronal Activity Related to Novelty, Familiarity and Reinforcement,* D.Phil. Thesis, Oxford University.

Wilson, F. A. W., and Rolls, E. T., 1985, Reinforcement-related unit activity in basal forebrain and amygdala, *Soc. Neurosci. Abstr.* **11:**160.1.

Wilson, F. A. W., Rolls, E. T., Yaxley, S., Thorpe, S. J., Williams, G. V., and Simpson, S. J., 1984, Responses of neurons in the basal forebrain of the behaving monkey, *Soc. Neurosci. Abstr.* **10:**37.8.

Wilson, F. A. W., Brown, M. W., and Riches, I. P., 1986, Response-related neuronal activity in the primate hippocampal formation, *Neurosci. Lett. Suppl.* **24:**S29.

Winocur, G., Oxbury, S., Roberts, R., Agnetti, A., and Davis, C., 1984, Amnesia in a patient with bilateral lesions to the thalamus, *Neuropsychologia* **22:**123–143.

Zola-Morgan, S., and Squire, L. R., 1984, Preserved learning in monkeys with medial temporal lesions: sparing of motor and cognitive skills, *J. Neurosci.* **4:**1072–1085.

Zola-Morgan, S., and Squire, L. R., 1985, Medial temporal lesions impair memory on a wide variety of tasks sensitive to human amnesia, *Behav. Neurosci.* **99:**22–34.

Zola-Morgan, S., Squire, L. R., and Amaral, D. G., 1984, Performance of monkeys with separate and combined lesions of the hippocampus and amygdala on delayed non-matching to sample, *Soc. Neurosci. Abstr.* **10:**116.7.

Zola-Morgan, S., Squire, L. R., and Amaral, D., 1986, Human amnesia and the medial temporal region: enduring memory impairment following a bilateral lesion limited to the CA1 field of the hippocampus, *J. Neurosci.* **6:**2950–2967.

31

Plasma Glucose Regulation of Memory Storage Processes

PAUL E. GOLD

1. INTRODUCTION

The findings of many studies performed during the past decade support the view that hormonal consequences of training exert important influences on memory storage (Gold and Zornetzer, 1983; McGaugh and Gold, 1986). These findings suggest that the neurobiological substrates of memory storage may be regulated in part by hormonal events that follow shortly after training. The now standard finding is that an animal trained with a weak footshock under conditions in which later retention performance is poor and in which hormonal responses are minimal will avoid as if it had been trained with a higher footshock if the appropriate hormones are injected immediately after training. In order to understand the normal roles of hormones in regulating memory, it is as important to assess the release of these hormones under different training conditions as it is to evaluate the pharmacological efficacy of hormonal treatments in modulating memory.

An important assumption underlying memory modulation research is that the hormones that act on memory do so by controlling the neurobiological processes that are the substrates of memory storage. There are many candidate mechanisms, several represented in this volume, for memory substrates; these mechanisms include anatomic, physiological, and biochemical changes that long outlast the training experience. We do not, of course, yet know which are necessary neurobiological correlates of memory, but the catalogue of correlates has expanded rapidly in recent years. In conjunction with completing this catalogue, it is important to begin also to evaluate the neurobiological processes that control the substrate mechanisms. This can be accomplished both by examining interactions between hormones and memory substrates (often using models of memory like long-term potentiation) and by using behavioral assessments of memory (probably engaging many memory substrate processes in parallel) to determine which hormones control memory storage and the neuroendocrine mechanisms by which they do so.

The strategy described here is analogous to that employed in embryology. The original elegant descriptions of the lineage and migration of cells have been greatly enriched by studies of the processes that regulate the timing and precise nature of cellular

PAUL E. GOLD • Department of Psychology, University of Virginia, Charlottesville, Virginia 22903.

differentiation. In a similar way, examinations of the neuroendocrine processes that regulate memory storage may eventually result in a more complete understanding of not only what events store the information but how these events are regulated, i.e., the factors that control the timing (when the substrate mechanisms are turned on and off) and the nature (which mechanisms are engaged) of the memory substrates. One implication of research on hormonal regulation of memory storage is that the biology of learning and memory requires attention not only to relevant brain mechanisms but also to peripheral hormonal events. Thus, the neurobiology of memory must eventually integrate peripheral regulators with memory storage mechanisms, producing a full organismic picture of the process by which experience modifies brain and behavior.

Just as one can catalogue neurobiological correlates of memory, it is also possible to catalogue the hormones and neurotransmitters that regulate memory. Not too surprisingly, the list is long, including peripheral and central catecholamines, opioids, pituitary hormones, as well as most central neurotransmitters and modulators (cf. Gold and Zornetzer, 1983). We decided several years ago to focus attention on one hormone in particular, selecting epinephrine as the hormone for major investigation on the basis of several pragmatic concerns: there is a wealth of information about the physiological actions of the hormone, there is a wide repertoire of agonists and antagonists that can be employed, and there are several very sensitive assays to use in assessing release of the hormone. These are advantages not available for most other neuroendocrine systems. For example, attempts to measure release of central neurotransmitters are difficult and generally indirect. In contrast, by using chronic catheter procedures, it is possible to measure plasma epinephrine at various times after training and get a firm assessment of the amount of epinephrine released. It is important to emphasize that the focus on epinephrine's roles in memory storage is largely based on practical concerns, not theoretical ones. By studying one hormonal regulator of memory storage in detail, we hope to determine the mechanism by which it acts on memory as well as to provide a conceptual framework that will make it easier to examine other neuroendocrine systems in the future.

2. EPINEPHRINE REGULATION OF MEMORY STORAGE PROCESSES

Epinephrine injected immediately after one-trial inhibitory avoidance training with mild footshock modulates later retention performance (Gold and van Buskirk, 1975; Gold *et al.*, 1977; Sternberg *et al.*, 1986). The effects of epinephrine on memory are dose dependent in an inverted-U manner; moderate doses enhance memory, and high doses impair memory. The effects are also time dependent; injections delayed by an hour or more after training do not affect later retention performance. Recent findings indicate that posttraining epinephrine injections also enhance memory for visual discrimination (Introini-Collison and McGaugh, 1986) and appetitive training as well (Sternberg *et al.*, 1985). Thus, the findings obtained with epinephrine—retrograde enhancement and retrograde amnesia—mirror the major effects of most treatments that affect memory consolidation and suggest the possibility that epinephrine actions might underlie the effects of many treatments on memory. Evidence consistent with this view comes from a series of studies that demonstrate that pretreatment with peripheral, but not central, injections of many adrenergic antagonists attenuates the effects on memory of many amnestic and memory-enhancing treatments (cf. Gold *et al.*, 1982a; Sternberg *et al.*, 1982). For example, supraseizure electrical stimulation of frontal cortex, subseizure stimulation of the amygdala, pentylenetetrazol, and cycloheximide all produce retrograde amnesia that can readily be blocked with peripheral injections of phenoxybenzamine or propranolol. Epinephrine release may therefore be an event common to these varied treatments and may be an important contributor to the effects of these treatments on memory.

To determine whether the epinephrine injections mimic the endogenous responses to footshock training, we examined the plasma epinephrine response to training (Gold and McCarty, 1981; McCarty and Gold, 1981). Plasma samples were obtained before and at several intervals after training. The results, shown in Fig. 1, indicate that a relatively strong footshock produces a transient elevation in plasma epinephrine, which peaks immediately (within 60 sec) after training. In the immediate posttraining samples, epinephrine levels increased from a base line of 100–300 pg/ml to 1200–1500 pg/ml. Using a similar design, we examined the effects of epinephrine injections on plasma epinephrine levels and found that the optimal memory-enhancing dose (0.1 mg/kg) was the dose that also most closely approximated the plasma epinephrine levels of untreated animals after strong footshock. Therefore, the epinephrine injections increase plasma levels to a physiologically relevant high level comparable to that seen after the high-footshock training.

Thus, epinephrine injections enhance memory storage; the posttraining epinephrine levels are comparable after high footshock or low footshock followed by a memory-enhancing dose of epinephrine, and peripherally administered adrenergic antagonists block the effects on memory of several classes of treatments that modulate memory. Epinephrine also enhances memory in juvenile rats that would otherwise forget rapidly (infantile amnesia) (Gold *et al.*, 1982b), in aged rodents (see below), and in animals that have memory deficits after amygdala damage (Bennett *et al.*, 1985; Liang, *et al.*, 1985). Furthermore, epinephrine enables classical conditioning to occur in animals that are deeply anesthetized (Fig. 2; Weinberger *et al.*, 1984; Gold *et al.*, 1986). Peripheral epinephrine injections also enhance the establishment of long-term potentiation of the perforant path–dentate gyrus system (Gold *et al.*, 1984).

3. MECHANISM OF EPINEPHRINE EFFECTS ON MEMORY: ROLE OF HYPERGLYCEMIA

Thus, there is substantial evidence that peripheral epinephrine regulates memory storage under a wide range of conditions. These results stand in apparent conflict with the findings that plasma epinephrine does not readily enter the central nervous system (Weil-Malherbe *et al.*, 1959). On the basis of these findings, it seems likely that a major portion of the mechanism by which epinephrine enhances memory must reside outside of the central nervous system; presumably such a mechanism would provide a "transduction" of the epinephrine effects to a process that acts directly on the central nervous system. We have therefore investigated several peripheral epinephrine actions that might mediate the hormone's effects on memory.

Recently, we began to examine the possibility that the hyperglycemic actions of epinephrine might contribute to the hormonal effects on memory storage. In the first experiment in this series (Gold, 1986), we found that glucose injections administered immediately, but not an hour, after training modulate memory storage. As shown in Fig. 3, the effects of glucose on memory are dose dependent in an inverted-U fashion. Thus, glucose has dose- and time-dependent effects on memory analogous to those observed with epinephrine. To determine further whether glucose might be an endogenous regulator of memory storage processes, we examined posttraining plasma glucose levels under

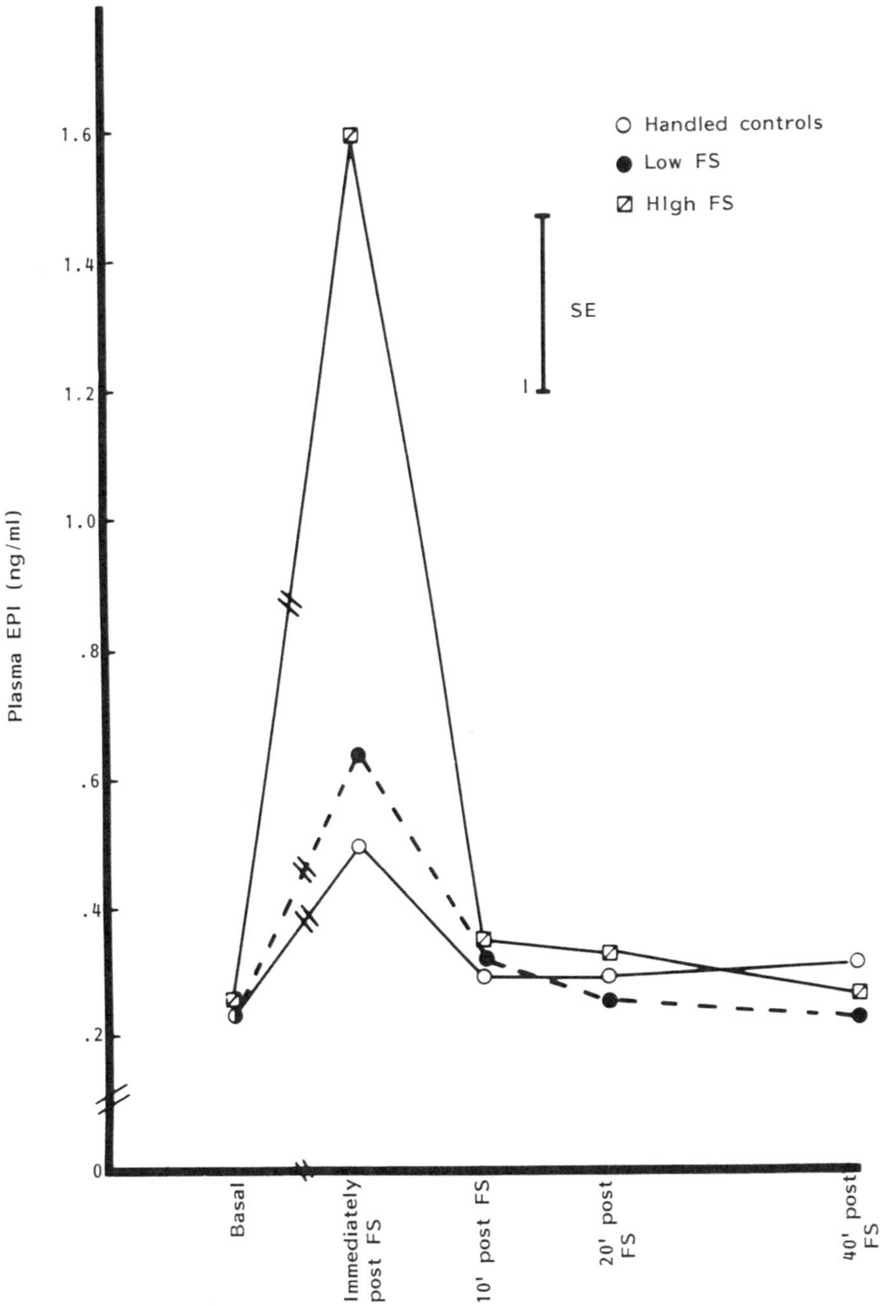

FIGURE 1. Plasma levels of epinephrine (EPI, ng/ml) under basal conditions and after handling or exposure to a training footshock. Values are means for six to seven animals per group, and the range of S.E.M.s is indicated. From McCarty and Gold, (1981), with permission.

FIGURE 2. Suppression ratio (duration of drinking during 1-min presentation of white noise divided by drinking during 1 min immediately preceding). White noise had been paired with cutaneous shock during ten classical conditioning trials presented while rats were deeply anesthetized. Rats that received peripheral administration of epinephrine exhibited conditioned fear on test trials 10 days after training; there was no evidence of conditioning under anesthesia in animals treated with saline. Additional controls (unpaired) received 0.01 mg/kg epinephrine prior to unpaired CS and US presentations or were neither anesthetized nor trained (untrained). From Weinberger *et al.* (1984) with permission.

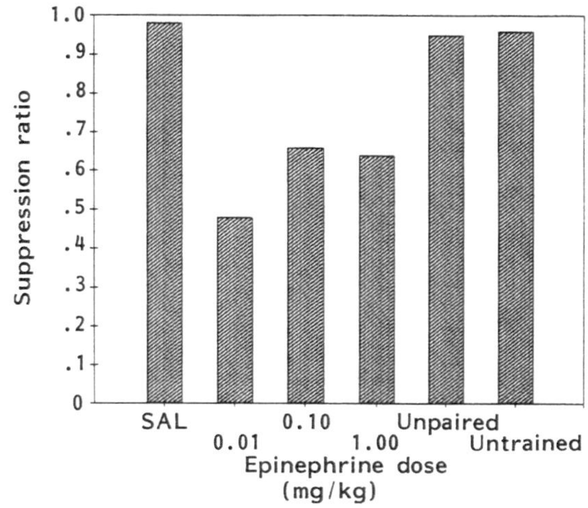

several conditions (Hall and Gold, 1986; Fig. 4). Like epinephrine, plasma glucose levels are sensitive to the specific training parameters; high but not low footshock results in a significant increase in plasma glucose levels, which is maximal (40% increase) 10 min after training. Moreover, memory-enhancing doses of epinephrine (0.1 mg/kg) and glucose (100 mg/kg) increase plasma glucose levels to an extent comparable to that observed after high-footshock training. Therefore, the posttraining glucose levels are comparably elevated under three conditions that result in good retention performance—high footshock, low footshock plus glucose, and low footshock plus epinephrine. Lower glucose levels (e.g., low footshock plus saline) predict relatively poor retention. Significantly higher glucose predicts poor retention, as seen after amnestic doses of glucose or epinephrine.

A study completed just recently examined the effects on memory of intracerebroventricular glucose injections (Lee *et al.*, 1987). Rats were trained on a one-trial inhibitory

FIGURE 3. Effects of a posttraining injection of glucose on 24-hr retention performance of inhibitory avoidance training. Note that glucose facilitated later retention at both the 10 and 100 mg/kg doses but did not affect retention performance at higher or lower doses. Also, glucose injected 1 hr after training had no effect on the test trial latencies. Thus, the effect of glucose on memory is an inverted-U function and is time dependent. From Gold (1987) with permission.

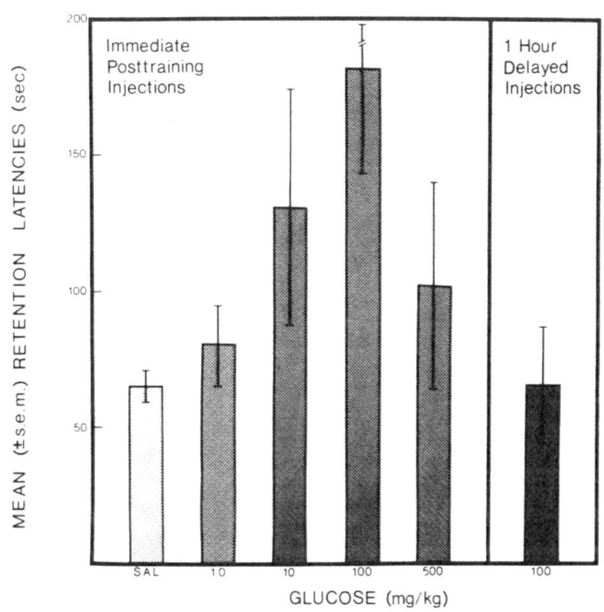

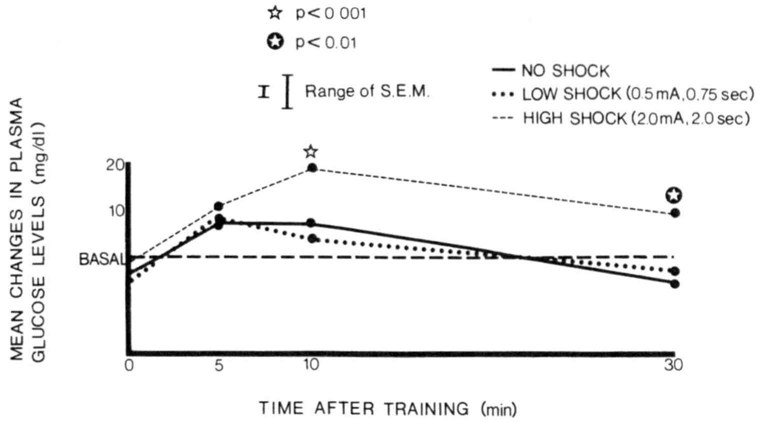

FIGURE 4. Mean plasma glucose difference scores (GLU, mg/dl) immediately and 5, 10, and 30 min after placement of rats into a test chamber and delivery of low (•, 0.5 mA, 0.75-sec duration) or a high (--, 2.0 mA, 2.0-sec duration) training footshock. Handled controls (—, SHAM) were placed in the chamber but did not receive footshock. Values are means, and the range of S.E.M.s is indicated. Note that high footshock elevated plasma glucose levels significantly above those of low footshock or handled groups. From Hall and Gold (1986) with permission.

avoidance task and received an immediate injection of artificial cerebrospinal fluid (1 μl) that contained 0–10 μg glucose. As shown in Fig. 5, the glucose injections enhanced memory in an inverted-U dose–response manner; delayed injections of the optimal dose (10 μg) had no effect on later retention performance. These findings are consistent with the possibility that plasma glucose may act on memory by directly affecting central nervous system functions; until additional evidence is obtained, however, it should be noted that the effects of peripheral and central glucose injections may have independent mechanisms of action by which they enhance memory.

As noted above, animals pretreated with peripherally administered adrenergic antagonists are resistant to the memory-modulating effects of a wide range of treatments. However, glucose (cf. Fig. 6) is an exception to this general rule. In this experiment (Gold *et al.*, 1986), glucose enhanced memory in animals pretreated with saline, phenoxybenzamine, or propranolol. These results are as predicted if glucose represents a step in memory modulation subsequent to the relevant peripheral adrenergic receptors.

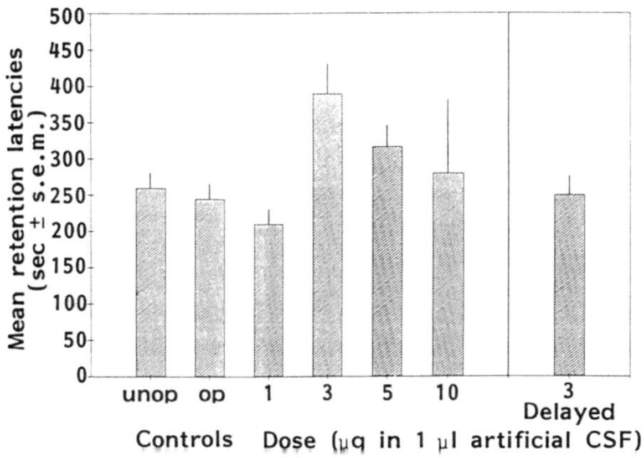

FIGURE 5. Effects on memory of posttraining intraventricular glucose injections. When injected immediately after training, 3 μg (in 1 μl) glucose injection significantly ($P \leq 0.02$) enhanced later retention performance in an inverted-U dose–response manner; a comparable injection administered 1 hr after training did not affect later retention. From Lee *et al.* (1986) with permission.

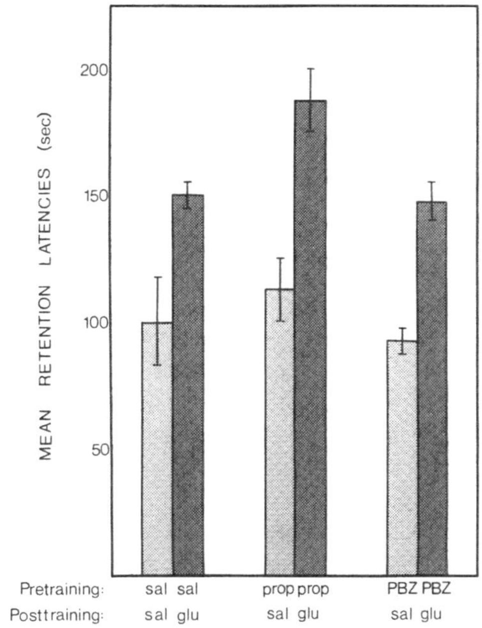

FIGURE 6. Retention performance of animals trained with low footshock in an inhibitory avoidance task after pretraining injections of saline, propranolol, or phenoxybenzamine and posttraining injections of saline or glucose. The adrenergic antagonists did not significantly affect retention performance, nor did the antagonists block the memory-enhancing effects of glucose. From Gold *et al.* (1986) with permission.

4. ADRENALECTOMY EFFECTS ON MEMORY AND PLASMA GLUCOSE: PARALLEL RECOVERY OF FUNCTION

The major stress-related increase in plasma glucose is accomplished by release of adrenomedullary epinephrine. An apparent conflict in this view of memory modulation is that adrenalectomy does not regularly impair learning and memory, and although there are some reports that adrenalectomy attenuates the effectiveness of memory-modulating treatments, these findings are also not always observed. A recent study by Borrell *et al.* (1983) provides a potential solution to this problem. In that study, rats trained within a few days of adrenalectomy had poor retention performance 24 hr later. In contrast, animals trained 1 week or more after surgery had retention performance comparable to that of intact controls. Thus, there appears to be some recovery of function within days of surgery, which allows memory to return to an impaired state. Adrenalectomized animals do not regain epinephrine, but the results obtained with glucose suggest that the animals might regain the capacity to regulate plasma glucose levels.

We recently obtained evidence consistent with this view (Hall *et al.*, 1987a). When plasma glucose levels are evaluated in rats 24 hr after adrenalectomy, basal values are significantly lower than those of control animals. Furthermore, plasma glucose levels in adrenalectomized rats do not exhibit the increase after a training footshock that is seen in intact animals. In contrast to these results, rats tested 1 week after adrenalectomy have basal levels that have recovered to values no longer statistically below those of controls. We do not yet know the mechanism by which the adrenalectomized animals regain glucose regulation. One possibility is that glucose comes under better neuronal control by the hypothalamus (Smythe *et al.*, 1984); alternatively, modifications in the control of glucagon and insulin might also accomplish glucose regulation in adrenalectomized animals. Whatever the mechanism, the results obtained with adrenalectomized rats fit very well with the general view that plasma glucose levels regulate memory storage. Soon after adrenalectomy, when basal glucose levels are low, memory is poor. A week later, when basal glucose levels have returned close to normal values, memory is no longer impaired.

5. AGING AND MEMORY: ENHANCEMENT WITH EPINEPHRINE AND GLUCOSE

The considerable increase in interest of late in cognitive changes that accompany aging has been accompanied by a wide range of neurobiological studies aimed at identifying factors that may underlie the behavioral changes. Among the list of neurolobiological changes that accompany aging in rodents, there is evidence that adrenomedullary release of epinephrine is reduced in response to footshock (McCarty, 1981). Because of the evidence that epinephrine release after footshock regulates memory storage in juvenile and young adult rodents, we have examined the effects of epinephrine on memory storage in aged rats and mice. Aged rodents exhibit rapid forgetting after training in several tasks, including both avoidance and appetitive procedures (Gold *et al.*, 1981; Sternberg *et al.*, 1985). For example, after training in an inhibitory avoidance task, 70-day-old rats retain the learned response for up to 6 weeks, 1-year-old rats show some evidence of forgetting at 1 week and substantial forgetting at 3 weeks, and 2-year-old rats exhibit forgetting within a day that is extensive after a 1-week training–test interval.

Although the specific time course varies with the particulars of the task, the general principle is that forgetting rate varies directly with age. We therefore examined retention performance in aged rats and mice that received injections of epinephrine immediately after training. The results (Fig. 7) indicate that the hormone injection significantly enhances later retention performance; i.e., forgetting is slowed in aged rats and mice that received epinephrine injections. Thus, the peripheral release of epinephrine after training is impaired in aged animals, and the animals forget rapidly. If epinephrine is experimentally administered at the time that younger animals would release the hormone, later retention performance is augmented.

Because of the substantial evidence that forebrain cholinergic dysfunction accompanies aging and, in particular, Alzheimer's disease (e.g., Coyle *et al.*, 1983), cholinergic antagonists such as scopolamine have been used frequently to model age-related memory deficits. Recently, we found that posttraining epinephrine or glucose injections reverse the amnesia that would otherwise be seen in animals pretreated with scopolamine (Croul *et al.*, 1986). We are currently examining the effects of scopolamine on peripheral release of epinephrine and glucose to determine whether these actions might mediate the amnesia produced by the cholinergic antagonist. The behavioral results obtained thus far are consistent with those of Flood and Cherkin (1986), who recently demonstrated that several noncholinergic drugs can reverse scopolamine-induced amnesia, suggesting that the central cholinergic deficit may not be the relevant action of the drug in impairing memory. Because some of the memory-enhancing treatments used in the Flood and Cherkin study and in our own studies to reverse scopolamine amnesia can effectively enhance memory in aged animals as well, it may be that scopolamine is a useful model of age-related memory impairments but that the drug actions responsible for the memory deficits may not primarily involve central cholinergic actions.

Finally, because glucose, unlike epinephrine, can be administered safely to human subjects, we examined the effects of glucose ingestion on memory in college students and in aged subjects (Hall *et al.*, 1987b). The elderly subjects were healthy volunteers, averaging 67 years of age. The experiment was performed on three consecutive mornings

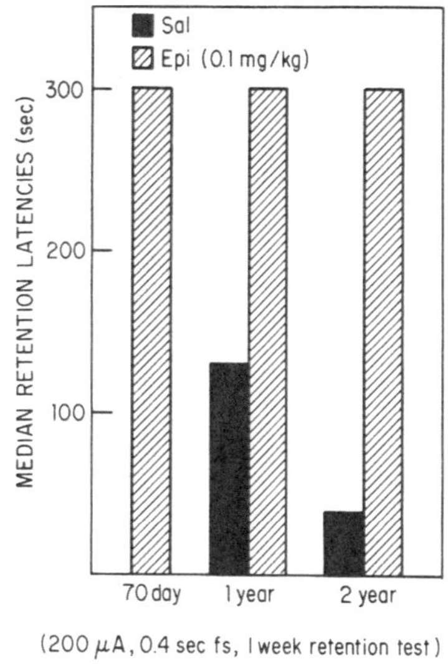

FIGURE 7. Retention performance of rats in a one-trial inhibitory avoidance task. Retention is expressed as latency to reenter the shock compartment 1 week following training. There is a significant decrease in retention performance in old rats. A post training injection of epinephrine enhanced retention performance of both 1- and 2-year-old rats. *N*s were 7–12 per group. From Sternberg *et al.* (1985) with permission.

using a blind, counterbalanced, crossover design in which each subject's performance under the glucose treatment could be compared to his or her performance under a saccharin control condition. The results, shown in Fig. 8, indicate that, compared to college students, the elderly subjects were impaired on several memory scores. Glucose ingestion significantly attenuated the impairment in elderly subjects, improving memory to levels approaching those of the college students. In particular, glucose significantly enhanced performance in the elderly subjects on the total memory score and on the logical memory individual test. The effects of glucose on memory in the young subjects were more restricted (only backward digit span). Additional tests are required of glucose efficacy in enhancing memory in young subjects.

In the course of this experiment, we evaluated blood glucose levels after the subjects ingested the glucose- or saccharin-flavored drink. The results indicated that the glucose treatment significantly elevated blood glucose levels as expected. However, these findings also revealed a most remarkable relationship between the glucose response, essentially a glucose tolerance test, and memory scores. The extent of the blood glucose response predicted the total memory score under glucose treatment (shown in Fig. 9). High blood glucose responses were related to poor memory performance. Of particular interest, high glucose changes (i.e., poor glucose control) predicted the memory scores under both glucose and saccharin conditions. Thus, the glucose tolerance test provided a means of segregating the population into those with good and poor memory scores. These results have the important implication that, in aged subjects, those individuals under relatively poor glucose control have impaired memory capacities.

We recently obtained an analogous correlation in 2-yr-old rats. The animals were trained and tested in a one-trial inhibitory avoidance task (as in Fig. 7). Blood glucose increases during the 2 hours after a glucose (500 mg/kg) injection were assessed 2–4 weeks after training and testing. The peak blood glucose increases were then related to the previous individual retention scores. The findings indicated that, as in human subjects, large increases in blood glucose levels (i.e., poor glucose regulation) were significantly correlated with low retention scores ($r = 0.9$).

In terms of the original research question, the animal models of aging, both 2-year-old animals and scopolamine-treated animals, were useful predictors of a treatment that

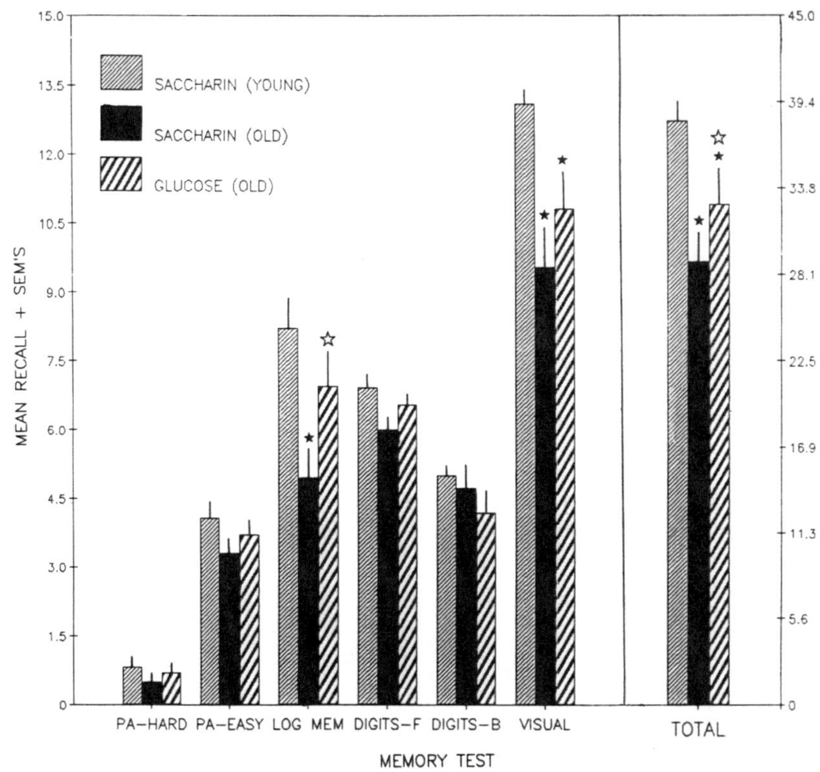

FIGURE 8. Effects of glucose ingestion on memory performance in elderly subjects. In a double-blind, counterbalanced, crossover design, subjects [college students and healthy volunteers (mean age 67 years)] ingested a fruit drink prepared with saccharin or glucose prior to learning and memory testing. Total memory scores and performance on the logical memory test were significantly enhanced under glucose versus saccharin treatment. Filled stars, $P < 0.05$ versus young; open stars, $P < 0.05$ versus elderly–saccharin. From Hall *et al.* (1986b) with permission.

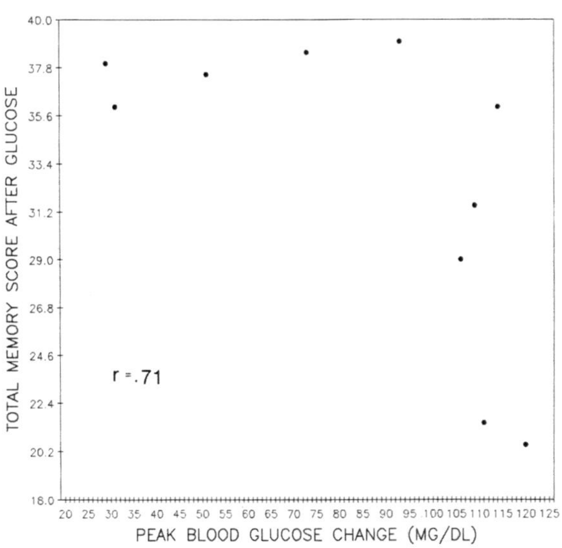

FIGURE 9. Peak blood glucose responses after glucose ingestion plotted against total memory scores obtained in elderly subjects on the glucose treatment day. Note that the subjects under poor blood glucose control (high levels) exhibited poor memory performance. The magnitude of the blood glucose responses to glucose ingestion similarly predicted the memory scores obtained during control (saccharin) sessions during which blood glucose levels were stable. This relationship was not observed in college students. From Hall *et al.* (1986b) with permission.

enhances memory in elderly humans. We do not yet know whether the effect of glucose on human memory is limited to aged populations. However, in conjunction with the results indicating that epinephrine and glucose have important roles in regulating memory storage in animals, the findings with human subjects support the broad task and species generality of epinephrine–glucose effects on memory.

6. CONCLUSIONS

The findings described here suggest that epinephrine enhances memory under a very wide range of conditions. The findings suggest further that the hyperglycemia in response to epinephrine release or injection may represent an important step in memory modulation subsequent to epinephrine release. Because glucose enters the brain readily, it is possible that glucose acts directly on the central nervous system to regulate memory storage. This view is supported by the recent findings that both peripheral and central injections of glucose enhance memory. We can now turn to investigations of the mechanism by which glucose controls the neuronal substrates of memory. There are several possible mechanisms that should be explored. One possibility is that plasma glucose regulates central intermediary metabolism, a possibility that seems somewhat unlikely since plasma glucose levels have little effect on overall central metabolism within physiological ranges; it is less certain whether metabolism in specific brain regions may be sensitive to glucose within normal ranges of plasma values. A second possibility is that plasma glucose acts on central glucoreceptors, generally studied in the hypothalamus (Oomura *et al.*, 1974, 1982) although such receptors may exist elsewhere as well. Related to these findings, and of interest to issues of neuroplasticity, there is evidence that amphetamine binding in the hypothalamus is regulated in a dynamic manner by plasma glucose levels (Hauger *et al.*, 1986). Additional evidence that plasma glucose has effects on central nervous system function is that, at physiologically relevant levels, plasma glucose may exert regulatory influences on central acetylcholine synthesis (Gibson and Blass, 1976; Blass and Gibson, 1979); if related to the effects of glucose on memory, such a mechanism would quite nicely integrate the peripheral adrenergic influences on memory with central cholinergic mechanisms implicated in age-related memory deficits. The goal of this research program is now to identify the next link in a sequence of events that include epinephrine and subsequent glucose increase in plasma to regulate neuronal changes that underlie memory.

ACKNOWLEDGMENTS. Supported by grants from the National Institute of Mental Health (MH 31141), the Office of Naval Research (N00014-85-K0472), the American Diabetes Association, and the University of Virginia Biomedical Research Support Program (2-S07-RR07094-20).

REFERENCES

Bennett, C., Liang, K. C., and McGaugh, J. L., 1985, Depletion of adrenal catecholamines alters the amnestic effect of amygdala stimulation, *Behav. Brain Res.* **15:**83–91.

Blass, J. P., and Gibson, G. E., 1979, Carbohydrates and acetylcholine synthesis: Implications for cognitive disorders, in: *Brain Acetylcholine and Neuropsychiatric Disease* (K. L. Davis and P. A. Berger, eds.), Plenum Press, New York, pp. 215–236.

Borrell, J., de Kloet, E. R., Versteeg, D. H. G., and Bohus, B., 1983, Inhibitory avoidance deficit following short-term adrenalectomy in the rat: The role of adrenal catecholamines, *Behav. Neural Biol.* **39:**241–258.

Coyle, J. T., Price, D. L., and DeLong, M. R., 1983, Alzheimer's disease: A disorder of cortical cholinergic innervation, *Science* **219:**1184–1190.

Croul, C., Stone, W. S., and Gold, P. E., 1986, Epinephrine and glucose attenuation of scopolamine-induced amnesia in mice, *Neurosci. Abstr.* **12:**709.

Flood, J. F., and Cherkin, A., 1986, Scopolamine effects on memory retention in mice: A model of dementia? *Behav. Neural Biol.* **45:**169–184.

Gibson, G. E., and Blass, J. P., 1976, Impaired synthesis of acetylcholine in brain accompanying mild hypoxia and hypoglycemia, *J. Neurochem.* **27:**37–42.

Gold, P. E., 1986, Glucose modulation of memory storage, *Behav. Neural Biol.* **45:**342–349.

Gold, P. E., and McCarty, R., 1981, Plasma catecholamines: Changes after footshock and seizure-producing frontal cortex stimulation, *Behav. Neural Biol.* **31:**247–260.

Gold, P. E., and van Buskirk, R. B., 1975, Facilitation of time-dependent memory processes with posttrial epinephrine injections, *Behav. Biol.* **13:**145–153.

Gold, P. E., and Zornetzer, S. F., 1983, The mnemon and its juices: Neuromodulation of memory processes, *Behav. Neural Biol.* **38:**151–189.

Gold, P. E., van Buskirk, R. B., and Haycock, J. W., 1977, Effects of posttraining epinephrine injections on retention of avoidance training in mice, *Behav. Biol.* **20:**197–204.

Gold, P. E., McGaugh, J. L., Hankins, L. L., Rose, R. P., and Vasquez, B. J., 1981, Age dependent changes in retention in rats, *Exp. Aging Res.* **8:**53–58.

Gold, P. E., McCarty, R., and Sternberg, D. B., 1982a, Peripheral catecholamines and memory modulation, in: *Neuronal Plasticity and Memory Formation* (C. Ajmone Marsan and H. Matthies, eds.), Raven Press, New York, pp. 327–338.

Gold, P. E., Murphy, J. M., and Cooley, S., 1982b, Neuroendocrine modulation of memory during development, *Behav. Neural Biol.* **35:**277–293.

Gold, P. E., Delanoy, R. L., and Merrin, J., 1984, Modulation of long-term potentiation by peripherally administered amphetamine and epinephrine, *Brain Res.* **305:**103–107.

Gold, P. E., Weinberger, N. M., and Sternberg, D. B., 1985, Epinephrine-induced learning under anesthesia: Retention performance at several training–testing intervals, *Behav. Neurosci.* **99:**1019–1022.

Gold, P. E., Vogt, J., and Hall, J. L., 1986, Posttraining glucose effects on memory: Behavioral and pharmacological characteristics, *Behav. Neural Biol.* **46:**145–155.

Hall, J. L., and Gold, P. E., 1986, The effects of training, epinephrine, and glucose injections on plasma glucose levels in rats, *Behav. Neural Biol.* **46:**156–167.

Hall, J. L., Croul, C., and Gold, P. E., 1987a, Recovery of plasma glucose regulation and memory after adrenalectomy, (in preparation).

Hall, J. L., Gonder-Frederick, L., Vogt, J., and Gold, P. E., 1987b, Memory enhancement in young adult and elderly humans: Effects of glucose ingestion, (in preparation).

Hauger, R., Hulihan-Giblin, B., Angel, I., Luu, M. D., Janowsky, A., Skolnick, P., and Paul, S. M., 1986, Glucose regulates [^{3}H](+)-amphetamine binding and Na^+K^+ ATPase activity in the hypothalamus: A proposed mechanism for the glucostatic control of feeding and satiety, *Brain Res. Bull.* **16:**281–288.

Introini-Collison, I. B., and McGaugh, J. L., 1986, Epinephrine modulates retention of an aversively-motivated discrimination task, *Behav. Neural Biol.* **45:**358–365.

Lee, M. K., Graham, S., and Gold, P. E., 1987, Memory enhancement with posttraining central glucose injections, *Behav. Neurosci.* (in press).

Liang, K. C., Bennett, C., and McGaugh, J. L., 1985, Peripheral epinephrine modulates the effects of post-training amygdala stimulation on memory, *Behav. Brain Res.* **15:**93–100.

McCarty, R., 1981, Aged rats: Diminished sympathetic–adrenal medullary response to acute stress, *Behav. Neural Biol.* **33:**204–212.

McCarty, R., and Gold, P. E., 1981, Plasma catecholamines: Effects of footshock level and hormonal modulators of memory storage, *Horm. Behav.* **15:**168–182.

McGaugh, J. L., and Gold, P. E., 1987, Hormonal modulation of memory, in: *Psychoendocrinology* (R. Brush and S. Levine, eds.), Academic Press, New York (in press).

Oomura, Y., Ooyama, H., Sujimori, M., Nakamura, T., and Yamada, Y., 1974, Glucose inhibition of the glucose-sensitive neurone on the rat lateral hypothalamus, *Nature* **247:**284–286.

Oomura, Y., Shimizu, N., Miyahara, S., and Hattori, K., 1982, Chemosensitive neurons in the hypothalamus: Do they relate to feeding behavior? in: *The Neural Basis of Feeding* (B. G. Hoebel and D. Novin, eds.), Haer Institute, Brunswick, Maine, pp. 551–566.

Smythe, G. A., Brunstein, H. S., Bradshaw, J. E., Nicholson, M. V., and Compton, P. J., 1984, Relationships between brain noradrenergic activity and blood glucose, *Nature* **308:**65–67.

Sternberg, D. B., Gold, P. E., and McGaugh, J. L., 1982, Noradrenergic sympathetic blockade: Lack of effect on memory or retrograde amnesia, *Eur. J. Pharmacol.* **81:**133–136.

Sternberg, D. B., Martinez, J., McGaugh, J. L., and Gold, P. E., 1985, Age-related memory deficits in rats and mice: Enhancement with peripheral injections of epinephrine, *Behav. Neural Biol.* **44:**213–220.

Sternberg, D. B., Korol, D., Novack, G. D., and McGaugh, J. L., 1986, Epinephrine-induced memory facilitation: Attenuation by adrenergic receptor antagonists, *Eur. J. Pharmacol.* **129:**189–193.

Weil-Malherbe, H., Axelrod, J., and Tomchick, R., 1959, Blood–brain barrier for adrenaline, *Science* **129:**1226–1227.

Weinberger, N. M., Gold, P.E., and Sternberg, D. B., 1984, Epinephrine enables Pavlovian fear conditioning under general anesthesia, *Science* **233:**605–607.

32

Behavioral Pharmacology of Memory

Opportunities for Cellular Explanations

ARTHUR CHERKIN and JAMES F. FLOOD

Dosis toxinum facit.

Paracelsus

When we try to pick out
anything by itself,
we find it hitched to
everything else in the universe.

John Muir

1. INTRODUCTION

Recent progress in understanding molecular and cellular mechanisms of conditioning make it timely to reevaluate interfaces between well-defined unitary cellular mechanisms and subtle, complex, multifactorial animal behaviors. The purpose of this chapter is to call attention to two robust phenomena observed during studies of the behavioral pharmacology of memory. We consider these phenomena to provide interesting possibilities for exploration at the cellular level. The first phenomenon is that pharmacological probes of memory in living organisms can have opposite effects. That is to say, a given drug in a given experimental paradigm can either enhance or impair memory, depending on dose. The second phenomenon is that several two-drug combinations exhibit powerful supraadditivity (as much as 20-fold) of memory-enhancing potency. The question is,

ARTHUR CHERKIN and JAMES F. FLOOD • Geriatric Research, Education and Clinical Center and Psychobiology Research Laboratory, Veterans Administration Medical Center, Sepulveda, California 91343; and Department of Psychiatry and Biobehavioral Sciences, University of California at Los Angeles School of Medicine, Los Angeles, California 90024.

can molecular and cellular approaches help to explain these two phenomena? The interfaces involve multiple interactions and functional linkages (Schmitt and Schneider, 1975). Atkinson (1975) has pointed out:

> Any discussion of biological processes must always involve oversimplification, because biomolecular reactions are always part of a larger whole. Biomolecular processes may be of some interest in their own right, but their biological significance relates to their effects on a larger system. . . .
>
> The black box approach, in which the intermediate mechanisms between a stimulus and its induced response are not considered, can lead, and has led, to much useful information in physiology and biophysics. It poses questions that a mechanistically oriented approach can then be used to resolve. However, this approach is limited and in a sense outmoded.
>
> In contrast, biochemists deal with small portions of large systems and frequently become so intrigued with the properties of these parts that they forget that they are, *in situ*, functioning elements within a complex interacting system. . . .
>
> These two basic approaches must be combined in order to obtain information that is biologically meaningful. Every aspect of an organism is designed by mutation and selection, but we must remember that while it is molecular detail that mutates, it is overall organismic function that is selected. Both the molecular and the black-box approaches must be used, but the area of real interest is in the linkage between them, the ways in which molecular detail is responsible for the overall response or for the living functioning organism.

The molecular structure of the acetylcholine receptor protein is known in great detail (Popot and Changeux, 1984). The cholinergic system is considered to play a major role in human memory and cognition (Davies, 1985). What are the linkages between these two facts? Can increased knowledge of either enrich our understanding of the other? Or, are intervening interactions so predominant that the cellular–behavioral linkage becomes obscured?

Altman (1985) expressed the challenge as follows:

> It is a serious problem for neurobiological research that the subject is studied on a number of different levels, ranging from the molecular constituents of membranes to whole animal behavior, and it is difficult to access one level from another. The lobster studies [of E. Kravitz and colleagues] are a good example of the power of neuropharmacology to provide a bridge between these levels. In order to understand how neuronal circuits work, it is now necessary to appreciate the way a chemical messenger can cause short- or long-term changes in receptors or channels, which in turn can influence a neuron's excitability, change its metabolism, modify transmitter release, or affect the synthesis of new protein, including receptors.
>
> We are only just beginning to understand these processes, but it seems that the ever-increasing list of neuroactive substances and the complexity of their actions may be the unifying factor so far lacking in neurobiology.

2. DUAL EFFECTS

2.1. On Memory *in Vivo*

Dual, opposite, dose-dependent effects of a single drug are well known in pharmacology. An old rule of thumb is that "low doses stimulate; high doses depress." Common experience teaches us that where two cocktails may stimulate an individual to animated conversation, six cocktails may induce stupor. It is plausible but incorrect to dismiss the depressant effects of high doses as simply reflecting toxic activity. As will be seen, we have observed dual effects in the absence of any symptom of "toxicity."

When a complete dose–response curve is developed for a drug that exhibits opposite

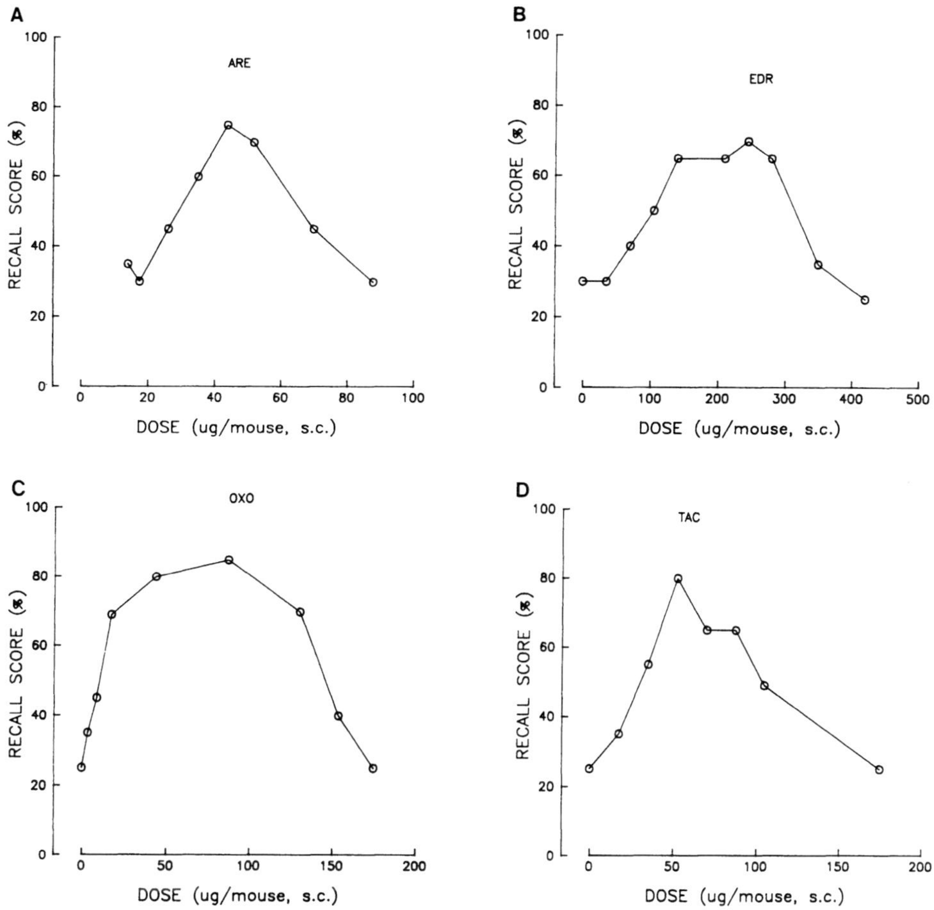

FIGURE 1. Inverted-U dose–response curves of four individual cholinergic drugs. (A) ARE, arecoline hydrobromide; (B) EDR, dedrophonium chloride; (C) OXO, oxotremorine sesquifumarate; (D) TAC, tetrahydroaminoacride hydrochloride sesquihydrate. Both ARE and OXO are muscarinic agonists; EDR and TAC are cholinesterase inhibitors.

effects, the curve has the shape of an inverted U (Fig. 1). Within the sensitive dose range, increasing doses induce increasing responses until a peak response is reached. Thereafter, increasing doses induce decreasing responses until the response falls to control levels.

To plot an inverted-U dose–response curve requires at least three data points within the sensitive range in order to detect (1) a small response to a low dose, (2) a peak response to an intermediate dose, and (3) a lower response to a higher dose. Published data do not always provide three such points. Logically, sufficiently low doses of a memory-active drug must fail to induce a measurable response. Even though doses lower than the peak dose have not been tested, we consider that an inverted U is indicated simply by a decline in response as the dose is increased beyond some peak response dose.

Davies (1985) considers that virtually all postsynaptically active cholinomimetic agents (muscarinic receptor agonists or cholinesterase inhibitors) will have an inverted-U dose–response curve. The even more ubiquitous nature of this phenomenon is indicated in Table I for 27 memory-enhancing compounds given by seven routes of administration

TABLE I. Dual Effect of Memory-Enhancing Compounds

Compound	Time of administration[b]	Species	Route	Dose[a] Peak (enhancing)	Dose[a] Higher (less enhancing)	Dose ratio (higher/peak)	Reference
Physostigmine	Post	Rat[c]	i.p.	0.50	0.75	1.5	Stratton and Petrinovich (1963)
	Post	Rat[d]	i.p.	0.75	1.00	1.3	Stratton and Petrinovich (1963)
	Post	Rat[e]	i.p.	0.03	0.06	2.0	Johns *et al.* (1985)
	Pre	Monkey	i.m.	0.01	0.04	4.0	Bartus (1979)
	Pre	Monkey	i.m.	0.032	0.056	2.0	Aigner and Mishkin (1986)
	Pre	Man	i.v.	0.014	0.028	2.0	Davis *et al.* (1978)
Cholinergics	Post	Mouse	i.c.v.				Flood *et al.* (1981)
Arecoline				0.1 μg/mouse	1.00 μg/mouse	10.0	
Choline				50.	200.	4.0	
Deanol				1.0	10.0	10.0	
DMPP				0.001	0.100	100.0	
Edrophonium				0.10	0.25	2.5	
Muscarine				0.05	0.10	2.0	
Oxotremorine				0.01	1.00	100.0	
Physostigmine				1.00	4.00	4.0	
Scopolamine	Pre	Mouse	s.c.	0.01	0.10	10.0	Flood and Cherkin (1986)
	Post	Mouse	i.p.	0.10	1.00	10.0	Flood and Cherkin (1986)
Etiracetam	Pre	Rat	i.p.	20.0	40.0	2.0	Wolthuis (1981)
Piracetam + choline	Post	Mouse	i.p.	50 + 50	100 + 100	2.0	Platel *et al.* (1984)
D-Amphetamine	Pre	Mouse	i.p.	0.63	1.25	2.0	Butler *et al.* (1981)
	Post	Mouse	i.p.	2.0	2.5	1.3	Krivanek and McGaugh (1969)
L-Amphetamine	Pre	Mouse	i.p.	5.0	10.0	2.0	Butler *et al.* (1981)
Norepinephrine	Post	Rat	i.c.	0.1	1.0	10.0	McGaugh *et al.* (1984)

Epinephrine	Post	Rat	s.c.	0.1	1.0	10.0	McGaugh *et al.* (1984)
Chlorpromazine	Pre	Mouse	i.p.	1.25	5.0	4.0	Butler *et al.* (1981)
Hydergine	Pre	Mouse	s.c.	2.5	3.0	1.2	Flood *et al.* (1985)
	Post		s.c.	15.0	20.0	1.3	Flood *et al.* (1985)
	Post		i.c.v.	15.0 μg	30.0 μgf	2.0	Flood *et al.* (1985)
Picrotoxin	Pre	Mouse	i.p.	1.25	5.0	4.0	Butler *et al.* (1981)
	Post	Mouse	i.p.	0.60	1.2	2.0	Butler *et al.* (1981)
Strychnine	Post	Mouse	i.p.	0.30	0.60	2.0	Bovet *et al.* (1966)
	Post	Mouse	i.p.	0.025	0.05	2.0	McGaugh and Krivanek (1970)
Pentylene tetrazol	Post	Mouse (Balb/c)	i.p.	5.0	10.0	2.0	Krivanek and McGaugh (1968)
	Post	Mouse (C57BL/6J)	i.p.	10.0	20.0	2.0	Krivanek and McGaugh (1968)
Ethosuximide	Pre	Mouse	i.p.	20.0	40.0	2.0	Butler *et al.* (1981)
Glucose	Post	Rat	s.c.	100.0	500.0	5.0	Gold (1986)
Glucose	Post	Rat	i.c.v.	3.0 μg	5.0 μg	1.7	Lee and Gold (1987)
$ACTH_{4-9}$ analogue	Post	Rat	s.c.	100.0 ng/rat	500.0 ng/rat	5.0	Fekete and De Wied (1982)
$ACTH_{4-10}$	Post	Rat	s.c.	100.0 μg/rat	1000.0 μg/rat	10.0	Fekete and De Wied (1982)
Naloxone	Post	Mouse	i.c.v.	5.0 μg	10.0 μg	2.0	Flood *et al.* (1985)
	Post	Chick	i.c.v.	16.0 μg	64.0 μg	4.0	Flood *et al.* (1985)
Naltrexone	Pre	Rat	i.p.	0.1	1.0	10.0	Bechara and van der Kooy (1985)
Flurothyl	Post	Chick	Inhaled	0.2% v/v	0.8% v/v	4.0	Cherkin *et al.* (1975)
	Post	Trout	In bath	25.0 μl/liter	200.0 μl/liter	8.0	Riege and Cherkin (1976)

[a] Doses are expressed in milligrams per kilogram unless otherwise specified. "Peak (enhancing)" doses gave maximal memory score. "Higher (less enhancing)" doses gave lower memory scores, often at control levels.
[b] Relative to time of training.
[c] Maze, bright rats; maze-learning paradigm.
[d] Maze, dull rats; maze-learning paradigm.
[e] Sham-operated rats; passive avoidance paradigm.

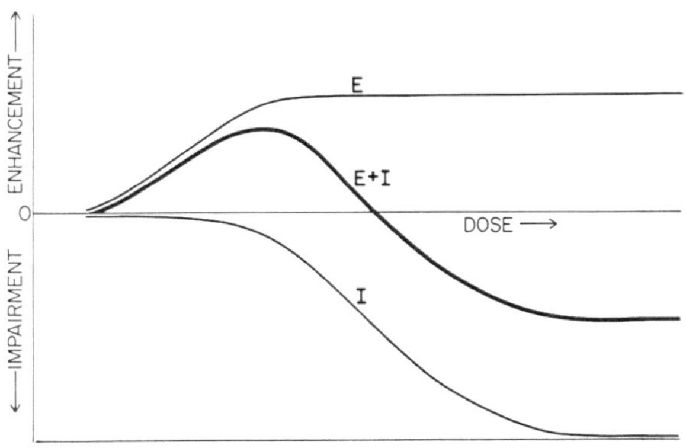

FIGURE 2. Hypothetical genesis of an inverted-U dose–response curve as the arithmetic sum of two ordinary dose–response curves.

to six organisms in 15 laboratories. The median ratio of higher dose to peak dose is 2.0. As more memory enhancers are tested over an adequate range of doses, the list of Table I will lengthen. For example, Butler *et al.* (1981) tested a series of 28 pyridine analogues for 1-week memory retention in mice. Of these, all nine compounds classed as active in enhancing retention gave inverted-U-shaped dose–response curves. In a later study of a series of 37 pyrrolidine acetamides, all 22 compounds that were active in reversing amnesia induced by electroconvulsive shock exhibited inverted-U-shaped dose–response curves (Butler *et al.*, 1984).

A simplistic conceptualization of how dual effects can produce inverted-U dose–response curves is illustrated in Fig. 2, adapted from Paalzow *et al.* (1985). The amnestic effect starts at a higher dose than the enhancing effect, and we assume that each separate effect reaches its own plateau. The separate plateaus cannot be observed behaviorally. Rather, the measured responses correspond to the summed responses of the two separate curves, resulting in the inverted-U dose–response curve actually observed. The problem with this conceptualization is that it explains nothing in molecular or cellular terms. We suggest, however, that it is not unreasonable to consider that a drug that modulates memory has at least two sets of influences at the cellular level. One set produces an enhancing effect. The other set produces an amnestic effect, starting at higher doses than the first set.

The descending limb of inverted-U curves has been attributed to nonspecific toxic effects and therefore considered to have little significance for mechanisms of action. We have established, however, that memory-active drugs can modulate memory up or down at doses far smaller than those causing behavioral toxicity. Also, certain drugs enhance memory despite concurrent behaviorally toxic effects. For example, oxotremorine improves memory retention scores in mice at doses that elicit tremors and other motor disturbances (Flood *et al.*, 1985). Furthermore, in the chick experiments listed in Table I, toxic and amnestic effects influence the measure of memory retention in opposite directions: toxic effects reduce pecking by chicks at a test target, whereas amnesia increases the peck rate.

Acceptance of the dual effect notion would require clear recognition that the "usual" pharmacological effect of a compound that modulates memory is probably not its exclusive functional effect. Indeed, an opposite functional effect can be revealed simply by increasing or decreasing the "usual" dose. Paracelsus was well aware of the critical importance of dose when he stated: "Dosis toxinum facit"; i.e., "The dose makes the poison."

2.2. Dual Effects *in Vitro*

Examples of inverted-U dose–response curves exist at the cellular level. The *in vitro* sodium-dependent high-affinity choline uptake into rat hippocampal synaptosomes after *in vivo* administration of memory modulators showed inverted-U dose responses for pramiracetam (Fig. 3) and 3,4-diaminopyridine (Shih and Pugsley, 1985). (We suggest that the five other memory modulators examined in that study might also have shown the inverted-U phenomenon had they been tested over a sufficient range of closely spaced doses.) Pramiracetam enhances cognitive functions in animals and man with the inverted-U dose–response curve in mice (Butler *et al.*, 1984) seen in Fig. 3.

A second *in vitro* example relates to regulation of rat myocardial adenylate cyclase activity by manganese ion. The dose–response curve of the activation is a typical inverted U, rising from an Mn^{2+} concentration of 10^{-5}M to a peak at about 3×10^{-4} M and falling toward base line at about 10^{-2}M (Steinberg *et al.*, 1986, Fig. 1b). The inverted-U dose effect was observed also with activation by gpp(NH)p. We are not implying a direct connection between adenylate cyclase activity and memory processing based on what may be a purely coincidental similarity of dose–response functions. We do suggest, however, that the U-shaped curve may be more general in biology than is now appreciated.

2.3. How Can Dual Effects Be Explained?

Several speculative interpretations of the dual effect of memory enhancers have been suggested, but none are satisfactory at the cellular level. They include:

1. Opposite responses to a given dose of a cholinergic drug at different times are related to variations in the CNS excitability of the test animals (Votava, 1967).
2. The dual effects of an analogue of $ACTH_{4-9}$ on retention of passive avoidance behavior (Fekete and De Wied, 1982) result from the induction at high doses of:
   a. Competitive action on ACTH-sensitive brain structures.
   b. Functional antagonistic action.

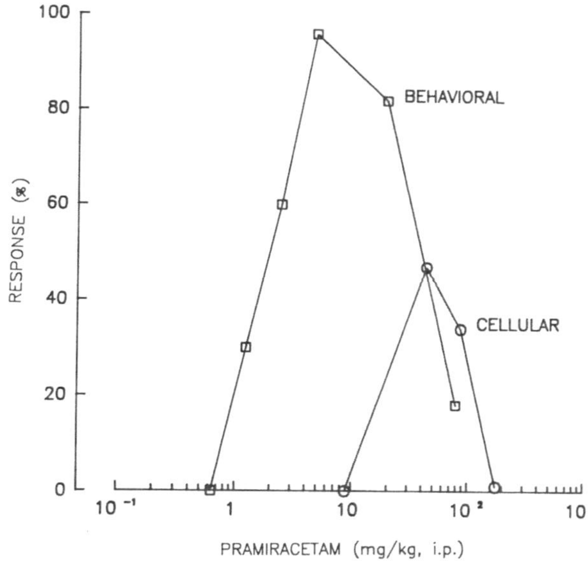

FIGURE 3. Inverted-U dose–response curves for pramiracetam at the cellular level (Shih and Pugsley, 1985) and the behavioral level (Butler *et al.*, 1984).

c. Interference with endogenous processes, preventing formation of endogenous neuropeptides related to ACTH.
d. Release of neuropeptide(s) with amnestic properties.

3. High doses of ACTH or peripheral catecholamines stimulate the release of memory-impairing β-endorphin in the brain (Izquierdo, 1984).
4. High doses of physostigmine (Davies, 1985) or muscarinic agonists (Wilson, 1986) cause deleterious nonspecific peripheral side effects.
5. Low doses of scopolamine bind preferentially to presynaptic receptors and enhance release of acetylcholine (and/or dopamine), enhancing memory processing. High doses bind also to postsynaptic receptors, reducing postsynaptic binding by acetylcholine and thus reducing cholinergic transmission (Flood and Cherkin, 1986).
6. Opiate compounds produce positive reinforcing effects in the brain but aversive effects in the periphery (Bechara and van der Kooy, 1985).
7. Impurities in drugs may induce increasingly toxic effects as higher doses are used (P. Scarpace, personal communication).

It may turn out that the behavioral paradigms employed in most pharmacological studies of mammalian memory are so complex and so heavily dependent on multiple interactions that the behavioral response is many removes from the underlying unitary cellular events. If so, elegant invertebrate models such as the sea snail, *Aplysia* (Kandel, 1985) or *Hermissenda* (Alkon *et al.*, 1985) may provide an opportunity for probing dual effects at the cellular level.

3. SUPRAADDITIVITY OF POTENCY

3.1. On Memory *in Vivo*

The second phenomenon of potential heuristic interest at the cellular level is the marked synergistic effect observed in appropriate two-drug combinations of memory-enhancing drugs. The paradigm we used (Flood *et al.*, 1983) involved three major steps. First, mice were trained to learn the safe arm of a T-maze by several trials during which entry into the opposite arm was punished by mild footshock. Second, the mice were injected with saline (control group) or with one of a series of graded doses of a drug or a drug combination within 2 min after training. Third, 1 week later, the mice were returned to the training apparatus and tested for retention of the initial training. All experiments were run blind to avoid experimenter bias. The experimental conditions (e.g., footshock intensity) were adjusted so that the saline control group had either a low memory-retention score or a high score. The former conditions were used for testing memory enhancement, and the latter for testing memory impairment.

3.1.1. Results: Intracerebroventricular Injections

Our initial experiments with arecoline hydrobromide (ARE), a cholinomimetic, edrophonium chloride (EDR: Tensilon®), an anticholinesterase, and their combination (Flood *et al.*, 1983) yielded a fortunate coincidence; namely, the entire dose–response curves for ARE and EDR proved to be identical (Fig. 4). Thus, a constant weight ratio of 1

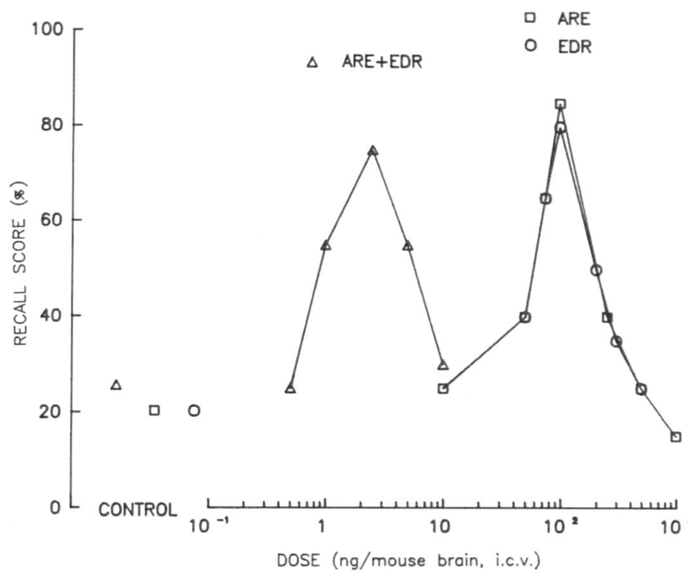

FIGURE 4. Dose–response curves of arecoline (ARE), of edrophonium (EDR), and of the combination (ARE + EDR). The equipotency of ARE and EDR is coincidental. Administration was by intraventricular injection within 1 min after training in a T-maze to avoid footshock applied 5 sec after sounding a warning buzzer. Retention of the trained avoidance was tested 1 week later. The ordinate represents the percentage of each group ($N = 20$) that met the criterion of memory retention. The abscissa indicates the dose of each drug; the total dose of the two equipotent drugs in combination is double this value. Note that the optimal total dose (2.5 + 2.5 = 5.0 ng) is 1/20 of the optimal dose (100 ng) of either drug alone. Note also the broadening of the inverted U for the combination.

ARE : 1 EDR could be utilized; i.e., 1 ng of ARE was equipotent with 1 ng of EDR at all doses. The ceiling response for memory retention (70–80% recall score) occurred at a dose of 100 ng of ARE or 100 ng of EDR. The same ceiling response for the two-drug combination occurred at 1/40 of these doses, i.e., at 2.5 ng of each drug (Fig. 4), which is a 95% reduction from the dose of either drug alone (Fig. 5).

The dose–response curves of ARE, EDR, OXO, and TAC (Fig. 1) proved to be

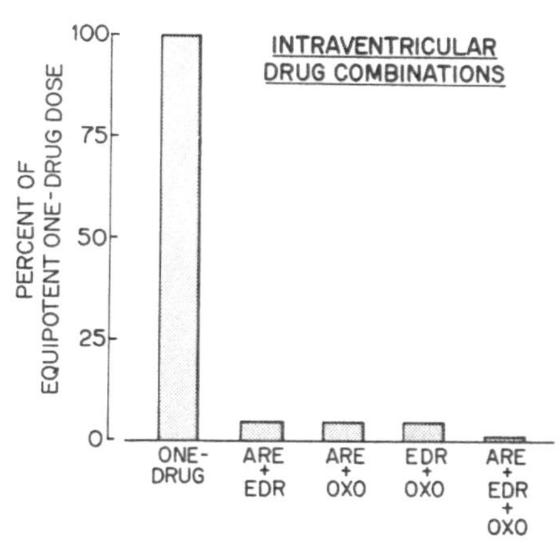

FIGURE 5. Reduced intraventricular doses of two-drug combinations required for peak response (75% or higher recall score). The doses in each optimal combination were converted to the equipotent dose of ARE (see text). The ordinate represents the percentage of this one-drug dose, normalized to 100%. The first bar refers to ARE, EDR, or OXO, each administered by itself.

typical of all the drugs subsequently tested in their inverted-U shape and also in the displacement of the drug-combination curves to much lower doses. The coincidental identity of the ARE and EDR dose–response curves was, however, unique to these two components. Therefore, for all other combinations, the dose of each drug was first converted to the equipotent dose of ARE required by mice to reach the maximal recall score. The reductions in the optimal intraventricular dose for maximal retention test performance (recall scores of 75–85%) are summarized in Fig. 5. The potentiation factor of the three-drug combination (ARE + OXO + EDR) was 67, compared with 20 for each of the three two-drug combinations (ARE + EDR; OXO + EDR; ARE + OXO). It is indeed remarkable that the interactions of the three-drug combination permitted the dose of each of its three components to be reduced to 1/200 of the equipotent dose of each component alone.

3.1.2. Results: Subcutaneous Injections

As shown in Fig. 6, the supraadditivity observed with intracerebral injection was also observed with subcutaneous injection, indicating the robustness of this phenomenon.

3.1.3. Potentiation of Efficacy

The marked potentiation described above relates to the increased potency of two or three drugs when used in combination. Another view relates to increased efficacy of a given dose of each drug in the combination, as was shown for choline plus piracetam in rats (Bartus *et al.*, 1981). We selected noneffective individual doses of ARE (5 ng) and OXO (0.5 ng), doses that have no greater effect than does saline. When we tested the three possible two-drug combinations with each drug at one-half of its noneffective dose, the retention test performance was raised to ceiling levels (75–85% recall). If the combined drugs had acted only additively, the recall scores would have been only 20–25%. Thus, the combinations showed marked supraadditivity of efficacy (Flood *et al.*, 1983).

3.2. Supraadditivity *in Vitro*

There are clear drawbacks to studying two-drug combinations at the cellular level, including the lack of a convincing rationale for two-drug studies, the difficulty of establishing the optimal ratio of the two drug doses or concentrations, the difficulty of defining

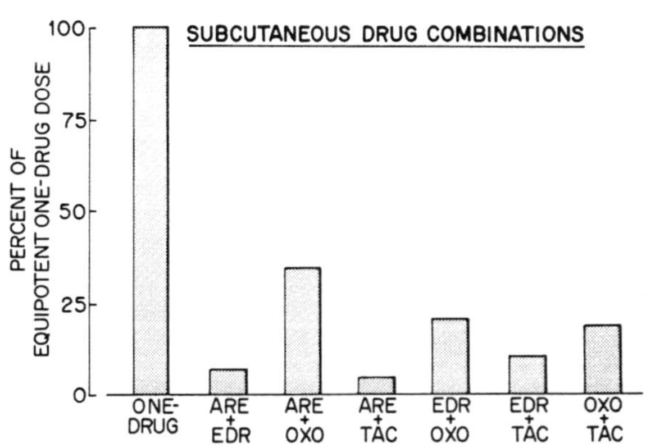

FIGURE 6. Reduced subcutaneous doses required for peak response (see legend to Fig. 5). The subcutaneous doses are much larger than the intraventricular doses (Fig. 5), but the potentiating effect of the drug combinations is similar.

the effective dose or concentration range, and, perhaps of greatest importance, the near impossibility of defining the precise mechanism of action of each individual drug. Since we have not carried out a systematic literature review, however, it is possible that such studies have escaped our attention. For example, Alkon *et al.* (1985) found that activation of two different kinases had a larger effect than either kinase alone in modulating potassium channels in *Hermissenda*.

4. CONCLUSION

We consider that two robust phenomena of memory processing observed at the behavioral level provide heuristic opportunities for explanation at the cellular level. The first phenomenon is the dual, opposite effects of memory modulators. The second is the marked supraadditivity ("synergism") of memory modulators when combined.

REFERENCES

Aigner, T. G., and Mishkin, M., 1986, The effects of physostigmine and scopolamine on recognition memory in monkeys, *Behav. Neural Biol.* **45:**81–87.

Alkon, D. L., Sakakibara, M., Forman, R., Harrigan, J., Lederhendler, I., and Farley, J., 1985, Reduction of two voltage-dependent K^+ currents mediates retention of a learned association, *Behav. Neural Biol.* **44:**278–300.

Altman, J., 1985, Tuning in to neurotransmitters, *Nature* **315:**537.

Atkinson, D. E., 1975, Allosteric interactions in enzyme systems, in: *Functional Linkage in Biomolecular Systems* (F. O. Schmitt and D. M. Schneider, eds.), Raven Press, New York, pp. 43–56.

Bartus, R. T., 1979, Physostigmine and recent memory: Effects in young and aged nonhuman primates, *Science* **206:**1087–1089.

Bartus, R. T., Dean, R. L., Sherman, K. A., Friedman, E., and Beer, B., 1981, Profound effects of combining choline and piracetam on memory enhancement and cholinergic function in aged rats, *Neurobiol. Aging* **2:**105–111.

Bechara, A., and van der Kooy, D., 1985, Opposite motivational effects of endogenous opioids in brain and periphery, *Nature* **314:**533–534.

Bovet, D., Bovet-Nitti, F., and Oliverio, A., 1966, Effects of nicotine on avoidance conditioning of inbred strains of mice, *Psychopharmacologia (Berl.)* **10:**1–5.

Butler, D. E., Poschel, B. P. H., and Marriott, J. G., 1981, Cognition-activating properties of 3-(aryloxy) pyridines, *J. Med. Chem.* **24:**346–350.

Butler, D. E., Nordin, I. C., L'Italien. Y. J., Zweisler, L., Poschel, P. H., and Marriott, J. G., 1984, Amnesia-reversal activity of a series of N-[(disubstituted-amino) alkyl]-2-oxo-1-pyrrolidineacetamides, including pramiracetam, *J. Med. Chem.* **27:**684–691.

Cherkin, A., Meinecke, R. O., and Garman, M. W., 1975, Retrograde enhancement of memory by mild flurothyl treatment in the chick, *Physiol. Behav.* **14:**151–158.

Davies, P., 1985, Is it possible to design rational treatments for the symptoms of Alzheimer's disease? *Drug Dev. Res.* **5:**69–76.

Davis, K. L., Mohs, R. C., Tinklenberg, J.R., Pfefferbaum, J. R., Hollister, L. E., and Kapell, B. S., 1978, Physostigmine: Improvement of long-term memory processes in normal humans, *Science* **210:**272–274.

Fekete, M., and De Wied, D., 1982, Dose-related facilitation and inhibition of passive avoidance behavior by the ACTH 4–9 analog (ORG 2766), *Pharmacol. Biochem. Behav.* **17:**177–182.

Flood, J. F., and Cherkin, A., 1986, Scopolamine effects on memory retention in mice: A model of dementia? *Behav. Neural Biol.* **45:**169–184.

Flood, J. F., Landry, D. W., and Jarvik, M. E., 1981, Cholinergic receptor interactions and their effects on long-term memory processing, *Brain Res.* **215:**177–185.

Flood, J. F., Smith, G. E., and Cherkin, A., 1983, Memory retention: Potentiation of cholinergic drug combinations in mice, *Neurobiol. Aging* **4:**37–43.

Flood, J. F., Smith, G. E., and Cherkin, A., 1985, Memory enhancement: Supraadditive effect of subcutaneous cholinergic drug combinations in mice, *Psychopharmacology* **86**:61–67.

Gold, P. E., 1986, Glucose modulation of memory storage, *Behav. Neural Biol.* **45**:342–349.

Izquierdo, I., 1984, Endogenous state dependency: Memory depends on the relation between the neurohumoral and hormonal states present after training and at the time of testing, in: *Neurobiology of Learning and Memory* (G. Lynch, J. L. McGaugh, and N. M. Weinberger, eds.), Guilford Press, New York, pp. 333–350.

Johns, C. A., Haroutunian, V., Greenwall, B. S., Mohs, R. G., Davis, B. M., Kanof, P., Horvath, T. B., and Davis, K. L., 1985, Development of cholinergic drugs for the treatment of Alzheimer's disease, *Drug Dev. Res.* **5**:77–96.

Kandel, E. R., 1985, Cellular mechanisms of learning and the biological basis of individuality, in: *Principles of Neural Science*, 2nd ed. (E. R. Kandel and J. H. Schwartz, ed.), Elsevier, New York, pp. 816–833.

Krivanek, J. A., and McGaugh, J. L., 1968, Effects of pentylenetetrazol on memory storage in mice, *Psychopharmacologia (Berl.)* **12**:303–321.

Krivanek, J. A., and McGaugh, J. L., 1969, Facilitating effects of pre- and posttrial amphetamine and discrimination learning, *Agents Actions* **1**:36–42.

Lee, M., and Gold, P. E., 1987, Memory enhancement and impairment with intracerebroventricular glucose injections (in preparation).

McGaugh, J. L., and Krivanek, J. A., 1970, Strychnine effects on discrimination learning in mice: Effects of dose and time of administration, *Physiol. Behav.* **5**:1437–1442.

McGaugh, J. L., Liang, K. C., Bennett, C., and Sternberg, D. B., 1984, Adrenergic influences on memory storage: Interaction of peripheral and central systems, in: *Neurobiology of Learning and Memory* (G. Lynch, J. L. McGaugh, and N. M. Weinberger, eds.), Guilford Press, New York, pp. 313–332.

Paalzow, L. K., Paalzow, G. H. M., and Tfelt-Hansen, P., 1985, Variability in bioavailability: Concentration versus effect, in: *Variability in Drug Therapy: Description, Estimation, and Control* (M. Rowland, L. B. Sheiner, and J. L. Steimer, eds.), Raven Press, New York, pp. 167–185.

Platel, A., Jalfre, M., Pawelec, C., Roux, S., and Porsolt, R. D., 1984, Habituation of exploratory activity in mice: Effects of combinations of piracetam and choline on memory processes, *Pharmacol. Biochem. Behav.* **21**:209–212.

Popot, J. L., and Changeux, J. P., 1984, Nicotinic receptor of acetylcholine: Structure of an oligomeric integral membrane protein, *Physiol. Rev.* **64**:1162–1239.

Riege, W. H., and Cherkin, A., 1976, Memory performance after flurothyl treatment in rainbow trout, *Psychopharmacologia* **46**:31–35.

Schmitt, F. O., and Schneider, D. M., eds., 1975, *Functional Linkage in Biomolecular Systems*, Raven Press, New York.

Shih, Y. H., and Pugsley, T. A., 1985, The effects of various cognition-enhancing drugs on *in vitro* rat hippocampal synaptosomal sodium dependent high affinity choline uptake, *Life Sci.* **36**:2145–2152.

Steinberg, S. F., Chow, Y. K., and Bilezikian, J. P., 1986, Regulation of rat heart membrane adenylate cyclase by magnesium and manganese, *J. Pharmacol. Exp. Ther.* **237**:764–772.

Stratton, L. O., and Petrinovich, L., 1963, Post-trial injections of an anticholinesterase drug and maze learning in two strains of rats, *Psychopharmacologia* **5**:47–54.

Votava Z., 1967, Pharmacology of the central cholinergic synapses, *Annu. Rev. Pharmacol.* **7**:233–240.

Wilson, C. A., 1986, Society for drug reasearch symposium: Senile dementia of the Alzheimer type, *Neurobiol. Aging* **7**:219–222.

Wolthuis, O. L., 1981, Behavioral effects of etiracetam in rats, *Pharmacol. Biochem. Behav.* **15**:247–255.

VI

Presynaptic and Postsynaptic Plasticity

33

Synaptic Efficacy Is Controlled by the Concentration of Transmitter in the Nerve Ending

BERNARD POULAIN, LADISLAV TAUC, GERARD BAUX, and PHILIPPE FOSSIER

1. INTRODUCTION

Plastic changes related to the quantity of transmitter released per impulse have been more or less directly correlated with induced changes in calcium influx (Kandel and Schwartz, 1982; Alkon, 1986; Krnjevic, 1986), which is known to trigger the transmitter release. As the cytosol of the presynaptic terminal is a site for other intensive events (which regulate the metabolic equilibrium of its molecular constituents), it seems legitimate to ask whether other processes than calcium concentration can influence the quantity of transmitter released. In particular, it was found that in stimulated nervous tissues the transmitter content varies, and it is tempting to consider a possible control of transmitter release by the presynaptic intracellular transmitter concentration. We attempted to obtain evidence for such an effect on a central cholinergic synapse.

Earlier studies suggested that the presynaptic content of acetylcholine (ACh) may be of importance in synaptic efficacy. Several authors reported a decrease of ACh release when its synthesis was depressed by the presence of hemicholinium-3, a choline uptake blocker (Elmqvist and Quastel, 1965; Jones and Kwanbunbumpen, 1970). In mammalian brain, enhancement of choline supply resulted in an increase in synthesis and release of ACh (Haubrich *et al.*, 1975). That the presynaptic ACh content is a factor affecting the quantity of transmitter released was also apparent from results in *Torpedo* electric organ (Dunant *et al.*, 1974; Corthay *et al.*, 1982) or in sympathetic ganglia (Birks, 1977), where ACh release had a parallel evolution to that of the presynaptic ACh content. In order to study the cellular basis of the relationship between presynaptic ACh content and its release, we took advantage of an identified *Aplysia* synapse where we could increase or decrease the concentration of ACh in the presynaptic neuron and measure the subsequent modifications of quantal parameters of the postsynaptic response (Tauc *et al.*, 1974; Tauc and Baux, 1982; Baux and Tauc, 1983; Poulain *et al.*, 1986a,b).

BERNARD POULAIN, LADISLAV TAUC, GERARD BAUX, and PHILIPPE FOSSIER • Laboratory of Cellular and Molecular Neurobiology, National Center for Scientific Research, 91190 Gif-sur-Yvette, France.

2. MATERIALS AND METHODS

In all experiments we used the now well-known inhibitory cholinergic synapse (Tauc *et al.*, 1974; Simonneau *et al.*, 1980; Baux and Tauc, 1983; Fossier *et al.*, 1983, 1986; Poulain *et al.*, 1986a; Baux *et al.*, 1986) located in the buccal ganglion of *Aplysia californica* (Pacific Biomarine, Venice, CA, USA). The somata of the pre- and postsynaptic cells are identified from one experiment to another and can be easily penetrated by several microelectrodes after removal of the connective tissue sheath covering the ganglion (Fig. 1). The postsynaptic inhibitory response evoked by a presynaptic action potential is associated with chloride permeability. As the postsynaptic neuron received huge numbers of presynaptic inputs, spontaneous miniature events related to a given presynaptic neuron cannot directly be determined from a recording. To obtain quantal parameters from postsynaptic events, we used a method that was previously elaborated for this same neuronal preparation (Simonneau *et al.*, 1980). Briefly, the pre- and postsynaptic neurons impaled with two 3-M KCl microelectrodes were simultaneously and respectively voltage clamped at −50 mV and −80 mV (Fig. 1). Because the distance of the synaptic terminal from the soma is short (300–500 μm), it was possible to induce a prolonged postsynaptic response by depolarizing the soma of the presynaptic neuron for several seconds in the presence of tetrodotoxin (TTX) to avoid spike generation. Simultaneously, recordings were made (Fig. 1) of the presynaptic membrane potential, of the presynaptic current, of the postsynaptic current, and its AC component. The long-duration induced postsynaptic current (LDIPSC) recording showed, on its top, typical noise resulting from the summation of discrete events representing miniature postsynaptic currents (MPSCs). One MPSC is the postsynaptic response evoked by the release of one quantum of ACh. The size of the MPSC was obtained from statistical analysis of the LDIPSC (see legend to Fig. 1). T_{min},

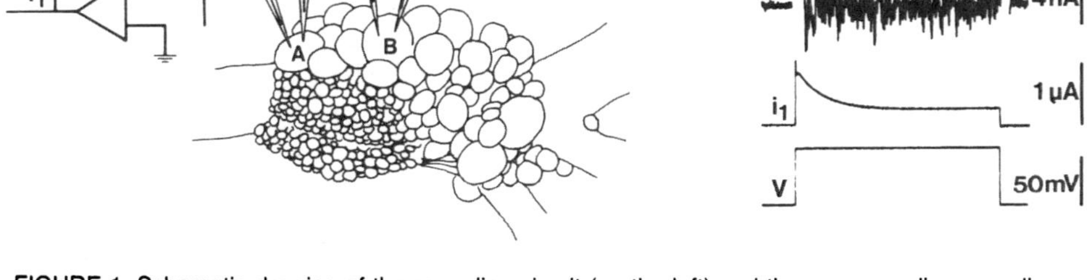

FIGURE 1. Schematic drawing of the recording circuit (on the left) and the corresponding recordings (on the right). One of the two presynaptic neurons (A) and one of the postsynaptic neurons (B) were simultaneously voltage clamped at, respectively, −50 and −80 mV. A virtual ground gives the sum, i_1, of both pre- and postsynaptic currents. i_2 is the postsynaptic current. The presynaptic current is then given by subtracting i_2 from i_1 (i_2 is negligible compared to i_1). Recordings (right panel): the long-duration induced postsynaptic current (LDIPSC, i_2) was obtained with sustained (3-sec) depolarization (V) of the presynaptic neuron. For analysis of the synaptic noise, i_2 was recorded at high gain as an AC trace (i_2A). The size of the MPSC is given by the function $2E^2/I_m$ where E^2 is the variance of the noise calculated from i_2A and I_m is the mean value of i_2.

the time constant of the MPSC decay, was calculated using a fast Fourier transform of the synaptic noise. Postsynaptic responses (evoked by a presynaptic action potential, LDIPSCs, and MPSCs) were recorded as currents in postsynaptic cells voltage clamped at -80 mV and sometimes converted into conductances, taking into account the driving force for Cl^- ions.

The quantal content (i.e., the number of quanta released by a stimulus) of the postsynaptic response was obtained by dividing the quantity of current of the LDIPSC by that of an MPSC. For more details see Baux *et al.* (1986).

The preparation was bathed with artificial sea water (ASW: NaCl, 460 mM; KCl, 10 mM; $CaCl_2$, 11 mM; $MgCl_2$, 25 mM; $MgSO_4$, 28 mM; Tris-HCl, 10 mM, pH 7.8) to which TTX was added at a final concentration of 10^{-4} M when the LDIPSC method was used.

For intracellular iontophoresis, acetylcholine chloride (ACh), choline chloride, or tetraethylammonium bromide (TEA) solutions were made at a concentration of 1 M in distilled water. Acetylcholinesterase (AChE, Worthington), at a concentration of 5% (w/v) in ASW, was injected into the presynaptic neuron by an air pressure system described earlier (Tauc *et al.*, 1974; Baux and Tauc, 1983). Choline oxidase (Sigma) was used either extracellularly at a concentration of 20 U/ml in ASW or intracellularly, injected by an air pressure system from a microelectrode containing a 1000 U/ml in ASW (Poulain *et al.*, 1986a).

Experiments were performed at room temperature (22°C).

3. RESULTS

3.1. Increase of Intraterminal ACh Concentration

To increase the ACh concentration in the presynaptic terminal, exogenous ACh or choline, its precursor, was injected into the presynaptic neuron by means of intracellular iontophoresis. In both cases, this led to an enhancement of the synaptic response that depended on the quantity of exogenous ACh or choline introduced (Fig. 2). This enhancement of the response because of increased transmitter release started rapidly after the beginning of injection (about 5 min, probably necessary for the accumulation of ACh in the soma and its diffusion and/or transport to the terminal) and lasted for several tens of minutes, and then, progressively, the response went back to its initial size. In all cases, the size of MPSCs (i.e., of the quantum) was unchanged (Fig. 3A,B). It was concluded that the increase of the postsynaptic response resulting from the increase of ACh concentration in the terminal was caused solely by a greater number of released quanta for the same presynaptic depolarization.

3.2. Decrease of Intracellular ACh Concentration

To do the opposite, we decreased the presynaptic ACh content by destroying it intracellularly or by preventing its synthesis.

3.2.1. Hydrolysis of Presynaptic ACh by Intracellular Injection of AChE

Intracellular ACh was destroyed by introducing exogenous AChE. Because the vesicular pool of ACh is protected against the injected AChE (Dunant *et al.*, 1972, 1974),

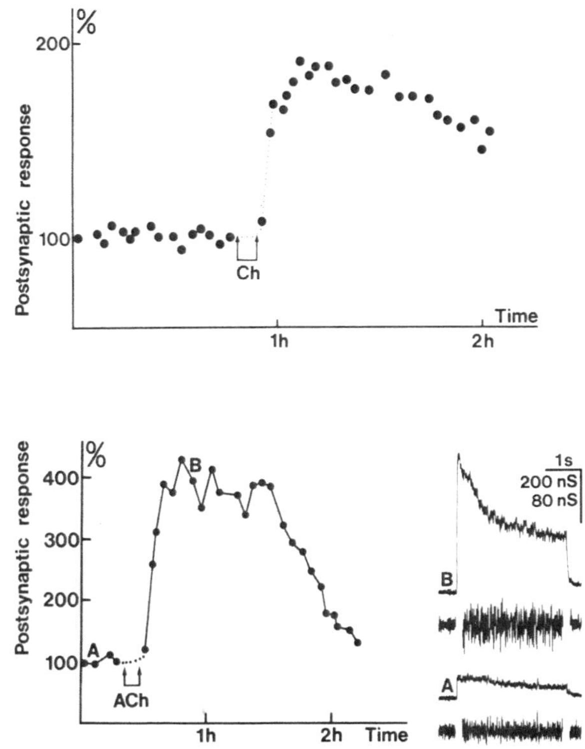

FIGURE 2. Enhancement of the postsynaptic response following intracellular injection of choline (upper graph) or ACh (lower graph). The iontophoretic injection of choline (50 nA for 7 min) or ACh (80 nA for 7 min) led to an increase of the size of the LDIPSC evoked by a constant presynaptic depolarization to +5 mV for 3 sec. On the right of the lower graph, the recordings show a control response (A) and a response after overloading the presynaptic neuron with exogenous ACh (B). Upper traces, DC recordings of the LDIPSC; lower traces, AC recordings of the LDIPSC. Modified from Poulain *et al.* (1986a) with permission.

only the cytoplasmic ACh was expected to be hydrolyzed. After a minimal delay of 90 min (corresponding to the diffusion/transport of the enzyme to the terminal), the size of the postsynaptic response began to decrease until a complete block of synaptic transmission took place (Fig. 4). It was ascertained that the decrease of the postsynaptic response was caused by the reduction of ACh release and that the release mechanism itself was not affected since the hydrolyzed ACh in the terminal could be replaced by a nonhydrolyzable analogue, carbachol, which was then released as false transmitter (Baux and Tauc, 1983). No decrease in the size of the MPSC was observed, and the depression of the postsynaptic response was exclusively related to a decrease in its quantal content.

3.2.2. Block of ACh Synthesis

In the absence of specific blockers of choline acetyltransferase, the enzyme responsible for ACh synthesis from choline and acetylcoenzyme A, we attempted to block ACh synthesis by antagonizing choline uptake. We could not use the classical choline uptake blocker hemicholinium-3 (HC-3) because of its complex effects at this synapse (Baux *et*

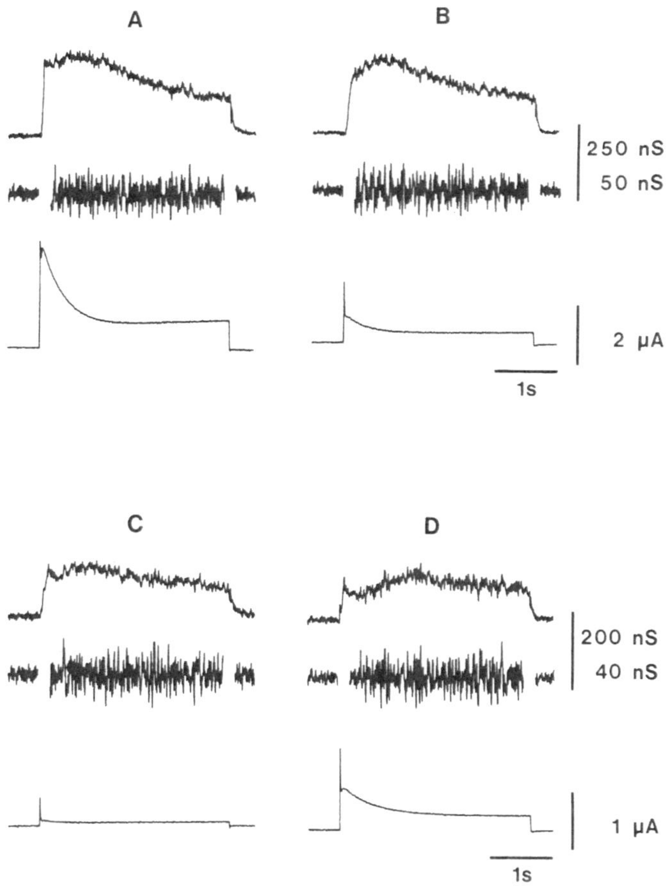

FIGURE 3. Comparison of LDIPSC of the same amplitude before (A) and after (B) presynaptic injection of ACh or before (C) and after (D) choline oxidase treatment. Because ACh injection led to an increase in the size of the LDIPSC for a given presynaptic stimulation, the presynaptic depolarization used in B (d0 mV for 3 sec) was lower than that used in the control, A (+10 mV for 3 sec) in order to obtain post-synaptic responses of the same amplitude. In contrasts choline oxidase treatment led to a decrease in the size of the LDIPSC. The presynaptic depolarization used in D (+10 mV for 3 sec) thus had to be higher than that used in the control, C (−5 mV for 3 sec). Upper trace, DC recordings; middle trace, AC recordings; bottom trace, presynaptic currrent. Whatever the treatment, the variance of the synaptic noise was identical to that of the controls (respectively, B to A and D to C). Since for each treatment the amplitude of the LDIPSC was the same, it could be concluded that the miniature was not changed when the ACh presynaptic content was modified. Modified from Poulain *et al.* (1986b) with permission.

al., 1986). We used choline oxidase instead, which, when present in the bath, oxidized extracellular choline (Fig. 5), making the latter unavailable for uptake by the presynaptic neuron. Subsequent intensive stimulation of the presynaptic neuron exhausted its ACh content. This was followed by a reduction in the size of the postsynaptic response (Fig. 6) solely, again, as a result of a decrease of the quantal content, as the MPSC size was unchanged (Fig. 3C,D). Similar results were obtained when choline oxidase was introduced into the presynaptic neuron by air pressure injection. The depression of synaptic transmission was effectively caused by a deficiency in available transmitter, since an

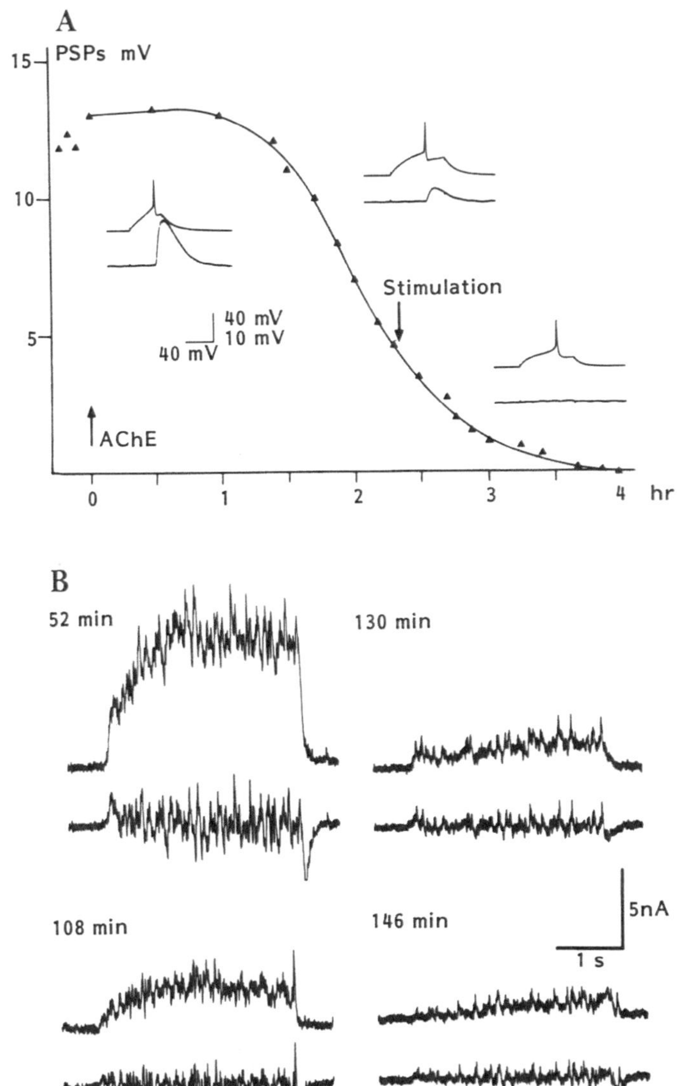

FIGURE 4. Decrease in presynaptic ACh content by intracellular injection of acetylcholinesterase (AChE). A: Decrease of the amplitude of the evoked postsynaptic potential (insets, lower traces) after intracellular injection of AChE (at time 0). The arrow indicates the time at which a high-frequency repetitive stimulation was applied; the rate of decrease in the postsynaptic response was not affected, indicating that stimulation did not act on an AChE-resistant ACh store. Insets show spikes and postsynaptic responses before and after AChE injection. The postsynaptic cell was current clamped to −80 mV. B: Depression of the LDIPSC after AChE injection. The enzyme was air-pressure injected into the presynaptic soma at time 0, and the response induced by a presynaptic depolarization to 0 mV for 3 sec at 52 min can be considered as control. Fluctuation analysis showed that the depression of the response with time results from a reduction in the quantal content, as the size of the individual miniatures remained unchanged. Upper trace, DC recordings; lower trace, AC recordings. From Tauc and Baux (1982) with permission.

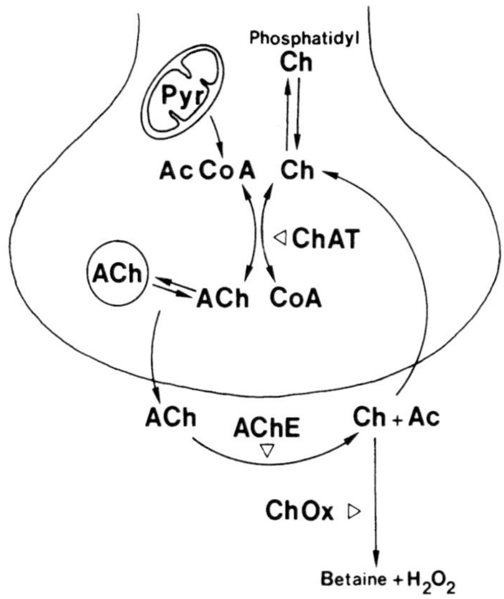

FIGURE 5. Schematic representation of the oxidation of extracellular choline by choline oxidase. Choline oxidase (ChOx) oxidizes choline (Ch) to peroxide (H_2O_2) and betaine. Thus, extracellularly applied choline oxidase prevents the supply and recycling of choline. Choline oxidase may also be introduced by an air-pressure injection into the presynaptic neuron in order to destroy endogenous choline. AcCoA, acetylcoenzyme A; ChAT, choline acetyltransferase; CoA, coenzyme A; ACh, acetylcholine; Pyr, pyruvate; AChE, acetylcholinesterase.

iontophoretic injection of exogenous ACh into the presynaptic neuron led to a recovery of synaptic efficacy (Fig. 6).

From all these experiments in which presynaptic ACh content was altered, it can be concluded that for a given Ca^{2+} influx, the number of quanta released is controlled by the presynaptic ACh content.

4. DISCUSSION

4.1. Potassium Currents, Free ACh, and Release

One may ask if changes in ACh release caused by its intraterminal modifications were actually related to a direct effect of its concentration on the release mechanism or to a trivial effect on presynaptic membrane properties that in turn changed Ca^{2+} influx into the terminal. Acetylcholine is a quaternary ammonium compound and has been demonstrated to block different voltage-dependent K^+ channels (Krjnevic, 1978; Yarom *et al.*, 1985). Indeed, when the intracellular content of ACh was increased by iontophoretic injection of exogenous ACh, the presynaptic outward current, mainly the late K^+ current, recorded in the voltage-clamped presynaptic neuron (Fig. 7, column a) during the depolarizing stimulus was reduced. In contrast, this current was increased when the intracellular ACh content was depressed by choline oxidase treatment (Fig. 7, column c). However, in our experimental situation, these changes in K^+ outward current did not significantly affect the release of ACh for a given presynaptic long-lasting depolarization.

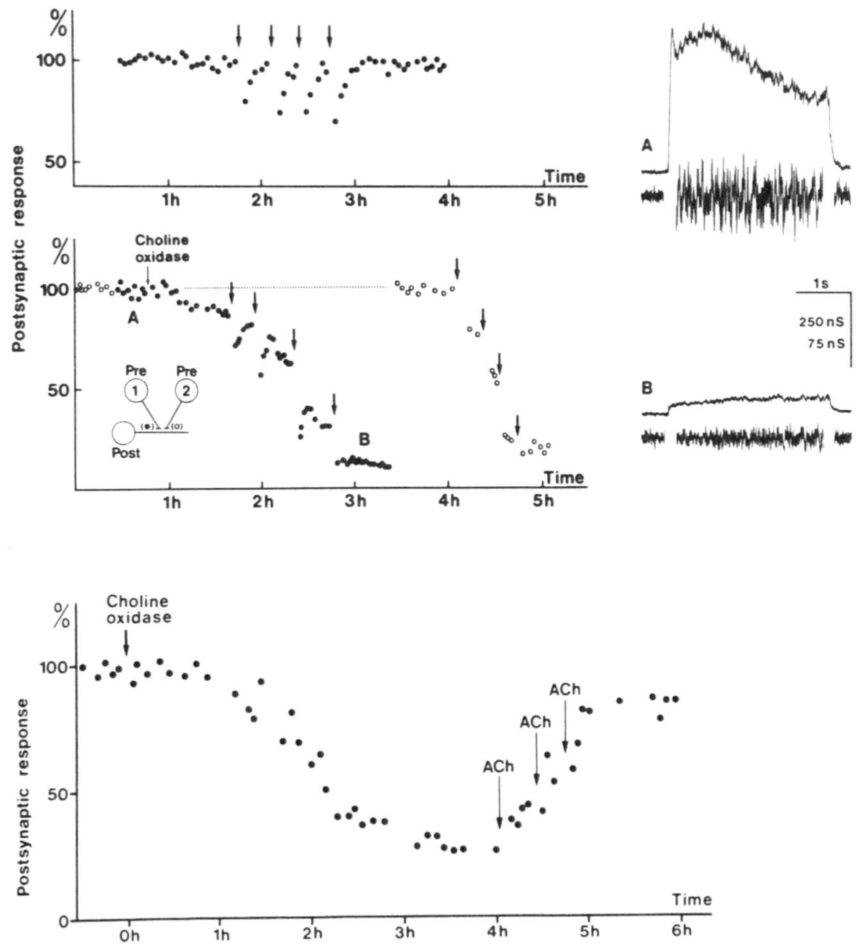

FIGURE 6. Decrease in presynaptic ACh content by choline oxidase treatment. Upper graph: evolution in the size of the postsynaptic response (LDIPSC) when the presynaptic neuron was submitted to periods of intense stimulation in the absence of choline oxidase in the bath. Each arrow, for both graphs, represents a series of four stimulations during which the presynaptic cell was depolarized to +10 mV for 1 min. The recovery of the control amplitude response was achieved in a few minutes, and the transient decrease of the postsynaptic response may be partly related to a transient decrease in calcium conductance. Middle graph: evolution in the size of the postsynaptic response when two equivalent presynaptic neurons (open and black circles; the inset shows a schematic drawing of the synaptic connections in the buccal ganglion) were submitted to periods of intense stimulation at different times of incubation in choline oxidase. First, control responses (expressed in percentage) were obtained for both neurons. After choline oxidase was bath applied, stimulation (arrows) of the first interneuron (black circles) led to a clear decrease in the response. Three hours after the application of choline oxidase, the LDIPSCs evoked by the second interneuron (open circle) were unchanged. However, intense stimulation (arrows) rapidly depressed the LDIPSC. The recordings (right panel) show LDIPSCs (for presynaptic depolarizations to +10 mV for 3 sec) recorded in A (control) and B (after choline oxidase treatment). The calculated MPSC expressed as a conductance was 1.26 + 0.29 nS for A and 1.27 + 0.32 nS for B. Bottom graph: effects of choline oxidase and acetylcholine (ACh) injected into the presynaptic neuron. After the introduction of choline oxidase, the postsynaptic response (LDIPSC) decreased after a delay that may result partly from the diffusion of the enzyme into the terminal. Subsequent injections of acetylcholine (arrows, 50 nA for 5 min) restored synaptic transmission. From Poulain *et al.* (1986a) with permission.

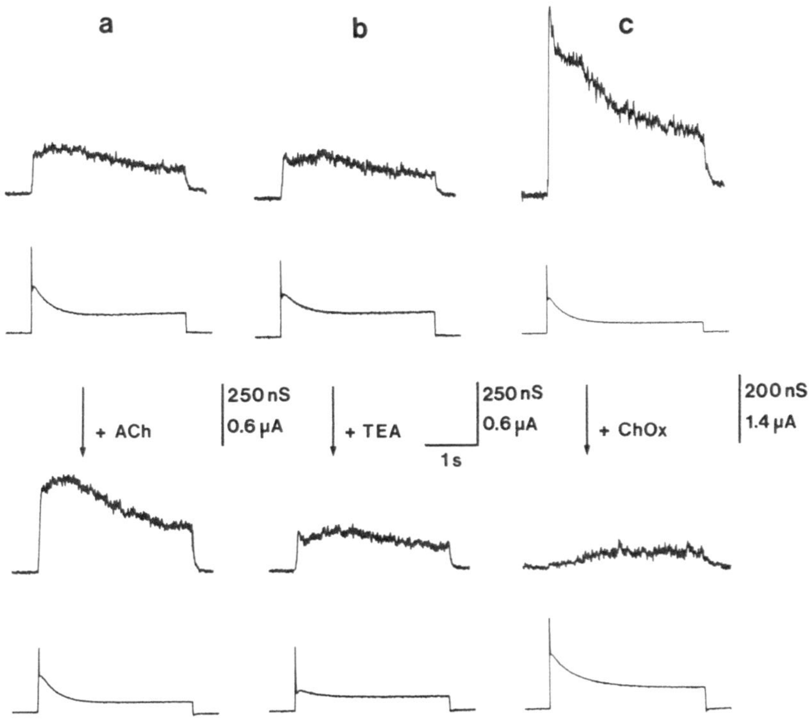

FIGURE 7. Potassium current as an experimental index of changes in presynaptic ACh concentration. For the three columns, the upper recordings show, for three different cells, control LDIPSCs induced by presynaptic depolarizations to +5 mV (in a) or to +10 mV (in b and c), and the lower line shows the control presynaptic currents (see Fig. 1). In the lower half of column a, after presynaptic injection of ACh (50 nA for 10 min), the size of the LDIPSC was enhanced because of an increased release of ACh. At the same time, because ACh is a quaternary ammonium, the presynaptic current (for the same presynaptic depolarization) was reduced. In the lower half of column b, a presynaptic injection of TEA (50 nA for 5 min) induced a similar decrease of the presynaptic current without modifying the amplitude of the LDIPSC. In the lower half of column c, bath-applied choline oxidase (4 hr) coupled with long presynaptic stimulations (six series of four stimulations of 1 min to +10 mV) led to a depression of the LDIPSC as a result of a lack of available presynaptic ACh. Note that the presynaptic current was increased at the same time. For all presynaptic current traces, the spike represents the early transitory K^+ current I_A.

When the presynaptic neuron was iontophoretically loaded with TEA, the decrease of the presynaptic current was similar to that induced by ACh (Fig. 7, column b), but no change in ACh release was observed. Also, when the presynaptic current was increased after choline oxidase treatment, the intracellular injection of TEA did not increase the release of transmitter (Poulain *et al.*, 1986a,b). Similar results were obtained when, instead of TEA, a quaternary ammonium ACh agonist, bethanechol, was injected into the presynaptic neuron (Tauc and Baux, 1982, 1985). We are thus confident that modifications in the size of the postsynaptic response following changes in presynaptic ACh content are closely related to the concentration of the transmitter available for release from the terminal. In addition, the changes in potassium conductance caused by choline oxidase treatment clearly indicate that a free cytoplasmic pool of ACh normally exists in the presynaptic neuron.

4.2. Mechanisms by Which Presynaptic ACh Controls Transmitter Release

When exogenous ACh was injected into the presynaptic neuron, the increase in the size of the postsynaptic response started less than 5 min after the start of the injection (Fig. 2). If we consider the time necessary for the somatically injected ACh to pile up and to reach the terminal situated some hundreds of micrometers away and for it to accumulate there at a significant concentration, the time remaining for an exchange between cytoplasmic and vesicular compartments can only be very short. It thus seems reasonable to conclude that the released ACh originated from the cytoplasmic pool unless very rapid exchange took place between the cytoplasmic and the vesicular ACh compartments.

Whether the presynaptic ACh content was experimentally increased or decreased, the size of the MPSC (i.e., that of the quantum of ACh in our experimental conditions showing no postsynaptic interference) remained constant. Thus, the size of the quantum is independent of the free ACh concentration. This result is in agreement with the original formulation of the quantal theory by Katz (1969): only full-sized quanta can be released. This indicates that the quantum release mechanism is a saturable structure. Such a role was attributed to synaptic vesicles (Katz, 1969). If we accept that released ACh quanta originate from the cytoplasmic pool, another structure for release must be involved. Various hypotheses have been proposed, such as the existence of "vesigates" (Tauc, 1982), membranous saturable structures offering intracellular presynaptic binding sites for cytoplasmic ACh (Tauc and Baux, 1982, 1985), or "mediatophores," proteins that translocate ACh across the presynaptic membrane (Israël and Manaranche, 1985; Birman *et al.*, 1986).

Taking into account the cytoplasmic origin of released ACh and the fact that release requires only saturated quanta, intraterminal ACh concentration would determine the probability of saturation of the structures forming the quanta and thus the number of releasable quanta. So, for a given presynaptic stimulus (i.e., the same Ca^{2+} influx), the number of quanta released is determined by the presynaptic concentration of ACh.

Undoubtedly such a dependence may play a nonnegligible role in the control of synaptic efficacy. In a stimulated synapse, the equilibrium between the loss of transmitter and the rate of its synthesis is not necessarily maintained. In addition, the rate of synthesis of the transmitter or the mechanism controlling its cytoplasmic concentration may be affected by the activity of presynaptic receptors or autoreceptors on the nerve terminal. We have recently reported (Baux *et al.*, 1986), on the same synapse used in the present study, evidence for the presence of presynaptic receptors, possibly of a muscarinic nature, the activation of which modulated ACh release, apparently without affecting the calcium influx.

So far, our evidence for the participation of the control of the cytoplasmic concentration of transmitter on its release concerns only cholinergic synapses. In order to generalize the existence of such a control, evidence should be obtained on synapses using other neurotransmitters. Progress in this direction will be more difficult, as it depends on the finding of a noncholinergic synaptic couple presenting the same favorable features as the cholinergic synapse in the *Aplysia* buccal ganglion.

ACKNOWLEDGMENTS. We are grateful to Dr. S. O'Regan for critical reading of the manuscript. This work was supported by grant no. 856021 from INSERM, no. 85/1177 from DRET, and ATP no. 6931 from CNRS.

REFERENCES

Alkon, D. L., 1986, Changes of membrane currents and calcium-dependent phosphorylation during associative learning, in: *Neural Mechanisms of Conditioning* (D. L. Alkon and C. D. Woody, eds.), Plenum Press, New York, pp. 3–18.

Baux, G., and Tauc, L., 1983, Carbachol can be released at a cholinergic ganglionic synapse as a false transmitter, *Proc. Natl. Acad. Sci. U.S.A.* **80:**5126–5128.

Baux, G., Poulain, B., and Tauc, L., 1986, Quantal analysis of action of hemicholinium-3 studied at a central cholinergic synapse of *Aplysia*, *J. Physiol. (Lond.)* **380:**209–226.

Birks, R. I., 1977, A long-lasting potentiation of transmitter release related to an increase in transmitter stores in a sympathetic ganglion, *J. Physiol. (Lond.)* **271:**847–862.

Birman, S., Israël, M., Lesbats, B., and Morel, N., 1986, Solubilization and partial purification of a presynaptic membrane protein ensuring calcium-dependent acetylcholine release from proteoliposomes, *J. Neurochem.* **47:**433–444.

Corthay, J., Dunant, Y., and Loctin, F., 1982, Acetylcholine changes underlying transmission of a single nerve impulse in the presence of 4-aminopyridine in *Torpedo*, *J. Physiol. (Lond.)* **325:**461–479.

Dunant, Y., Gautron, J., Israël, M., Lesbats, B., and Manaranche, R., 1972, Les compartiments d'acétylcholine de l'organe électrique de la torpille et leurs modifications par la stimulation, *J. Neurochem.* **19:**1987–2002.

Dunant, Y., Gautron, J., Israël, M., Lesbats, B., and Manaranche, R., 1974, Evolution de la décharge de l'organe électrique de la torpille et variations simultanées de l'acétylcholine au cours de la stimulation, *J. Neurochem.* **23:**635–643.

Elmqvist D., and Quastel, D. M. J., 1965, Presynaptic action of hemicholinium at the neuromuscular junction, *J. Physiol. (Lond.)* **177:**463–482.

Fossier, P., Baux, G., and Tauc, L., 1983, Possible role of acetylcholinesterase in regulation of postsynaptic receptor efficacy at a central inhibitory synapse of *Aplysia*, *Nature* **301:**710–712.

Fossier, P., Baux, G., and Tauc, L., 1986, Acetylcholinesterase and synaptic efficacy, in: *Neural Mechanisms of Conditioning* (D. L. Alkon and C. D. Woody, eds.), Plenum Press, New York, pp. 341–354.

Haubrich, D. R., Wang, P. F. L., Clody, D. E., and Wedeking, P. W., 1975, Increase in rat brain acetylcholine induced by choline or deanol, *Life Sci.* **17:**975–980.

Israël, M., and Manaranche, R., 1985, The release of acetylcholine from a cellular towards a molecular mechanism, *Biol. Cell* **55:**1–14.

Jones, S. F., and Kwanbunbumpen, S., 1970, Some effects of nerve stimulation and hemicholinium on quantal transmitter release at the mammalian neuromuscular junction, *J. Physiol. (Lond.)* **207:**51–61.

Kandel, E. R., and Schwartz, J. H., 1982, Molecular biology of learning: Modulation of transmitter release, *Science* **218:**433–443.

Katz, B., 1969, *The Release of Neural Substances*, Liverpool University Press, Liverpool.

Krnjevic, K., 1978, Intracellular actions of a transmitter, in: *Iontophoresis and Transmitter Mechanisms in the Mammalian Central Nervous System* (R. W. Ryall and J. S. Kelly, eds.), Elsevier, New York and Amsterdam, pp. 155–157.

Krnjevic, K., 1986, Role of calcium ions in learning, in: *Neural Mechanisms of Conditioning* (D. L. Alkon and C. D. Woody, eds.), Plenum Press, New York, pp. 251–259.

Poulain, B., Baux, G., and Tauc, L., 1986a, Presynaptic transmitter content controls the number of quanta released at a neuro-neuronal cholinergic synapse, *Proc. Natl. Acad. Sci. U.S.A.* **83:**170–173.

Poulain, B., Baux, G., and Tauc, L., 1986b, The quantal release at a neuro-neuronal synapse is regulated by the content of acetylcholine in the presynaptic cell, *J. Physiol. (Paris.)* **81:**270–277.

Simonneau, M., Tauc, L., and Baux, G., 1980, Quantal release of acetylcholine examined by current fluctuation analysis at an identified neuro-neuronal synapse of *Aplysia*, *Proc. Natl. Acad. Sci. U.S.A.* **77:**1661–1665.

Tauc, L., 1982, Nonvesicular release of neurotransmitter, *Physiol. Rev.* **62:**857–893.

Tauc, L., and Baux, G., 1982, Are there intracellular acetylcholine receptors in the cholinergic synaptic nerve terminals? *J. Physiol. (Paris.)* **78:**366–372.

Tauc, L., and Baux, G., 1985, Mechanisms of acetylcholine release at neuroneuronal synapses, in: *Non-vesicular Transport* (S. S. Rothman and J. L. Ho, eds.), John Wiley & Sons, New York, pp. 253–269.

Tauc, L., Hoffmann, A., Tsuji, S., Hinzen, D. H., and Faille, L., 1974, Transmission abolished on a cholinergic synapse after injection of acetylcholinesterase into the presynaptic neurone, *Nature* **250:**496–498.

Yarom, Y., Bracha, O., and Werman, R., 1985, Intracellular injection of acetylcholine blocks various potassium conductances in vagal motoneurons, *Neuroscience* **16:**739–752.

34

Short-Term and Long-Term Plasticity Mediated by Changes in Responding Synapses at Crustacean Neuromuscular Junctions

H. L. ATWOOD, J. M. WOJTOWICZ, and F. W. Y. TSE

1. PLASTICITY AT CRUSTACEAN NEUROMUSCULAR JUNCTIONS

Neuromuscular junctions of crustaceans have been frequently employed as model systems to investigate mechanisms of synaptic plasticity (reviews by Atwood, 1976, 1982). The large size of the individual motor neurons and their limited number permit investigations of synaptic physiology and morphology at the level of individual nerve terminals and even of individual synapses. Furthermore, crustacean neuromuscular junctions display physiological properties akin to those of synapses in the central nervous systems of both vertebrate and invertebrate species. Several forms of short-term and long-term synaptic plasticity have been described, and cellular mechanisms of these are under investigation.

There are six major forms of synaptic plasticity described to date at crustacean neuromuscular junctions.

1. Short-term facilitation. Following one impulse or a brief train, the probability of release of transmitter by a subsequent impulse is enhanced for 1 to several seconds (reviewed by Zucker, 1982). The predominant mechanism for this phenomenon is thought to be residual calcium from a preceeding impulse or train of impulses (Katz and Miledi, 1968; Parnas *et al.*, 1982; Zucker and Lara-Estrella, 1983). Residual calcium increases the likelihood of transmitter release by the facilitated nerve ending.
2. Presynaptic inhibition. The reverse of short-term facilitation is presynaptic inhibition. Synaptic transmission at neuromuscular junctions of motor neurons to certain crustacean limb muscles can be decreased during activity of peripheral inhibitory axons, which form axoaxonal synapses with the excitatory motor axons (Dudel and Kuffler, 1961; Atwood and Morin, 1970). The effect is mediated through a GABA-operated chloride conductance channel in the terminal mem-

H. L. ATWOOD, J. M. WOJTOWICZ, and F. W. Y. TSE • Department of Physiology, University of Toronto, Toronto, Ontario, Canada M5S 1A8. *Present address of F.W.Y.T.*: Department of Medical Physiology, University of Calgary, Calgary, Alberta, Canada T2N 4N1.

brane (Takeuchi and Takeuchi, 1966a,b; Fuchs and Getting, 1980). Increased chloride conductance alters the electrical properties of nerve terminals, thus decreasing transmitter release.

3. Synaptic depression. Repetitive activity in phasic motor neurons such as the "fast" motor axons of limb muscles (Hoyle and Wiersma, 1958) and the "motor giant" axon to crayfish fast abdominal flexor muscles leads to synaptic depression. Two types are observed: low-frequency depression, which occurs at low stimulus repetition rates and cannot be explained by depletion of transmitter stores (Zucker and Bruner, 1977), and high-frequency depression, which appears with impulse repetition rates above 4–5 Hz, often after a transient period of short-term facilitation. This form of depression may involve depletion of transmitter stores at least in part (Bryan and Atwood, 1981).
4. Long-term facilitation. Tonic motor axons of crustacean limb muscles show gradual progressive enhancement of transmission when subjected to maintained stimulation at frequencies greater than 4–5 Hz (Sherman and Atwood, 1971; Atwood *et al.*, 1975; Jacobs and Atwood, 1981). Once established, the enhancement of transmission persists for a long time, perhaps as long as 24 hr. Phasic motor axons, even though their transmission may be depressed by such stimulation, nevertheless exhibit a prolonged aftereffect of enhanced transmission that persists for 1–24 hr (Lnenicka and Atwood, 1985b). Long-term facilitation is presynaptic in origin (Wojtowicz and Atwood, 1986) and requires the presence of extracellular sodium and calcium ions for full expression (Wojtowicz and Atwood, 1985).
5. Long-term adaptation. Increasing or decreasing the daily level of impulse activity in a phasic motor neuron leads to adaptive changes in synaptic transmission that persist for weeks once established (Lnenicka and Atwood, 1985a,b). Increased daily motor activity leads to a decrease in initial output of transmitter and to less high-frequency synaptic depression during maintained stimulation. (The decrease in initial output of transmitter is the opposite effect to long-term facilitation.) Morphological changes associated with adaptation in motor nerve terminals include an increased mitochondrial content (Atwood *et al.*, 1985). Long-term adaptive changes appear to be mediated by the neuronal cell body (see Lnenicka, 1987, for recent experimental evidence).
6. Neurohormonal modulation. Circulating neurohormones (in particular, serotonin and octopamine) accentuate synaptic transmission through presynaptic and postsynaptic actions. At the neuromuscular junction, serotonin exerts a strong presynaptic effect, increasing transmitter release (Dudel, 1965a). Extracellular calcium is not required for this effect (Glusman and Kravitz, 1982), but at the crayfish neuromuscular junction, extracellular sodium ion is required (Dixon and Atwood, 1985). The effect of octopamine is primarily postsynaptic (enhancement of muscular contraction), but an activity-dependent presynaptic enhancement of transmission has been described at the crustacean neuromuscular junction (Florey and Rathmayer, 1978; Breen and Atwood, 1983). Central as well as peripheral effects of octopamine are known in crustaceans (review by Kravitz *et al.*, 1985).

All of the above phenomena entail modifications of transmitter-releasing properties of the crustacean nerve terminal. The question arises: Are these modifications brought about through changes in the number of individual responding synapses, or are they attributable to variation in release probability of a fixed number of synapses? In the following discussion, we present relevant evidence on long-term facilitation and presynaptic inhibition to illustrate that changes in the number of responding synapses play a role in plastic changes at the crustacean neuromuscular junction.

2. MORPHOLOGICAL SUBSTRATE FOR PLASTICITY

Electron microscopic examination of crustacean neuromuscular junctions reveals that motor nerve terminals are generously endowed with morphological synapses (Fig. 1). Typically, each transverse section through a motor nerve ending contains several structures that would normally be classed as synapses. The "complete" synapse is distinguished by electron-dense pre- and postsynaptic membranes with uniform separation (about 20 nm), a cluster of synaptic vesicles near the presynaptic membrane, and one or more presynaptic "dense bodies" (the morphological equivalent of the vertebrate "active zone"). Recent evidence from freeze-fracture studies suggests that exocytosis of synaptic vesicles occurs at the perimeters of the dense bodies in lobster neuromuscular junctions (Pearce *et al.*, 1986).

Serial sectioning of nerve terminals has shown that total synaptic complement amounts to one to five morphological synapses per micrometer of terminal length (Govind *et al.*, 1982; Atwood and Marin, 1983). The number of synapses per micrometer of terminal length is apparently lower for terminals releasing large amounts of transmitter for a single impulse than for those releasing small amounts of transmitter (Atwood and Marin, 1983). This observation by itself indicates that transmitter output at a nerve terminal is not a simple function of the number of available synapses: additional factors are involved.

Inspection of individual synapses in serial sections has shown that they differ in morphological detail. Some lack presynaptic dense bodies and associated vesicle clusters (Jahromi and Atwood, 1974). Some have only one presynaptic dense body, whereas others have more than one; the size and shape of these structures are not uniform. These nonuniformities suggest corresponding variation in physiological function.

Statistical analyses of transmitter release at individual nerve endings, as measured

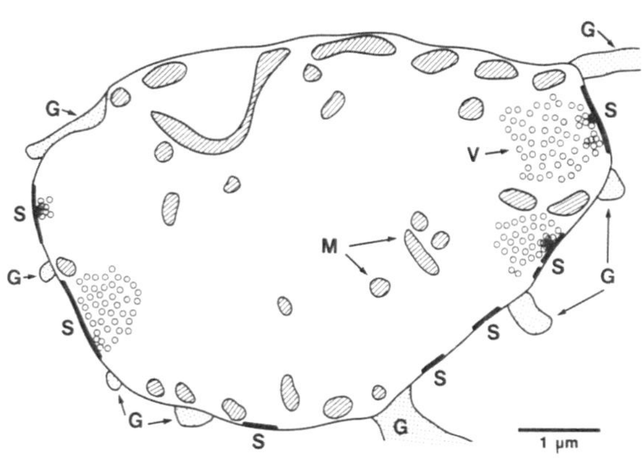

FIGURE 1. Morphological features of the crustacean neuromuscular junction (after Fritz *et al.*, 1980). Transverse section of a lobster (*Homarus americanus*) nerve terminal showing numerous synapses (S), some of which are associated with well-defined clusters of synaptic vesicles (V) near presynaptic dense projections. Numerous mitochondrial profiles (M) are also seen in the nerve terminal. Glial cell processes (G) intervene between adjacent synaptic contacts.

by focal extracellular microelectrodes, have provided good fits to a binomial model of transmitter release in which *n* responding units release with average probability *p*. Quantal content of transmission per impulse, *m*, is the product of *n* and *p*. At low frequencies of stimulation, low values of *m* occur at individual nerve endings, and estimates of *n* are also low (typically in the range of 1–5: Wernig, 1972; Zucker, 1973; Smith, 1983). It has been postulated that *n* represents the number of responding synapses that can release transmitter (Zucker, 1973). This idea has been strengthened by the work of Korn *et al.* (1982) on inhibitory inputs to the Mauthner cell of fish. They showed a morphological correlate for the binomial parameter *n* in the number of individual synaptic boutons associated with a particular inhibitory axon. If the general hypothesis that the synapse functions as an all-or-nothing unit applies at the crustacean neuromuscular junction, then there is clearly a discrepancy between the morphologically defined synapses of a motor nerve terminal and those detected by statistical analysis; the latter are too few in number to match the former.

By the same token, if some of the morphological synapses are inactive at low frequencies of stimulation, they may form a "reserve" that can be drawn on as physiological conditions change. If that is the case, and if the general statistical model of transmitter release is valid, it should be possible to detect changes in *n* as transmission is altered either in the short term (short-term facilitation, presynaptic inhibition) or in the long term (long-term facilitation and adaptation).

We have conducted tests of this hypothesis by measuring statistical changes in transmission associated with plasticity.

3. PRESYNAPTIC INHIBITION

It has long been known that reduced quantal output at presynaptically inhibited crustacean motor nerve endings is associated with a GABA-mediated increase in chloride conductance in the motor terminal membrane (Dudel, 1965b; Takeuchi and Takeuchi, 1966b; Fuchs and Getting, 1980). The inhibitory conductance increase often changes the membrane potential of the nerve terminal, but this change is probably not the cause of reduced transmitter output, because presynaptic inhibition occurs regardless of whether the motor terminal membrane is depolaried or hyperpolaried (Baxter and Bittner, 1981). Reduction in amplitude of the presynaptic action potential as a result of increased chloride conductance could account for some reduction of transmitter release. However, the small degree of spike reduction (5–10%) measured in presynaptic terminals (Baxter and Bittner, 1981) may not be enough to account for the rather large reductions of transmitter release often observed, particularly at crab neuromuscular junctions (Atwood and Bittner, 1971).

Possibly, the reduction of spike amplitude is larger in boutonlike terminal endings on which inhibitory synapses act (Atwood *et al.*, 1984). Local effects of this sort would not be detected by a microelectrode impaling a presynaptic nerve ending even a short distance away. It is even possible that synapses on particular boutons may be entirely inactivated by inhibitory action, should the reduction in spike amplitude be large enough. Inactivation of sets of synapses (Fig. 2) would be expected to produce a reduction in *n* of the binomial statistical model.

Our measurements of quantal release at individual nerve endings in a crab muscle have fulfilled this prediction (Tse, 1986; Tse and Atwood, 1986). Endings that normally release several quantal units for each impulse with little trial-to-trial fluctuation ("high-output" synapses) can be modeled by a binomial distribution in which 10–20 responding

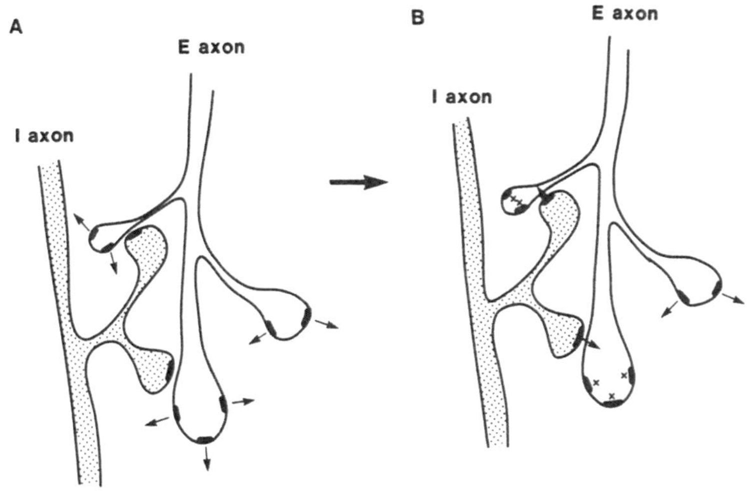

FIGURE 2. Possible inactivation of excitatory synapses on boutonlike structures during presynaptic inhibition (based on morphological work and modelling of Atwood *et al.*, 1984). (A) Before inhibition, several E synapses transmitting. (B) Inhibition in progress, most E synapses inactivated. Inhibitory synapses attenuate the excitatory fibers' action potential severely in boutons; transmitter release in such locations may be almost completely suppressed.

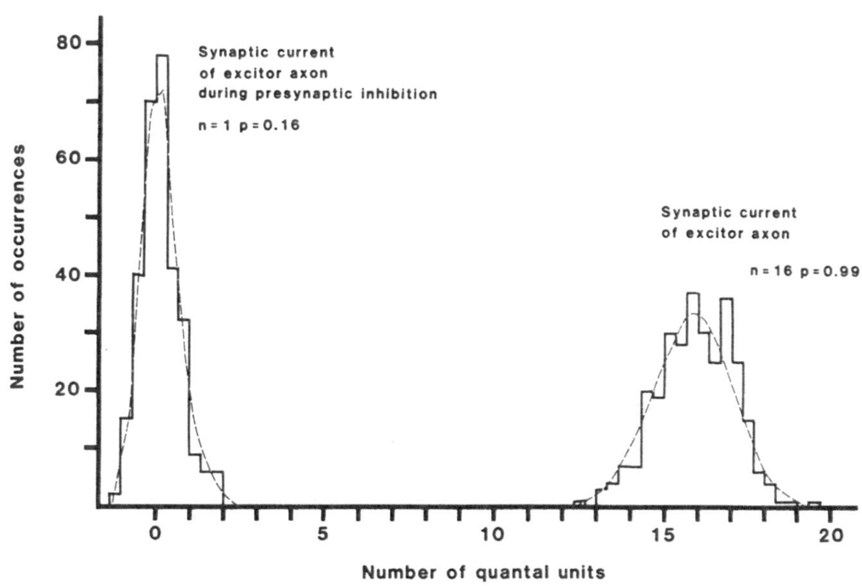

FIGURE 3. Changes in binomial parameters during presynaptic inhibition at a crab neuromuscular junction (from work of Tse, 1986; see Tse and Atwood, 1986). Analysis of transmitter release indicates conversion from a condition in which 16 responding units release with high probability to one in which only one responding unit releases with fairly low probability. Thus, the majority of available units (synapses?) have been removed from the responding population by presynaptic inhibition.

responding units each release with very high probability (Fig. 3). On application of presynaptic inhibition, both *n* and *p* are strongly reduced. Many of the resulting distributions cannot be well fitted by a binomial distribution; trial-to-trial conditions may vary because of local changes in inhibitory effectiveness. In any event, it is clear that some of the originally reliable synapses become either completely inactivated or marginal during presynaptic inhibition.

Morphological studies in which sites of presynaptic inhibition were focally labeled by activity-dependent uptake of horseradish peroxidase (Tse, 1986; Tse *et al.*, 1987) have shown that the number of excitatory synapses seen in serial reconstruction greatly exceeds the binomial *n*. Furthermore, some sites experiencing strong presynaptic inhibition may have relatively few axoaxonal synapses. These findings reinforce the hypothesis of a "reserve" of inactive excitatory synapses and lend further credence to the idea of a "gating" function for axoaxonal synapses.

4. LONG-TERM FACILITATION

Following a relatively short period of stimulation at 10–20 Hz, excitatory synapses of the crayfish opener muscle show a persistent enhancement of transmitter release that is not caused by changes in quantal unit amplitude or by changes in membrane potential or spike amplitude of the presynaptic terminal (Fig. 4). Changes in the metabolism of the nerve terminal mediated by sodium and calcium ions are likely involved (Wojtowicz and Atwood, 1985, 1986). We envision possible recruitment of previously inactive synapses to the responding pool (Fig. 5).

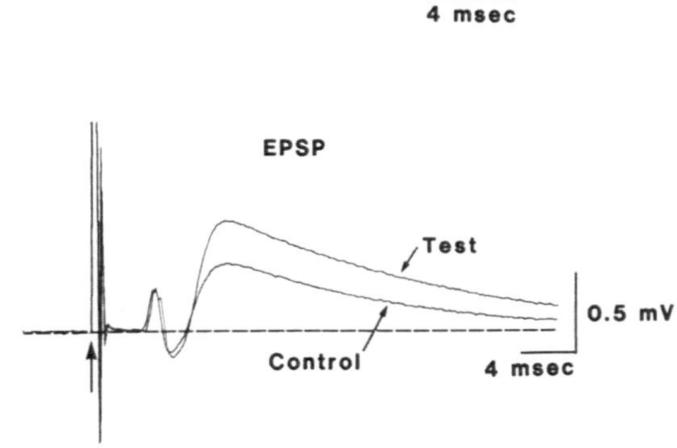

FIGURE 4. Long-term facilitation in the crayfish opener muscle (from Wojtowicz and Atwood, 1986). Intracellular records of postsynaptic potentials (lower traces) recorded at a frequency of 5 Hz before ("Control") and after ("Test") induction of long-term facilitation by a 20-Hz tetanus. Intracellular records from the presynaptic nerve branch (top traces) showed no long-lasting changes in either the resting membrane potential or the action potential.

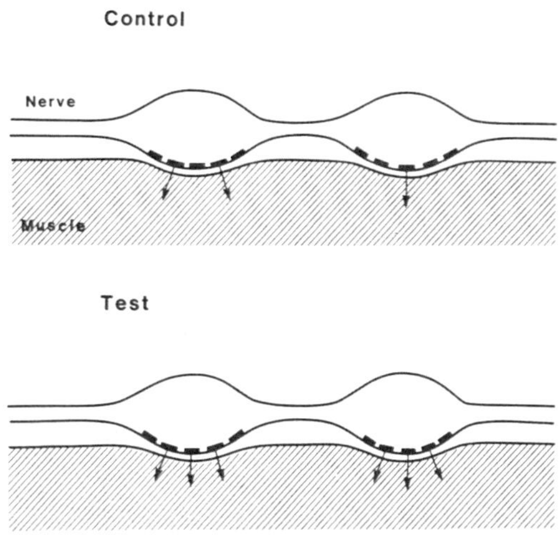

FIGURE 5. Synaptic recruitment model for long-term facilitation (from Wojtowicz and Atwood, 1986). The percentage of responding synapses is increased after induction of long-term facilitation.

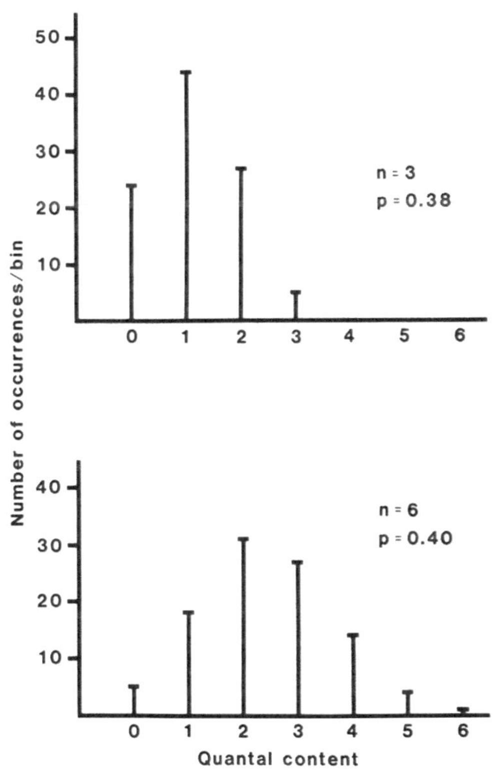

FIGURE 6. Binomial parameters determined for a crayfish muscle fiber before and after long-term facilitation (from Wojtowicz and Atwood, 1986). The histograms show distributions of quantal units released by 100 impulses. The number of responding units (n) increases from 3 to 6, but the average probability of response (p) is little altered.

Our statistical analyses of transmitter release for individual small muscle fibers of the crayfish opener muscle have shown that binomial distributions can often be matched to the amplitude distribution of synaptic potentials (Wojtowicz and Atwood, 1986). Following induction of long-term facilitation, binomial parameter n increases, although p shows little change (Fig. 6). Thus, the hypothesis of synaptic recruitment is borne out by the available evidence.

A problem with the analyses conducted to date is that probability of response is assumed to be uniform across the population of responding units. This is almost certainly an oversimplification (see Hirst *et al.*, 1981; Jack *et al.*, 1981). Further analysis must consider the likelihood of nonuniform response probabilities.

5. CONCLUSION

Both presynaptic inhibition and long-term facilitation have been clearly established as presynaptic in origin on the basis of rigorous criteria. It is probable that the more slowly occurring but longer-lasting phenomenon of long-term adaptation is also presynaptic (Lnenicka and Atwood, 1985a,b). Thus, the crayfish neuromuscular junction presents a situation in which presynaptic mechanisms of synaptic plasticity can be investigated at the cellular level in an accessible preparation.

Although it is well known that the simple binomial model of transmitter release is not entirely adequate for the process of transmitter release at the neuromuscular junction and elsewhere because of nonuniformities in p (Brown *et al.*, 1976; Silinsky, 1985; Lustig *et al.*, 1986), nevertheless, it provides a first approximation that is useful as an indication of what may be happening in the nerve terminal. Our estimates show substantial changes in n for both presynaptic inhibition and long-term facilitation. A possible interpretation is that the number of responding synapses can change in response to physiological conditions. Morphological studies suggest a large pool of available synapses on crustacean nerve terminals. Therefore, it is not unreasonable to postulate that individual synapses can readily be switched on and off according to functional demands. This theme is explored in more detail by Atwood and Wojtowicz (1986).

ACKNOWLEDGMENTS. Research support was received from the Medical Research Council of Canada and NSERC, Canada. Technical assistance was provided by Dr. Leo Marin and Ms. Marianne Hegström-Wojtowicz.

REFERENCES

Atwood, H. L., 1976, Organization and synaptic physiology of crustacean neuromuscular systems, *Prog. Neurobiol.* **7**:291–391.

Atwood, H. L., 1982, Synapses and neurotransmitters, in: *The Biology of Crustacea*, Vol. 3 (H. L. Atwood, and D. C. Sandeman, eds.), Academic Press, New York, pp. 105–150.

Atwood, H. L., and Bittner, G. D., 1971, Matching of excitatory and inhibitory inputs to crustacean muscle fibers, *J. Neurophysiol.* **34**:157–170.

Atwood, H. L., and Marin, L., 1983, Ultrastructure of synapses with different transmitter-releasing characteristics on motor axon terminals of a crab, *Hyas areneas*, *Cell Tissue Res.* **231**:103–115.

Atwood, H. L., and Morin, W. A., 1970, Neuromuscular and axoaxonal synapses of the crayfish opener muscle, *J. Ultrastruct. Res.* **32**:351–369.

Atwood, H. L., and Wojtowicz, J. M., 1986, Short-term and long-term plasticity at crustacean motor synapses, *Int. Rev. Neurobiol.* **28**:275–362.

Atwood, H. L., Swenarchuk, L. E., and Gruenwald, C. R., 1975, Long-term synaptic facilitation during sodium accumulation in nerve terminals, *Brain Res.* **100:**198–204.

Atwood, H. L., Stevens, J. K., and Marin, L., 1984, Axoaxonal synapse location and consequences for presynaptic inhibition in crustacean motor axon terminals, *J. Comp. Neurol.* **225:**64–74.

Atwood, H. L., Lnenicka, G. A., and Marin, L., 1985, Morphological responses to conditioning stimulation in a phasic motor axon of crayfish (*Procambarus clarkii*), *J. Physiol.* (*Lond.*) **365:**26P.

Baxter, D. A., and Bittner, G. D., 1981, Intracellular recordings from crustacean motor axons during presynaptic inhibition, *Brain Res.* **223:**422–428.

Breen, C., and Atwood, H. L., 1983, Octopamine—a neurohormone with presynaptic activity-dependent effects at crayfish neuromuscular junctions, *Nature* **303:**716–718.

Brown, H. B., Perkel, D. H., and Feldman, M. W., 1976, Evoked neurotransmitter release: Statistical effects of nonuniformity and nonstationarity, *Proc. Natl. Acad. Sci. U.S.A.* **73:**2913–2917.

Bryan, J. S., and Atwood, H. L., 1981, Two types of synaptic depression at synapses of a single crustacean motor axon, *Mar. Behav. Physiol.* **8:**99–121.

Dixon, D., and Atwood, H. L., 1985, Crayfish motor nerve terminal's response to serotonin examined by intracellular microelectrode, *J. Neurobiol.* **16:**409–424.

Dudel, J., 1965a, Facilitatory effects of 5-hydroxy-tryptamine on the crayfish neuromuscular junction, *Naunyn Schmiedebergs Arch. Exp. Pathol. Pharmacol.* **249:**515–528.

Dudel, J., 1965b, The mechanism of presynaptic inhibition at the crayfish neuromuscular junction, *Pflugers Arch.* **284:**66–80.

Dudel, J., and Kuffler, S. W., 1961, Presynaptic inhibition at the crayfish neuromuscular junction, *J. Physiol.* (*Lond.*) **155:**543–562.

Florey, E., and Rathmayer, M., 1978, The effects of octopamine and other amines on the heart and on neuromuscular transmission in decapod crustaceans: Further evidence for a role as neurohormone, *Comp. Biochem. Physiol.* **61C:**229–237.

Fritz, L. C., Atwood, H. L., and Jahromi, S. S., 1980, Lobster neuromuscular junctions treated with black widow spider venom: Correlation between ultrastructure and physiology, *J. Neurocytol.* **9:**699–721.

Fuchs, P. A., and Getting, P. A., 1980, Ionic basis of presynaptic inhibitory potentials at crayfish claw opener, *J. Neurophysiol.* **43:**1547–1557.

Glusman, S., and Kravitz, E. A., 1982, The action of serotonin on excitatory nerve terminals in lobster nerve–muscle preparations, *J. Physiol.* (*Lond.*) **325:**223–241.

Govind, C. K., Meiss, D. E., and Pearce, J., 1982, Differentiation of identifiable lobster neuromuscular synapses during development, *J. Neurocytol.* **11:**235–247.

Hirst, G. D. S., Redman, S. J., and Wong, K., 1981, Post-tetanic potentiation and facilitation of synaptic potentials evoked in cat spinal motoneurones, *J. Physiol.* (*Lond.*) **321:**97–109.

Hoyle, G., and Wiersma, C. A. G., 1958, Excitation at neuromuscular junctions in crustacea, *J. Physiol.* (*Lond.*) **143:**403–425.

Jack, J. J. B., Redman, S. J., and Wong, K., 1981, Modifications to synaptic transmission at group Ia synapses on cat spinal motoneurones by 4-aminopyridine, *J. Physiol.* (*Lond.*) **321:**111–126.

Jacobs, J. R., and Atwood, H. L., 1981, Long term facilitation of tension in crustacean muscle and its modulation by temperature, activity and circulating amines, *J. Comp. Physiol.* **144:**335–343.

Jahromi, S. S., and Atwood, H. L., 1974, Three-dimensional ultrastructure of the crayfish neuromuscular apparatus, *J. Cell Biol.* **63:**599–613.

Katz, B., and Miledi, R., 1968, The role of calcium in neuromuscular facilitation, *J. Physiol.* (*Lond.*) **195:**481–492.

Korn, H., Mallet, A., Triller, A., and Faber, D. S., 1982, Transmission at a central inhibitory synapse. II. Quantal description of release, with a physical correlate for binomial *n*, *J. Neurophysiol.* **48:**679–707.

Kravitz, E. A., Beltz, B., Glusman,, S., Goy, M., Harris-Warrick, R., Johnston, M., Livingstone, M., and, Schwarz, T., 1985, The well-modulated lobster: The roles of serotonin, octopamine, and proctolin in the lobster nervous system, in: *Model Neural Networks and Behavior* (A. I. Selverston ed.), Plenum Press, New York, pp. 339–360.

Lnenicka, G. A., and Atwood, H. L., 1985a, Age-dependent long-term adaptation of crayfish phasic motor axon synapses to altered activity, *J. Neurosci.* **5:**459–467.

Lnenicka, G. A., and Atwood, H. L., 1985b, Long-term facilitation and long-term adaptation at synapses of a crayfish phasic motoneuron, *J. Neurobiol.* **16:**97–110.

Lnenicka, G. A., and Atwood, H. L., 1987, Long-term changes in neuromuscular synapses with altered sensory input to a crayfish mononeuron, *Exp. Neurol.* (in press).

Lustig, C., Parnas, H., and Segel, L. A., 1986, On the quantal hypothesis of neurotransmitter release: An explanation for the calcium dependence of the binomial parameters, *J. Theor. Biol.* **120:**205–213.

Parnas, H., Dudel, J., and Parnas, I., 1982, Neurotransmitter release and its facilitation in crayfish. 1. Saturation kinetics of release, and of entry and removal of calcium, *Pflugers Arch.* **393:**1–14.

Pearce, J., Govind, C. K., and Shivers, R. R., 1986, Intramembranous organization of lobster excitatory neuromuscular synapses, *J. Neurocytol.* **15:**241–252.

Sherman, R. G., and Atwood, H. L., 1971, Synaptic facilitation: Long-term neuromuscular facilitation in crustaceans, *Science* **171:**1248–1250.

Silinsky, E. M., 1985, The biophysical pharmacology of calcium-dependent acetylcholine secretion, *Pharmacol. Rev.* **37:**81–132.

Smith, D. O., 1983, Variable activation of synaptic release sites at the neuromuscular junction, *Exp. Neurol.* **80:**520–528.

Takeuchi, A., and Takeuchi, N., 1966a, A study of the inhibitory action of gamma-aminobutyric acid on the neuromuscular transmission in the crayfish, *J. Physiol.* (*Lond.*) **183:**418–432.

Takeuchi, A., and Takeuchi, N., 1966b, On the permeability of the presynaptic terminal of crayfish neuromuscular junction during synaptic inhibition and the action of gamma-aminobutyric acid, *J. Physiol.* (*Lond.*) **183:**433–449.

Tse, F. W., 1986, *Presynaptic Inhibition of Crustacean Neuromuscular Synapses*, Ph.D. thesis, University of Toronto.

Tse, F. W., and Atwood, H. L., 1986, Presynaptic inhibition at the crustacean neuromuscular junction and elsewhere, *News Physiol. Sci.* **1:**47–50.

Tse, F. W., Marin, L., and Atwood, H. L., 1987, Focal labeling of terminals with active synapses recorded by an extracellular macro-patch electrode, *J. Neurosci. Methods* **21:**17–29.

Wernig, A., 1972, Changes in statistical parameters during facilitation at the crayfish neuromuscular junction, *J. Physiol.* (*Lond.*) **226:**751–759.

Wojtowicz, J. M., and Atwood, H. L., 1985, Correlation of presynaptic and postsynaptic events during establishment of long term facilitation at the crayfish neuromuscular junction, *J. Neurophysiol.* **54:**220–230.

Wojtowicz, J. M., and Atwood, H. L., 1986, Long-term facilitation alters transmitter releasing properties at the crayfish neuromuscular junction, *J. Neurophysiol.* **55:**484–498.

Zucker, R. S., 1973, Changes in the statistics of transmitter release during facilitation, *J. Physiol.* (*Lond.*) **229:**787–810.

Zucker, R. S., 1982, Processes underlying one form of synaptic plasticity: Facilitation, in: *Conditioning* (C. D. Woody, ed.), Plenum Press, New York, pp. 249–264.

Zucker, R. S., and Bruner, J., 1977, Long-lasting depression and the depletion hypothesis at crayfish neuromuscular junctions, *J. Comp. Physiol.* **121:**223–240.

Zucker, R. S., and Lara-Estrella, L. O., 1983, Post-tetanic decay of evoked and spontaneous transmitter release and a residual-calcium model of synaptic facilitation at crayfish neuromuscular junctions, *J. Gen. Physiol.* **81:**355–372.

Postsynaptic Events Associated with Long-Lasting Activity-Induced Changes in Excitability of Neocortical Neurons

Studies in the Anesthetized Rat and in Slices *in Vitro*

LYNN J. BINDMAN, TIM MEYER, and CLIVE A. PRINCE

1. INTRODUCTION

It is well established that changes in excitability of neocortical neurons that persist, undiminished, for tens of minutes can be induced by increasing the firing rate for just a few minutes. Any of a number of experimental procedures can be used to increase firing, including depolarizing current, synaptic activation, and iontophoresis of glutamate or acetylcholine (Burns, 1957; Bindman *et al.*, 1964; Bindman and Boisacq-Schepens, 1966; McCabe, 1973; Woody *et al.*, 1978). The site of the underlying change has been localized to the neocortex in experiments carried out a neuronally isolated slabs of cortex *in vivo* (Bliss *et al.*, 1968). Both increases and decreases in excitability of different neurons were observed following stimulation in a number of these studies.

Subsequently, long-lasting activity-induced increases in synaptic transmission were produced in the hippocampus (Bliss and Lomo, 1973), and there is much current work, and controversy, on the contributions of pre- and postsynaptic mechanisms to the induction of long-term potentiation in hippocampal neurons (see Malinow and Miller, 1986; Wigström *et al.*, 1986). Postsynaptic mechanisms have been strongly implicated in the production of activity-induced long-lasting increases in excitability in neocortical neurons by Bindman *et al.* (1979, 1982) in experiments in which synaptic transmission was blocked.

We argued that a long-lasting postsynaptic change in excitability might well be associated with a persistent alteration in membrane properties that could be detected using intracellular recording and measurement. For example, Woody *et al.* (1978) had produced persistent increases in excitability of neocortical neurons following iontophoretic application of acetylcholine in combination with depolarization-induced discharge; they found

LYNN J. BINDMAN, TIM MEYER, and CLIVE A. PRINCE • Department of Physiology, University College London, London WC1E 6BT, England.

that increased excitability was associated with increased apparent input resistance. In fact, Bindman and Prince (1986) found that following several minutes of increased firing brought about by intracellular application of depolarizing currents, prolonged changes in excitability were associated with prolonged changes in apparent input resistance. However, limitations on the stability of intracellular recordings and on the manipulations and measurements that can be carried out *in vivo* pointed to the use of isolated slices of neocortex for further investigations. We here report (1) comparative data we obtained of electrical properties of neurons *in vivo* and in the slice preparation and (2) observations from our *in vivo* studies and preliminary data from neurons in slices suggesting that the maintenance of prolonged changes in excitability, induced either by intracellular stimulation with depolarizing current or by synaptic activation, was associated in different cells with different modifications of cellular properties.

2. METHODS

2.1. Preparations

Methods used for rats anesthetized with urethane were reported in Bindman and Prince (1986). Methods used to prepare slices incorporating sensorimotor cortex can be obtained from Bindman *et al.* (1985). Coronal slices 400 or 450 μm thick were placed on lens tissue supported by nylon mesh in a recording chamber with recirculating artificial CSF at 35–37°C below the slice and warm humidified 95% O_2/5% CO_2 above it. Composition of artificial CSF was NaCl, 132mM when $CaCl_2$ was 2.5mM and 134mM when $CaCl_2$ was 1.1mM; $NaHCO_3$, 19mM; KCl, 2mM; KH_2PO_4, 1.2mM; $MgSO_4$, 1.1mM; and D-glucose, 10mM.

2.2. Intracellular Recordings and Measurements

Identical recording techniques and methods of measurement were used for both the anesthetized rat and for slices *in vitro* (see Bindman and Prince, 1986). Current was injected via a bridge circuit into the microelectrode, and bridge balance was adjusted throughout the recording to compensate for electrode resistance. Capacitance compensation for the electrode ensemble was adjusted to be slightly undercompensated.

Measurements for comparative data were made from neurons from which a stable spike amplitude and resting potential could be recorded for at least 10 min (mean 63 min) after recovery from any initial injury following penetration. Spike amplitudes were $\geqslant 72$ mV ($\geqslant 61$ mV threshold to peak) and varied by $\leqslant 4$ mV in the stable period of recording; resting membrane potentials were $\geqslant 60$ mV and varied by $\leqslant 4$ mV/10 min. Measurements of long-lasting changes in excitability were made in cells from which stable recordings were obtained for at least 30 min that also met the criteria listed above during prestimulation periods.

Excitability was measured as one or more of the following: spontaneous firing rate *in vivo,* rate of firing in response to depolarizing constant-current pulses, or latency of firing on the depolarizing current pulse.

Input resistance was measured using depolarizing and hyperpolarizing constant-current pulses. For the comparative data we used the linear range of voltage responses to hyperpolarizing current pulses (approximately 0 to -0.9 nA for neurons *in vivo* and

0 to −0.4 nA for neurons in slice). Averages of eight successive responses to each current step were obtained, and the peak amplitude was plotted against applied current. The slope of the linear portion in the hyperpolarizing range was calculated. For some neurons *in vivo,* slope resistances were obtained for membrane potential versus current (Bindman and Prince, 1986).

The averaged voltage response to hyperpolarizing constant-current pulses was also used to obtain the membrane time constant. Since the mean apparent input resistance in slices was twice that *in vivo,* the time constants in slices were measured with approximately half the current (mean values −0.37 and −0.65 nA, respectively) to keep the sag of voltage low and comparable in the two situations. Semilogarithmic plots were made of the averaged voltage response against time during charging of the cell. The charging response was well fitted by a single exponential for most cells; the membrane time constant was calculated as the negative reciprocal of the best-fit line. The r^2 value for the line was 0.99 or greater for all neurons. The voltage tended to sag after about 15 msec in some cells, and the points after the sag were omitted. A few neurons in slices showed a second-order time constant, i.e., a transient early response reflecting a nonisopotential region of the neuron (Stafstrom *et al.,* 1984). The early response was also omitted from the calculation of the membrane time constant. The membrane capacity (C_m) was estimated from the membrane time constant (τ_m) and the input resistance (R_{in}); i.e., $C_m = \tau_m/R_{in}$.

3. SIGNIFICANT DIFFERENCES IN ELECTRICAL PROPERTIES OF NEOCORTICAL NEURONS IN SLICES COMPARED WITH THOSE *IN VIVO*

There have been many studies of electrical properties of neocortical neurons in intact brain and in isolated brain slices, but comparisons have been made across laboratories using different techniques. Since neurons in slices are in abnormal condition in several respects, we compared measurements made of electrical properties of 19 neocortical neurons from the anesthetized rat with 27 neurons from slices of the same area of neocortex *in vitro.* More than 80% of penetrations into neurons comprising the two samples were made at depths greater than 800 μm below the pial surface, i.e., in layers V and VI. The data are summarized in Table I.

The significant differences were in spike amplitude measured from the resting potential (item 3 in Table I): firing threshold (4 and 5), input resistance (6), time constant (7), and membrane capacity (8). Although mean spike amplitude measured from the resting potential (3) was significantly larger in neurons from slices, the increased amplitude was partly attributable to the larger mean resting potential; the remaining mean increase in spike overshoot potential was not significant.

We found the mean firing threshold (5) to be significantly raised in slices by 5 mV ($P = 0.037$, one-tailed t test). The raised threshold is predictable from the elevated divalent cation concentrations (Frankenheuser, 1957) used in the slice-bathing medium by us and most workers maintaining slices near body temperature. With Ca^{2+} at 2.5 mM and Mg^{2+} at 1.1 mM (as for all the slice data in Table I), we obtained stable recordings in the majority of experiments. However, when we used Ca^{2+} and Mg^{2+} concentrations (each at 1.1 mM) close to those in rat CSF, we could not obtain stable intracellular recordings.

A major and consistent difference commented on by previous authors is that neurons

TABLE I. Comparison of Properties of Neocortical Neurons *in Vivo* and in Slices *in Vitro*[a]

	In vivo (mean ± S.E.)	In slices (mean ± S.E.)
1. Resting potential (mV)	−80.9 ± 2.0	−84.3 ± 2.3
2. Spike amplitude (mV, threshold to peak)	78.8 ± 1.7	78.1 ± 1.2
3. Spike amplitude (mV from rp)	103.7 ± 2.7	111.7 ± 1.7
4. Spike threshold (mV from rp)	+24.8 ± 1.6	+33.5 ± 1.3
5. Spike threshold (membrane potential, mV)	−56.2 ± 2.0	−51.4 ± 1.7
6. Input resistance[b] (MΩ)	18.4 ± 1.6	35.6 ± 2.4
7. Time constant[c] (τ_m) (msec)	6.8 ± 0.44	9.7 ± 0.57
8. Membrane capacity (nF)	0.43 ± 0.04	0.29 ± 0.02

[a] Stable recordings maintained in each neuron for at least 10 min. *In vivo* sample n = 19 except for 7 and 8, when n = 17. Slice sample n = 27 except for 1 and 5, when n = 25.
[b] Resistance measured in negative linear range of *V–I* relationship.
[c] Time constant measured with negative pulses in linear range of *V–I* relationship.

in slices have on average a higher input resistance than those from intact brain; in our comparative data, the mean apparent input resistance (item 6 in Table I) in slice cells was twice that in the anesthetized rat ($P < 0.001$, two-tailed t test). Histological measurements of the length of dendritic branches in neocortical cells in sensory cortex of the rat gave a mean radius of 185 μm for basal dendrites and 161 μm for apical dendrites of layer V neurons (Winkelmann *et al.*, 1973). Obviously many neuronal processes will be lopped off in the slicing procedure. We attempted to assess the extent to which loss of surface membrane might contribute to the higher mean input resistance in neurons from slices by estimating membrane capacities. Table I item 8 shows that the mean membrane capacity of the slice sample was significantly smaller than the mean of the *in vivo* sample ($P < 0.005$) suggesting that the neurons in the slices had less surface membrane.

The smaller mean surface area of neurons in our slice sample could have arisen from loss of dendrites in slicing or, alternatively, from the greater stability of slice preparations, enabling stable intracellular recordings to be made from smaller cells in slices than *in vivo*. A smaller variance of membrane capacities would be expected in the slice sample if dendrites had been lost, whereas a larger variance would be expected if recordings had been made from both large and small cells. We found a smaller variance of membrane capacities in our slice sample compared with our *in vivo* sample (0.013 versus 0.025), which is consistent with loss of dendrites.

However, a smaller mean surface area of neurons could not have been the only factor determining the higher mean input resistance of the slice sample. When we considered a subgroup of neurons with membrane capacities of < 0.42 nF, an arbitrarily chosen value, then the mean membrane capacity of these cells was not significantly different in the two preparations (0.28 ± 0.03 nF, n = 8 *in vivo* versus 0.27 ± 0.02 nF, n = 25 in slices), but the mean input resistance of the subgroup of neurons from slices was still significantly greater than that of the *in vivo* subgroup (37.1 ± 2.4 MΩ and 21.5 ± 2.6 MΩ, respectively, $P < 0.001$). Therefore factors other than size, such as an increase in mean specific input resistance or a smaller current leak around the electrode or uncoupling of cells, must have contributed to the raised mean input resistance of the neurons in slices.

What are the implications of the differing electrical properties of neurons in slices from those *in vivo* for the study of long-term changes? First, the raised firing threshold reinforces the importance of examining changes in subthreshold currents or voltages (Stafstrom *et al.*, 1982, 1984a,b, 1985; Crill, 1986) and spike latency, even when firing rate is not altered in slice preparations. Secondly, the large increase in apparent input resistance in slices may be caused partly by abnormal extraneuronal ion concentrations and partly by loss of transmitter and modulator influences on receptors or voltage-sensitive channels (e.g., Halliwell and Adams, 1982). The differences are unlikely to arise from the anesthetic, since measurements of input resistance made in the awake cat by Woody and colleagues (Woody and Gruen, 1978; Woody *et al.*, 1985) are closer to those in our *in vivo* sample than to those in slices. The extent to which the abnormal conditions in slices may affect the production and maintenance of long-term changes in membrane properties of neocortical neurons is at present unknown and needs investigation.

4. POSTSYNAPTIC EVENTS ASSOCIATED WITH PROLONGED CHANGES OF EXCITABILITY OF NEOCORTICAL NEURONS *IN VIVO* AND IN SLICES

The studies of Bindman and Prince (1986) have been extended to include 14 neurons *in vivo* in which excitability and apparent input resistance were continually monitored before and after a few minutes of increased firing induced by intracellular current stimulation. Prolonged changes in excitability were produced in all these cells. We have also observed long-lasting activity-induced changes following stimulation in nine neurons in slices.

4.1. Prolonged Increases in Excitability

Nine persisting increases in excitability were observed following stimulation with intracellular depolarizing current in eight cells in the anesthetized rat; each increase was statistically significant. The increased excitability following stimulation lasted for as long as we monitored the aftereffect (mean duration after stimulation, 21 min, range 5 to 55 min). The stable prestimulation periods for these nine persisting increases had a mean duration of 15 min (range 5 to 31 min). In an additional cell, the firing returned to prestimulation levels by 3 min after stimulation.

In each case increased neuronal excitability was associated with a significant increase in apparent input resistance of the cell. We suspected that different mechanisms might be involved in different cells in the production of increases in excitability and input resistance because (1) there were variable membrane potential shifts following stimulation, with depolarization accompanying increased excitability in some cells but hyperpolarization or no change in others, and (2) the increased apparent input resistance was found over different ranges of membrane potential (relative to firing threshold) in different cells.

We have also been able to produce long-lasting increases in excitability in neurons in slices (up to 55 min following stimulation before monitoring ended). Firing rate was increased for a few minutes during stimulation either by exciting neurons synaptically or by intracellular injection of depolarizing current. We have observed increased excitability associated with increased apparent input resistance over a wide range of membrane potential (e.g., from firing threshold to > 30 mV negative to firing threshold) but have also found increased excitability without any change in apparent input resistance as for the unit illustrated in Figs. 1 to 4.

Figure 1 shows examples of responses of the neuron to constant depolarizing current pulses (0.55 nA) 10 min before the start of stimulation (above) and 6.5 min following stimulation (below). Stimulation consisted of 9.5 min of depolarizing pulses superimposed on a steady depolarization, which caused the neuron to fire at a mean rate of *ca.* 20 Hz (*ca.* 37 Hz for 0.55 sec each sec). Following stimulation the latency and interspike interval were reduced; there was no significant change in membrane potential. Figure 2 shows a plot of the interval between the two spikes before and after stimulation, where each point is the mean of ten successive measurements. The interspike interval was consistently reduced for 15 min following stimulation, at which point the neuron depolarized and analysis was abandoned.

Pulse plots were made at times indicated by PP on Fig. 2 to determine apparent

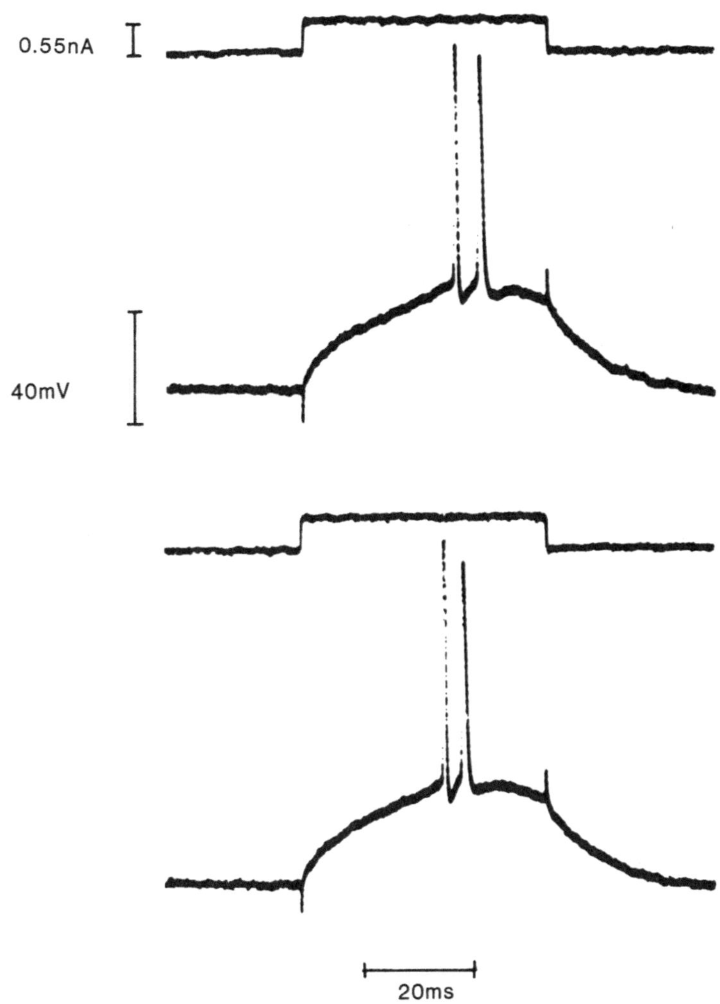

FIGURE 1. Examples of firing of layer VI neuron in slice preparation in response to 0.55-nA pulses applied intracellularly. Each pair of traces consists of current record above and voltage below. Upper records, 10 min before onset of stimulation; lower records, 6.5 min after stimulation ended. Resting potential about −80 mV; apparent input resistance 42 MΩ in linear range of voltage response to negative current pulses.

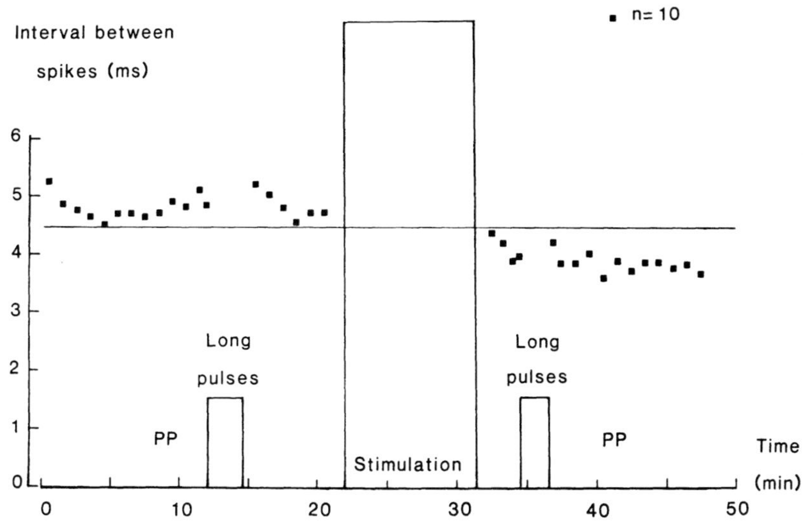

FIGURE 2. Graph of interval between first and second spikes elicited by constant-current pulses (as in Fig. 1) applied at 1 Hz. Each point is the mean of ten successive measurements made at indicated time. For 1 min immediately following stimulation, there is a posttetanic depression with hyperpolarization of the membrane, during which the cell did not fire. Subsequently, the interspike interval was consistently decreased whether the membrane potential was depolarized or hyperpolarized by about 2 mV.

input resistance over a range of membrane potentials. No change in apparent input resistance was detected between pre- and poststimulation pulse plots with hyperpolarizing or subthreshold depolarizing pulses (i.e., for membrane potentials ranging from 47 mV more hyperpolarized to 24 mV more depolarized than resting potential). The reduced latency of the first spike was probably caused by a slight reduction in firing threshold with no significant change in membrane potential.

Figure 3 shows examples of responses to 250-msec depolarizing pulses of 1 nA applied 7.5 min before (A and B) and 5 min after stimulation (A′ and B′). A and A′ show six consecutive traces superimposed, whereas B and B′ are single traces. Following stimulation, the latency of the first spike and most interspike intervals were significantly reduced. The afterhyperpolarization of each spike was also reduced (measured from spike threshold to peak hyperpolarization, as indicated by horizontal lines after the last spike in the train shown on B and B′). We made comparisons of afterhyperpolarization amplitudes for current pulses applied at resting potentials slightly hyperpolarized to the prestimulation value (B and B′) as well as slightly depolarized: the reduced interspike intervals and reduced afterhyperpolarizations were found in both situations and were not consequent on altered membrane potential following stimulation.

The graphs in Fig. 4 show interburst intervals expressed as frequency of firing for t_1 and t_2 for increasing levels of current. (Inset shows responses to +0.94-nA pulses with t_1, t_2, and t_3 labeled.) Each point is the mean frequency of six successive responses. Standard error bars are shown where they exceed the size of the symbol. The firing frequency is significantly increased for t_1 and t_2 but not for t_3 (not graphed).

The activity-induced changes in the cell illustrated could be detected only around firing threshold, in contrast to other cells in which increased apparent input resistance was seen over membrane potentials tens of millivolts negative to firing threshold.

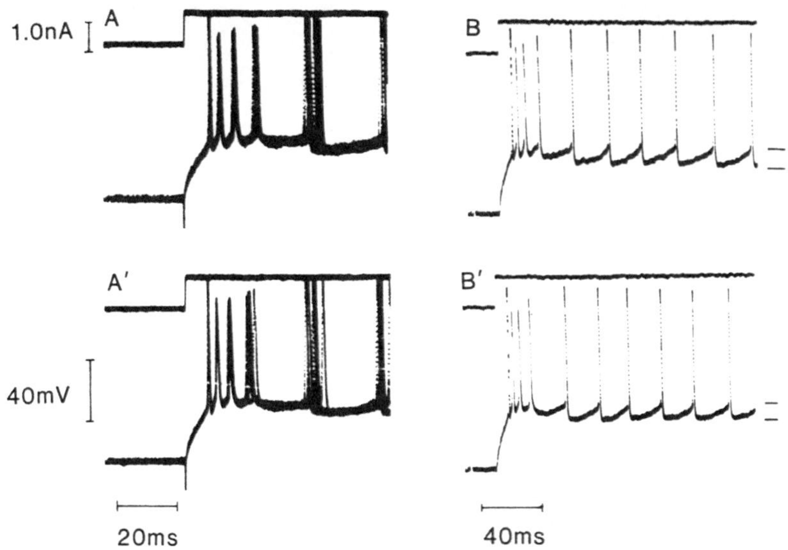

FIGURE 3. Responses of same neuron as in Fig. 2 to long depolarizing pulses of 1 nA before (A and B) and after (A' and B') stimulation, at times shown on graph of Fig. 2 by columns labeled long pulses. Note slight bridge imbalance in B. For details see text.

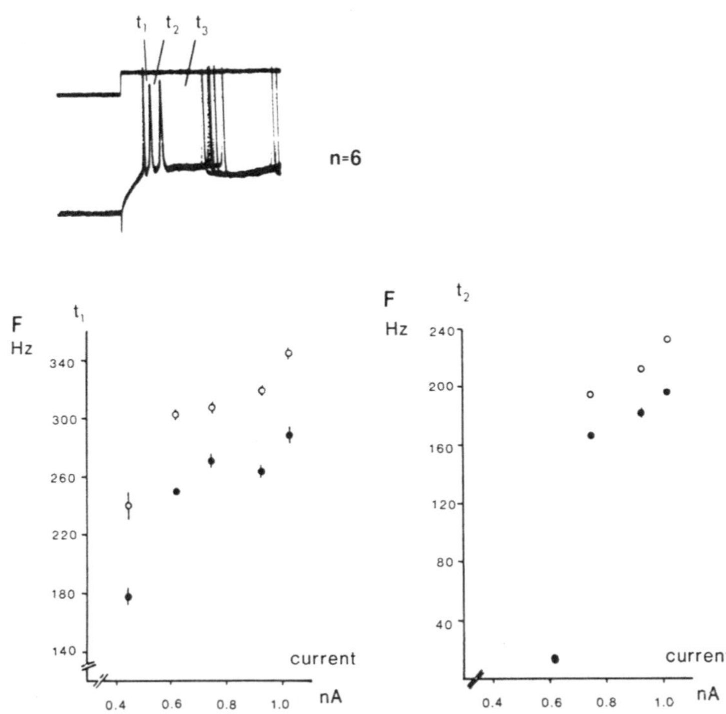

FIGURE 4. Graphs showing frequency of firing of same neuron to long (250-msec) pulses at different depolarizing current levels, with t_1 plotted on left and t_2 at right side. Inset shows six superimposed responses to 0.94 nA, with t_1, t_2, and t_3 labeled. Bars show ± 1 S.E. where it is larger than the symbols. Filled circles before and open circles after stimulation at times indicated by long pulses on Fig. 2.

4.2. Persistent Decreases in Excitability

We obtained decreases in excitability in neurons in both the anesthetized rat and in slices but do not discuss persistent decreases in this account.

5. DISCUSSION

We have found that increases in excitability lasting for tens of minutes could be produced in neocortical neurons by increasing the firing rate for a few minutes using intracellularly injected depolarizing current. We could detect changes in neuronal properties in association with the persistent increase in excitability, in both anesthetized rats and in slices *in vitro*. Increased apparent input resistance was found in most neurons when excitability was increased, but we also found increased excitability manifest as reduced interspike intervals and reduced afterhyperpolarizations, with no accompanying change in apparent input resistence. Thus, there is probably more than one postsynaptic mechanism supporting long-lasting increases in neocortical excitability produced with this method.

Both the changes in postsynaptic properties (i.e., increased input resistance and decreased afterhyperpolarizations) would confer a nonselective alteration of excitability upon the neuron, in that all synaptic inputs to it would be made more effective by an increased input resistance, and all excitatory inputs would evoke a higher frequency of firing when afterhyperpolarizations were reduced.

5.1. Mechanisms Involved in the Production and Maintenance of Nonselective Long-Lasting Changes in Postsynaptic Excitability

Among possible means for inducing persisting changes in postsynaptic conductances are (1) the action of excitatory transmitter on the postsynaptic membrane in combination with depolarization and firing of the cell and (2) depolarization and/or firing alone.

5.1.1. Transmitters

Extracellular iontophoresis of acetylcholine combined with cell firing induced by intracellularly applied depolarizing pulses, was shown to initiate a prolonged increase in excitability and input resistance in neocortical neurons (Woody *et al.*, 1978). Persisting increases in input resistance were also found by Baranyi and Feher (1981) in half the neocortical neurons in which an enhancement of EPSPs was produced by pairing of synaptic inputs with antidromic firing. It is not clear in the other half of the sample, whether the important events occurred in neurons presynaptic to the ones impaled or whether intracellular events other than increased input resistance sustained the EPSP enhancement (see Section 5.2). The transmitters involved in the experiments are not identified.

It is known that a number of transmitters modify various postsynaptic conductances in the short term; for example, acetylcholine reduces a voltage-sensitive potassium current (I_m) in hippocampal neurons (Halliwell and Adams, 1982). Increased firing frequency associated with reduced afterhyperpolarizations, with no accompanying change in input resistance (as in the cell of Figs. 3 and 4), has been brought about in hippocampal neurons by phorbol esters, drugs that, like some transmitters, activate protein kinase C intracellularly (Malenka *et al.*, 1986). However, it is not known if such transmitter-induced modifications in I_m or afterhyperpolarizations are persistent.

5.1.2. Cellular Processes Initiated by Depolarization and/or Firing

One of the methods of stimulation we used was chosen to bypass synaptic inputs. This method was intracellularly applied depolarizing current adequate to produce repetitive firing. While the activation of recurrent excitatory circuits by this method cannot be ruled out *in vivo,* we have not seen any recurrent excitatory potentials in slices. Moreover, recurrent excitation could not have accounted for the prolonged increases in excitability in cat neocortical neurons produced by repetitive, antidromic firing in the presence of synaptic block by Mg^{2+} (Bindman *et al.*, 1979). We consider that the prolonged excitation we induced by intracellular depolarizing current could have initiated persisting changes in postsynaptic conductances without the mediation of transmitter.

It is known that the entry of calcium ions into a neuron can initiate a delayed decrease in potassium conductance reflected by a persistent increase in input resistance and depolarization (Alkon, 1984). In the experiments in cat neocortex, in which long-lasting increases in excitability were produced in the presence of sufficient Mg^{2+} to block synaptic transmission, and hence presumably to block entry of Ca^{2+} via voltage-sensitive channels into the neurons, a raised intracellular $[Ca^{2+}]$ could have occurred secondary to a raised intracellular $[Na^{+}]$ (Lev-tov and Rahamimoff, 1980; Gustafsson and Wigstrom, 1983).

While not excluding the participation of transmitters or modulators in the production of long-lasting changes in excitability in naturally functioning cortex, our experiments demonstrate that adequate postsynaptic excitation can trigger long-lasting increases in excitability associated with maintained changes in postsynaptic events.

5.2. Selective Increases in Excitability of Particular Synaptic Inputs

Postsynaptic control of the induction of long-term potentiation (LTP) of synaptic transmission at hippocampal CA1 neurons has been demonstrated by Wigstrom *et al.* (1986), Wigstrom and Gustafsson (1986), and Kelso *et al.* (1986). Using a Hebbian model of associative learning, they produced LTP by repetitive pairing of intracellularly applied depolarizing current with a stimulus to an afferent pathway, such that the responses occurred within a few tens of milliseconds of one another. A second synaptic response to stimulation of a different input pathway which was timed to occur 400 msec after the intracellular depolarization, was not potentiated. Bindman *et al.* (1987) repeated the experimental paradigm in neocortical neurons, and found that LTP could be induced in about 20% of neurons by pairing postsynaptically applied depolarizing current with a synaptic response. They also confirmed that there was no change in an unpaired synaptic EPSP, nor was there any change in input resistance of the cells when LTP of one input had been elicited.

Thus, the long-lasting increase in efficacy of synaptic transmission was selective in that it was confined to the EPSP that was paired temporally with the intracellular depolarization and firing. Bindman *et al.* (1987) found that increased EPSPs were detectable after about ten pairings (intracellular depolarizing pulses were 100 to 200 msec, and the neurons fired 6 to 10 times/pulse) although in fact they used 25 to 50 pairings in the production of LTP that persisted, undiminished, for up to 50 min. The postsynaptic depolarization giving rise to LTP was therefore of much shorter duration than in the experiments that gave rise to postsynaptic changes in input resistance or afterhyperpolarizations.

Although a postsynaptic depolarization can be used to *induce* LTP, the evidence is against there being nonselective changes in postsynaptic properties associated with the *maintenance* of LTP, in either hippocampus or neocortex. This finding contrasts with those from the experiments described in this chapter, in which a more prolonged postsynaptic depolarization gave rise to long-lasting increases in excitability that were associated with maintained changes in postsynaptic membrane properties.

ACKNOWLEDGMENTS. We thank the Wellcome Trust for its support of T. M. and J. J. B. Jack for helpful advice.

REFERENCES

Alkon, D. L., 1984, Calcium-mediated reduction of ionic currents: A biophysical memory trace, *Science* **226**:1037–1045.

Baranyi, A., and Feher, O., 1981, Synaptic facilitation requires paired activation of convergent pathways in the neocortex, *Nature* **290**:413–415.

Bindman, L. J., and Boisacq-Schepens, N., 1966, Persistent changes in the rate of firing of single, spontaneously active cortical cells in the rat produced by peripheral stimulation, *J. Physiol. (Lond.)* **185**:14–17P.

Bindman, L. J., and Prince, C. A., 1986, Persistent changes in excitability and input resistance of cortical neurons in the rat, in: *Neural Mechanisms of Conditioning* (D. L. Alkon and C. D. Woody, eds.), Plenum Press, New York and London, pp. 291–305.

Bindman, L. J., Lippold, O. C. J., and Redfearn, J. W. T., 1964, The action of brief polarizing currents on the cerebral cortex of the rat, (1) during current flow, and (2) in the production of long-lasting after-effects, *J. Physiol. (Lond.)* **172**:369–382.

Bindman, L. J., Lippold, O. C. J., and Milne, A. R., 1979, Prolonged changes in excitability of pryamidal tract neurons in the cat, *J. Physiol. (Lond.)* **286**:457–477.

Bindman, L. J., Lippold, O. C. J., and Milne, A. R., 1982, A postsynaptic mechanism underlying long-lasting changes in the excitability of pryamidal tract neurones in the anaesthetized cat, in: *Conditioning* (C. D. Woody, ed.), Plenum Press, New York, pp. 171–178.

Bindman, L. J., Meyer, T., and Pockett, S., 1987, Long-term potentiation in rat neocortical neurones in slices, produced by repetitive pairing of an afferent volley with intracellular depolarizing current, *J. Physiol. (Lond.)* **386**:90P.

Bindman, L. J., Meyer, T., and Prince, C. A., 1985, Intracellular measurements of the electrical properties of neurones in the cerebral cortex of the rat: Comparative data from slices *in vitro* and from the anaesthetized animal, *J. Physiol. (Lond.)* **341**:7–8P.

Bliss, T. V. P., and Lomo, T., 1973, Long-lasting potentiation of synaptic transmission in the dentate area of the anaesthetized rabbit following stimulation of the perforant path, *J. Physiol. (Lond.)* **232**:331–356.

Bliss, T. V. P., Burns, B. D., and Uttley, A. M., 1968, Factors affecting the conductivity of pathways in the cerebral cortex, *J. Physiol. (Lond.)* **195**:339–367.

Burns, B. D., 1957, Electrophysiologic basis of normal and psychotic function, in: *Psychotropic Drugs* (S. Garattini and V. Ghetti, eds.), Elsevier, Amsterdam, pp. 177–184.

Crill, W. E., 1986, A critique of modeling population responses for mammalian central neurons, in: *Neural Mechanisms of Conditioning* (D. L. Alkon and C. D. Woody, eds.), Plenum Press, New York, pp. 307–310.

Frankenheuser, B., 1957, The effect of calcium on the myelinated nerve fibre, *J. Physiol. (Lond.)* **137**:245–260.

Gustafsson, B., and Wigstrom, H., 1983, Hyperpolarization following long-lasting tetanic activation of hippocampal pyramidal cells, *Brain Res.* **275**:159–163.

Halliwell, J. V., and Adams, P. R., 1982, Voltage clamp analysis of muscarinic excitation in hippocampal neurons, *Brain Res.* **250**:71–92.

Kelso, S. R., Ganong, A. H., and Brown, T. H., 1986, Hebbian synapses in hippocampus, *Proc. Natl. Acad. Sci. U.S.A.* **83**:5326–5330.

Lev-Tov, A., and Rahamimoff, R., 1980, A study of tetanic and post-tetanic potentiation of miniature end-plate potentials at the frog neuromuscular junction, *J. Physiol. (Lond.)* **309**:247–273.

Malenka, R. C., Madison, D. V., Andrade, R., and Nicoll, R. A., 1986, Phorbol esters mimic some cholinergic actions in hippocampal pyramidal neurons, *J. Neurosci.* **6**:475–480.

Malinow, R., and Miller, J. P., 1986, Postsynaptic hyperpolarization during conditioning reversibly blocks induction of long-term potentiation, *Nature* **320:**529–530.

McCabe, B. J., 1973, *Production of Prolonged Changes in Cortical Neuronal Activity by Iontophoresis of* L-Glutamate in Anaesthetized and Unanaesthetized Rats, Ph.D. thesis, University of London.

Stafstrom, C. E., Schwindt, P. C., and Crill, W. E., 1982, Negative slope conductance due to a persistent subthreshold sodium current in cat neocortical neurons *in vitro, Brain Res.* **236:**221–226.

Stafstrom, C. E., Schwindt, P. C., Flatman, J. A., and Crill, W. E., 1984a, Properties of subthreshold response and action potential recorded in layer V neurons from cat sensorimotor cortex *in vitro, J. Neurophysiol.* **52:**244–263.

Stafstrom, C. E., Schwindt, P. C., and Crill, W. E., 1984b, Cable properties of layer V neurons from cat sensorimotor cortex *in vitro, J. Neurophysiol.* **52:**278–289.

Stafstrom, C. E., Schwindt, P. C., Chubb, M. C., and Crill, W. E., 1985, Properties of persistent sodium conductance and calcium conductance of layer V neurons from cat sensorimotor cortex *in vitro, J. Neurophysiol.* **53:**153–170.

Wigström, H., and Gustafsson, B., 1986, Postsynaptic control of hippocampal long-term potentiation, *J. Physiol. (Paris)* **81:**228–236.

Wigström, H., Gustafsson, B., Huang, Y.-Y. and Abraham, W. C., 1986, Hippocampal long-term potentiation is induced by pairing single afferent volleys with intracellularly injected depolarizing current pulses, *Acta Physiol. Scand.* **126:**317–319.

Winkelmann, A., Kunz, G., Winkelmann, E., Kirsche, W., Neumann, H., and Wenzell, J., 1973, Quantitative Untersuchungen an Dendriten der grossen pyramidenzellen der Lamin V des sensorichsen Cortex der Ratte, *J. Hirnforsch.* **14**:137–149.

Woody, C.D., and Gruen, E., 1978, Characterization of electrophysiological properties of intracellularly recorded neurons in the neocortex of awake cats: A comparison of the response to injected current in spike overshoot and undershoot neurons, *Brain Res.* **158**:343–357.

Woody, C.D., Swartz, B., and Gruen, E, 1978, Effects of acetylcholine and cyclic GMP on input resistance of cortical neurons in awake cats, *Brain Res.* **158**:373–395.

Woody, C.D., Bindman, L.J., Gruen, E., and Betts, B., 1985, Two different mechanisms control inhibition of spike discharge in neurons of cat motor cortex after stimulation of the pyramidal tract, *Brain Res.* **332**:369–375.

36

Postsynaptic Activity-Dependent Facilitation of Excitatory Synaptic Transmission in the Neocortex

ATTILA BARANYI and MAGDOLNA B. SZENTE

1. INTRODUCTION

Heterosynaptic facilitation (HF), analyzed first by Kandel and Tauc (1965), is an activity-dependent, lasting amplification of synaptic transmission in convergent neural pathways. As a cellular associative mechanism, HF can account for the temporal specificity and stimulus–response specificity of classical conditioning in *Aplysia* (Kandel and Tauc, 1965; Kandel, 1976). Heterosynaptic facilitation is a presynaptic phenomenon in molluscs, and action potential generation in the postsynaptic neuron is neither necessary nor sufficient for HF induction in these species (Wurtz *et al.*, 1967; Carew *et al.*, 1984). This finding is at odds with Hebb's postulate (Hebb, 1949) proposing that synaptic transmission is enhanced only when the presynaptic stimulus occurs in conjunction with action potential generation in the postsynaptic cell. Apart from *Aplysia,* plastic synaptic changes in agreement with the Hebb model have been documented in vertebrate ganglion cells (Schulman and Weight, 1976; Kumamoto and Kuba, 1983; Mochida and Libet, 1985), in the hippocampus (McNaughton *et al.*, 1978; Levy and Steward, 1979; Lynch *et al.*, 1983; Kuhnt, 1984; Scharfman and Sarvey, 1985; Wigstrom and Gustafsson, 1985; Wigstrom *et al.*, 1986; Kelso *et al.*, 1986; Malinow and Miller, 1986), and in the visual cortex (Hubel and Wiesel, 1965; Rauschecker and Singer, 1981; Bienenstock *et al.*, 1983; Fregnac and Imbert, 1984; Bear and Singer, 1986).

In earlier intracellular microelectrode studies on acute cats, we first demonstrated heterosynaptic facilitation of identified EPSPs in the mammalian neocortex, and a postsynaptic locus for responsible modifications was suggested (Baranyi and Feher, 1978, 1981a,c). It was also shown that intensive postsynaptic activation of cat neocortical neurons in itself could modify membrane excitability (McCabe, 1976; Bindman *et al.*, 1979; Tzebelikos and Woody, 1979; Baranyi and Feher, 1978, 1981b). In some studies antidromic activation was successfully applied as an unconditioned stimulus in Pavlovian conditioning situation (Black-Cleworth *et al.*, 1975; O'Brien *et al.*, 1977; Martin *et al.*,

ATTILA BARANYI and MAGDOLNA B. SZENTE • Department of Comparative Physiology, Attila Jozsef University of Sciences, Szeged, Hungary. *Present address of A.B.:* Neuropsychiatric Institute, University of California at Los Angeles Medical Center, Los Angeles, California 90024.

1980; Quinn and O'Brien, 1983). In this chapter we demonstrate that conjunctive pre- and postsynaptic firing activity induces synaptic facilitation via calcium-dependent postsynaptic processes in the motor cortex of the cat. A preliminary account of a portion of these results has already been reported (Baranyi and Szente, 1985).

2. INDUCTION OF HETEROSYNAPTIC FACILITATION IN THE NEOCORTEX BY A PAIRING PARADIGM

Experiments were performed on the exposed motor cortex of cats anesthetized with 35 mg/kg pentabarbital. The surgical methods and details of stimulation and recording procedures have been reported elsewhere (Baranyi and Feher, 1981a–d). Intracellular recording of neuronal activity and simultaneous intra- and juxtacellular iontophoretic drug injections were performed with multibarrel microelectrodes (Baranyi and Chase, 1984). Conventional intracellular techniques were used to measure the apparent input resistance (R_m) or to inject drugs into the recorded cell. Standard iontophoretic methods were employed for juxtacellular drug injections.

Since it is possible to maintain intracellular recordings for only a few hours in the neocortex *in vivo,* functional plasticity on only a short time scale could be monitored. In the motor cortex, large layer V pyramidal tract (PT) neurons as well as other cells (non-pyramidal-tract, nPT) have well-identified mono- and oligosynaptic inputs from the specific thalamic nucleus (VL) as well as from the callosal and somatosensory system or from other PT and nPT cells via the recurrent axon collateral system. Thus, these neurons are well suited for studying neuronal and synaptic plasticity. Another advantage is that the localization of synaptic endings from different sources is well known (at least in the case of PT cells; see Peters and Jones, 1984).

2.1. Relationship to Conjunctive Postsynaptic Firing

To induce plastic changes in neuronal excitability and synaptic transmission that can be related to conjunctive pre- and postsynaptic activity, we applied an EPSP–action potential pairing sequence (Baranyi and Feher, 1978, 1981; Fig. 1). This paradigm was based originally on classical behavioral conditioning of type I (Kandel and Tauc, 1965). A similar procedure was recently used to evoke plastic changes in the hippocampal slice (Kelso *et al.,* 1986; Gustafsson and Wigstrom, 1986). The stimulus parameters of the test pathway were set to produce an EPSP (conditioned stimulus, CS). The EPSP was then paired with postsynaptic action potentials (unconditioned stimulus, US; Kandel and Tauc, 1965). Our complete conditioning procedure consisted of the following steps. (1) In a habituation period, the test EPSP was elicited 150 times at a frequency of 0.1–1 Hz. The monosynaptic character of the EPSP was verified by its ability to follow 100-Hz stimulation with a short and constant latency. Thereafter, action potentials of the recorded cell were induced by heterosynaptic orthodromic stimulation (via different synaptic pathways) or by direct stimulation using 1 to 5-nA cathodal current pulses of 10–50 msec duration via the recording microelectrode; antidromic action potentials as USs were obtained in PT cells following stimulation of the bulbar pyramidal tract. (2) In a pseudoconditioning period, the test EPSP and the action potentials were given at 0.1–1 Hz, but without a constant interstimulus interval (ISI) for a total of 150 trials. (3) The

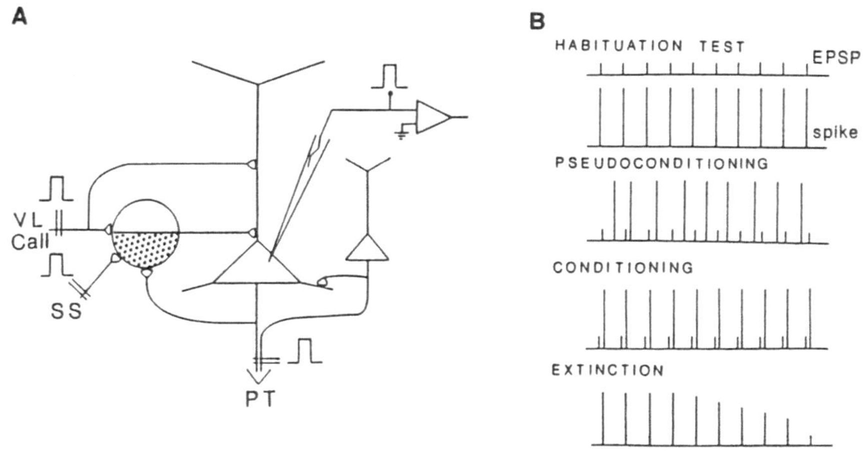

FIGURE 1. A: Schema of a single neural network in the cat motor cortex that illustrates the possibilities for heterosynaptic interactions. Stimulation of the thalamic ventrolateral nucleus (VL), the callosal (Call) and somatosensory system (SS), and the pyramidal tract (PT) ellicits mono- or oligosynaptic responses in PT and nPT cells (triangle and circle, respectively). Shaded part of the circle cell indicates that the nPT cell can be an excitatory or inhibitory interneuron. B: Conditioning paradigm for inducing plastic changes. For more details see text.

conditioning period consisted of test EPSP and action potential (heterosynaptic, antidromic, or current-induced) stimulus pairs applied 150 times with a fixed ISI (0–200 msec) at a frequency of 0.1–1 Hz. In the final extinction period, responses to the test stimuli were evoked every 1–10 sec until the enhanced synaptic response returned to the conditioning level. Records shown in Table I confirmed our previous findings on a relatively large sample of neurons (n = 961 cells).

TABLE I. Synaptic Facilitation in PT and nPT Neurons during Different Conditioning Circumstances[a]

Conditioned stimulus	Unconditioned stimulus: Antidromic action potential	Direct action potential	Synaptic action potential	PT cells	nPT cells
VL EPSP	46(165)	28(86)	53(198)	85(292)	42(157)
	28%	33%	27%	29%	27%
SS EPSP	22(70)	19(49)	56(174)	80(211)	17(82)
	31%	39%	32%	38%	21%
Call EPSP	9(34)	6(22)	15(52)	21(71)	9(37)
	26%	27%	29%	30%	24%
PT EPSP	—	6(37)	17(74)	—	23(111)
		16%	23%		21%
Total	77(269)	59(194)	141(498)	186(574)	91(387)
	29%	30%	28%	32%	24%

[a] Groups of 961 neurons conditioned with EPSP–action potential stimulus pairs. Test EPSPs were paired with heterosynaptic action potentials [pyramidal tract (PT), callosal (Call), or somatosensory (SS) activation from the contralateral radial nerve] in 498 cells. In 141 of these cells, HF could be observed (double framed group). The numbers indicate successful pairings leading to synaptic facilitation, and the values are also given in percentage of the total tested cases (in parentheses).

Based on these observations, we can draw the following conclusions. In the case when different CSs and USs were used, 16–39% of PT and nPT cells showed augmented synaptic responses to test stimuli following EPSP–action potential pairings. In the remaining cells, identical pairings produced no changes in the test response. Heterosynaptic facilitation was detected in 141 cells (28% of 498). According to these data, a higher percentage of PT cells than nPT cells showed synaptic facilitation for all test EPSP categories. In the case of randomized US presentations (pseudoconditioning) or after repetitive CS or US applications (habituation test) that excluded the regular coincidence of EPSPs and postsynaptic spikes, no lasting facilitation of the test response was detected. In addition, when a single action potential train (10–200 Hz for 1–10 sec) served as the US without pairing, no lasting enhancement of comparably long duration was observed. For example, a PT stimulus train of 100 Hz for 5 sec potentiated the VL EPSP for only 97 ± 14 sec (n = 41 PT neurons).

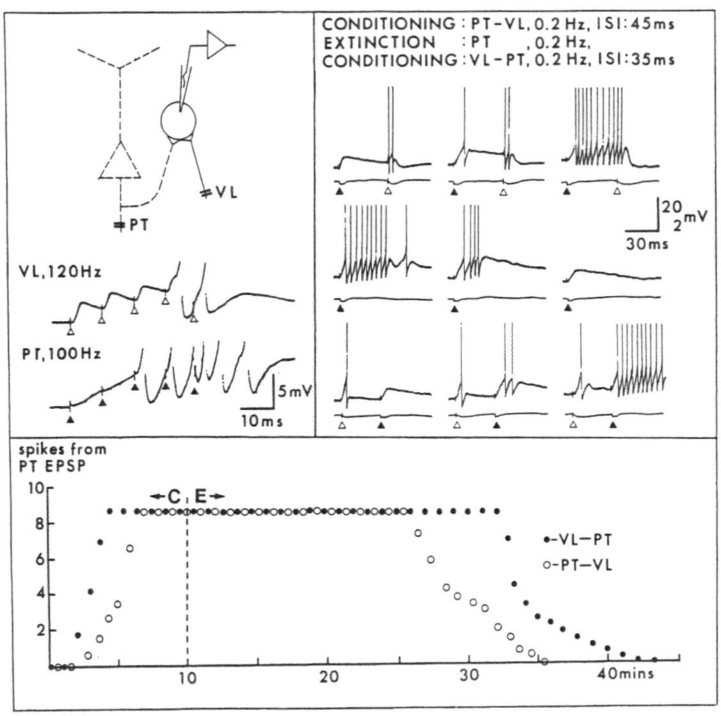

FIGURE 2. Heterosynaptic facilitation (HF) of the PT EPSP in an nPT neuron by EPSP–action potential (forward) and action potential–EPSP (backward) sequences. Upper left part illustrates the recording situation and monosynaptic responses of the nPT neuron (solid circle) to ventrolateral thalamic (VL) and pyramidal tract (PT) stimulation. Upper right part shows records during PT–VL pairings (denoted by upward dark and empty triangles, respectively) at the first, 50th, and 100th stimulus pairs. Upper traces are intracellular microelectrode records; lower traces are surface field potential records taken at a nearby cortical point. Large HF of the test EPSP appeared and resulted in an action potential burst to a single PT pulse. During the extinction period (middle traces), the HF of the PT response gradually declined. Records were taken at the tenth, 30th, and 40th minutes. Reconditioning with a backward sequence (VL spike–PT EPSP) resulted again in HF of the PT response. Records were taken at the fifth, 25th, and 50th pairings. Diagram on the bottom illustrates the increased efficiency of PT responses lasting for up to 43 min.

2.2. Stimulus Composition and Sequence in the Efficient Pairing Paradigm

In the case of 136 cells (out of 463), paired antidromic spikes as well as current-induced firing were as effective as orthodromic heterosynaptic activation in the induction of synaptic enhancement. This observation suggests a role of postsynaptic firing in the synaptic facilitation independently of the induction mode of spikes, i.e., whether induction was antidromic, current induced, or orthodromic. Both action potential–EPSP and EPSP-action potential sequences induced synaptic facilitation for a mean period of 22 ± 7 min (range of 10–60 min in 127 cells; Fig. 2). In the case of the action potential–EPSP sequence (backward pairing), ISIs in the range of 0 to 200 msec were effective. If the US triggered three to five spikes (burst) under pairing, synaptic facilitation occurred even at ISI values of 250 msec. In the case of the forward sequence (EPSP–spike), ISI values over 50–100 msec reduced the efficacy of the pairing procedure. It should be mentioned that the duration of oligosynaptic EPSPs in the neocortex is generally shorter than 50 msec (Takahashi, 1967; Purpura, 1972; Baranyi and Feher, 1981; Deschenes *et al.*, 1982).

3. MECHANISM OF POSTSYNAPTIC CHANGES UNDERLYING PERSISTENT SYNAPTIC FACILITATION

Enhancement of the test EPSP was the only consequence of pairing procedures in 62% of successfully conditioned PT and 60% of nPT neurons. In other cells an increase in input resistance (7.2 ± 2.1 to 10.4 ± 1.9 mΩ, n = 70 PT cells), membrane depolarization (5.6 ± 2.1 mV), and an elevation of background firing activity (182 ± 6.1% of the preconditioning value) accompanied synaptic enhancement. In 34 PT cells only an increase in input resistance paralleled the synaptic facilitation, without membrane depolarization or an increase in firing activity.

3.1. Is Synaptic Facilitation Exclusively Dependent on Time-Locked Postsynaptic Firing Activity?

The EPSP-spike stimulus pairs without constant ISI never induced persistent facilitation in the test response. In addition, no synaptic facilitation could be induced in the following tests: (1) when orthodromic activation of the US pathway was set to elicit only EPSPs (heterosynaptic pairings, tested in case of 62 cells), (2) when depolarizing current pulses were used as USs and provoked only membrane depolarization without firing (in 40 cells), and (3) when hyperpolarizing pulses were used as USs (in 40 cells).

3.2. Postsynaptic Hyperpolarization Prevents the Induction of Synaptic Facilitation

Involvement of postsynaptic activity-dependent processes was demonstrated by a series of experiments (Fig. 3) in which paired action potential generation in the recorded neuron was inhibited by intracellular anodal current pulses (n = 47 neurons; see also Baranyi and Feher, 1981a–c) and juxtasomatic GABA injections (n = 33 cells). Neither synaptic facilitation nor HF could be detected, although similar pairing procedures in the same cell were highly efficient before injections. Similar observations were recently

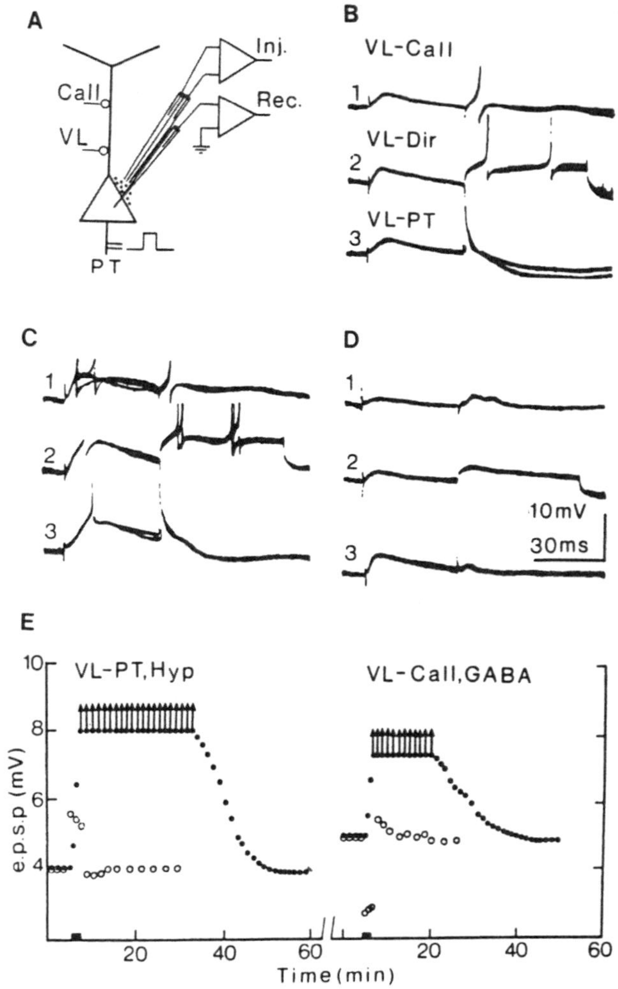

FIGURE 3. Inhibition of conditioned facilitation of monosynaptic VL EPSPs in fast PT neurons by preventing postsynaptic firing. A: Stimulation and recording arrangements. Intracellular recordings (Rec.) were made simultaneously with juxtasomatic iontophoretic GABA injections (Inj.). The VL stimulus parameters were set to induce monosynaptic EPSPs in the PT cells while callosal stimuli (Call) provoked orthodromic action potentials. B: Control responses in three different PT cells at the beginning of the pairing procedure (third through fifth pairings). Test EPSPs were paired (150 times at a frequency of 0.2 Hz) with heterosynaptic action potentials (VL-Call., B 1), with current-induced spikes (1.2 nA, 50 msec, VLd-Ddir., B 2), and with antidromic spikes (B 3). C: Responses to the 108–110th pairings, when facilitated VL responses were seen in all PT cells. D: Juxtacellular GABA injections (−20-nA, 200-msec pulses were given simultaneously with pairings, D 1,2) or anodal current (−2-nA, 2-msec pulses, D 3) prevented the neurons from firing action potentials in the stimulus pairs. Pairing procedures under these conditions failed to induce synaptic facilitation. Action potentials in all records are truncated. E: Time course of changes in the amplitude of test EPSPs during repeated conditioning and extinction series. Black circles show changes during the first conditioning period in the third (VL-PT) and first (VL-Call.) neurons. Upward-going arrows illustrate that test EPSPs attained the firing threshold during conditioning. Empty circles represent EPSP changes when postsynaptic firing in the stimulus pairs was prevented. The circles represent the average of ten individual data points. Conditioning periods are indicated by thick lines on the time scale.

published by Scharfman and Sarvey (1985), Malinow and Miller (1986), and Kelso *et al.* (1986) on CA1 cells of the *in vitro* hippocampal slice.

3.3. Intracellular EGTA and Colchicine Injections Incapacitate Neurons to Undergo Plastic Changes

In 81 PT neurons, we examined some details of postsynaptic processes that can be related to pairing-induced modifications. Basically, the role of two factors was considered: (1) the postsynaptic firing-induced somatic and synaptic transmission (mostly dendritic) induced Ca^{2+} entry into the postsynaptic cells, and (2) the significance of the fast neurotubular transport system.

Recent studies in *Aplysia* (Byrne, 1985; Goelet *et al.*, 1986), *Hermissenda* (Alkon, 1984), vertebrate sympathetic ganglion cells (Kumamoto and Kuba, 1983; Libet, 1984), hippocampal neurons (Lynch *et al.*, 1983; Lynch and Baudry, 1984; Gustaffson and Wigstrom, 1986; Kelso *et al.*, 1986), and cells of the motor cortex of cats (Woody *et al.*, 1984, 1986) indicate that alteration in a calcium-dependent potassium current via second messenger systems and protein phophorylation (Nestler *et al.*, 1984) can be related to the plastic changes of synaptic and membrane functions.

Considering the extensive dendritic arborization of neocortical neurons, the observation that sodium-mediated action potentials (which are generated at the initial segment) could induce specific alterations in EPSPs generated far away from the cell body raises the question: What mechanism establishes a functional relationship between the spatially separated, synaptically activated, and calcium-dependent dendritic processes (Lynch and Baudry, 1984; Pumain and Heinemann, 1985; Wigstrom and Gustaffson, 1985; Bear and Singer, 1986) and the intrasomatic biochemical processes underlying postsynaptic firing. The neurotubular fast transport system with a speed for 1 μm/sec (Kreutzberg, 1981) is a factor that seems to be relevant. It is able to carry newly synthesized membrane and synaptic constituents from the Golgi apparatus of the perikaryon (Fambrugh and Devreotes, 1978) to remote dendritic branches within minutes. For these reasons, we investigated the action of intrasomatic microinjections of EGTA (calcium chelator) and colchicine (intracellular transport blocker) on conditioned plastic changes (Fig. 4). Both drugs are known to be inhibitors of learning or, in simpler systems, to block plastic alterations and modify synaptic responsiveness (Baux *et al.*, 1981; Sutula *et al.*, 1982; Carlson *et al.*, 1983; Lynch *et al.*, 1983).

Injections of EGTA prevented both the induction of synaptic facilitation and concomitant changes in the input resistance (in 24 cells out of 31); EGTA also blocked the slow conductance changes underlying spike and depolarizing pulse afterhyperpolarizations. All of the above actions attained their maximum values within 2–3 min and were reversible within 20–30 min. The effect of colchicine (2 min of injection) culminated within 3–5 min following the intrasomatic injection and was restricted only to a blockage of synaptic facilitation (in 36 cells out of 50). This drug did not alter the membrane potential or input resistance. Changes induced by colchicine proved to be irreversible over 3 hr of recording in 11 cells. Electron microscopic autoradiography of 17 recovered PT cells (together with F. Joo) using [^{3}H]colchicine injections demonstrated that the drug remained inside the soma and exerted its effect on the postsynaptic side, similarly to that observed in *Aplysia* (Baux *et al.*, 1981).

Since both EGTA and colchicine act inside injected cells, the firing-induced elevation of intracellular calcium and the intact neurotubular transport system proved to be vital components of conditioned synaptic and neuronal excitability changes in the motor cortex.

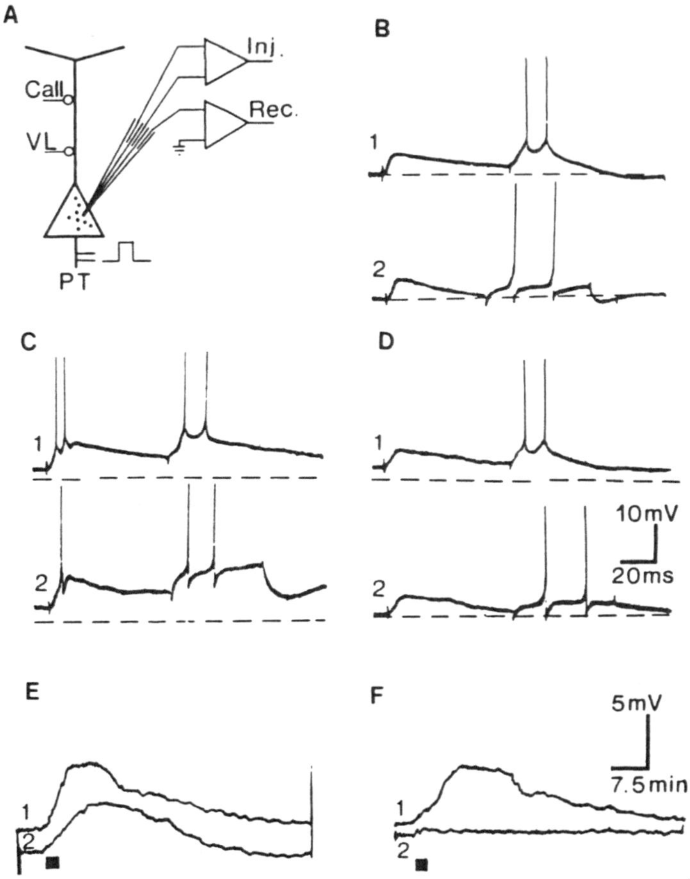

FIGURE 4. Inhibition of conditioned facilitation of monosynaptic VL EPSPs by intrasomatic EGTA and colchicine microinjections. A: Schematic drawing of experimental arrangement for intracellular drug application and recording. B: Control responses at the beginning of pairings (VL-Call., B 1 and VL-Dir., B 2). Records taken at 120th pairing when the test responses exhibited large heterosynaptic facilitation (C 1) or synaptic facilitation (C 2). D: Intracellular EGTA (0.2 M potassium EGTA, 0.5 nA for 5 min, D 2) and colchicine (1 nA for 5 min, D 1) injections before the second conditioning period prevented synaptic facilitation in both neurons. E: Slow pen records of membrane potentials during first conditioning–extinction series reveals moderate membrane depolarizations concomitant with synaptic facilitations. F: Membrane potential records of the second conditioning period. The synaptic facilitation was abolished by colchicine, but membrane potential change occurred again (1). The EGTA injection prevented both the synaptic facilitation and the membrane depolarization (2). Thick black lines indicate conditioning periods at the bottom.

4. DISCUSSION

In the motor cortex of the cat, the cellular learning process manifested itself as a reversible increase in the amplitude of a test EPSP paired with conjunctive action potentials. The memory was represented by a pairing-specific, persistent (up to 60 min) facilitated state of the test response. In addition, in many cells the conditioned synaptic enhancement was concomitant with an increase in neuronal excitability reflected in an elevation of input resistance, membrane depolarization, and an increased background firing activity.

4.1. Role of Heterosynaptic Interactions in Formation of Elementary Memory Trace

In the process by which the local subthreshold analogue response, the EPSP, converts into an all-or-none action potential, the synaptic connections of the neuron are modified and a new pathway is established for signal transmission. Under natural circumstances, the cortical neuron can be considered to be a coincidence detector of associative events in which repetitive conjunction of EPSPs and action potentials of different origins (via activation of other synaptic pathways) results in the heterosynaptic facilitation of the EPSP. This may be the basic process of associative learning in the neocortex at a cellular level (Hebb, 1949; Brindley, 1969; Marr, 1970; Gardner-Medwin, 1976, 1978; Olds, 1980; Palm, 1982; Fregnac and Imbert, 1984).

4.2. Additional Requirements for Hebb's Theory

In 28% of the total conditioned cell population, we observed a specific facilitation of synaptic responses as a result of consequent pre- and postsynaptic firing, as predicted by Hebb. In addition, we provided direct evidence that the conditioned alteration is a postsynaptic phenomenon, since intracellular anodal currents and juxtasomatic GABA injections prevented the development of synaptic facilitation, as did intracellular injection of the transport blocker colchicine and the calcium chelator EGTA.

The question arises, why did only 28% of the stimulated cells show plastic changes?

The observation can be interpreted in two ways. (1) Considering a test pathway, the pairing procedure induced changes with certain probability in only some of the connections. Probably, other connections of the same stimulated pathways were already affected in an earlier learning process of the adult brain and attained a maximal functional state. (The proportion of plastic neurons is supposedly higher in a critical period of immature brain.) In the view of the observer, the formal condition of conjunctive pre- and postsynaptic firing (Hebb's postulate) in the majority of the afferent pathways of the adult brain is insufficient in itself. (2) Another possible explanation (see also Kelso *et al.*, 1986; Wigstrom *et al.*, 1986) is that the postsynaptic sodium action potential could not in itself provide an adequate postsynaptic condition. Rather, other factors such as postsynaptic depolarizations high enough to result in firing activity (and a concomitant calcium entry into the cell), changes in input resistance and in the cable properties of the cell (Woody *et al.*, 1977; Holmes and Woody, 1986), or reduction of different outward currents (A current, calcium-dependent potassium current) should be considered. In addition, according to our hypothesis, proper synchronization of the synaptic input signals and the postsynaptic intracellular traffic of different substances is also required.

Further extension of the current experiments may disclose more details of postsynaptic mechanisms.

REFERENCES

Alkon, D. L., 1984, Calcium mediated reduction of ionic currents: A biophysical memory trace, *Science* **226:**1037–1045.

Baranyi, A., and Chase, M. H., 1984, Ethanol-induced modulation of the membrane potential and synaptic activity of trigeminal motoneurons during sleep and wakefulness, *Brain Res.* **307:**233–245.

Baranyi, A., and Feher, O., 1978, Conditioned changes of synaptic transmission in the motor cortex of the cat, *Exp. Brain Res.* **33:**283–298.

Baranyi, A., and Feher, O., 1981a, Selective facilitation of synapses in the neocortex by heterosynaptic activation, *Brain Res.* **212:**164–168.

Baranyi, A., and Feher, O., 1981b, Intracellular studies on cortical synaptic plasticity: Conditioning effect of antidromic activation of test-EPSPs, *Exp. Brain Res.* **41:**124–134.

Baranyi, A., and Feher, O., 1981c, Synaptic facilitation requires paired activation of convergent pathways in the neocortex, *Nature* **290:**413–415.

Baranyi, A., and Feher, O., 1981d, Long-term facilitation of excitatory synaptic transmission in single motor cortex neurons of the cat produced by repetitive pairing of synaptic potentials and action potentials following intracellular stimulation, *Neurosci. Lett.* **23:**303–308.

Baranyi, A., and Szente, M., 1986, The effect of intracellular colchicine and EGTA injection on the heterosynaptic facilitation of cortical neurons, *Acta Physiol. Hung.* **68:**353(Absr.).

Baux, G., Simonneau, M., and Tauc, L., 1981, Action of colchicine on membrane currents and synaptic transmission in *Aplysia* ganglion cells, *J. Neurobiol.* **12:**75–85.

Bear, M. F., and Singer, W., 1986, Modulation of visual cortical plasticity by acetylcholine and noradrenaline, *Nature* **320:**172–176.

Bienenstock, E. L., Fregnac, Y., and Thorpe, S., 1983, Iontophoretic clamp of activity in visual cortical neurons in the cat: A test of Hebb's hypothesis, *J. Physiol.* (*Lond.*) **345:**123P.

Bindman, L. J., Lippold, O. C. J., and Milne, A. R., 1979, Prolonged changes in excitability of pyramidal tract neurons in the cat: A postsynaptic mechanism, *J. Physiol.* (*Lond.*) **286:**457–477.

Black-Cleworth, P., Woody, C. D., and Nieman, J. A., 1975, Conditioned eyeblink obtained by using electrical stimulation of the facial nerve as unconditioned stimulus, *Brain Res.* **90:**44–56.

Brindley, G. S., 1969, Nerve net models of plausible size that perform many simple learning tasks, *Proc. R. Soc.* (*Lond.*) **174:**173–191.

Byrne, J. H., 1985, Neuronal and molecular mechanisms underlying information storage in Aplysia: Implications for learning and memory, *Trends Neurosci.* **9:**478–482.

Carew, T. J., Hawkins, R. D., Abrams, T. W., and Kandel, E. R., 1984, A test of Hebb's postulate at identified synapses which mediate classical conditioning in *Aplysia, J. Neurosci.* **4:**1217–1224.

Carlson, N. R., Laxer, K. D., and Mason, M. A., 1983, Intracerebral infusions of colchicine abolish kindled epileptogenic foci in cats, *Soc. Neurosci. Abstr.* **9:**764.

Deschenes, M., Landry, P., and Clercq, M., 1982, Reanalysis of the ventrolateral input in slow and fast pyramidal tract neurons of the cat motor cortex, *Neuroscience* **7:**2149–2157.

Fambrough, M. D., and Devreotes, P. M., 1978, Newly synthetized acetylcholine receptors are located in the Golgi apparatus, *J. Cell. Biol.* **76:**237–244.

Fregnac, Y., and Imbert, M., 1984, Development of neuronal selectivity in primary visual cortex of cat, *Physiol. Rev.* **64:**325–434.

Gardner-Medwin, A. R., 1976, The recall of events through the learning of associations between their parts, *Proc. R. Soc.* (*Lond.*) [Biol.] **194:**375–402.

Gardner-Medwin, A. R., 1978, The possible significance for learning of some different types of synaptic modification, *Biochem. Soc. Trans.* **6:**841–844.

Goelet, P., Castellucci, V. F., Schacher, S., and Kandel, E. R., 1986, The long and the short of long-term memory—a molecular framework, *Nature* **322:**419–422.

Gustafsson, B., and Wigstrom, H., 1986, Hippocampal long-lasting potentiation produced by pairing single volleys and brief conditioning tetani evoked in separate afferents, *J. Neurosci.* **6:**1575–1582.

Hebb, D. O., 1949, *The Organization of Behavior,* John Wiley & Sons, New York.

Holmes, W. R., and Woody, C. D., 1987, Effects of uniform and nonuniform synaptic "activation-distributions" on the cable properties of modeled cortical pyramidal neurons, *J. Neurophysiol.* (in press).

Hubel, D. H., and Wiesel, T. N., 1965, Binocular interaction in striate cortex of kittens reared with artificial squint, *J. Neurophysiol.* **28:**1041–1049.

Kandel, E. R., 1976, *Cellular Basis of Behavior,* W. H. Freeman, San Francisco.

Kandel, E. R., and Tauc, L., 1965, Mechanism of heterosynaptic facilitation in the giant cell of the abdominal ganglion of *Aplysia depilans, J. Physiol.* (*Lond.*) **181:**28–47.

Kelso, S. R., Ganong, A. H., and Brown, T. H., 1986, Hebbian synapses in hippocampus, *Proc. Natl. Acad. Sci. U.S.A.* **83:**5326–5330.

Kreutzberg, G. W., 1981, Parameters of dendritic transport, *Neurosci. Res. Prog. Bull.* **20:**45–55.

Kuhnt, U., 1984, Long-lasting changes of synaptic excitability induced by repetitive intracellular current injection in hippocampal neurons, *Neurosci. Lett. Suppl.* **18:**S27.

Kumamoto, E., and Kuba, K., 1983, Sustained rise in ACh sensitivity of sympathetic ganglion cell induced by postsynaptic electrical activities, *Nature* **305:**145–146.

Levy, W. B., and Steward, O., 1979, Synapses as associative elements in the hippocampal formation, *Brain Res.* **175**:233–245.

Libet, B., 1984, Heterosynaptic interaction at a sympathetic neuron as a model for induction and storage of a postsynaptic memory trace, in: *Neurobiology of Learning and Memory* (G. Lynch, J. L. McGaugh, and N. M. Weinberger, eds.), Guilford Press, New York, pp. 405–430.

Lynch, G., and Baudry, T., 1984, The biochemistry of memory: A new and specific hypothesis, *Science* **224**:1057–1063.

Lynch, G., Larson, J., Kelso, S., Barrionuevo, G., and Shottler, F., 1983, Intracellular injections of EGTA block induction of hippocampal long-term potentiation, *Nature* **305**:719–721.

Malinow, R., and Miller, J. P., 1986, Postsynaptic hyperpolarization during conditioning reversibly block induction of long-term potentiation, *Nature* **320**:529–530.

Marr, D. A., 1970, A theory for cerebral neocortex, *Proc. R. Soc. Lond.* [*Biol.*] **176**:164–234.

Martin, G. K., Land, T., and Thompson, R. F., 1980, Classical conditioning of the rabbit (*Oryctolagus cuniculus*) nicticating membrane response with electrical brain stimulation, *J. Comp. Physiol. Psychol.* **94**:216–226.

McCabe, B., 1976, An after-effect of local stimulation of neurons in the cerebral cortex of the unanaesthetized rat, *J. Physiol.* (*Lond.*) **263**:140–141.

McNaughton, B. L., Douglas, P. M., and Goddard, G. V., 1978, Synaptic enhancement in fascia dentata: Cooperativity among coactive afferents, *Brain Res.* **157**:277–293.

Mochida, S., and Libet, B., 1985, Synaptic long-term enhancement (LTE) induced by heterosynaptic neuronal input, *Brain Res.* **39**:360–363.

Nestler, E. J., Walaas, J. S., and Greengard, P., 1984, Neuronal phosphoproteins. Physiological and clinical implications, *Science* **225**:1357–1365.

O'Brien, J. H., Wilder, M. B., and Stevens, C. D., 1977, Conditioning of cortical neurons in cats with antidromic activation as the unconditioned stimulus, *J. Comp. Physiol. Psychol.* **91**:918–929.

Olds, J., 1980, Thoughts on cerebral function: The cortex as an action system, in: *Biology of Reinforcement, Facets of Brain-Stimulation Reward* (A. Routtenberg, ed.), Academic Press, New York, pp. 149–167.

Palm, G., 1982, *Neural Assemblies: An Alternative Approach to Artificial Intelligence Studies of Brain Function*, Vol. 7, Spinger-Verlag, Berlin, Heidelberg, New York.

Peters, A., and Jones, E. G., 1984, *Cellular Components of the Cerebral Cortex*, Plenum Press, New York, London.

Pumain, R., and Heinemann, H., 1985, Stimulus and aminoacid-induced calcium and potassium changes in rat neocortex, *J. Neurophysiol.* **53**:1–16.

Purpura, D. P., 1972, Intracellular studies on synaptic organizations in mammalian brain, in: *Structure and Function of Synapses* (D. P. Purpura and G. D. Pappas, eds.), Raven Press, New York, pp. 253–302.

Quinn, K. J., and O'Brien, J. H., 1983, Cortical motor neuron activity in the cat during classical conditioning with central stimulation as the CS and US, *Behav. Neurosci.* **97**:28–41.

Rauschecker, J. P., and Singer, W., 1981, The effects of early visual experience and their possible explanation by Hebb synapses, *J. Pysiol.* (*Lond.*) **310**:215–239.

Scharfman, H., and Sarvey, J. M., 1985, Postsynaptic firing during repetitive stimulation is required for long-term potentiation in hippocampus, *Brain Res.* **331**:267–274.

Schulman, J. A., and Weight, F. F., 1976, Synaptic transmission: Long-lasting potentiation by a postsynaptic mechanism, *Science* **194**:1434–1439.

Sutula, T., Goldsmith, R., and Steward, O., 1982, Effect of colchicine on synaptic transmission and long-term potentiation in the dentate gyrus, *Soc. Neurosci. Absr.* **9**:240.

Takahashi, K., Kubota, K., and Uno, M., 1967, Recurrent facilitation in cat pyramidal tract cells, *J. Neurophysiol.* **30**:22–34.

Tzebelikos, E., and Woody, C. D., 1979, Intracellularly studies on excitability changes in coronal–pericruciate neurons following low-frequency stimulation of the corticobulbar tract, *Brain Res. Bull.* **4**:635–641.

Wigstrom, H., and Gustaffson, B., 1985, On long-lasting potentiation in the hippocampus: A proposed mechanism for its dependence on coincident pre- and postsynaptic activity, *Acta Physiol. Scand.* **123**:519–522.

Wigstrom, H., Gustafsson, B., Huang, Y. Y., and Abraham, W. C., 1986, Hippocampal long-term potentiation is induced by pairing single afferent valleys with intracellularly injected depolarizing current pulse, *Acta Physiol. Scand.* **126**:317–319.

Woody, C. D., Buerger, J. A., Ungar, R. A., and Levine, D. S., 1977, Modelling aspects of learning by altering biophysical properties of a simulated neuron, *Biol. Cybernet.* **23**:73–82.

Woody, C. D., Alkon, D. L., and Hay, B., 1984, Depolarization-induced effects of Ca^{2+}-calmodulin-dependent protein kinase injection, *in vivo,* in single neurons of cat motor cortex, *Brain Res.* **321:**192–197.

Woody, C. D., Bartfai, T., Gruen, E., and Nairn, A. C., 1986, Intracellular injection of cGMP-dependent protein kinase results in increase input resistance in neurons of the mammalian motor cortex, *Brain Res.* **386:**379–385.

Wurtz, R. H., Castellucci, V. F., and Nusrala, J. M., 1967, Synaptic plasticity: The effect of the action potential in the postsynaptic neuron, *Exp. Neurol.* **18:**350–368.

37

The Role of Neuronal Activity in the Long-Term Regulation of Synaptic Performance at the Crayfish Neuromuscular Junction

GREGORY A. LNENICKA

1. INTRODUCTION

Long-term changes in synapses produced by altered impulse activity may be an important mechanism underlying forms of learning. The long-term effects of impulse activity on the physiology and morphology of well-defined synapses can be studied at the neuromuscular junction (Robbins, 1980). This is particularly true for crustacean neuromuscular synapses, where muscles are innervated by relatively few motoneurons. In crustacean muscles the activity of identified motoneurons can be monitored and selectively altered for extended periods *in vivo*. Identified neuromuscular synapses can be examined physiologically and morphologically. In addition, many of the physiological and morphological features of crustacean neuromuscular synapses are similar to synapses in the central nervous system (Atwood, 1982).

The naturally occurring differences in the physiology and morphology of phasically and tonically active neuromuscular synapses are discussed. Evidence is presented that the different impulse activity levels in phasic and tonic motoneurons may influence the differentiation of their neuromuscular synapses.

2. PHASIC AND TONIC MOTONEURONS

Crustacean motoneurons have a broad range of impulse activity levels with corresponding differences in neuromuscular synaptic performance. These differences are most pronounced when comparisons are made between tonically and phasically active motoneurons. Phasic motoneurons produce rapid transient movements and are generally inactive. Tonic motoneurons produce slower, more sustained movement and are often active for prolonged periods.

GREGORY A. LNENICKA • Department of Physiology, University of Toronto, Toronto, Ontario, Canada M5S 1A8. *Present address:* Neurobiology Research Center, Department of Biological Sciences, State University of New York, Albany, New York 12222.

The neuromuscular synapses of the motoneurons are generally well adapted to their normal pattern of activity. The phasic motor synapses release maximal amounts of transmitter at the low frequencies at which they normally operate. Prolonged stimulation at moderate frequencies (5 Hz) normally results in synaptic depression. Tonic motor synapses release maximal amounts of transmitter during high-frequency activation because of a build-up of short-term facilitation. This allows the tonic motor synapses to grade transmitter release throughout the broad range of frequencies over which they normally operate. Prolonged stimulation does not normally produce synaptic fatigue. Similar although less pronounced differences exist between vertebrate phasic and tonic neuromuscular synapses (Gertler and Robbins, 1978).

Phasic and tonic motor terminals have ultrastructural differences that could be responsible for some of the observed differences in physiology. In particular, invertebrate tonic motor terminals have a greater cross-sectional area and contain a greater volume of mitochondria and more synaptic vesicles than phasic motor terminals (Atwood and Jahromi, 1978; Atwood and Johnston, 1968; Hill and Govind, 1981; Titmus, 1981). These differences may be responsible for the greater fatigue resistance of tonic motor synapses.

A good example of these corresponding differences in activity and synaptic properties is provided by the phasically active fast closer excitor (FCE) and the tonically active slow closer excitor (SCE), which innervate the claw closer muscle of crayfish (Fig. 1). The impulse activity of the FCE and SCE can be monitored by recording closer muscle myograms. The FCE produces a large myogram at low frequencies of stimulation, which is easily distinguished from the smaller SCE myogram. *In vivo* recordings from freely moving animals show that the SCE averages approximately 6000 impulses per hour and the FCE fires an average of one impulse per hour (Pahapill *et al.*, 1985). Most of the impulse activity in the SCE occurs at frequencies of 10–20 Hz with some activity as high as 200 Hz.

The SCE neuromuscular synapses generally produce small EPSPs at low frequencies of stimulation, which facilitate at high frequencies of stimulation (Fig. 1). The FCE produces large EPSPs at low stimulation frequencies, which show depression at higher

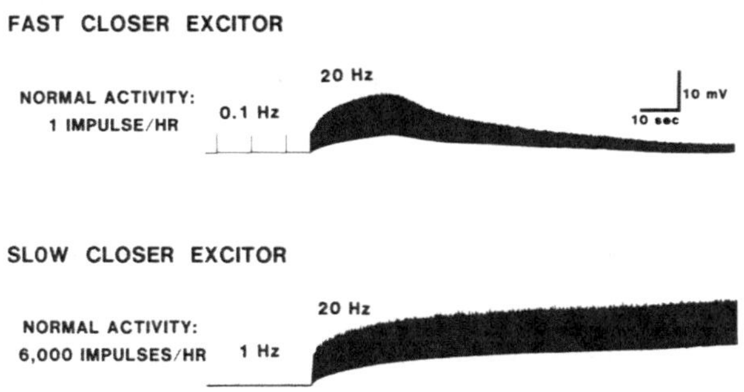

FIGURE 1. Differentiation of activity and synaptic performance in the two excitatory motoneurons innervating the crayfish claw closer-muscle. The mean impulse activity of the fast closer excitor (FCE) and the slow closer excitor (SCE) was determined from closer-muscle myograms recorded in freely moving crayfish. Excitatory postsynaptic potentials were recorded from closer-muscle fibers during stimulation of the FCE and SCE. The normally inactive FCE produces large EPSPs at low frequencies of stimulation (0.1 Hz), which depress rapidly during high-frequency stimulation. The more active SCE produces small EPSPs (<0.5 mV) at low frequencies of stimulation (1 Hz); however, during high-frequency stimulation dramatic facilitation occurs, and no synaptic depression is observed.

frequencies of activation. The differences in EPSP amplitude produced by phasic and tonic neuromuscular synapses are caused by differences in transmitter release (Atwood, 1976).

3. NEUROMUSCULAR SYNAPSES ADAPT TO ALTERED NEURONAL ACTIVITY

3.1. Long-Term Alteration of Motoneuron Impulse Activity

To determine if the different levels of activity in the motoneurons played a role in the differentiation of synaptic properties, I altered the impulse activity of the FCE over

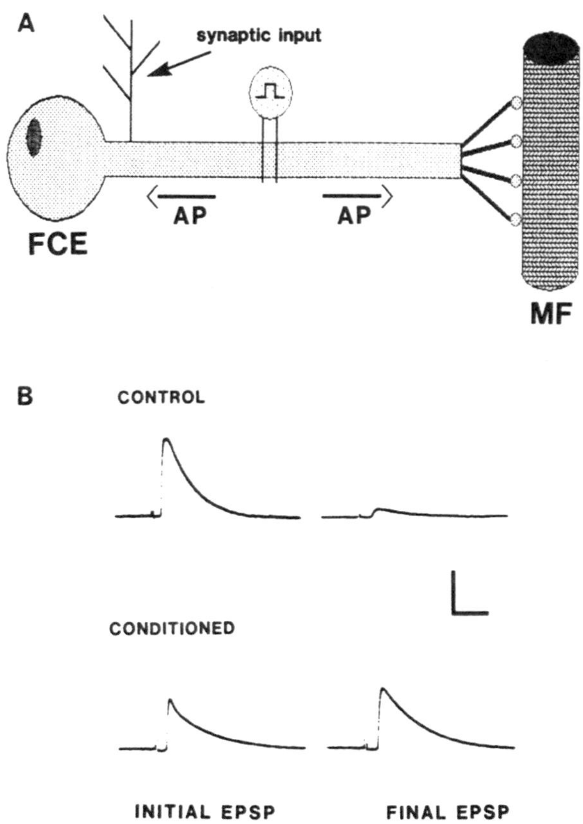

FIGURE 2. The effect of chronic stimulation of the FCE on its neuromuscular synaptic properties. A: The motor axon of the FCE was stimulated *in vivo* for 2 hr per day at 5 Hz. After 2 weeks of stimulation, the nueromuscular synaptic properties were compared for the conditioned and contralateral control FCEs. B: Representative EPSPs produced by the conditioned and control FCE neuromuscular synapses. Initial EPSP was recorded during 0.01-Hz stimulation, and the final EPSP was recorded after 30 min of stimulation at 5 Hz. The conditioned neuromuscular synapses produce a smaller initial EPSP and show greater fatigue resistance than the contralateral control. These changes in EPSP amplitude are caused by changes in transmitter release. Calibration: 20 msec; 4 mV.

a period of days. An increase in impulse activity was produced by *in vivo* electrical stimulation of the FCE. Stimulating electrodes were implanted near the claw nerve, and recording electrodes were implanted over the closer muscle. *In vivo* stimulation of the FCE was monitored by recording the FCE myograms from the closer muscle (Lnenicka and Atwood, 1985a). The FCE motor axon was stimulated *in vivo* at 5 Hz for 2 hr per day in one of the paired claws (Fig. 2A). Acute studies of synaptic physiology and morphology were performed 1 day following the final *in vivo* stimulation period in both conditioned and control claws. The FCE impulse activity was decreased by immobilizing one of the paired claws in the closed position for 1 to 8 weeks. This procedure resulted in an eightfold decrease in FCE impulse activity (Pahapill *et al.*, 1985).

3.2. Adaptation of Synaptic Physiology

The *in vivo* stimulation regimen resulted in an adaptation of the neuromuscular synapses to the more tonic lifestyle. Conditioning of the FCE produced a 50% reduction in the low-frequency EPSP amplitude and greater resistance of the neuromuscular synapses to fatigue during high-frequency stimulation (Fig. 2B). These changes, which are referred to as long-term adaptation (LTA), result from a reduction in initial transmitter release and greater transmitter release during prolonged stimulation (Lnenicka and Atwood, 1985a). The changes persist for at least 10 days after the final *in vivo* conditioning period.

Subsequent experiments have shown that 3 days of stimulation are required to produce a detectable adaptation of synaptic properties. The alteration in synaptic physiology persists for at least a week after only 3 days of stimulation (Lnenicka and Atwood, 1985b).

Immobilization produces a change in synaptic properties opposite to that produced by imposed activity: increased initial EPSP amplitude and decreased fatigue resistance (Pahapill *et al.*, 1985). This effect is first detected after 1 week of immobilization and becomes fully expressed after 4 weeks of immobilization.

3.3. Adaptation of Synaptic Ultrastructure

Adaptive changes in motor terminal ultrastructure are observed after the *in vivo* stimulation regimen. The ultrastructure of conditioned FCE motor terminals was quantified through reconstruction of serial electron micrographs of the terminals. These measurements were compared to those obtained from control FCE and SCE terminals. All changes occurring in FCE motor terminal ultrastructure during conditioning altered the terminal morphology to resemble more closely that of the SCE (Lnenicka *et al.*, 1986).

The most dramatic changes involved an increase in the size of the mitochondria and a reorganization of the terminal (Fig. 3). The mitochondrial volume per unit length of terminal more than doubled because of an increase in the cross-sectional area of individual mitochondria. Control FCE motor terminals were of uniform diameter, with the mitochondria and synapses randomly distributed along its length. After stimulation the terminals became more varicose, with the synapses and mitochondria localized in the varicosities.

The increased mitochondrial size and proximity to the synapses could provide a greater energy supply to the synapses, resulting in their increased fatigue resistance. There were no changes in synaptic morphology that could directly account for the decrease in initial transmitter release. It may be that a decrease in the density of calcium channels

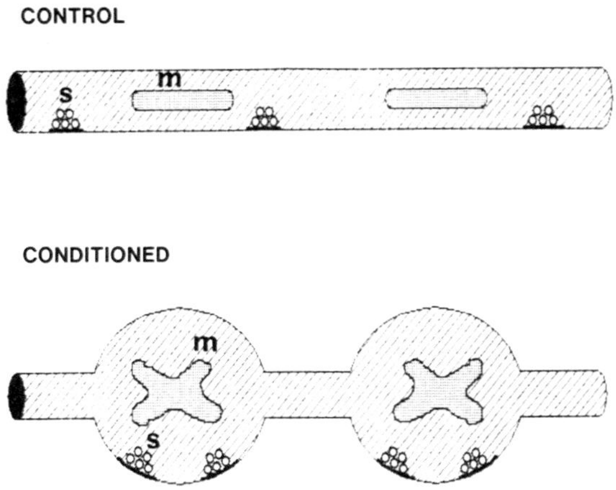

FIGURE 3. Changes in the morphology of FCE motor terminals produced by long-term tonic stimulation. The control FCE motor terminals were uniform in diameter and contained small mitochondria. The mitochondria (m) and synapses (s) were randomly distributed along the terminal. The conditioned FCE motor terminals contained large mitochondria, often with numerous side branches. The terminals were varicose, with the mitochondria (m) and synapses (s) localized in the varicosities.

at the neuromuscular synapses could be responsible for this change (Walrond and Reese, 1985; Atwood and Lnenicka, 1986).

4. CENTRAL REGULATION OF NEUROMUSCULAR SYNAPTIC PROPERTIES

The cell body is required to produce the alterations in synaptic physiology associated with LTA. After section of a crustacean motor axon, the motor terminals remain functional even though they are decentralized from the cell body (Bittner, 1973). *In vivo* stimulation of decentralized terminals produces long-term facilitation (LTF; see Atwood *et al.*, Chapter 34) but not LTA (Lnenicka and Atwood, 1985). Thus, it appears that factors from the cell body are necessary for production of LTA. The gradual onset and relative permanence of LTA suggest that changes in protein synthesis could be involved. In addition, since the majority of mitochondrial proteins are synthesized in the cytoplasm (Schatz and Mason, 1974), an increase in mitochondrial mass should require changes in cytoplasmic protein synthesis. The restriction of cytoplasmic protein synthesis to the cell body in neurons (Lasek and Brady, 1982) suggests the involvement of the cell body in these alterations in neuromuscular synaptic morphology and physiology.

Recent experiments demonstrate that not only is the cell body required for the expression of LTA, but an increase in the central electrical activity of the motoneuron is sufficient to produce adaptation of the neuromuscular synapses. The FCE axon was selectively stimulated antidromically at 5 Hz for 2 hr a day (Fig. 4A). This was accomplished by blocking the nerve locally, distal to the site of stimulation. The local block was produced by placing a small gelatin cube containing TTX against the nerve during the stimulation period. Removal of the cube after stimulation resulted in recovery of impulse traffic within 1 hr.

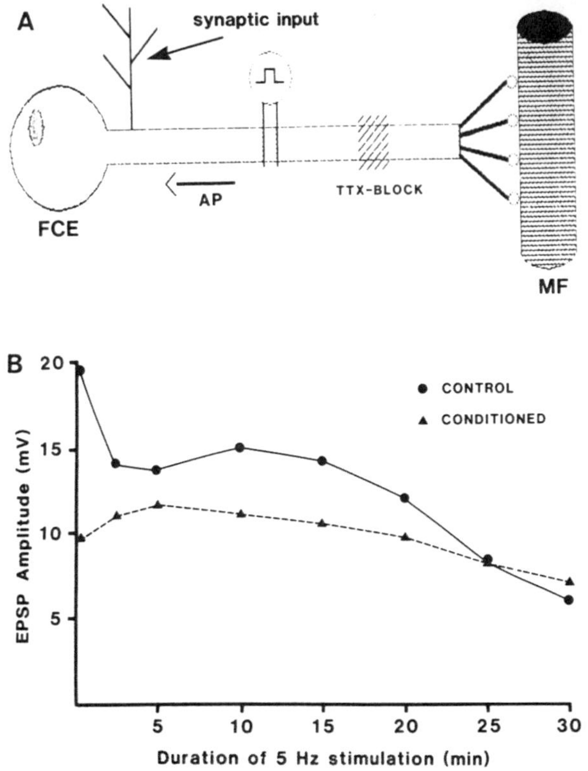

FIGURE 4. The effect of selective antidromic stimulation of the FCE on neuromuscular synaptic properties. A: The FCE motor axon was stimulated *in vivo* at 5 Hz for 2 hr per day. During stimulation, a reversible local TTX block was applied to the motor axon distal to the site of stimulation. This procedure produced antidromic impulses without increasing the activity of the neuromuscular synapses. After 3 days of stimulation the neuromuscular synaptic properties were tested in the conditioned and contralateral control FCEs. B: Representative EPSP amplitudes measured at conditioned and control FCE neuromuscular synapses. Central stimulation of the FCE motoneuron produces a decrease in the initial EPSP amplitude and reduces synaptic fatigue.

Selective antidromic stimulation of the FCE for 3 days produces LTA of its neuromuscular synapses. The initial EPSP amplitude is reduced approximately 50%, and the synapses are more fatigue resistant (Fig. 4B). It is not yet known if changes in terminal morphology accompany these changes in physiology.

5. CONCLUSIONS

The normally occurring differences in the physiology and morphology of neuromuscular synapses may be largely related to the differences in motoneuron activity. The initial transmitter release and fatigue properties of the neuromuscular synapses adapt to the activity levels of the motoneuron. This adaptation may be a general property, as shown by similar results for the vertebrate neuromuscular junction (Robbins and Fischbach, 1971) and spinal sensory synapses (Gallego *et al.*, 1979).

These changes in neuromuscular synaptic properties can be produced by increasing the central impulse activity of the motoneuron in the absence of changes in synaptic activity. Calcium influx affects the synthesis and/or degradation of various skeletal muscle proteins, including oxidative and glycolytic enzymes (Lawrence and Salsgiver, 1983) and acetylcholine receptors (Rubin, 1985). It may be that central impulse activity regulates the production of synaptic proteins responsible for LTA through changes in Ca^{2+} influx.

REFERENCES

Atwood, H. L., 1976, Organization and synaptic physiology of crustacean neuromuscular systems, *Prog. Neurobiol.* **7**:291–391.

Atwood, H. L., 1982, Synapses and neurotransmitters, in: *The Biology of Crustacea,* Vol. 3 (H. L. Atwood and D. C. Sandeman, eds.), Academic Press, New York, pp. 105–150.

Atwood, H. L., and Jahromi, S. S., 1978, Fast axon synapses of a crab leg muscle, *J. Neurobiol.* **9**:1–15.

Atwood, H. L., and Johnston, H. S., 1968, Neuromuscular synapses of a crab motor axon, *J. Exp. Zool.* **167**:457–470.

Atwood, H. L., and Lnenicka, G. A., 1986, Structure and function in synapses: Emerging correlations, *Trends Neurosci.* **9**(6):248–250.

Bittner, G. D., 1973, Degeneration and regeneration in crustacean neuromuscular systems, *Am. Zool.* **13**:379–408.

Gallego, R., Kuno, M., Nunez, R., and Snider, W. D., 1979, Disuse enhances synaptic efficacy in spinal motoneurons, *J. Physiol. (Lond.),* **291**:191–205.

Gertler, R. A., and Robbins, N., 1978, Differences in neurotransmission in red and white muscles, *Brain Res.* **142**:160–164.

Hill, R. H., and Govind, C. K., 1981, Comparison of fast and slow synaptic terminals in lobster muscle, *Cell Tissue Res.* **221**:303–310.

Lasek, R. J., and Brady, S. T., 1982, The axon: A prototype for studying expressional cytoplasm, *Cold Spring Harbor Symp. Quant. Biol.* **46**:113–124.

Lawrence, J. C., and Salsgiver, W. J., 1983, Levels of enzymes of energy metabolism are controlled by activity of cultured rat myotubes, *Am. J. Physiol.* **244**:C348–C355.

Lnenicka, G. A., and Atwood, H. L., 1985a, Age-dependent long-term adaptation of crayfish phasic motor axon synapses to altered activity, *J. Neurosci.* **5**(2):459–467.

Lnenicka, G. A., and Atwood, H. L., 1985b, Long-term facilitation and long-term adaptation at synapses of a crayfish phasic motoneuron, *J. Neurobiol.* **16**:97–110.

Lnenicka, G. A., Atwood, H. L., and Marin, L., 1986, Morphological transformation of synaptic terminals of a phasic motoneuron by long-term tonic stimulation, *J. Neurosci.* **6**:2252–2258.

Pahapill, P. A., Lnenicka, G. A., and Atwood, H. L., 1985, Asymmetry of motor impulses and neuromuscular synapses produced in crayfish claws by unilateral immobilization, *J. Comp. Physiol.* **157**(4):461–467.

Robbins, N., 1980, Plasticity at the mature neuromuscular junction, *Trends Neurosci.* **3**:120:122.

Robbins, N., and Fischbach, G. D., 1971, Effect of chronic disuse of rat soleus neuromuscular junctions on presynaptic function, *J. Neurophysiol.* **34**:570–578.

Rubin, L. L., 1985, Increases in muscle Ca^{++} mediate changes in acetylcholinesterase and acetylcholine receptors caused by muscle contraction, *Proc. Natl. Acad. Sci. U.S.A.* **82**:7121–7125.

Schatz, G., and Mason, T. L., 1974, The biosynthesis of mitochondrial proteins, *Annu. Rev. Biochem.* **43**:51–87.

Titmus, M. J., 1981, Ultrastructure of identified fast excitatory, slow excitatory and inhibitory neuromuscular junctions in the locust, *J. Neurocytol.* **10**:363–385.

Walrond, J. P., and Reese, T. S., 1985, Structure of axon terminals and active zones at synapses on lizard twitch and tonic muscle fibers, *J. Neurosci.* **5**:1118–1131.

38

Long-Term Effects of Firing on Excitability of Hippocampal CA1 Neurons

S. POCKETT and O. C. J. LIPPOLD

1. INTRODUCTION

Backfiring a hippocampal CA1 cell by stimulation of its axon can, under certain circumstances, cause it to become less excitable to synaptic input (Dunwiddie and Lynch, 1978). We have investigated the basis of this lowered excitability.

A reasonable hypothesis is that the cell's excitability is lowered through the activation of recurrent inhibitory pathways such as those described by Andersen *et al.* (1964). We have tested this hypothesis by backfiring the cells in the presence of sufficient Mg^{2+} to block synaptic transmission completely, thus eliminating any effect of recurrent inhibitory pathways. We find the lowering of excitability still occurs and suggest possible alternative mechanisms.

2. METHODS

Hippocampal slices approximately 400 μm thick were prepared from adult rats using a tissue chopper. They were stored at room temperature on a net in a holding chamber superfused with fluid containing (mM) Na^+ 151.25, K^+ 3, Cl^- 135, HCO_3^- 26, $H_2PO_4^-$ 1.25, glucose 10, Ca^{2+} 2, Mg^{2+} 2 and bubbled with 95% O_2, 5% CO_2. After a period of about 3 hr, the slices were transferred to a submerged-slice recording chamber superfused with the same salt solution at room temperature. Extracellular recordings were made from the CA1 region using 2 to 10-MΩ glass microelectrodes, which were filled with the superfusing salt solution to eliminate any possible ionic effects of fluid leakage from the microelectrode. Monopolar stimulation was delivered either to the Schaeffer collateral/commissural pathway or to the efferent fibers of the CA1 cells through sharpened tungsten wire electrodes insulated to the tip with lacquer. Stimuli were biphasic 0.1-msec

S. POCKETT • Department of Physiology, University of Auckland, Auckland, New Zealand. O. C. J. LIPPOLD • Department of Human Physiology, Royal Holloway and Bedford New College, Egham TW20 OEX, England.

pulses of 2–10 V. Responses were recorded on an oscilloscope, and the amplitude of population spikes was recorded on a chart recorder using an analogue peak height detector, which put out long voltage pulses of amplitude proportional to the amplitude of the population spikes. Selected responses were also digitized at 10 kHz and written out on another recorder using a Franklin Ace microcomputer.

Test pulses were delivered every 5 sec to the Schaeffer collateral/commissural pathway. Because slices that were allowed to become dry or anoxic gave unstable responses to the test pulses, the amplitude of the population spike usually increasing with time, all slices were stimulated once every 5 sec for at least 30 min before any treatment designed to change the population spike was applied, and only slices in which the responses to test pulses remained constant during this 30-min period were used. Test pulses were bipolar 0.1-msec pulses at a voltage producing a submaximal population spike (usually 3–4 V). Conditioning stimuli comprised six bursts, at 10-sec intervals, of 50 bipolar 0.1-msec pulses, at 200 sec^{-1}, at a voltage producing a supramaximal population spike (usually 6–7 V). Conditioning pulses delivered to the Schaeffer collateral pathway produced LTP in that pathway in four out of six slices.

3. RESULTS

Control experiments showed that it was possible, in eight out of eight slices tested, to apply 25 mM Mg^{2+} to abolish synaptic transmission completely and then wash out the Mg^{2+} and regain the same responses to Schaeffer-collateral test pulses as before Mg^{2+} application (see Fig. 1).

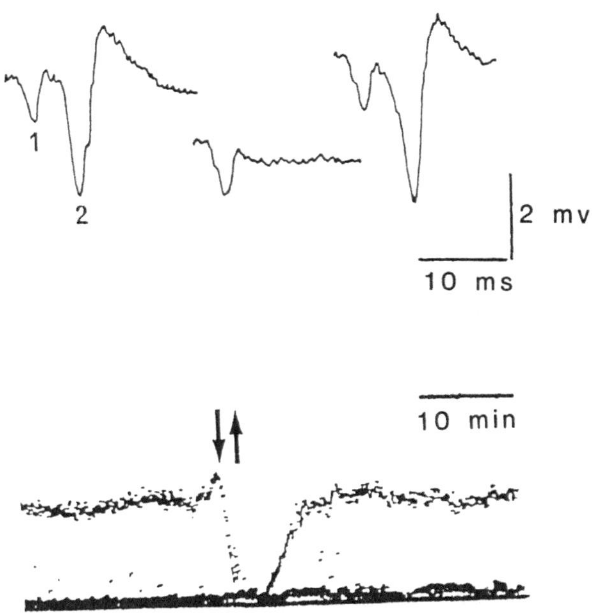

FIGURE 1. Field potentials in CA1 cells before, during, and after superfusion with high-Mg^{2+} solution. Top records show typical field potentials recorded from CA1 in response to a standard Schaeffer collateral/commissural test pulse, before, during, and after 25 mM Mg^{2+}. 1 indicates presynaptic volley, which is still evident in high-Mg^{2+} solution. 2 indicates population spike. Lower record is a dottogram of population spike amplitudes through the course of an experiment. Vertical arrows indicate beginning of Mg^{2+} superfusion (↓) and beginning of washout (↑).

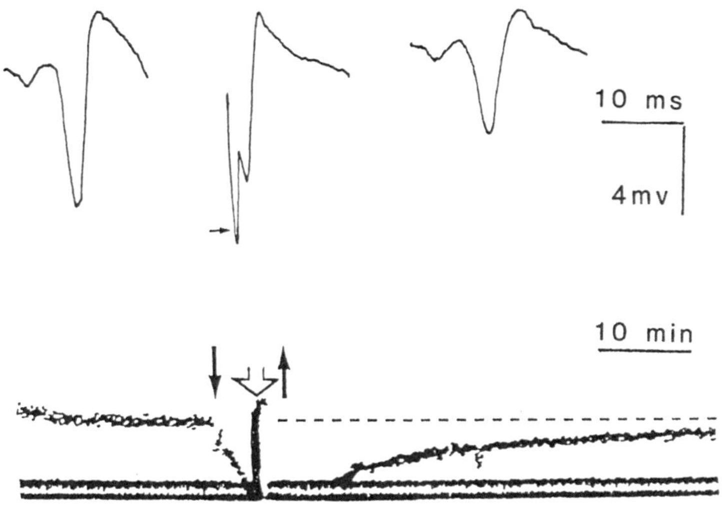

FIGURE 2. Field potentials in CA1 cells in normal solution, during antidromic stimulation in presence of high-Mg^{2+} solution, and after washout of Mg^{2+}. Top records show typical field potentials recorded from CA1 in reponse to standard Schaeffer collateral/commissural test pulses before (left record) and after (right record) 25 mM Mg^{2+}. Middle record shows field potential recorded during antidromic conditioning stimulation of CA1 cells in presence of high Mg^{2+}. Small horizontal arrow indicates stimulus artifact (not shown in other records). Lower record is a dottogram of population spike amplitudes throughout the whole experiment. Solid vertical arrows indicate period of Mg^{2+} superfusion as in Fig. 1. Open vertical arrow shows antidromic conditioning pulses.

However, if conditioning pulses were applied to the axonal outflow of the CA1 cells during the presence of 25 mM Mg^{2+}, the responses to Schaeffer collateral test pulses were generally smaller after Mg^{2+} washout than before Mg^{2+} application. In eight out of eight slices, there was total depression of the response for 8–10 min after the conditioning stimulus, i.e., 3–5 min longer than it took for washout of Mg^{2+}. In two of the eight slices this total depression continued for the duration of the experiments, 30 min and 2 hr, respectively. In another two of the eight slices, the 8- to 10-min period of total depression was followed by recovery of the response to equal the preconditioning response. In the other four slices, the 8 to 10-min period of total depression was followed by a prolonged period of 30–50% depression; i.e., the responses were 70% to 50% of the amplitude of the preconditioning response. One of these four slices produced the records in Fig. 2.

4. DISCUSSION

Since antidromic stimulation of CA1 cells delivered in the complete absence of synaptic transmission produced a depression of cellular excitability, it must be concluded that the lowered excitability was a direct result of the firing of the CA1 cells and not a consequence of activation of inhibitory synapses onto these cells.

How might this direct effect occur? One possibility is that the firing of action potentials in the cell changes the characteristics of K^+ channels in the cell membrane. For example, if the voltage dependence of M channels (Halliwell and Adams, 1982) or Ca^{2+}-dependent K^+ channels (Zbicz and Weight, 1985) were altered so that these channels were more open at the cells' resting potential, cellular excitability would be lowered.

Likewise, a change in the characteristics of Cl^- channels could lower cellular excitability. These possibilities are under investigation.

ACKNOWLEDGMENTS. This work was supported by the Isaacs Trust, the New Zealand Lottery Board Medical Research Distribution Committee, and the New Zealand Medical Research Council.

REFERENCES

Andersen P., Eccles, J. C., and Løyning, Y., 1964, Pathway of postsynaptic inhibition in the hippocampus, *J. Physiol. (Lond.)* **27**:608–619.

Dunwiddie, T., and Lynch, G., 1978, Long-term potentiation and depression of synaptic responses in the rat hippocampus: Localization and frequency dependency, *J. Physiol. (Lond.)* **276**:353–367.

Halliwell, J. V., and Adams, P. R., 1982, Voltage-clamp analysis of muscarinic excitation in hippocampal neurons, *Brain Res.* **250**:71–92.

Zbicz, K. L., and Weight, F. F., 1985, Transient voltage and calcium-dependent outward currents in hippocampal CA3 pyramidal neurons, *J. Neurophysiol.* **53**:1038–1058.

Locus-Coeruleus-Induced Enhancement of the Perforant-Path Evoked Potential

Effects of Intradentate β Blockers

CAROLYN W. HARLEY and SUZANNE EVANS

1. INTRODUCTION

Norepinephrine (NE) iontophoresed in the dentate gyrus of anesthetized rat (Neuman and Harley, 1983) or superfused on the hippocampal slice (Lacaille and Harley, 1985; Stanton and Sarvey, 1986) produces a significant and reliable enhancement of the perforant-path-evoked population spike. This NE-induced enhancement of population spike amplitude can last for many minutes or hours.

Pharmacological studies using the hippocampal slice preparation have implicated β receptors as the critical mediators of the NE-induced enhancement. Either timolol (Lacaille and Harley, 1985), propranolol, or metoprolol (Stanton and Sarvey, 1986) superfused on the hippocampal slice will prevent NE induction of both short- and long-lasting population spike enhancement. Isoproterenol, a β agonist, can mimic the enhancing effects of NE on population spike amplitude, and the α antagonist phentolamine is ineffective in preventing enhancement (Lacaille and Harley, 1985). The α agonist phenylephrine does not induce enhancement of population spike amplitude, and significant depressive effects of this α agonist on population spike amplitude have been observed (Lacaille and Harley, 1985).

Although the effects of α and β agents in the *in vitro* studies appear relatively straightforward, a recent *in vivo* investigation suggests that the pharmacological characterization of adrenergic effects in the dentate is more complex and may critically vary with drug placement location. Winson and Dahl (1985) have reported that iontophoresis of the β antagonist sotalol at the cell body layer reliably enhances perforant-path-evoked population spike amplitude, that iontophoresis of the β agonist isoproterenol reduces population spike amplitude, that the α agonist phenylephrine increases both the EPSP slope and population spike amplitude, and that NE, itself, has no reliable effect unless

CAROLYN W. HARLEY and SUZANNE EVANS • Psychology Department and Faculty of Medicine, Memorial University of Newfoundland, St. Johns, Newfoundland, Canada A1B 3X9.

applied for 5 min or longer. Long pulses of NE can produce the long-lasting enhancement already mentioned.

When iontophoresed in the middendritic region, α and β agonists and antagonists as well as NE reduce the population EPSP slope recorded at that level without any effects on the EPSP slope measured at the cell body layer, although the population spike was reduced at the cell body recording site.

In a related paper, Dahl and Winson (1985) demonstrated that some of the short-term effects they observed with adrenergic agents in the dentate could also be produced by electrical activation of the locus coeruleus (LC). With LC stimulation, the perforant-path-evoked population spike amplitude recorded at the cell body layer was enhanced without observable changes in EPSP slope at that site, while the EPSP slope, monitored simultaneously at the dendritic level, was significantly diminished.

The LC or A6 NE cell group is the sole source of central NE in the dentate gyrus, and we have also been concerned with comparing the effects of direct NE application in the dentate with the effects of LC activation. In particular, we have been interested in whether the physiological route of NE release could induce long-lasting enhancement of perforant path input as well as whether the pharmacology of LC effects in the dentate would resemble that of the *in vitro* preparation.

Enhancement of the perforant-path-evoked potential with electrical stimulation had been reported earlier (Bliss and Wendlandt, 1977; Assaf *et al.*, 1979), and using similar stimulation parameters but longer periods of perforant path–LC pairing, we were able to produce clear long-lasting enhancement of population spike amplitude (Harley *et al.*, 1982). However, the latencies of the earliest LC stimulation effects observable in the dentate suggested that more rapidly conducting pathways might also be affected by the stimulation.

In order to improve the specificity of LC stimulation, we elected to use glutamate ejections in the LC with the aim of selectively activating cell bodies while minimizing involvement of fibers of passage. Using 50 to 150-nl glutamate ejections within 300 μm of LC cell bodies, we were able reliably and consistently to enhance the perforant path population spike amplitude. The enhancement could be long-lasting, and systemic injections of propranolol attenuated or blocked the LC-induced enhancement (Harley and Milway, 1986).

The present series of experiments investigates further the role of β receptors in mediating LC-induced enhancement of the perforant-path-evoked potential using intradentate administration of the β blockers timolol and propranolol.

2. METHOD

The perforant-path-evoked potential was monitored in the dentate gyrus of Sprague–Dawley female rats (280–320 g) anesthetized with urethane (1.5 g/kg, i.p.). The perforant path was stimulated with a monophasic square wave at 0.1 Hz, and measurements of the population EPSP slope, population spike amplitude, and population spike onset latency were collected on line from the digitized evoked potential (1 point/75 μsec). These values were average for each 30 sec by a 16-bit microcomputer, and a two-tailed 95% confidence interval was generated for each parameter during a 5- to 10-min control period prior to any experimental manipulation.

In control experiments ($n = 9$), saline (pH 5.5), phosphate-buffered saline (pH 7.3), timolol (200 μM in phosphate-buffered saline), or propranolol (200 μM) was ejected in the dentate gyrus at various times during evoked potential recording.

In the LC activation experiments ($n = 14$), dentate gyrus ejections of the same substances were followed within 1–3 min by LC activation using glutamate (0.25 M or 0.5 M) ejection from a micropipette placed in or near the LC. Locus coeruleus activation effects were also evaluated prior to any drug ejection.

The drug ejection micropipettes were typically fixed parallel to the recording pipette (3 M NaCl) so that a 1.5-mm distance occurred between the tips ($n = 17$), with the drug pipette tip lateral and posterior to, but slightly above, the recording pipette tip. In some experiments a double-barrel pipette was used with a single tip for recording and drug ejection ($n = 3$), or the drug pipette tip was placed 300 μm above the recording tip at approximately the same site ($n = 3$).

A BH Neurophore II pressure ejection unit delivered pulses of 5–30 psi with 100- to 500-msec pulse widths to the glutamate pipette in the LC and the drug ejection pipette in the dentate gyrus. Ejection parameters were chosen to give drop volumes in the range of 50–150 nl for a single ejection cycle. Pipette tip sizes varied from 10 to 30 μm. At the end of each experiment, the drug and glutamate pipettes were filled with 1% pyronin Y and ejected to mark the site of solution delivery. Alternate Nissl-stained and unstained 40-μm sections were examined to locate all pipettes.

A detailed description of methodology is available in Harley and Milway (1986). One change in the present study is the use in most of the experiments ($n = 17$) of two points separated by 300 μsec on the linear portion of the EPSP slope to provide the EPSP slope measurement. Previously the peak of the EPSP slope had been used as one of the points in determining slope.

3. RESULTS

3.1. Direct Effects of Saline and Drug Ejections on the Perforant-Path-Evoked Potential

Saline ejections generally did not affect evoked potential parameters (18/24 ejections). Four ejections did produce reduction in or loss of the evoked potential, and in one animal, two ejections produced brief (less than 5 min) increases in population spike amplitude. Both kinds of effects appeared within 1 min of drug ejection. The four decreases were associated either with the highest ejection pressure used (30 psi) or with placement of the ejection pipette near the hippocampal fissure.

The direct effects of timolol and propranolol ejections were similar to those of saline. Typically, no significant change in evoked potential parameters occurred (see also Figs. 1 and 2), although with multiple ejections transient effects were occasionally noted (see Fig. 2).

As with the saline ejections, loss of the evoked potential with timolol occurred in one experiment following ejection in the hippocampal fissure and in another experiment using a 30-psi, 500-msec ejection. A significant decrease in evoked potential also occurred in one experiment employing a double-barrel pipette. These effects all appeared within 1 min following ejection.

In two animals timolol ejection dorsal to the buried blade of the dentate produced

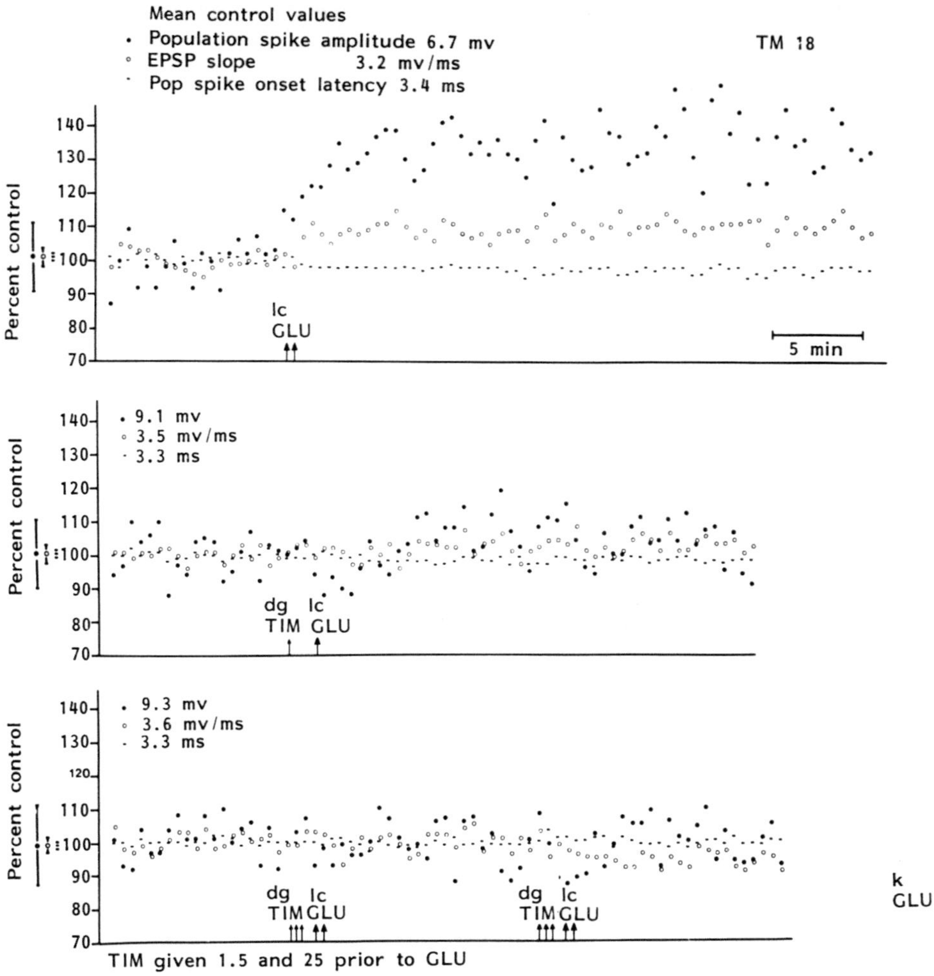

FIGURE 1. Effects of an intradentate timolol ejection on the enhancement of the perforant-path-evoked potential induced by glutamate ejection in the locus coeruleus.

a clear and significant increase in population spike amplitude lasting 5–15 min; however, this effect could not be produced again with subsequent timolol ejections.

3.2. Locus-Coeruleus-Induced Enhancement of the Perforant-Path-Evoked Potential

In the 14 experiments with glutamate ejections into the LC, a significant enhancement of the population spike occurred within 1.5 min of glutamate ejection (the peak increases ranged from 115% to 200%). The EPSP slope was also significantly increased in ten of the 14 experiments (peak increases, 105% to 115%), although the effect on EPSP slope was typically briefer than effects on population spike amplitude. Onset latency was generally unaffected by LC activation.

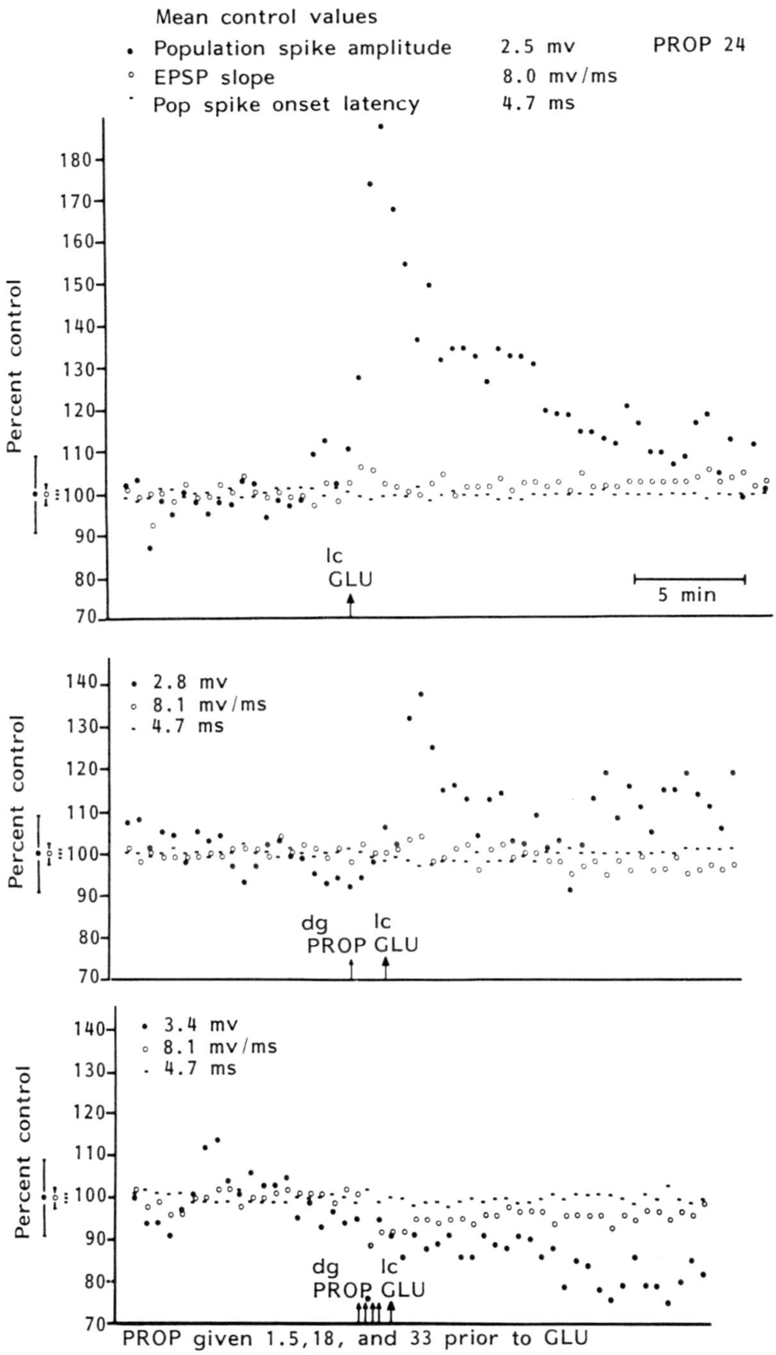

FIGURE 2. Effects of an intradentate propranolol ejection on the enhancement of the perforant-path-evoked potential induced by glutamate ejection in the locus coeruleus.

In six experiments in which glutamate enhancement effects were not blocked, replicable and consistent increases in population spike amplitude were seen with six to ten successive ejections of LC glutamate.

Long-term increases in population spike amplitude, i.e., lasting longer than 30 min or for the duration of the experiment, occurred in seven of the 14 experiments. In four experiments, several long-lasting increases were initiated; that is, there were cumulative stepwise increases in amplitude during the experiments.

3.3. Saline and Drug Effects on LC-Induced Enhancement

Of the 14 experiments with LC-induced enhancement, 13 involved β-blocker drug ejections in the dentate area; in one experiment saline alone was ejected in the dentate. Saline ejection, as expected, did not affect LC-induced enhancement.

One of the drug ejection experiments could not be interpreted since the initial ejection of glutamate in the LC induced a long-lasting and progressive increase in population spike amplitude similar to those illustrated by Neuman and Harley (1983) and Winson and Dahl (1985). There was no attenuation of the progressive increase by timolol.

In the other 12 experiments, effective attenuation of LC-induced enhancement was seen after the first drug ejection in six experiments, and effective attenuation was not seen after the first drug ejection in the remaining six. These two groups are described separately.

3.3.1. Effective Attenuation of LC-Induced Enhancement by β-Blocker Ejection

Both timolol ($n = 4$) and propanolol ($n = 2$) produced effective attenuation of the population spike enhancement effect. Data from both an effective timolol and an effective propanolol experiment are presented in Figs. 1 and 2, respectively.

In the timolol experiment illustrated, three glutamate ejections were made prior to the first drug ejection. All three produced increases of similar magnitude, and the third ejection (illustrated) initiated a long-lasting increase in both EPSP slope and population spike amplitude starting 1 min after glutamate ejection.

Glutamate ejections ($n = 3$) following timolol failed to produce any comparable enhancement. After the first drug ejection (see Fig. 1), a significant increase is seen more than 5 min after the glutamate ejection. The delayed effect may or may not be related to LC activation but does indicate that a significant enhancement is possible despite the long-lasting elevation of population spike amplitude that had occurred prior to drug ejection. No other increases were observed with successive drug–glutamate pairings. On the other hand, no reduction in the previously initiated long-lasting enhancement occurred with drug delivery either, and this was the case in all experiments in which long-lasting effects and effective blocking occurred ($n = 3$).

In the propranolol experiment of Fig. 2, two glutamate ejections were given prior to propranolol delivery. Both produced enhancement effects of similar magnitude (220%, 188%), but only the second is shown. Following the initial propranolol ejection, a relative attenuation of the peak enhancement occurs, although a long-lasting increase in population spike amplitude follows. With repeated drug delivery no enhancement of population spike amplitude can be induced, although a significant depression is initiated with LC activation. A further glutamate ejection (not shown) increased the depression of the population spike amplitude, and recovery was not seen prior to terminating the experiment.

In another three effective blocking experiments, LC-induced depression appeared after repeated drug ejections, and even in Fig. 1, transient depressive effects can be seen with some glutamate ejections. In all six experiments attenuation of enhancement was seen as soon as the first drug ejection, but it was also the case that attenuation became more pronounced or LC-induced depression appeared with further drug–glutamate pairings, suggesting cumulative or more complete blocking effects.

In five experiments, LC activation significantly increased EPSP slope as well as affecting the population spike amplitude; however, in three experiments β blockers were more effective in preventing the population spike amplitude increases than in attenuating the EPSP slope increases.

3.3.2. Experiments without Attenuation of Enhancement

In five drug experiments, a single episode of drug ejection was followed by repeated glutamate ejections in the LC in order to monitor the changes in glutamate-induced effects over time. No significant attenuation of enhancement occurred for either the initial or successive glutamate ejections after the single drug delivery in these experiments.

In the remaining experiment a three-cycle ejection of timolol prior to LC glutamate did not affect enhancement, but a six-cycle ejection did attenuate the enhancement peak from 150% to 130% with recovery occuring 35 min after drug ejection.

3.3.3. Histological Results

Locus coeruleus placements of the glutamate pipette tip in the enhancement experiments were in or within 200 μm of the LC. In the effective blocking experiments, four of the six placements were similar. The recording site was in the exposed blade of the dentate gyrus, and dye clearly spread from the drug pipette medially over the molecular layer of the exposed blade. In the other two effective experiments, ejections occurred from a double-barrel pipette at the recording site in the cell layer of the buried blade.

In the drug experiments in which blockade was not seen, drug pipette placements were either in the hilar region with the recording tip in the cell layer of the exposed or buried blade of the dentate or the drug tips were dorsal to the buried blade cell layer with ejection on the lateral margin of the blade and with recording on the medial slope of the blade.

4. DISCUSSION

These results suggest (1) that timolol and propranolol can block the enhancement effects of LC activation, (2) that previously induced long-lasting enhancement is not attenuated by administering β blockers, (3) that LC activation may also produce a depression of population spike amplitude in the dentate that is unmasked in the presence of β receptor blockade and may be mediated by release of NE onto α receptors, and (4) that placement of β antagonists within the dentate gyrus may be critical for determining their effectiveness; in particular, β antagonists ejected in the hilar region appear less effective than ejections that spread to the molecular layer.

These conclusions must be tempered by our failure to chart a consistent temporal pattern of blockade and recovery as might be expected to occur with reversible blockers and by the relatively small number of experiments with a given histological placement of drug tip ejection.

The LC-induced enhancement effects we observed here were very similar to those observed in our earlier study (Harley and Milway, 1986). The blocking effects we observed here are also consistent with those reported in the slice by Lacaille and Harley (1985) and by Stanton and Sarvey (1986) in that a blockade of presumed NE enhancement effects occurs with no significant alteration of base-line evoked potentials.

Systemic injections of propranolol had previously been shown to produce attenuation of LC-induced population spike enhancement with less clear-cut attenuation of EPSP slope increases (Harley and Milway, 1986), and these results are similar to the blocking effects seen in the present study. However, systemic propranolol consistently and significantly increased the spike onset latency, whereas intradentate blockers *in vivo,* like those administered *in vitro,* rarely produced latency changes. This supports our suggestion that the latency effects of systemic propranolol are of extrahippocampal origin.

These results contrast with the enhancement effects seen with iontophoresis of sotalol (Winson and Dahl, 1985; personal observations). Although enhancement effects did occur twice with the ejection of timolol, this was not a typical result. These differences with pressure ejection may be ascribable either to differences in the mode of delivery and quantity of antagonist applied or to the differences in antagonist. Mueller *et al.* (1982) have reported differential effects of pressure ejecting timolol, which selectively blocked excitatory NE effects on units in CA1, as compared with sotalol, which affected both inhibition and excitation. Mueller *et al.* also report that pressure-ejected NE produces a different profile of cellular effects than that observed with iontophoresed NE. A direct comparison of the two methods and the two blockers in the dentate would be of interest.

The apparent depression of population spike amplitude induced by LC stimulation after repeated ejection of β blockers suggests that α receptors in the dentate may normally act to depress population spike amplitude. Such contrasting effects of α and β receptors have been extensively documented in hippocampal CA1 by Mueller and collaborators (e.g., Mueller *et al.,* 1982). It would be interesting to know if intradentate pressure ejection of α blockers could lead to an increase in LC enhancement effects.

It has been reported that α-receptor distribution corresponds more closely to the pattern of NE innervation in the dentate and that α receptors in particular are densest in the infragranular zone, whereas β receptors exhibit a sparser and more uniform distribution (Crutcher and Davis, 1980). Nonetheless, most of the functional effects of NE activation in the dentate known at present appear to be mediated by β receptors. More specifically, the present results suggest that β receptors in the molecular layer or at the cell body layer may be of more importance in mediating LC activation effects than those in the hilar region or infragranular zone. A similar conclusion may be drawn from the NE-induced enhancement of glutamate release in the dentate (Lynch and Bliss, 1986). This NE effect can be blocked by propranolol but not by α blockers and is presumably mediated by presynaptic β receptors on perforant path terminals in the molecular layer.

REFERENCES

Assaf, S. Y., Mason, S. T., and Miller, J. J., 1979, Noradrenergic modulation of neuronal transmission between the entorhinal cortex and the dentate gyrus of the rat, *J. Physiol. (Lond.)* **292:**52P.

Bliss, T. V. P., and Wendlandt, S., 1977, Effects of stimulation of locus coeruleus on synaptic transmission in the hippocampus, *Proc. XIII Cong. IUPS* **225:**81.

Crutcher, K. A., and Davis, J. N., 1980, Hippocampal α- and β-adrenergic receptors. Comparison of [^{3}H]dihydroalprenolol and [^{3}H]WB4101 binding with noradrenergic innervation in the rat, *Brain Res.* **182:**107–117.

Dahl, D., and Winson, J., 1985, Action of norepinephrine in the dentate gyrus I. Stimulation of locus coeruleus, *Exp. Brain Res.* **59:**491–496.

Harley, C. W., and Milway, J. S., 1986, Glutamate ejection in the locus coeruleus enhances the perforant path-evoked population spike in the dentate gyrus, *Exp. Brain Res.* **63:**143–150.

Harley, C. W., Lacaille, J. C., and Milway, S., 1982, Potentiation of the perforant path evoked potential in the dentate gyrus by locus coeruleus stimulation, *Neurosci. Abstr.* **8:**483.

Lacaille, J. C., and Harley, C. W., 1985, The action of norepinephrine in the dentate gyrus: Beta-mediated facilitation of evoked potentials *in vitro, Brain Res.* **358:**210–220.

Lynch, M. A., and Bliss, T. V. P., 1986, Noradrenaline modulates the release of [^{14}C]glutamate from dentate but not from CA1/CA3 slices of rat hippocampus, *Neuropharmacology* **25:**493–498.

Mueller, A. L., Palmer, M. R., Hoffer, B. J., and Dunwiddie, T. V., 1982, Hippocampal noradrenergic responses *in vivo* and *in vitro,* characterization of alpha and beta components, *Naunyn Schmiedebergs Arch. Pharmacol.* **318:**259–266.

Neuman, R. S., and Harley, C. W., 1983, Long-lasting potentiation of the dentate gyrus population spike by norepinephrine, *Brain Res.* **273:**162–165.

Stanton, P. K., and Sarvey, J. M., 1986, Blockade of norepinephrine-induced long-lasting potentiation in the hippocampal dentate gyrus by an inhibitor of protein synthesis, *Brain Res.* **361:**276–283.

Winson, J., and Dahl, D., 1985, Action of norepinephrine in the dentate gyrus. II. Iontophoretic studies, *Exp. Brain Res.* **59:**497–506.

40

Persistent Changes of Single-Cell Responses in Kitten Striate Cortex Produced by Pairing Sensory Stimulation with Iontophoretic Application of Neurotransmitters and Neuromodulators

JOACHIM M. GREUEL, HEIKO J. LUHMANN, and WOLF SINGER

1. INTRODUCTION

During a critical period of early postnatal life, the development of the kitten visual cortex depends to a substantial degree on sensory experience. In normally reared animals, most neurons respond to stimulation of both eyes and to movement of contours in either direction (Hubel and Wiesel, 1962). Responses to all orientations of visual stimuli are equally represented. Dramatic changes in this organization of the visual cortex can be obtained, however, by suturing one eye closed (Wiesel and Hubel, 1963) or by rearing kittens in an artificial environment (Blakemore and Cooper, 1970; Hirsch and Spinelli, 1970; Cynader and Chernenko, 1976; Singer, 1976). After monocular deprivation, the majority of cortical cells can no longer be driven from the deprived eye. Similarly, changes in orientation preferences are observed after rearing kittens in a visual environment containing only contours of a single orientation (Blakemore and Cooper, 1970; Hirsch and Spinelli, 1970).

There is evidence suggesting that these adaptive processes in the visual cortex (1) require postsynaptic activity (Wilson *et al.*, 1977; Singer *et al.*, 1977; Rauschecker and Singer, 1979), (2) have a threshold (Singer, 1982; Greuel *et al.*, 1987), (3) require the integrity of the noradrenergic and/or cholinergic modulatory systems (Pettigrew and Kasamatsu, 1978; Bear and Singer, 1986), and (4) may require the activation of postsynaptic Ca^{2+} fluxes (Geiger and Singer, 1986; Kleinschmidt *et al.*, 1986). Virtually all the evidence on plastic changes in the visual cortex had been obtained in chronic experiments

JOACHIM M. GREUEL, HEIKO J. LUHMANN, and WOLF SINGER • Max Planck Institute for Brain Research, Department of Neurophysiology, 6000 Frankfurt 71, Federal Republic of Germany.

by manipulating the kitten's environment for days or weeks and analyzing the response properties of a population of neurons thereafter. There are only a few reports on changes in single-cell response properties that could be obtained in an acute experiment (Tsumoto and Freeman, 1981; Frégnac *et al.*, 1984; Geiger and Singer, 1986).

Taking into account the known requirements for plasticity, we wanted to induce persistent changes in single cells while recording from them. We tried to mimic the conditions that promote plastic modifications by pairing a visual stimulus with iontophoretic application of the neuromodulators acetylcholine (ACh) and norepinephrine (NE), the putative neurotransmitter L-glutamate (GLU), and/or the amino acid receptor agonist N-methyl-D-aspartate (NMDA). Glutamate was chosen because its depolarizing action promotes postsynaptic activation; NMDA, since the NMDA receptor is linked to an ionophore that allows the passage of Ca^{2+} ions (Flatman *et al.*, 1983; McDermott *et al.*, 1986), and ACh and NA were applied because of their established involvement in cortical plasticity (Bear and Singer, 1986).

2. METHODS

2.1. Preparation

For the experiments, 28 kittens were used. Their ages varied between 4 and 6 weeks. Anesthesia was initiated with 5 mg/kg sodium hexobarbital i.p. and 10 mg/kg xylazine hydrochloride (5%) i.m. After the animal was placed in a stereotactic head holder,the anesthesia was maintained by artificial ventilation with nitrous oxide (70% N_20, 30% O_2) supplemented by an i.v. infusion of 6% pentobarbital (1.8 mg/kg per hr) combined with 0.7 mg/kg per hr hexacarbacholine bromide. In later experiments the pentobarbital was replaced by 0.2–0.4% halothane. The end-tidal CO_2 concentration in the expired air was adjusted between 3.5% and 4%, and the body temperature was maintained at 37.5°C. In addition, EEG and heart rate were monitored continuously. Loss of fluid was compensated with a glucose–Ringer solution applied via an orally inserted gastric catheter. The nictitating membranes were retracted with phenylephrine, the pupils dilated with atropine, and the corneas were covered with contact lenses. The refractive state was determined by a refractometer, and appropriate spectacle lenses were selected to focus the retina on a tangent screen. Retinal landmarks were plotted on the screen with the help of a fundus camera. After exposure of the visual cortex, multibarreled piggy-back electrodes were advanced in 2-μm steps into area 17 until a single unit was isolated.

2.2. Drugs and Electrodes

The drugs used were acetylcholine chloride (ACh, 2M, pH 4.5), norepinephrine hydrochloride (NE, 0.2 M, pH 3.8), L-glutamic acid (GLU, 0.1 M, pH 8.5), N-methyl-D-aspartic acid (NMDA, 50 mM, pH 8), and γ-aminobutyric acid (GABA, 1 M, pH 3.5), all obtained from Sigma Chemical Company. The pH was adjusted with either HCl or NaOH. The drugs were filled into a seven-barreled micropipette with a tip diameter of 2–4 μm (Fig. 1). One barrel contained a 2 M NaCl solution for balance currents to avoid current artifacts. DC resistances ranged between 10 MΩ and 400 MΩ. The recording pipette was bent and then glued to the multibarreled pipette, the tip of the recording electrode (tip diameter <1 μm) protruding by 30–50 μm. This arrangement reduced

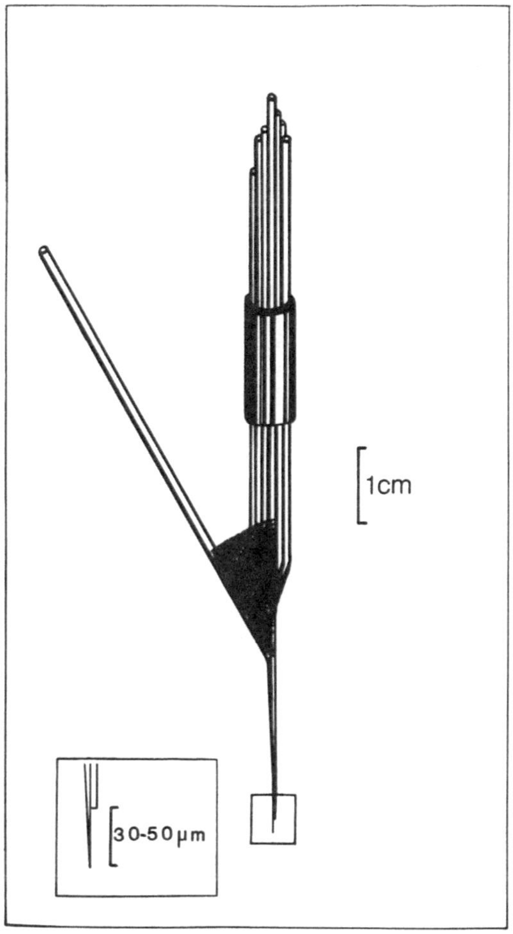

FIGURE 1. Drawing of a multibarreled micropipette. See text for further details.

current artifacts and had the effect that the drugs were applied close to the apical dendrites rather than close to the soma when the electrode was perpendicular to the cortical lamination. The recording electrode was filled with 1.5 M potassium citrate, and the DC resistance varied between 10 and 30 MΩ. The retaining and ejecting currents used for controlling the release of the drugs were: ACh, −8 nA (retaining) and +20 to +50 nA (ejecting); NE, −8 nA (retaining) and +20 to +70 nA (ejecting); GLU, +10 nA (retaining) and −5 to −20 nA (ejecting); NMDA, +12 nA (retaining) and −2 to −18 nA (ejecting); GABA −9 nA (retaining) and +1 to +10 nA (ejecting).

2.3. Cell Classification and Conditioning Procedure

After isolation of a single unit, the receptive fields were first analyzed with hand-held light stimuli consisting of slits whose length, width, and orientation could be varied. Then a moving light bar of appropriate size was delivered by an optomechanical stimulator, and peristimulus time histograms (PSTHs) were compiled with a laboratory computer and subsequently displayed as the sum of 10–15 epochs. Data were collected from the responses elicited by the optimal ocular dominance/orientation preference and by at least one suboptimal visual stimulus. The responses of the cells to the optimal stimulus of each eye were classified according to their vigor in a four-group rating scale. Weak responses were rated in class 1. Clear and reproducible responses that could vary in time were rated in class 2. Clear, reproducible, and stable responses were assigned to class 3 or, if they were particularly vigorous, to class 4.

Before conditioning, histograms of the responses to the optimal and suboptimal stimuli were compiled again after a delay of 10 min to control for spontaneous fluctuations of response properties. The conditioning was performed subsequently by pairing the suboptimal stimulus with iontophoretic ejection of ACh, NE, NMDA, and/or GLU (Fig. 2). In a few cases GABA was applied during the time when no pairing occurred to suppress spontaneous activity. Pairing was performed once every 20 sec. This interval was chosen in order to obtain a reasonable number of pairings in an acceptable time

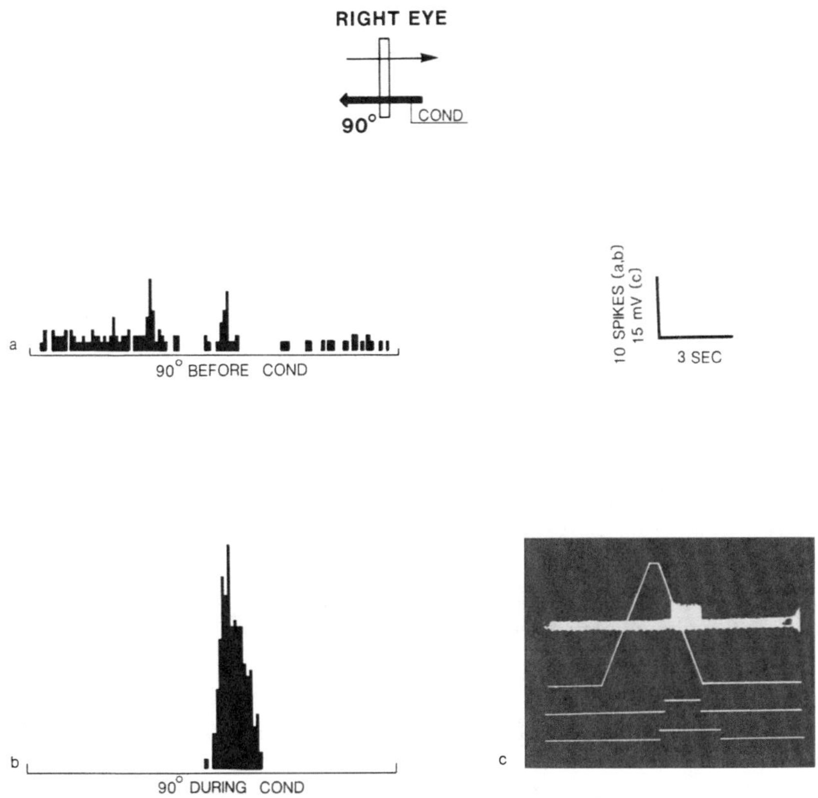

FIGURE 2. Conditioning procedure. The bar at the top indicates a bar-shaped light stimulus of appropriate orientation that was projected onto a tangent screen placed in front of the kitten. Arrows indicate direction of movement. a: PSTH compiled out of ten sweeps. b: Conditioning of the backwards-directed movement was performed by pairing the visually evoked response with iontophoretic ejection of ACh, NA, NMDA, and GLU. Note the marked increase in activity. Here, in addition, the spontaneous activity was suppressed with low-current application of GABA. c: Photograph taken from the oscilloscope showing a single pairing of visual stimulation with iontophoresis. First trace, recorded potential; second trace, movement of the bar, indicated by the positive (forward) and negative (backward) slope; third trace, pulse of ionophoretic current; fourth trace, opening of a shutter to allow visual stimulation.

while avoiding marked desensitization or drug accumulation. For most cells investigated, the conditioning period lasted 30 min (~100 pairings).

2.4. Comparison of Responses

Changes in the response properties were assessed (1) by comparing the PSTHs before and after conditioning and (2) by comparing the PSTHs compiled with suboptimal and

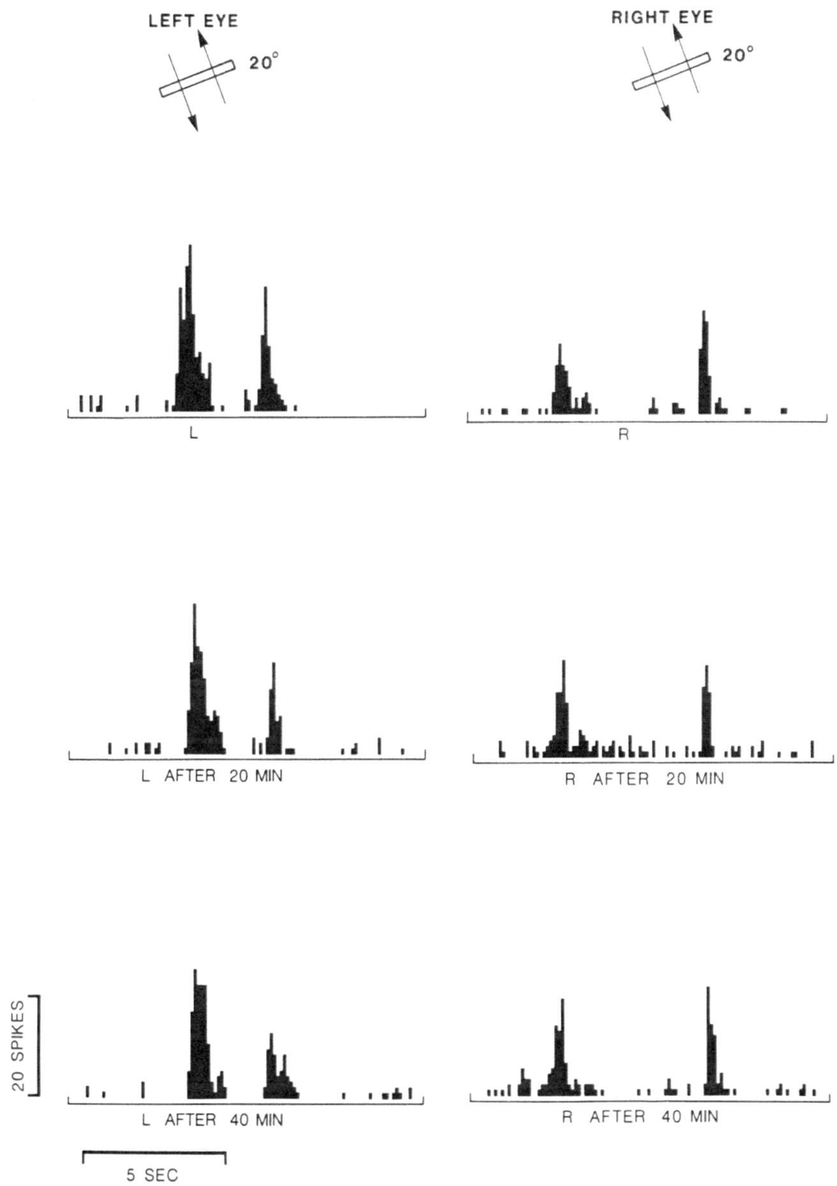

FIGURE 3. Poststimulus time histograms of one cell to demonstrate stability of response properties. Samples of representative responses taken at the times indicated are shown.

optimal stimulation. For 62% of the conditioning procedures (47 out of 76), the analysis was supplemented by a quantitative evaluation: we determined the number of spikes in the evoked response and corrected the data for spontaneous activity. Finally, we calculated the percentages of response changes after conditioning and determined the total percentage of change by subtracting the percentages obtained for the optimal response from the percentages of the conditioned response. If the differences exceeded the 80% threshold, the changes in response properties were regarded as induced by conditioning.

3. RESULTS

3.1. Stability of Response Properties

In order to obtain data on the spontaneous fluctuations of response properties, we performed a series of control experiments. We presented a sequence of optimal (e.g., to the left eye) and suboptimal (e.g., to the right eye) stimuli and compiled PSTHs once every 10 min. The PSTHs were collected in this way for up to 60 min. In 31 out of the 36 cells that were assigned to response classes 3 or 4, the spontaneously occurring changes in response amplitudes were less than ±40% around a mean (Fig. 3). Even when fluctuations occurred, they always affected both the optimal and suboptimal responses in a similar way. The results were essentially the same when the responses to different orientations of light bars were compared. Our results are in line with evidence obtained by Macy *et al.* (1982) showing that there is no spontaneous change in ocular dominance and with Henry *et al.* (1973) and Heggelund and Albus (1978), who report that the orientation preference also remains invariant over time. Therefore, we accepted changes following conditioning as specific when they were selective and exceeded the 80% threshold.

3.2. Changes Obtained by the Conditioning Procedure

For the conditioning procedure different combinations of drugs were used, and only cells rated in response classes 3 or 4 were considered. We were able to induce changes in ocular dominance, orientation preference, and direction selectivity (DS) in 47% (36 out of 76) of the conditioning procedures. Four different types of changes were found.

1. Homosynaptic potentiation. A change is called homosynaptic potentiation when a cell's response to the conditioned stimulus improves whereas the response to the unconditioned stimulus remains unchanged. A example is shown in Fig. 4. The conditioning was unidirectional and lasted for 20 min. The visually evoked response was paired with iontophoretic ejection of ACh, NE, and NMDA. After the conditioning there was an increase in both the unconditioned and the conditioned response. The general increase of responsiveness is probably a consequence of an unselective increase of excitability as reflected by the enhanced spontaneous activity. However, the improvement of the conditioned response is above what would be expected from a general excitability increase. The differential change exceeds the 80% threshold and is therefore regarded as induced by the conditioning procedure.

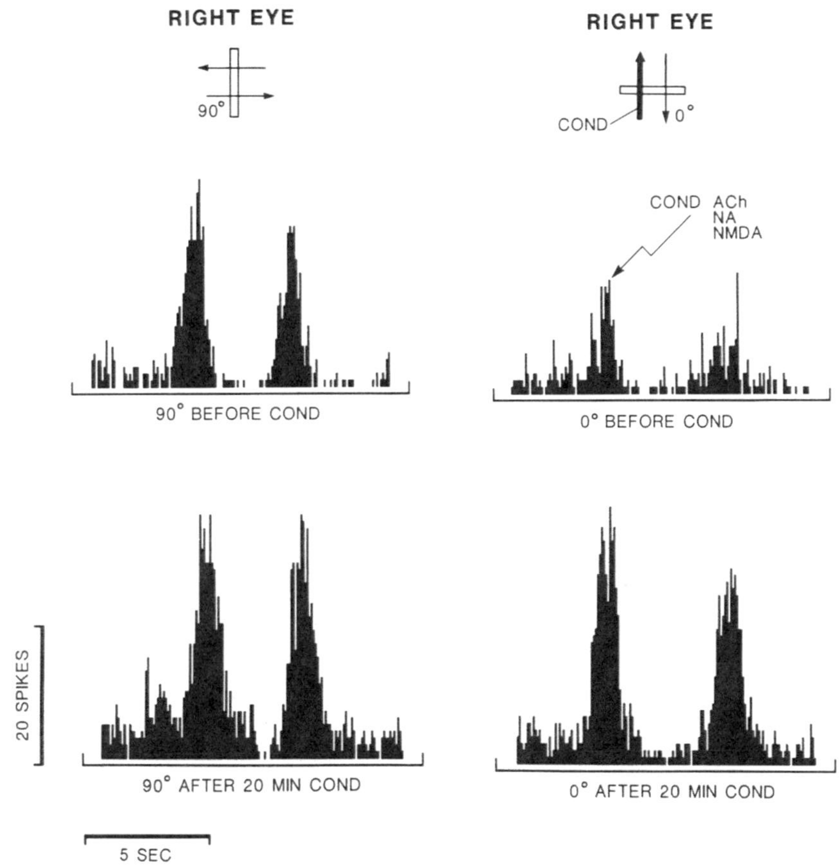

FIGURE 4. Homosynaptic potentiation. Significant increase of the conditioned response, here with a nonspecific change of the unconditioned response (see text). The arrow marked COND points to the conditioned response.

2. Heterosynaptic depression. A change is called heterosynaptic depression when the response to the conditioned stimulus remains unchanged and the response to the unconditioned stimulus is depressed. The cell shown in Fig. 5 responded slightly better to a horizontal bar (0°) than to an oblique bar (135°). After 25 min of conditioning with NMDA and GLU, there was only a small improvement of the conditioned response to the oblique bar but a pronounced depression of the unconditioned response to the horizontal bar.
3. Homosynaptic depression. We call a response modification homosynaptic depression when only the conditioned response is reduced. The neuron shown in Fig. 6 was conditioned with ACh for 30 min. Both eyes gave roughly equal responses before the conditioning. During conditioning the conditioned response decreased markedly, and this depression persisted for more than 10 min (longest period tested for this cell). The fact that the responses to both directions were reduced indicates that afferents from bidirectionally responding cells lost their potency in driving the neuron to the conditioned stimulus. The unconditioned response decreased only slightly; therefore, we did not attribute this change to our conditioning procedure.

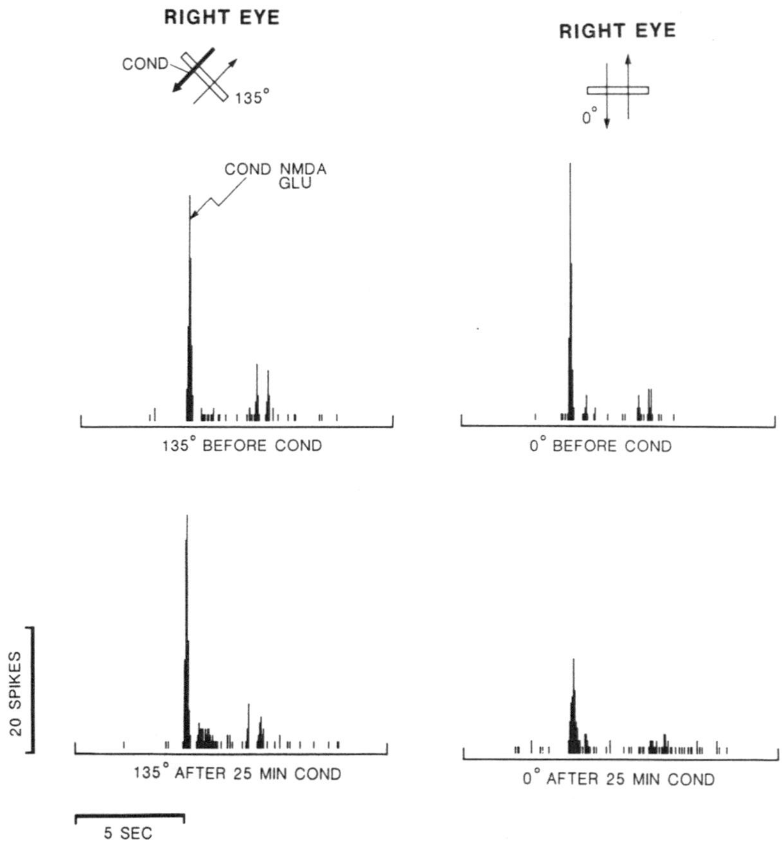

FIGURE 5. Heterosynaptic depressing. The conditioned response remained mostly unchanged; the unconditioned response decreased.

4. Combined changes. Changes are called "combined" when the response to the conditioned stimulus increases and the response to the unconditioned stimulus decreases. A decrease of the response to the conditioned stimulus and a simultaneous increase of the unconditioned response was never observed. Figure 7 shows the responses of a neuron that was conditioned with NMDA and GLU for 30 min. Before conditioning there were only slight variations of the response amplitudes. After conditioning, the response to the previously optimal stimulus had dropped close to zero, while the conditioned response showed a marked potentiation. The potentiation persisted for at least 15 min (longest period tested for this cell). The previously optimal response remained reduced, although a slight recovery could be observed after 15 min.

We never observed a heterosynaptic potentiation, i.e., a selective improvement of the unconditioned response, and we also never found covariations (potentiations or depressions) of both the conditioned and unconditioned responses that were large enough to be attributable to conditioning. The direction of the modifications was not predictable from the combination of the applied drugs. The changes in response properties could persist for more than 40 min (longest period tested). All types of modifications occurred with about equal frequency (Table I).

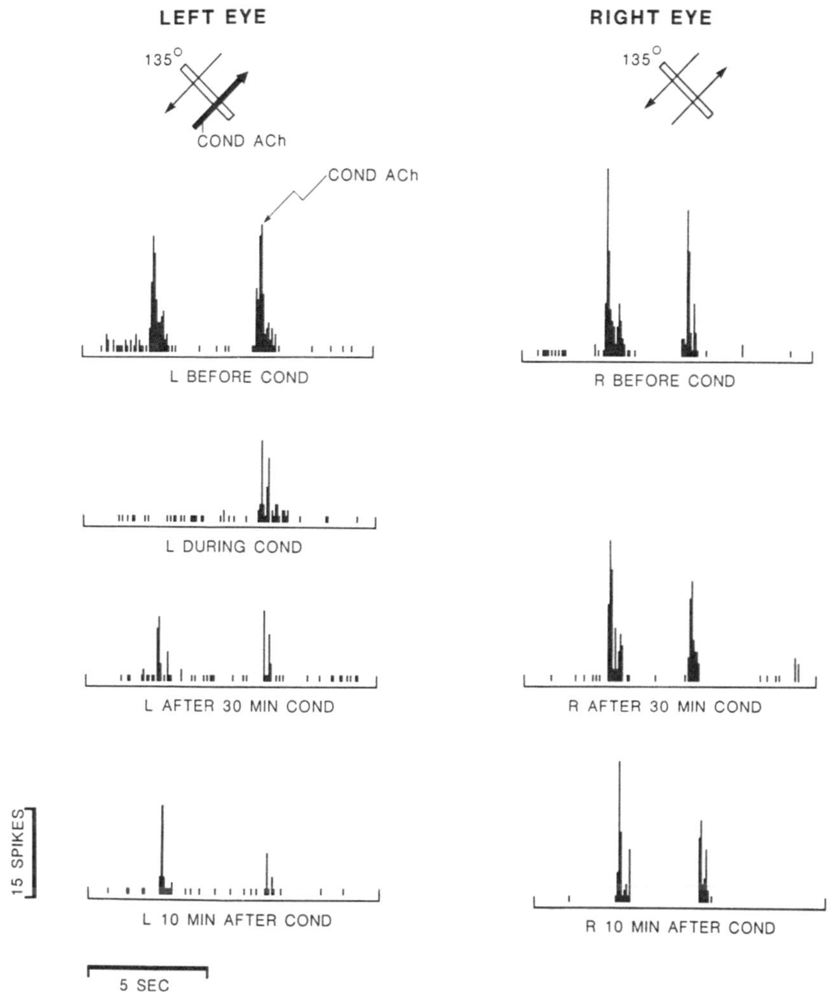

FIGURE 6. Homosynaptic depression. The conditioned response decreased markedly; the unconditioned response did not change significantly. The effect could still be seen 10 min after the termination of the conditioning.

4. DISCUSSION

Persistent changes in response properties of cortical neurons could be induced by pairing a visual stimulus with iontophoretic ejection of ACh, NE, NMDA, and/or GLU. We found four types of modifications: homosynaptic potentiation, heterosynaptic depression, homosynaptic depression, and combined changes (potentiation of the conditioned response and depression of the unconditioned response). These modifications cannot be attributed to spontaneous changes, since the fluctuations of response properties were very small and the modifications obtained were specific. It is well established that repetitive visual stimulation alone is not appropriate to alter the response properties in anesthetized animals (Freeman and Bonds, 1979; Singer and Rauschecker, 1982); this kind of stimulation is either insufficient to produce a change or may lead to adaptation effects (Saul

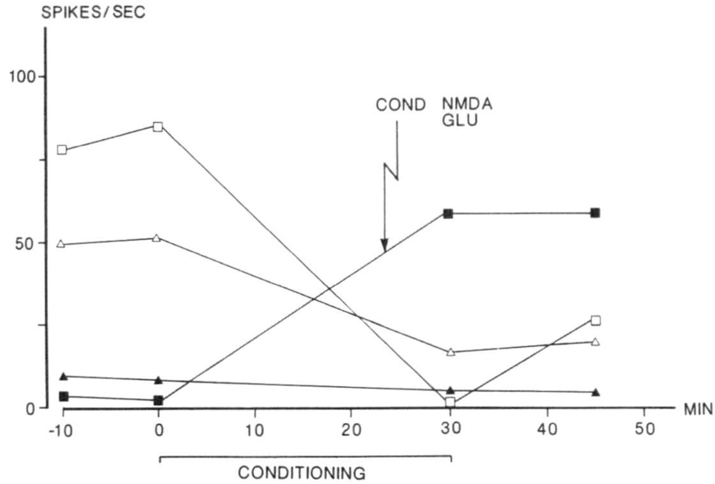

FIGURE 7. Combined change. Potentiation of the conditioned response and depression of the unconditioned response. The response amplitude is expressed as spikes/sec (sum of ten sweeps). Open symbols, responses to optimal stimulation; filled symbols; responses to suboptimal stimulation (same eye, different orientations). Squares, responses to forward-directed movements; triangles, responses to backwards-directed movements. Note the persistence of the effect and the slight recovery of the response to the previously optimal stimulus 15 min after termination of the conditioning.

and Daniels, 1985; Marlin and Cynader, 1986; J.M. Greuel, H. J. Luhmann, and W. Singer, unpublished observations). However, pairing visual stimulation with iontophoresis can produce persistent changes in response properties of cortical neurons; it is conceivable that these changes are based on the same mechanisms as those that occur during a period of deprivation or in kittens reared in an artificial environment.

Interestingly, each combination of drugs could produce a variety of response modifications. As one possible explanation, it may be assumed that all drugs act on a final common pathway, thereby inducing a cascade of intracellular events. Acetylcholine, NE, and GLU are known to promote the breakdown of phosphatidylinositol 4,5-bisphosphate (Berridge and Irvine, 1984; García-Sainz, 1985; Sladeczek *et al.*, 1985; Nicoletti *et al.*, 1986). One product, inositol trisphosphate, releases Ca^{2+} from intracellular stores (for review see Berridge and Irvine, 1984) and therefore contributes to an increase of intracellular free Ca^{2+}, as does NMDA by opening a Ca^{2+}- permeable ionophore (McDermott *et al.*, 1986). Calcium ions are necessary for the induction of long-term potentiation (Lynch *et al.*, 1983) and may also be responsible for the induction of long-term depression (Ekerot and Kano, 1985).

Acetylcholine and NA are known to activate second messenger systems. These

TABLE I. Occurence of Different Types of Response Modifications[a]

	OD	OP	DS
Homosynaptic potentiation	2	5	9
Heterosynaptic depression	2	3	0
Homosynaptic depression	2	4	2
Combined effects	3	3	1

[a] OD, ocular dominance; OP, orientation preference; DS, direction selectivity.

second messenger systems may contribute to biochemical events specific for long-term modifications; alternatively, ACh and NE could promote plasticity by reducing potassium conductivities (McCormick and Prince, 1986; Nakajima *et al.*, 1986; Madison and Nicoll, 1982; Sah *et al.*, 1985). A decrease of potassium conductivity would lead to enhanced depolarizations, which in turn might either promote Ca^{2+} influx via high-threshold voltage-dependent Ca^{2+} channels or reduce the Mg^{2+} block of the NMDA-receptor-coupled ionophore.

5. SUMMARY

Long-term changes of neuronal response properties were induced in the visual cortex of anesthetized kittens by pairing visual stimulation with iontophoretic ejection of acetylcholine, norepinephrine, N-methyl-D-aspartate, and/or L-glutamate. Four types of response modifications were observed with about equal frequency: homosynaptic potentiation (increase of conditioned response and no change of control response), heterosynaptic depression (no change of conditioned response and decrease of control response), homosynaptic depression (decrease of conditioned response and no change of control response), and combined effects (increase of conditioned response and decrease of control response). No correlation was found between the applied drug combination and the resulting modifications. The changes could persist for more than 40 min (longest period tested). We conclude that the drugs used promote long-term modifications of response properties and hypothesize that mechanisms similar to long-term potentiation/long-term depression may account for the observed changes.

ACKNOWLEDGMENTS. We would like to thank Mrs. C. Ziegler, Mrs. R. Ruhl-Völsing, and Mrs. G. Trauten for excellent technical assistance. J. M. Greuel was supported by a grant from the Studienstiftung des deutschen Volkes.

REFERENCES

Bear, M. F., and Singer, W., 1986, Modulation of visual cortical plasticity by acetylcholine and noradrenaline, *Nature* **320:**172–176.

Berridge, M. J., and Irvine, R. F., 1984, Inositol trisphosphate: A novel second messenger in cellular signal transduction, *Nature* **312:**315–321.

Blakemore, C., and Cooper, G. F., 1970, Development of the brain depends on the visual environment, *Nature* **228:**477–478.

Cynader, M., and Chernenko, G., 1976, Abolition of direction selectivity in the visual cortex of the cat, *Science* **193:**504–505.

Ekerot, C.-F., and Kano, M., 1985, Long-term depression of parallel fibre synapses following stimulation of climbing fibres, *Brain Res.* **342:**357–360.

Flatman, J. A., Schwindt, P. C., Crill, W. E., and Stafstrom, C. E., 1983, Multiple actions of N-methyl-D-aspartate on cat neocortical neurons *in vitro, Brain Res.* **266:**169–173.

Freeman, R. D., and Bonds, A. B., 1979, Cortical plasticity in monocularly deprived immobilized kittens depends on eye movement, *Science* **206:**1093–1095.

Frégnac, Y., Thorpe, S., Shulz, D., and Bienenstock, E., 1984, Modification of function in cat visual cortical neurones induced by control of the correlation between postsynaptic activity and visual input, *Soc. Neurosci. Abstr.* **10:**1078.

García-Sainz, J. A., 1985, α_1-Adrenergic and M_1-muscarinic actions and signal propagation, *Trends. Pharmacol. Sci.* **6:**349–350.

Geiger, H., and Singer, W., 1986, A possible role of Ca^{++}-currents in developmental plasticity, *Exp. Brain Res., Ser.* **14:**256–270.

Greuel, J. M., Luhmann, H. J., and Singer, W., 1987, Evidence for a threshold in experience-dependent long-term changes of kitten visual cortex, *Dev. Brain Res.* **34:**141–149.

Heggelund, P., and Albus, K., 1978, Response variability and orientation discrimination of single cells in striate cortex of cat, *Exp. Brain Res.* **32:**197–211.

Henry, G. H., Bishop, P. O., Tupper, R. M., and Dreher, B., 1973, Orientation specificity and response variability of cells in the striate cortex, *Vis. Res.* **13:**1771–1779.

Hirsch, H. V. B, and Spinelli, D. N., 1970, Visual experience modifies distribution of horizontally and vertically oriented receptive fields in cats, *Science* **168:**869–871.

Hubel, D. H., and Wiesel, T. N., 1962, Receptive fields, binocular interaction and functional architecture in the cat's visual cortex, *J. Physiol. (Lond.)* **160:**106–154.

Kleinschmidt, A., Bear, M. F., and Singer, W., 1986, Effects of the NMDA-receptor antagonist APV on visual cortical plasticity in monocularly deprived kittens, *Neurosci. Lett. Suppl.* **26:**S58.

Lynch, G., Larson, J., Kelso, S., Barrionuevo, G., and Schottler, F., 1983, Intracellular injections of EGTA block induction of hippocampal long-term potentiation, *Nature* **305:**719–721.

Macy, A., Ohzawa, I., and Freeman, R. D., 1982, A quantitative study of the classification and stability of ocular dominance in the cat's visual cortex, *Exp. Brain Res.* **48:**401–408.

Madison, D. V., and Nicoll, R. A., 1982, Noradrenaline blocks accommodation of pyramidal cell discharge in the hippocampus, *Nature* **299:**636–638.

Marlin, S. G., and Cynader, M. S., 1986, Direction selective adaptation of cat striate cortex cells, *Invest. Ophthalmol. Vis. Sci. Suppl.* **27:**244.

McCormick, D. A., and Prince, D. A., 1986, Mechanisms of action of acetylcholine in the guinea pig cerebral cortex *in vitro, J. Physiol. (Lond.)* **375:**169–194.

McDermott, A. B., Mayer, M. L., Westbrook, G. L., Smith, S. J., and Barker, J. L., 1986, NMDA-receptor activation increases cytoplasmic calcium concentration in cultured spinal cord neurones, *Nature* **321:**519–522.

Nakajima, Y., Nakajima, S., Leonard, R. J., and Yamaguchi, K., 1986, Acetylcholine raises excitability by inhibiting the fast transient potassium current in cultured hippocampal neurons, *Proc. Natl. Acad. Sci. U.S.A.* **83:**3022–3026.

Nicoletti, F., Meck, J. L., Iadarola, M. J., Chuang, D. M., Roth, B. L., and Costa, E., 1986, Coupling of inositol phospholipid metabolism with excitatory amino acid recognition sites in rat hippocampus, *J. Neurochem.* **46:**40–46.

Pettigrew, J. D., and Kasamatsu, T., 1978, Local perfusion of noradrenaline maintains visual cortical plasticity, *Nature* **271:**761–763.

Rauschecker, J. P., and Singer, W., 1979, Changes in the circuitry of the kitten visual cortex are gated by postsynaptic activity, *Nature* **280:**58–60.

Sah, P., French, C.R., and Gaye, P. W., 1985, Effects of noradrenaline on some potassium currents in CA1 neurones in rat hippocampus, *Neurosci. Lett.* **60**:295–300.

Saul, A. B., and Daniels, J. D., 1985, Adaptation effects from conditioning area 17 cortical units in kittens during physiological recording, *Soc. Neurosci. Abstr.* **11:**461.

Singer, W., 1976, Modification of orientation and direction selectivity of cortical cells in kittens with monocular vision, *Brain Res.* **118:**460–468.

Singer, W., 1982, Central core control of developmental plasticity in the kitten visual cortex: I. Diencephalic lesions, *Exp. Brain Res.* **47:**209–222.

Singer, W., 1985, Central control of developmental plasticity in the mammalian visual cortex, *Vis. Res.* **25:**389–396.

Singer, W., and Rauschecker, J. P., 1982, Central core control of developmental plasticity in the kitten visual cortex: II. Electrical activation of mesencephalic and diencephalic projections, *Exp. Brain Res.* **47:**223–233.

Singer, W., Rauschecker, J., and Werth, R., 1977, The effect of monocular exposure to temporal contrasts on ocular dominance in kittens, *Brain Res.* **134:**568–572.

Singer, W., Tretter, F., and Yinon, U., 1982, Central gating of developmental plasticity in kitten visual cortex, *J. Physiol. (Lond.)* **324:**221–237.

Sladeczek, F., Pin, J.-P., Récasens, M., Bockaert, J., and Weiss, S., 1985, Glutamate stimulates inositol phosphate formation in striatal neurones, *Nature* **317:**717–719.

Tsumoto, T., and Freeman, R. D., 1981, Ocular dominance in kitten cortex: Induced changes of single cells while they are recorded, *Exp. Brain Res.* **44:**347–351.

Wiesel, T. N., and Hubel, D. H., 1963, Single-cell responses in striate cortex of kittens deprived of vision in one eye, *J. Neurophysiol.* **26:**1003–1017.

Wilson, J. R., Webb, S. V., and Sherman, S. M., 1977, Conditions for dominance of one eye during competitive development of central connections in visually deprived cats, *Brain Res.* **136:**277–287.

41

Norepinephrine-Dependent Neuronal Plasticity in Kitten Visual Cortex

T. KASAMATSU and T. SHIROKAWA

1. INTRODUCTION

Higher mammals such as cats, monkeys, and humans have a pair of frontal eyes. Visual scientists want to understand this simple fact in evolution as a necessary condition for having stereopsis, or the three-dimensional depth sensation in vision. We look at a single object in space using the two eyes. An image of the object in visual space is encoded into a sequence of electrical impulses at the two retinas, and the coded information is sent upward to the visual centers through the two deliberately separated channels in the subcortical structures. A set of impulses impinge on a single cell for the first time in the occipital area of the neocortex and have somehow to be integrated and decoded in order to create the three-dimensional visual sensation of the object at which we are looking. This is thought to be the neuronal scenario of having stereopsis.

Therefore, by placing a recording microelectrode in the primary visual cortex of cats, for example, one can pick up single-unit activity that is excited by presentation of a proper visual target in front of the left or right eye or both (mapping of visual receptive fields). If we pay particular attention to the extent of overlap of binocular fields in single cells, the profile of the normal visual cortex may be shown in the form of a frequency histogram that is compiled of cells having different levels of binocular overlap (ocular dominance (OD) histogram). This OD histogram is usually made following the seven-group scheme originally proposed by Hubel and Wiesel (1962). Briefly, group 4 cells receive well-balanced excitatory input from the two eyes, and group 1 and group 7 cells are excited by stimulation of only the contralateral and ipsilateral eye, respectively. Cells in the remaining four groups receive input from the two eyes with various degrees of binocular interaction.

Wiesel and Hubel (1963) also carried out an ingenious experiment. They found that the OD distribution in young kittens is not fixed but is modifiable by surgically closing the eyelids of one eye. After monocular lid suture for a brief period, about 1 week, virtually all visual cortical cells change their ocularity and become responsive to stimulation of only the nondeprived eye. The deprived eye is functionally disconnected from

T. KASAMATSU • The Smith–Kettlewell Eye Research Foundation at Pacific Presbyterian Medical Center, San Francisco, California 94115. T. SHIROKAWA • Department of Neurophysiology, Institute of Higher Nervous Activity, Osaka University Medical School, Kita, Osaka 530, Japan.

the visual center. This is primarily an age-dependent phenomenon. It is generally accepted that this change in OD is caused primarily by changes in synaptic connectivity within the visual cortex. Since these findings, the monocular deprivation paradigm has been extensively used to study OD plasticity. It should be pointed out that we invariably obtain a U- or W-shaped histogram as an intermediate before the shift in OD has been finalized following monocular deprivation. What are the likely mechanisms underlying the OD plasticity? Unfortunately, not much is known at the moment about the cellular and chemical basis of this type of synaptic plasticity present in the developing neocortex.

2. REVIEW OF EARLY FINDINGS

More than 10 years ago, we started to look closely at this problem. Kasamatsu and Pettigrew (1976) proposed that the catecholamine (CA)-containing system in the brain is necessary for maintaining OD plasticity at the high level in the visual cortex of developing kittens. Lately, as is shown below, this proposal has been extended along the cascades of biochemical events possibly involved. Detailed accounts of the marriage between OD plasticity and the central CA system have appeared elsewhere (e.g., Kasamatsu, 1983). Therefore, we briefly review here only a few results from our early studies in this area.

We first wanted to study whether OD plasticity disappears from the kitten visual cortex in which CA terminals have been destroyed. The presence of rich CA terminal fields was shown in cat visual cortex by both CA histochemistry and electron microscopy (Itakura *et al.*, 1981). Since the locus coeruleus from which the ascending norepinephrine (NE) fibers originate is not the compact nucleus in cats that it is in rats, we decided to utilize the chemical lesion of CA terminal fields with a CA-related neurotoxin, 6-hydroxydopamine (6-OHDA) (e.g., Kostrzewa and Jacobowitz, 1974; Jonsson, 1980). The results from the first experiment in the series were quite intriguing; we found many binocularly driven cells despite monocular deprivation (Kasamatsu and Pettigrew, 1976, 1979). We favored over other alternatives an interpretation that OD plasticity disappeared because of the lack of CA terminals in the 6-OHDA-treated brain. However, there are a few serious shortcomings in this study: (1) the lack of specificity in the site of drug action, since 6-OHDA was repeatedly injected intraventricularly, and (2) high doses of 6-OHDA (>10 mg total).

Next we tried to localize the plasticity-suppressing effect of 6-OHDA within the visual cortex by directly infusing the cortex with 4 mM 6-OHDA (a total of 170 μg) in the acidic vehicle solution. We used the continual microinfusion method (Kasamatsu *et al.*, 1981). We again found many binocular cells, despite monocular deprivation, in the cortex that had been locally infused with 6-OHDA (Kasamatsu *et al.*, 1979). The results were satisfactory, and obviously their interpretation became much easier than before. However, everything seems to hinge on a single point: is the lesion by directly infused 6-OHDA confined within the visual cortex, and, if so, how specific is it? To answer these crucial issues, we carried out a series of histochemical and biochemical studies in a visual cortex that had been locally infused with 4 mM 6-OHDA for a week, a condition comparable to that in physiological studies. The following three methods were used: (1) elucidation of the 6-OHDA-affected area by a glyoxylic-acid-induced CA histofluorescence method (Kasamatsu *et al.*, 1981), (2) a high-pressure liquid chromatography (HPLC) assay of endogenous CAs remaining in the 6-OHDA-affected area (Kasamatsu *et al.*, 1981), and (3) delineation of the size of CA lesion by *in vitro* uptake of [^{3}H]norepinephrine by cortical slices obtained from the CA-affected area (Nakai and Kasamatsu, 1984).

Taken together, the histochemical and biochemical studies strongly suggested that in our paradigm of localized infusion, 6-OHDA destroyed primarily NE terminals within the CA-affected area except for a small area of nonspecific damage at the center of cannulation (Nakai *et al.*, 1981, 1987). We therefore thought that the lack of NE terminals in the 6-OHDA-treated cortex was a primary cause of the disappearance of OD plasticity. We next tested this interpretation more rigorously by modifying the design of our experiments.

We asked whether the plasticity that had been lost to the preceding 6-OHDA treatment could be restored by exogenous NE. Only in the hemisphere locally infused with NE did we obtain a shift in ocular dominance in response to monocular deprivation (Pettigrew and Kasamatsu, 1978; Kasamatsu *et al.*, 1979). This made a sharp contrast with the nearly normal OD distribution obtained from the other hemisphere of the same animal. As expected, we noted the presence of a concentration–effect relationship between exogenous NE in an osmotic minipump and the extent of changes in ocular dominance. In a separate series of studies, we determined the threshold concentration of NE that was needed to restore the plasticity to the kitten cortex that had become aplastic because of the 6-OHDA treatment. The threshold was at about 50 μM in an osmotic minipump and was calculated as 3×10^{-7} M at the recording site (Kasamatsu *et al.*, 1981). In summary, it seems to be very likely that the central NE system is deeply involved in the regulation of OD plasticity that is typically seen in the developing visual cortex.

3. RECENT DEVELOPMENTS

Recently, however, the above conclusion has been challenged by a few "negative" results. In addition to the original methods of repeated injections of 6-OHDA into the lateral ventricle and direct infusion of the visual cortex with 6-OHDA, various other methods have been utilized to reduce endogenous NE in the kitten brain. They include a few injections of 6-OHDA into the peritoneal space of neonatal kittens (Bear and Daniels, 1983; Bear *et al.*, 1983), multiple injections of 6-OHDA into the locus coeruleus area (Adrien *et al.*, 1985), massive electrolytic lesion of the ascending NE projection path at the lateral hypothlamus (Daw *et al.*, 1984), and intraventricular as well as systemic injections of DSP-4, another type of NE-related neurotoxin (Daw *et al.*, 1985a). What is common in all these seemingly straightforward experiments is the following: in spite of substantial reduction of endogenous NE in the visual cortex, which is proven by either an HPLC or enzymatic assay, the physiological effect of monocular lid suture was sustained, suggesting that a decrease in endogenous NE is not necessarily correlated with the low level of OD plasticity. Despite these confusions, however, it is worthwhile to note that two other laboratories independently confirmed our early results when they recorded from kitten visual cortex that had been directly infused with 4 mM 6-OHDA using the continuous microperfusion method (Daw *et al.*, 1983; Bear *et al.*, 1983; Paradiso *et al.*, 1983). Critical reviews of these "negative" results and their interpretations have appeared elsewhere (Kasamatsu *et al.*, 1984; Kasamatsu and Shirokawa, 1985; Shirokawa and Kasamatsu, 1986; Kasamatsu, 1987). Different views of the issue have also been expressed (Daw *et al.*, 1985b; Frégnac and Imbert, 1984).

Faced with these unexpected and puzzling findings, some authors proposed that the failure to block the shift in OD might be attributable to an unspecified compensatory process counteracting the effects of 6-OHDA outside the NE system (Bear *et al.*, 1983). Others even said that the "positive" results obtained from the kitten cortex directly infused with 6-OHDA were caused by "nonspecific" effects of 6-OHDA (Daw *et al.*, 1984;

1985a; Trombley *et al.*, 1986), thus implying strongly that our interpretation of the early findings is not right. The situation is certainly perplexing. Nevertheless, we would like to point out that restoration of OD plasticity to the aplastic kitten cortex that has been pretreated with 6-OHDA deserves particular attention in its own right. The plasticity-enhancing effect of NE is irrefutable independent of the specificity of 6-OHDA's action. Furthermore, there is a corroborative report that the expected shift in ocular dominance following monocular deprivation was prevented from occurring, at least partly, by systemic injections of an α-adrenergic agonist clonidine (Nelson *et al.*, 1985).

There are basically two ways, we think, to clarify this conceptual confusion: (1) to examine directly whether the logic behind the main theme of our NE hypothesis for OD plasticity is totally ungrounded and (2) to devise experiments that may decide one way or the other which interpretation of the "conflicting" results is incorrect or at least insufficient. Knowing that the two approaches are both needed and not mutually exclusive of each other, we decided at this time to address the first of the above two alternatives. We believe that the strategy we took seems to be more urgent, constructive, and rewarding as well.

Let us now expand the original NE hypothesis further to include β adrenoreceptors in this matter. The reason for this is obvious; when we think of how NE works either after released from the NE terminals or infused externally, it is impossible to escape from the presence of NE-related receptors within the cortex, including β adrenoreceptors. In what follows, we describe the results obtained from three studies, all of which strongly indicate the unique role of β adrenoreceptors in the regulation of OD plasticity.

First, we studied the normal ontogeny of both endogenous monoamines and related receptors in cat neocortex. We used an HPLC method for the former and *in vitro* binding assays with [^{3}H]-labeled β receptor antagonist dihydroalprenolol for the latter. As illustrated in Fig. 1, the number of specific β-adrenoreceptor binding sites was higher than the adult value between 5 and 11 weeks postnatally, whereas endogenous NE increases continually toward adulthood (Jonsson and Kasamatsu, 1983).

These biochemical results strongly suggest to us that if the central NE system is involved in the regulation of OD plasticity, it is not the absolute level of endogenous NE by itself but rather the number of specific β-adrenoreceptor binding sites that should be correlated with the extent of the plasticity.

In the next series of experiments, we used adult cats as a model of the aplastic visual cortex, since under usual conditions the adult cortex is not susceptible to monocular deprivation. What we wanted to show is significant changes in the OD distribution in

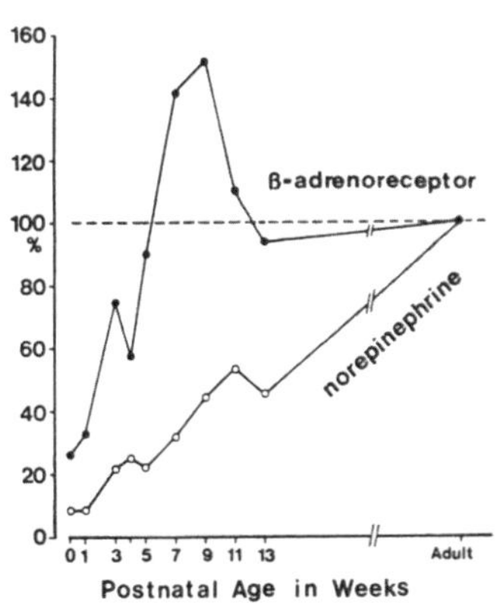

FIGURE 1. Postnatal ontogeny of endogenous NE (open circles) and β-adrenoreceptor binding (closed circles) in the visual cortex of normal cats. Each point is an average of three to five determinations. The data are expressed as percentages of the respective adult values. (Reproduced with minimum modification from Jonsson and Kasamatsu, 1983.)

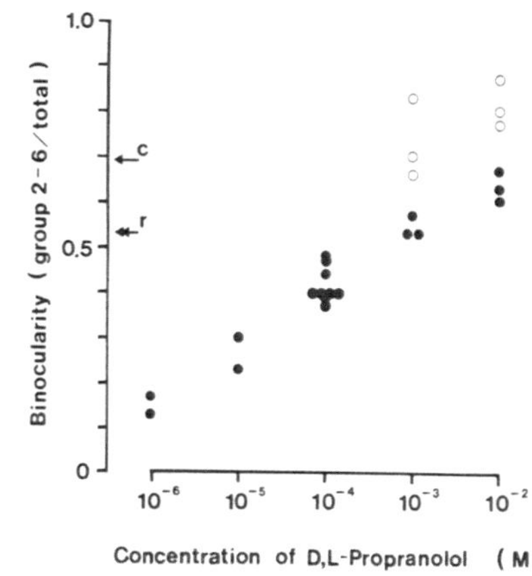

FIGURE 2. Concentration-dependent suppression of ocular dominance plasticity. The proportion of binocular cells obtained in either the monocularly deprived cortex (filled circles, $N = 19$) or the drug-infused, otherwise normal cortex (open circles, $N = 6$) is plotted against the logarithmic concentration of propranolol in an osmotic minipump. The lower end of the 95% confidence limits (*C*, 0.69) and that of the 5% rejection limits (*r*, 0.53) of the mean binocularity and its standard deviation (0.75 ± 0.10) are indicated on the ordinate. See Kasamatsu *et al*. (1985) for the source of these values. (Reproduced with minimum modification from Shirokawa and Kasamatsu, 1986.)

the mature visual cortex by electrical stimulation of the NE cells in the locus coeruleus complex. We found a significant decrease in binocularity when brief monocular exposure was accompanied by electrical stimulation of the locus coeruleus (Kasamatsu *et al.*, 1985). The OD distribution was W-shaped, indicating the partial restoration of the plasticity to the mature brain. This result is consistent with our early finding of a significant decrease in binocularity in the NE-infused cortex of monocularly deprived adult cats (Kasamatsu *et al.*, 1979).

The final topic for discussion is the direct test of the involvement of β adrenoreceptors in OD plasticity. Using the continual microinfusion method, we directly infused the kitten visual cortex with β adrenoreceptor antagonists such as propranolol and sotalol at various concentrations (10^{-6} to 10^{-2} M in an osmotic minipump). We found a concentration-dependent suppression of the expected shift in ocular dominance that usually follows monocular lid suture (Shirokawa and Kasamatsu, 1986). These results are summarized in Fig. 2. The half-maximal suppression was obtained at $\sim 10^{-4}$ M propranolol (in an osmotic minipump). In the same study we examined the spatial spread of locally and continually infused [^{3}H]propranolol, thus enabling us to calculate the likely concentration of propranolol at a given recording site, which was usually ~2 mm away from the center of the continual infusion (5.8×10^{-7} M).

The results with sotalol were essentially the same as those obtained with propranolol except for a minor difference: the effect of sotalol seemed to be saturated at the concentration at least tenfold lower than that for propranolol. Reasons for this were not known. Under the comparable condition, α_1 adrenoreceptor antagonists such as phenoxybenzamine and piperoxan did not block the shift in ocular dominance. The cortical infusion with 10^{-2} M metoprolol (β_1 antagonist), however, resulted in the effects comparable to those produced by propranolol. The present results with adrenoreceptor antagonists strongly suggest that activation of β adrenoreceptors is crucially involved in the regulation of OD plasticity.

4. CONCLUSION

What we have described above is a summary of the current state of the NE hypothesis for OD plasticity and its extension. Our preliminary results further suggest that as a next step adenosine cyclic monophosphate (cAMP) is involved in enhancing the plasticity

(Kasamatsu, 1980, 1986). It is plausible that cAMP serves as a key step in the regulation of OD plasticity. If we are allowed to speculate along this line, a cartoon in Fig. 3 shows one possible mechanism, which is primarily a postsynaptic story. Recently, Aoki and Siekevitz (1985) showed that the *in vitro* phosphorylation of microtubule-associated protein 2 (MAP2) was kept low in the visual cortex of dark-reared cats and that the phosphorylatability quickly increased to normal values on brief exposure to light. It is quite intriguing that the metabolism of cytoskeletal proteins within the visual cortex can be altered through an animal's visual experience. Since MAP2 immunoreactivity is confined to the somatodendritic membrane of neurons in the brain (Bloom *et al.*, 1984; Huber and Matus, 1984; De Camilli *et al.*, 1984; Caceres *et al.*, 1984), an increase in phosphorylated MAP2 *in vivo* may serve as a prerequisite to altering the distribution pattern of visual receptors (whose nature is unknown currently), which are usually trapped at the subsynaptic site on the postsynaptic membrane of a visual cortical cell (see 2 in Fig. 3), thus eventually causing drastic changes in the effciency of a synapse under study. In this model we need not expect reduction of the total number of visual receptors per cell as the primary cause of a decrease in synaptic efficiency; a change in their distribution may be sufficient.

The NE system appears to be one of the necessary mechanisms in the regulation of OD plasticity. This does not exclude, however, other possibilities, which include various transmitters, peptides, and hormones other than the NE system. We neuroscientists are archeologists of brain evolution: removing one layer of function/structure in the brain, we usually encounter beneath it other layers of the unknown. Therefore, we are ready

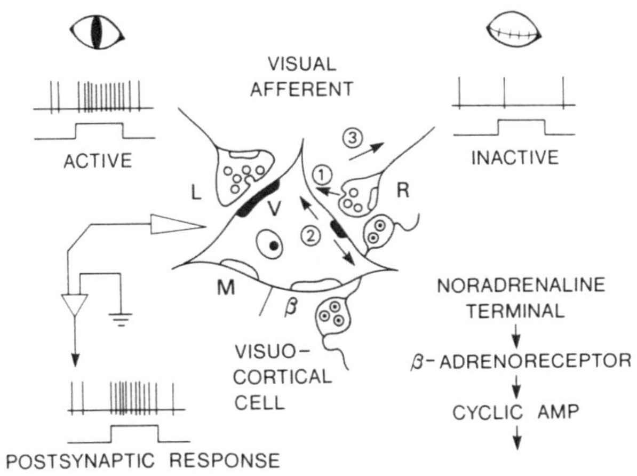

FIGURE 3. A possible scheme to explain a decrease in synaptic efficiency at an inactive synaptic site. A visual terminal (R), which used to receive a patterned visual afferent, becomes quiescent following monocular lid suture. Subsequently, release of transmitters, which are unknown currently, should be less (1) than normal and incoherent with postsynaptic firing. This may trigger the dispersion of receptor molecules (V), which are also currently unidentified, from the subsynaptic membrane of a visuocortical cell (2). Another terminal (L) driven by bombardment from an active afferent can maintain the existing synaptic contact. The sustained changes in the subsynaptic membrane may eventually lead to retraction of the inactive terminal (3). The hypothetized process (2) may be facilitated by the higher level of *in vivo* phosphorylation of cytoskeletal proteins such as MAP2, which are thought to anchor various receptor molecules inserted into the cell membrance. The central norepinephrine (NE) system may be crucially involved in this process by increasing intracellular cAMP following the activation of β adrenoreceptors (β) by NE. Muscarinic cholinergic receptors (M) may also contribute to the scheme.

to be happily surprised at any unexpected findings, only if they are "positive" and thus testable by other methods. Such an example is seen in a recent proposal by Bear and Singer (1986) that both NE and cholinergic inputs to the visual cortex must be destroyed to decrease OD plasticity in kittens. We decided to take this proposal seriously and started to examine it in detail, paying particular attention to the likely interaction of the NE and the cholinergic systems in this function (Imamura and Kasamatsu, 1986).

ACKNOWLEDGMENTS. We are grateful for efficient secretarial service by Ms. N. Wilson. Ms. C. Dang also helped us with the English grammar. These studies were supported by USPHS grant EY-06733 to T.K. and BRSG RR05981.

REFERENCES

Adrien, J., Blanc, G., Buisseret, P., Frégnac, Y., Gary-Bobo, E., Imbert, M., Tassin, J.-P., and Trotter, Y., 1985, Noradrenaline and functional plasticity in kitten visual cortex: A re-examination, *J. Physiol. (Lond.)* **367**:73–98.

Aoki, C., and Siekevitz, P., 1985, Ontogenic changes in the cyclic adenosine 3′,5′-monophosphate stimulatable phosphorylation of cat visual cortex proteins, particularly of microtubule-associated protein 2 (MAP2): Effects of normal and dark rearing and of the exposure to light, *J. Neurosci.* **5**:2465–2483.

Bear, M. F., and Daniels, J. D., 1983, The plastic response to monocular deprivation persists in kitten visual cortex after chronic depletion of norepinephrine, *J. Neurosci.* **3**:407–416.

Bear, M. F., and Singer, W., 1986, Modulation of visual cortical plasticity by acetylcholine and noradrenaline, *Nature* **320**:172–176.

Bear, M. F., Paradiso, M. A., Schwartz, M., Nelson, S. B., Carnes, K. M., and Daniels, J. D., 1983, Two methods of catecholamine depletion in kitten visual cortex yield different effects on plasticity, *Nature* **302**:245–247.

Bloom, G. S., Schoenfeld, T. A., and Vallee, R. B., 1984, Widespread distribution of the major polypeptide component of MAP1 (microtubule-associated protein 1) in the nervous system, *J. Cell Biol.* **98**:320–330.

Caceres, A., Binder, L. I., Payne, M. R., Bender, P., Rebhun, L., and Steward, O., 1984, Differential subcellular localization of tubulin and the microtubule-associated protein MAP2 in brain tissue as revealed by immunocytochemistry with monoclonal hybridoma antibodies, *J. Neurosci.* **4**:394–410.

Daw, N. W., Rader, R. K., Robertson, T. W., and Ariel, M., 1983, Effects of 6-hydroxydopamine on visual deprivation in kitten striate cortex, *J. Neurosci.* **3**:907–914.

Daw, N. W., Robertson, T. W., Rader, R. K., Videen, T. O., and Coscia, C. J., 1984, Substantial reduction in cortical noradrenaline by lesion of adrenergic pathway does not prevent effects of monocular deprivation, *J. Neurosci.* **4**:1354–1360.

Daw, N. W., Videen, T. O., Parkinson, D., and Rader, R. K., 1985a, DSP-4 [N-(2-chloroethyl)-N-ethyl-2-bromobenzylamine] depletes noradrenaline in kitten visual cortex without altering the effects of monocular deprivation, *J. Neurosci.* **5**:1925–1933.

Daw, N. W., Videen, T. O., Robertson, T., and Rader, R. K., 1985b, An evaluation of the hypothesis that noradrenaline affects plasticity in the developing visual cortex, in: *The Visual System* (A. Fein, ed.), Alan R. Liss, New York, pp. 133–144.

Daw, N. W., Videen, T. O., Rader, R. K., Robertson, T. W., and Coscia, C. J., 1985c, Substantial reduction of noradrenaline in kitten visual cortex by intraventricular injection of 6-hydroxydopamine does not always prevent ocular dominance shift after monocular deprivation, *Exp. Brain Res.* **59**:30–35.

De Camilli, P., Milles, P. E., Navone, F., Theurkauf, W. E., and Vallee, R. B., 1984, Patterns of MAP2 distribution in the nervous system studied by immunofluorescence, *Neuroscience* **11**:819–846.

Frégnac, Y., and Imbert, M., 1984, Development of neuronal selectivity in primary visual cortex of cat, *Physiol. Rev.* **64**:325–434.

Hubel, D. H., and Wiesel, T. N., 1962, Receptive fields, binocular interaction and functional architecture in the cat's visual cortex, *J. Physiol. (London.)* **160**:106–154.

Huber, G., and Matus, A., 1984, Differences in the cellular distributions of two microtubule-associated proteins, MAP1 and MAP2, in rat brain, *J. Neurosci.* **4**:151–160.

Imamura, K., and Kasamatsu, T., 1986, Noradrenergic and cholinergic interaction in ocular dominance plasticity, *Soc. Neurosci. Abstr.* **12**:1372.

Itakura, T., Kasamatsu, T., and Pettigrew, J. D., 1981, Norepinephrine-containing terminals in kitten visual cortex: Laminar distribution and ultrastructure, *Neuroscience* **6**:159–175.

Jonsson, G., 1980, Chemical neurotoxins as denervation tools in neurobiology, *Annu. Rev. Neurosci.* **3**:169–187.

Jonsson, G., and Kasamatsu, T., 1983, Maturation of monoamine neurotransmitters and receptors in cat occipital cortex during postnatal critical period, *Exp. Brain Res.* **50**:449–458.

Kasamatsu, T., 1980, A possible role for cyclic nucleotides in plasticity of visual cortex, *Soc. Neurosci. Abstr.* **6**:494.

Kasamatsu, T., 1983, Neuronal plasticity maintained by the central norepinephrine system in the cat visual cortex, in: *Progress in Psychobiology and Physiological Psychology*, Vol. 10 (J. M. Sprague and A. N. Epstein, eds.), Academic Press, New York, pp. 1–112.

Kasamatsu, T., 1986, Changes in ocular dominance of adult cats following monocular lid suture: The effects of directly infused forskolin, *Invest. Ophthalmol. Vis. Sci. Suppl.* **27**:153.

Kasamatsu, T., 1987, Norepinephrine hypothesis for visual cortical plasticity: Thesis, antithesis, and recent development, in: *Current Topics in Developmental Biology, Vol. 21* (R. K. Hunt, A. A. Moscona, and A. Monroy, eds.), Academic Press, Orlando, pp. 367–389.

Kasamatsu, T., and Pettigrew, J. D., 1976, Depletion of brain catecholamine: Failure of ocular dominance shift after monocular occlusion in kittens, *Science* **194**:206–209.

Kasamatsu, T., and Pettigrew, J. D., 1979, Preservation of binocularity after monocular deprivation in the striate cortex of kittens treated with 6-hydroxydopamine, *J. Comp. Neurol.* **185**:139–162.

Kasamatsu, T., and Shirokawa, T., 1985, Involvement of β-adrenoreceptors in the shift of ocular dominance after monocular deprivation, *Exp. Brain Res.* **59**:507–514.

Kasamatsu, T., Pettigrew, J. D., and Ary, M., 1979, Restoration of visual cortical plasticity by local microperfusion of norepinephrine, *J. Comp. Neurol.* **185**:163–182.

Kasamatsu, T., Itakura, T., and Jonsson, G., 1981, Intracortical spread of exogenous catecholamines: Effective concentrations for modifying cortical plasticity, *J. Pharmacol. Exp. Ther.* **217**:841–850.

Kasamatsu, T., Itakura, T., Jonsson, G., Heggelund, P., Pettigrew, J. D., Nakai, K., Watabe, K., Kuppermann, B. D., and Ary, M., 1984, Neuronal plasticity in cat visual cortex: A proposed role for the central noradrenaline system, in: *Monoamine Innervation of Cerebral Cortex* (L. Descarries, T. Reader, and H. H. Jasper, eds.), Alan R. Liss, New York, pp. 301–319.

Kasamatsu, T., Watabe, K., Heggelund, P., and Schöller, E., 1985, Plasticity in cat visual cortex restored by electrical stimulation of the locus coeruleus, *Neurosci. Res.* **2**:365–386.

Kostrzewa, R. M., and Jacobowitz, D. M., 1974, Pharmacological action of 6-hydroxydopamine, *Pharmacol. Rev.* **26**:199–288.

Nakai, K., and Kasamatsu, T., 1984, Accelerated regeneration of central catecholamine fibers in cat occipital cortex: Effects of substance P, *Brain Res.* **323**:364–379.

Nakai, K., Jonsson, G., and Kasamatsu, T., 1981, Regrowth of central catecholaminergic fibers in cat visual cortex following localized lesion with 6-hydroxydopamine, *Soc. Neurosci. Abstr.* **7**:675.

Nakai, K., Jonsson, G., and Kasamatsu, T., 1987, Norepinephrinergic reinnervation of cat occipital cortex following localized lesions with 6-hydroxydopamine, *Neurosci. Res.* **4**:433–453.

Nelson, S. B., Schwartz, M., and Daniels, J. D., 1985, Clonidine and cortical plasticity: Possible evidence for noradrenergic involvement, *Dev. Brain Res.* **23**:39–50.

Paradiso, M. A., Bear, M. F., and Daniels, J. D., 1983, Effects of intracortical infusion of 6-hydroxydopamine on the response of kitten visual cortex to monocular deprivation, *Exp. Brain Res.* **51**:413–422.

Pettigrew, J. D., and Kasamatsu, T., 1978, Local perfusion of norepinephrine maintains visual cortical plasticity, *Nature* **271**:761–763.

Shirokawa, T., and Kasamatsu, T., 1986, Concentration-dependent suppression by β-adrenergic antagonists of the shift in ocular dominance following monocular deprivation in kitten visual cortex, *Neuroscience* **18**:1035–1046.

Trombley, P., Allen, E. E., Soyke, J. Blaha, C. D., Lane, R. F., and Gordon, B., 1986, Doses of 6-hydroxydopamine sufficient to deplete norepinephrine are not sufficient to decrease plasticity in the visual cortex, *J. Neurosci.* **6**:266–273.

Wiesel, T. N., and Hubel, D. H., 1963, Single-cell responses in striate cortex of kittens deprived of vision in one eye, *J. Neurophysiol.* **26**:1003–1017.

42

Possible Indications of Noradrenergic Involvement in Behavioral Plasticity

TREVOR ARCHER

1. INTRODUCTION

The modifiability of neuronal substrates by early postnatal experience, as in the case of early visual experience (Hubel and Weisel, 1965), stimulated much interest in behavioral plasticity. The early hypothesis concerning central catecholamine (Carlsson *et al.*, 1962) involvement in plasticity was demonstrated in experiments depleting catecholamines by use of the catecholamine neurotoxin 6-hydroxydopamine (Kasamatsu and Pettigrew, 1976). Subsequent investigations indicated that administration of the putative neurotransmitter norepinephrine (NE) restored plasticity abolished by catecholamine depletion after 6-OHDA (Pettigrew and Kasamatsu, 1978). In the present account, two possible indications of behavioral plasticity implicating noradrenergic modulation are described: (1) the improvement of rats' ability to perform in the Hebb–Williams maze task following post-weaning housing in a complex environment is abolished by postnatal NE depletion, and (2) central and spinal depletion of NE antagonizes the analgesia induced by the 5-hydroxytryptamine (5-HT) agonist 5-methoxy-N,N-dimethyltryptamine, and the intrathecal administration of NE restores the analgesic effects of the 5-HT agonist.

2. COMPLEX HOUSING CONDITIONS IMPROVE HEBB–WILLIAMS MAZE PERFORMANCE: ROLE OF NOREPINEPHRINE

The involvement of NE in cognitive processes has been established using both instrumental (Archer, 1982a,b; Archer *et al.*, 1983b) and classical conditioning (Archer *et al.*, 1982, 1983a, 1986a) tasks. To date, an impressive volume of evidence has been accumulated to indicate an important role of NE terminals in mediating behavioral plasticity in a variety of procedures (e.g., Kasamatsu and Pettigrew, 1979; Kasamatsu *et al.*, 1979; Davies *et al.*, 1985). In this context, it has been shown that housing and rearing conditions modulate to varying degrees both behavioral plasticity and neurochemical

TREVOR ARCHER • R & D Laboratories, Astra Alab AB, S-151 85 Södertälje, Sweden.

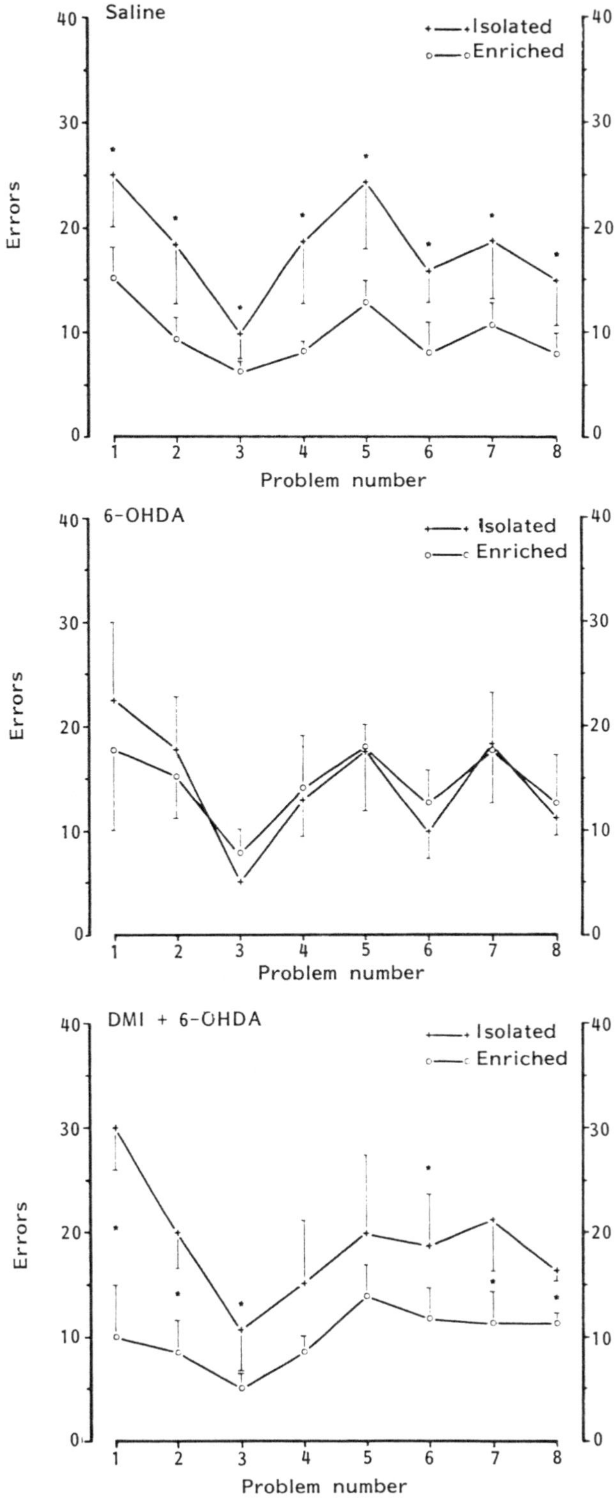
Saline
Isolated
Enriched
Errors
Problem number
6-OHDA
Isolated
Enriched
Errors
Problem number
DMI + 6-OHDA
Isolated
Enriched
Errors
Problem number

changes in the central nervous system (Diamond *et al.*, 1964; Diamond, 1967; Rosenzweig *et al.*, 1971; Smith, 1972). Furthermore, some recent evidence seems to implicate NE involvement in this respect (O'Shea *et al.*, 1983). The experiment described below demonstrates that complex rearing conditions affect the performance of rats in the Hebb–Williams maze and that depletion of NE abolishes the improvement observed in the animals reared in a complex environment (Mohammed *et al.*, 1986).

Neonatal male rat pups were treated at birth with either 6-hydroxydopamine (6-OHDA, 100 μg/g s.c.), vehicle (saline containing 0.1% ascorbic acid), or desipramine (DMI, 25 μg/g s.c.) 30 min before 6-OHDA. On day 21 (after weaning), the rats in each treatment condition were placed in either a complex or isolated housing environment for 35 days. For the complex condition, the rats were group housed in large wire mesh cages and had access to a variety of objects that they could manipulate, whereas for the isolated condition rats were housed singly in the normal Plexiglas™ cages. On day 56 all the rats moved to the Plexiglas™ cages and housed singly throughout. On days 71 to 78 (after birth), all the rats were tested on the first eight problems of the Hebb–Williams maze task, and on days 81 and 82, spontaneous motor activity was measured. Following this, all the rats were sacrificed, and neurochemical analysis of forebrain regions was performed.

Each problem in the Hebb–Williams maze confronted the rats with a task of varying complexity so that for each animal a particular pattern of performance was obtained over all eight problems. Figure 1 presents the performance pattern by the complex and isolated groups in each of the three treatment conditions. For the saline and DMI plus 6-OHDA treatment conditions, the complex housing condition caused a significant overall improvement in performance, whereas no such improvement was obtained for the 6-OHDA treatment condition. Rats from the isolated housing condition demonstrated significantly greater spontaneous motor activity than those from the complex housing condition, independent of the treatment (see Fig. 2). Thus, NE depletion appears to antagonize the cognitive (maze learning) but not the hyperactivity aspect of the behavioral changes resulting from complex or isolated housing conditions. The neurochemical analysis indicated that the neonatal 6-OHDA treatment caused a selective depletion of NE and the NE metabolite 3,4-dihydroxyphenylethylene glycol (DOPEG). The neurochemical effects of 6-OHDA treatment were antagonized by the DMI pretreatment. The neurochemical data (see Table I) appear to confirm NE involvement in the behavioral plasticity induced by the differential postweaning housing conditions.

3. SPINAL NOREPINEPHRINE DEPLETION ABOLISHES 5-METHOXY-N,N-DIMETHYLTRYPTAMINE-INDUCED ANALGESIA: RESTORATION BY INTRATHECAL NOREPINEPHRINE ADMINISTRATION

It was recently found that the analgesia induced by either 5-methoxy-N,N-dimethyltryptamine (5-MeODMT), 5-hydroxytryptamine (5-HT), or 5-HT agonists was abolished

FIGURE 1. Median (± quartiles) number of errors per problem in the Hebb–Williams maze for 6-OHDA-, DMI- plus 6-OHDA-, and saline-treated rats housed in an isolated (IC) or a complex environment (EC) for 35 days following weaning. Saline- and DMI- plus 6-OHDA-treated rats raised in EC made significantly fewer errors than those raised in ICs (Mann–Whitney *U*-tests, $p < 0.01$). Both 6-OHDA groups made more errors than the complex environment groups in the saline and DMI-plus 6-OHDA conditions.

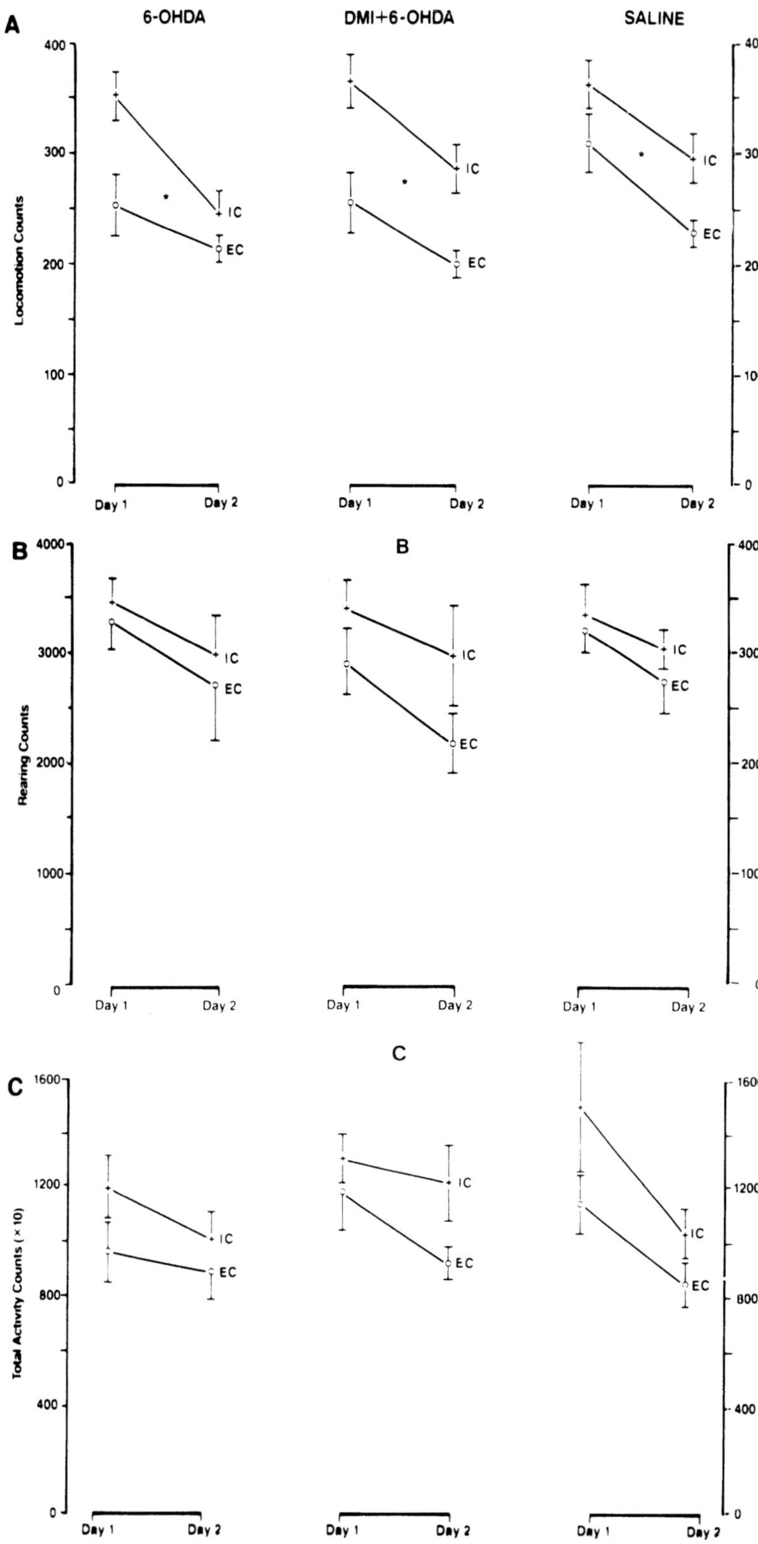

FIGURE 2. Mean counts for locomotion, rearing, and total activity for the 6-OHDA-, DMI-plus 6-OHDA-, and saline-treated rats housed in an isolated (IC) or a complex enviroment (EC) for 35 days following weaning. Motor activity was measured on two consecutive days. $^{*}P < 0.01$, Students *t*-test.

TABLE I. Catecholamine and Catecholamine Metabolite Levels in the Frontal Cortex of 6-OHDA-, DMI- Plus 6-OHDA-, and Saline-Treated Rats Housed in a Complex (EC) or Isolated (IC) Environment[a]

Treatment	NE	DOPEG	DOPEG/NE	DOPA	A	DA	DOPAC	DOPAC/DA
NaCl–IC	346 ± 18	27 ± 1.6	0.078	12 ± 1.2	2.2 ± 0.5	52 ± 1.5	41 ± 1.9	0.79
NaCl–EC	372 ± 9.1	25 ± 0.9	0.067	9.8 ± 0.3	1.1 ± 0.2	35 ± 2.6	35 ± 1.6	0.60
6-OHDA–IC	154 ± 61***	13 ± 0.9***	0.084	6.3 ± 0.3***	1.3 ± 0.5	72 ± 2.2***	40 ± 1.6	0.56
	(45)	(48)	(108)	(53)	(59)	(138)	(98)	(71)
6-OHDA–EC	193 ± 33***	17 ± 2***	0.088	7.4–0.5***	1.2 ± 0.3	69 ± 2.5**	38 ± 2.0	0.55
	(52)	(68)	(131)	(76)	(109)	(119)	(109)	(92)
DMI + 6-OHDA–IC	367 ± 50	24 ± 2.5	0.065	8.7 ± 0.6**	2.5 ± 0.7	63 ± 3.8	38 ± 2.2	0.60
	(106)	(89)	(83)	(73)	(114)	(121)	(93)	(76)
DMI + 6-OHDA–EC	347 ± 23	26 ± 2.1	0.070	8.5 ± 0.5*	1.7 ± 0.5	53 ± 3.9	33 ± 2.5	0.62
	(101)	(96)	(104)	(87)	(155)	(91)	(94)	(103)

[a] Values are expressed as mean ± S.E.M., in nanograms per gram wet weight, assayed according to Durkin *et al.* (1985). Percentage of appropriate control (IC or EC) in parentheses. Significance according to Student's *t*-test: $^{***}P < 0.001$, $^{**}P < 0.01$, $^{*}P < 0.05$.

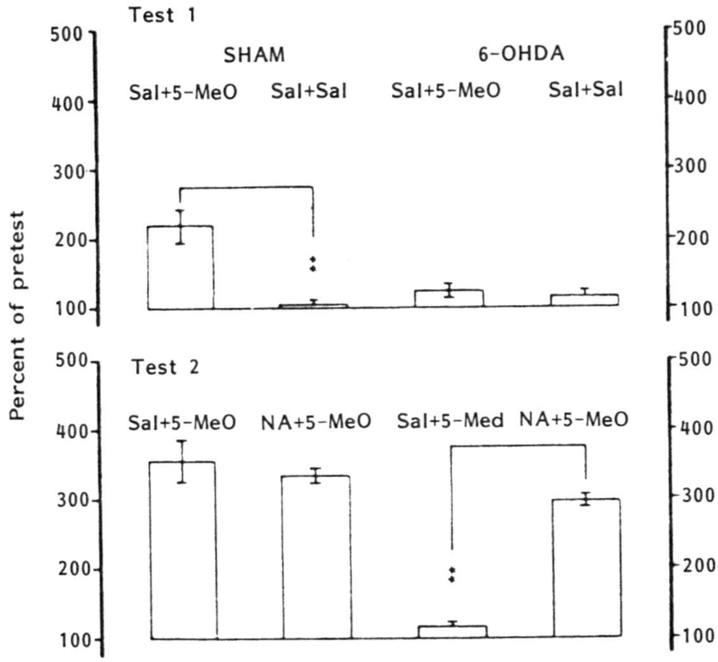

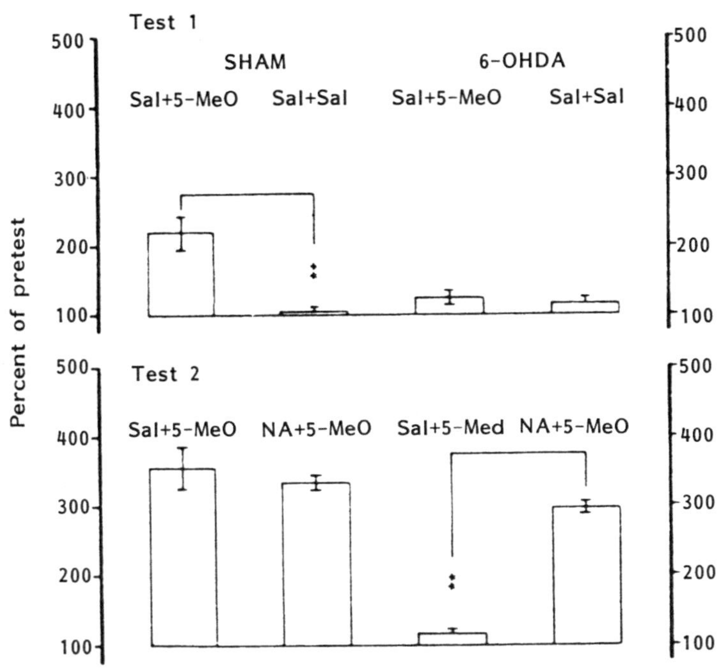

FIGURE 3. Blockade of MeODMT-induced analgesia by intrathecal 6-OHDA treatment and reinstatement of 5-MeODMT-induced analgesia by intrathecal NE administration. Test 1: Intrathecal saline was administered immediately before either 5-MeODMT (1 mg/kg) or saline. Test 2: Intrathecal NE (2 μg) or saline was administered immediately before 5-MeODMT (1 mg/kg). Test 2 was performed 1 week after test 1. Tail-flick data are shown in A, hot-plate in B, and shock-titration in C. In each case tail-flick, hot-plate, and shock-titration tests were performed 15 min after the drug administrations. **$p < 0.01$, Tukey HSD test (Redrawn from Minor *et al.*, 1987.)

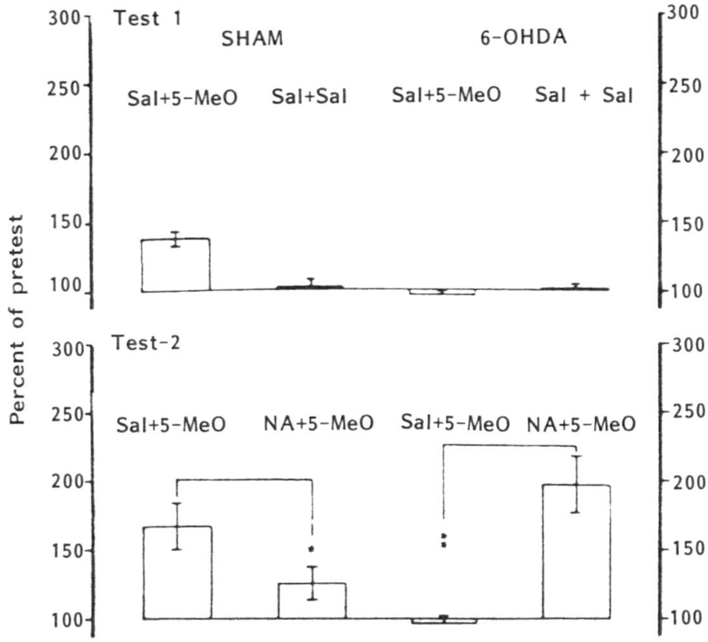

FIGURE 3. (*Continued*)

by treatments that induced spinal depletions of NE (Archer *et al.*, 1985a,b, 1986b; Minor *et al.*, 1985; Post *et al.*, 1985; Danysz *et al.*, 1986). The experiment described below demonstrated a complete blockade of 5-MeODMT-induced analgesia following the intrathecal administration of 6-OHDA, which causes a selective depletion of NE spinally. Later, the analgesia induced by 5-MeODMT was restored by the intrathecal administration of NE, which was administered in two doses (1 and 2 μg) immediately prior to the systemic administration of 5-MeODMT (Minor *et al.*, 1987).

Three different tests of analgesia were used: tail-flick, hot-plate, and shock titration, as described previously (Archer *et al.*, 1985a,b). Male Sprague–Dawley rats aged 90–95 days were injected intrathecally with either 6-OHDA (20 μg in 10 μl vehicle) or vehicle (15 μl), and the indwelling catheters were fixed with the tip of the catheter in the lumbar subarachnoidal space (Yaksh and Rudy, 1976). The 6-OHDA- and vehicle-treated rats were administered 5-MeODMT (1 mg/kg s.c.) 2 weeks after the operations, and the analgesia tests were performed 15 min later. The analgesia induced by 5-MeODMT was abolished completely by prior 6-OHDA treatment (see Figs. 3 and 4) in all three tests of nociception. One week later, the 6-OHDA- and vehicle-treated rats were administered NE (1 or 2 μg) intrathecally just prior to the injection of 5-MeODMT (1 mg/kg s.c.), and nociception testing was performed 15 min later. The results are straightforward: intrathecal NE administration restored 5-MeODMT-induced analgesia (see Figs. 3 and 4). This result cannot be explained on the basis of a supersensitivity effect as a result of NE denervation in the spinal cord, since in each case this effect was carefully tested for. The monoamine assays of the rats are presented in Table II. Intrathecal administration of 6-OHDA caused severe NE depletion (2% of control values) in both the lumbar and thoracic regions of the spinal cord, whereas dopamine (DA) concentrations were reduced to a much smaller extent although still significantly (77 and 65% of control values).

Intrathecal NE administration restored the analgesia induced by 5-MeODMT in 6-OHDA-treated rats (Minor *et al.*, 1987), and this evidence appears to be consistent with

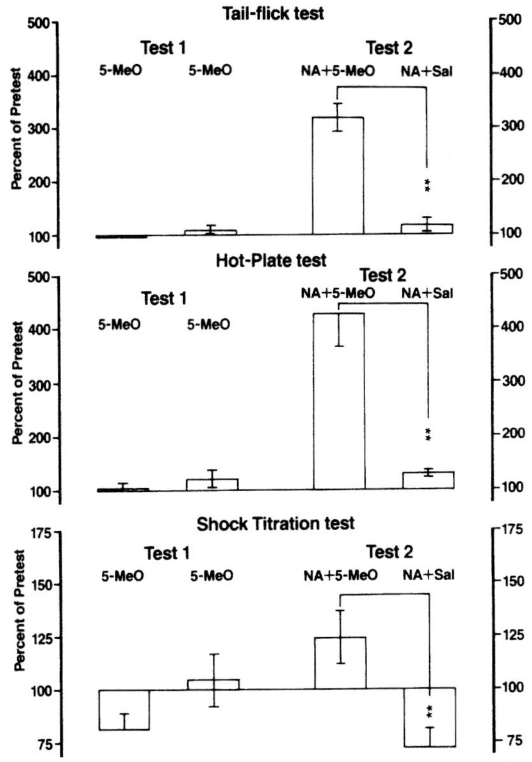

FIGURE 4. Reinstatement of 5-MeODMT-induced analgesia by intrathecal NE administration. All the rats were administered 6-OHDA (20 μg in 10 μl) intrathecally 2 weeks before testing. Test 1: 5-MeODMT (1 mg/kg) was administered to both groups. Test 2: intrathecal NE (1 μg) was administered immediately prior to either 5-MeODMT (1 mg/kg:) or saline. Test 2 was performed 1 week after test 1. Tail-flick, hot-plate, and shock-titration testing were performed 15 min after the drug administration. $^{**}p < 0.01$, Tukey HSD test. (Redrawn from Minor *et al.*, 1987.)

TABLE II. Monoamine Assays on the Lumbar and Thoracic Spinal Cords of Intrathecally 6-OHDA- and Vehicle-Treated Rats[a]

		Monoamines (μg/g wet weight tissue)	
		Lumbar	Thoracic
Norepinephrine	6-OHDA	5.5 ± 4** (2)	5.6 ± 3** (2)
	Saline	301 ± 38	296 ± 22
Dopamine	6-OHDA	102 ± 4* (77)	9.8 ± 3.5* (65)
	Saline	13.2 ± 4.5	15 ± 2

[a] The 6-OHDA (20 μg m 10μl) was injected intrathecally 10 min after pargyline (20 mg/kg i.p.). Values are expressed as means ± S.E.M. of six animals. Percentage of control value in parentheses. Significance: $^{**}P < 0.001$, $^{*}P < 0.01$, Student's *t*-test.

other findings. Hammond *et al.* (1985) demonstrated that stimulation of the nucleus raphe magnus and the nucleus reticularis paragigantocellularis increased the efflux of NE but not 5-HT following pretreatment with fluoxetine and desipramine. Further, Sagen and Proudfit (1984) indicated that activation of nucleus raphe magnus neurons induced hypoalgesia partially mediated by activation of spinally projecting NE neurons. Taken together, these diverse findings serve to demonstrate a noteworthy plasticity in NE–5-HT interactions even at a spinal location.

The two experiments (Mohammed *et al.*, 1986; Minor *et al.*, 1987) presented above seem to offer further evidence of noradrenergic involvement in behavioral plasticity. The effects of a diverse environmental enrichment are quite general (e.g., Cummins and Livesey, 1979; Bhide and Bedi, 1984a), and recent investigations indicate that an 80-day exposure to an enriched environment produces widespread morphological changes and alterations of neuronal and synaptic characterization in the forebrains of both well-fed and undernourished rats (Bhide and Bedi, 1984b,c). Differential environmental housing was also shown to produce marked differences in the cerebral anatomy of rats, most especially in the thickness of the occipital cortex (Katz and Davies, 1984), which is most interesting in view of the finding that rats reared in an isolated environment maintain base-line levels of bar pressing for intensities of contingent light that were clearly aversive to rats reared under enriched environmental conditions (Rose *et al.*, 1986).

At present it is not known whether NE depletion affects any of the morphological, neuronal, and synaptic changes brought about by an enriched environment. It is to be noted that in many cases of differential environmental housing, this period was considerably more extended than in the study described above. It is to be noted also that supracritical postsynaptic activation as well as the presence of both NE and acetylcholine were found to be necessary for changes in plasticity (Bear and Singer, 1986). These investigators found that although ocular dominance of neurons in the cat striate cortex could be reduced by 6-OHDA infusion during the period of monocular deprivation, the action of acetylcholine on cortical neurons was interfered with. The results suggest that intracortical administration of 6-OHDA affects plasticity by altering both noradrenergic and cholinergic neurotransmission (Bear and Singer, 1986). At present there appears to be tentative support for a functional role of central noradrenergic connections (e.g., Björklund *et al.*, 1979) in forebrain synaptogenesis underlying the early development of processes mediating visual and cognitive experience.

REFERENCES

Archer, T., 1982a, DSP4-(*n*-2-chloroethyl-N-ethyl-2-bromobenzylamine), a new noradrenaline neurotoxin, and the stimulus conditions affecting acquisition of two-way active avoidance, *J. Comp. Physiol. Psychol.* **96:**476–490.

Archer, T., 1982b, Signalled and unsignalled avoidance impairments following noradrenaline depletion with DSP4: An hypothesis incorporating an associative and a non-associative factor, *Scand. J. Psychol.* **24:**75–87.

Archer, T., Cotic, T., and Järbe, T. U. C., 1982, Attenuation of the context effect and lack of unconditioned stimulus-preexposure effect in taste-aversion learning following treatment with DSP4, the selective noradrenaline neurotoxin, *Behav. Neural Biol.* **25:**159–173.

Archer, T., Mohammed, A. K., and Järbe, T. U. C., 1983a, Latent inhibition following systemic DSP4: Effects due to presence and absence of contextual cues in taste-aversion learning, *Behav. Neural Biol.* **38:**287–306.

Archer, T., Mohammed, A. K., Ross, S. B., and Söderberg, U., 1983b, T-maze learning, spontaneous activity and food intake recovery following systemic administration of the noradrenaline neurotoxin, DSP4, *Pharmacol. Biochem. Behav.* **19:**121–130.

Archer, T., Arwesträm, E., Jonsson, G., Minor, B. G., and Post, C., 1985a, Complete blockade and attentuation of 5-hydroxytryptamine induced analgesia following NA depletion in rats and mice, *Acta Pharmacol. Toxicol.* **57**:255–261.

Archer, T., Minor, B. G., and Post, C., 1985b, Blockade and reversal of 5-methoxy-N,N-dimethyltryptamine-induced analgesia following noradrenaline depletion, *Brain. Res.* **333**:55–61.

Archer, T., Cotic, T., and Järbe, T. U. C., 1986a, Noradrenaline and sensory preconditioning in the rat, *Behav. Neurosci.* **100**:704–711.

Archer, T., Jonsson, G., Minor, B. G., and Post, C., 1986b, Noradrenergic–serotonergic interactions and nociception in the rat, *Eur. J. Pharmacol.* **120**:295–308.

Bear, M. F., and Singer, W., 1986, Modulation of visual cortical plasticity by acetylcholine and noradrenaline, *Nature* **320**:172–176.

Bhide, P. G., and Bedi, K. S., 1984a, Effects of environmental diversity on well fed and previously undernourished rats: Neuronal and glial cell measurements in the visual cortex (area 17), *J. Anat.* **138**:447–461.

Bhide, P. G., and Bedi, K. S., 1984b, The effects of a lengthy period of environmental diversity on well-fed and previously undernourished rats. I. Neurons and glial cells, *J. Comp. Neurol.* **227**:269–304.

Bhide, P. G., and Bedi, K. S., 1984c, The effects of a lengthy period of environmental diversity on well-fed and previously undernourished rats. II. Synapse-to-neuron ratios, *J. Comp. Neurol.* **227**:305–310.

Björklund, A., Segal, M., and Stenevi, U., 1979, Functional reinnervation of rat hippocampus by locus coeruleus inplants, *Brain Res.* **170**:409–426.

Carlsson, A., Falck, B., and Hillarp, N.-Å., 1962, Cellular localization of brain monoamines, *Acta Physiol. Scand.* [*Suppl.*] **196**:1–28.

Cummins, R. A., and Livesey, P. J., 1979, Enrichment–isolation, cortex length and the rank order effect, *Brain Res.***178**:89–98.

Danysz, W., Jonsson, G., Minor, B. G., Post, C., and Archer, T., 1986, Spinal and locus coeruleus noradrenergic lesions abolish the analgesic effects of 5-methoxy-N,N-dimethyltryptamine, *Behav. Neural Biol.***46**:71–86.

Davies, D. C., Horn, G., and McCabe, B. J., 1985, Noradrenaline and learning: Effects of the noradrenaline neurotoxin DSP4 and imprinting in the domestic chick, *Behav. Neurosci.* **99**:652–660.

Diamond, M. C., 1967, Extensive cortical depth measures and neuron size increases in the cortex of environmentally enriched rats, *J. Comp. Neurol.* **131**:357–364.

Diamond, M. C., Krech, D., and Rosenzweig, M. R., 1964, The effects of enriched environments on the histology of rat cerebral cortex, *J. Comp. Neurol.* **123**:111–119.

Durkin, T. A., Caliguri, E. J., Mefford, I. N., Lake, D. M., McDonald, I. A., Sundström, E., and Jonsson, G., 1985, Determination of catecholamines in tissue and body fluids using microbore and high pressure liquid chromatography with amperometric detection, *Life Sci.* **37**:1803–1810.

Hammond, D. L., Tyce, G. M., and Yaksh, T. L., 1985, Efflux of 5-hydroxytryptamine and noradrenaline into spinal cord superfusates during stimulation of the rat medulla, *J. Physiol.* (*Lond.*) **359**:151–162.

Hubel, D. H., and Wiesel, T. N., 1965, Binocular interaction in striate cortex of kittens reared with artificial squint, *J. Neurophysiol.* **28**:1041–1059.

Kasamatsu, T., and Pettigrew, J. D., 1976, Depletion of brain catecholamines: Failure of ocular dominance shift after monocular occlusion in kittens, *Science* **194**:206–209.

Kasamatsu, T., and Pettigrew, J. D., 1979, Preservation of binocularity after monocular deprivation in the striate cortex of kittens treated with 6-hydroxydopamine, *J. Comp. Neurol.* **185**:139–162.

Kasamatsu, T., Pettigrew, J. D., and Ary, M., 1979, Restoration of visual cortical plasticity by local microperfusion of norepinephrine, *J. Comp. Neurol.* **185**:163–182.

Katz, H. B., and Davies, C. A., 1984, Effects of differential environments on the cerebral anatomy of rats as a function of previous and subsequent housing conditions, *Exp. Neurol.* **83**:274–287.

Minor, B. G., Post, C., and Archer, T., 1985, Blockade of intrathecal 5-hydroxytryptamine-induced antinociception in rats by noradrenaline depletion, *Neurosci. Lett.* **54**:39–44.

Minor, B. G., Persson, M. L., Post, C., Jonsson, G., and Archer, T., 1987, Intrathecal noradrenaline restores 5-methoxy-N,N-dimethyltryptamine induced antinociception abolished by intrathecal 6-hydroxydopamine, *J. Neurol Transm.* (in press).

Mohammed, A. K., Jonsson, G., and Archer, T., 1986, Selective lesioning of forebrain noradrenaline neurons at birth abolishes the improved maze learning performance induced by rearing in complex environment, *Brain Res.* **398**:6–10.

O'Shea, L., Saari, M., Pappas B. A., Ings, R., and Stange, K., 1983, Neonatal 6-hydroxydopamine attenuates the neural and behavioral effects of enriched rearing in the rat, *Eur. J. Pharmacol.* **92**:43–47.

Pettigrew, J. D., and Kasamatsu, T., 1978, Local perfusion of norepinephrine maintains visual cortical plasticity, *Nature* **271**:761–763.

Post, C., Minor, B. G., Davies, M., and Archer, T., 1985, Analgesia induced by 5-hydroxytryptamine receptor agonists is blocked or reversed by noradrenaline depletion in rats, *Brain Res.* **363**:18–27.

Rose, D. F., Love, S., and Dell, P. A., 1986, Differential reinforcement effects in rats reared in enriched and impoverished environments, *Physiol. Behav.* **36**:1139–1145.

Rosenzweig, M. R., Bennet, E. L., and Diamond, M. C., 1971, Chemical and anatomical plasticity of brain: Replications and extensions, in: *Macromolecules and Behavior* (J. Gaito, ed.), Appleton Century Crofts, New York, pp. 205–278.

Sagen, J., and Proudfit, H. K., 1984, Effect of intrathecally administered noradrenergic antagonists on nociception in the rat, *Brain Res.* **310**:295–301.

Smith, H. W., 1972, Effects of environmental enrichment on open-field activity and Hebb–Williams problem solving in rats, *J. Comp. Physiol. Psychol.* **80**:163–186.

Yaksh, T. L., and Rudy, T. A., 1976, Chronic catheterization of spinal subarachnoidal space, *Physiol. Behav.* **17**:1031–1036.

VII

Biophysical Considerations and New Approaches

43

Digital Imaging of Ca^{2+} Levels in CNS Neurons under Conditions That Induce Facilitating Increases in Ca^{2+} Levels and Sustained Ca^{2+} Elevation

JOHN A. CONNOR and PHILIP E. HOCKBERGER

1. INTRODUCTION

It is generally recognized that intracellular calcium ions (Ca^{2+}) are important in the events that underlie cellular forms of neuronal conditioning. The experimental evidence supporting this view is clearest where there are correlations between electrophysiological and direct optical measurements of calcium fluxes, for example, in invertebrate neurons (Connor and Alkon, 1984; Boyle *et al.*, 1984; Connor *et al.*, 1986). The much smaller size and poor accessibility of vertebrate central nervous system (CNS) neurons have made a cellular analysis of conditioning mechanisms in these systems more difficult. The evidence suggesting that Ca^{2+} ions are involved in changes underlying vertebrate learning has been primarily circumstantial (see, e.g., Lynch and Baudry, 1984). Recently, the development of membrane-permeable Ca^{2+}-sensitive dyes (Tsien, 1980; Grynkiewicz *et al.*, 1985) and high-resolution digital imaging technology have made it possible to measure spatially-resolved free Ca^{2+} changes in mammalian neurons (Connor, 1986; Connor *et al.*, 1987). At the present time the measurements are optimized using tissue-cultured neurons, where experimental difficulties such as cell inaccessibility, nonspecific optical absorbance, and scattering artifacts can be minimized. However, in the future it should be possible to extend this technology to other preparations, e.g., brain slices.

We report here some of the first measurements of Ca^{2+} changes that have been made on mammalian neurons (also see MacDermott *et al.*, 1986). This study has examined the response of cultured cerebellar granule and Purkinje cells from rat using two stimulus paradigms: multiple exposures to high-potassium salines for granule cells and glutamate application for Purkinje cells. Each paradigm caused changes in the Ca^{2+}-regulating processes of these cells. With repeated K^+-induced depolarizations, there was facilitation

JOHN A. CONNOR and PHILIP E. HOCKBERGER • Department of Molecular Biophysics, AT&T Bell Laboratories, Murray Hill, New Jersey 07974.

of the Ca^{2+} response in granule cells. In Purkinje cells there was sustained elevation of $[Ca^{2+}]$ following focal application of glutamate, which lasted several minutes. These effects on the Ca^{2+}-regulating processes may shed light on the role that Ca^{2+} ions play in affecting synaptic efficacy between these cells (cf. Ito, 1982; Ekerot and Kano, 1985).

2. RESULTS

2.1. Characterization of Cerebellar Cells in Culture

An important issue that must be addressed in a study of this nature is what similarities the cells analyzed in tissue culture bear to neurons *in vivo*. Several laboratories have shown that cultured cerebellar granule cells display morphological and biochemical properties similar to cells *in vivo* (Lasher and Zagon, 1972; Messer, 1977; Gallo *et al.*, 1982; Levi *et al.*, 1984). Likewise, numerous studies have analyzed the electrical properties of cultured Purkinje cells (Hild and Tasaki, 1962; Nelson and Peacock, 1972; Geller and Woodward, 1974; Gahwiler, 1976; Marshall *et al.*, 1980; Moonen *et al.*, 1982; Gruol, 1983). We have also used electrophysiological as well as immunocytochemical methods to show that both granule and Purkinje cells differentiate in culture. Furthermore, the timing of this differentiation process under our culture conditions is consistent with similar events *in vivo*.

We have cultured cerebella from embryonic and postnatal rats using procedures that have been described in detail elsewhere (Hockberger *et al.*, 1987a). Intracellular and voltage-clamp measurements were performed using whole-cell patch-recording techniques (Hamill *et al.*, 1981). Cells were grown on poly-D-lysine-coated glass coverslips, and recordings were made in either culture medium or Krebs saline under phase-contrast optics.

We have found four stages of granule cell development in explant cultures, which are summarized in Table I: neuroblast, immature, intermediate, and mature cells. Neuroblasts appear first and predominate during days 1–3 *in vitro* (DIV). The cell bodies are located within 50 μm of the explant, extend two or three long thin processes, are inexcitable, and staining for neuron-specific enolase (NSE) is absent. These cells probably correspond to bipolar and tripolar neuroblasts in the external granular layer of the developing cerebellum (Cajal, 1911; del Cerro and Snyder, 1972), which are also NSE-negative (Schmechel *et al.*, 1980).

The second stage, immature granule cells, appears by the end of the first week *in vitro*. The cell bodies are located 50–140 μm from the explant, are still inexcitable, but now stain lightly for NSE (Fig. 1). Small voltage-dependent currents can be elicited under voltage clamp, and application of the transmitters GABA and glutamate hyperpolarize and depolarize the membrane potential, respectively. These neurons may correspond to granule cells that are migrating *in vivo* from the external to the internal granular layer between postnatal days 5 and 20. Migrating granule cells are occasionally NSE-positive *in vivo*, but the intensity of staining increases after most cells have reached the internal granular layer (Schmechel *et al.*, 1980; Marangos *et al.*, 1980).

The third stage is intermediate granule cells and is characterized by further movement away from the explant as well as by the first appearance of soma excitability (Fig. 2A–D) and more intense staining for NSE. Large currents can be evoked under voltage clamp from these cells (Fig. 3), and GABA but not glutamate responses are recorded from the

TABLE I. Classification of the Four Stages of Granule Cell Development in Postnatal Day-4 Explant Cultures

Cell type	First appearance (days *in vitro*)	Locate (μm from explant)	Anti-NSE reactivity	Excitability	Soma currents	GABA response	Glutamate response
Neuroblast	1–3	0–50	–	No	I_{leak}	–	–
Immature	3–7	50–140	+	No	Small I_A, I_K, I_{Na}, I_{Ca}	+	+
Intermediate	11–18	140–500	+ +	Soma spikes	Large I_A, I_K, I_{Na}, I_{Ca}	+ +	–
Mature	>30	140–500	+ + +	Axonal spikes	I_{leak}	+ +	–

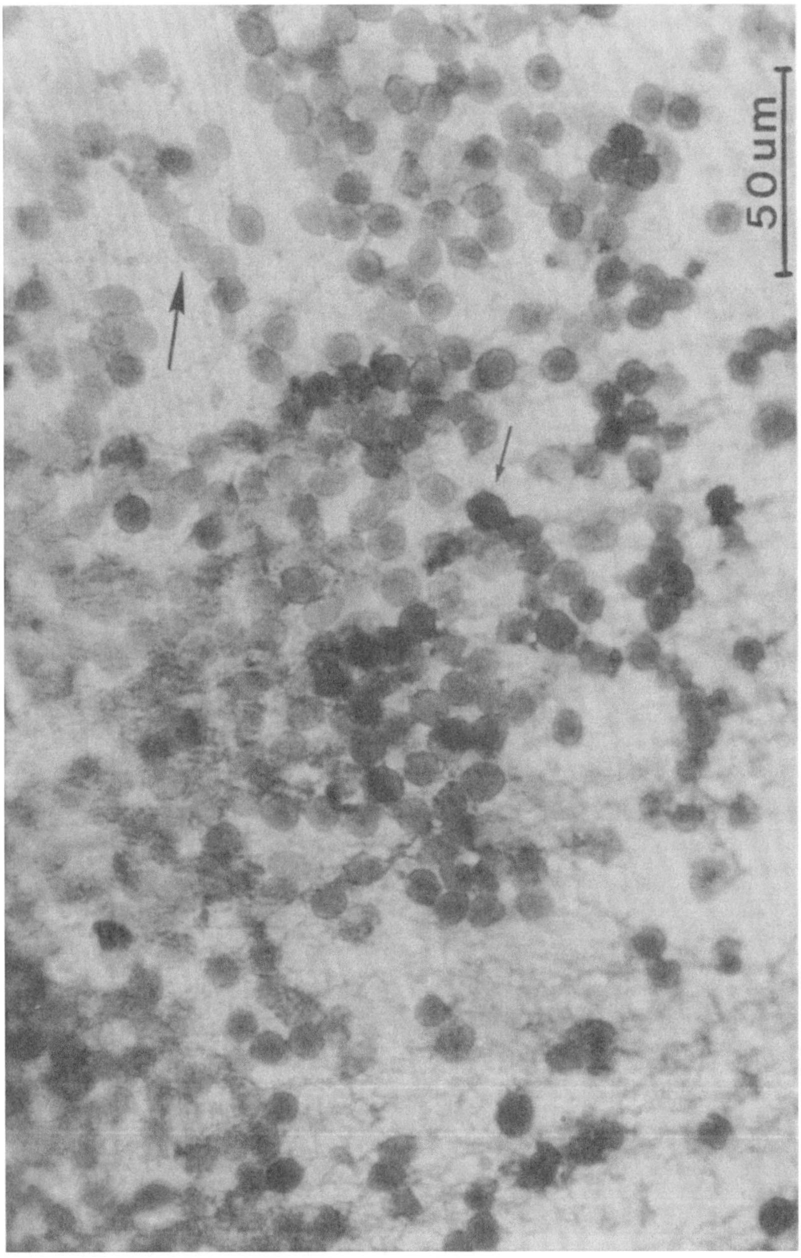

FIGURE 1. Brightfield photomicrograph of a relatively uniform population of cultured cerebellar granule cells (5 DIV) stained with the peroxidase–antiperoxidase method using antibodies directed against neuron-specific enolase (NSE). Cells that contained this marker of differentiation displayed a dark precipitation reaction (small arrow). Granule cell neuroblasts were NSE-negative and were visualized using the counterstain hematoxylin (large arrow).

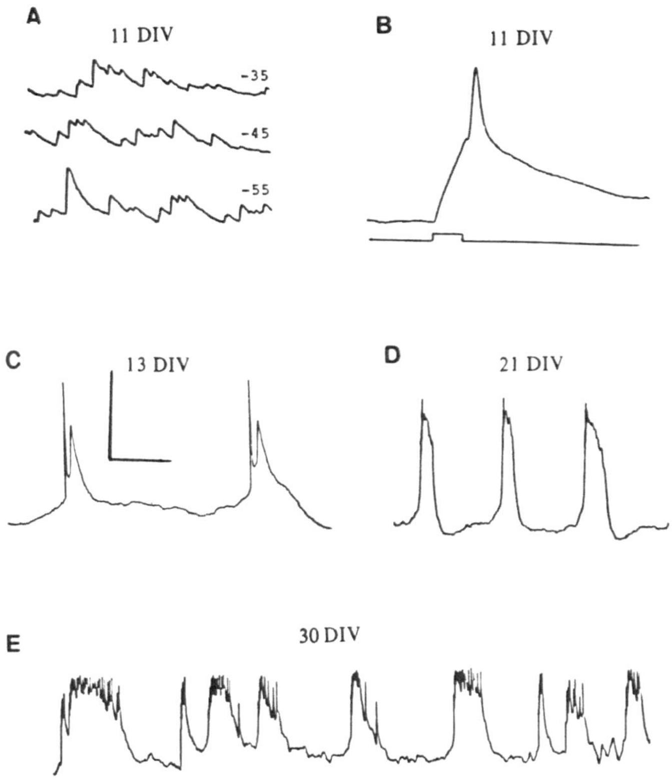

FIGURE 2. Intracellular recordings from the somata of cultured granule cells after different lengths of time *in vitro*. During the second week, excitatory synaptic potentials (A) and small action potentials (B) were recorded, sometimes exhibiting rhythmic firing patterns (C). The duration of the depolarizing phase of the rhythmic firing increased over time in culture (D), and by the end of 1 month only small axonal spikes were elicited during the depolarizing phase (E). The calibration bars are 30 mV throughout and 400 msec (A), 100 msec (B), 200 msec (C), and 1 sec (D,E). Reprinted from Hockberger *et al.*, 1987a.

soma. We have performed a complete analysis of the voltage-dependent conductances of these cells (cf. Hockberger *et al.*, 1987a), and the types are summarized in Table I. These include inward sodium and calcium currents and outward potassium (transient and delayed rectifier) currents. Granule cells at this stage are intermediate in character between immature and mature granule cells (cf. Fig. 3), and this type is most likely found within the internal granular layer during postnatal days 10–30.

The final stage of development in culture includes cells that stain intensely for NSE and still respond to GABA, but the soma no longer displays voltage-dependent currents or glutamate responses. Axonal currents are sometimes present, and membrane potential oscillations contain spikes that may result from action potentials originating from axonal regions (Fig. 2E). The absence of somatic action potentials is consistent with observations in adult cat (Eccles *et al.*, 1966) and turtle cerebellum (Walsh *et al.*, 1974), but evidence is lacking in the rat. Action potentials in the axons, i.e., parallel fibers, first appear around postnatal day 7 in the rat, although most parallel fibers are still maturing at this stage (Shimono *et al.*, 1976). Synaptic connections between granule cells and the other neurons in the cerebellum *in vivo* are functional starting on postnatal days 10–12, and

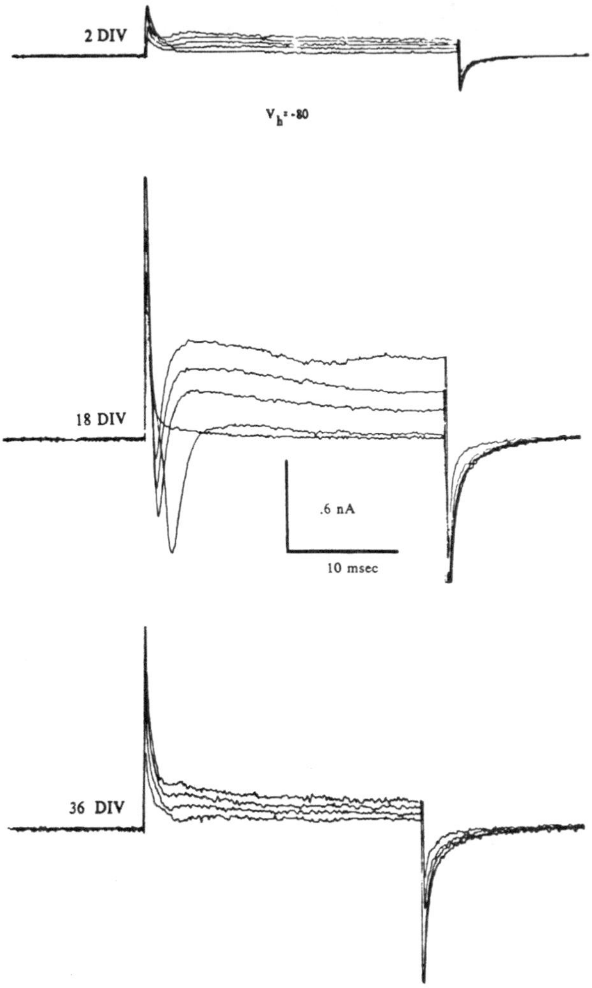

FIGURE 3. Current traces evoked from granule cell somata under voltage clamp changed progressively over time in culture. Neuroblasts (2 DIV) did not display any measurable voltage-dependent conductances. Intermediate granule cells (18 DIV) had inward and outward currents that were voltage-independent, and mature cells (36 DIV) lacked these conductances. All records were obtained from cells held at −80 mV while voltage was stepped positively to equivalent potentials.

complete development takes several more weeks (Woodward *et al.*, 1971; Shimono *et al.*, 1976). Thus, granule cell maturation in culture takes approximately 1 month and involves several identifiable stages of differentiation. Each of these stages *in vitro* corresponds to a discrete phase in the development of granule cells *in vivo*.

Our analysis of the differentiation process for cultured Purkinje cells is less complete. A major difficulty in this analysis has been in identifying these cells in culture at early stages of development. Purkinje cells *in vivo* form a monolayer about 3 days after birth in the rat, and a thick apical dendrite develops several days later (Addison, 1911; Altman, 1972a). The cell bodies double in size to about 20 μm during the first week, with secondary and tertiary apical branches forming by the end of the second postnatal week (Altman, 1972b, 1982). By the time Purkinje cells can be identified in culture using these char-

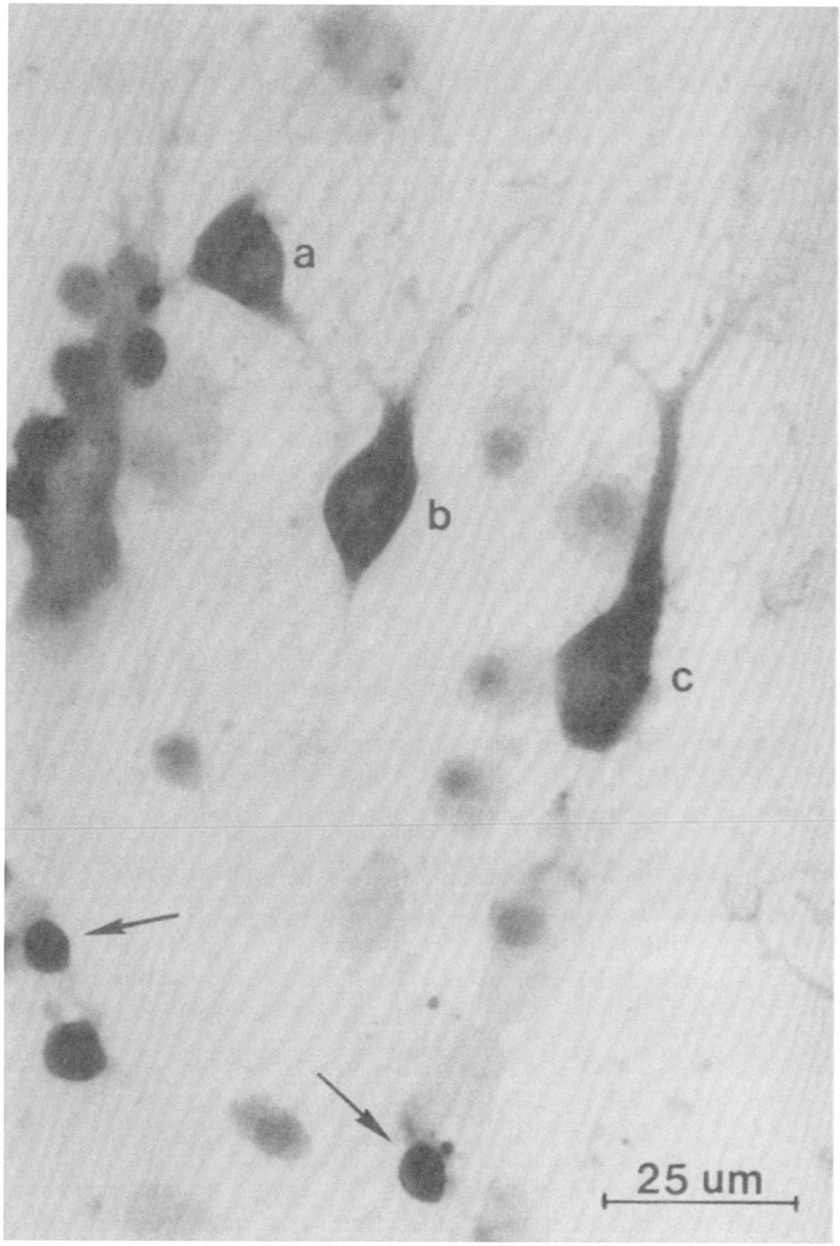

FIGURE 4. Brightfield photomicrograph of cultured Purkinje cells at three different stages of morphological development after 1 week *in vitro*. During the first stage (a), the soma is round (*ca.* 15 μm) and multipolar. As the cell develops, a thick apical dendrite begins to form from one end of the soma, and the axon from the other (b). With continued development, the apical dendrite elongates and branches (c). Most Purkinje cells require 2–3 weeks in culture before the latter stage develops. In this slide Purkinje and granule cells (arrows) were stained immunocytochemically for NSE.

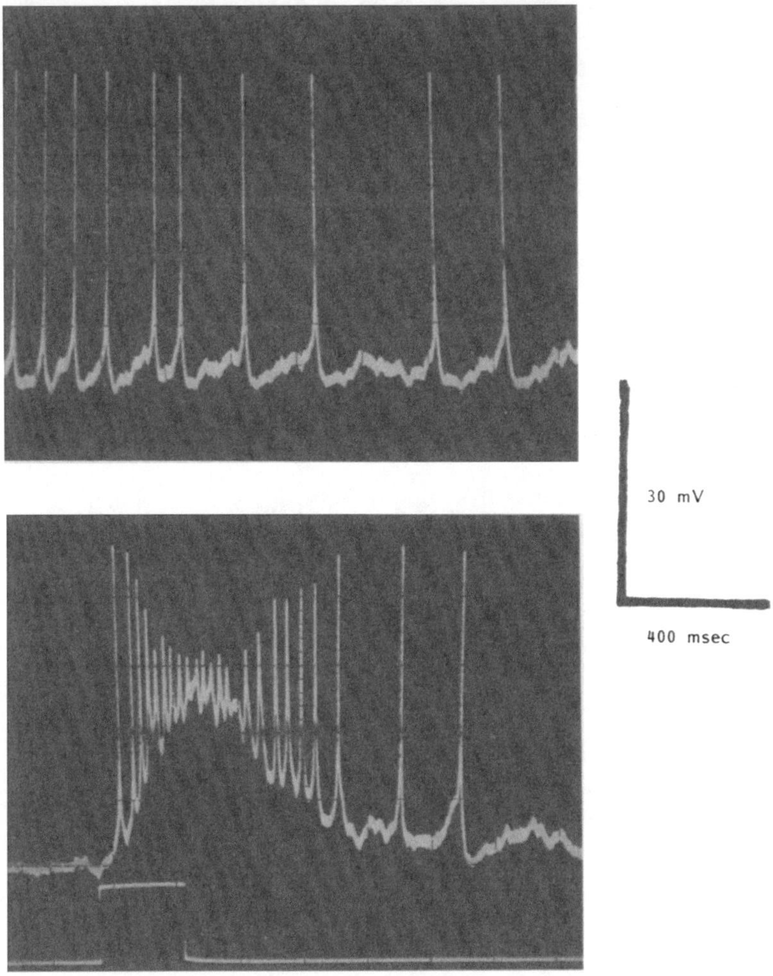

FIGURE 5. A: Intracellular recordings from a Purkinje cell after 29 DIV before (top) and during iontophoretic application of glutamate onto the cell (bottom). B: Current traces evoked from this cell under voltage clamp at −60 mV with positive voltage steps.

acteristics (see Fig. 4), we have found that the cells are already excitable. They exhibit a variety of voltage-dependent conductances as well as membrane potential responses to GABA, glutamate, kainate, and NMDA (Hockberger *et al.*, 1987b). These results are consistent with reports that rat Purkinje cells *in vivo* are excitable at birth and can respond to several different neurotransmitters (Woodward *et al.*, 1969, 1971).

Figure 5A shows intrasomatic recordings from a cultured Purkinje cell displaying spontaneous action potentials and a burst of activity after iontophoretic application of glutamate. Records like these were routinely obtained from cells after 1 month in culture and are similar to recordings obtained by Gruol (1983) under similar *in vitro* conditions. The response to glutamate was a slow depolarization with attenuated action potentials during the peak of the response. A similar response was evoked with a depolarizing

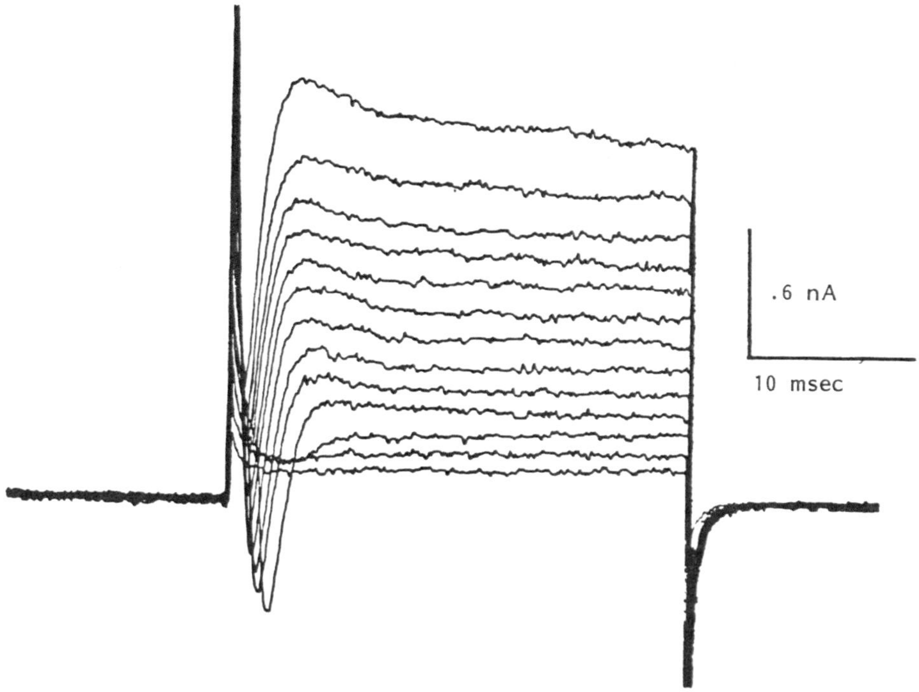

FIGURE 5. (*Continued*)

current stimulus, and its shape is reminiscent of intracellular waveforms obtained from adult Purkinje cells in slice preparations (Llinas and Sugimori, 1980; Crepel *et al.*, 1981).

We have not as yet performed a complete analysis of the voltage-dependent conductances in cultured Purkinje cells. However, our preliminary evidence suggests that these cells have several different types of conductances. Figure 5B shows typical current traces elicited under voltage clamp with membrane potential held at −60 mV. Positive voltage steps were applied in 5-mV increments, and a small inward current was apparent at −45 mVB. Greater steps activated larger inward as well as outward currents. The inward current is at least partly reduced by tetrodotoxin (TTX), indicating the presence of a Na^+ current (data not shown). Calcium-imaging data (see below) suggest that a large Ca^{2+} current also contributes to the inward current. Performing voltage-clamp steps from various holding potentials has also demonstrated the presence of three different outward currents, which we believe are similar to the delayed rectifier, transient, and calcium-activated potassium currents (also see Gruol, 1984).

In spite of this impressive electrophysiological differentiation, the cultured Purkinje cells do not display the beautiful dendritic arborization so characteristic of these cells *in vivo* (Cajal, 1911). The latter apparently requires the presence of parallel fibers in the correct orientation (Altman and Anderson, 1972; Berry and Bradley, 1976). Nevertheless, using immunocytochemical methods, we have found that cultured Purkinje cells express several antigenic determinants that are also expressed *in vivo*: Thy-1, Leu-4, and NSE (for *in vivo* results see Barclay, 1979; Garson *et al.*, 1982; Schmechel *et al.*, 1978). Figure 4 shows three NSE-positive Purkinje cells in culture at different stages of morphological development. For the experiments described below on intracellular Ca^{2+} mea-

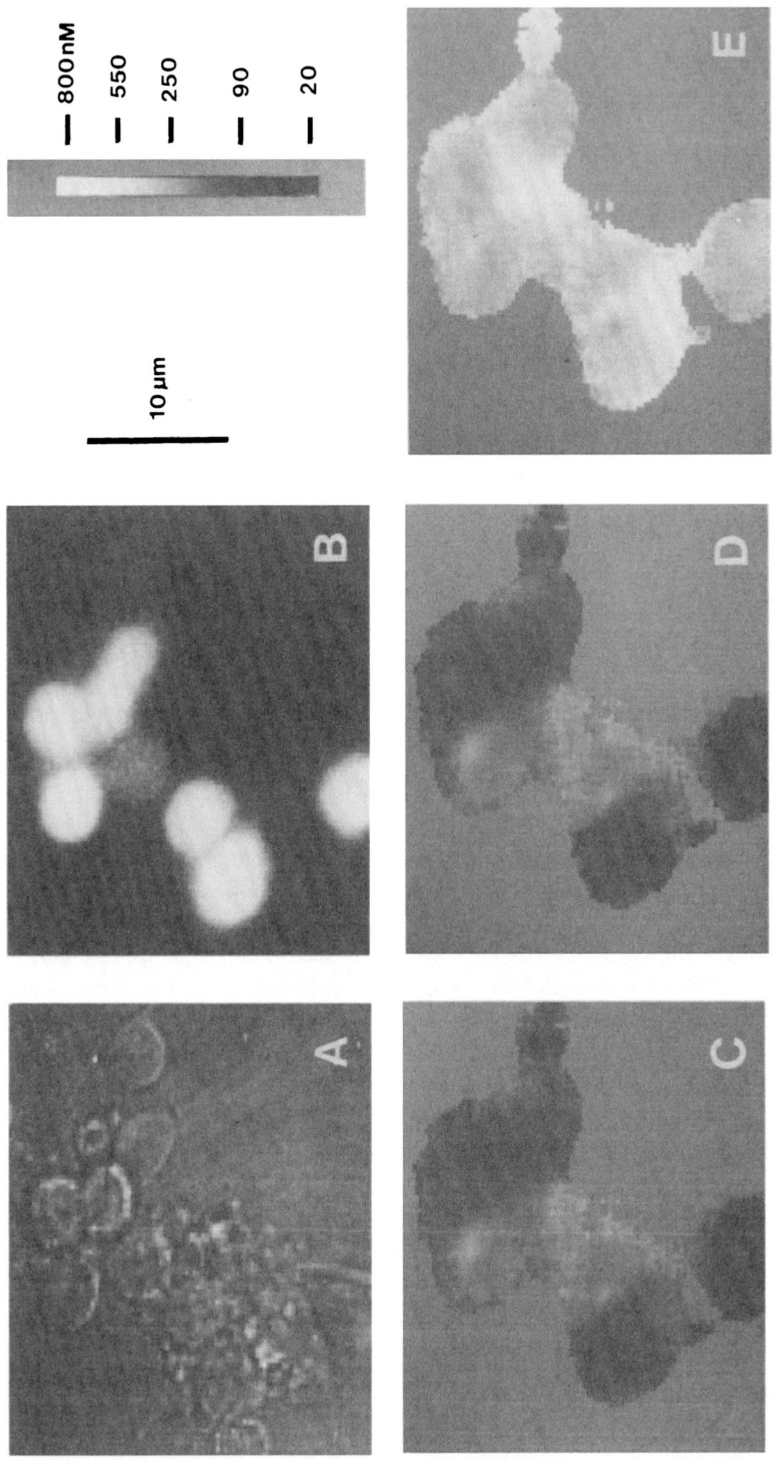
800nM
550
250
90
20
10 μm
A
B
C
D
E

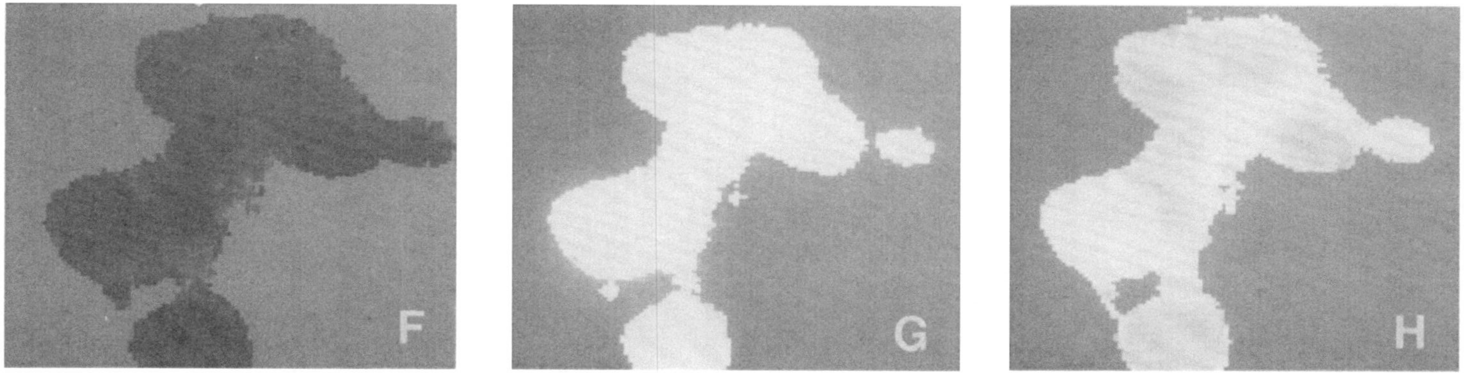

FIGURE 6. Images of granule cells in early, immature stage of development taken with CCD camera. A: Nomarski image of eight granule neurons (6 DIV). B: Fluorescence image of small field, 380-nm excitation. C: Fluorescence ratio image (340/380 nm) of same field, with cells bathed in normal saline. The ratio values correspond to Ca^{2+} level (see text). The Ca^{2+} levels are displayed on an eight-bit (256 level) gray scale with low values being mapped to dark shades and high values bright. Approximate Ca^{2+} levels are indicated on the shading bar. D: Cells in 25 mM K for 75 sec displayed ratios not significantly changed. E: Cells in high-K^+saline 2 min 15sec. F: Four-minute recovery in normal saline. G: Second pass in high-K^+ saline, 75sec as in B; Ca^{2+} levels increased to nearly 1 μM in two of the neurons. Cells in high-K^+ saline for 2 min 15sec (as in E) on second pass. Calibration bar, 10 μm. Reprinted from Hockberger *et al.*, 1987a.

surements, we used only cells that displayed the more differentiated appearance: flask-shaped somata and thick apical dendrite with some secondary and tertiary branching. We have not yet attempted to combine Ca^{2+} imaging and immunocytochemistry on the same cell.

2.2. Facilitating Ca^{2+} Increases in Granule Cells

Intracellular Ca^{2+} levels have been monitored using fura-2 fluorescence and digital image processing. In this method, the fluorescence of the indicator trapped inside cells is imaged at two different excitation wavelengths, 340 and 380 nm. Each of these images is corrected for backround fluorescence and camera dark current. This gives a field of numbers, each corresponding to the absolute fluorescence intensity at a point in the image. Each point size, or picture element (pixel), represents an area approximately 0.5 μm on a side depending on camera setting and microscope objective power. A ratio image is then formed by dividing each pixel value in the 340-nm excitation image by its counterpart in the 380-nm image. This ratio is a function of the $[Ca^{2+}]$ in equilibrium with the indicator, and it is relatively insensitive to factors such as cell geometry or indicator concentration. The charge-coupled-device (CCD) camera used in these experiments is a linear photometer unlike silicon-intensified target (SIT) video cameras. Details of both methods for imaging intracellular Ca^{2+} ions have been reported (Williams *et al.*, 1985; Connor, 1986).

Tissue culture explants growing on #1 glass coverslips were loaded with fura-2 by 30- to 50-min exposures to the membrane-permeant form of the molecule, fura-2/AM, at a nominal concentration of 4–5 μM in a defined medium. Following this step, cells were rinsed in the medium and incubated for at least 1 hr to allow deesterification of the indicator (Tsien, 1981; Tsien *et al.*, 1982). Coverslips were mounted on an inverted microscope, and measurements were made at 32–35°C as described previously (Connor, 1986).

Resting Ca^{2+} concentrations in granule neurons were generally in the range of 50–80 nM when measured in Krebs saline. Values were somewhat higher when measured in culture medium, generally between 100 and 150 nM, or in media with five times normal potassium (high K^+) where the Ca^{2+} levels were in excess of 150 nM. It has been our experience, and that of others (Lasher and Zagon, 1972; Thangnipon *et al.*, 1983; Kingsbury *et al.*, 1985), that survivability and outgrowth of cultured granule cells are enhanced by high-K^+ medium. These observations indicate not only that cells can survive better with prolonged Ca^{2+} elevation but that growth and differentiation are enhanced under such conditions (cf. Connor, 1986; Hockberger *et al.*, 1987).

We have used transient exposure to high-K^+ saline to examine the Ca^{2+} levels of granule cells. Figure 6 shows a series of digitally processed images, both Nomarski (A) and fura-2 fluorescence (B–H) images, from a field of nine granule cells (6 days in culture). Figure 6B is a single-wavelength (380 nm) fluorescence image of the field showing that eight of the neurons had trapped a sufficient amount of the indicator to enable us to detect the dye's presence. Figure 6C–H are ratio images (340 nm/380 nm) that reflect internal Ca^{2+} levels displayed on an eight-bit gray scale (256 level) where a brighter image indicates higher Ca^{2+}. The cells were given three successive exposures to high-K^+ (25 mM) saline in Krebs saline with 4-min recovery periods in normal-K^+ (5 mM) saline in between. Figure 6C shows Ca^{2+} levels in the cells in normal Krebs saline, and the data of Fig. 6D were taken 75 sec after high-K^+ saline had begun to reach the cells. The Ca^{2+} levels were not measurably different between the images in 6C and 6D. Images of Fig. 6E were taken 1 min later and show elevated Ca^{2+} levels in the cells. Following a 4-min recovery, Ca^{2+} levels were at or below initial values (Fig. 6F).

A second exposure of these cells to high K^+ led to a qualitatively different behavior. At the 75-sec point the Ca^{2+} levels had increased dramatically in most of the cells (Fig. 6G), and 1 min later values were reduced partway back to base line. The latter were maintained for the duration of the high-K^+ exposure (Fig. 6H). Data pooled from five of the neurons are plotted in Fig. 7 for all three high-K^+ runs and show that the second and third passes produced nearly the same Ca^{2+} change, each much larger than the first. Where individual exposure times to high K^+ were longer than the 2–3 min shown here, Ca^{2+} levels settled to a plateau that was maintained for at least 10 min.

The facilitating Ca^{2+} responses shown in Figs. 6 and 7 are typical of those seen in over 80 neurons observed at the immature stage of development, i.e., before they had reached full excitability (cf. Table I). The facilitation was also present in intermediate neurons; however, in these cells exposure to high-K^+ saline generally produced a period of intense firing that drove Ca^{2+} levels briefly into the micromolar range (near the upper limit of the indicator measurement range). Facilitation was not easily observed with subsequent exposures. Blocking spike activity by bathing the cells in 600 nM TTX, though, reduced the K^+-induced Ca^{2+} changes to approximately the same range as that observed in the immature neurons and revealed the facilitating response.

Figure 8 shows high-K^+ stimulation data pooled from two intermediate neurons and illustrates the reduction of the Ca^{2+} transient brought about by blocking spike activity with TTX. Given the slow nature of changing external solutions and the limited speed of reading and storing the CCD data, it is probable that the peak of the Ca^{2+} transient was missed in most cells. Even so, peak fluorescence ratios of 6.5 to 7 were observed in high-K^+ saline (Fig. 8A), corresponding to Ca^{2+} levels of 1 to 2 μM. This peak

FIGURE 7. Graphic representation of the pooled data from four of the neurons illustrated in Fig. 6 showing the mean Ca^{2+} levels (with S.D. bar) of the neurons during three consecutive exposures to 25 mM K saline. During the third pass TTX was present in the bathing solution (0.3 μM). Measurements were made only at the times indicated by the data points, which were connected by straight lines for illustration. The Ca^{2+} levels were not followed during recovery periods (interrupted traces) but were generally 80–90% complete within 1 to 2 min. Reprinted from Hockberger *et al.*, 1987a.

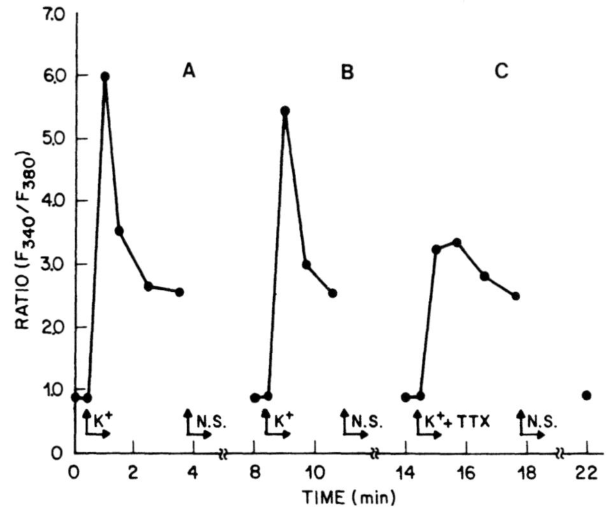

FIGURE 8. Pooled numerical data from two granule cells responding to K^+ stimulation before and after exposure to TTX. Cells were at a stage of development at which somatic action potentials were generated by depolarization (intermediate stage). Two high-K^+ runs were made before TTX was added to the bath. Reprinted from Hockberger *et al.*, 1987a.

rapidly decayed to a maintained plateau. After a large initial response such as this, subsequent Ca^{2+} elevations were either the same, as illustrated in the figure, or smaller. When cells at this stage were exposed to TTX at the beginning of an experiment so that the Ca^{2+} transient was severely damped, facilitation similar to that shown in Fig. 6 and 7 was displayed. The facilitation decreased measurably over 10-min measurement periods, but the recovery time course has not been reliably assessed at present.

The slow time course of onset and recovery of Ca^{2+} changes results primarily from the bath-exchange method of applying high potassium. When a high-K^+ saline is ejected from a micropipette 30–40 μm from the cells, the Ca^{2+} increase is maximal within the time required to obtain the data (*ca.* 2 sec), and recovery is complete within 60 to 90 sec. This type of stimulation has not been used to test facilitation because of the difficulty in obtaining quantifiably reproducible ejections.

We have not yet performed simultaneous electrophysiology and Ca^{2+} imaging because of technical difficulties, but we have monitored companion populations of granule cells using whole-cell patch recording. Exposure to 25 mM K^+ saline depolarizes both classes of granule cell to approximately −30 mV (i.e., the inexcitable early-stage cells and the older, TTX-poisoned cells). The depolarizations are identical from stimulus to stimulus and completely recover when the K^+ level is restored to normal. That is, there is no increase in the degree of membrane depolarization, nor does the depolarization outlast the exposure to high-K^+ saline.

2.3. Sustained Elevation of Ca^{2+} in Purkinje Cells

Glutamate, when applied to the cell body and proximal neurites of Purkinje cells, produced a large and occasionally long-lasting change in internal Ca^{2+} levels. This response was present as soon as the cells could be identified by morphological criteria (about 2 weeks in culture) and has been followed as far out as day 50. The long-lasting responses were most reliably observed with the cells bathed in their growth medium

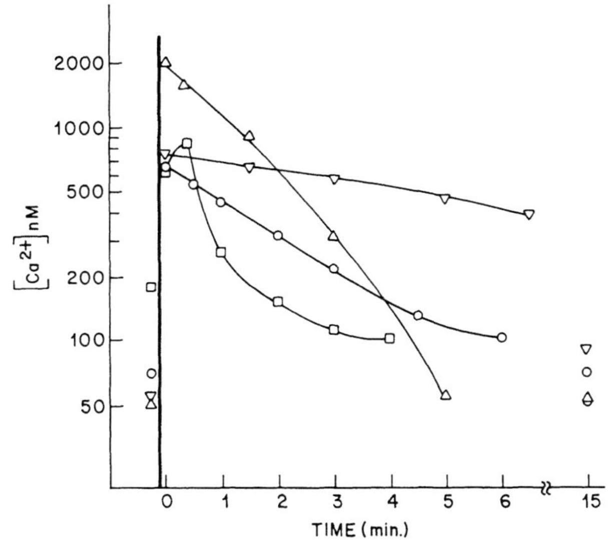

FIGURE 9. Numerical data from five individual Purkinje neurons showing the time course of Ca^{2+} concentration changes following the application of four iontophoretic pulses of glutamate to their somata. The pulses (300 nA, 0.5 sec) were delivered within a 3-sec interval.

rather than Krebs saline. Therefore, both electrical and optical measurements have been made in medium.

Iontophoretic application of glutamate (three or four pulses of 0.5 sec duration, 300 nA amplitude) produced Ca^{2+} concentration changes that lasted several minutes. Figure 9 plots the time course of Ca^{2+} changes measured at the soma in four neurons in response to such a stimulus. It is clear from the records that there is considerable cell-to-cell variability, but in all cases the time course of the Ca^{2+} change persisted for a number of minutes. During the long Ca^{2+} responses, the bathing medium (0.3–0.4 ml) was generally replaced once or twice.

The duration of the long-lasting Ca^{2+} response does not simply reflect slow Ca^{2+} regulation within the cells. Cells that displayed a prolonged Ca^{2+} response to glutamate showed 90% recovery to focally applied high-K^+ media ($n = 8$) within 1.5 min. The records of Fig. 10 show one such experiment in which a 4-sec pressure pulse of high-

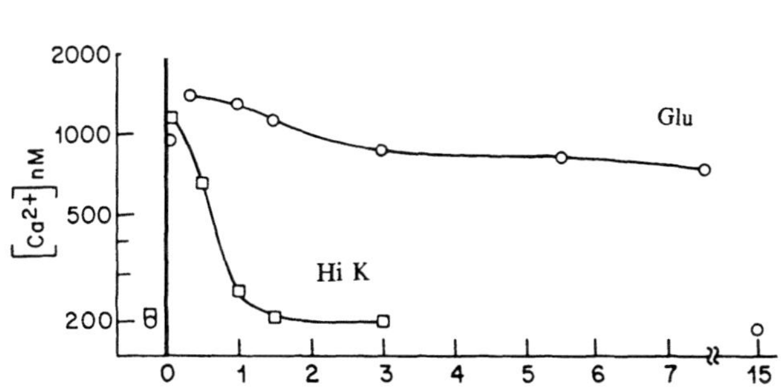

FIGURE 10. Comparison of the Ca^{2+} response to glutamate and high K^+. In this example the high K^+ was administered after the glutamate; however, the order of delivery made no difference to the relative persistence of the two responses.

K^+ medium was administered following the recovery from two previous responses to glutamate. The time course of the Ca^{2+} recovery is much faster than that of the recovery following glutamate application by iontophoresis.

Electrical recordings were carried out without medium change or bath circulation during a particular measurement so that local transmitter persistence may have been a greater problem there. Nevertheless, the response to small doses of glutamate was a depolarization that recovered in a matter of seconds (see Fig. 11A). Larger doses of glutamate, as in Fig. 11B, induced a longer-lasting depolarization, but we have not as yet identified the conditions necessary to obtain this effect.

Under conditions in which the stimulus was large and included a greater area of the cells, as by pressure ejection of 100 μM glutamate, the Ca^{2+} elevation persisted for even longer periods. Figure 12 illustrates the response when glutamate was applied to the fork region of the dendrite by a 5-sec pressure pulse. Figure 12B shows the intracellular Ca^{2+} levels before the glutamate application. The soma level was approximately 60 nM, while that in the dendritic region was significantly higher, 150 to 180 nM. This regionalization was often, but not always, observed in Purkinje cells as well as in other cultured cell types (cf. Connor, 1986). Immediately after the stimulus, Ca^{2+} levels rose to approximately 1 μM throughout the neuron and remained elevated for 3–4 min. During this time it was impossible to obtain a good ratio image of the dendrites of this particular cell because of the faint fluorescence at 380-nm excitation. Figure 12C shows the Ca^{2+}

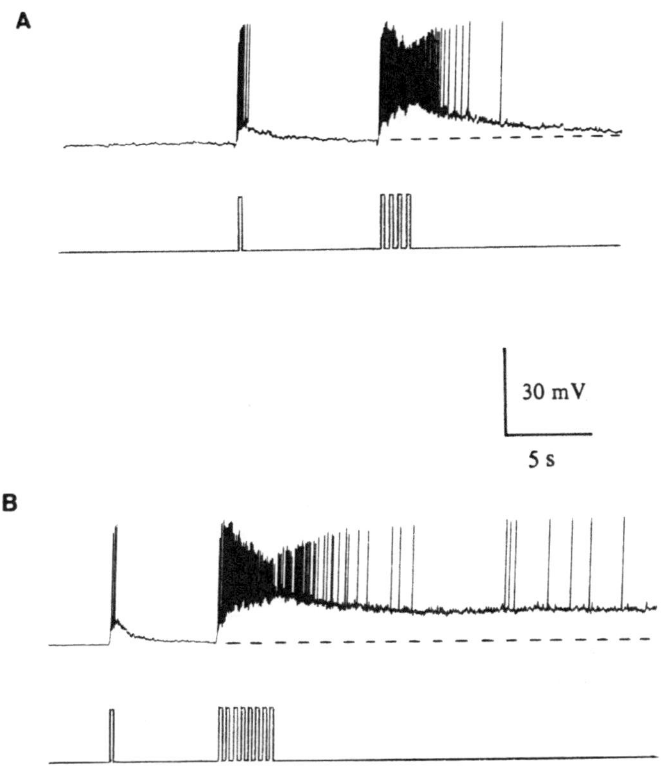

FIGURE 11. Membrane potential responses of a Purkinje cell (30 DIV) following different iontophoretic doses of glutamate. A: The membrane potential depolarization (upper trace) was transient with one or four pulses of glutamate (lower trace), recovering after several seconds. B: With a larger dose, the depolarization lasted several minutes.

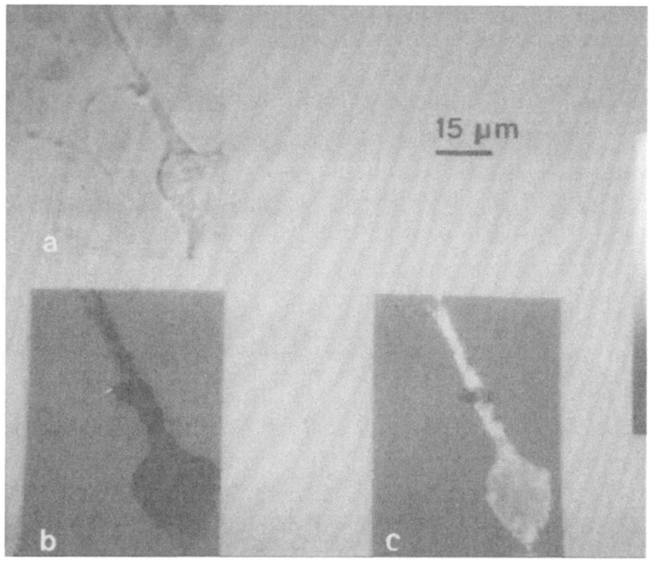

FIGURE 12. Nomarski image of Purkinje cell 30 days in culture. Branching processes and terminals from other cells can be seen in the left-hand portion of the picture. B: The Ca^{2+} levels in this neuron before pressure application of 100 µM glutamate to the branching region of the upper neurite. C: The Ca^{2+} levels 8 min after glutamate application. Duration of the pressure pulse was 5 sec. See text for Ca^{2+} levels. Calibration bar, 15 µm.

distribution 8 min after the stimulus, when the levels were still high but resolvable; soma was 250 nM, and dendrite was 650 nM. Recovery required about 12 min.

After Ca^{2+} levels recovered from such large stimuli, we have noted in a number of instances that subsequent applications of glutamate are either ineffective or much reduced. For example, in the cell of Fig. 12, a second application produced a transient Ca^{2+} increase that was nearly as large as the first but recovered in 2 min. At the present time we have not analyzed enough cells to interpret this complicated phenomenon, but such behavior may be related to long-term depression described in Purkinje cells *in vivo* (cf. Ekerot and Kano, 1985).

3. DISCUSSION

We have presented observations on the time course of Ca^{2+} changes following stimulation in two types of cerebellar neurons: granule and Purkinje cells. Although these results were obtained using cultured cells, many morphological, electrical, and immunostaining properties were similar to granule and Purkinje cells *in vivo*. We emphasize that many of the electrical and optical measurements reported here are simply not possible *in vivo* at the present time. It remains to be seen whether these measurements are indicative of the *in vivo* condition.

The facilitating effect of high K^+ is extremely interesting in that it provides cultured granule cells with a mechanism for implementing use dependence. For example, it is quite plausible that the first of a train of stimuli might give a negligible response and that subsequent stimuli would produce disproportionately larger responses. Such a mechanism could contribute to changes in parallel fiber–Purkinje cell transmission following conditioning procedures that involve cerebellar circuitry (see Steinmetz *et al.*, Chapter 14).

The critical difference between the Ca^{2+} facilitation observed here and conditions analogous to the "residual Ca^{2+} hypothesis" for synaptic transmitter release (see Katz and Miledi, 1968; Rahamimoff, 1968) is that the cytoplasmic Ca^{2+} concentration returns completely to base line between stimuli rather than remaining at an elevated level. Such a condition of residual Ca^{2+} has been shown, for example, during posttetanic potentiation in *Aplysia* neuron L-10 (Connor *et al.*, 1986b). At present we do not know whether the basis of the facilitation resides in modification of the Ca^{2+} channel population with repeated use, as shown for adrenal chromaffin cells (Hoshi *et al.*, 1984), or whether the granule cells have a moderately low buffer capacity for Ca^{2+} that takes relatively long periods (over 10 min) to empty.

The long-lasting elevation of Ca^{2+} in response to glutamate may be important for understanding modulation of Purkinje cell characteristics. A rather brief exposure to the transmitter can create a modified state with regard to Ca^{2+}-activated processes for some time thereafter. The data are less informative than they might be because both electrical activity and Ca^{2+} changes were not tracked simultaneously in the same neuron. For example, it is unclear whether the sustained response is caused by sustained electrical activity or possibly mediated by second messenger systems. Extracellular recordings of electrical activity have been made while simultaneously monitoring Ca^{2+} responses, and these records have not shown long-persisting spike activity. Such a recording technique would fail, however, to detect steady shifts in membrane potential. Invasive loading of the cells with indicator and intracellular recording is being attempted at present, but whole-cell recording techniques often tend to dialyze out factors in the cytoplasm that regulate some aspect or another of Ca^{2+} transport (see Hockberger *et al.*, 1987a).

4. SUMMARY

Most neurons respond to stimulation with a transient influx of calcium ions through voltage-gated channels, which raises the concentration of intracellular free calcium [Ca^{2+}]. Among the cellular processes affected by [Ca^{2+}] elevation are transmitter release, ion conductance activation, axoplasmic transport, and receptor-mediated events. Facilitation of Ca^{2+} influx has been suggested as a possible mechanism underlying various types of synaptic plasticity including homosynaptic and heterosynaptic facilitation. Sustained elevation of [Ca^{2+}] has been implicated in short-term and long-term potentiation of synaptic transmission. Direct measurements of [Ca^{2+}] changes in most of these cases, however, have not been possible.

We have recently developed an imaging system for examining Ca^{2+} levels in mammalian CNS neurons *in vitro* (Connor, 1986). This system employs a CCD camera and the fluorescent indicator fura-2 for monitoring spatially-resolved changes in [Ca^{2+}] on a time scale of 2 to 3 sec. Cerebellar granule and Purkinje cells were studied in culture, and electrical correlates of the responses were assessed using whole-cell patch recording. We have made two observations that may be important for understanding the interaction between granule cells and Purkinje cells: (1) facilitating increases in [Ca^{2+}] levels were detected in granule cells with repeated potassium-induced depolarizations; and (2) sustained elevation of [Ca^{2+}] was measured in Purkinje cells following focal application of glutamate, which lasted several minutes.

ACKNOWLEDGMENTS. We gratefully acknowledge the technical assistance of Hsiu-Yu Tseng in developing the culture methods and performing the immunostaining. That work not sponsored by AT&T Bell Laboratories was funded by a grant from the Air Force Office of Scientific Research (F49620).

REFERENCES

Addison, W., 1911, The development of the Purkinje cells and of the cortical layers in the cerebellum of the albino rat, *J. Comp. Neurol.* **21**:459–487.

Altman, J., 1972a, Postnatal development of the cerebellar cortex in the rat, I. The external germinal layer and the transitional molecular layer, *J. Comp. Neurol.* **145**:353–398.

Altman, J., 1972b, Postnatal development of the cerebellar cortex in the rat, II. Phases in the maturation of Purkinje cells and of the molecular layer, *J. Comp. Neurol.* **145**:399–464.

Altman, J., 1982, Morphological development of the rat cerebellum and some of its mechanisms, *Exp. Brain Res.* **6**(suppl.):8–46.

Altman, J., and Anderson, W., 1972, Experimental reorganization of the cerebellar cortex. I. Morphological effects of elimination of all microneurons with prolonged X-irradiation started at birth, *J. Comp. Neurol.* **146**:355–406.

Barclay, A., 1979, Localization of the Thy-1 antigen in the cerebellar cortex of rat brain by immunofluorescence during postnatal development, *J. Neurochem.* **32**:1249–1257.

Berry, M., and Bradley, P., 1976, The growth of the dendritic trees of Purkinje cells in irradiated agranular cerebellar cortex, *Brain Res.* **116**:361–387.

Boyle, M., Klein, M., Smith, S., and Kandel, E., 1984, Serotonin increases intracellular Ca^{2+} transients in voltage-clamped sensory neurons of *Aplysia californica*, *Proc. Natl. Acad. Sci. U.S.A.* **81**:7642–7646.

Cajal, R., 1911, *Histologie du Systeme Nerveux del l'Homme et des Vertebre*, Maloine, Paris, reprinted by Consejo Superior de Investigaciones Cientificas, Madrid, 1972.

Connor, J. A., 1986, Digital imaging of free calcium changes and of spatial gradients in growing processes in single, mammalian central nervous system cells, *Proc. Natl. Acad. Sci. U.S.A.* **83**:6179–6183.

Connor, J., and Alkon, D. L., 1984, Light- and voltage-dependent increases of calcium ion concentration in molluscan photoreceptors, *J. Neurophysiol.* **81**:745–752.

Connor, J. A., Kretz, R., and Shapiro, E., 1986, Ca levels measured in a presynaptic neuron of *Aplysia* under conditions that modulate transmitter release, *J. Physiol. (Lond.)* **375**:625–642.

Connor, J. A., Tseng H., and Hockberger, P., 1987, Depolarization and transmitter induced changes in intracellular Ca of rat cerebellar granule cells in explant cultures, *J. Neurosci.* **7**:1384–1400.

Crepel, F., Dhanjal, S., and Garthwaite, J., 1981, Morphological and electrophysiological characteristics of rat cerebellar slices maintained *in vitro*, *J. Physiol. (Lond.)* **316**:127–138.

del Cerro, M., and Snider, R., 1972, Studies on the developing cerebellum. II. The ultrastructure of the external granular layer, *J. Comp. Neurol.* **144**:131–164.

Eccles, J., Llinaś, R., and Sasaki, K., 1966, The excitatory synaptic action of climbing fibres on the Purkinje cells of the cerebellum, *J. Physiol. (Lond.)* **182**:268–296.

Ekerot, C., and Kano, M., 1985, Long-term depression of parallel fibre synapses following stimulation of climbing fibres, *Brain Res.* **342**:357–360.

Gahwiler, B., 1976, Spontaneous bioelectric activity of cultured Purkinje cells during exposure to glutamate, glycine and strychnine, *J. Neurobiol.* **7**:97–107.

Gallo, V., Ciotti, M., Coletti, A., Aloisi, F., and Levi, G., 1982, Selective release of glutamate from granule cells differentiating in culture, *Proc. Natl. Acad. Sci. U.S.A.* **79**:7919–7923.

Garson, J., Beverley, P., Coakham, H., and Harper, E., 1982, Monoclonal antibodies against human T lymphocytes label Purkinje neurones of many species, *Nature* **298**:375–377.

Geller, H., and Woodward, D., 1974, Responses of cultured cerebellar neurons to iontophoretically applied amino acids, *Brain Res.* **74**:67–80.

Gruol, D., 1983, Cultured cerebellar neurons: Endogenous and exogenous components of Purkinje cell activity and membrane response to putative transmitters, *Brain Res.* **263**:223–241.

Gruol, D., 1984, Intracellular and single channel analysis of voltage-sensitive ionic mechanisms in the somal and dendritic membranes of cultured cerebellar Purkinje neurons, *Soc. Neurosci. Abstr.* **10**:939.

Grynkiewicz, G., Poenie, M., and Tsien, R. Y., 1985, A new generation of Ca indicators with greatly improved fluorescence properties, *J. Biol. Chem.* **260**:3440–3450.

Hamill, O., Marty, A., Neher, E., Sakmann, B., and Sigworth, F., 1981, Improved patch-clamp techniques for high-resolution current recording from cells and cell-free membrane patches, *Pflugers Arch.* **391**:85–100.

Hild, W., and Tasaki, I., 1962, Morphological and physiological properties of neurons and glial cells in tissue culture, *J. Neurophysiol.* **25**:277–304.

Hockberger, P., Tseng, H., and Connor, J. A., 1987a, Immunocytochemical and electrophysiological differentiation of rat cerebellar granule cells in explant cultures, *J. Neurosci.* **7**:1370–1383.

Hockberger, P., Tseng, H., and Connor, J. A., 1987b, Electrophysiological properties of cerebellar Perkinje cells after dissociation from late embryonic and early postnatal rats, *Soc. Neurosci. Abstr.* **13**:1119.

Hoshi, T., Rothlein, J., and Smith, S., 1984, Facilitation of Ca channel currents in bovine adrenal chromaffin cells, *Proc. Natl. Acad. Sci. U.S.A.* **81**:5871–5875.

Ito, M., 1982, Mechanisms of motor learning, in: *Competition and Cooperation in Neural Nets* (S. Amari and M. Arbib, eds.), Springer, New York, pp. 418–429.

Katz, B., and Miledi, R., 1968, The role of calcium in neuromuscular facilitation, *J. Physiol.* (*Lond.*) **195**:481–492.

Kingsbury, A., Gallo, V., Woodhaus, P., and Balazs, R., 1985, Survival, morphology and adhesion properties of cerebellar interneurones cultured in chemically defined and serum-supplemented medium, *Dev. Brain Res.* **17**:17–25.

Lasher, R., and Zagon, I., 1972, The effect of potassium on neuronal differentiation in cultures of dissociated newborn rat cerebellum, *Brain Res.* **41**:482–488.

Levi, G., Aloisi, F., Ciotti, M., and Gallo, V., 1984, Autoradiographic localization and depolarization-induced release of acidic amino acids in differentiating cerebellar granule cell cultures, *Brain Res.* **290**:77–86.

Llinas, R., and Sugimori, M., 1980, Electrophysiological properties of *in vitro* Purkinje cell dendrites in mammalian cerebellar slices, *J. Physiol.* (*Lond.*) **305**:171–195.

Lynch, G., and Baudry, M., 1984, The biochemistry of memory: A new and specific hypothesis, *Science* **224**:1057–1063.

MacDermott, A., Mayer, M., Westbrook, G., Smith, S., and Burker, J., 1986, NMDA-receptor activation increases cytoplasmic calcium concentration in cultured spinal cord neurones, *Nature* **32**:519–522.

Marangos, P. Schmechel, D., Parma, A., and Goodwin, F., 1980, Developmental profile of neuron-specific (NSE) and non-neuronal (NNE) enolase, *Brain Res.* **190**:185–193.

Marshall, K., Wojotwicz, J., and Hendleman, W., 1980, Patterns of functional connections in organized cultures of cerebellum, *Neuroscience* **5**:1847–1857.

Messer, A., 1977, The maintenance and identification of mouse cerebellar granule cells in monolayer culture, *Brain Res.* **130**:1–12.

Moonen, G., Neale, E., MacDonald, R., Gibbs, W., and Nelson, P., 1982, Cerebellar macroneurons in microexplant cell culture. Methodology, basic electrophysiology, and morphology after horseradish peroxidase injection, *Dev. Brain Res.* **5**:59–73.

Nelson, P., and Peacock, J., 1972, Electrical activity in dissociated cell cultures from fetal mouse cerebellum, *Brain Res.* **61**:163–164.

Rahamimoff, R., 1968, A dual effect of calcium ions on neuromuscular facilitation, *J. Physiol.* (*Lond.*) **195**:471–480.

Schmechel, D., Marangos, P., Zis, A., Brightman, M., and Goodwin, F., 1978, Brain enolases as specific markers of neuronal and glial cells, *Science* **199**:313–315.

Schmechel, D., Brightman, M., and Marangos, P., 1980, Neurons switch from non-neuronal enolase to neuron-specific enolase during differentiation, *Brain Res.* **190**:195–214.

Shimono, T., Nosaka, S., and Sasaki, K., 1976, Electrophysiological study on the postnatal development of neuronal mechanisms in the rat cerebellar cortex, *Brain Res.* **108**:279–294.

Thangnipon, W., Kingsbury, A., Webb, M., and Balazs, R., 1983, Observations on rat cerebellar cells *in vitro*: Influence of substratum, potassium concentration, and relationship between neurones and astrocytes, *Dev. Brain Res.* **11**:177–189.

Tsien, R., 1980, New calcium indicators and buffers with high selectivity against magnesium and protons: Design, synthesis, and properties of prototype structures, *Biochemistry* **19**:2396–2404.

Tsien, R. Y., 1981, A non-disruptive technique for loading calcium buffers and indicators into cells, *Nature* **280**:527–528.

Tsien, R. Y., Pozzan, T., and Rink, T. J., 1982, Calcium homeostasis in intact lymphocytes; cytoplasmic free calcium monitored with a new, intracellularly trapped fluorescent indicator, *J. Cell Biol.* **94**:325–334.

Walsh, J., Houk, J., and Mugnaini, E., 1974, Identification of unitary potentials in turtle cerebellum and correlations with structures in granular layer, *J. Neurophysiol.* **7**:30–47.

Williams, D. A., Fogarty, K. E., Tsien, R. Y., and Fay, F. S., 1985, Calcium gradients in single smooth muscle cells revealed by the digital imaging microscope using fura-2, *Nature* **318:**558–561.

Woodward, D., Hoffer, B., and Lapham, L., 1969, Postnatal development of electrical and enzyme histochemical activity in Purkinje cells, *Exp. Neurol.* **23:**120–139.

Woodward, D., Hoffer, B., Siggins, G., and Bloom, F., 1971, The ontogenetic development of synaptic junctions, synaptic activation and responsiveness to neurotransmitter substances in rat cerebellar Purkinje cells, *Brain Res.* **34:**73–97.

44

A Method to Investigate a Metabolic Process in a Single Neuron and Its Utilization in the Study of Fast Axonal Transport of Acetylcholine in a Cholinergic Neuron of *Aplysia*

HIROYUKI KOIKE, YOSHITOMO UMITSU, and HIROKO MATSUMOTO

1. INTRODUCTION

Intracellular recording techniques have been used successfully to analyze the electrical activities of specific single neurons among numerous heterogeneous populations of cells in the central nervous system. We have pursued development of a technique for studying the metabolic processes of single neurons by intracellular injection of radioactive substances via double-barreled Pyrex® capillary micropipettes (Koike *et al.*, 1972; Koike and Tsuda, 1979, 1980; Koike and Matsumoto, 1985). This simple idea met several technical difficulties. (1) How could sufficient amounts of radioactive markers be applied into the soma of a single neuron? (2) How could we know whether all the materials injected were utilized for the metabolism in the neuron or not? (3) How could we eliminate inherent artifacts for this method such as leakage from the cell and reuptake of the leakage into nearby cells as well as intracellular diffusion of the injected substance. Now most of these difficulties have been overcome except for an inherent problem of time-consuming analyses resulting from single injections.

2. THE METHOD OF INTRACELLULAR INJECTION

A double-barreled Pyrex® capillary was used for intracellular injection (Fig. 1). One barrel was filled with an electrolyte (usually 3 M KCl) for intracellular recording from the target neuron. The other barrel was filled with the injecting solution. The tip of the Pyrex® capillary was gently broken to make its outer diameter about 1 μm and the solution

HIROYUKI KOIKE, YOSHITOMO UMITSU, and HIROKO MATSUMOTO • Department of Neurophysiology, Tokyo Metropolitan Institute for Neurosciences, Fuchu City, Tokyo 183, Japan.

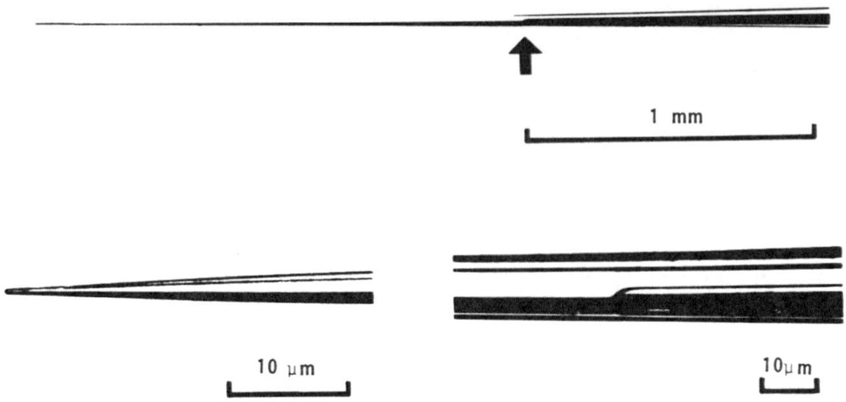

FIGURE 1. Photographs of a double-barreled capillary. Injecting solution was filled from the tip to the position indicated by the upward arrow (Koike and Tsuda, 1979).

was introduced from the tip by capillary action aided by a weak vacuum. The filling procedure was monitored under a binocular microscope, and the filled volume was usually of the order of nanoliters (Fig. 1). The capillary was used to penetrate a target neuron soma, and after the neuron was identified, the solution was injected into the cell body by air pressure. Usually the injected volume corresponded to 10–50% of the cell volume, and the injection caused a significant cell volume increase and membrane expansion. To avoid damage to the cell by the injection, the membrane potential was monitored throughout the course of injection and was used as an index for activating a pressure-release valve in order to control the injection (Fig. 2).

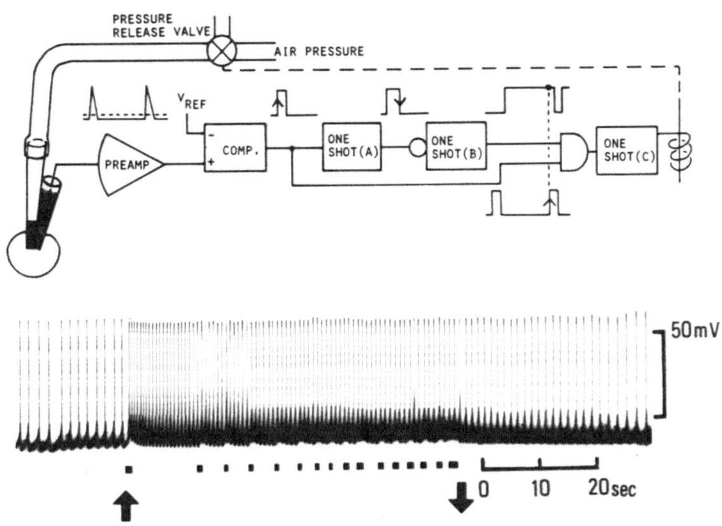

FIGURE 2. A circuit diagram for controlled-pressure injection. As soon as the membrane depolarizes or spike discharges are produced faster than a predetermined rate, the applied pressure is released. Lower record indicates a membrane potential change during the injection that began at the upward arrow and terminated at the downward arrow. The periods when pressure was actually applied are indicated by bars (Koike, 1979).

3. THE PREPARATION FOR THE INJECTION

A giant cholinergic neuron (R2) in the abdominal ganglion of *Aplysia kurodai* was used for injection. The neuron was easily identified by its location and by antidromic activation of the cell by electrical stimulation of either the right connective or the branchial nerve. Its large size enabled intracellular injection to be done relatively easily. However, by using a controlled-pressure injection device as mentioned above, we also succeeded in injecting sufficient amounts of various substances into smaller neurons: abdominal neurons of lobsters, stretch receptor cells of crayfish (Koike and Tsuda, 1979), and photoreceptor cells of barnacles (Koike and Tsuda, 1980).

4. THE RADIOACTIVE SUBSTANCES INJECTED

Radioactive substances of high specific activity were selected for injection. They were dissolved in distilled water and concentrated further under filtered nitrogen gas. Those used for the present report were [^{14}C]acetylcholine (ACh) (labeled in the acetyl moiety) and [^{3}H]leucine. Acetylcholine was a transmitter substance of the injected neuron, and leucine was a precursor of a protein known to be transported axonally at a fast rate. The latter served as a control for axonal transport of ACh.

5. THE ANALYSIS OF RADIOACTIVE SUBSTANCES ALONG THE AXON

Following injection of a mixture of [^{14}C]ACh and [^{3}H]leucine into a neuron, the nervous tissues were cultured for several hours, then rapidly frozen under powdered dry ice. The frozen tissues were cut (usually) at 1-mm lengths along the neuron's axon. Each cut fragment of tissue was separately homogenized in 5% trichloroacetic acid (TCA) solution, kept in a refrigerator overnight, and then centrifuged. The radioactivity derived from the centrifuged pellet indicated [^{3}H]protein and bound [^{14}C]ACh in the tissue, whereas that from the supernatant was free [^{3}H]leucine and soluble [^{14}C]ACh. The radioactivities were counted using a liquid scintillation counter. Because of very low levels of radioactivity just above the background, each scintillation sample was counted at least five times for 5 min each time. The resulting counts were converted to disintegration numbers.

6. THE INTRACELLULAR SYNTHESIS OF RADIOACTIVE PROTEIN

The cellular synthesis of protein from injected [^{3}H]leucine was detected by the appearance of ^{3}H radioactivity in the TCA precipitate from a segment that contained the cell body of the injected neuron. The protein radioactivity was detected as early as several minutes after injection and increased to about 50% of the total radioactivity recovered from the cell body segment within an hour. It became almost constant thereafter.

Radioactivity from ^{14}C was also detected from the TCA precipitate of the cell body and increased with time to about 40% after 5 hr, slower than the time course of the formation of [^{3}H]protein. This fraction of ACh would correspond to a bound form of ACh, which was expected to be present in synaptic vesicles. The remaining ACh recovered from the TCA supernatant provided an indication of the soluble ACh. No other ^{14}C chemical compounds formed from the injected ACh (labeled in the acetyl moiety) were detected by paper electrophoresis (Koike and Nagata, 1979) or by gel filtration analysis (not shown). Instead, the radioactive ACh was recovered from the synaptic vesicle fraction by analysis of sucrose density gradient ultracentrifugation (Koike, 1984). The exogenous ACh was likely to be loaded into synaptic vesicles or some equivalent structures in the cholinergic neuron, and amino acid was converted to protein within the cell.

7. THE INTRAAXONAL MIGRATION OF RADIOACTIVITIES

With time, the radioactivities migrated into the nervous tissue in which the axons of the injected neuron were running. Examples of the distributions of radioactivities are shown in Figs. 3–6. Following 5 hr of culturing after the injection (Fig. 3), the radioactivities were detected as far as 12–13 mm from the cell body. The protein fraction of radioactivity (TCA precipitate), indicated by open circles in the upper part of the figure, showed a plateaulike distribution, whereas the free leucine (TCA soluble fraction) simply showed a decreasing tendency away from the cell body. In the lower part of the figure, ^{14}C radioactivity of soluble ACh (TCA soluble) showed a decreasing tendency, whereas the bound ACh (TCA precipitate) showed a small hump or peak near the cell body. The profiles of proteins, free leucine, soluble ACh, and bound ACh were more clearly differentiated from each other by longer culturing following injection (Fig. 4). Proteins moved more than 30 mm with a broad plateau, and free leucine showed a decrease again. Both forms of ACh showed a hump or shoulder, but they were less distant from the cell than the proteins.

8. THE INTRAAXONAL DIFFUSION OF FREE LEUCINE AND SOLUBLE ACh

The appearance of free leucine along the axon of the injected neuron was explained by intraaxonal diffusion (Koike and Matsumoto, 1985). The thick curve in the upper figure indicates the theoretical diffusion profile of a substance whose diffusion constant in the axoplasm is 6.0×10^{-6} cm^{2}/sec. The profiles of free leucine reasonably coincided with the theoretical curve (Figs. 3 and 4), whereas those of soluble ACh overscaled the diffusion curve of 5.7×10^{-6} cm^{2}/sec. These diffusion constants were derived from values obtained experimentally in the same axoplasm (Koike and Nagata, 1979). The passive characteristic of the migration of free leucine was demonstrated when the nervous preparation was cultured in a medium that contained an antimitotic drug, colchicine, at 10 mM (Fig. 5). At this extraordinarily high concentration, colchicine blocked the fast axonal transport of protein and ACh (see also Koike and Nagata, 1979). The resulting appearance of radioactivities from the TCA-soluble fraction under colchicine agreed well with the theoretical diffusion profiles both in free leucine and soluble ACh (Fig. 5). This indicates that radioactive protein, bound ACh, and part of the soluble ACh were moved along the axon by an active process of axonal transport.

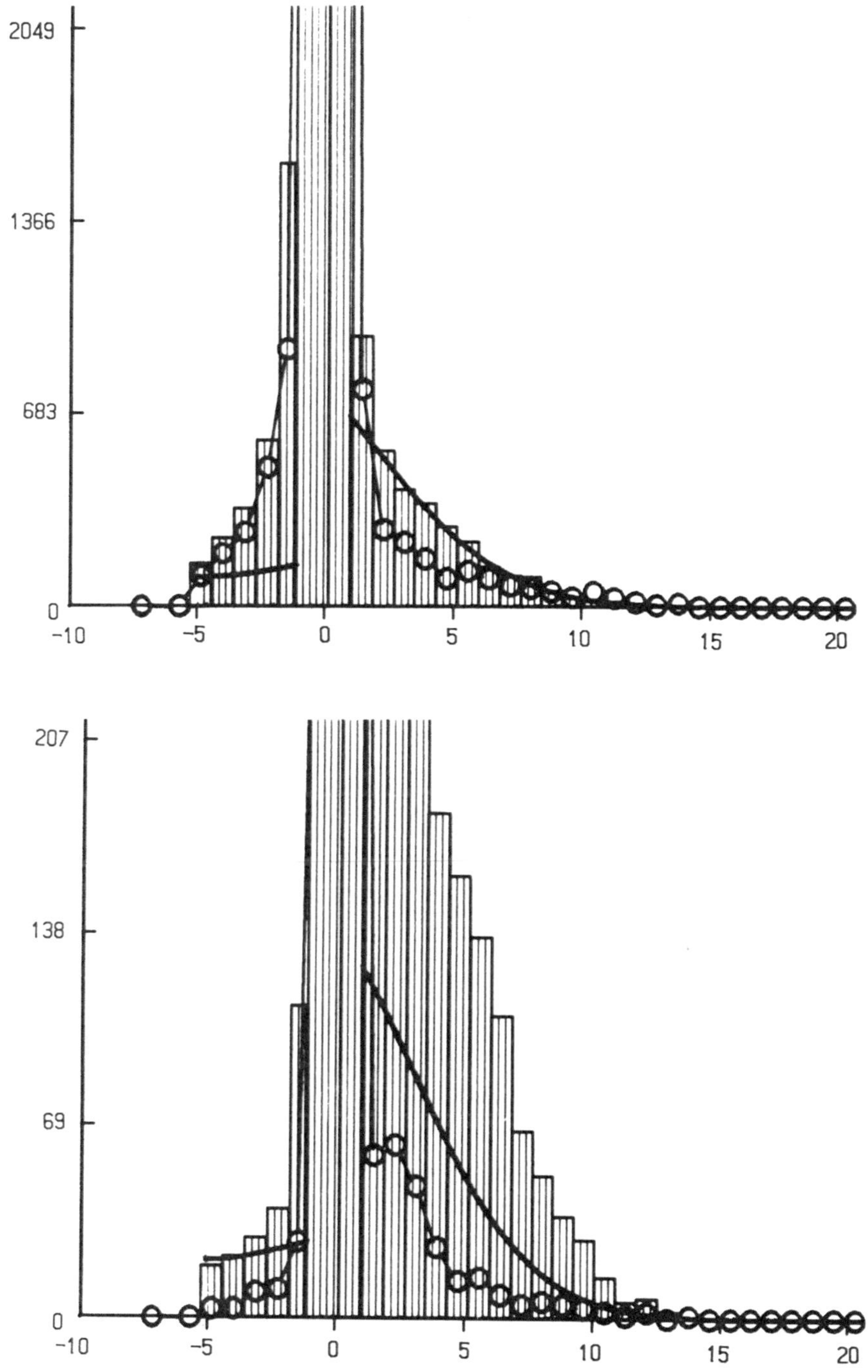

FIGURE 3. Following 5 hr of culturing after injection of [^{3}H]leucine and [^{14}C]acetylcholine (Ach), the profiles of radioactivities were plotted along the nerves in which axons of the injected neuron were running (main axon to the right and minor axon to the left from the cell body at 0). The TCA precipitates are shown by open circles, and TCA-soluble activities by shaded bars. Solid curves indicate the theoretical diffusion profiles of leucine and ACh in the axoplasm. Upper figure from ^{3}H radioactivity and lower from ^{14}C.

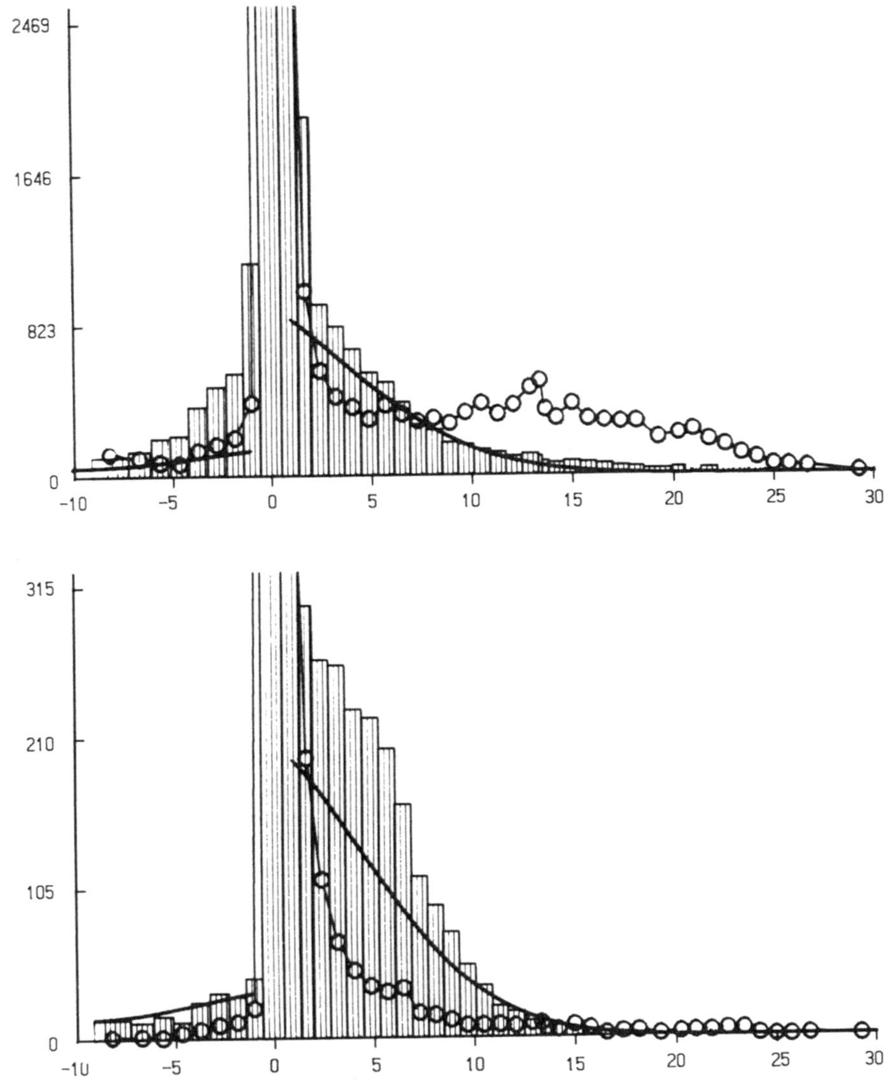

FIGURE 4. Ten hours of culturing after injection.

9. DIFFERENCE IN INTRAAXONAL TRANSPORT BETWEEN PROTEINS AND ACETYLCHOLINE

9.1. Transport Velocities

The farthest detectable position of the radioactive protein (moving front) against the culturing period following injection gave an axonal transport velocity of about 2.5 mm/hr on average. Similarly, bound ACh traveled about 1.8 mm/hr. These values were obtained at 25°C and are in the range of fast axonal transport. Thus, between protein and ACh there was a small but significant difference in the rate of transport.

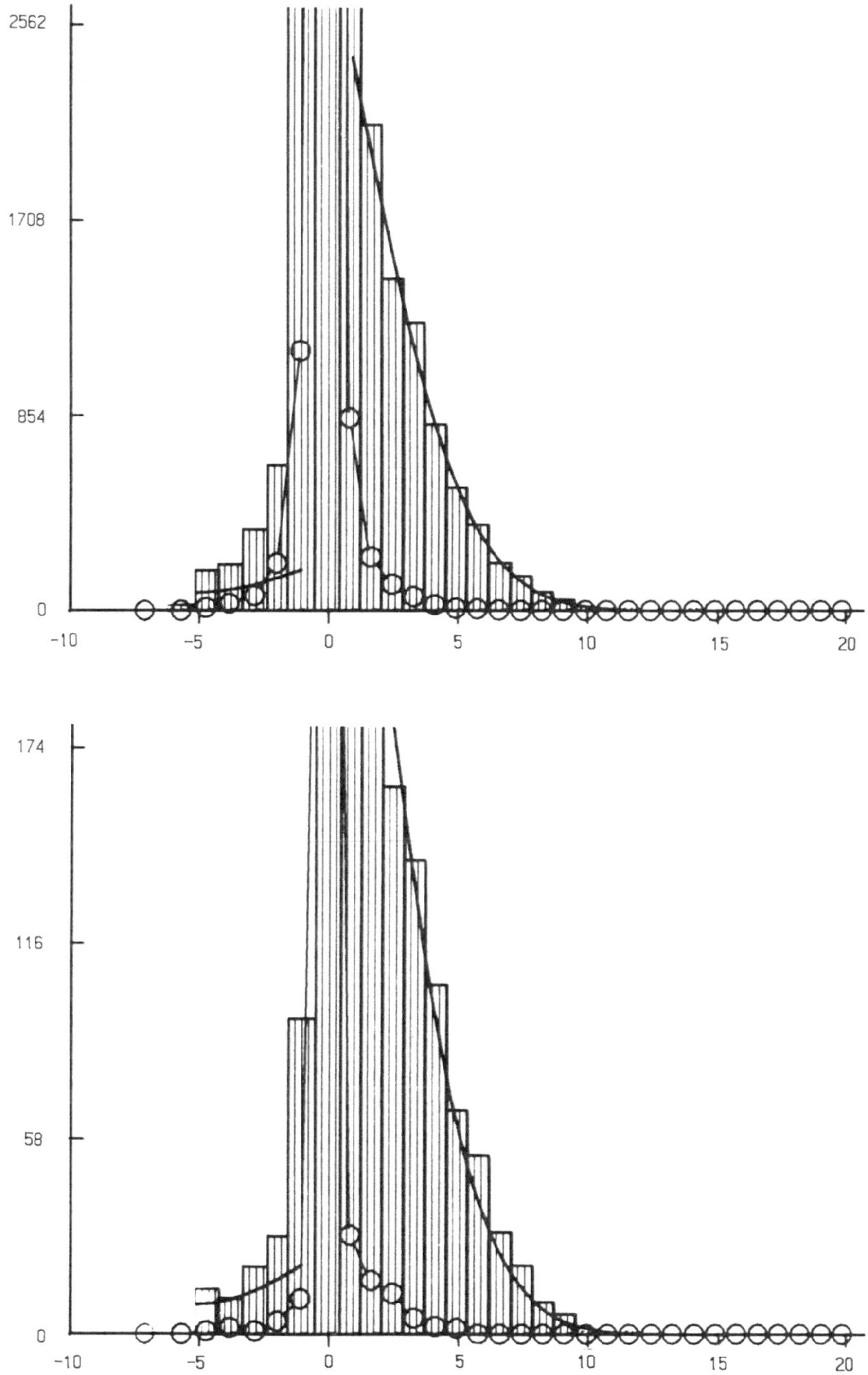

FIGURE 5. The preparation was cultured in 10 mM colchicine for 1 hr prior to injection and for an additional 3 hr; then it was frozen. Diffusion curves fit the profiles of soluble components.

9.2. Pharmacological Effects

Colchicine at 10 mM concentration blocked axonal transport of protein as well as ACh. However, at moderate concentration of 1 mM, only the movement of protein was blocked. That of ACh was not significantly affected (Fig. 6) (Koike, 1983). When the nerve was perfused separately, and the cell body of the injected neuron and its vicinity were perfused by a standard medium while the remote nerve was perfused with the colchicine (1–5 mM)-containing medium, axonal transport of the protein was stopped at the site where colchicine was perfused, but ACh moved continuously into the colchicine perfusion area (Koike *et al.*, 1985). In contrast to the selective action of colchicine, local cooling (4–8°C) blocked movements of both substances simultaneously.

10. TRANSMITTER SPECIFIC MOVEMENT

Axonal transport of ACh was also examined using a noncholinergic neuron (R1) that sends its axon into the same right connective (Fig. 7). Except for a small movement of bound ACh, injected ACh was not transported axonally as in the cholinergic neuron. The diffusion of soluble ACh was noted by the coincidence of the result with the theoretical diffusion curve. The fast axonal transport of ACh observed in the cholinergic neurons of *Aplysia* indicates that it represents a transmitter-specific system.

Axonal transport of neuron-specific transmitter other than ACh has been reported earlier (Dahlström, 1968; Banks *et al.*, 1971; Goldman *et al.*, 1980; Price and McAdoo, 1981; McLean and Lewis, 1984; and so on). It is generally accepted that the synthesis site for ACh is solely in the nerve terminal because of the existence of the synthesizing enzyme choline acetyltransferase and availability of acetyl-CoA and choline in the axon terminal. However, high concentrations of endogenous ACh have also been detected from the somas of cholinergic neurons (McCaman *et al.*, 1973). The cellular supply of the particular transmitter substance, ACh, was not likely to be an experimental artifact caused by exogenous application of ACh. When we injected radioactive choline instead of ACh, the cell synthesized radioactive ACh in the cell body and successively sent ACh down the axon (Koike *et al.*, 1972). Axonally transported ACh was released from the axon terminal in response to the neuron's activity, but not diffused soluble ACh (Koike *et al.*, 1974).

Axonal transport of ACh was a little slower in velocity than that of membrane protein observed in the same axon, and it was less sensitive to application of the antimitotic drug colchicine. Recently, a video-enhanced, contrast-differential interference contrast microscopy method (AVEC DIC) has shown unambiguously that microtubules are the main cellular structure to support organelle axonal transport (Allen *et al.*, 1982), and two different soluble factors have been shown to function as a motor for orthodromic and retrograde organelle movements, respectively (Vale *et al.*, 1985a,b). These investigations also noted a difference in velocity among endogenous organelles and exogenous latex beads along a single microtubule. We observed a slower velocity of ACh axonal transport, but this may not have been caused by an exogenous origin of ACh, because after injection of radioactive choline, even the cellularly synthesized ACh moved with identical speed to that observed here.

Axonal transport of ACh was selectively less sensitive to colchicine. We observed some microtubules in the axon even after 5 hr of culturing in 5 mM colchicine (Koike

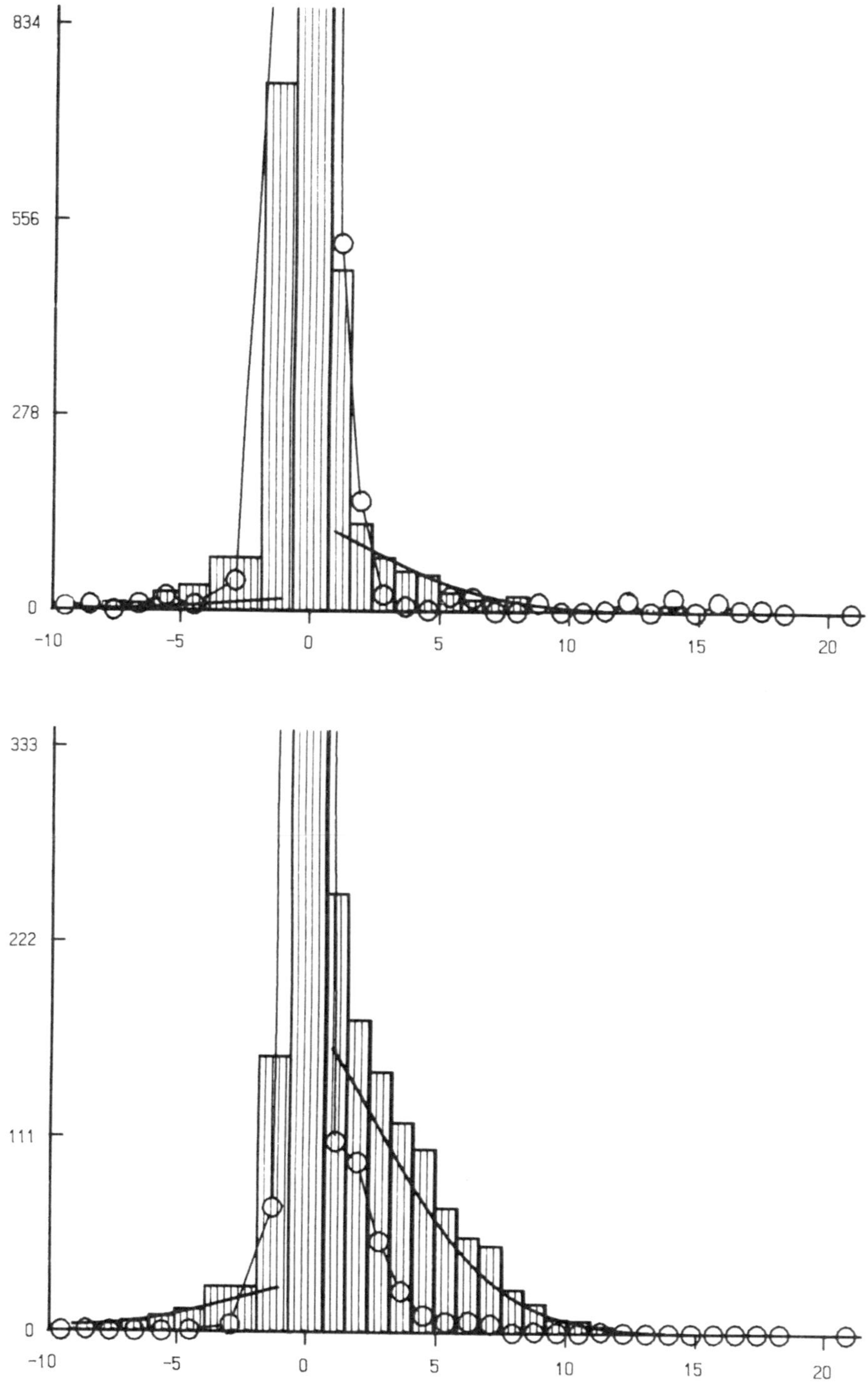

FIGURE 6. The preparation was cultured in 1 mM colchicine for 1 hr prior to injection and then for an additional 5 hr. The profile of ACh does not coincide with the theoretical curve.

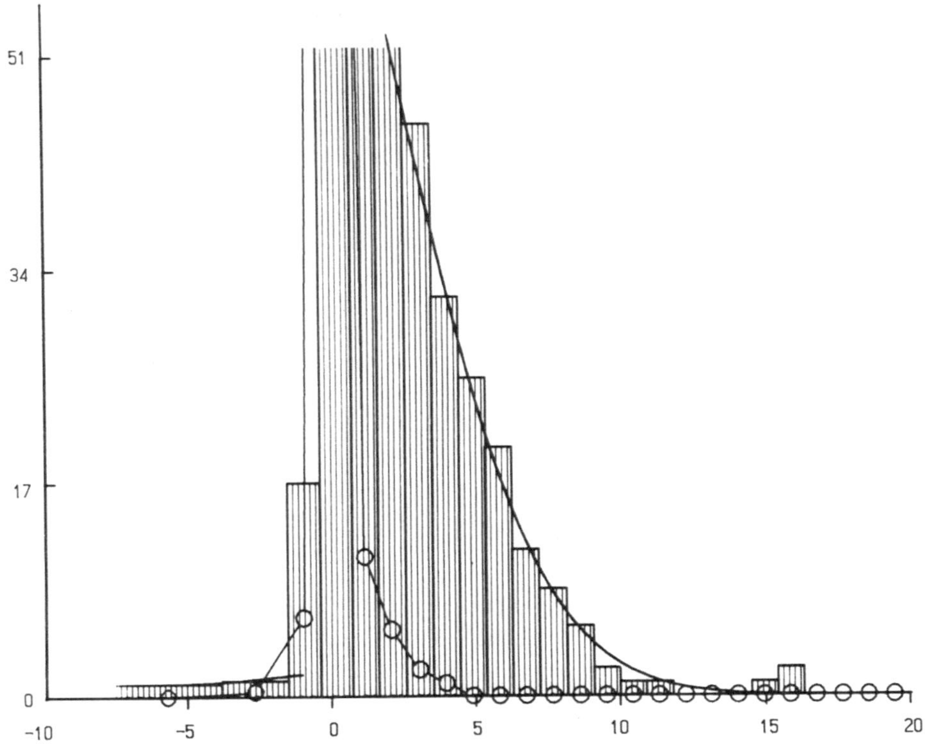

FIGURE 7. [^{14}C]Acetylcholine was injected into a noncholinergic neuron and cultured for 5 hr. The neuron has an axon to the right.

et al., 1985). Therefore, it is more likely that the microtubules that transport transmitter ACh and protein are not identical, but rather, a particular group of colchicine-resistant microtubules prefer ACh as a transport partner. Alternatively, the environments surrounding each group of microtubules are different and cause preference for ACh and resistance to colchicine at the same time. The latter possibility was suggested by the finding that organelle transport velocities were different between cellular domains of microtubule-rich regions and those of actin-rich, few-microtubule regions, perhaps because of fewer possibilities for attaching microtubules and organelles and also because of the higher resistance against movement (Bridgman *et al.*, 1986). Acetylcholine supplied from the cell body might not be a major source for the transmitter released at the axon terminal but might have some trophic functions for synaptic plasticity.

REFERENCES

Allen, R. D., Metuzals, J., Tasaki, I., Brady, S. T., and Gilbert, S. P., 1982, Fast axonal transport in squid giant axon, *Science* **218:**1127–1129.

Banks, P., Mayor, D., Mitchell, M., and Tomlinson, D., 1971, Studies on the translocation of noradrenaline-containing vesicles in post ganglionic sympathetic neurones *in vitro*. Inhibition of movement by colchicine and vinblastine and evidence for the involvement of axonal microtubules, *J. Physiol.* (*Lond.*) **216:**625–639.

Bridgman, P. C., Kacher, B., and Reese, T. S., 1986, The structure of cytoplasm in directly frozen cultured cells. II. Cytoplasmic domains associated with organelle movements, *J. Cell Biol.* **102:**1510–1521.

Dahlström, A., 1968, Observations on the accumulation of noradrenaline in the proximal and distal parts of peripheral adrenergic nerves after compression, *J. Anat.* **99**:677–689.

Goldman, J. E., Kim, K. S., and Schwartz, J. H., 1980, Axonal transport of [^{3}H]serotonin in an identified neuron of *Aplysia californica*, *J. Cell Biol.* **70**:304–318.

Koike, H., 1979, A new device for controlled intracellular injection using pneumatic pressure, *Integr. Control Funct. Brain* **2**:39–41.

Koike, H., 1983, Transmitter specific axonal transport of acetylcholine in a neuron of *Aplysia*: Colchicine resistant and cytochalasin sensitive, *Proc. Int. Union Physiol. Soc.* **15**:297.

Koike, H., 1984, Evidence of axonal transport of vesicular acetylcholine in a cholinergic neuron of *Aplysia*, *Neurosci. Lett. Suppl.* **17**:S3.

Koike, H., and Matsumoto, H., 1985, Fast axonal transport of membrane protein and intra-axonal diffusion of free leucine in a neuron of *Aplysia*, *Neurosci. Res.* **2**:281–285.

Koike, H., and Nagata, Y., 1979, Intra-axonal diffusion of [^{3}H]acetylcholine and [^{3}H]gamma-aminobutyric acid in a neurone of *Aplysia*, *J. Physiol.* (*Lond.*) **295**:397–417.

Koike, H., and Tsuda, K., 1979, Intracellular acetylcholine synthesis and GABA synthesis in some crustacean neurons, in: *Neurobiology of Chemical Transmission* (M. Otsuka and Z. Hall, eds.), John Wiley & Sons, New York, pp. 65–76.

Koike, H., and Tsuda, K., 1980, Cellular synthesis and axonal transport of gamma-aminobutyric acid in a photoreceptor cell of the barnacle, *J. Physiol.* (*Lond.*) **305**:125–138.

Koike, H., Eisenstadt, M., and Schwartz, J. H., 1972, Axonal transport of newly synthesized acetylcholine in an identified neuron of *Aplysia*, *Brain Res.* **37**:152–159.

Koike, H., Kandel, E. R., and Schwartz, J. H., 1974, Synaptic release of radioactivity after intrasomatic injection of ^{3}H-choline into an identified cholinergic interneuron of *Aplysia californica*, *J. Neurophysiol.* **37**:815–827.

Koike, H., Umitsu, Y., and Matsumoto, H., 1985, Fast axonal transport of membrane protein and acetylcholine in an *Aplysia* cholinergic neuron: Its modification by local colchicine perfusion or lowering temperature, *J. Physiol. Soc. Jpn.* **47**:375.

McCaman, R. E., Weinreich, D., and Borys, H., 1973, Endogenous levels of acetylcholine and choline in individual neurons of *Aplysia*, *J. Neurochem.* **21**:473–476.

McLean, D. B., and Lewis, S. F., 1984, Axoplasmic transport of somatostatin and substance P in the vagus nerve of the rat, guinea pig, and cat, *Brain Res.* **307**:135–145.

Price, C. H., and McAdoo, D. J., 1981, Localization of axonally transported [^{3}H]glycine in vesicles of identified neurons, *Brain Res.* **219**:307–315.

Vale, R. D., Reese, T. S., and Sheetz, M. P., 1985a, Identification of a novel force-generating protein, kinesin, involved in microtubule-based motility, *Cell* **42**:39–50.

Vale, R. D., Schnapp, B. J., Mitchison, T., Steuer, E., Reese, T. S., and Sheetz, M. P., 1985b, Different axoplasmic proteins generate movement in opposite directions along microtubules *in vitro*, *Cell* **43**:623–632.

45

Modulation of the Gaba- and Pentobarbital-Gated Cl Current by Intracellular Calcium in Frog Sensory Neurons

TOORU MARUYAMA, YUTAKA OOMURA, JUNICHI SADOSHIMA, NAOFUMI TOKUTOMI, JAN BEHRENDS, and NORIO AKAIKE

1. INTRODUCTION

Voltage-dependent Ca^{2+} currents are well known to play an important role in the regulation of a vast number of neuronal functions. The transient increase in intracellular free calcium resulting from Ca^{2+}-channel activation has been shown to trigger various ionic currents in the soma of a variety of nerve cells, i.e., calcium-dependent potassium, chloride, or unselective anion conductances. In addition to this well-established interaction with ionic channels, Ca^{2+} has been demonstrated to modulate desensitization kinetics of ACh receptors in the neuromuscular end plate (Miledi, 1980) and to inhibit a nicotinic chloride response in *Aplysia* neurons (Chemeris *et al.*, 1982) as well as the nicotinic cation response in frog sympathetic ganglion cells (Morita *et al.*, 1979). These findings point to a possible involvement of Ca^{2+} ions not only in the regulation of pure ionic channels but also in the modulation of transmitter-operated receptor–ionophore complexes.

In the present experiments, we examined possible interactions between voltage-dependent Ca^{2+} currents (resulting in an increase of intracellular free calcium) and Cl^- currents elicited by activating GABA receptors in the internally perfused frog sensory neuron.

TOORU MARUYAMA, JUNICHI SADOSHIMA, NAOFUMI TOKUTOMI, JAN BEHRENDS, and NORIO AKAIKE • Department of Physiology, Faculty of Medicine, Kyushu University, Fukuoka 812, Japan. YUTAKA OOMURA • Department of Physiology, Faculty of Medicine, Kyushu University, Fukuoka 812, Japan; and Department of Biological Control Systems, National Institute for Physiological Sciences, Okazaki 444, Japan.

2. METHODS

2.1. Preparation

Dorsal root ganglia dissected from decapitated bullfrogs (*Rana catesbeiana*) were used in all experiments. The thick connective tissue surrounding the ganglion was carefully stripped off using a pair of microforceps, and the capsules enveloping the ganglion masses were digested in 10 ml normal Ringer solution containing 0.3% collagenase and 0.05% trypsin at pH 7.4 for about 15 min at 37°C. During enzyme treatment, the preparation was gently shaken by bubbling 99.9% O_2. Thereafter, single cells were isolated from the ganglion mass with finely polished pins under binocular observation and left overnight in a culture medium consisting of equal parts of Ringer solution and isotonic Eagle's MEM (Nissui, Japan) at room temperature (about 22°C).

2.2. Solution

The ionic compositions of the standard solutions were (mM): internal, CsCl 95, Cs-aspartate 10, TEA-Cl 25, EGTA 0.5 [Ca^{2+} buffer ($[Ca^{2+}]_i = 3 \times 10^{-8}$ M)]; external, Tris-Cl 89, CsCl 2, $CaCl_2$ 2, TEA-Cl 25, glucose 5. The pH of all solutions was adjusted to 7.4 with Tris base or HEPES.

2.3. Suction Pipette

A suction pipette technique was used for voltage clamp and internal perfusion (Hattori *et al.*, 1984; Ishizuka *et al.*, 1984). A part of an individual neuron was aspirated by applying a negative pressure of about 3 cm Hg through the suction pipette. The aspirated membrane either broke spontaneously or was ruptured by application of 5- to 20-nA square-wave pulses of depolarizing current (10–50 msec). Thereafter, the neuron was internally perfused at a constant flow rate of 1 ml/min. Adequacy of internal perfusion was evaluated by determining how close the reversal potential for GABA-induced Cl^- responses was to the Cl^- equilibrium potential, which is at +4 mV as calculated from the Nernst equation based on the Cl^- activities in the internal and external solutions.

2.4. Electrical Measurements

The membrane potential was measured through an Ag/AgCl wire in a Ringer–agar plug mounted on the suction pipette holder. The reference electrode was also an Ag/AgCl wire in a Ringer–agar plug. Resistance between the suction pipette filled with standard internal solution and the reference electrode was 200 to 300 kΩ. Both electrodes were connected to a voltage-clamp circuit, and membrane potential was controlled by a single-electrode voltage-clamp system switching at a frequency of 10 kHz and passing current for 36% of the cycle (Ishizuka *et al.*, 1984). In this system, the suction electrode could carry time-averaged currents exceeding 100 nA at a switching frequency of 10 kHz. Both current and voltage were monitored on a digital storage oscilloscope (National, type VP-57300A) and stored on a magnetic data recorder (Teac, type MR-30).

2.5. External Solution Application Method

The "concentration-clamp" method was used for extremely rapid application of an external solution within 2 to 4 msec, with or without an agonist (Akaike *et al.*, 1986)

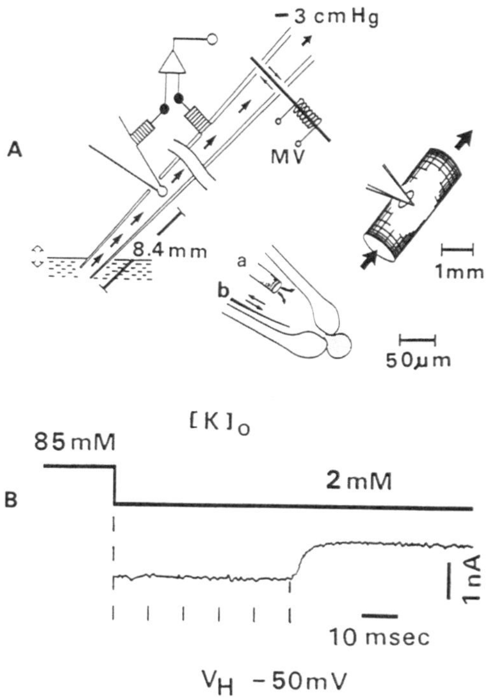

FIGURE 1. A: Schematic illustration of the "concentration-clamp" technique, which combines internal perfusion via a suction pipette with a method for rapid exchange of the external solution (Akaike *et al.*, 1986). Left: Cell-attached tip of the suction electrode in the solution-exchange tube, which is exposed to external solution at the lower end. Diagram illustrating the positions of voltage-recording and reference Ag/AgCl wires. Lower middle: Schematic illustration of suction pipette tip with the single neuron attached to it: a, internal solution inlet; b, puncture wire, used mainly to clean the electrode tip. Left: Illustration of the dimensions of pipette tip and solution-exchange tube. For details see Section 2.5. B: Measurement of the speed of solution exchange. A cell was clamped to −50 mV in an external solution containing 85 mM K^+. On exchange of the solution to 2 mM K^+, an outward current could be observed with a latency of 50 msec. The rising phase of this outward current had a time constant of about 2 to 3 msec, indicating the speed of complete exchange of external solution around the cell.

(Fig. 1). The cell-attached tip of the suction pipette was inserted into a plastic tube through a circular hole approximately 500 μm in diameter. The lower end of this tube could be directly exposed to external solutions by moving up a stage on which drug-containing dishes were placed. Suction (−3 cm Hg) applied to the upper end of the tube was controlled by an electromagnetic valve driven by 24 V DC. The power supply was switched on for the desired duration by a stimulator (Nihon Koden, type SEN-7103). Results are expressed as mean ± S.E.M.

3. RESULTS AND DISCUSSION

3.1. Suppression of Voltage-Dependent Ca^{2+} Currents by Activation of GABA-Gated Cl^- Conductance

Suppression of a large, slowly inactivating Ca^{2+} inward current (I_{Ca}) by application of 3×10^{-6} M GABA is shown in Fig. 2A. The cell was voltage clamped to a holding

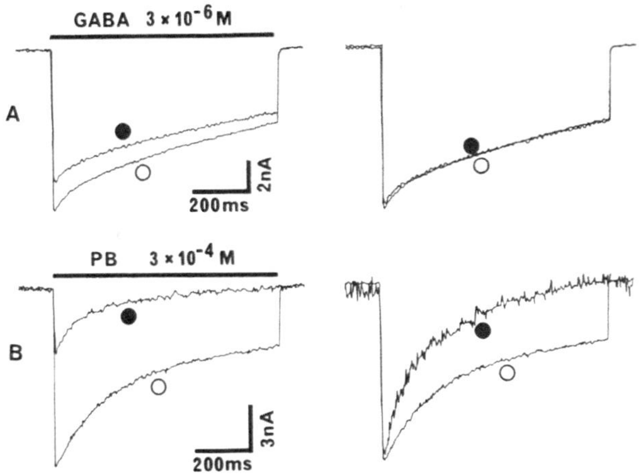

FIGURE 2. Suppression of calcium currents (I_{Ca}) by GABA and pentobarbital (PB). Parts A and B (left) show control I_{Ca} (open circles) and reduced I_{Ca} elicited after application of the respective agonist. Currents were evoked by application of 800-msec depolarizing voltage command pulses from the holding potential of −50 to −20 mV. They were corrected for leakage currents by subtraction of the response to a hyperpolarizing pulse of the same amplitude. On the right-hand side, peak-normalized current traces are shown to illustrate the kinetic differences between GABA and PB action on I_{Ca}. For discussion see text.

potential of −50 mV, and I_{Ca} was elicited by +30-mV command pulses of 800 msec duration. GABA effectively reduced I_{Ca} amplitude without affecting kinetics of either activation or inactivation. Figure 2B shows a similar action of pentobarbital (PB) (3×10^{-4} M), at which concentration PB also induced I_{Cl}. Note, however, that in the case of PB, the I_{Ca} inactivation process is greatly enhanced, indicating a considerably different mode of action. See the normalized current traces on the right-hand side.

These findings are in accordance with those of Deisz and Lux (1985), who found that GABA depressed I_{Ca} in a similar manner in chick sensory neurons in culture. It should be noted, however, that these authors determine their GABA receptor to be of the $GABA_B$ type, whereas in our case, sensitivity to bicuculline (not shown) clearly indicates that the GABA receptor in frog sensory neurons is pharmacologically a $GABA_A$ receptor. Also, unlike Deisz and Lux (1985), we were not able to mimic GABA effects on I_{Ca} with baclofen. Regardless of the receptor type, however, our most recent experiments have shown that bicuculline does not block this GABA effect on I_{Ca}. Further, I_{Ca} was also affected by GABA in the presence of picrotoxin. These observations suggest that this particular GABA action is not mediated through either the GABA receptor or the Cl^- channels but rather is a conventional channel-blocking action from the outside of the cell membrane. This could also explain Deisz and Lux's findings, rendering differences in receptor type unimportant. However, the lack of effect of baclofen in our preparation then becomes a pivotal issue.

3.2. Inhibition of GABA-Receptor-Mediated Cl^- Currents by Elevated Intracellular Calcium

Figure 3 (inset) shows a typical response to rapid application of 10^{-5} M GABA. The cell was voltage clamped to a holding potential of −50 mV. When an I_{Ca} was elicited

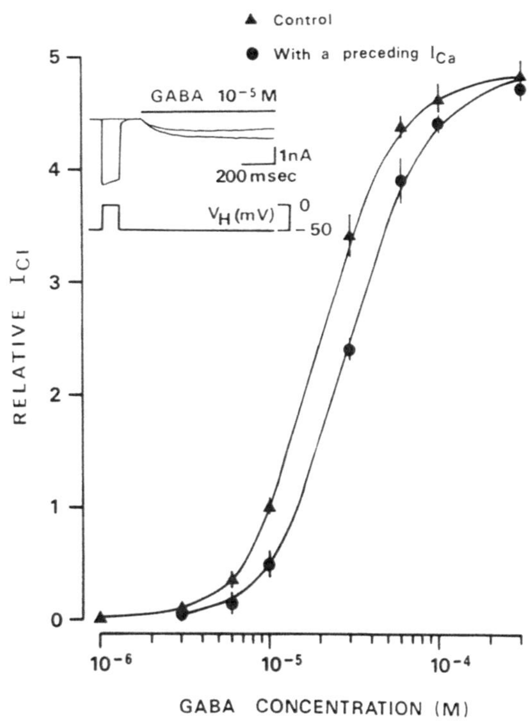

FIGURE 3. Effect of a preceding I_{Ca} on the dose–response curve for GABA. Ordinate: Cl^- current (I_{Cl}) amplitude relative to that of a control response to 10^{-5} M GABA. Abscissa: GABA concentrations applied (logarithmic). Data points are means ± SEM of six experiments. The inset shows an original record obtained from a cell clamped to −50 mV. When GABA application was preceded by I_{Ca} elicited by a voltage command step to 0 mV for 100 msec (see voltage trace), the control response to 10^{-5} M GABA was markedly reduced. The cell was perfused with 0.5 mM EGTA.

by a 100-msec depolarizing voltage command pulse to 0 mV prior to the application of GABA, the GABA-gated I_{Cl} was markedly reduced. This reduction of GABA responses could be enhanced by prolonging the duration of the command pulse (not shown).

A number of observations suggest that in this suppression of GABA-gated I_{Cl} by a preceding I_{Ca}, a transient elevation of intracellular free Ca^{2+} ($[Ca^{2+}]_i$) plays the central role. In cells perfused with 0.5 mM EGTA–Ca buffer ($[Ca^{2+}]_i = 3 \times 10^{-8}$ M), an I_{Ca} elicited by a 100-msec depolarizing command pulse of +50 mV reduced responses to 10^{-5} M GABA by 50%, whereas in cells perfused with 2.5 mM EGTA alone, this reduction was only by 25% (not shown). Further evidence arises from the observation that the degree of I_{Cl} inhibition achieved by an I_{Ca} of constant amplitude and duration also depended on the delay between Ca^{2+} pulse and rapid GABA-gated I_{Cl} application, increasing with shorter delays. With very short delays (10 msec), an apparent early desensitization of the Cl^- current could be observed, which can be accounted for by a time-dependent activation process of the mechanism that mediates the Ca^{2+}-dependent depression of GABA-evoked Cl^- current (not shown).

Therefore, to quantify the relationship between the amount of voltage-dependent Ca^{2+} influx and the reduction in GABA-gated I_{Cl}, we elicited I_{Ca}s of various amplitudes during a steady-state response to rapid application of GABA-receptor agonists (Fig. 4),

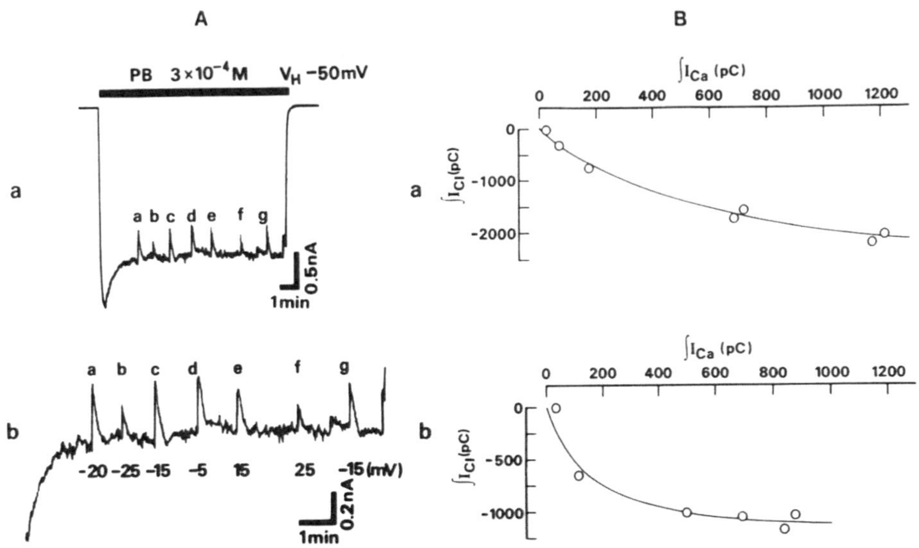

FIGURE 4. A: Steady-state I_{Cl} to rapid application of 3 × 10^{-4} M PB, shown in its entirety in a and on a faster time scale in b. The I_{Ca} was elicited several times during the steady-state responses by 300-msec command pulses to different potentials, which are indicated for each response to I_{Ca}. Reduction in I_{Cl} is represented by an apparent outward current hump. B: Graphs a and b display the integral of the reduction in steady-state I_{Cl} (ordinate) plotted against the integral of the I_{Ca} evoked during the steady-state response (abscissa). Graph a uses the data of the recording shown in A (a and b), whereas graph b is from a like recording during a steady-state response to 3 × 10^{-6} M GABA, the original record of which is not shown.

thus avoiding the kinetic distortions brought about by the mixing of I_{Cl} kinetics with the time course of the transient increase of $[Ca^{2+}]_i$, which was a major drawback of the former protocol (see Fig. 3).

Figure 4A shows a steady-state response to 3 × 10^{-4} M PB in a cell voltage clamped to −50 mV. During the steady-state plateau, I_{Ca} was elicited several times (a–g) by command pulses of constant duration but varying amplitudes. It is clear from the records that the I_{Ca}-induced inhibition of the steady-state I_{Cl} reaches a maximum following the command pulse to −15 mV (c) and then decreases on further increase of command voltage amplitudes. This is because at voltages positive to *ca.* −10 mV, the total amount of Ca^{2+} influx reaches a maximum. Thus, although the peak current amplitude still grows, premature inactivation of I_{Ca} yields an overall decrease in calcium flux into the cell. The finding that this decrease in current amount leads to a weaker GABA response suppression lends further support to the notion that an increase in $[Ca^{2+}]_i$ rather than the current itself is pivotal to the GABA-response-blocking action of I_{Ca}.

Thus, the graphs shown in Fig. 4B display the integral of the reduction in steady-state Cl^- current (I_{Cl}) plotted as a function of the integral of the Ca^{2+} influx through the membrane (I_{Ca}) for, respectively, the steady-state response to PB shown above (a) and a like response to 3 × 10^{-6} M GABA, of which the original record is not shown. Both graphs illustrate a clear dependence of GABA response block on I_{Ca} amount.

To gain information about the possible mechanism by which $[Ca^{2+}]_i$ suppresses the GABA-activated I_{Cl}, we studied its effect on the GABA dose–response curve. The result is shown in Fig. 3. The suppression of GABA-induced I_{Cl} decreased with increasing GABA concentrations; i.e., the dose-response curve was shifted to the right without a

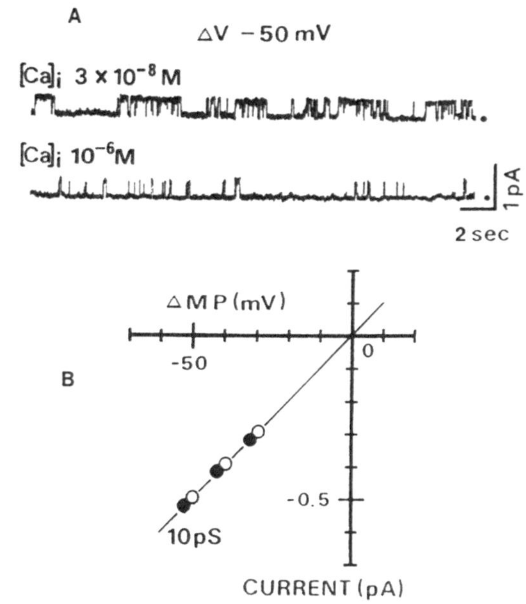

FIGURE 5. A: Patch-clamp recording obtained using the inside-out configuration. External and internal solutions were identical to those used in whole-cell clamp experiments. The patch-clamp pipette contained 3×10^{-6} M GABA; $[Ca^{2+}]_i$ was changed from 3×10^{-8} M (upper trace) to 10^{-6} M. Driving force (ΔV) for I_{Cl} was -50 mV. B: Single-channel current amplitude plotted as a function of driving force under control conditions and in the presence of elevated $[Ca^{2+}]_i$ (open and closed circles, respectively). The values remain unchanged and fit a straight line that corresponds to a unitary conductance of 10 pS.

change in the maximum GABA response. This competitive mode of inhibition suggests a Ca^{2+}-mediated decrease in receptor affinity for GABA rather than a block of the Cl^- channel. However, our recent patch-clamp data, obtained using the inside-out configuration, clearly show a decrease in open time and a prolongation of closed time intervals of the GABA-gated single-channel current as $[Ca^{2+}]_i$ increased (Fig. 5A), while unitary conductance remained unchanged (Fig. 5B). This indicates the possibility of direct interaction of $[Ca^{2+}]_i$ with the gating mechanisms of the Cl^- channel. Further investigation must clarify this issue by close kinetic analysis.

ACKNOWLEDGMENTS. This research was supported by a Grant-in-Aid to N. Akaike from the Japanese Ministry of Education, Science, and Culture, No. 6148109.

REFERENCES

Akaike, N., Inoue, M., and Krishtal, O. A., 1986, "Concentration-clamp" study of γ-aminobutyric-acid-induced chloride current kinetics in frog sensory neurones, *J. Physiol. (Lond.)* **379:**171–185.

Chemeris, N. K., Kazachenko, V. N., Kislov, A. N., and Kurchikov, A. L., 1982, Inhibition of acetylcholine responses by intracellular calcium in *Lymnaea stagnalis* neurones, *J. Physiol. (Lond.)* **323:**1–19.

Deisz, R. A., and Lux, H. D., 1985, γ-Aminobutyric acid-induced depression of calcium currents of chick sensory neurons, *Neurosci. Lett.* **56:**205–210.

Hattori, K., Akaike, N., Oomura, Y., and Kuraoka, S., 1984, Internal perfusion studies demonstrating GABA-induced chloride responses in frog primary afferent neurons, *Am. J. Physiol.* **246:**C259–265.

Ishizuka, S., Hattori, K., and Akaike, N., 1984, Separation of ionic currents in the somatic membrane of frog sensory neurons, *J. Membr. Biol.* **78:**19–28.
Miledi, R., 1980, Intracellular calcium and desensitization of acetylcholine receptors, *Proc. R. Soc. Lond.* **B209:**447–452.
Morita, K., Kato, E., and Kuba, K., 1979, A possible role of intracellular Ca^{2+} in the regulation of the ACh receptor–ion channel complex of the sympathetic ganglion cell, *Kurume Med. J.* **26:**371–376.

46

Contrasting Roles of a Brain-Specific Protein Kinase C Substrate

Has Protein F1 Evolved a New Function in CNS of Higher Vertebrates?

ROBERT B. NELSON and ARYEH ROUTTENBERG

1. INTRODUCTION

Increased phosphorylation of the neuronal membrane-bound protein F1 and translocation of its kinase, Ca^{2+}- and phospholipid-stimulated protein kinase C (PKC), have been related to long-term increases in adult synaptic efficacy in a number of reports from our laboratory (Routtenberg *et al.*, 1985; Lovinger *et al.*, 1985; Akers *et al.*, 1986).

Recently, in collaboration with several other laboratories, we have reported that protein F1 appears to be identical to three other proteins reported in the literature. The first of these, pp46, is the major phosphoprotein in a purified fraction of neuronal growth cones (Nelson *et al.*, 1986; Katz *et al.*, 1985). Growth cones are the terminal enlargements of extending neurites and occur only during nervous system development prior to synaptogenesis. Synthesis of pp46 is very high during nervous system development but drops dramatically in adult brain (Simkowitz *et al.*, 1986). The second protein, termed GAP43, shows dramatic increases in synthesis and fast axonal transport following optic nerve crush, prompting its designation as a "growth-associated protein" (Snipes *et al.*, 1986; Skene and Willard, 1981a–c). Increased synthesis of GAP43 following nerve crush occurs only in fiber pathways capable of regeneration, suggesting that the presence of this protein might be crucial in the regeneration process. Third, the B50 protein, which again appears to be identical to protein F1, is brain specific and enriched in presynaptic versus postsynaptic membranes (Gispen *et al.*, 1985a, 1986; Kristjansson *et al.*, 1982). Initial evidence suggesting a role for the B50 protein in phosphatidylinositol turnover has been presented (Gispen *et al.*, 1985b). We refer to this multiply designated protein as protein F1 in this chapter unless directly referring to findings from another laboratory.

The coidentification of protein F1 with proteins that appear to be important in development and regeneration and the implication of PKC in growth-related processes (as the primary receptor for tumor-promoting phorbol esters; Castagna *et al.*, 1982) have

ROBERT B. NELSON • Neuroscience Group, The DuPont Company, Wilmington, Delaware 19898.
ARYEH ROUTTENBERG • Cresap Neuroscience Laboratory, Northwestern University, Evanston, Illinois 60201.

led us to propose that protein F1 might participate in neural plasticity by mediating presynaptic terminal growth (Routtenberg, 1985). There are a number of dichotomies, however, between the roles described for protein F1 in development and regeneration and its role in adult synaptic plasticity. In the present review, we further compare these roles and, on the basis of these comparisons, propose that protein F1 may have acquired the function of regulating synaptic plasticity in mature CNS of higher vertebrates, replacing in that neural system an evolutionarily older function that involves regeneration of neuronal processes.

2. PROTEIN F1 AND ADULT NEURAL PLASTICITY

A high-frequency train of stimulation applied to any of several fiber pathways in the hippocampus or other specific fiber pathways in brain results in long-term potentiation (LTP), a prolonged change in transsynaptic communication (measured physiologically as an enhanced population spike or population EPSP in the postsynaptic cells; Bliss and Lomo, 1973). Several properties of LTP—including the briefness of stimulation necessary to induce it, its persistence (measured for months in chronically recorded animals; Douglas and Goddard, 1975), its similarities to neuronal events measured during learning (Berger, 1984), and its property of associativity (i.e., in some instances LTP can only be induced if concomitant activation of nearby synapses occurs; Levy and Steward, 1983; Larson and Lynch, 1986)—have made LTP an attractive model for use in studying how information storage might occur.

For theoretical reasons reviewed elsewhere (Routtenberg, 1982), we have chosen to study changes in protein phosphorylation following induction of LTP in an effort to uncover biochemical events that may accompany information storage. Following high-frequency stimulation of the perforant path (an afferent fiber system of the hippocampal formation), we assayed different rat brain regions and found a selective increase in the *in vitro* phosphorylation of a single substrate termed protein F1 (M_r 47 kDa, pI 4.5) in animals sacrificed 5 min but not 1 min after the LTP-inducing stimulation (Routtenberg *et al.*, 1985; Nelson and Routtenberg, 1985). This increase in phosphorylation at 5 min was found only in dorsal hippocampal formation, the region containing the potentiated synapses, and did not appear after low-frequency control stimulation (which does not produce LTP). The degree of LTP achieved (measured as the increase in population spike amplitude) was directly correlated with the extent of increase in protein F1 phosphorylation across animals, suggesting a direct relationship between protein F1 phosphorylation and LTP.

The absence of an effect on phosphorylation at 1 min following high-frequency stimulation suggested that protein F1 phosphorylation was related to later aspects of LTP (perhaps its maintenance) but not to its initiation. We therefore investigated whether the increase in protein F1 phosphorylation following high-frequency stimulation would persist in parallel with the much longer durations noted for LTP. We found the elevation in protein F1 phosphorylation *in vitro* still occurred 1 hr following induction of LTP in anesthetized rats (Lovinger *et al.*, 1987). Extending the poststimulation period by using chronically implanted, freely moving animals, we then found that increased protein F1 phosphorylation in dorsal hippocampus was detectable *in vitro* 3 days following induction of LTP (Lovinger *et al.*, 1985). In these latter two studies, a direct correlation was found between protein F1 phosphorylation and the persistence of the potentiated response (i.e., the final spike amplitude divided by the spike amplitude measured immediately after high-frequency stimulation). These results demonstrated that a net increase of phosphate incorporation into protein F1 can still be detected days after the precipitating event and suggested that protein F1 phosphorylation might be associated more with maintenance of the potentiated response than with initial enhancement of the response. This hypothesis is further supported by the observation that protein F1 phosphorylation had no direct relationship with the magnitude of potentiation seen immediately after high-frequency stimulation.

3. PROTEIN KINASE C TRANSLOCATION FOLLOWING INDUCTION OF LONG-TERM POTENTIATION

As a logical next step in our research, we sought to understand the role that enzymes involved in protein F1 phosphorylation might have in LTP. Since our evidence suggested that protein F1 is phosphorylated by the Ca^{2+}- and phospholipid-stimulated protein kinase C (PKC; Kikkawa *et al.*, 1982; Akers and Routtenberg, 1985), we chose to examine whether LTP was accompanied by an increase in membrane PKC activity. Some alternatives to an increase in kinase activity, such as decreased phosphatase activity or increased substrate availability, have been discussed in detail elsewhere (Routtenberg, 1982).

Protein kinase C is normally distributed in both cytosol and membrane fractions. Recent evidence has shown that tumor-promoting phorbol esters, as well as elevated Ca^{2+} levels, can activate PKC by translocating it from the cytosolic to the membrane-associated state, where cofactors for PKC activity reside (Kraft and Andersen, 1983; Wolf *et al.*, 1985). Because protein F1 is a membrane-bound protein (Nelson and Routtenberg, 1985), we were intrigued by the idea that translocation of PKC to the membrane might be responsible for the increase in protein F1 phosphorylation we observed following LTP. In order to test this hypothesis, we induced LTP in the dentate gyrus of anesthetized animals and assayed for PKC activity (using exogenous histone as the phosphate acceptor) in both cytosolic and membrane fractions from dorsal hippocampus after LTP.

In agreement with the changes we found in protein F1 phosphorylation following LTP, we detected a decrease in cytosolic PKC activity and a corresponding increase in membrane-associated PKC activity at 1 hr but not 1 min following the high-frequency stimulation (Akers *et al.*, 1986). The sum of PKC activity in the two fractions did not change significantly after induction of LTP, suggesting that PKC was physically transferred from the cytosol to the membrane. Thus, increased phosphorylation of membrane-bound protein F1 following induction of LTP may be caused by activation of PKC through its movement to the membrane.

4. PROTEIN KINASE C SUBSTRATE PHOSPHORYLATION IN THE PRIMATE VISUAL PROCESSING SYSTEM

Since phosphorylation of protein F1 and translocation of PKC have been related to neural plasticity through our LTP studies, an increased expression of protein kinase C activity and protein F1 phosphorylation might be expected in areas of the brain that have been implicated in the storage of environmental information through neurobehavioral studies. A test of this hypothesis would also provide an alternative means for relating protein F1 to expression of neural plasticity.

In collaboration with M. Mishkin, we chose to explore this hypothesis by examining protein phosphorylation in the occipitotemporal pathway of rhesus monkey cerebral cortex (Jones and Powell, 1970; Kuypers *et al.*, 1965; Pandya and Kuypers, 1969). This pathway consists of a series of corticocortical relays extending from striate cortex, or primary visual cortex, to inferotemporal cortex. Electrophysiological, 2-deoxyglucose, and lesion studies indicate that this pathway is involved primarily if not exclusively in the processing of visual information (Mishkin, 1982; Macko *et al.*, 1982). Neurobehavioral evidence suggests that temporal areas of this pathway participate in the storage of visual representations (Mishkin, 1982; Mishkin and Ungerleider, 1982). Lesions of striate cortex, on the other hand, can form scotomata in portions of the visual field but do not impair acquisition and memory of tasks gained through intact portions of the visual field. Moreover, the receptive fields of neurons in the occipital regions are extremely resistant to modifying influences following the critical period of development, making these neurons unlikely candidates for information storage (Weisel and Hubel, 1963; Hubel *et al.*, 1977). We predicted, therefore, that protein F1 phosphorylation would be higher in temporal than in occipital areas of this pathway because of the apparent importance of temporal cortex in information storage.

On characterizing phosphoproteins in rhesus monkey cerebral cortex, we found a 50-kDa protein that resembled protein F1 in rat (Nelson *et al.*, 1986). Coelectrophoresis of 50 kDa and protein F1 on two-dimensional gels revealed that the two proteins share identical isoelectric points and microheterogeneity but that 50 kDa and protein F1 have different migration rates (corresponding to 3 kDa). Two-dimensional electrophoresis of fragments of 50 kDa and protein F1 obtained after limited proteolysis with *S. aureus* V8 protease yielded identical phosphopeptide maps, suggesting extensive homology between the two proteins.

We also found that the kinase specificity for 50 kDa was similar to that of protein F1. Protein kinase C added to monkey brain homogenate yielded a dose-dependent increase in the phosphorylation of 50 kDa and a second PKC substrate of 81 kDa. Cyclic AMP, on the other hand, had no effect on phosphorylation of 50 kDa, a result previously found for protein F1. Thus, despite the slight difference in migration, the many similarities between 50 kDa and protein F1 strongly suggest that the two proteins are species-specific homologues probably sharing closely related functions.

When we assayed endogenous phosphorylation of proteins from regions along the occipitotemporal pathway, we found a topographical gradient of ^{32}P incorporation into both 50 kDa/protein F1 and the 81-kDa protein (Nelson *et al.*, 1986). The values were found lowest in striate cortex and showed a progressive increase moving rostrally through adjacent regions to temporal cortex.

The specificity of this regional heterogeneity in phosphorylation was striking: whereas ^{32}P incorporation into protein F1 and the 81-kDa protein was ten- and 13-fold greater, respectively, in temporal cortical areas than in occipital cortical areas, no other phosphoproteins displayed significant differences in phosphorylation across regions in the range analyzed (30–100 kDa). In particular, neither the phosphorylation of synapsin Ia/Ib, well-characterized synaptic proteins, nor that of the α subunit of pyruvate dehydrogenase, a key mitochondrial enzyme, increased along this pathway, suggesting that the increases in phosphorylation of 50 kDa/protein F1 and 81 kDa were not caused by differences in the regional density of synapses or differences in energy utilization among the cortical regions assayed. We are currently investigating what components of the PKC phosphorylation system were responsible for the changes in substrate phosphorylation observed. These results suggest the hypothesis that phosphorylation of protein F1, which has already been implicated in the LTP model of information storage, and possibly phosphorylation of other PKC substrates may underlie regional participation in visual information storage in primate cerebral cortex.

5. PROTEIN F1 AS A MAJOR GROWTH CONE PHOSPHOPROTEIN (pp46)

In collaboration with K. Pfenninger, we found that protein F1 appears to be the same as a major growth cone phosphoprotein termed pp46 (Katz *et al.*, 1985) on the basis of identical molecular weight, isoelectric point, microheterogeneity on two-dimensional gels, phosphorylation by PKC, cAMP-independent endogenous phosphorylation, membrane enrichment, and two-dimensional phosphopeptide maps following limited proteolysis with *S. aureus* V8 protease (Nelson *et al.*, 1985). Potentially more interesting than this coidentification are the quantitative differences in protein F1 between growth cone and synaptosomal preparations.

The endogenous phosphorylation of protein F1 was 20-fold higher in growth cones than in a crude adult synaptosome preparation. In fact, protein F1 was the most highly labeled phosphoprotein detected in growth cones in the range of phosphoproteins assayed (30–100 kDa), in contrast to adult brain where there are many phosphoproteins more prominent than protein F1. This higher phosphorylation in growth cones is related in part to a higher ratio of protein F1 to total protein in the growth cones versus crude adult synaptosomes as measured qualitatively by protein staining of two-dimensional gels.

Of four major growth cone phosphoproteins detectable in the 30- to 100-kDa range, three, including protein F1, could be phosphorylated by exogenously added purified protein kinase C, suggesting that PKC activity might be relatively more important in growth cone function than activity of other protein kinases. Evidence from other laboratories has also suggested that PKC might play an important role in neurite outgrowth and synaptogenesis. Protein kinase C can be bound with high affinity by 12-O-tetradecanoylphorbol-13-acetate (TPA), a potent tumor promoter (Niedel *et al.*, 1983). When labeled TPA has been used to determine PKC levels in particular tissue types, the highest levels of TPA binding have been found in growth-cone-rich regions of fetal rat brain (Murphy *et al.*, 1983). The role that PKC might occupy in neuronal functions has also been explored using TPA, since this substance, besides directly binding to PKC, is known to stimulate PKC activity strongly (Castagna *et al.*, 1982). In cultured dorsal root ganglia cells or cultured neuroblastoma, TPA induces neurite outgrowth (Hsu *et al.*, 1984), suggesting that PKC and its substrates in growth cones play a crucial role in growth.

6. PROTEIN F1 AS A GROWTH-ASSOCIATED PROTEIN LINKED TO AXONAL REGENERATION (GAP43)

In collaboration with J. Freeman, we have coidentified F1 with GAP43 (Skene and Willard, 1981a–c; Benowitz and Lewis, 1983) on the basis of identical molecular weight, isoelectric point, microheterogeneity on two-dimensional gels, phosphorylation by PKC, membrane enrichment, and immunologic cross reactivity of protein F1 with an antibody raised against GAP43 (Snipes *et al.*, 1986). GAP43 synthesis and subsequent fast axonal transport are greatly increased during the axonal sprouting that follows nerve crush or axotomy. This increase in GAP43 synthesis typically persists until axons have reached their targets and only occurs in nervous systems where successful nerve regeneration occurs, i.e., the CNS and PNS of anamniotic creatures such as fish and amphibians but only the PNS of amniotes such as reptiles, birds, and mammals (Skene, 1984).

GAP43 synthesis is elevated in CNS of higher vertebrates during axonal development prior to synaptogenesis, and expression of pp46 is high in growth cones. One therefore might predict that CNS of higher vertebrates would be capable of axonal regeneration during nervous system development to the extent that expression of the gene for this protein is important in the capacity for axonal regeneration. This prediction is confirmed by studies demonstrating regrowth of CNS fiber pathways in neonatal rats for a limited period following birth, after which the same pathways will fail to regenerate (Kalil and Reh, 1979; Bernstein and Stelzner, 1983). Thus, in regard to induction of protein F1 synthesis, the regenerative state of the neuron recapitulates initial axonal development. Given our evidence for PKC phosphorylation of protein F1 during LTP, it will be of interest in future studies to see whether PKC has a modulatory influence on protein F1 activity during development and regeneration. The high levels of pp46 phosphorylation in growth cones would indicate that such a role, at least in development, is quite likely.

7. PROTEIN F1 AS A PUTATIVE REGULATOR OF PHOSPHOLIPID KINASE ACTIVITY AND ITS LOCALIZATION (B50)

Protein kinase C is a multifunctional enzyme with a long list of putative physiological substrates (Nishizuka, 1986). This raises the issue of how PKC might play a specific role in adult synaptic plasticity without simultaneously altering a host of other cellular events. One means of achieving this specificity of action would be to isolate changes in PKC activity related to adult synaptic plasticity to a defined cellular compartment and to limit the distribution of PKC substrates important for synaptic plasticity to this same compartment.

Localization studies on the B50 protein, which is cross-reactive with and most likely identical to protein F1 on the basis of its physical properties (Gispen *et al.*, 1986; Nelson and Routtenberg, 1985), have shown that both phosphorylation and immunoreactivity of the protein are restricted to brain (Kristjansson *et al.*, 1982; Oestreicher *et al.*, 1986). Within the nervous system, the protein has a heterogeneous distribution, being highest in such areas as septum, hippocampus, and cerebral cortex. Within the neuron itself, the evidence to date, both from ultrastructural immunocytochemical studies and from our studies of the growth cone, suggests that protein F1 concentration is highest in the presynaptic terminal (Gispen *et al.*, 1985; Nelson *et al.*, 1985). Finally, within the terminal, protein F1 is tightly associated with the membrane (Skene and Willard, 1981b; Nelson and Routtenberg, 1985). It is of interest in this regard that PKC activity, despite its wide distribution among tissues and across species, has its highest enrichment in neural plasma membranes (Kikkawa *et al.*, 1982), strongly suggesting that it has an important role in regulating protein F1 function and suggesting more generally that it has an important role in functions specific to nervous system.

The only evidence suggesting a specific biochemical role for protein F1 to date comes from the Gispen laboratory's studies of the B50 protein. Their data indicate that B50 may regulate the activity of a kinase that phosphorylates phosphatidylinositol-4-phosphate to phosphatidylinositol-4,5-diphosphate (Gispen *et al.*, 1985b), although this role has yet to be shown definitively using either fully purified B50 or the fully purified lipid kinase.

The phosphatidylinositol turnover cycle has recently attracted much attention because a number of protooncogene-derived proteins have been found to act at different points of this cycle, implicating phosphatidylinositol turnover as an important component of both normal and neoplastic growth. Of particular relevance to the Gispen findings is that at least two protooncogene products, derived from *c-ros* and *c-src*, act as kinases in the conversion of phosphatidylinositol to phosphatidylinositol-4,5-diphosphate (Macara, 1985). Thus, the role the Gispen group proposes for protein F1 (B50) is potentially a direct regulatory control over a protooncogene-derived protein. The occupation of such a biochemical function by protein F1 is intriguing in light of the evidence implicating this protein in growth processes, i.e., as GAP43 in axonal regeneration and as pp46 in growth cone function and neural development. It is also of interest in this regard that a neuron-specific form of the *c-src*-derived protein is expressed at high levels during neurite outgrowth (Lynch *et al.*, 1986), coinciding with the developmental window of increased protein F1 synthesis.

8. CONTRASTING THE ROLES OF PROTEIN F1: A HYPOTHESIS

Protein F1 has thus been implicated in multiple nervous system functions on the basis of its coidentification with other proteins in the literature, suggesting that a common molecular mechanism underlies all of these functions. Three of these roles—one in neuronal development, a second in neuronal regeneration, and a third as the putative regulator of a protooncogene product—suggest that protein F1 may participate in growth, especially growth of neuronal processes. We have previously proposed, on the basis of these growth-related roles, that protein F1's function in synaptic plasticity may also involve growth, in this case growth of existing presynaptic terminals or sprouting of new terminals (Routtenberg, 1985).

Although many commonalities exist in the roles suggested for protein F1, there are also important distinctions among these roles, in particular, (1) whether an increase in protein F1 synthesis might be involved and (2) what types of growth might be involved. For example, both the role of pp46 in development and the role of GAP43 in regeneration involve extensive axonal formation and elongation. Such a phenomenon does not occur normally in the mature CNS of higher vertebrates, however. Since protein F1 is studied in this neural system, the role of protein F1 in synaptic plasticity probably does not involve extensive growth of neuronal processes. In a similar vein, the roles of GAP43 and pp46 involve large increases in protein synthesis during development and regeneration, indicating an important genetic component in the regulation of GAP43/pp46 activity. Although changes in protein F1 synthesis during adult synaptic plasticity have not been studied, two observations suggest that the role of protein F1 synthesis in adult synaptic plasticity versus development and regeneration may be quite different: (1) increased GAP43 synthesis does not occur in response to axonal injury in the CNS of higher vertebrates; and (2) the induction and early maintenance of LTP and changes in protein F1 phosphorylation occur well before increased synthesis and transport of protein F1 could take place. Although this latter observation does not rule out a role for protein F1 synthesis in more prolonged maintenance of LTP, it does indicate that the full physiological manifestation of LTP occurs prior to protein F1 synthesis, something that does not appear to be true of regeneration.

On the basis of their coidentification, it is still likely that protein F1, GAP43, pp46, and B50 all share the same molecular function and that protein F1 participates in some type of growth. Limited forms of growth have been reported in mature CNS of higher vertebrates, including terminal and paraterminal sprouting and alterations in morphology of existing presynaptic terminals. Of particular relevance to the proposal that synaptic plasticity involves protein-F1-mediated growth are reports of increased synapse number and altered synaptic morphology following induction of LTP (Lee *et al.*, 1980; Desmond and Levy, 1983; Van Harreveld and Fifkova, 1975). If such growth involves protein F1, it would probably depend on already existing pools of protein F1 in the presynaptic terminals, since this growth is first detectable at poststimulation intervals too brief for protein synthesis and transport to occur.

The site of protein F1's action in mature CNS of higher vertebrates appears to have moved from the tips of extending neurites (during development and regeneration) to mature presynaptic terminals (during adult synaptic plasticity). As mentioned above, it appears that the mechanisms for controlling protein F1 activity in this neural system may be altered as well. A shift in the types of control exerted on protein F1 activity at the protein synthetic level versus the posttranslational level is logical from a functional standpoint: genetic control of protein F1 activity, although appropriate for controlling elongation of a single axon for an uninterrupted duration of time, would be inefficient and probably impractical in the primary control of protein F1 activity at individual synaptic terminals. The briefness with which synaptic efficacy changes, the signals necessary to target protein F1 to individual terminals, and the cumbersome distances over which chemically mediated communication between nucleus and terminal occur are all problems that would have to be surmounted. Posttranslational modification of protein F1 would provide a more practical means for rapidly altering protein F1 activity at individual terminals. Evidence for such control emerges from our LTP experiments and appear to involve phosphorylation of protein F1 by PKC. Thus, an existing pool of protein F1 in each synapse, quickly mobilized by PKC-mediated phosphorylation, could alter terminal morphology or establish new terminals and thereby alter synaptic efficacy.

The primary regulation of protein F1 activity by posttranslational means during adult synaptic plasticity would allow for a steadier and more limited rate of protein F1 synthesis in replenishing presynaptic terminal stores. However, the mechanisms providing for this limited expression of protein F1 might be incompatible with mechanisms regulating the full-scale activation of protein F1 synthesis during development or regeneration. Such a shift in the factors governing protein F1 gene expression would provide one explanation for the curious loss of capacity for axonal regeneration in mature CNS of higher vertebrates, since regeneration appears to involve much greater protein F1 levels than are constitutively expressed in this neural system.

If the capacity for axonal regeneration in mature CNS of higher vertebrates was lost through the emergence of protein F1's role in synaptic plasticity, the theory of natural selection would predict that this new mechanism of synaptic plasticity conferred a greater reproductive fitness to higher vertebrates—presumably through an improved behavioral adaptivity to the environment—than did the capacity for CNS regeneration together with phylogenetically older mechanisms for synaptic plasticity. Thus, we postulate that the loss of capacity for CNS regeneration observed across a broad spectrum of higher vertebrates may be the result of an evolutionary tradeoff in which protein F1 acquired a role in adult synaptic plasticity incompatible with—and more adaptive than—its role in regeneration.

REFERENCES

Akers, R. F., and Routtenberg, A., 1985, Kinase C phosphorylates a protein involved in synaptic plasticity, *Brain Res.* **334**:147–151.

Akers, R. F., Lovinger, D., Colley, P., Linden, D., and Routtenberg A., 1986, Translocation of protein kinase C activity may mediate hippocampal long term potentiation, *Science* **231**:587–589.

Benowitz, L. I., and Lewis, E. R., 1983, Increased transport of 44,000- to 49,000-dalton acidic proteins during regeneration of the goldfish optic nerve: A two-dimensional gel analysis, *J. Neurosci.* **3:**2153–2163.

Berger, T. W., 1984, Long-term potentiation of hippocampal synaptic transmission affects rate of behavioral learning, *Science* **224:**627–630.

Bernstein, E., and Stelzner, D., 1983, Plasticity of the corticospinal tract following mid-thoracic spinal injury in post-natal rat, *J. Comp. Neurol.* **221:**382–400.

Bliss, T. V. P., and Lomo, T., 1973, Long lasting potentiation of synaptic transmission in the dentate area of the anesthetized rabbit following stimulation of the perforant path, *J. Physiol. (Lond.)* **232:**357–374.

Castagna, M., Takai, Y., Kaibuchi, K., Sano, K., Kikkawa, U., and Nishizuka, Y., 1982, Direct activation of calcium-activated, phospholipid-dependent protein kinase by tumor-promoting phorbol esters, *J. Biol. Chem.* **257:**7847–7851.

Desmond, N. L., and Levy, W. B., 1983, Synaptic correlates of associative potentiation/depression: An ultrastructural study in the hippocampus, *Brain Res.* **265:**21–30.

Douglas, R. M., and Goddard, G. V., 1975, Long-term potentiation of the perforant path–granule cell synapse in the rat hippocampus, *Brain Res.* **86:**205–215.

Gispen, W. H., Leunissen, J. L. M., Oestreicher, A. B., Verkleij, A. J., and Zwiers, H., 1985a, Presynaptic localization of 50 phosphoprotein: The (ACTH)-sensitive protein kinase substrate involved in rat brain polyphosphoinositide metabolism, *Brain Res.* **328:**381–385.

Gispen, W. H., Van Dongen, C. J., De Graan, P. N. E., Oestreicher, A. B., and Zwiers, H., 1985b, The role of phosphoprotein B50 in phosphoinositide metabolism in brain synaptic plasma membranes, in: *Inositol and Phosphoinositides* (J. E. Bleasdale, G. Hauser, and J. Eichberg, ed.), Humana Press, Dallas, pp. 399–414.

Gispen, W. H., De Graan, P. N. E., Chan, S. Y., and Routtenberg, A., 1986, Comparison between the neural acidic proteins B50 and F1, in: *Progress in Brain Research, Volume 69* (W. H. Gispen and A. Routtenberg, eds.), Elsevier, Amsterdam, pp. 383–386.

Hsu, L., Natyzak, D., and Laskin, J. D., 1984, Effects of the tumor promoter 12-O-tetradecanoylphorbol-13-acetate on neurite outgrowth from chick embryonic sensory ganglia, *Cancer Res.* **44:**4607–4614.

Hubel, D. H., Weisel, T. N., and LeVay, S., 1977, Plasticity of ocular dominance columns in monkey striate cortex, *Phil. Trans. R. Soc. Lond. [Biol.]* **278:**377–409.

Jones, E. G., and Powell, T. P. S., 1970, An anatomical study of converging sensory pathways within the cerebral cortex of the monkey, *Brain* **503:**793–820.

Kalil, K., and Reh, T., 1979, Regrowth of severed axons in the neonatal CNS; establishment of normal connections, *Science* **205:**1158–1161.

Katz, F., Ellis, L., and Pfenninger, K. H., 1985, Nerve growth cones isolated from fetal rat brain: Calcium dependent protein phosphorylation, *J. Neurosci.* **5:**1402–1411.

Kikkawa, U., Takai, Y., Minakuchi, R., Inohara, S., and Nishizuka, Y., 1982, Calcium-activated, phospholipid-dependent protein kinase from rat brain, *J. Biol. Chem.* **257:**13341–13348.

Kraft, A. S., and Andersen, W. B., 1983, Phorbol esters increase the amount of calcium, phospholipid-dependent protein kinase associated with the plasma membrane, *Nature* **301:**621–623.

Kristjansson, G. I., Zwiers, H., Oestricher, A. B., and Gispen, W. H., 1982, Evidence that the synaptic phosphoprotein B50 is localized exclusively in nerve tissues, *J. Neurochem.* **39:**371–378.

Kuypers, H. G. J., Szwarcbart, M. K., Mishkin, M., and Rosvold, H. E., 1965, Occipitotemporal corticocortical connections in the rhesus monkey, *Exp. Neurol.* **11:**245–262.

Larson, J., and Lynch, G., 1986, Induction of synaptic potentiation in hippocampus by patterned stimulation involves two events, *Science* **232:**985–988.

Lee, K. S., Schottler, F., Oliver, M., and Lynch, G., 1980, Brief bursts of high-frequency stimulation produce two types of structural change in rat hippocampus, *J. Neurophysiol.* **44:**247–258.

Levy, W. B., and Steward, O., 1983, Temporal contiguity requirements for long-term associative potentiation/depression in the hippocampus, *Neuroscience* **8:**791–797.

Lovinger, D. M., Akers, R. F., Nelson, R. B., Barnes, C. A., McNaughton, B. L., and Routtenberg, A., 1985, A selective increase in the phosphorylation of protein F1, a protein kinase C substrate, directly related to three day growth of long term synaptic enhancement, *Brain Res.* **343:**137–143.

Lovinger, D. M., Colley, P. A., Akers, R. F., Nelson, R. B., and Routtenberg, A., 1986, Direct relation of long-duration synaptic potentiation to phosphorylation of membrane protein F1: A substrate for membrane protein kinase C, *Brain Res.* **399:**205–211.

Lynch, S. A., Brugge, J. S., and Levine, J. M., 1986, Induction of an altered *c-src* protein accompanies the neural differentiation of an embryonal cell line, *Soc. Neurosci. Abstr.* **12:**216.

Macara, I. G., 1985, Oncogenes, ions, and phospholipids, *Am. J. Physiol.* **248:**C3–C11.

Macko, K. A., Jarvis, C. D., Kennedy, C., Miyaoka, M., Shinohara, M., Sokoloff, L., and Mishkin, M., 1982, Mapping the primate visual system with 2-[^{14}C]deoxyglucose, *Science* **218:**394–397.

Matus, A. I., Ng, M. L., and Mazat, J. P., 1980, Protein phosphorylation in synaptic membranes: Problems of interpretation, in: *Protein Phosphorylation and Bio-Regulation* (G. Thomas, E. J. Podesta, and J. Gorson, eds.), Karger, Basel, pp. 25–36.

Mishkin, M., 1982, A memory system in the monkey, *Phil. Trans. R. Soc. Lond. [Biol.]* **298:**85–95.

Mishkin, M., and Ungerleider, L. G., 1982, Contributions of striate inputs to the visuospatial functions of parietopreoccipital cortex in monkeys, *Behav. Brain Res.* **6:**57–77.

Murphy, K. M. M., Gould, R. J., Oster-Granite, M. L., Gearheart, J. D., and Snyder, S. H., 1983, Phorbol esters receptors: Autoradiographic identification in the developing rat, *Science* **222:**1036–1038.

Nelson, R. B., and Routtenberg, A., 1985, Characterization of the 47kD protein F1 (pI 4.5), a kinase C substrate directly related to neural plasticity, *Exp. Neurol.* **89:**213–224.

Nelson, R. B., Routtenberg, A., Hyman, C., and Pfenninger, K. H., 1985, A phosphoprotein, F1, directly related to neuronal plasticity in adult rat brain may be identical to a major growth cone membrane protein, *Soc. Neurosci. Abstr.* **11:**927.

Nelson, R. B., Friedman, D. P., O'Neill, J. B., Mishkin, M., and Routtenberg, A., 1986, Protein kinase C substrate phosphorylation in primate cerebral cortex (e.g., protein F1) is increased in those stages of the occipitotemporal visual processing pathway important for information storage, *Soc. Neurosci. Abstr.* **12:**1168.

Niedel, J. E., Kuhn, L. J., and Vandenbark, G. R., 1983, Phorbol diester receptor copurifies with protein kinase C, *Proc. Natl. Acad. Sci. U.S.A.* **80:**36–40.

Nishizuka, Y., 1986, Studies and perspectives of protein kinase C, *Science* **233:**305–312.

Oestreicher, A. B., Dekker, L. V., and Gispen, W. H., 1986, A radioimmunoassay for the phosphoprotein B50: Distribution in rat brain, *J. Neurochem.* **46:**1366–1369.

Pandya, D. N., and Kuypers, H. G. J. M., 1969, Cortico-cortical connections in the rhesus monkey, *Brain Res.* **13:**13–36.

Routtenberg, A., 1982, Memory formation as a posttranslational modification of brain proteins, in: *Mechanisms and Models of Neural Plasticity. Proceedings VIth International Neurobiology IBRO Symposium on Learning and Memory* (C. A. Marsden and H. Matthies, eds.), Raven Press, New York, pp. 17–24.

Routtenberg, A., 1985, Protein kinase C activation leading to protein F1 phosphorylation may regulate synaptic plasticity by presynaptic terminal growth, *Behav. Neural Biol.* **44:**186–200.

Routtenberg, A., Lovinger, D., and Steward, O., 1985, Selective increase in the phosphorylation of a 47kD protein (F1) directly related to long-term potentiation, *Behav. Neural Biol.* **43:**3–11.

Simkowitz, P., Ellis, L., and Pfenninger, K. H., 1987, Developmentally regulated membrane proteins of nerve growth cones and synaptic endings, (submitted for publication).

Skene, J. H. P., 1984, Growth-associated proteins and the curious dichotomies of nerve regeneration, *Cell* **37:**697–700.

Skene, J. H. P., and Willard, M., 1981a, Changes in axonally transported proteins during axon regeneration in toad retinal ganglion cells, *J. Cell Biol.* **89:**86–95.

Skene, J. H. P., and Willard, M., 1981b, Axonally transported proteins associated with axon growth in rabbit central and peripheral nervous system, *J. Cell Biol.* **89:**96–103.

Skene, J. H. P., and Willard, M., 1981c, Characteristics of growth-associated polypeptides in regenerating toad retinal ganglion cell axons, *J. Neurosci.* **1:**419–426.

Snipes, J., Freeman, J. A., Costello, B., Chan, S., and Routtenberg, A., 1986, A growth associated protein, GAP43, is immunologically and structurally related to the plasticity associated protein, protein F1, *Soc. Neurosci. Abstr.* **12:**500.

Van Harreveld, A., and Fifkova, E., 1975, Swelling of dendritic spines in the fascia dentata after stimulation of the perforant fibers as a mechanism of post-tetanic stimulation, *Exp. Neurol.* **49:**736–739.

Weisel, T. N., and Hubel, D. H., 1963, Single-cell responses in striate cortex of kittens deprived of vision in one eye, *J. Neurophysiol.* **26:**1003–1117.

Wolf, M., Cuatrecasas, P., and Sahyoun, N., 1985, Interaction of protein kinase C with membranes is regulated by Ca^{++}, phorbol esters, and ATP, *J. Biol. Chem.* **260:**15718–15722.

47

Evolutionary Origin of Electrical Excitability

BERTIL HILLE

1. INTRODUCTION

Other chapters in this volume invoke synapses, transmitters, channels, and second messengers as cellular elements of learning. This chapter considers the evolutionary origins of these elements, particularly ionic channels.* Apparently each of the elements is present in the first animals with a nervous system, and we are as yet not aware of major improvements made through subsequent evolution. Thus, if these are indeed the prerequisites of learning, we can expect that mechanisms of learning have broad similarities across taxonomic boundaries. More complex accounts of this subject have been given in Hille (1984, 1987), where full references for many of the statements made here may be found.

2. AN OUTLINE OF EVOLUTION

An evolutionary tree of living organisms (Fig. 1) shows several stages of astounding architectural revolution followed by radiating diversification. Because of the long time separating us from these events, the exact relationships among major taxa are still subjects of debate and hypothesis.

The most primitive cellular kingdom, the prokaryotes, is typified by bacteria. They are unicellular and surrounded by a conventional cytoplasmic membrane containing unglycosylated but otherwise architecturally conventional membrane proteins. Bacteria have active transport of ions, and, like eukaryotes, their cytoplasm has a typically high K^+ content and extremely low free Ca^{2+}. They have a negative membrane potential set up by electron transport, which generates their energy-storing protomotive force across the cytoplasmic membrane. Many bacteria have no requirement for for Na^+ or Cl^- and can be grown indefinitely in media lacking these ions. In general, bacteria have no intracellular vesicles or organelles.

* This chapter is dedicated in the memory of the late Roger Eckert. His investigations of excitability gave important insights into channel biology.

BERTIL HILLE • Department of Physiology and Biophysics, University of Washington Medical School, Seattle, Washington 98195.

FIGURE 1. Possible evolutionary relationships of living organisms with an emphasis on taxa whose channels have received electrophysiological study. (From Hille, 1984.)

The transitions from prokaryote to eukaryote entailed a major overhaul and increase in complexity of the cellular blueprint. Eukaryotes have for the first time a cell nucleus and more than one chromosome. The chromosomes are linear and segregate in a spindle at cell division. Eukaryotes carry out their major electron-transport energy coupling in intracellular mitochondria instead of on the cytoplasmic membrane. Membrane and secretory proteins are synthesized on the rough endoplasmic reticulum, glycosylated and packaged in the Golgi, and delivered to the cell surface by endocytosis. Cytoskeletal filaments and tubules participate in establishing cell shape, shape change, and intracellular transport of membranous organelles. Calcium and calmodulin serve as an intracellular signaling and regulatory couple. Protein phosphorylation under control of cyclic nucleotides and of diacylglycerol modulate intracellular activities. These advances made possible four modern kingdoms: the unicellular protists and the multicellular plants, fungi, and animals.

The first surviving animal experiment, the sponges, apparently have no axons or nervous system. However, in the next surviving step, the cnidaria (coelenterates) and ctenophores (comb jellies) have axons, chemical synapses, gap junctions, sense organs, rapid muscular movements, and other components of typical animal neuromuscular systems (Anderson and Schwab, 1982). Not much successful neurochemistry has been done on representatives of this evolutionary stage. There is contradictory evidence for biogenic amines and peptides as neurotransmitters (Martin and Spencer, 1983). By the time one gets to the nematodes (roundworms), however, there are cells large enough for classical electrophysiology (*Ascaris*) and yet organisms of such short generation time as to expedite neurogenetics (*Caenorhabditis elegans*). Here there are behavioral mutants with defects in acetylcholine, dopamine, 5-hydroxytryptamine, and octopamine transmission (Chalfie, 1984). Electrophysiology provides evidence for excitatory nicotinic cholinergic transmission and for inhibitory (GABAergic transmission). Thus, the familiar themes seem to be in place at this stage.

After the nematodes there are many order-of-magnitude increases of complexity of the wiring of nervous systems and vast improvements in sensory and motor organs. However, the major classes of molecules involved and the principles of rapid cellular signaling may be the same in all the higher animals whether vertebrate, arthropod, annelid, or mollusk.

3. DISTRIBUTION OF VOLTAGE-GATED CHANNELS

To discuss evolution of ionic channels requires knowing where they can be found, a question best approached for now by electrophysiological methods. Classical voltage-clamp methods applied to large axons readily demonstrate tetrodotoxin-sensitive Na^+ channels in chordates, mollusks, arthropods, and annelids (Fig. 2). Sodium currents activate and inactivate with similar (but clearly not identical) time courses, and even the peak current densities are similar among phyla except in vertebrate myelinated axons, where densities of Na^+ channels in nodes of Ranvier are ten- to 50-fold higher than in other giant axons. This overriding similarity of Na^+ channel function does not mean that channel structure has been invariant. Indeed, the amino acid sequences of Na^+ channels are surprisingly diverse even within the vertebrates. From cDNA clones we know that there is 38% amino acid sequence dissimilarity between the major Na^+ channel polypeptide of a rat brain Na^+ channel and an electric eel electric organ Na^+ channel (Noda *et al.*, 1986).

Although Na^+-permeable and even Na^+-selective channels have been reported elsewhere, no Na^+ channels with voltage-gated properties even partly like those in animals have been reported outside the animal kingdom. The lowest phyla with known sodium spikes are the cnidaria and ctenophores (Anderson and Schwab, 1982), where there is, however, no tetrodotoxin sensitivity. At the level of the platyhelminths (flatworms) we find block by tetrodotoxin (Koopowitz and Keenan, 1982). Apparently, among surviving animal phyla the Na^+ channel is present in the first animals to have a nervous system. Indeed, it seems likely that the evolution of voltage-gated Na^+ channels is one of the several innovations that made it possible to have animal life with large, many-celled organisms that retain the millisecond response time characteristic of the quicker unicellular organisms.

Despite an absence of voltage-gated Na^+ channels, higher plants, algae, and protists have well-documented electrical excitability. In the case of higher plants, slow action potentials can propagate a considerable distance from excitation initiated at mechanoreceptors. This is how news of the jiggling of a fly is spread to the leaf-closing cells of the Venus's-flytrap and the alarm of contact with a browser is spread over an entire

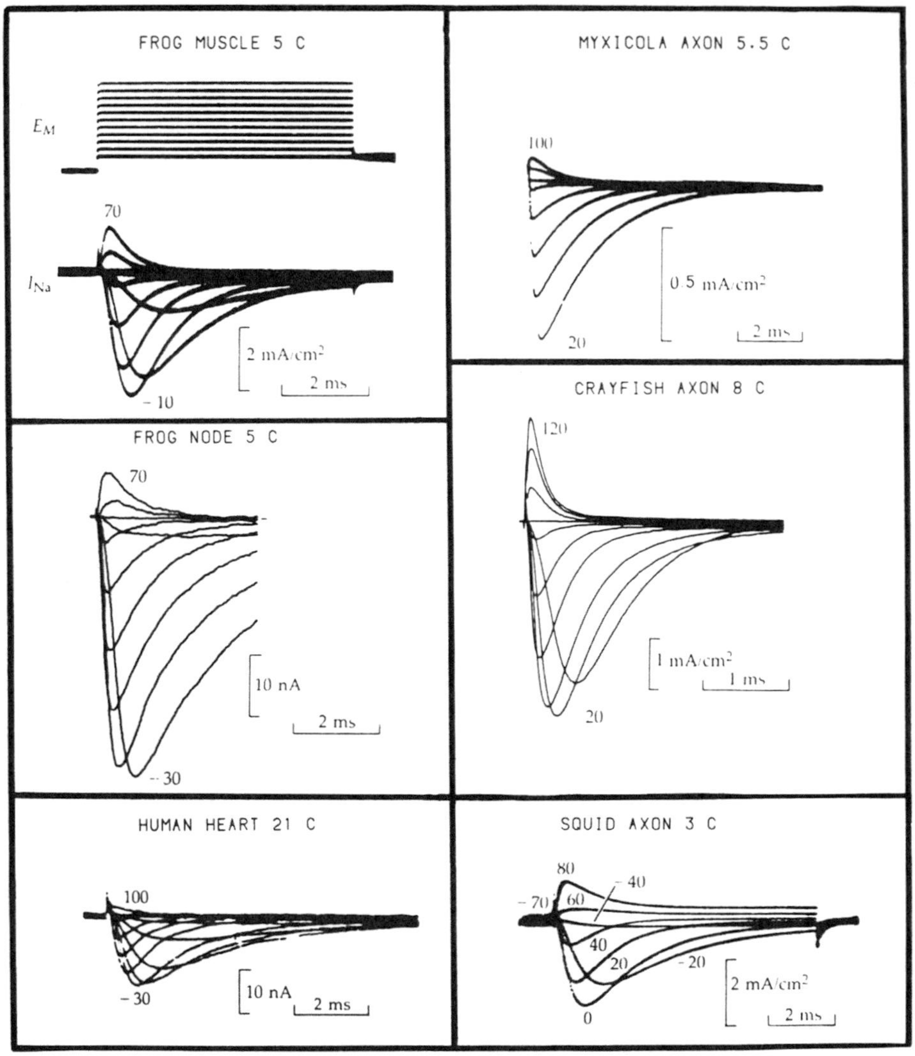

FIGURE 2. Voltage-clamp evidence for voltage-gated Na^+ channels in the four most advanced animal phyla. In each frame a similar family of depolarizing clamp pulses is applied to a preparation whose K^+ currents are blocked pharmacologically. The membranes are frog semitendinosus muscle (5°C), frog sciatic node of Ranvier (5°C), dissociated human atrial cells (21°C), giant axons of *Myxicola* (annelid) ventral cord (5.5°C), crayfish (arthropod) ventral cord (8°C), and squid (mollusk) giant axon (3°C). References are given in Hille (1984).

branch of the *Mimosa* sensitive plant. The action potentials propagate through cytoplasmic bridges from cell to cell—a form of electrical coupling without gap junctions. Voltage-gated K^+ channels are also said to mediate the large K^+ fluxes underlying opening and closing of respiratory stomata of plant leaves (Schroeder *et al.*, 1987). Voltage clamp reveals TEA-blockable, voltage-gated K^+ channels in all eukaryotic kingdoms (Fig. 3). The action potentials of protozoa are definitely Ca^{2+} spikes, and those of plants seem to be Ca^{2+} and/or Cl^- spikes. Thus, it is probable that various voltage-gated K^+ and Ca^{2+} channels are present in every eukaryote.

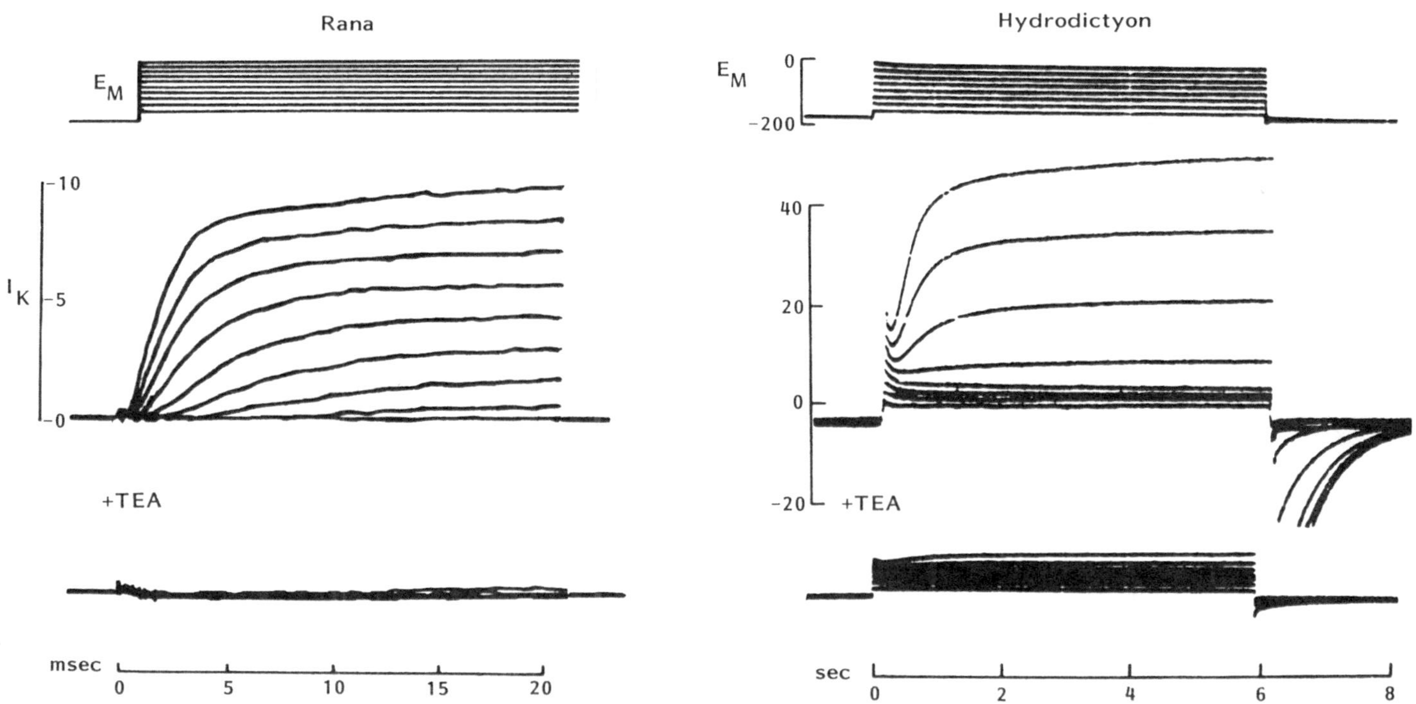

FIGURE 3. Voltage-clamp comparison of voltage-gated K^+ currents in frog and a green alga, *Hydrodictyon*. Depolarizations elicit delayed K^+ currents that are sensitive to tetraethylammonium (TEA) ions applied in the bath. Data from tetrodotoxin-treated node of Ranvier ±60 mM TEA (After Hille, 1967) and from a *Hydrodictyon* coenocyte in artificial pond water ±10 mM TEA. (After Findlay and Coleman, 1983.)

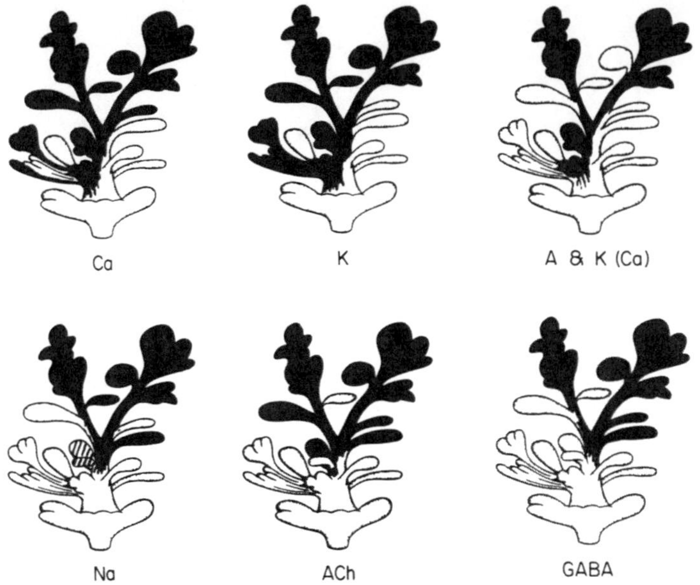

FIGURE 4. Reports of ionic channels in phylogeny. The phylogenetic trees are drawn as in Fig. 1 and shaded where electrophysiological observations give evidence for voltage-gated Ca^{2+} channels, delayed K^+ channels, A (transient) and K(Ca) channels (which have the same distribution), Na^+ channels, excitatory ACh receptors, and inhibitory GABA receptors. Full references are given in Hille (1984, 1987). (Modified from Hille, 1987.)

Figure 4 summarizes where in phylogeny different channel types have been clearly identified by electrophysiological methods. The ones that seem to be new to the animals are the voltage-gated Na^+ channel and perhaps channels that respond to the binding of neurotransmitters.

4. AN EVOLUTIONARY SPECULATION

Electricity is a convenient way to spread signals throughout a cell more quickly than could be done with chemical diffusion. When electrical signals reach their destination, however, the message must be translated into another form to produce whatever biological response is appropriate. Broadly, this translation takes only one form so far as we know: a depolarizing signal eventually leads to an increased Ca^{2+} permeability in some cell membrane, resulting in an increased intracellular free calcium concentration. Conversely, a hyperpolarizing signal shuts down Ca^{2+} permeability, resulting in a decreased cytoplasmic calcium. The calcium signal then is translated into changed enzyme activities, contraction, secretion, modulation of channel gating, and other outputs. Thus, electrical signals ultimately exert their influence on Ca^{2+}-sensitive cytoplasmic processes.

I have proposed (Hille, 1984) that if electrical signals serve Ca^{2+}-regulated processes, they may have arisen in parallel with the evolution of Ca^{2+}-sensitive biological processes. So far as we know, bacteria use neither Ca^{2+} signals nor electrical signals, yet all eukaryotes do. Thus, around 1400 million years ago, as the eukaryotic cell plan was being formulated, calmodulinlike molecules to control cytoplasmic processes and ionic channels to control cell Ca^{2+} may have been developing in parallel (Fig. 5). The opportunity to begin to exploit Ca^{2+} or K^+ gradients for electrical signaling may have arisen when the protoeukaryote received symbiotic mitochondria so that the cytoplasmic membrane was no longer essential for storing all of the protonmotive free energy of the cell. What the original Ca^{2+}-permeable channels were we do not know. They might have been sensitive to voltage, osmotic stretch, other mechanical forces, or environmental chemicals. However, still in this early period they apparently gave rise to a highly ion-selective and steeply voltage-sensitive prototype that diversified into a variety of subtypes of K^+ and Ca^{2+} channels. This variety of subtypes was then passed on to all modern eukaryotes, where they have been adapted for many kinds of cell irritability and responsiveness.

The animals are the only kingdom to commit themselves to having a high Na^+ concentration in their interstitial fluid. Certainly there are many marine forms in other kingdoms, but these groups have freshwater forms as well, and they do not create a high-Na^+ interstitium. The commitment of the animals means that all animal cells have a Na^+–K^+ pump (probably original with the animals) and all animals can use Na^+-based action potentials and Na^+-coupled cotransport for amino acids, sugars, neurotransmitters, and some inorganic ions as well as Na^+–Ca^{2+} exchange. These Na^+-oriented devices evolved early in metazoan phylogeny some 700 million years ago.

The Na^+-based action potential can propagate far more rapidly than a Ca^{2+}-based one. This is not because Na^+ channels are faster, for surely the Ca^{2+} channel proteins could easily be modified by evolution to operate at whatever speed is adaptive, and some Ca^{2+} channels are quite fast. Rather, it is because the Na^+ ion does not itself have the messenger role that Ca^{2+} ions have. Therefore Na^+ channel densities and Na^+ fluxes 100 times larger than are found for Ca^{2+} are quite tolerable, and it is current density that makes speed. Large Ca^{2+} fluxes raising the cytoplasmic free Ca^{2+} to millimolar levels would severely compromise the messenger role of Ca^{2+} and would probably be toxic to many other cellular processes.

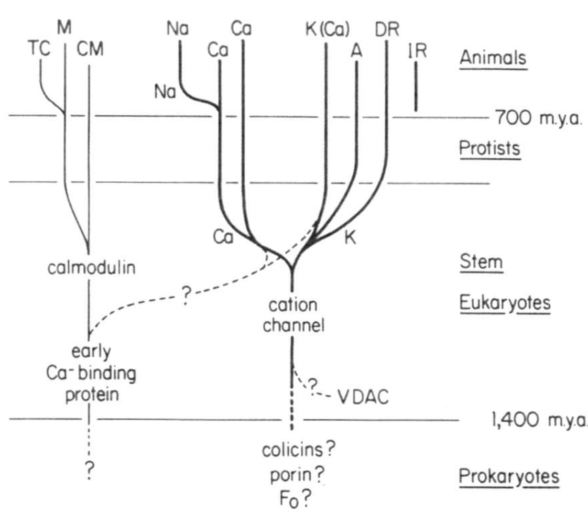

FIGURE 5. Proposed descent of voltage-gated channels from a stem eukaryotic cation channel. Abbreviations: TC, troponin C; M, myosin light chain; CM, calmodulin; Na, Na^+ channel; Ca, Ca^{2+} channel; K(Ca), A, DR, IR, several types of K^+ channel. (After Hille, 1987.)

5. CONCLUSION

Voltage-gated channels are an evolutionary innovation dating back at least to the origin of the eukaryotes. Exocytosis and most second messenger systems date this far back as well. Except for mechanoreceptors, we do not yet know of any sensory or ligand-gated channels outside of the animal kingdom. In the low animal phyla, most of the subcellular elements of the nervous system already seem present, and in the higher phyla (mollusk, arthropod, annelid, and vertebrate) these elements are mature and relatively indistinguishable. Thus, apparent differences in "intelligence" and ability to learn among animal phyla should be attributed largely either to the evolution of more complex wiring or to molecules that we have not considered here. The mechanisms used in learning could be basically the same in all phyla.

REFERENCES

Anderson, P. A. V., and Schwab, W. E., 1982, Recent advances and model systems in coelenterate neurobiology, *Prog. Neurobiol.* **19:**213–236.

Chalfie, M., 1984, Neural development in *Caenorhabditis elegans*, *Trends Neurosci.* **7:**197–202.

Findlay, G. P., and Coleman, H. A., 1983, Potassium channels in the membrane of *Hydrodictyon africanum*, *J. Membr. Biol.* **75:**241–251.

Hille, B., 1967, The selective inhibition of delayed potassium currents in nerve by tetraethylammonium ion, *J. Gen. Physiol.* **50:**1287–1302.

Hille, B., 1984, *Ionic Channels in Excitable Membranes*, Sinauer Associates, Sunderland, MA.

Hille, B., 1987, Evolutionary origins of voltage-gated channels and synaptic transmission, in: *Synaptic Function* (G. M. Edelman, W. E. Gall, and W. M. Cowan, eds.), Neurosciences Research Foundation/John Wiley & Sons, New York, pp. 163–176.

Koopowitz, H., and Keenan, L., 1982, The primitive brains of platyhelminthes, *Trends Neurosci.* **5:**77–79.

Martin, S. M., and Spencer, A. N., 1983, Neurotransmitters in coelenterates, *Comp. Biochem. Physiol.* **74C:**1–14.

Noda, M., Ikeda, T., Suzuki, H., Takeshima, H., Takahashi, T., Kuno, M., and Numa, S., 1986, Expression of functional sodium channels from cloned cDNA, *Nature* **322:**826–828.

Schroeder, J. L., Raschke, K., and Neher, E., 1987, Voltage-sensitive K^+ channels in guard cell protoplasts, *Proc. Natl. Acad. Sci. U.S.A.* **84:**4108–4112.

Contributors

NORIO AKAIKE
Department of Physiology
Faculty of Medicine
Kyushu University
Fukuoka 812, Japan

DANIEL L. ALKON
Section on Neural Systems
Laboratory of Biophysics
IRP
National Institute of Neurological and Communicative Disorders and Stroke
National Institutes of Health at the Marine Biological Laboratory
Woods Hole, Massachusetts 02543

VAHE E. AMASSIAN
Department of Physiology
State University of New York Health Science Center at Brooklyn
Brooklyn, New York 11203

SHUJI AOU
Department of Biological Control Systems
National Institute for Physiological Sciences
Okazaki 444, Japan

TREVOR ARCHER
R&D Laboratories
Astra Alab AB
S-151 85 Södertälje, Sweden

H. L. ATWOOD
Department of Physiology
University of Toronto
Toronto, Ontario, Canada M5S 1A8

FRANK H. BAKER
Department of Neurology
Johns Hopkins University
Baltimore, Maryland 21205

PAVEL M. BALABAN
Laboratory of Conditioned Reflexes and Physiology of Emotions
Institute of Higher Nervous Activity and Neurophysiology
USSR Academy of Sciences
Moscow, USSR

ATTILA BARANYI
Department of Comparative Physiology
Attila Jozsef University of Sciences
Szeged, Hungary
Present address:
Neuropsychiatric Institute
University of California at Los Angeles Medical Center
Los Angeles, California 90024

PAULA BARRETT
Departments of Internal Medicine and Cell Biology
School of Medicine and School of Nursing
Yale University
New Haven, Connecticut 06510

GERARD BAUX
Laboratory of Cellular and Molecular Neurobiology
National Center for Scientific Research
91190 Gif-sur-Yvette, France

JAN BEHRENDS
Department of Physiology
Faculty of Medicine
Kyushu University
Fukuoka 812, Japan

LYNN J. BINDMAN
Department of Physiology
University College London
London WC1E 6BT, England

M. W. BROWN
Department of Anatomy
The Medical School
University of Bristol
Bristol BS8 1TD, England

ARTHUR CHERKIN
Geriatric Research, Education and Clinical Center and Psychobiology Research Laboratory
Veterans Administration Medical Center
Sepulveda, California 91343
and Department of Psychiatry and Biobehavioral Sciences
University of California at Los Angeles School of Medicine
Los Angeles, California 90024

JOHN CLEMENTS
Experimental Neurology Unit
John Curtin School of Medical Research
Australian National University
Canberra, A.C.T., Australia

JOHN A. CONNOR
Department of Molecular Biophysics
AT & T Bell Laboratories
Murray Hill, New Jersey 07974

DOUGLAS A. COULTER
Laboratory of Biophysics
National Institute of Neurological and Communicative Disorders and Stroke
National Institutes of Health at the Marine Biological Laboratory
Woods Hole, Massachusetts 02543

MAHLON R. DELONG
Departments of Neuroscience and Neurology
Johns Hopkins University
Baltimore, Maryland 21205

JOHN F. DISTERHOFT
Department of Cell Biology and Anatomy
Northwestern University Medical School
Chicago, Illinois 60611
and
Laboratory of Biophysics
National Institute of Neurological and Communicative Disorders and Stroke
National Institutes of Health at the Marine Biological Laboratory
Woods Hole, Massachusetts 02543

ROBERT W. DOTY
Center for Brain Research
University of Rochester
Rochester, New York 14642

JOHN C. ECCLES
Contra CH6611, Switzerland

SUZANNE EVANS
Psychology Department and Faculty of Medicine
Memorial University of Newfoundland
St. Johns, Newfoundland, Canada A1B 3X9

JAMES F. FLOOD
Geriatric Research, Education and Clinical Center and Psychobiology Research Laboratory
Veterans Administration Medical Center
Sepulveda, California 91343
and
Department of Psychiatry and Biobehavioral Sciences
University of California at Los Angeles School of Medicine
Los Angeles, California 90024

PHILIPPE FOSSIER
Laboratory of Cellular and Molecular Neurobiology
National Center for Scientific Research
91190 Gif-sur-Yvette, France

MASAJI FUKUDA
Department of Physiology
Faculty of Medicine
Toyama Medical and Pharmaceutical University
Sugitani, Toyama 930-01, Japan

PAUL E. GOLD
Department of Psychology
University of Virginia
Charlottesville, Virginia 22903

KAMIL A. GRAJSKI
Graduate Group in Biophysics
University of California at Berkeley
Berkeley, California 94720
Present address:
Coleman and Epstein Memorial Laboratories
University of California, San Francisco
San Francisco, California 94143

JOACHIM M. GREUL
Max Planck Institute for Brain Research
Department of Neurophysiology
6000 Frankfurt 71, Federal Republic of Germany

BENGT GUSTAFSSON
Department of Physiology
University of Göteborg
S-400 33 Göteborg, Sweden

CAROLYN W. HARLEY
Psychology Department and Faculty of Medicine
Memorial University of Newfoundland
St. Johns, Newfoundland, Canada A1B 3X9

U. HEINEMANN
Institute for Normal and Pathological Physiology
University of Cologne
5000 Cologne 41, Federal Republic of Germany

BERTIL HILLE
Department of Physiology and Biophysics
University of Washington Medical School
Seattle, Washington 98195

TOSHITSUGU HIRANO
Department of Psychology
Faculty of Letters
Kyoto University
Sakyoku, Kyoto 606, Japan

PHILIP E. HOCKBERGER
Department of Molecular Biophysics
AT & T Bell Laboratories
Murray Hill, New Jersey 07974

WILLIAM R. HOLMES
Mathematical Research Branch
National Institute of Diabetes and Digestive and Kidney Diseases
National Institutes of Health
Bethesda, Maryland 20892

GERT HOLSTEGE
Department of Anatomy II
Medical Faculty
Erasmus University Rotterdam
3000 DR Rotterdam, The Netherlands

YAN-YOU HUANG
Department of Physiology
University of Göteborg
S-400 33 Göteborg, Sweden
Sabbatical from:
Shanghai Institute of Physiology
Academia Sinica
People's Republic of China

CARLOS ISALES
Departments of Internal Medicine and Cell Biology
School of Medicine and School of Nursing
Yale University
New Haven, Connecticut 06510

HERBERT H. JASPER
Center for Research in Neurological Sciences
Department of Physiology
University of Montreal
Montreal, Quebec, Canada H3C 3J7
and The Montreal Neurological Institute
McGill University
Montreal, Quebec, Canada H32 1E7

T. KASAMATSU
The Smith–Kettlewell Eye Research Foundation at Pacific Presbyterian Medical Center
San Francisco, California 94115

A. HARRY KLOPF
Avionics Laboratory
Air Force Wright Aeronautical Laboratories
Wright-Patterson Air Force Base, Ohio 45433

HIROYUKI KOIKE
Department of Neurophysiology
Tokyo Metropolitan Institute for Neurosciences
Fuchu City, Tokyo 183, Japan

BORIS I. KOTLYAR
Department of Higher Nervous Activity
Lomonosov Moscow State University
Moscow, USSR

K. KOYANO
Department of Physiology
Saga Medical School
Saga 840-01, Japan

K. KUBA
Department of Physiology
Saga Medical School
Saga 840-01, Japan

E. KUMAMOTO
Department of Physiology
Saga Medical School
Saga 840-01, Japan

LÁSZLÓ LÉNÁRD
Department of Biological Control Systems
National Institute for Physiological Sciences
Okazaki 444, Japan

JEFFREY D. LEWINE
Center for Brain Research
University of Rochester
Rochester, New York 14642

O. C. J. LIPPOLD
Department of Human Physiology
Royal Holloway and Bedford New College
Egham TW20 0EX, England

GREGORY A. LNENICKA
Department of Physiology
University of Toronto
Toronto, Ontario, Canada M5S 1A8
Present address:
Neurobiology Research Center
Department of Biological Sciences
State University of New York
Albany, New York 12222

CHRISTINE G. LOGAN
Department of Psychology
University of Southern California
Los Angeles, California 90089

F.H. LOPES DA SILVA
Department of General Zoology
University of Amsterdam
1098 SM Amsterdam, The Netherlands

HEIKO J. LUHMANN
Max Planck Institute for Brain Research
Department of Neurophysiology
6000 Frankfurt 71, Federal Republic of Germany

TOORU MARUYAMA
Department of Physiology
Faculty of Medicine
Kyushu University
Fukuoka 812, Japan

HIROKO MATSUMOTO
Department of Neurophysiology
Tokyo Metropolitan Institute for Neurosciences
Fuchu City, Tokyo 183, Japan

B. P. C. MELCHERS
Department of General Zoology
University of Amsterdam
1098 SM Amsterdam, The Netherlands

TIM MEYER
Department of Physiology
University College
London WC1E 6BT, England

S. MINOTA
Department of Physiology
Saga Medical School
Saga 840-01, Japan

SUSAN J. MITCHELL
Department of Neurology
Johns Hopkins University
Baltimore, Maryland 21205

FUJIO MURAKAMI
Department of Biophysical Engineering
Faculty of Engineering Science
Osaka University
Toyonaka, Osaka 560, Japan

KIYOMI NAKAMURA
Department of Physiology
Faculty of Medicine
Toyama Medical and Pharmaceutical University
Sugitani, Toyama 930-01, Japan

YASUHIKO NAKANO
Department of Biological Control Systems
National Institute for Physiological Sciences
Okazaki 444, Japan

ROBERT B. NELSON
Neuroscience Group
The DuPont Company
Wilmington, Delaware 19898

HISAO NISHIJO
Department of Physiology
Faculty of Medicine
Toyama Medical and Pharmaceutical University
Sugitani, Toyama 930-01, Japan

HITOO NISHINO
Department of Biological Control Systems
National Institute for Physiological Sciences
Okazaki 444, Japan

YOICHI ODA
Department of Biophysical Engineering
Faculty of Engineering Science
Osaka University
Toyonaka, Osaka 560, Japan

TAKETOSHI ONO
Department of Physiology
Faculty of Medicine
Toyama Medical and Pharmaceutical University
Sugitani, Toyama 930-01, Japan

YUTAKA OOMURA
Department of Physiology
Faculty of Medicine
Kyushu University
Fukuoka 812, Japan
and
Department of Biological Control Systems
National Institute for Physiological Sciences
Okazaki 444, Japan

S. POCKETT
Department of Physiology
University of Auckland
Auckland, New Zealand

BERNARD POULAIN
Laboratory of Cellular and Molecular Neurobiology
National Center for Scientific Research
91190 Gif-sur-Yvette, France

CLIVE A. PRINCE
Department of Physiology
University College London
London WC1E 6BT, England

NING QIAN
Department of Biophysics
Johns Hopkins University
Baltimore, Maryland 21218

WILFRID RALL
Mathematical Research Branch
National Institute of Diabetes and Digestive and Kidney Diseases
National Institutes of Health
Bethesda, Maryland 20892

HOWARD RASMUSSEN
Departments of Internal Medicine and Cell Biology
School of Medicine and School of Nursing
Yale University
New Haven, Connecticut 06510

STEPHEN REDMAN
Experimental Neurology Unit
John Curtin School of Medical Research
Australian National University
Canberra, A.C.T., Australia

RUSSELL T. RICHARDSON
Department of Neurology
Johns Hopkins University
Baltimore, Maryland 21205

I. P. RICHES
Department of Anatomy
The Medical School
University of Bristol
Bristol BS8 1TD, England

JAMES L. RINGO
Center for Brain Research
University of Rochester
Rochester, New York 14642

ARYEH ROUTTENBERG
Cresap Neuroscience Laboratory
Northwestern University
Evanston, Illinois 60201

JUNICHI SADOSHIMA
Department of Physiology
Faculty of Medicine
Kyushu University
Fukuoka 812, Japan

JOHN M. SARVEY
Department of Pharmacology
Uniformed Services University of the Health Sciences
Bethesda, Maryland 20814-4799

ALMUT SCHÜZ
Max Planck Institute for Biological Cybernetics
7400 Tübingen, Federal Republic of Germany

IDAN SEGEV
Mathematical Research Branch
National Institute of Diabetes and Digestive and Kidney Diseases
National Institutes of Health
Bethesda, Maryland 20892
Present address:
Department of Neuroscience
Institute of Life Sciences
The Hebrew University
Jerusalem, Israel

TERRENCE J. SEJNOWSKI
Department of Biophysics
Johns Hopkins University
Baltimore, Maryland 21218

T. SHIROKAWA
Department of Neurophysiology
Institute of Higher Nervous Activity
Osaka University Medical School
Kita, Osaka 530, Japan

WOLF SINGER
Max Planck Institute for Brain Research
Department of Neurophysiology
6000 Frankfurt 71, Federal Republic of Germany

P. K. STANTON
Department of Biophysics
Johns Hopkins University
Baltimore, Maryland 21218

JOSEPH E. STEINMETZ
Department of Psychology
Indiana University
Bloomington, Indiana 47405

MAGDOLNA B. SZENTE
Department of Comparative Physiology
Attila Jozsef University of Sciences
Szeged, Hungary

NORIKO TAKUWA
Departments of Internal Medicine and Cell Biology
School of Medicine and School of Nursing
Yale University
New Haven, Connecticut 06510

YOH TAKUWA
Departments of Internal Medicine and Cell Biology
School of Medicine and School of Nursing
Yale University
New Haven, Connecticut 06510

JOEP TAN
Department of Anatomy II
Medical Faculty
Erasmus University Rotterdam
3000 DR Rotterdam, The Netherlands

K. TANAKA
Department of Physiology
Saga Medical School
Saga 840-01, Japan

LADISLAV TAUC
Laboratory of Cellular and Molecular Neurobiology
National Center for Scientific Research
91190 Gif-sur-Yvette, France

RICHARD F. THOMPSON
Department of Psychology
University of Southern California
Los Angeles, California 90089

NATALYA O. TIMOFEEVA
Department of Higher Nervous Activity
Lomonosov Moscow State University
Moscow, USSR

NAOFUMI TOKUTOMI
Department of Physiology
Faculty of Medicine
Kyushu University
Fukuoka 812, Japan

F. W. Y. TSE
Department of Physiology
University of Toronto
Toronto, Ontario, Canada M5S 1A8
Present address:
Department of Medical Physiology
University of Calgary
Calgary, Alberta, Canada T2N 4N1

S. TSUJI
Department of Physiology
Saga Medical School
Saga 840-01, Japan

NAKAAKIRA TSUKAHARA
Department of Biophysical Engineering
Faculty of Engineering Science
Osaka University
Toyonaka, Osaka 560, Japan

YOSHITOMO UMITSU
Department of Neurophysiology
Tokyo Metropolitan Institute for Neurosciences
Fuchu City, Tokyo 183, Japan

JACQUELINE J. VAN HAM
Department of Anatomy II
Medical Faculty
Erasmus University Rotterdam
3000 DR Rotterdam, The Netherlands

W. J. WADMAN
Department of General Zoology
University of Amsterdam
1098 SM Amsterdam, The Netherlands

CHRISTIAN WERTENBAKER
Departments of Ophthalmology and Neurology
Albert Einstein College of Medicine of Yeshiva University
Bronx, New York 10461

HOLGER WIGSTRÖM
Department of Physiology
University of Göteborg
S-400 33 Göteborg, Sweden

F. A. W. WILSON
Section of Neuroanatomy
School of Medicine
Yale University
New Haven, Connecticut 06510

J. M. WOJTOWICZ
Department of Physiology
University of Toronto
Toronto, Ontario, Canada M5S 1A8

CHARLES D. WOODY
Mental Retardation Research Center
Brain Research Institute
University of California at Los Angeles
Los Angeles, California 90024

IGOR S. ZAKHAROV
Laboratory of Conditioned Reflexes and Physiology of Emotions
Institute of Higher Nervous Activity and Neurophysiology
USSR Academy of Sciences
Moscow, USSR

WALTER ZAWALICH
Departments of Internal Medicine and Cell Biology
School of Medicine and School of Nursing
Yale University
New Haven, Connecticut 06510

L. J. ZIJP
Department of General Zoology
University of Amsterdam
1098 SM Amsterdam, The Netherlands

Index

FSC
www.fsc.org
MIX
Papier | Fördert gute Waldnutzung
FSC® C083411

Kolibri 360 GmbH
Ferdinand-Jühlke-Straße 7
99095 Erfurt, Deutschland
produktsicherheit@kolibri360.de